DATE DUE

VITAMINS AND HORMONES

VOLUME 67

Editorial Board

―――――――――

TADHG P. BEGLEY

ANTHONY R. MEANS

BERT W. O'MALLEY

LYNN RIDDIFORD

ARMEN H. TASHJIAN, JR.

―――――――――

VITAMINS AND HORMONES

ADVANCES IN RESEARCH AND APPLICATIONS

TRAIL
(TNF–Related Apoptosis-Inducing Ligand)

Editor-in-Chief

GERALD LITWACK

Department of Biochemistry and Molecular Pharmacology
Jefferson Medical College
Thomas Jefferson University
Philadelphia, Pennsylvania and
Toluca Lake, California

VOLUME 67

ELSEVIER
ACADEMIC
PRESS

AMSTERDAM • BOSTON • HEIDELBERG • LONDON
NEW YORK • OXFORD • PARIS • SAN DIEGO
SAN FRANCISCO • SINGAPORE • SYDNEY • TOKYO

Academic Press is an imprint of Elsevier

Elsevier Academic Press
525 B Street, Suite 1900, San Diego, California 92101-4495, USA
84 Theobald's Road, London WC1X 8RR, UK

This book is printed on acid-free paper. ∞

Copyright © 2004, Elsevier Inc. All Rights Reserved.

No part of this publication may be reproduced or transmitted in any form or by any means, electronic or mechanical, including photocopy, recording, or any information storage and retrieval system, without permission in writing from the Publisher.

The appearance of the code at the bottom of the first page of a chapter in this book indicates the Publisher's consent that copies of the chapter may be made for personal or internal use of specific clients. This consent is given on the condition, however, that the copier pay the stated per copy fee through the Copyright Clearance Center, Inc. (www.copyright.com), for copying beyond that permitted by Sections 107 or 108 of the U.S. Copyright Law. This consent does not extend to other kinds of copying, such as copying for general distribution, for advertising or promotional purposes, for creating new collective works, or for resale.
Copy fees for pre-2004 chapters are as shown on the title pages. If no fee code appears on the title page, the copy fee is the same as for current chapters.
0083-6729/2004 $35.00

Permissions may be sought directly from Elsevier's Science & Technology Rights Department in Oxford, UK: phone: (+44) 1865 843830, fax: (+44) 1865 853333, e-mail: permissions@elsevier.com.uk. You may also complete your request on-line via the Elsevier homepage (http://elsevier.com), by selecting "Customer Support" and then "Obtaining Permissions."

For all information on all Academic Press Publications visit our Web site at www.academicpress.com

ISBN: 0-12-709867-4

PRINTED IN THE UNITED STATES OF AMERICA
04 05 06 07 08 9 8 7 6 5 4 3 2 1

Former Editors

ROBERT S. HARRIS
Newton, Massachusetts

JOHN A. LORRAINE
*University of Edinburgh
Edinburgh, Scotland*

PAUL L. MUNSON
*University of North Carolina
Chapel Hill, North Carolina*

JOHN GLOVER
*University of Liverpool
Liverpool, England*

GERALD D. AURBACH
*Metabolic Diseases Branch
National Institute of Diabetes
and Digestive and Kidney Diseases
National Institutes of Health
Bethesda, Maryland*

KENNETH V. THIMANN
*University of California
Santa Cruz, California*

IRA G. WOOL
*University of Chicago
Chicago, Illinois*

EGON DICZFALUSY
*Karolinska Sjukhuset
Stockholm, Sweden*

ROBERT OLSEN
*School of Medicine
State University of New York
at Stony Brook
Stony Brook, New York*

DONALD B. MCCORMICK
*Department of Biochemistry
Emory University School of Medicine
Atlanta, Georgia*

Contents

Contributors XVII
Preface XXIII

1

Specificity of Molecular Recognition Learned from the Crystal Structures of TRAIL and the TRAIL:sDR5 Complex

Sun-Shin Cha, Young-Lan Song, and Byung-Ha Oh

I. Introduction: The TNF and TNF Receptor Superfamilies 2
II. TRAIL and its Receptors 3
III. Selective Antitumor Activity and Biologic Functions of TRAIL 4
IV. Overall Structure of TRAIL 5
V. Zinc Binding and its Critical Role 5
VI. Unique Frame Insertion in the AA″ Loop 8
VII. Structure of TRAIL:sDR5 Complex 9
VIII. Determinants Conferring Specificity of Receptor Recognition 12
IX. Conclusions and Perspectives 13
References 14

2

CRYSTAL STRUCTURE OF RANK LIGAND INVOLVED IN BONE METABOLISM

SHUICHIRO ITO AND TADASHI HATA

I. Introduction   20
II. The OPG/RANKL/RANK System   21
III. Three-Dimensional Structure of RANKL   25
IV. Conclusion   29
   References   30

3

PROMOTER OF TRAIL-R2 GENE

TATSUSHI YOSHIDA AND TOSHIYUKI SAKAI

I. Introduction   36
II. Basic Structure of TRAIL-R2 Promoter   37
III. Regulator of TRAIL-R2   40
IV. Possibility of Cancer Therapy Using TRAIL-R2   43
   References   45

4

TRANSCRIPTIONAL REGULATION OF THE TRAIL-R3 GENE

CARMEN RUIZ DE ALMODÓVAR, ABELARDO LÓPEZ-RIVAS, JUAN MIGUEL REDONDO, AND ANTONIO RODRÍGUEZ

I. Introduction: TRAIL and its Receptors   52
II. Decoy Receptors   54
III. TRAIL-R3   56
IV. Future Directions   60
   References   60

5

MONOCLONAL ANTIBODIES AGAINST TRAIL

NINA-BEATE LIABAKK AND TERJE ESPEVIK

I. Introduction   66
II. Overview of Monoclonal Antibodies Against TRAIL   67
III. Applications of Monoclonal Antibodies Against TRAIL   70
IV. Physiologic Significance of TRAIL   74
V. Future Aspects   76
References   76

6

MODULATION OF TRAIL SIGNALING COMPLEX

CHUNHAI HAO, JIN H. SONG, UROSH VILIMANOVICH, AND NORMAN M. KNETEMAN

I. TRAIL: A New Member of TNF Family   82
II. TRAIL Death Receptors: DR4 and DR5   83
III. TRAIL Decoy Receptors: DcR1, DcR2, and OPG   84
IV. TRAIL-Induced DISC: Cleavage of Apoptosis Initiating Protease Caspase-8   85
V. TRAIL-Induced Bid Cleavage: Activation of Mitochondrial Pathways   87
VI. TRAIL-Induced Cell Proliferation: RIP-Mediated NF-κB Activation   87
VII. TRAIL-Induced c-FLIP Recruitment: DISC Modulation   88
VIII. TRAIL-Induced PED Recruitment: Differential Modulation of the DISC   90
IX. TRAIL Physiologic Functions: Tumor Surveillance   91
X. Recombinant TRAIL: A Novel Cancer Therapeutic Agent   91
References   93

7

TRAIL AND NFκB SIGNALING—A COMPLEX RELATIONSHIP

HARALD WAJANT

I. Introduction   102
II. Regulation of TRAIL-Induced Apoptosis by NFκB   113
III. TRAIL Activates NFκB   119
    References   121

8

CARDINAL ROLES IN APOPTOSIS AND NFκB ACTIVATION

LISA BOUCHIER-HAYES AND SEAMUS J. MARTIN

I. Introduction   134
II. Regulation of NFκB Activation   134
III. CARD Proteins   135
IV. CARD Proteins and NFκB Activation   135
V. Functions of CARDINAL   138
VI. CARDINAL as a Negative Regulator of Diverse NFκB Activation Pathways   138
VII. CARDINAL and Apoptosis   140
VIII. CARDINAL and Caspase-1 Activation   140
IX. The Inflammasome   141
X. CARDINAL Contradictions   142
XI. Conclusions   143
    References   143

9

TRAIL IN THE AIRWAYS

NOREEN M. ROBERTSON, MARY ROSEMILLER, ROCHELLE G. LINDEMEYER,
ANDRZEJ STEPLEWSKI, JAMES G. ZANGRILLI, AND GERALD LITWACK

I. Introduction   150
II. TRAIL: Synthetic and Signaling Pathways   150
III. Effects of TRAIL in the Airways   153
IV. Conclusions   162
    References   162

10

TRAIL DEATH RECEPTORS, BCL-2 PROTEIN FAMILY, AND ENDOPLASMIC RETICULUM CALCIUM POOL

M. SAEED SHEIKH AND YING HUANG

I. Introduction   170
II. The Endoplasmic Reticulum   170
III. Endoplasmic Reticulum Ca^{2+} Pools and Regulation of Apoptosis   172
IV. Concluding Remarks   183
    References   185

11

FLIP PROTEIN AND TRAIL-INDUCED APOPTOSIS

WILFRIED ROTH AND JOHN C. REED

I. Introduction   190
II. FLIP: Structure and Mechanisms of Action   190
III. FLIP Function in Normal Physiology and Disease   194
IV. TRAIL and FLIP   197
    References   199

12

EPIDERMAL GROWTH FACTOR AND TRAIL INTERACTIONS IN EPITHELIAL-DERIVED CELLS

SPENCER BRUCE GIBSON

I. Epidermal Growth Factor Receptor–Mediated Cell Survival 208
II. Epidermal Growth Factor Signaling Pathways 209
III. Transcriptional Activation 213
IV. Antiapoptotic Proteins 214
V. Targeting Epidermal Growth Factor Receptor Signaling for Treatment of Cancer 216
VI. TRAIL Death Receptors 217
VII. TRAIL as a Cancer Treatment 218
VIII. Convergence of the Epidermal Growth Factor and TRAIL Signaling Pathways 219
References 222

13

TRAIL AND CERAMIDE

YONG J. LEE AND ANDREW A. AMOSCATO

I. Introduction 230
II. TRAIL 231
III. Sphingolipids 234
IV. TRAIL and Ceramide 243
References 244

14

TRAIL AND VIRAL INFECTION

JÖRN STRÄTER AND PETER MÖLLER

I. Introduction 258
II. Apoptosis as a Process of Degradation of Host Cell and Viral Constituents 259
III. Death Receptor/Ligand Systems 260
IV. The Cytotoxic Activity of TRAIL Against Virus-Infected Cells 262
V. Viral Strategies to Circumvent TRAIL-Induced Apoptosis 264

- VI. TRAIL in Virus-Induced Immunosuppression 265
- VII. Viruses and TRAIL in Malignant Disease 266
- VIII. Conclusions 268
     - References 269

15

Modulation of TRAIL Signaling for Cancer Therapy

Simone Fulda and Klaus-Michael Debatin

- I. Introduction 276
- II. The Core Apoptotic Machinery 277
- III. TRAIL and its Receptors 278
- IV. TRAIL Signaling 278
- V. Defective TRAIL Signaling in Cancers 279
- VI. TRAIL and Cancer Therapy 281
- VII. Conclusions 284
     - References 285

16

Interferon-Gamma and TRAIL in Human Breast Tumor Cells

Carmen Ruiz de Almodóvar, Abelardo López-Rivas, and Carmen Ruiz-Ruiz

- I. Introduction 292
- II. Interferon-γ 293
- III. TRAIL System 297
- IV. Regulation by IFN-γ of TRAIL-Induced Apoptosis in Breast Tumor Cells 303
- V. Conclusions 308
     - References 309

17

RETINOIDS AND TRAIL: TWO COOPERATING ACTORS TO FIGHT AGAINST CANCER

LUCIA ALTUCCI AND HINRICH GRONEMEYER

I. Introduction   320
II. Origin of Retinoids   322
III. Mechanism of Retinoid Action   322
IV. Anticancer Activity of Retinoic Acids   325
V. Retinoid Action on Human Cancers   328
VI. APO2L/TRAIL and its Receptors   330
VII. Preventive and Therapeutic Potential of Retinoids: The TRAIL Connection   334
VIII. Future Directions   336
References   338

18

POTENTIAL FOR TRAIL AS A THERAPEUTIC AGENT IN OVARIAN CANCER

TOURAJ ABDOLLAHI

I. Introduction   348
II. Ovarian Cancer   348
III. Tumor Necrosis Factor–Related Apoptosis-Inducing Ligand   351
IV. Interleukin-8   354
V. Role of p38 MAPK in Apoptosis   356
VI. Summary   357
References   358

19

TRAIL AND CHEMOTHERAPEUTIC DRUGS IN CANCER THERAPY

XIU-XIAN WU, OSAMU OGAWA, AND YOSHIYUKI KAKEHI

I. Introduction   366
II. Receptors for TRAIL   367
III. Signaling Pathways of TRAIL Receptors   368
IV. Bioactivity of TRAIL   369
V. Synergistic Effect of TRAIL and Chemotherapeutic Drugs   372
VI. Molecular Mechanisms of the Synergistic Effect   372
VII. Conclusions and Prospects   376
References   377

20

ADDITIVE EFFECTS OF TRAIL AND PACLITAXEL ON CANCER CELLS: IMPLICATIONS FOR ADVANCES IN CANCER THERAPY

CHRISTINE ODOUX AND ANDREAS ALBERS

I. Introduction   386
II. Tumor Necrosis Factor–Related Apoptosis-Inducing Ligand   388
III. Paclitaxel   392
IV. Biologic and Clinical Benefits of TRAIL and Paclitaxel Combination Therapy   397
V. Conclusions   402
References   403

21

REGULATION OF SENSITIVITY TO TRAIL BY THE PTEN TUMOR SUPPRESSOR

YOUNG E. WHANG, XIU-JUAN YUAN, YUANBO LIU, SAMARPAN MAJUMDER, AND TERRENCE D. LEWIS

I. Introduction   411
II. Regulation of TRAIL-Induced Apoptosis   411

III. Regulation of the Phosphatidylinositol-3 Kinase
     Pathway by PTEN   414
IV. Modulation of TRAIL Sensitivity by the Phosphatidylinositol-3
     Kinase/PTEN/Akt Pathway   416
V. Downstream Targets of the Phosphatidylinositol-3
     Kinase/PTEN/Akt Pathway   417
VI. Strategies for Overcoming TRAIL Resistance   419
VII. Conclusion   421
     References   422

22

TRAIL AND MALIGNANT GLIOMA

CHRISTINE J. HAWKINS

I. Malignant Glioma   428
II. Apoptosis in Cancer   430
III. Apoptosis Pathways Overview   433
IV. TRAIL   435
V. TRAIL and Glioma   437
VI. Future Directions   444
    References   444

23

REGULATION OF TRAIL-INDUCED APOPTOSIS BY ECTOPIC EXPRESSION OF ANTIAPOPTOTIC FACTORS

BHARAT B. AGGARWAL, UDDALAK BHARDWAJ, AND YASUNARI TAKADA

I. Introduction   454
II. Negative Regulation of TRAIL-Induced Apoptosis   458
III. Inhibitors of Antiapoptotic Factors as Therapeutic Agents   471
IV. Effect of TRAIL on Normal Cells   473
V. Conclusions   474
   References   475

INDEX   485

CONTRIBUTORS

Numbers in parenthesis indicate the pages on which the authors' contributions begin.

Touraj Abdollahi (347) Department of Biochemistry and Molecular Pharmacology, Jefferson Medical College, Thomas Jefferson University, Philadelphia, Pennsylvania 19107.

Bharat B. Aggarwal (453) Cytokine Research Section, Department of Bioimmunotherapy, The University of Texas, M. D. Anderson Cancer Center, Houston, Texas 77030.

Andreas Albers (385) Department of Pathology, Otolaryngology, and Tumor Immunology, University of Pittsburgh Cancer Institute, Research Pavilion at The Hillman Cancer Center, University of Pittsburgh Medical School, Pittsburgh, Pennsylvania 15213.

Lucia Altucci (319) Dipartimento di Patologia Generale, Seconda Università degli Studi di Napoli 80138, Napoli Italy.

Andrew A. Amoscato (229) Department of Pathology, University of Pittsburgh, Pittsburgh, Pennsylvania 15213.

Uddalak Bhardwaj (453) Cytokine Research Section, Department of Bioimmunotherapy, The University of Texas, M. D. Anderson Cancer Center, Houston, Texas 77030.

Lisa Bouchier-Hayes (133) Division of Cellular Immunology, La Jolla Institute for Allergy and Immunology, San Diego, California 92121.

Sun-Shin Cha (1) Beamline Division, Pohang Accelerator Laboratory, Pohang, Kyungbuk 790-784, Korea.

Klaus-Michael Debatin (275) University Children's Hospital, D-89075 Ulm, Germany.

Terje Espevik (65) Institute of Cancer Research and Molecular Medicine, Norwegian University of Science and Technology, N-7489 Trondheim, Norway.

Simone Fulda (275) University Children's Hospital, D-89075 Ulm, Germany.

Spencer Bruce Gibson (207) Department of Biochemistry and Medical Genetics, Manitoba Institute of Cell Biology, University of Manitoba, Winnipeg, Manitoba R3E 0V9, Canada.

Hinrich Gronemeyer (319) Department of Cell Biology and Signal Transduction, Institut de Génétique et de Biologie Moléculaire et Cellulaire/Centre National de la Recherche Scientifique/Institut National de la Santé et de la Recherche Médicale/University Louis Pasteur, 67404 Illkirch Cedex, C. U. de Strasbourg, France.

Chunhai Hao (81) Department of Laboratory Medicine & Pathology, University of Alberta, Edmonton, Alberta T6G 2B7, Canada.

Tadashi Hata (19) Biomedical Research Laboratories, Sankyo Co., Ltd., Tokyo 140-8710, Japan.

Christine J. Hawkins (427) Murdoch Children's Research Institute, Department of Haematology and Oncology, Royal Children's Hospital, Department of Paediatrics, University of Melbourne, Parkville, Victoria 3052, Australia.

Ying Huang (169) Department of Pharmacology, State University of New York Upstate Medical University, Syracuse, New York 13210.

Shuichiro Ito (19) Biomedical Research Laboratories, Sankyo Co., Ltd., Tokyo 140-8710, Japan.

Yoshiyuki Kakehi (365) Department of Urology, Kagawa University, Kagawa 761-0793, Japan.

Norman M. Kneteman (81) Department of Surgery, University of Alberta, Edmonton, Alberta T6G 2B7, Canada.

Yong J. Lee (229) Department of Surgery and Pharmacology, University of Pittsburgh, Pittsburgh, Pennsylvania 15213.

Terrence D. Lewis (409) Lineberger Comprehensive Cancer Center, Departments of Medicine and Pathology and Laboratory Medicine, University of North Carolina School of Medicine, Chapel Hill, North Carolina 27599-7295.

Nina-Beate Liabakk (65) Institute of Cancer Research and Molecular Medicine, Norwegian University of Science and Technology, N-7489 Trondheim, Norway.

Rochelle G. Lindemeyer (149) Department of Pediatric Dentistry, School of Dental Medicine, University of Pennsylvania, Pennsylvania 19104.

Gerald Litwack (149) Department of Biochemistry and Molecular Pharmacology, Jefferson Medical College, Thomas Jefferson University, Philadelphia, Pennsylvania 19107.

Yuanbo Liu (409) Lineberger Comprehensive Cancer Center, Departments of Medicine and Pathology and Laboratory Medicine, University of North Carolina School of Medicine, Chapel Hill, North Carolina 27599-7295.

Abelardo López-Rivas (51, 291) Department of Cellular Biology and Immunology, Instituto de Parasitología y Biomedicina, Consejo Superior de Investigaciones Cientificas, Granada E-18001, Spain.

Samarpan Majumder (409) Lineberger Comprehensive Cancer Center, Departments of Medicine and Pathology and Laboratory Medicine, University of North Carolina School of Medicine, Chapel Hill, North Carolina 27599-7295.

Seamus J. Martin (133) Molecular Cell Biology Laboratory, Department of Genetics, The Smurfit Institute, Trinity College, Dublin 2, Ireland.

Peter Möller (257) Department of Pathology, University Hospital of Ulm, D-89081 Ulm, Germany.

Christine Odoux (385) Division of Hematology/Oncology, University of Pittsburgh Cancer Institute, Research Pavilion at The Hillman Cancer Center, University of Pittsburgh Medical School, Pittsburgh, Pennsylvania 15213.

Osamu Ogawa (365) Department of Urology, Graduate School of Medicine, Kyoto University, Kyoto 606-8507, Japan.

Byung-Ha Oh (1) Center for Biomolecular Recognition, Department of Life Science and Division of Molecular and Life Science, Pohang University of Science and Technology, Pohang, Kyungbuk 790-784, Korea.

Juan Miguel Redondo (51) Centro de Biología Molecular 'Severo Ochoa,' Universidad Autónoma de Madrid, Madrid E-28049, Spain, and Centro Nacional de Investigaciones Cardiovasculares, Tres Cantos E-28760, Spain.

John C. Reed (189) The Burnham Institute, La Jolla, California 92037.

Noreen M. Robertson (149) Department of Biochemistry and Molecular Pharmacology, Jefferson Medical College, Thomas Jefferson University, Philadelphia, Pennsylvania 19107.[*]

Antonio Rodríguez (51) Centro de Biología Molecular 'Severo Ochoa,' Universidad Autónoma de Madrid, Madrid E-28049, Spain, and Centro Nacional de Investigaciones Cardiovasculares, Tres Cantos E-28760, Spain.

---

[*]Current address: Department of Biochemistry, College of Medicine, Drexel University, Philadelphia, Pennsylvania 19102-1192.

Mary Rosemiller (149) Department of Biochemistry and Molecular Pharmacology, Jefferson Medical College, Thomas Jefferson University, Philadelphia, Pennsylvania 19107.

Wilfried Roth (189) The Burnham Institute, La Jolla, California 92037.

Carmen Ruiz de Almodóvar (51, 291) Department of Cellular Biology and Immunology, Instituto de Parasitología y Biomedicina, Consejo Superior de Investigaciones Cientificas, Granada E-18001, Spain.

Carmen Ruiz-Ruiz (291) Facultad de Medicina, Universidad de Granada, Granada E-18071, Spain.

Toshiyuki Sakai (35) Department of Molecular-Targeting Cancer Prevention, Graduate School of Medical Science, Kyoto Prefectural University of Medicine, Kyoto 602-8566, Japan.

M. Saeed Sheikh (169) Department of Pharmacology, State University of New York Upstate Medical University, Syracuse, New York 13210.

Young-Lan Song (1) Center for Biomolecular Recognition, Department of Life Science and Division of Molecular and Life Science, Pohang University of Science and Technology, Pohang, Kyungbuk 790-784, Korea.

Jin H. Song (81) Department of Laboratory Medicine & Pathology, University of Alberta, Edmonton, Alberta T6G 2B7, Canada.

Andrzej Steplewski (149) Department of Biochemistry and Molecular Pharmacology, Jefferson Medical College, Thomas Jefferson University, Philadelphia, Pennsylvania 19107.

Jörn Sträter (257) Department of Pathology, University Hospital of Ulm, D-89081 Ulm, Germany.

Yasunari Takada (453) Cytokine Research Section, Department of Bioimmunotherapy, The University of Texas, M. D. Anderson Cancer Center, Houston, Texas 77030.

Urosh Vilimanovich (81) Department of Laboratory Medicine & Pathology, University of Alberta, Edmonton, Alberta T6G 2B7, Canada.

Harald Wajant (101) Department of Molecular Internal Medicine, Medical Polyclinic, University of Würzburg, D-97070 Würzburg, Germany.

Young E. Whang (409) Lineberger Comprehensive Cancer Center, Departments of Medicine and Pathology and Laboratory Medicine, University of North Carolina School of Medicine, Chapel Hill, North Carolina 27599-7295.

Xiu-Xian Wu (365) Department of Urology, Kagawa University, Kagawa 761-0793, Japan.

Tatsushi Yoshida (35) Department of Molecular-Targeting Cancer Prevention, Graduate School of Medical Science, Kyoto Prefectural University of Medicine, Kyoto 602-8566, Japan.

Xiu-Juan Yuan (409) Lineberger Comprehensive Cancer Center, Departments of Medicine and Pathology and Laboratory Medicine, University of North Carolina School of Medicine, Chapel Hill, North Carolina 27599-7295.

James G. Zangrilli (149) Division of Critical Care, Pulmonary, Allergic, and Immunologic Diseases, Thomas Jefferson University, Philadelphia, Pennsylvania 19107.

PREFACE

The discovery of TRAIL (TNF–Related Apoptosis-Inducing Ligand) also referred to as Apo-2, ushered in an era of intense research because TRAIL induces many cancer cells to undergo programmed cell death (apoptosis) while having virtually no effect on normal cells. This important protein deserves extensive review at a formative time in the development of our knowledge concerning its mechanism of action and the ways in which it can be used as a cancer chemotherapeutic agent. Consequently, this volume reviews the current status of research on TRAIL. The contents range from structure, biology, and biochemistry to the foundations for the use of TRAIL as a chemotherapeutic agent.

The volume begins with: "Specificity of Molecular Recognition Learned from the Crystal Structures of TRAIL and the TRAIL:sDR5 Complex" by S.-S. Cha, Y.-L. Song, and B.-H. Oh. This is followed by another structural analysis of a related protein: "Crystal Structure of RANK Ligand Involved in Bone Metabolism" by S. Ito and T. Hata. At the DNA level, T. Yoshida and T. Sakai discuss: "Promoter of TRAIL-R2 Gene." "Transcriptional Regulation of the TRAIL-R3 gene" is next, presented by C. R. de Almodóvar, A. López-Rivas, J. M. Redondo, and A. Rodríguez.

Consideration of the biochemical aspects of TRAIL begins with a paper by N.-B. Liabakk and T. Espevik on: "Monoclonal Antibodies Against TRAIL." C. Hao, J. H. Song, U. Vilimanovich, and N. M. Kneteman write on: "Modulation of TRAIL Signaling Complex." Consistent with the relationship of TRAIL to NFκB, H. Wajant introduces "TRAIL and NFκB Signaling—A Complex Relationship," and L. Bouchier-Hayes and S. J. Martin introduce: "CARDINAL Roles in Apoptosis and NFκB

Activation." "TRAIL in the Airways" is reviewed next by N. M. Robertson, M. Rosemiller, R.G. Lindemeyer, A. Steplewski, J. G. Zangrilli, and G. Litwack. "TRAIL Death Receptors, Bcl-2 Protein Family, and Endoplasmic Reticulum Calcium Pool" is covered by M. S. Sheikh and Y. Huang. Next, "FLIP Protein and TRAIL-Induced Apoptosis" is presented by W. Roth and J. C. Reed. "Epidermal Growth Factor and TRAIL Interactions in Epithelial-Derived Cells" is provided by S. B. Gibson. Y. J. Lee and A. A. Amoscato review: "TRAIL and Ceramide." Finally, in this section, J. Sträter and P. Möller discuss: "TRAIL and Viral Infection."

In relation to chemotherapeutic approaches using TRAIL, S. Fulda and K.-M. Debatin review; "Modulation of TRAIL Signaling for Cancer Therapy." C. R. de Almodóvar, A. López-Rivas, and C. Ruiz-Ruiz present: "Interferon-Gamma and TRAIL in Human Breast Tumor Cells." "Retinoids and TRAIL: Two Cooperating Actors to Fight Against Cancer" by L. Altucci and H. Gronemeyer follows. T. Abdollahi discusses: "Potential for TRAIL as a Therapeutic Agent in Ovarian Cancer." "TRAIL and Chemotherapeutic Drugs in Cancer Therapy" appears by X.-X. Wu, O. Ogawa, and Y. Kakehi. C. Odoux and A. Albers report: "Additive Effects of TRAIL and Paclitaxel on Cancer Cells: Implications for Advances in Cancer Therapy." "Regulation of Sensitivity to TRAIL by the PTEN Tumor Suppressor" is reviewed by Y. E. Whang, X.-J. Yuan, Y. Liu, S. Majumder, and T. D. Lewis. Two contributions complete the volume; they are: "TRAIL and Malignant Glioma" by C. J. Hawkins and "Regulation of TRAIL-Induced Apoptosis by Ectopic Expression of Antiapoptotic Factors" by B. B. Aggarwal, U. Bhardwaj, and Y. Takada.

Gerald Litwack
Toluca Lake, California
September, 2003

Specificity of Molecular Recognition Learned from the Crystal Structures of TRAIL and the TRAIL:sDR5 Complex

Sun-Shin Cha,[*] Young-Lan Song,[†] and Byung-Ha Oh[†]

[*]Beamline Division, Pohang Accelerator Laboratory
[†]Center for Biomolecular Recognition, Department of Life Science and Division of Molecular and Life Science, Pohang University of Science and Technology
Pohang, Kyungbuk 790-784, Korea

I. Introduction: The TNF and TNF Receptor Superfamilies
II. TRAIL and its Receptors
III. Selective Antitumor Activity and Biologic Functions of TRAIL
IV. Overall Structure of TRAIL
V. Zinc Binding and its Critical Role
VI. Unique Frame Insertion in the AA″ Loop
VII. Structure of TRAIL:sDR5 Complex
VIII. Determinants Conferring Specificity of Receptor Recognition
IX. Conclusions and Perspectives
References

TRAIL is a member of the tumor necrosis factor (TNF) superfamily. TRAIL has drawn a lasting attention because of its selectivity and efficacy in inducing apoptosis in a variety of cancer cells but not in normal cells. The structures of both TRAIL and the protein in complex with the extracellular domain of death receptor 5 (sDR5) were elucidated. Because each factor of the ligand family and the receptor family is large, it poses an intriguing question of how recognition between cognate ligands and receptors is achieved in a highly specific manner without cross interactions. This review focuses on the unique properties of TRAIL and molecular strategies for the specific recognition between the two family members primarily based on the crystal structures of TRAIL and the TRAIL:sDR5 complex. © 2004 Elsevier Inc.

I. INTRODUCTION: THE TNF AND TNF RECEPTOR SUPERFAMILIES

Tumor necrosis factor (TNF) and its corresponding TNF receptor (TNFR) superfamilies play pivotal roles in regulating the immune system, organogenesis, and homeostasis (Locksley *et al.*, 2001). So far, 18 TNF family ligands and 30 TNFR family receptors have been identified (Ashkenazi, 2002; Bodmer *et al.*, 2002). The receptors belonging to the TNFR family are type I transmembrane proteins with several exceptions that contain an extracellular domain only. The extracellular domain of these receptors is characterized by the pseudorepeats of cysteine-rich domain (CRD) (Naismith *et al.*, 1996). The structure of TNF-α (Jones *et al.*, 1989), the first structure for the TNF family, showed that the protein forms a bell-shaped homotrimer. The structure of TNF-β in complex with the extracellular domain of TNF-R55 (sTNF-R55) revealed that three receptor molecules bind to the three identical interfaces between the ligand subunits (Banner *et al.*, 1993). Subsequent structure determination of four other TNF family members (Cha *et al.*, 1999; Hymowitz *et al.*, 2000; Ito *et al.*, 2002; Karpusas *et al.*, 1995, 2002; Lam *et al.*, 2001; Liu *et al.*, 2002; Oren *et al.*, 2002) and two of these in complex with a respective receptor (Cha *et al.*, 2000; Hymowitz *et al.*, 1999; Kim *et al.*, 2003; Liu *et al.*, 2002; Mongkolsapaya *et al.*, 1999) have established that the functional unit of TNF-like ligands is a homotrimer and that the binding of the trimeric ligands to the receptors induces the trimerization of receptors, leading to an initiation of transmembrane signaling. However, a variation on this theme is found in the structures of the complex between BAFF/TALL-1/BlyS and its receptor BAFF-R or BCMA, which shows a virus-like assembly composed of 20 copies of the 3:3 ligand:receptor complex (Kim et al., 2003).

II. TRAIL AND ITS RECEPTORS

TNF-related apoptosis-inducing ligand (TRAIL), also known as Apo2L, was identified based on the sequence homology to FasL/Apo1L and TNF (Pitti et al., 1996; Wiley et al., 1995). Full-length TRAIL is a type II membrane protein composed of four parts: an extracellular TNF-like domain, an extracellular stalk, a transmembrane helix, and a small cytoplasmic domain. The extracellular TNF-like domain of TRAIL (referred to as TRAIL throughout the text) can be released from the cell surface by a proteolytic cleavage at the stalk region (Wiley et al., 1995). Unusually, TRAIL interacts with five different receptors that belong to the TNFR superfamily: DR4 (Pan et al., 1997b), DR5 (Pan et al., 1997a; Sheridan et al., 1997; Walczak et al., 1997), decoy receptor 1 (DcR1) (Degli-Esposti et al., 1997b; Pan et al., 1997a; Sheridan et al., 1997), DcR2 (Degli-Esposti et al., 1997a; Marsters et al., 1997), and osteoprotegerin (OPG) (Emery et al., 1998). Two membrane-bound death receptors, DR4 and DR5, contain extracellular CRDs, a transmembrane helix, and a cytoplasmic death domain. The two receptors are capable of transducing apoptotic signal. The decoy receptors, DcR1 and DcR2, contain extracellular CRDs, which exhibit significant homology to those of DR4 and DR5 (54–58% sequence identity). However, DcR2 contains a truncated, nonfunctional cytoplasmic death domain, and DcR1 is devoid of any transmembrane or cytoplasmic residues. Both the decoy receptors therefore can bind TRAIL but do not mediate apoptosis, acting antagonistically to the two signaling receptors. The fifth TRAIL receptor OPG, which binds to a TNF family member RANKL/TRANCE and inhibits osteoclastogenesis (Simonet et al., 1997), was found to interact with TRAIL (Emery et al., 1998). OPG exhibits the weakest binding affinity for TRAIL among the five TRAIL receptors, with the dissociation constant K_D of 400 nM as compared to 2, 70, and 200 nM for the binding of TRAIL to DR5, DR4, and DcR1, respectively (Truneh et al., 2000). While physiologic effects of the relatively weak interaction between TRAIL and OPG need to be firmly established, OPG was shown to be a survival factor for human prostate cancer cells (Holen et al., 2002) and to potentially function as a paracrine survival factor for human myeloma cells (Shipman and Croucher, 2003). Recent reviews detail the pathways of death signaling stimulated by TRAIL (Almasan and Ashkenazi, 2003; LeBlanc and Ashkenazi, 2003), which involves FADD and caspase-8 and -10 as downstream effector molecules (Kischkel et al., 2000; Sprick et al., 2000, 2002).

III. SELECTIVE ANTITUMOR ACTIVITY AND BIOLOGIC FUNCTIONS OF TRAIL

Perhaps, TRAIL is the only known cytokine that effectively and specifically kills tumor cells. TRAIL induces apoptosis in a variety of tumor cells and some virally infected cells but does not affect normal cells *in vitro* (Lawrence *et al.*, 2001; Pitti *et al.*, 1996; Wiley *et al.*, 1995). Repeated injections of TRAIL actively suppressed implanted human tumors in mice (Walczak *et al.*, 1999) and caused no detectable cytotoxic effect on normal tissues and organs in a nonhuman primate model (Ashkenazi *et al.*, 1999). To explain the selective antitumor activity of TRAIL, a decoy hypothesis was put forth; preferential expression of the decoy receptors in normal cells compared with tumor cells interferes with the TRAIL action (Marsters *et al.*, 1997; Mongkolsapaya *et al.*, 1998; Pan *et al.*, 1997a; Sheridan *et al.*, 1997). However, examination of a panel of human tumor cell lines failed to demonstrate any obvious correlation between the mRNA expression level of the TRAIL receptors and the level of sensitivity to TRAIL (Griffith and Lynch, 1998; Lincz *et al.*, 2001). Furthermore, not all tumor cells are sensitive to the cytotoxic effects of TRAIL (Griffith and Lynch, 1998). Although the mechanisms for sensitivity or resistance of cells to TRAIL-induced apoptosis remain unclear and need further investigation, intracellular apoptosis regulators, such as FLICE-inhibitory proteins, inhibitors of apoptosis proteins, BCL-2 family proteins, and inhibitors of apoptosis-associated factor-1, appear to contribute to the resistance to TRAIL-induced cell death, besides the decoy receptors, as discussed in a recent review (Almasan and Ashkenazi, 2003).

What is the physiologic function of TRAIL? Takeda *et al.* identified the *in vivo* function of TRAIL as a tumor suppressor by using neutralizing monoclonal antibody against TRAIL that significantly increased liver metastases of several tumor cell lines in an interferon-γ–dependent manner (Takeda *et al.*, 2001, 2002). Consistently, in TRAIL gene-deficient mice in comparison with the wild-type control mice, liver metastases and fibrosarcoma induction by a chemical carcinogen increased (Cretney *et al.*, 2002; Sedger *et al.*, 2002). The gene knockout mice were viable and exhibited no sign of hematologic defects. These studies demonstrate that TRAIL plays an important role in immune surveillance against tumor initiation and metastasis but it does not play an essential role in the developmental process. The other *in vivo* role of TRAIL lies in antiviral immune surveillance; cytomegaloviral infection upregulated the expression of TRAIL, DR4, and DR5 potentiated by interferon-γ, which sensitizes the infected cells to apoptosis by TRAIL through upregulation of the TRAIL gene (Sedger *et al.*, 1999).

IV. OVERALL STRUCTURE OF TRAIL

Like other TNF superfamily members whose structures are available (TNF-α: Cha et al., 1998; Eck and Sprang, 1989; Jones et al., 1989; TNF-β: Eck et al., 1992; CD40L: Karpusas et al., 1995; TRAIL: Cha et al., 1999; Hymowitz et al., 2000; RANKL/TRANCE: Ito et al., 2002; Lam et al., 2001; and BAFF/TALL-1/BlyS: Karpusas et al., 2002; Liu et al., 2002; Oren et al., 2002), the TRAIL monomer contains two antiparallel β-pleated sheets with the Greek-key topology that form a β-sandwich as a core scaffold, and interacts with the adjacent subunits in a head-to-tail fashion to form a bell-shaped homotrimer (Fig. 1A). One end of the trimer is wider (called 'bottom') than the other end (called 'top'). Conventionally, the secondary structural elements of TNF family members are described according to the notation used for TNF-β (Eck et al., 1992). In this notation, the β-strands are assigned in the order from N to C terminus of the amino acid sequence except that the first two short strands are labeled as A and A″ and the following two short strands are labeled as B and B′. The β-strands A″, A, H, C, and F form the inner β-sheet involved in the intersubunit contacts for the trimer formation, while the β-strands B′, B, G, D, and E form the outer β-sheet (Fig. 1A). In the intersubunit interactions, in addition to the packing at the trimeric interface, one edge of the β-sandwich (strands E and F) in each subunit is packed against the face of the inner sheet of the neighboring subunit. Eight aromatic residues (His125, Phe163, Tyr183, Tyr185, Tyr189, Tyr243, Phe274, and Phe278), which are present on the surface of the inner sheet, form a tight hydrophobic core at the trimeric interface (Fig. 1B) or provide a hydrophobic platform for the edge-to-face interactions between adjacent subunits.

V. ZINC BINDING AND ITS CRITICAL ROLE

So far, TRAIL is the only member of TNF family that contains a metal ion. TRAIL contains one cysteine residue, Cys230, which is located at the trimer interface (Fig. 1A). Three cysteinyl sulfur atoms of the TRAIL trimer, related by the molecular threefold axis, tetrahedrally coordinate a zinc ion in conjunction with the fourth coordination arm, which is presumably a chloride ion (Cha et al., 2000; Hymowitz et al., 1999, 2000). No zinc atom was reported in the two available structures of TRAIL that was obtained by refolding from inclusion bodies (Cha et al., 1999; Mongkolsapaya et al., 1999), indicating that the Zn^{2+} is not necessary for the folding of TRAIL. However, the importance of the zinc coordination for the stability and the

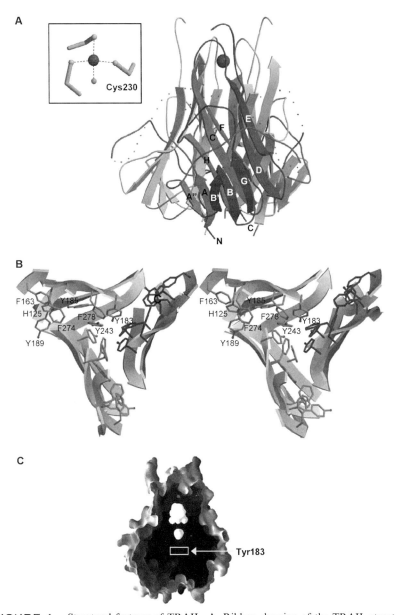

FIGURE 1. Structural features of TRAIL. A, Ribbon drawing of the TRAIL structure (PDB code: 1DG6). Each subunit is represented by different darknesses. The β-strands for one subunit are labeled. The inner β-strands (A″, A, H, C, and F) and the outer β-strands (B′, B, G, D, and E) are shown in light and dark gray, respectively. The disordered residues 131–141 of the loop are indicated by small spheres. The bound Zn^{2+} is shown. The inset shows the coordination by the three Cys230 residues and a putative chloride ion. B, The intersubunit interactions. Eight aromatic residues, which are engaged in hydrophobic edge-to-face or

bioactivity of TRAIL was demonstrated by the point mutations of Cys230 to Ala, Gly, or Ser. These mutants gave rise to a monomeric or dimeric form of TRAIL that is biologically inactive (Bodmer et al., 2000; Seol and Billiar, 2000; Trabzuni et al., 2000). Furthermore, the gradual conversion of active trimer into inactive smaller species was observed with the refolded TRAIL that contained no Zn^{2+} (Cha et al., 1999). The biologically active refolded TRAIL was eluted as virtually a single peak from a size exclusion column used for the protein purification. However, after the purified TRAIL was stored for some time at 4 °C, it was eluted as two separate peaks. These observations strongly suggest that the metal coordination is built to preserve the trimeric conformation of TRAIL, which is critical for the bioactivity of the protein.

The Zn^{2+}-coordination appears to affect the cytotoxicity of TRAIL toward human hepatocytes. A recombinant TRAIL with an N-terminal polyhistidine tag (called TRAIL-His) induced apoptosis of hepatocytes (Jo et al., 2000), while TRAIL that lacks an exogenous sequence tag did not (Lawrence et al., 2001). Because the flexible N-terminal region of TRAIL is just a linker that connects the extracellular structured domain to the transmembrane helix, the polyhistidine tag would not alter the TRAIL structure. Thus, the disparity in the bioactivity between the two TRAIL versions is quite difficult to grasp. However, there is an interesting difference between the two, which deserves attention. Metal analysis showed 0.46 mol zinc/mol trimer for the hepatotoxic TRAIL-His and 0.92 mol zinc/mol trimer for the TRAIL with no tag (Lawrence et al., 2001). Accordingly, the two recombinant TRAIL versions displayed different composition of oligomers; the hepatotoxic TRAIL-His was composed of 79% trimer and 21% dimer, whereas the TRAIL with no tag was homogeneous, with 99% trimer. The tagged version has a low solubility and tends to aggregate. It could be possible that the multimerization of the tagged TRAIL may result in higher-order clustering of the DR4 and DR5 receptors, leading to a signal that surpasses the high threshold for apoptosis activation of normal cells (Almasan and Ashkenazi, 2003).

The investigation of the subunit interactions of TRAIL along the molecular three fold axis provides insights into why the metal binding is essential for the stabilization of the trimeric conformation. From the bottom to the midway of the molecule, a cluster of nine aromatic residues (Tyr183,

---

trimeric core interactions, are shown in ball-and-sticks. The view is looking down from the top of the molecule along the molecular threefold axis. The hydrogen bonds between the hydroxyl groups of the tyrosine residues are omitted for clarity. C, Cavities at the trimer interface. The molecular 3-fold axis of the TRAIL trimer is roughly on the face of the paper. The cavities shown in white were calculated by the *GRASP* program. The bound Zn^{2+} is shown in the circle. The approximate positions of the Tyr183 residues are indicated by a rectangular box.

Tyr243, and Phe278 from each subunit) are involved in tight hydrophobic and hydrogen-bonding interactions (Fig. 1B) leaving no vacant space. However, a large cavity is present at the trimer interface of TRAIL between Tyr183 and Cys230 (Fig. 1C). The presence of the cavity indicates a loose subunit association in this region. Furthermore, the segment (residues 231–236) located between Cys230 and the top of the molecule is disordered in the crystals (PDB code: 1D2Q), unless it is held by symmetry-related molecules (PDB code: 1DG6). Consequently, the Zn^{2+} coordination is the only source of interactions in the large region from the midway to the top of the trimer interface. Therefore, the loss of the Zn^{2+} would be unfavorable for the stability of the TRAIL trimer. Noticeably, TNF-α has a large void space open to the bulk solvent in the upper half region of the trimeric interface, which is a similar feature observed in the structure of TRAIL. It was shown that the TNF-α trimer converts into inactive monomeric forms under physiologic conditions (Schuchmann *et al.*, 1995). These observations indirectly underscore the importance of Zn^{2+} binding for the stability of TRAIL.

VI. UNIQUE FRAME INSERTION IN THE AA″ LOOP

TRAIL has an extraordinarily elongated AA″ loop due to an insertion of 12–16 amino acids (depending on the proteins) in the loop compared with other TNF family members (Fig. 2). The AA″ loop in TRAIL traverses one entire subunit structure twice, whereas the corresponding loops are substantially shorter in the structures of other TNF family members (Fig. 3). The first half of the AA″ loop of TRAIL (residues 145–160) sits on top of the middle part of the β-sandwich and is involved in many favorable interactions (Fig. 3A), while the other half of the loop (residues 130–144) sits on top of the upper part of the molecule and is involved in only a few interactions. As a consequence, the second half of the loop is disordered (PDB code: 1DG6) or only partly observed (PDB code: 1D2Q). Four aromatic residues (Tyr209, Tyr211, Tyr213, Phe257), clustered on the surface of the outer β-sheet, interact with five residues on the AA″ loop (Lys145, Ala146, Leu147, Gly148, Lys150) that are absent in all the other TNF family members (Figs. 2, 3). None of the four aromatic acids is a conserved residue. Presumably, they have been selected for the favorable interactions with the unusually elongated loop. These structural observations suggested a functional importance of the loop as discussed later.

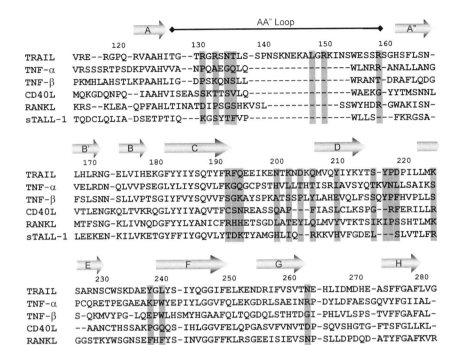

FIGURE 2. Sequence alignments of six TNF family proteins. The secondary structure assignment of TRAIL, in the order from N to C terminus of the amino acid sequence, is shown at the top of the sequence. The residue numbers correspond to the amino acid sequence of TRAIL. The AA″ loop of TRAIL is also marked. The receptor binding residues, identified on the basis of the structure of the TRAIL:sDR5 complex (PDB code: 1DU3), are darkly shaded.

VII. STRUCTURE OF TRAIL:sDR5 COMPLEX

TRAIL and sDR5 form a tight 3:3 symmetric complex. The elongated sDR5 molecules bind to the grooves between neighboring subunits of TRAIL (Fig. 4). The top of TRAIL faces the plasma membrane upon binding to the receptor molecules. The binding interface can be divided into three areas: lower contact region (close to the membrane), a central region, and upper contact region. Rather than detailing the mode of interactions in each region, we only highlight the interactions in the central contact region, which is distinguished from those in the structures of the TNF-β:sTNF-R55 complex and of BAFF/TALL-1/BlyS in complex with the extracellular domain of BAFF-R or sBCMA. While a part of the AA″ loop (residues 131–135) of TRAIL penetrates into the central binding interface and participates in the interaction with sDR5 upon the complex formation (Cha *et al.*, 2000),

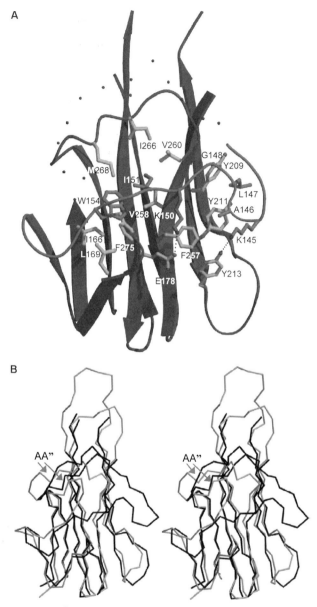

FIGURE 3. The AA″ loop of TRAIL. A, Interactions of the AA″ loop with the core scaffold of TRAIL (PDB code: 1D2Q). The AA″ loop is shown with the disordered residues on the loop represented by small spheres. The dotted lines represent hydrogen bonds. B, Stereo view of the superimposed Cα traces of TRAIL (black) and TNF-α (gray). Although not shown here, the other available structures of the TNF family members exhibit that the AA″ loops of these are similarly short as that of TNF-α.

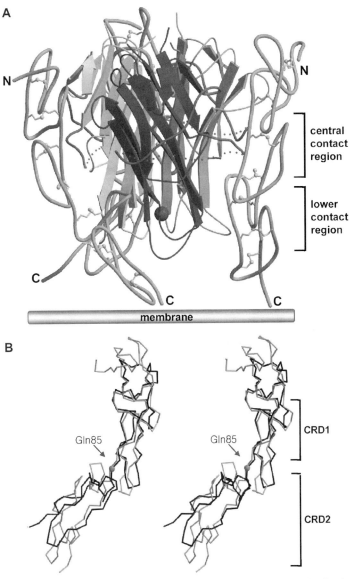

FIGURE 4. Structure of the TRAIL:sDR5 complex. A, The ribbon and tube drawing of the TRAIL:sDR5 complex (PDB code: 1DU3). The AA″ loop is shown with the disordered residues 136–145 on the loop represented by small spheres. The three receptor molecules are represented in tubes. The lines with two small spheres indicate disulfide bonds. The bound Zn^{2+} is represented as a sphere. B, Stereo view of the superimposed Cα traces of sDR5 (black) and sTNF-R55 (gray). The Cα atom of Gln85, which is located at the connection between CRD1 and CRD2, is rendered as a sphere. A torsional angle change of ~30 degrees occurs at Gln85 of sDR5 compared with sTNFR-55.

the other two complexes lack such ligand:receptor interactions in the corresponding region (Banner *et al.*, 1993; Kim *et al.*, 2003; Liu *et al.*, 2003). Consistent with the interactions of the AA″ loop with sDR5 observed in the crystal structure, a deletion of residues 137–152 or replacing residues 135–153 by the Ser-Leu-Leu sequence to mimic the short AA″ loop of TNF-β resulted in a significantly reduced apoptotic activity (Cha *et al.*, 1999) or affinity for sDR5 (Mongkolsapaya *et al.*, 1999). These data demonstrated that the frame insertion in the AA″ loop of TRAIL plays an important role in the binding of the cognate receptors.

VIII. DETERMINANTS CONFERRING SPECIFICITY OF RECEPTOR RECOGNITION

The structures of TRAIL and TRAIL:sDR5 complex, along with those of other related members provide valuable information on how TNF family members recognize cognate receptors specifically. Four different molecular strategies controlling the binding specificity between the two family members can be deduced. One obvious determinant is the variation of amino acids at the binding interface. The amino acid similarity among TNF family members is largely confined to internal residues constituting the structural scaffold. The external surfaces of these molecules, especially the receptor-binding sites show little sequence similarity (Fig. 2). As a result, despite the shared structural scaffold, each member in TNF family displays a characteristic surface property in the receptor binding site. The TNFR superfamily members should have coevolved to accommodate the changes in the receptor-binding sites of the corresponding ligands. The sequence variation at the receptor-binding surface of the ligand molecules and the complementary replacement of residues in the receptor molecules is definitely the major determinant controlling the binding specificity between the two family members. The frame insertion, such as that observed in the AA″ loop of TRAIL, is a drastic means used to alter the geometric and electrostatic property of the receptor binding surface without changing the structural scaffold. In the TRAIL structure, the β-turn of the AA″ loop penetrates into and delineates the receptor-binding site, drastically altering the common receptor binding surface. As mentioned previously, the loop makes unique interactions with sDR5 at the central contact region in the TRAIL:sDR5 structure (Cha *et al.*, 2000). Conformational flexibility of receptors appears to play a general role in compensating for geometric alterations due to sequence variations in the receptor-binding sites of the TNF family ligands. The structure of sTNF-R55 revealed a built-in plasticity of the molecule that adapts its structure to changes in solvent conditions (Naismith *et al.*, 1996). The conformation of the free receptor is different not only from that of the receptor in complex with TNF-β, but also from that of sDR5 in complex with TRAIL. Furthermore, the conformations of sDR5 and sTNF-R55, both in the ligand-bound form, are different from each other. The difference can be described as a domain movement primarily as a result of torsional angle changes at the connection between CRD1 and CRD2 (Fig. 4B) (Cha *et al.*, 2000). The interdomain flexibility is probably a common feature of the receptor family. Finally, structural variation of receptor molecules appears to be an important determinant. The structure of BAFF-R, one of the receptors for BAFF/TALL-1/BlyS consists of only a partial CRD and, therefore, is a highly focused interaction module. It is analogous to the first half of the canonical CRD but is stabilized by an additional noncanonical disulfide bond (Gordon *et al.*, 2003). It was shown that this disulfide bridge is critical for determining the binding specificity of BAFF-R to TALL-1 instead of APRIL, which shares the highest sequence homology with TALL-1 (Kim *et al.*, 2003). The minimal module binds to the intersubunit cleft of the ligand at the lower contact region only and interacts with an adjacent receptor molecule in a head-to-head fashion (Gordon *et al.*, 2003; Kim *et al.*, 2003; Liu *et al.*, 2003). The unique contacts, never seen in the structures of related complexes, could contribute to the stabilization of the interactions between the BAFF/TALL-1/BlyS trimers. These intertrimeric interactions mediate the virus-like assembly of the ligand molecules (Kim *et al.*, 2003; Liu *et al.*, 2003), which is critical for the physiologic activity of the cytokine (Liu *et al.*, 2002).

IX. CONCLUSIONS AND PERSPECTIVES

Most of genes have been generated by gene duplication and diversification from common ancestral genes. Although a group of proteins belonging to a same family are related to each other in the primary sequence, the functions of individual members are usually different from each other. TNF and TNFR families are no exception to this notion. Because the two family members exert their functions through protein:protein interactions, they are an excellent system to study for probing into the molecular determinants that provide specificity of recognition between cognate partners. We discussed four different underlying molecular strategies enabling the highly specific protein:protein interactions between the two families. Later, unexpected mode of interactions could be found; the already found higher-order clustering of the BAFF-R by the virus-like assembly of the BAFF/TALL-1/BlyS trimer was a total surprise. The structures of TRAIL and TRAIL:sDR5 complex provide important information that could be extremely useful for developing TRAIL as a safe therapeutic agent for the treatment of cancers. Previous tests for the efficacy of TRAIL in animal models required injection of a large amount of the protein. By site-directed mutagenesis, it would be possible to improve the trimeric stability of TRAIL and to enhance the affinity of the protein for DR4 and DR5. Such TRAIL mutants may overcome this shortcoming.

ACKNOWLEDGMENTS

Supported by Creative Research Initiatives of the Korean Ministry of Science and Technology.

REFERENCES

Almasan, A., and Ashkenazi, A. (2003). Apo2L/TRAIL: Apoptosis signaling, biology, and potential for cancer therapy. *Cytokine Growth Factor Rev.* **14,** 337–348.

Ashkenazi, A. (2002). Targeting death and decoy receptors of the tumour-necrosis factor superfamily. *Nat. Rev. Cancer* **2,** 420–430.

Ashkenazi, A., Pai, R. C., Fong, S., Leung, S., Lawrence, D. A., Marsters, S. A., Blackie, C., Chang, L., McMurtrey, A. E., Hebert, A., DeForge, L., Koumenis, I. L., Lewis, D., Harris, L., Bussiere, J., Koeppen, H., Shahrokh, Z., and Schwall, R. H. (1999). Safety and antitumor activity of recombinant soluble Apo2 ligand. *J. Clin. Invest.* **104,** 155–162.

Banner, D. W., D'Arcy, A., Janes, W., Gentz, R., Schoenfeld, H. J., Broger, C., Loetscher, H., and Lesslauer, W. (1993). Crystal structure of the soluble human 55 kd TNF receptor-human TNF beta complex: Implications for TNF receptor activation. *Cell* **73,** 431–445.

Bodmer, J. L., Meier, P., Tschopp, J., and Schneider, P. (2000). Cysteine 230 is essential for the structure and activity of the cytotoxic ligand TRAIL. *J. Biol. Chem.* **275,** 20632–20637.

Bodmer, J. L., Schneider, P., and Tschopp, J. (2002). The molecular architecture of the TNF superfamily. *Trends Biochem. Sci.* **27,** 19–26.

Cha, S. S., Kim, J. S., Cho, H. S., Shin, N. K., Jeong, W., Shin, H. C., Kim, Y. J., Hahn, J. H., and Oh, B.-H. (1998). High resolution crystal structure of a human tumor necrosis factor-alpha mutant with low systemic toxicity. *J. Biol. Chem.* **273,** 2153–2160.

Cha, S. S., Kim, M. S., Choi, Y. H., Sung, B. J., Shin, N. K., Shin, H. C., Sung, Y. C., and Oh, B. H. (1999). 2.8 Å resolution crystal structure of human TRAIL, a cytokine with selective antitumor activity. *Immunity* **11,** 253–261.

Cha, S. S., Sung, B. J., Kim, Y. A., Song, Y. L., Kim, H. J., Kim, S., Lee, M. S., and Oh, B. H. (2000). Crystal structure of TRAIL-DR5 complex identifies a critical role of the unique frame insertion in conferring recognition specificity. *J. Biol. Chem.* **275,** 31171–31177.

Cretney, E., Takeda, K., Yagita, H., Glaccum, M., Peschon, J. J., and Smyth, M. J. (2002). Increased susceptibility to tumor initiation and metastasis in TNF-related apoptosis-inducing ligand-deficient mice. *J. Immunol.* **168,** 1356–1361.

Degli-Esposti, M. A., Dougall, W. C., Smolak, P. J., Waugh, J. Y., Smith, C. A., and Goodwin, R. G. (1997a). The novel receptor TRAIL-R4 induces NF-kappaB and protects against TRAIL-mediated apoptosis, yet retains an incomplete death domain. *Immunity* **7,** 813–820.

Degli-Esposti, M. A., Smolak, P. J., Walczak, H., Waugh, J., Huang, C. P., DuBose, R. F., Goodwin, R. G., and Smith, C. A. (1997b). Cloning and characterization of TRAIL-R3, a novel member of the emerging TRAIL receptor family. *J. Exp. Med.* **186,** 1165–1170.

Eck, M. J., and Sprang, S. R. (1989). The structure of tumor necrosis factor-alpha at 2.6 Å resolution. Implications for receptor binding. *J. Biol. Chem.* **264,** 17595–17605.

Eck, M. J., Ultsch, M., Rinderknecht, E., de Vos, A. M., and Sprang, S. R. (1992). The structure of human lymphotoxin (tumor necrosis factor-beta) at 1.9-A resolution. *J. Biol. Chem.* **267,** 2119–2122.

Emery, J. G., McDonnell, P., Burke, M. B., Deen, K. C., Lyn, S., Silverman, C., Dul, E., Appelbaum, E. R., Eichman, C., DiPrinzio, R., Dodds, R. A., James, I. E., Rosenberg, M., Lee, J. C., and Young, P. R. (1998). Osteoprotegerin is a receptor for the cytotoxic ligand TRAIL. *J. Biol. Chem.* **273,** 14363–14367.

Gordon, N. C., Pan, B., Hymowitz, S. G., Yin, J., Kelley, R. F., Cochran, A. G., Yan, M., Dixit, V. M., Fairbrother, W. J., and Starovasnik, M. A. (2003). BAFF/BLyS receptor 3 comprises a minimal TNF receptor-like module that encodes a highly focused ligand-binding site. *Biochemistry* **42,** 5977–5983.

Griffith, T. S., and Lynch, D. H. (1998). TRAIL: A molecule with multiple receptors and control mechanisms. *Curr. Opin. Immunol.* **10,** 559–563.

Holen, I., Croucher, P. I., Hamdy, F. C., and Eaton, C. L. (2002). Osteoprotegerin (OPG) is a survival factor for human prostate cancer cell. *Cancer Res.* **62,** 1619–1623.

Hymowitz, S. G., Christinger, H. W., Fuh, G., Ultsch, M., O'Connell, M., Kelley, R. F., Ashkenazi, A., and de Vos, A. M. (1999). Triggering cell death: The crystal structure of Apo2L/TRAIL in a complex with death receptor 5. *Mol. Cell* **4,** 563–571.

Hymowitz, S. G., O'Connell, M. P., Ultsch, M. H., Hurst, A., Totpal, K., Ashkenazi, A., de Vos, A. M., and Kelley, R. F. (2000). A unique zinc-binding site revealed by a high-resolution X-ray structure of homotrimeric Apo2L/TRAIL. *Biochemistry* **39,** 633–640.

Ito, S., Wakabayashi, K., Ubukata, O., Hayashi, S., Okada, F., and Hata, T. (2002). Crystal structure of the extracellular domain of mouse RANK ligand at 2.2-A resolution. *J. Biol. Chem.* **277,** 6631–6636.

Jo, M., Kim, T. H., Seol, D. W., Esplen, J. E., Dorko, K., Billiar, T. R., and Strom, S. C. (2000). Apoptosis induced in normal human hepatocytes by tumor necrosis factor-related apoptosis-inducing ligand. *Nat. Med.* **6,** 564–567.

Jones, E. Y., Stuart, D. I., and Walker, N. P. (1989). Structure of tumour necrosis factor. *Nature* **338,** 225–228.

Karpusas, M., Cachero, T. G., Qian, F., Boriack-Sjodin, A., Mullen, C., Strauch, K., Hsu, Y. M., and Kalled, S. L. (2002). Crystal structure of extracellular human BAFF, a TNF family member that stimulates B lymphocytes. *J. Mol. Biol.* **15,** 1145–1154.

Karpusas, M., Hsu, Y.-M., Wang, J., Thompson, J., Lederman, S., Chess, L., and Thomas, D. (1995). 2 Å crystal structure of an extracellular fragment of human CD40 ligand. *Structure* **3,** 1031–1039.

Kim, H. M., Yu, K. S., Lee, M. E., Shin, D. R., Kim, Y. S., Paik, S. G., Yoo, O. J., Lee, H., and Lee, J. O. (2003). Crystal structure of the BAFF-BAFF-R complex and its implications for receptor activation. *Nat. Struct. Biol.* **10,** 342–348.

Kischkel, F. C., Lawrence, D. A., Chuntharapai, A., Schow, P., Kim, K. J., and Ashkenazi, A. (2000). Apo2L/TRAIL-dependent recruitment of endogenous FADD and caspase-8 to death receptors 4 and 5. *Immunity* **12,** 611–620.

Lam, J., Nelson, C. A., Ross, F. P., Teitelbaum, S. L., and Fremont, D. H. (2001). Crystal structure of the TRANCE/RANKL cytokine reveals determinants of receptor-ligand specificity. *J. Clin. Invest.* **108,** 971–979.

Lawrence, D., Shahrokh, Z., Marsters, S., Achilles, K., Shih, D., Mounho, B., Hillan, K., Totpal, K., DeForge, L., Schow, P., Hooley, J., Sherwood, S., Pai, R., Leung, S., Khan, L., Gliniak, B., Bussiere, J., Smith, C. A., Strom, S. S., Kelley, S., Fox, J. A., Thomas, D., and Ashkenazi, A. (2001). Differential hepatocyte toxicity of recombinant Apo2L/TRAIL versions. *Nat. Med.* **7,** 383–385.

LeBlanc, H. N., and Ashkenazi, A. (2003). Apo2L/TRAIL and its death and decoy receptors. *Cell Death Differ.* **10,** 66–75.

Lincz, L. F., Yeh, T. X., and Spencer, A. (2001). TRAIL-induced eradication of primary tumour cells from multiple myeloma patient bone marrows is not related to TRAIL receptor expression or prior chemotherapy. *Leukemia* **15**, 1650–1657.

Liu, Y., Hong, X., Kappler, J., Jiang, L., Zhang, R., Xu, L., Pan, C. H., Martin, W. E., Murphy, R. C., Shu, H. B., Dai, S., and Zhang, G. (2003). Ligand-receptor binding revealed by the TNF family member TALL-1. *Nature* **423**, 49–56.

Liu, Y., Xu, L., Opalka, N., Kappler, J., Shu, H. B., and Zhang, G. (2002). Crystal structure of sTALL-1 reveals a virus-like assembly of TNF family ligands. *Cell* **108**, 383–394.

Locksley, R. M., Killeen, N., and Lenardo, M. J. (2001). The TNF and TNF receptor superfamilies: Integrating mammalian biology. *Cell* **104**, 487–501.

Marsters, S. A., Sheridan, J. P., Pitti, R. M., Huang, A., Skubatch, M., Baldwin, D., Yuan, J., Gurney, A., Goddard, A. D., Godowski, P., and Ashkenazi, A. (1997). A novel receptor for Apo2L/TRAIL contains a truncated death domain. *Curr. Biol.* **7**, 1003–1006.

Mongkolsapaya, J., Cowper, A. E., Xu, X. N., Morris, G., McMichael, A. J., Bell, J. I., and Screaton, G. R. (1998). Lymphocyte inhibitor of TRAIL (TNF-related apoptosis-inducing ligand): A new receptor protecting lymphocytes from the death ligand TRAIL. *J. Immunol.* **160**, 3–6.

Mongkolsapaya, J., Grimes, J. M., Chen, N., Xu, X. N., Stuart, D. I., Jones, E. Y., and Screaton, G. R. (1999). Structure of the TRAIL-DR5 complex reveals mechanisms conferring specificity in apoptotic initiation. *Nat. Struct. Biol.* **6**, 1048–1053.

Naismith, J. H., Devine, T. Q., Kohno, T., and Sprang, S. R. (1996). Structures of the extracellular domain of the type I tumor necrosis factor receptor. *Structure* **4**, 1251–1262.

Oren, D. A., Li, Y., Volovik, Y., Morris, T. S., Dharia, C., Das, K., Galperina, O., Gentz, R., and Arnold, E. (2002). Structural basis of BLyS receptor recognition. *Nat. Struct. Biol.* **9**, 288–292.

Pan, G., Ni, J., Wei, Y. F., Yu, G., Gentz, R., and Dixit, V. M. (1997a). An antagonist decoy receptor and a death domain-containing receptor for TRAIL. *Science* **277**, 815–818.

Pan, G., O'Rourke, K., Chinnaiyan, A. M., Gentz, R., Ebner, R., Ni, J., and Dixit, V. M. (1997b). The receptor for the cytotoxic ligand TRAIL. *Science* **276**, 111–113.

Pitti, R. M., Marsters, S. A., Ruppert, S., Donahue, C. J., Moore, A., and Ashkenazi, A. (1996). Induction of apoptosis by Apo-2 ligand, a new member of the tumor necrosis factor cytokine family. *J. Biol. Chem.* **271**, 12687–12690.

Schuchmann, M., Hess, S., Bufler, P., Brakebusch, C., Wallach, D., Porter, A., Riethmuller, G., and Engelmann, H. (1995). Functional discrepancies between tumor necrosis factor and lymphotoxin alpha explained by trimer stability and distinct receptor interactions. *Eur. J. Immunol.* **25**, 2183–2189.

Sedger, L. M., Glaccum, M. B., Schuh, J. C., Kanaly, S. T., Williamson, E., Kayagaki, N., Yun, T., Smolak, P., Le, T., Goodwin, R., and Gliniak, B. (2002). Characterization of the in vivo function of TNF-alpha-related apoptosis-inducing ligand, TRAIL/Apo2L, using TRAIL/Apo2L gene-deficient mice. *Eur. J. Immunol.* **32**, 2246–2254.

Sedger, L. M., Shows, D. M., Blanton, R. A., Peschon, J. J., Goodwin, R. G., Cosman, D., and Wiley, S. R. (1999). IFN-gamma mediates a novel antiviral activity through dynamic modulation of TRAIL and TRAIL receptor expression. *J. Immunol.* **163**, 920–926.

Seol, D. W., and Billiar, T. R. (2000). Cysteine 230 modulates tumor necrosis factor-related apoptosis-inducing ligand activity. *Cancer Res.* **60**, 3152–3154.

Sheridan, J. P., Marsters, S. A., Pitti, R. M., Gurney, A., Skubatch, M., Baldwin, D., Ramakrishnan, L., Gray, C. L., Baker, K., Wood, W. I., Goddard, A. D., Godowski, P., and Ashkenazi, A. (1997). Control of TRAIL-induced apoptosis by a family of signaling and decoy receptors. *Science* **277**, 818–821.

Shipman, C. M., and Croucher, P. I. (2003). Osteoprotegerin is a soluble decoy receptor for tumor necrosis factor-related apoptosis-inducing ligand/Apo2 ligand and can function as a paracrine survival factor for human myeloma cells. *Cancer Res.* **63**, 912–916.

Simonet, W. S., Lacey, D. L., Dunstan, C. R., Kelley, M., Chang, M. S., Luthy, R., Nguyen, H. Q., Wooden, S., Bennett, L., Boone, T., Shimamoto, G., DeRose, M., Elliott, R., Colombero, A., Tan, H. L., Trail, G., Sullivan, J., Davy, E., Bucay, N., Renshaw-Gegg, L., Hughes, T. M., Hill, D., Pattison, W., Campbell, P., Sander, S., Van, G., Tarpley, J., Derby, P., Lee, R., and Boyle, W. J. (1997). Osteoprotegerin: A novel secreted protein involved in the regulation of bone density. *Cell* **89,** 309–319.

Sprick, M. R., Rieser, E., Stahl, H., Grosse-Wilde, A., Weigand, M. A., and Walczak, H. (2002). Caspase-10 is recruited to and activated at the native TRAIL and CD95 death-inducing signalling complexes in a FADD-dependent manner but can not functionally substitute caspase-8. *EMBO J.* **21,** 4520–4530.

Sprick, M. R., Weigand, M. A., Rieser, E., Rauch, C. T., Juo, P., Blenis, J., Krammer, P. H., and Walczak, H. (2000). FADD/MORT1 and caspase-8 are recruited to TRAIL receptors 1 and 2 and are essential for apoptosis mediated by TRAIL receptor 2. *Immunity* **12,** 599–609.

Takeda, K., Hayakawa, Y., Smyth, M. J., Kayagaki, N., Yamaguchi, N., Kakuta, S., Iwakura, Y., Yagita, H., and Okumura, K. (2001). Involvement of tumor necrosis factor-related apoptosis-inducing ligand in surveillance of tumor metastasis by liver natural killer cells. *Nat. Med.* **7,** 94–100.

Takeda, K., Smyth, M. J., Cretney, E., Hayakawa, Y., Kayagaki, N., Yagita, H., and Okumura, K. (2002). Critical role for tumor necrosis factor-related apoptosis-inducing ligand in immune surveillance against tumor development. *J. Exp. Med.* **195,** 161–169.

Trabzuni, D., Famulski, K. S., and Ahmad, M. (2000). Functional analysis of tumour necrosis factor-alpha-related apoptosis-inducing ligand (TRAIL): Cysteine-230 plays a critical role in the homotrimerization and biological activity of this novel tumoricidal cytokine. *Biochem. J.* **350,** 505–510.

Truneh, A., Sharma, S., Silverman, C., Khandekar, S., Reddy, M. P., Deen, K. C., McLaughlin, M. M., Srinivasula, S. M., Livi, G. P., Marshall, L. A., Alnemri, E. S., Williams, W. V., and Doyle, M. L. (2000). Temperature-sensitive differential affinity of TRAIL for its receptors: DR5 is the highest affinity receptor. *J. Biol. Chem.* **275,** 23319–23325.

Walczak, H., Degli-Esposti, M. A., Johnson, R. S., Smolak, P. J., Waugh, J. Y., Boiani, N., Timour, M. S., Gerhart, M. J., Schooley, K. A., Smith, C. A., Goodwin, R. G., and Rauch, C. T. (1997). TRAIL-R2: A novel apoptosis-mediating receptor for TRAIL. *EMBO J.* **16,** 5386–5397.

Walczak, H., Miller, R. E., Ariail, K., Gliniak, B., Griffith, T. S., Kubin, M., Chin, W., Jones, J., Woodward, A., Le, T., Smith, C., Smolak, P., Goodwin, R. G., Rauch, C. T., Schuh, J. C., and Lynch, D. H. (1999). Tumoricidal activity of tumor necrosis factor-related apoptosis-inducing ligand in vivo. *Nat. Med.* **5,** 157–163.

Wiley, S. R., Schooley, K., Smolak, P. J., Din, W. S., Huang, C. P., Nicholl, J. K., Sutherland, G. R., Smith, T. D., Rauch, C., and Smith, C. A. (1995). Identification and characterization of a new member of the TNF family that induces apoptosis. *Immunity* **3,** 673–682.

2

Crystal Structure of RANK Ligand Involved in Bone Metabolism

Shuichiro Ito and Tadashi Hata

*Biomedical Research Laboratories, Sankyo Co., Ltd.
Tokyo 140-8710, Japan*

I. Introduction
II. The OPG/RANKL/RANK System
   A. OPG (The Decoy Receptor, Belonging to the TNF Receptor Family)
   B. RANKL (The Ligand, Belonging to the TNF Ligand Family)
   C. RANK (The Receptor, Belonging to the TNF Receptor Family)
   D. Adaptor Proteins in RANK Signaling
III. Three-Dimensional Structure of RANKL
   A. Overall Structure
   B. Receptor Binding Sites
   C. Conserved Interaction with the Receptor
   D. Residue Specific to TNF-β, but Nonspecific to TRAIL
   E. Specific Interaction to both TNF-β and TRAIL
   F. Mutagenesis Studies
IV. Conclusion
   References

Bone remodeling involves the resorption of bone by osteoclasts and the synthesis of bone matrix by osteoblasts. Recently, an essential cytokine system for osteoclast biology has been identified and extensively characterized. This system consists of a ligand, receptor activator of NF-κB ligand (RANKL), a receptor, RANK, and its soluble decoy receptor, osteoprotegerin (OPG). RANKL, a member of the tumor necrosis factor (TNF) family, triggers osteoclastogenesis by forming a complex with RANK, a member of the TNF receptor family. Because members of the TNF family have the same topology and the extracellular domains of the TNF receptor family members also adopt the same structural scaffold, in addition to their rapid increase in the number, this poses an intriguing question of how recognition between cognate ligands and receptors is achieved in a highly specific manner. Structural studies on the mouse RANKL extracellular domain showed that the RANKL is trimeric, and each subunit has a β-strand jellyroll topology like the other members of the TNF family. A comparison of RANKL with TNF-β and TNF-related apoptosis-inducing ligand (TRAIL), whose structures were determined to be in the complex form with their respective receptor, revealed conserved and specific features of RANKL in the TNF superfamily. Residues important for receptor binding and activation have also been confirmed by mutagenesis experiments. Further structural and mutational studies on the RANKL/RANK/OPG system will provide useful information for developing drug candidates that inhibit osteoclastogenesis and mediate problems of bone metabolism. © 2004 Elsevier Inc.

I. INTRODUCTION

Bone is a dynamic organ that is continuously degraded (resorbed) and reconstructed (synthesized). Bone remodeling involves the resorption of bone by osteoclasts and the synthesis of bone matrix by osteoblasts (Suda et al., 1992). Osteoclasts and osteoblasts arise from distinct cell lineages and maturation processes. Osteoclasts differentiate from hematopoietic precursors of the monocyte/macrophage lineage, whereas osteoblasts arise from mesenchymal stem cells. Osteoclasts are formed *in vitro* from osteoclast progenitors by coculturing with osteoblasts or bone marrow stromal cells in the presence of stimulators, including colony-stimulating factor-1 (CSF-1 or M-CSF), interleukin (IL)-6, IL-11, parathyroid hormone, prostaglandin E2, and vitamin D3. The activation of osteoclasts is regulated by osteoblasts/stromal cells through a mechanism of cell-to-cell interaction with osteoclast progenitors. An imbalance of osteoclast and osteoblast activity results in certain diseases, including osteoporosis.

An essential cytokine system in osteoclast differentiation and activation has been identified and characterized. This system consists of members belonging to the tumor necrosis factor (TNF) ligand family and TNF receptor family: a ligand, receptor activator of NF-κB ligand (RANKL); a receptor, RANK; and a soluble decoy receptor, osteoprotegerin (OPG). RANKL induces osteoclastogenesis by sending signals to osteoclast progenitors. The interaction between RANKL expressed by osteoblasts/stromal cells and its receptor, RANK, expressed on osteoclast precursors is essential for osteoclastogenesis. OPG is a decoy receptor of RANKL and inhibits osteoclast maturation. RANKL, RANK, and OPG play critical roles in bone metabolism.

The TNF ligand family binds to the extracellular domain of the TNF receptor family and triggers the receptor activation and the intracellular cascade (Locksley *et al.*, 2001). Despite diversity in the primary structure, TNF ligand family members are thought to adopt a generally similar tertiary fold, namely a β-strand jellyroll topology. Furthermore, based on sequence homology, as all the TNF receptor family members have cysteine-rich domains (CRDs) in their extracellular domains, they also thought to adopt the same structural scaffold. Despite the same structural scaffold, the recognition between cognate ligands and receptors is achieved in a highly specific manner with typical dissociation constants in the nanomolar to picomolar range. The emerging complexity of the TNF and TNF receptor families raises questions, at the molecular level, of how recognition between cognate ligands and receptors is achieved in a highly specific manner. The mechanism by which these molecules achieve specificity cannot be understood without knowledge of their three-dimensional structures.

In this chapter, we describe the roles of the three proteins, RANKL, RANK, and OPG, in bone metabolism. Then, we discuss the conserved and specific interactions between the ligands and the receptors based on the crystal structure of RANKL and the comparison with the other ligands, TNF-β and TNF-related apoptosis-inducing ligand (TRAIL) in complex with their respective receptors.

II. THE OPG/RANKL/RANK SYSTEM

A. OPG (THE DECOY RECEPTOR, BELONGING TO THE TNF RECEPTOR FAMILY)

OPG (also known as osteoclastogenesis inhibitory factor [OCIF]) was isolated and cloned independently by two groups (Simonet *et al.*, 1997; Tsuda *et al.*, 1997; Yasuda *et al.*, 1998a). OPG was found to be a novel member of the TNF receptor family. Transgenic mice overexpressing OPG and animals injected with OPG (Simonet *et al.*, 1997; Yasuda *et al.*, 1998a) demonstrated that this protein increases bone mass and suppresses bone resorption associated with osteoclast development. In addition, OPG knockout mice showed severe osteoporosis due to enhanced osteoclastogenesis (Bucay et al., 1998; Mizuno et al., 1998). These results suggest that OPG is a physiologically important inhibitor in osteoclastic bone resorption.

OPG is initially synthesized as a 401-amino acid peptide, with a 21-amino acid propeptide that is cleaved, resulting in a mature protein of 380 amino acids (Simonet et al., 1997; Yasuda et al., 1998a). Sequence identity between human OPG and mouse OPG is 86%. In contrast to all other TNF receptor superfamily members, OPG lacks transmembrane and cytoplasmic domains and is secreted as a soluble protein. OPG contains four CRDs (CRD1–CRD4) in the N-terminal region. The C-terminal region contains two death domain homologous (DDH1 and DDH2) regions, a heparin binding domain, and a cysteine residue (Cys400) necessary for dimerization (Simonet et al., 1997; Yamaguchi et al., 1998; Yasuda et al., 1998a). OPG can exist as a monomer or dimer as a result of the formation of a disulfide bond with Cys400. The mutation of Cys400 to Ser does not affect the osteoclastogenesis-inhibiting activity of OPG, suggesting that dimerization of OPG is not necessary for its activity. Furthermore, analysis of the domain-deletion mutants of OPG revealed that CRDs but not the DDH regions are essential for inducing biologic activity *in vitro*. When the transmembrane domain of Fas was inserted between the CRDs and DDH regions and the mutant protein was expressed, apoptosis was induced in the transfected cells (Yamaguchi et al., 1998). The physiologic significance of the DDH region in OPG is still not known.

B. RANKL (THE LIGAND, BELONGING TO THE TNF LIGAND FAMILY)

Using the decoy receptor, OPG, as a probe, the ligand was isolated and cloned independently by two groups and named osteoprotegerin ligand (OPGL) or osteoclast differentiation factor (ODF) (Lacey et al., 1998; Yasuda et al., 1998b). Molecular cloning of the ligand also revealed this protein to be identical to RANKL, which was first cloned as a regulatory factor of dendritic cells (Anderson et al., 1997) and to TNF-related activation-induced cytokine (TRANCE), which was cloned as a regulatory factor of T cells (Wong et al., 1997), and thus considered a novel member of the TNF ligand family.

The committee for standard nomenclature adopted the name of RANKL for the ligand (The American Society for Bone and Mineral Research President's Committee on Nomenclature, 2000). RANKL, belonging to the TNF ligand family, is a type II transmembrane protein with a short cytoplasmic tail and transmembrane region that is linked to the extracellular domain by a stalk region. Human and mouse RANKL have 317 and 316 amino acid residues, respectively, and share 87% identity.

The major role of RANKL in bone is the stimulation of osteoclast differentiation (Lacey et al., 1998; Malyankar et al., 2000), enhancement of the activity of mature osteoclasts (Lacey et al., 1998), and inhibition of osteoclast apoptosis (Fuller et al., 1998). Indeed, in the presence of low levels of macrophage-colony stimulating factor (M-CSF), RANKL appears to be both necessary and sufficient for the complete differentiation of osteoclast precursor cells into mature osteoclasts (Lacey et al., 1998; Yasuda et al., 1998b). In addition, RANKL has a number of effects on immune cells, including induction of cluster formation by dendritic cells, and effects on cytokine-activated T cell proliferation (Anderson et al., 1997). RANKL knockout mice exhibited typical osteopetrosis and lacked osteoclasts but had normal osteoblast stromal cells (Kong et al., 1999). These facts suggest that RANKL is an absolute requirement for osteoclast development.

C. RANK (THE RECEPTOR, BELONGING TO THE TNF RECEPTOR FAMILY)

The receptor for RANKL is RANK, a member of the TNF receptor family. The interaction between RANKL expressed by osteoblasts/stroma cells and its receptor RANK expressed by osteoclast precursors is essential for osteoclastogenesis. RANK knockout mice exhibited osteopetrosis due to an absence of osteoclasts (Li et al., 2000). Human RANK is a 616-amino acid peptide, with a 28-amino acid signal peptide, an N-terminal extracellular domain, a short transmembrane domain of 21 amino acids, and a large C-terminal cytoplasmic domain (Anderson et al., 1997). Like OPG, RANK contains four extracellular CRDs.

D. ADAPTOR PROTEINS IN RANK SIGNALING

Recent studies indicate that the cytoplasmic tail of RANK interacts with TNF receptor-associate factor 1 (TRAF1), TRAF2, TRAF3, TRAF5, and TRAF6 (Galibert et al., 1998; Kim et al., 1999). Among the TRAF proteins, TRAF6 is critical for RANK signaling in osteoclasts because a knockout resulted in compromised differentiation and defective activation of osteoclasts (Naito et al., 1999). TRAF6 is a unique TRAF protein that participates in the signal transduction through both the TNF receptor family and the IL-1 receptor/Toll-like receptor family (Cao et al., 1996a; Cao et al., 1996b). Crystal structures of TRAF6 alone and in complexes with TRAF6-binding peptides from CD40 and RANK have been determined by X-ray crystallography, showing that the peptides from CD40 and RANK bind to the same crevice in TRAF6 (Ye et al., 2002). CD40 also binds to TRAF2 (McWhirter et al., 1999), but in an orientation

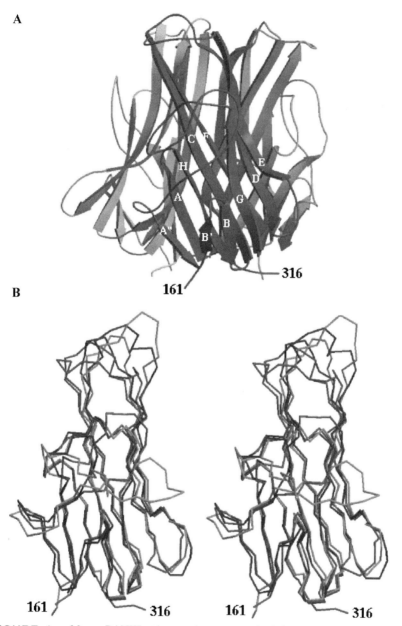

FIGURE 1. Mouse RANKL trimer and monomer. A, Schematic view of the mouse RANKL trimer shown with strands as arrows and labeled by the notation of Eck and Sprang, 1989. In this orientation, the cell surface containing mouse RANKL is located at the bottom of

differing by about 40 degrees from TRAF6, which explains the marked structural differences among receptor recognition by TRAF6 and other TRAFs (Ye *et al.*, 2002).

The first step in signaling by members of the TNF receptor family was thought to involve ligand-induced trimerization of the receptor. However, recent reports demonstrate that TNF receptor and Fas in unstimulated cells are constitutively oligomerized through a self-association of domain, CRD1, termed the preligand assembly domain, before ligand binding and that the preassembled receptors are nonsignaling (Chan *et al.*, 2000; Siegel *et al.*, 2000). It is possible that ligand binding induces conformational changes of the oligomerized subunits and/or the formation of oligomers different from those present in the unstimulated state. Currently, the detailed structural mechanism of receptor activation by ligand remains unclear.

Recent reviews (Boyle *et al.*, 2003; Khosla, 2001; Lee and Kim, 2003; Suda *et al.*, 1999; Theill *et al.*, 2002) contain more details and complete references on the four classes of proteins involved in this system.

III. THREE-DIMENSIONAL STRUCTURE OF RANKL

A. OVERALL STRUCTURE

The crystal structure of the mouse RANKL extracellular domain has been determined by X-ray crystallography (Ito *et al.*, 2002; Lam *et al.*, 2001) and deposited in the Protein Data Bank (PDB, *www.rcsb.org/*) as codes 1JTZ and 1IQA. In the crystal structure, RANKL is a trimer (Fig. 1A), which was also shown by analytical ultracentrifugation (Ito *et al.*, 2002; Willard *et al.*, 2000). Like the other members of the TNF-β ligand family, the RANKL subunit has a β-strand jellyroll topology (Fig. 1A). The RANKL monomer is composed of two β-sheets; one is the inner β-sheet composed of the β-strands A″, A, H, C, and F, and the other is the outer β-sheet composed of β-strands B′, B, G, D, and E. The inner β-sheet is involved in the intersubunit contacts.

Despite the low sequence identity, the overall structure of RANKL is similar to that of the other TNF family members (Fig. 1B). The extracellular domain of RANKL has 23% sequence identity with TNF-β and 33% with TRAIL. In comparing the structure of RANKL to those of TNF-β (PDB

---

the figure. B, Stereoview of the superposition of the Cα traces for mouse RANKL, TNF-β, and TRAIL monomers. The figures were prepared with the programs Molscript (Kraulis, 1991) and Raster3D (Merrit and Murphy, 1994). (Adapted from Ito *et al.*, 2002.)

code 1TNR [Banner *et al.*, 1993]) and TRAIL (PDB code 1DV4 [Mongkolsapaya *et al.*, 1999]) in the complex forms with their respective receptors, the root mean square deviations of RANKL are 1.17 Å with TNF-β (126 equivalent Cα atoms) and 1.27 Å with TRAIL (131 equivalent Cα atoms). The trimeric interface consists of a 7440 Å^2 area (2480 Å^2/subunit) and has a highly hydrophobic character with a high proportion of aromatic residues. The trimeric interface is well conserved in the TNF family, whereas the conformations of the surface loops are extremely variable (Fig. 1B).

The loops that have large differences are the AA" loop, linking the β-strands of A and A" and, likewise, the DE loop and EF loop. The largest difference in the sequence alignment is the length of the AA" loop; that of RANKL is seven residues longer than that of TNF-β and nine residues shorter than that of TRAIL.

B. RECEPTOR BINDING SITES

The bindings of RANKL to its receptor RANK and decoy receptor OPG are expected to be similar to those of the TNF-β–TNF receptor-1 (TNFR1) and TRAIL-DR5 complexes. The two crystal structures of the TNF-β–TNFR1 and TRAIL-DR5 complexes show that the receptors bind to the ligand diagonally along the crevices in the subunit-subunit interface of the ligands (Fig. 2A) (Banner *et al.*, 1993; Cha *et al.*, 1999; Hymowitz *et al.*, 1999; Mongkolsapaya *et al.*, 1999). Residues of the ligands that are involved in receptor binding are in the AA" loop, the C-terminal region of the C strand, the N terminus of the D strand, the DE loop, the N terminus of the F strand, and the GH loop. Members of the TNF receptor family are characterized by extracellular CRD repeats containing three disulfide bridges with a cysteine-knot topology. The number of repeats in TNFR1 is four, three in DR5, and four in RANK and OPG. Among the repeats, the second and third CRDs (CRD2 and CRD3) are known to be important for ligand binding (Fig. 2B, 2C).

Three important residues for the ligand-receptor interaction were proposed (Ito *et al.*, 2002), and mutagenesis of selected residues in the loops significantly modulated receptor activation (Lam *et al.*, 2001). The proposed residues in mouse RANKL are Ile248 in the DE loop for a conserved interaction, Gln236 in the N-terminal region of the D strand, and Lys180 in the AA" loop for a specific interaction. All of these three residues are conserved in human RANKL. In addition to the ligand, two loops in the receptors of RANK and OPG can also be considered important for ligand interaction. They are the loop in the CRD2 for a conserved interaction and the loop in the CRD3 for a specific interaction.

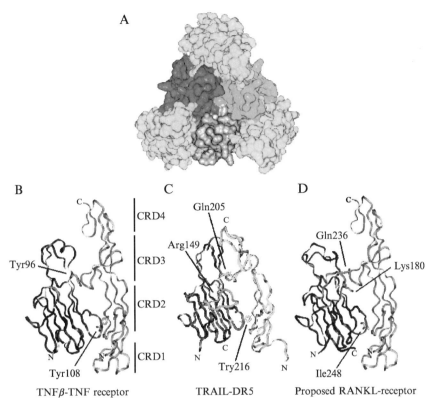

FIGURE 2. Interactions between the TNF and TNF receptor families. A, Surface representation of the crystal structure of the TNF-β–TNFR1 complex. The three monomers of TNF-β located at the center are depicted in different shades of gray. The three TNFR1 monomers are situated in the outside of the TNF-β trimer so that they are inserted between the two monomers of TNF-β. B, Ribbon rendering of the TNF-β and TNFR1 complex. TNFR1 is at the right side and TNF-β is at the left side. Tyr-96 in the N-terminal region of the D strand and Tyr-108 in the DE loop are shown as stick models. The approximate boundaries of CRD1, CRD2, CRD3, and CRD4 of TNFR1 are marked by bars. Shown is the side view of A. C, Ribbon rendering of the TRAIL and DR5 complex. Arg-149 in the AA″ loop, Gln-205 in the N-terminal region of the D strand, and Tyr-216 in the DE loop are shown as stick models. D, Ribbon rendering of the proposed mRANKL and receptor complex. The figure is generated after a superposition of mRANKL with TNF-β of the TNF-β–TNFR1 complex and shows mRANKL with TNFR1. mRANKL is located at the left side, and TNFR1 is located at the right side. Lys-180 in the AA″ loop, Gln-236 in the N-terminal region of the D strand, and Ile-248 in the DE loop are shown as stick models.

C. CONSERVED INTERACTION WITH THE RECEPTOR

Although the charge distributions on the surface of mouse RANKL, TNF-β, and TRAIL are very different, the DE loop region of mouse RANKL has a hydrophobic character as seen in TNF-β and TRAIL. These loop regions of TNF-β and TRAIL bind to their respective receptors mainly through hydrophobic interactions, suggesting that RANKL also interacts with OPG and RANK in the same manner. Mutagenesis studies show that Tyr-108 in TNF-β and Tyr-216 in TRAIL are critical for receptor binding (Hymowitz et al., 2000; Van Ostade et al., 1994). In the ligand-receptor complex structures, Tyr-108 in the DE loop of TNF-β directly interacts with residues of the receptor, and Tyr-216 in TRAIL also interacts with residues in the hydrophobic groove of the receptor surface in CRD2 (Fig. 2B, 2C). Both mouse RANKL and human RANKL have a hydrophobic residue, Ile (Ile-248 in mouse RANKL and Ile-249 in human RANKL), at an equivalent position to Tyr-108 in TNF-β and Tyr-216 in TRAIL. Despite its hydrophobic character, the DE loop is exposed to the solvent. The DE loop appears to be flexible in the free state but upon binding of the receptor through hydrophobic interactions, it becomes ordered in the complex form.

D. RESIDUE SPECIFIC TO TNF-β, BUT NONSPECIFIC TO TRAIL

Another critical residue of mouse RANKL for receptor binding is considered to be Gln-236 (Gln-237 in human RANKL), which lies in the N-terminal region of the D strand. Gln-205 in TRAIL forms hydrogen bonds to the backbone of Glu-151 and Met-152 of DR5 in the TRAIL-DR5 complex (Cha et al., 2000; Hymowitz et al., 1999; Mongkolsapaya et al., 1999). Mutating Gln-205 in TRAIL to Ala decreases its binding activity (Hymowitz et al., 2000). In contrast, TNF-β has a tyrosine residue (Tyr-96) at this position and binds to TNFR1 through van der Waals interaction. The residues of RANK and OPG equivalent to Glu-151 and Met-152 of DR5 are also acidic and hydrophobic, respectively (Glu-126 and Cys-127 in mouse RANK, Glu-125 and Cys-126 in human RANK, Glu-116 and Phe-117 in mouse OPG and human OPG, respectively). Gln-236 of mouse RANKL may form hydrogen bonds to RANK and OPG in the same manner as that of TRAIL in the TRAIL-DR5 complex.

Cross-reactivity of TRAIL to DR5 and OPG (the decoy receptor for RANKL) has been observed (Emery et al., 1998; Truneh et al., 2000), although the physiologic relevance of this observation is still unclear. TRAIL binds to DR5 with an affinity of $K_D = 2$ nM and also binds to OPG with an affinity of $K_D = 400$ nM, which were determined at 37°C by isothermal titration microcalorimetry (Truneh et al., 2000). The explanation for this cross-reactivity may lie in the identity of the residue in the N-terminal region of the D strand, Gln-236 in mouse RANKL, Gln-237 in human RANKL, and Gln-205 in TRAIL, accompanied by the similar sequence of the receptor loop.

E. SPECIFIC INTERACTION TO BOTH TNF-β AND TRAIL

The most dramatic difference among the three ligands is the length of the AA″ loop. Arg-149 of TRAIL in the AA″ loop, which has the longest insertion, makes a salt bridge with Glu-147 in DR5 (Fig. 2C) (Cha *et al.*, 2000; Hymowitz *et al.*, 1999; Mongkolsapaya *et al.*, 1999), whereas the shorter AA″ loop of TNF-β has no such interaction in the TNF-β–TNFR1 complex (Fig. 2B). The deletion mutant of TRAIL, which lacks the AA″ loop, had a completely abolished binding activity (Cha *et al.*, 1999; Mongkolsapaya *et al.*, 1999). The residue in RANKL, at a similar position to Arg-149 of TRAIL, is a lysine residue (Lys-180 of mouse RANKL and Lys-181 of human RANKL). Furthermore, residues at a similar position to Glu-147 of DR5 are Asp-124 in mouse RANK, Asp-123 in human RANK, and Glu-114 in mouse OPG and human OPG. Superposition of mouse RANKL with TNF-β in complex with TNFR1 shows that Lys-180 in the long AA″ loop is able to form a possible interaction with the receptor, presumably through a salt bridge with Asp in RANK and Glu in OPG. (Fig. 2D). In TRAIL, there were alternate conformations in several surface loops that appear to be a consequence of receptor binding. Among those loops, the AA″ loop is disordered in the free TRAIL structure but is ordered in the complex structure (Cha *et al.*, 2000; Mongkolsapaya *et al.*, 1999). The AA″ loop of mouse RANKL is also likely to undergo a conformational change upon receptor binding.

F. MUTAGENESIS STUDIES

Structure-based mutational experiments in mouse RANKL have been reported (Lam *et al.*, 2001). In these studies, three mutations were generated to affect the biologic function of RANKL: a single amino acid substitution in the DE loop, replacing Ile248 with Asp (I248D); a deletion of the AA″ loop, from Gly177 to Leu183 (AA″ loop deletion); and a replacement of the AA″ loop of RANKL (residues 177–183) with that of TNF-β (AA″ loop swap). A single amino acid substitution in the DE loop significantly lowered the potency of RANKL and the mutants with a deletion or substitution of the AA″ loop failed to induce osteoclast precursors to differentiate *in vitro* (Lam *et al.*, 2001).

IV. CONCLUSION

RANKL, a member of the TNF family, triggers osteoclastogenesis by forming a complex with RANK, a member of the TNF receptor family. OPG is a decoy receptor of RANKL and inhibits osteoclast maturation. RANKL, RANK, and OPG play critical roles in bone metabolism. The crystal structure of the mouse RANKL extracellular domain showed that RANKL is trimeric, and each subunit adopts a β-strand jellyroll topology like the other members of the TNF family (Ito et al., 2002; Lam et al., 2001). A comparison of RANKL with TNF-β and TRAIL revealed conserved and specific features of RANKL in the TNF superfamily. Structure-based mutagenesis experiments confirmed the key regions for receptor binding and activation: deletion or swap of the AA" loop completely abolished the osteoclastogenesis activity and the point mutation in the DE loop decreased the affinity by eightfold (Lam et al., 2001).

Inhibition of the RANKL-RANK binding is expected to provide a useful drug to mediate bone-related disorders including osteoporosis. Currently, the decoy receptor OPG is under investigation for human therapy as an osteoclastogenesis inhibitor and mediator of the bone metabolism problems. Based on the structural and mutational studies, the two major binding sites between RANKL and RANK may be separated by a distance, at most, of 30 Å. Therefore, one implication of these studies is that it may be difficult to produce small molecules such as RANKL mimetics. However, further structural and mutational studies on the RANKL/RANK/OPG system will help generate small molecules as drug candidates, that will inhibit osteoclastogenesis and mediate disorders of bone metabolism.

REFERENCES

American Society for Bone and Mineral Research President's Committee on Nomenclature (2000). Proposed standard nomenclature for new tumor necrosis factor family members involved in the regulation of bone resorption. *J. Bone Miner. Res.* **12**, 2293–2296.

Anderson, D. M., Maraskovsky, E., Billingsley, W. L., Dougall, W. C., Tometsko, M. E., Roux, E. R., Teepe, M. C., DuBose, R. F., Cosman, D., and Galibert, L. (1997). A homologue of the TNF receptor and its ligand enhance T-cell growth and dendritic-cell function. *Nature* **390**, 175–179.

Banner, D. W., D'Arcy, A., Janes, W., Gentz, R., Schoenfeld, H. J., Broger, C., Loetscher, H., and Lesslauer, W. (1993). Crystal structure of the soluble human 55 kd TNF receptor–human TNF beta complex: Implications for TNF receptor activation. *Cell* **73**, 431–445.

Boyle, W. J., Simonet, W. S., and Lacey, D. L. (2003). Osteoclast differentiation and activation. *Nature* **423**, 337–342.

Bucay, N., Sarosi, I., Dunstan, C. R., Morony, S., Tarpley, J., Capparelli, C., Scully, S., Tan, H. L., Xu, W., Lacey, D. L., Boyle, W. J., and Simonet, W. S. (1998). Osteoprotegerin-deficient mice develop early onset osteoporosis and arterial calcification. *Genes Dev.* **12**, 1260–1268.

Cao, Z., Henzel, W. J., and Gao, X. (1996a). IRAK: A kinase associated with the interleukin-1 receptor. *Science* **271**, 1128–1131.

Cao, Z., Xiong, J., Takeuchi, M., Kurama, T., and Goeddel, D. V. (1996b). TRAF6 is a signal transducer for interleukin-1. *Nature* **383**, 443–446.

Cha, S. S., Kim, M. S., Choi, Y. H., Sung, B. J., Shin, N. K., Shin, H. C., Sung, Y. C., and Oh, B. H. (1999). 2.8 Å resolution crystal structure of human TRAIL, a cytokine with selective antitumor activity. *Immunity* **11**, 253–261.

Cha, S. S., Sung, B. J., Kim, Y. A., Song, Y. L., Kim, H. J., Kim, S., Lee, M. S., and Oh, B. H. (2000). Crystal structure of TRAIL-DR5 complex identifies a critical role of the unique frame insertion in conferring recognition specificity. *J. Biol. Chem.* **275,** 31171–31177.

Chan, F. K., Chun, H. J., Zheng, L., Siegel, R. M., Bui, K. L., and Lenardo, M. J. (2000). A domain in TNF receptors that mediates ligand-independent receptor assembly and signaling. *Science* **288,** 2351–2354.

Eck, M. J., and Sprang, S. R. (1989). The structure of tumor necrosis factor-α at 2.6 Å resolution: Implications for receptor binding. *J. Biol. Chem.* **264,** 17595–17605.

Emery, J. G., McDonnell, P., Burke, M. B., Deen, K. C., Lyn, S., Silverman, C., Dul, E., Appelbaum, E. R., Eichman, C., DiPrinzio, R., Dodds, R. A., James, I. E., Rosenberg, M., Lee, J. C., and Young, P. R. (1998). Osteoprotegerin is a receptor for the cytotoxic ligand TRAIL. *J. Biol. Chem.* **273,** 14363–14367.

Fuller, K., Wong, B., Fox, S., Choi, Y., and Chambers, T. J. (1998). TRANCE is necessary and sufficient for osteoblast-mediated activation of bone resorption in osteoclasts. *J. Exp. Med.* **188,** 997–1001.

Galibert, L., Tometsko, M. E., Anderson, D. M., Cosman, D., and Dougall, W. C. (1998). The involvement of multiple tumor necrosis factor receptor (TNFR)-associated factors in the signaling mechanisms of receptor activator of NF-kappaB, a member of the TNFR superfamily. *J. Biol. Chem.* **273,** 34120–34127.

Hymowitz, S. G., Christinger, H. W., Fuh, G., Ultsch, M., O'Connell, M., Kelley, R. F., Ashkenazi, A., and de Vos, A. M. (1999). Triggering cell death: The crystal structure of Apo2L/TRAIL in a complex with death receptor 5. *Mol. Cell* **4,** 563–571.

Hymowitz, S. G., O'Connell, M. P., Ultsch, M. H., Hurst, A., Totpal, K., Ashkenazi, A., de Vos, A. M., and Kelley, R. F. (2000). A unique zinc-binding site revealed by a high-resolution X-ray structure of homotrimeric Apo2L/TRAIL. *Biochemistry* **39,** 633–640.

Ito, S., Wakabayashi, K., Ubukata, O., Hayashi, S., Okada, F., and Hata, T. (2002). Crystal structure of the extracellular domain of mouse RANK ligand at 2.2-Å resolution. *J. Biol. Chem.* **277,** 6631–6636.

Khosla, S. (2001). Minireview: The OPG/RANKL/RANK system. *Endocrinology* **142,** 5050–5055.

Kim, H. H., Lee, D. E., Shin, J. N., Lee, Y. S., Jeon, Y. M., Chung, C. H., Ni, J., Kwon, B. S., and Lee, Z. H. (1999). Receptor activator of NF-κB recruits multiple TRAF family adaptors and activates c-Jun N-terminal kinase. *FEBS Lett.* **443,** 297–302.

Kong, Y. Y., Yoshida, H., Sarosi, I., Tan, H. L., Timms, E., Capparelli, C., Morony, S., Oliveira-dos-Santos, A. J., Van, G., Itie, A., Khoo, W., Wakeham, A., Dunstan, C. R., Lacey, D. L., Mak, T. W., Boyle, W. J., and Penninger, J. M. (1999). OPGL is a key regulator of osteoclastogenesis, lymphocyte development and lymph-node organogenesis. *Nature* **397,** 315–323.

Kraulis, P. J. (1991). MOLSCRIPT: A program to produce both detailed and schematic plots of protein structures. *J. Appl. Crystallogr.* **24,** 946–950.

Lacey, D. L., Timms, E., Tan, H. L., Kelley, M. J., Dunstan, C. R., Burgess, T., Elliott, R., Colombero, A., Elliott, G., Scully, S., Hsu, H., Sullivan, J., Hawkins, N., Davy, E., Capparelli, C., Eli, A., Qian, Y. X., Kaufman, S., Sarosi, I., Shalhoub, V., Senaldi, G., Guo, J., Delaney, J., and Boyle, W. J. (1998). Osteoprotegerin ligand is a cytokine that regulates osteoclast differentiation and activation. *Cell* **93,** 165–176.

Lam, J., Nelson, C. A., Ross, F. P., Teitelbaum, S. L., and Fremont, D. H. (2001). Crystal structure of the TRANCE/RANKL cytokine reveals determinants of receptor-ligand specificity. *J. Clin. Invest.* **108,** 971–979.

Lee, Z. H., and Kim, H. H. (2003). Signal transduction by receptor activator of nuclear factor kappa B in osteoclasts. *Biochem. Biophys. Res. Commun.* **305,** 211–214.

Li, J., Sarosi, I., Yan, X. Q., Morony, S., Capparelli, C., Tan, H. L., McCabe, S., Elliott, R., Scully, S., Van, G., Kaufman, S., Juan, S. C., Sun, Y., Tarpley, J., Martin, L., Christensen, K., McCabe, J., Kostenuik, P., Hsu, H., Fletcher, F., Dunstan, C. R., Lacey, D. L., and Boyle,

W. J. (2000). RANK is the intrinsic hematopoietic cell surface receptor that controls osteoclastogenesis and regulation of bone mass and calcium metabolism. *Proc. Natl. Acad. Sci. USA* **97,** 1566–1571.

Locksley, R. M., Killeen, N., and Lenardo, M. J. (2001). The TNF and TNF receptor superfamilies: Integrating mammalian biology. *Cell* **104,** 487–501.

Malyankar, U. M., Scatena, M., Suchland, K. L., Yun, T. J., Clark, E. A., and Giachelli, C. M. (2000). Osteoprotegerin is an $a_v\beta_3$-induced, NF-κB-dependent survival factor for endothelial cells. *J. Biol. Chem.* **275,** 20959–20962.

McWhirter, S. M., Pullen, S. S., Holton, J. M., Crute, J. J., Kehry, M. R., and Alber, T. (1999). Crystallographic analysis of CD40 recognition and signaling by human TRAF2. *Proc. Natl. Acad. Sci. USA* **96,** 8408–8413.

Merrit, E. A., and Murphy, M. E. P. (1994). Raster3D version 2.0, a program for photorealistic molecular graphics. *Acta Crystallogr.* **D50,** 869–873.

Mizuno, A., Amizuka, N., Irie, K., Murakami, A., Fujise, N., Kanno, T., Sato, Y., Nakagawa, N., Yasuda, H., Mochizuki, S., Gomibuchi, T., Yano, K., Shima, N., Washida, N., Tsuda, E., Morinaga, T., Higashio, K., and Ozawa, H. (1998). Severe osteoporosis in mice lacking osteoclastogenesis inhibitory factor/osteoprotegerin. *Biochem. Biophys. Res. Commun.* **247,** 610–615.

Mongkolsapaya, J., Grimes, J. M., Chen, N., Xu, X. N., Stuart, D. I., Jones, E. Y., and Screaton, G. R. (1999). Structure of the TRAIL-DR5 complex reveals mechanisms conferring specificity in apoptotic initiation. *Nat. Struct. Biol.* **6,** 1048–1053.

Naito, A., Azuma, S., Tanaka, S., Miyazaki, T., Takaki, S., Takatsu, K., Nakao, K., Nakamura, K., Katsuki, M., Yamamoto, T., and Inoue, J. (1999). Severe osteopetrosis, defective interleukin-1 signalling and lymph node organogenesis in TRAF6-deficient mice. *Genes Cells* **4,** 353–362.

Siegel, R. M., Frederiksen, J. K., Zacharias, D. A., Chan, F. K., Johnson, M., Lynch, D., Tsien, R. Y., and Lenardo, M. J. (2000). Fas preassociation required for apoptosis signaling and dominant inhibition by pathogenic mutations. *Science* **288,** 2354–2357.

Simonet, W. S., Lacey, D. L., Dunstan, C. R., Kelley, M., Chang, M. S., Luthy, R., Nguyen, H. Q., Wooden, S., Bennett, L., Boone, T., Shimamoto, G., DeRose, M., Elliott, R., Colombero, A., Tan, H. L., Trail, G., Sullivan, J., Davy, E., Bucay, N., Renshaw-Gegg, L., Hughes, T. M., Hill, D., Pattison, W., Campbell, P., Sander, S., Van, G., Tarpley, J., Derby, P., Lee, R., Amgen EST Program, and Boyle, W. J. (1997). Osteoprotegerin: A novel secreted protein involved in the regulation of bone density. *Cell* **89,** 309–319.

Suda, T., Takahashi, N., and Martin, T. J. (1992). Modulation of osteoclast differentiation. *Endocr. Rev.* **13,** 66–80.

Suda, T., Takahashi, N., Udagawa, N., Jimi, E., Gillespie, M. T., and Martin, T. J. (1999). Modulation of osteoclast differentiation and function by the new members of the tumor necrosis factor receptor and ligand families. *Endocr. Rev.* **20,** 345–357.

Theill, L. E., Boyle, W. J., and Penninger, J. M. (2002). RANK-L and RANK: T cells, bone loss, and mammalian evolution. *Annu. Rev. Immunol.* **20,** 795–823.

Truneh, A., Sharma, S., Silverman, C., Khandekar, S., Reddy, M. P., Deen, K. C., McLaughlin, M. M., Srinivasula, S. M., Livi, G. P., Marshall, L. A., Alnemri, E. S., Williams, W. V., and Doyle, M. L. (2000). Temperature-sensitive differential affinity of TRAIL for its receptors. DR5 is the highest affinity receptor. *J. Biol. Chem.* **275,** 23319–23325.

Tsuda, E., Goto, M., Mochizuki, S., Yano, K., Kobayashi, F., Morinaga, T., and Higashio, K. (1997). Isolation of a novel cytokine from human fibroblasts that specifically inhibits osteoclastogenesis. *Biochem. Biophys. Res. Commun.* **234,** 137–142.

Van Ostade, X., Tavernier, J., and Fiers, W. (1994). Structure-activity studies of human tumour necrosis factors. *Protein Eng.* **7,** 5–22.

Willard, D., Chen, W. J., Barrett, G., Blackburn, K., Bynum, J., Consler, T., Hoffman, C., Horne, E., Iannone, M. A., Kadwell, S., Parham, J., and Ellis, B. (2000). Expression, purification, and characterization of the human receptor activator of NF-κB ligand (RANKL) extracellular domain. *Protein Expr. Purif.* **20**, 48–57.

Wong, B. R., Rho, J., Arron, J., Robinson, E., Orlinick, J., Chao, M., Kalachikov, S., Cayani, E., Bartlett, F. S. 3rd, Frankel, W. N., Lee, S. Y., and Choi, Y. (1997). TRANCE is a novel ligand of the tumor necrosis factor receptor family that activates c-Jun N-terminal kinase in T cells. *J. Biol. Chem.* **272**, 25190–25194.

Yamaguchi, K., Kinosaki, M., Goto, M., Kobayashi, F., Tsuda, E., Morinaga, T., and Higashio, K. (1998). Characterization of structural domains of human osteoclastogenesis inhibitory factor. *J. Biol. Chem.* **27**, 5117–5123.

Yasuda, H., Shima, N., Nakagawa, N., Mochizuki, S. I., Yano, K., Fujise, N., Sato, Y., Goto, M., Yamaguchi, K., Kuriyama, M., Kanno, T., Murakami, A., Tsuda, E., Morinaga, T., and Higashio, K. (1998a). Identity of osteoclastogenesis inhibitory factor (OCIF) and osteoprotegerin (OPG): A mechanism by which OPG/OCIF inhibits osteoclastogenesis in vitro. *Endocrinology* **139**, 1329–1337.

Yasuda, H., Shima, N., Nakagawa, N., Yamaguchi, K., Kinosaki, M., Mochizuki, S., Tomoyasu, A., Yano, K., Goto, M., Murakami, A., Tsuda, E., Morinaga, T., Higashio, K., Udagawa, N., Takahashi, N., and Suda, T. (1998b). Osteoclast differentiation factor is a ligand for osteoprotegerin/osteoclastogenesis-inhibitory factor and is identical to TRANCE/RANKL. *Proc. Natl. Acad. Sci. USA* **95**, 3597–3602.

Ye, H., Arron, J. R., Lamothe, B., Cirilli, M., Kobayashi, T., Shevde, N. K., Segal, D., Dzivenu, O. K., Vologodskaia, M., Yim, M., Du, K., Singh, S., Pike, J. W., Darnay, B. G., Choi, Y., and Wu, H. (2002). Distinct molecular mechanism for initiating TRAF6 signalling. *Nature* **418**, 443–447.

3

Promoter of TRAIL-R2 Gene

Tatsushi Yoshida and Toshiyuki Sakai

Department of Molecular-Targeting Cancer Prevention
Graduate School of Medical Science, Kyoto Prefectural University of Medicine
Kyoto 602-8566, Japan

I. Introduction
II. Basic Structure of TRAIL-R2 Promoter
    A. Core Promoter Region of TRAIL-R2 *Gene*
    B. *Transcription Factor Binding Sequences in TRAIL-R2 Promoter*
    C. *Hypermethylation of TRAIL-R2 Promoter*
    D. *Comparison of TRAIL-R2 Promoter with TRAIL-R1 Promoter*
    E. *Comparison of TRAIL-R2 Promoter with Mouse Homologue MK Promoter*
III. Regulator of TRAIL-R2
    A. *p53 Activates* TRAIL-R2 *Gene*
    B. *ER* Ca^{2+} *Pool Depletion Activates TRAIL-R2*
    C. *Other Regulators*
IV. Possibility of Cancer Therapy Using TRAIL-R2
References

TRAIL-R2 promoter does not have a typical TATA-box but two functional Sp1-binding sites. TRAIL-R2 promoter belongs to the class of TATA-less and GC-box–containing promoters. The minimal promoter element is contained in the region spanning −198 to −116 upstream of translational initiation codon ATG. Computer analysis shows putative transcription factor binding sites such as c-Ets, AML-1a, c-Myb, Sp1, and GATA-1 in TRAIL-R2 promoter. Hypermethylation of TRAIL-R2 is not frequent compared with that of TRAIL-R3 and TRIAL-R4. There are no potential transcription factor binding sites in highly homologous regions between TRAIL-R2 promoter and TRAIL-R1 promoter, or between TRAIL-R2 promoter and mouse homologue mouse killer (MK) promoter. TRAIL-R2 is known to be a downstream gene of p53, a tumor-suppressor gene, and a p53-binding site in TRAIL-R2 intron 1 is responsible for p53-dependent transcription. Thapsigargin, endoplasmic reticulum Ca^{2+}-ATPase inhibitor calcium releaser, upregulates TRAIL-R2 expression via the promoter region. Many regulators of TRAIL-R2 have been reported. However, it has not been demonstrated whether they regulate TRAIL-R2 via the promoter region. Here, we show a list of these regulators. Finally, we demonstrate the possibility of cancer therapy using regulation of TRAIL-R2 promoter. © 2004 Elsevier Inc.

I. INTRODUCTION

Tumor necrosis factor–related apoptosis-inducing ligand-receptor 2 (TRAIL-R2) is a member of the TNF receptor family. TRAIL-R2 has also been called Death Receptor 5 (DR5), Apo2, TRAIL receptor inducer of cell killing 2 (TRICK2), or KILLER/DR5 (MacFarlance et al., 1997; Pan et al., 1997a; Screaton et al., 1997; Sheridan et al., 1997; Walczak et al., 1997; Wu et al., 1997). Four human TRAIL receptors have been identified: TRAIL-R1 (also called DR4) (Pan et al., 1997b), TRAIL-R2, TRAIL-R3 (DcR1 or TRID) (Degli-Esposti et al., 1997a; Mongkolsapaya et al., 1998; Sheridan et al., 1997), and TRAIL-R4 (DcR2 or TRUNDD) (Degli-Esposti et al., 1997b; Marsters et al., 1997; Pan et al., 1998). TRAIL-R1 and TRAIL-R2 are proapoptotic receptors containing an intracellular death domain. These receptors associate their death domain with the same domain in adapter proteins such as FADD (Ashkenazi and Dixit, 1998; Kischkal et al., 2000; Sprick et al., 2000). Subsequently, adapter proteins signal apoptosis via activation of caspases. TRAIL-R3 and TRAIL-R4 are decoy receptors. TRAIL-R3 lacks the death domain and TRAIL-R4 contains a nonfunctional death domain. TRAIL-R3 and TRAIL-R4 antagonize TRAIL-inducing apoptosis by competing with TRAIL-R1 and TRAIL-R2 for ligand binding. TRAIL-R2 overexpression induces apoptosis of human cancer cell lines and prevents colony formation by cancer cells (Wu et al., 1997). Many drugs and genes upregulate TRAIL-R2 expression following apoptosis. It is unknown whether most of these mechanisms depend on promoter regulation. Therefore, it is important to define these mechanisms in detail.

We present the basic structure of TRAIL-R2 promoter in Section II and regulators that have been identified to control TRAIL-R2 expression via promoter in Section IIIA and B. In Section IIIC and Table I, we also describe other regulators that had not previously been clarified as regulating at the promoter level. Finally, we raise the possibility of cancer therapy using TRAIL-R2 promoter.

II. BASIC STRUCTURE OF TRAIL-R2 PROMOTER

A. CORE PROMOTER REGION OF *TRAIL-R2* GENE

We cloned 5'-flanking region of the *TRAIL-R2* gene and analyzed the promoter region (Yoshida *et al.*, 2001). Transcription start sites were shown by RNase protection assay. There are two major transcription start sites around -122 and -137 upstream of the translational initiation codon ATG (Fig. 1). About 2.5 kbp of the 5'-flanking region of the *TRAIL-R2* gene demonstrated authentic promoter activity when this region was subcloned into luciferase assay vector (Fig. 2). In the region shorter than -605, promoter activities gradually decreased. Moreover, the region shorter than -198 did not demonstrate promoter activities. The minimal promoter element was thought to be contained in the region spanning -198 to -116. Mutation analysis showed that two Sp1 sites were involved in transcription activity of the region spanning -198 to -116 (Fig. 3). TRAIL-R2 promoter does not have a typical TATA-box. Ubiquitously expressed genes often lack a TATA-box but have GC-boxes such as the Sp1 site (Melton *et al.*, 1986; Ohbayashi *et al.*, 1996). TRAIL-R2 promoter belongs to the class of TATA-less and GC-box–containing promoters.

B. TRANSCRIPTION FACTOR BINDING SEQUENCES IN TRAIL-R2 PROMOTER

We searched for putative transcription factor binding sequences in the TRAIL-R2 promoter region by TFSEARCH (http://pdap1.trc.rwcp.or.jp) and the Web signal scan program (http://www.dna.affrc.go.jp/sigscan/signal1.pl) (Yoshida *et al.*, 2001). There are potential regulatory factor binding sites such as c-Ets2, AML-1a, c-Myb, Sp1, and GATA-1 (Fig. 1). It has not yet been elucidated whether these factors regulate TRAIL-R2 expression, and further analyses are needed.

FIGURE 1. Structure of TRAIL-R2 promoter. Arrowheads show transcription start sites. Circles and squares show putative transcription factor binding sites. The box indicates Exon 1. Restriction enzyme recognition sites are described.

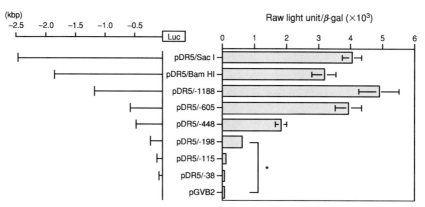

FIGURE 2. Basal activity of TRAIL-R2 promoter. Reporter plasmids containing various sizes of 5'-deleted human TRAIL-R2 promoter and luciferase genes were transfected into MCF7 breast cancer cells. Structures of relevant plasmids are shown on the left. $*P < .01$.

C. HYPERMETHYLATION OF TRAIL-R2 PROMOTER

Van Noesel et al. (2002) analyzed promoter hypermethylation of TRAIL-Rs. They studied a group of pediatric tumor cell lines (nine neuroblastoma and three peripheral primitive neuroectodermal tumors [PNETs]) and three adult tumor cell lines. Lack of expression of TRAIL-R3 and TRAIL-R4 was shown in 13 of the 15 cell lines and 10 of 15, respectively. In the nonexpressing cell lines, hypermethylation was found in nine of 13 cell lines and in nine of 10 cell lines, respectively. However, TRAIL-R1 and TRAIL-R2 were not expressed in seven of 15 and in three of 15 cell lines, respectively, and were methylated in five of seven and in two of three nonexpressing cell lines, respectively. Moreover a panel of 28 fresh neuroblastoma tumor samples lacked expression of TRAIL-R3 and TRAIL-R4 in 23 of 28 tumors and in 20 of 28 tumors, respectively. Hypermethylation of TRAIL-R3 and TRAIL-R4 was observed in six of 28 tumors and seven of 28 tumors. TRAIL-R1 and TRAIL-R2 were both expressed in 22 of 28 tumors, and there was no promoter methylation observed. Taken together, hypermethylation of TRAIL-R3 and TRAIL-R4 promoters may be important in the downregulation of each expression. In

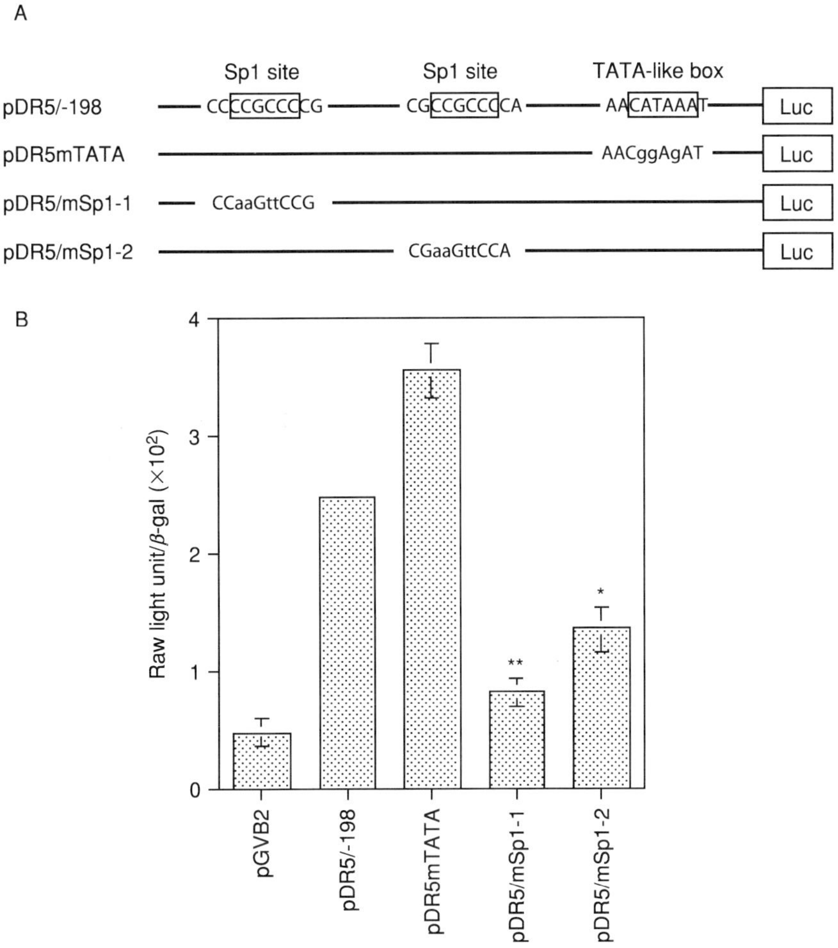

FIGURE 3. Two Sp1 sites are involved in transcription activation of the TRAIL-R2 promoter. A, Structure of reporter plasmids used in this luciferase assay. Two Sp1 sites and TATA-like box are indicated by the box. Substituted nucleotides are in lowercase letters. B, The reporter plasmids containing TRAIL-R2 promoter and luciferase gene were transfected into MCF7 cells. **$P < .002$; *$P < .025$ against the activity by pDR5/-198.

contrast, hypermethylation of TRAIL-R1 and TRAIL-R2 is not frequent compared with that of TRAIL-R3 and TRAIL-R4.

D. COMPARISON OF TRAIL-R2 PROMOTER WITH TRAIL-R1 PROMOTER

All TRAIL receptors have been mapped to the same chromosomal locus 8p21–22, suggesting that they have evolved as a result of gene duplication. TRAIL-R1 is another proapoptotic receptor like TRAIL-R2 (Pan *et al.*,

1997b). Promoter regions of TRAIL-R2 and TRAIL-R1 are shown in Fig. 4. First methionine of TRAIL-R2 corresponds to isoleucine (45th amino acid from the first methionine) of TRAIL-R1. Therefore, −1000 to first codon ATG of TRAIL-R2 was compared with −868 to +135 of TRAIL-R1. The promoter sequence of TRAIL-R2 shares 43.6% homology with that of TRAIL-R1. There was no potential transcription factor binding sequence in the highly homologous regions. TRAIL-R2 is expressed ubiquitously in many tissues; however, TRAIL-R1 is expressed strongly in peripheral blood leukocytes, small intestine, spleen and thymus, and weakly in other tissues. This difference in distributions of TRAIL-R1 and TRAIL-R2 in many tissues may reflect the low homology to sequences of the promoter regions. However, common mechanisms inducing both TRAIL-R1 and TRAIL-R2 seem to exist, because both TRAIL-R1 and TRAIL-R2 are most strongly expressed in peripheral blood leukocytes and are simultaneously induced by etoposide. To elucidate these mechanisms, promoter analyses are required.

E. COMPARISON OF TRAIL-R2 PROMOTER WITH MOUSE HOMOLOGUE MK PROMOTER

Wu *et al.* (1999a) reported cloning of the mouse homologue of TRAIL-R2 (MK). Like TRAIL-R2, MK induces apoptosis, inhibits colony formation, and is upregulated by p53 or DNA-damaging agents. MK has more similarity to the human TRAIL receptor TRAIL-R2 than to TRAIL-R1 (79% versus 76% at the protein level, respectively). MK promoter sequences are described on GenBank (Accession: AB031082). Sequence of TRAIL-R2 promoter shares 41.9% homology with MK promoter (Fig. 5). There are no potential transcription factor binding sequences in highly homologous regions. The expression of MK is particularly abundant in the heart, lung, and kidney; however, TRAIL-R2 is expressed in most tissues. Regulation of these promoters may be different.

III. REGULATOR OF TRAIL-R2

A. P53 ACTIVATES *TRAIL-R2* GENE

The tumor suppressor p53 is the most frequently mutated gene in human malignant tumors and induces a number of proapoptotic genes and cell-cycle arrest genes. Wu *et al.* (1997, 1999b) showed that TRAIL-R2 is a p53-downstream gene. Takimoto and El-Deiry (2000) investigated DNA fragments containing high-affinity p53-binding sites from a human TRAIL-R2 genomic locus using an immunoselection protocol. Three p53-binding sites were identified in the TRAIL-R2 genomic locus located −0.82 kb upstream of the translational codon ATG, within intron 1 (+0.25 kb downstream of

FIGURE 4. Alignment of comparison between the 5′-flanking region of TRAIL-R1 and TRAIL-R2. The 5′-flanking region containing the TRAIL-R2 promoter is compared with that of another family gene TRAIL-R1. This alignment was analysed by CLUSTAL W program.

```
TRAIL-R2  -----TATCATGTTTTATTTCCTAGACTAGTGACAA-ATGAAAGCTAAGTGTAGCAAGGG
MK        TGGAGAAGAGAGATTGATTGATTGAGATTGTGAAATTACCAAAATGCACTGATGTGGGTA
          ------*----*-**-***---*----*-****-*--*--**----*-**--*---*--

TRAIL-R2  TGCAGGGACACAGGCAC-----ATTTGTGGACTAGGTGTGAGTGTAAGC-TGGGTTCGAT
MK        GATGAGGTCATCTATTCTTTTTGTTTGTTTGTGTTTTGTTGGTTTAGGGGTGTGTGTGTG
          -----**-**---*------*------******-------***--**-**--*--**--

TRAIL-R2  GGTCTTTTGGCCAACATA-GTGAACCCCTGTGTCTACTAAAAATACAAAAATTAGCCAGG
MK        TGTGTGTGTGTGTGTGTGTGTGTGTGTGTGTGTGTGTGTTTAAAG-ACATGGTTT--CTCTG
          -**-*-*---*------*---**------****-*---*--***--**----*--*---*

TRAIL-R2  CGTGGTGGTG-CAGGCCTGTAGTCCCAGCTACATGGGAGGCTG------AGGTGGGAGTA
MK        TGTAGCCCTGGCTGTCCTGAAATTCACTCTGTAGACCAGGCTGGACTCTAACTCAGAGAC
          -**-*----**-*-*------*------*---********-------------*--**--

TRAIL-R2  TCGCTTGAACCTGGGAG-ACGGAAGTTGCAGTGAGCCGGGATCACACCAC----CGTTCA
MK        CGGCCTGCCTCTGCTCGTGCTAGGATTGAAGTGCTTTACCACCACGGCTAAGTGTGTTCA
          --**-**---**-----*------***-****--------*-***--*--------*****

TRAIL-R2  CCAATCTGAGC--CACAGAGAGACTGTCTCAAAAAATAAACCACAAGGAAGGGAGGGAGG
MK        GTGTTCCAAACGTTACAGAGACACGGATTGATTTCGCTTCTGTTACTGGGGTGAGACCAT
          ----**-*----*****--**--*-------*------*--*--*-*---**-----

TRAIL-R2  GGGAGGGGGAGGGAGGGAGGAAAGAGAAAGAGAGAAAGGAAGGAAAGAGAAAGCAGGAAG
MK        CCATTCTATCATTTCCTTTTCTGCGTTCTCAGATACTGCCTGGTAGCATACAACTGTAAC
          ----------------------***-*---*----**-*---*-*-*-*-*-**-

TRAIL-R2  GAAGGAAAGAAGAAGAAAGAAGACGAAAGAACGAAAGAAAAGGAAAGAAGAGAGGAGAGA
MK        AAAAACAAATTCAAGAA-GAGGAAGAGAAGAAGAAGGAGAGATGAAGGAGACAGAAGA
          -**----**----*****-***--**--**-*--**-*-**--***--*--****----***

TRAIL-R2  ACAGAAGGGGCAGGTGCCCCTGGGAAGGGGAGAAGATCAAGACG-CGCCTGGAAAG-CGG
MK        TGAGGAAGGAGGAGTACTCACTACTACAG-ACATCTTCAAGTCGGCTCCTCCAGGGACAC
          --**--*-**----**-*-*-------*---*-*-*----*****--**-*-***--*---*--

TRAIL-R2  ACTCTGAACCTCAAGACCCTGTTCACAGCCAAGCGCGCGACCCCGGGAGGCGTCAACTCC
MK        AAGCTGAAACTCAAAGGTGAATATCATATTAGTCAAAATATGAAGGGACAGG-CAGACCT
          *--*****-*****----------*---------**--*-------***----*-----

TRAIL-R2  CCAAGTGCCTCCCTCAACTCATTTCCCCCAAGTTTCGGTGCCTGTCCTGGCGCGGACAGG
MK        TTGAGTATTACGAGTTGGTTTTTTTTTTTAATAAGCATGGGCGTCTC----CAACCACG
          ---***----*-------*--***-------*-*------*-----**--------*---*-*

TRAIL-R2  ACCCAGAAACAA--ACCACAGCCCGGGGCGCAGCCGCCAGGGCGAAGGTTAGTTCCGGTC
MK        ACCTGACAATTATAATAAAAGTGGTGTGATTCTTTAAAAAAATAAAAGTT-GTTCAGATT
          ***--****----*--*---*---*--*----*--*-----*-----*****-*--*--*-

TRAIL-R2  CCTTCCC-CTCCCCTCCCCACTTGGACGCGCTTGCGGAGGATTGCGTTGACGAGACTCTT
MK        CTGTCTGGCGACTGCTCCCAAGTTTTCATA-TTAGGTAGGATCGGGGTT---GGACGCGT
          *--**---*--*----****--*---*-----**--*-*----**--*-*-*---**

TRAIL-R2  ATTTATTGTCACCAACCTGTGGTGGAATTTGCAGTTGCACATTGGATCTGATTCGCCCCG
MK        AATCA--AGCTCCTCCCAAGGGTGGACTTCCCCACCCCACCCCACCCCACCCCGGCCTG
          *-*-*-----*-**--*----**--****-----*------*-----------*--**-*

TRAIL-R2  CCCCGAATGACGCCTGCCCGGAGGCAGTGAAAGTACAGCC--GCGCCGCCCCAAGTCAGC
MK        GTGCGC---GCGCAAGCCCGCGGAGGTTGGAGGTACCACTAAGCGTCAAGTCGCCGCGAC
          ---**------*----***-*----*-*-****--*----***-*----***-*

TRAIL-R2  CTGGACACATAAATCAGCACGCGGCCGGAGAACCCCGCAATCTTTGCGCCCACAAAATAC
MK        AAGAATCCAGAACTTTTCTGGGAGTGAGGAAATCCAGAGAACTTTTTTAGGAGTGAGGGG
          --*-*---**-**-*----*--*---*----**-*--*----****--------*-----

TRAIL-R2  ACCGACGATGCCCGATCTACTTTAAGGGCTGAAACCCACGGGCCTGAGAGACTATAAGAG
MK        ACAGCC-ATCCTTCGTGGCTTTTGGGAGCTGAAGCCGCAGGGTTTCGGATGAGCTGACAC
          **-*-*-**-----*----***--*-******-**---**--*--*------*-*-*-

TRAIL-R2  CGTTCCCTACCGCCATG
MK        CATG-------------
          *-*-------------
```

FIGURE 5. Alignment of comparison between the 5′-flanking of TRAIL-R2 and MK. The 5′-flanking region containing the TRAIL-R2 promoter is compared with that of mouse homolog gene MK. This alignment was analysed by the CLUSTAL W program.

the ATG), and within intron 2 (+1.25 kb downstream of the ATG). Luciferase assays with constructs containing deletion or mutation indicated the p53-binding site within intron 1 was responsible for p53-dependent transactivation of TRAIL-R2. The other two p53-binding sites can bind to p53 *in vitro* but seem not to be involved in promoter activation by p53.

B. ER Ca^{2+} POOL DEPLETION ACTIVATES TRAIL-R2

Thapsigargin (TG) induces perturbation in Ca^{2+} homeostasis and induces apoptosis. He *et al.* (2002a) showed that TG-induced apoptosis is coupled with upregulation of TRAIL-R2. The TG-mediated induction of TRAIL-R2 mRNA and protein occurs in a p53-independent manner because TG modulates their levels in both p53-positive and p53-negative cells. They carried out luciferase assays and show that the construct containing the 5′-proximal promoter region and part of exon 1 was upregulated by TG. The critical transcriptional element on the region, however, has not been determined. He *et al.* also demonstrated that TG upregulated TRAIL-R2 by mRNA stability as well as by enhancing TRAIL-R2 promoter activity.

C. OTHER REGULATORS

Many regulators of TRAIL-R2 have been reported. However, it has not been demonstrated whether they regulate TRAIL-R2 via the promoter region. Regulators reported to date are shown in Table I. In future studies, analysis of whether the regulations occur through TRAIL-R2 promoter and identification of the transcription factors regulating TRAIL-R2 expression are needed.

IV. POSSIBILITY OF CANCER THERAPY USING TRAIL-R2

TRAIL induces apoptosis preferentially in tumor cells (Ashkenazi *et al.*, 1999; Lawrence *et al.*, 2001; Walczak *et al.*, 1999). TRAIL in combination with chemotherapy using CDDP, etoposide, or CPT-11 resulted in synergistic cell death and significantly extended the survival of mice bearing tumor xenografts compared with either TRAIL or drug alone (Nagane *et al.*, 2000; Naka *et al.*, 2002). Treatments with these drugs were accompanied by upregulation of TRAIL-R2. Therefore, a drug upregulating TRAIL-R2 expression in combination with TRAIL appears to lead to an effective therapy for cancer (Fig. 6A). It is important to understand the mechanism of TRAIL-R2 expression including regulation of the promoter and to find more effective drugs.

TABLE I. Regulators of TRAIL-R2 Expression

Drug, gene, or condition	Protein level	mRNA level	Reference
Ara-C	↑	ND	Wen et al. (2000)
ATRA	ND	↑	Altucci et al. (2001)
Betulinic acid	↑	↑	Meng and El-Deiry (2001)
Bile acid	↑	↑	Higuchi et al. (2001)
			Higuchi and Gores (2003)
			Sun et al. (1999a)
CD437	ND	↑	Sun et al. (1999b)
			Sun et al. (2000a)
			Sun et al. (2000b)
CDDP	ND	↑	Nagane et al. (2000)
Cholestasis	↑	↑	Higuchi et al. (2002)
CMV infection	↑	↑	Sedger et al. (1999)
CPT-11	↑	ND	Naka et al. (2002)
			Wu et al. (1997)
Doxorubicin	↑	↑	Sheikh et al. (1998)
			Wen et al. (2000)
			Mitsiades et al. (2001)
			Gibson et al. (2000)
Etoposide	↑	↑	Nagane et al. (2000)
			Wu et al. (1997)
Glucocorticoid	↑	↑	Meng et al. (2001)
Indole-3-carbinol	↑	↑	Jeon et al. (2003)
INF-α	↑	↑	Shigeno et al. (2003)
INF-γ	↑	↑	Meng and El-Deiry (2001)
			Wu et al. (1997)
IR	ND	↑	Sheikh et al. (1998)
			Gong and Almasan (2000)
β-Lapachone	↑	↑	Meng and El-Deiry (2001)
2-Methoxyestradiol	↑	↑	LaVallee et al. (2003)
MMS	ND	↑	Sheikh et al. (1998)
NF-κB	↑	↑	Ravi et al. (2001)
			Chen et al. (2003)
			Wu et al. (1997)
p53	↑	↑	Wu et al. (1999b)
			Takimoto and El-Deiry (2000)
SC′-236	↑	↑	He et al. (2002b)
Sulindac sulfide	↑	↑	Huang et al. (2001)
			He et al. (2002b)
Taxol	↑	No effect	Nimmanapalli et al. (2001)
Thapsigargin	↑	↑	He et al. (2002a)
			Yamaguchi et al. (2003)
TNF-α	ND	↑	Sheikh et al. (1998)

ND: not determined; ↑: upregulation.

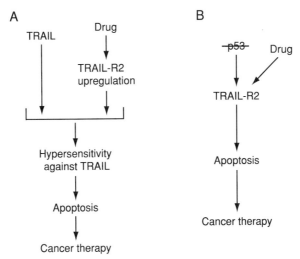

FIGURE 6. Models for cancer therapy using TRAIL-R2. A, A model of synergistic effect using TRAIL together with a drug that upregulate TRAIL-R2. B, A model of gene-regulating chemotherapy.

From another perspective, TRAIL-R2 is regulated by p53, which is a tumor-suppressor gene that is inactivated in approximately half of human malignancies. Overexpression of TRAIL-R2 can induce apoptosis of cancer cell lines and inhibits colony formation. We proposed that methods of upregulating p53 target genes would be useful for cancer therapy, and termed this method "gene-regulating chemotherapy" (Fig. 6B) (Sakai, 1996; Sowa and Sakai, 2000). Therefore, chemical compounds that can upregulate TRAIL-R2 may lead to new cancer therapy drugs. For this purpose, further analysis of the TRAIL-R2 promoter is required.

ACKNOWLEDGMENTS

This work was supported by a Grant-in-Aid from the Japanese Ministry of Education, Culture, Sports, Science and Technology; a Grant-in-Aid from Japan Society for the Promotion of Science; and SRF Grant for Biomedical Research.

REFERENCES

Altucci, L., Rossin, A., Raffelsberger, W., Reitmair, A., Chomienne, C., and Gronemeyer, H. (2001). Retinoic acid-induced apoptosis in leukemia cells is mediated by paracrine action of tumor-selective death ligand TRAIL. *Nature Med.* **7,** 680–686.
Ashkenazi, A., and Dixit, V. M. (1998). Death receptors: Signaling and modulation. *Science* **281,** 1305–1308.

Ashkenazi, A., Pai, R. C., Fong, S., Leung, S., Lawrence, D. A., Marsters, S. A., Blackie, C., Chang, L., McMurtrey, A. E., Hebert, A., DeForge, L., Koumenis, I. L., Lewis, D., Harris, L., Bussiere, J., Koeppen, H., Shahrokh, Z., and Schwall, R. H. (1999). Safety and antitumor activity of recombinant soluble Apo2 ligand. *J. Clin. Invest.* **104,** 155–162.

Chen, X., Kandasamy, K., and Srivastava, R. K. (2003). Differential roles of RelA (p65) and c-Rel subunits of nuclear factor κB in tumor necrosis factor-related apoptosis-inducing ligand signaling. *Cancer Res.* **63,** 1059–1066.

Degli-Esposti, M. A., Smolak, P. J., Walczak, H., Waugh, J., Huang, C. P., DuBose, R. F., Goodwin, R. G., and Smith, C. A. (1997a). Cloning and characterization of TRAIL-R3, a novel member of the emerging TRAIL receptor family. *J. Exp. Med.* **186,** 1165–1170.

Degli-Esposti, M. A., Dougall, W. C., Smolak, P. J., Waugh, J. Y., Smith, C. A., and Goodwin, R. G. (1997b). The novel receptor TRAIL-R4 induces NF-κB and protects against TRAIL-mediated apoptosis, yet retains an incomplete death domain. *Immunity* **7,** 813–820.

Gibson, S. B., Oyer, R., Spalding, A. C., Anderson, S. M., and Johnson, G. L. (2000). Increased expression of death receptors 4 and 5 synergizes the apoptosis response to combined treatment with etoposide and TRAIL. *Mol. Cell. Biol.* **20,** 205–212.

Gong, B., and Almasan, A. (2000). Apo2 ligand/TNF-related apoptosis-inducing ligand and death receptor 5 mediate the apoptotic signaling induced by ionizing radiation in leukemic cells. *Cancer Res.* **60,** 5754–5760.

He, Q., Lee, D. I., Rong, R., Yu, M., Luo, X., Klein, M., El-Deiry, W. S., Huang, Y., Hussain, A., and Sheikh, M. S. (2002a). Endoplasmic reticulum calcium pool depletion-induced apoptosis is coupled with activation of the death receptor 5 pathway. *Oncogene* **21,** 2623–2633.

He, Q., Luo, X., Huang, Y., and Sheikh, M. S. (2002b). Apo2L/TRAIL differentially modulates the apoptotic effects of sulindac and a COX-2 selective non-steroidal anti-inflammatory agent in Bax-deficient cells. *Oncogene* **21,** 6032–6040.

Higuchi, H., Bronk, S. F., Takikawa, Y., Werneburg, N., Takimoto, R., El-Deiry, W., and Gores, G. J. (2001). The bile acid glycochenodeoxycholate induces TRAIL-receptor2/DR5 expression and apoptosis. *J. Biol. Chem.* **276,** 38610–38618.

Higuchi, H., Bronk, S. F., Taniai, M., Canbay, A., and Gores, G. J. (2002). Cholestasis increases tumor necrosis factor-related apoptosis-inducing ligand (TRAIL)-R2/DR5 expression and sensitizes the liver to TRAIL-mediated cytotoxicity. *J. Phamacol. Exp. Ther.* **303,** 461–467.

Higuchi, H., and Gores, G. J. (2003). Bile acid regulation of hepatic physiology: IV. Bile acids and death receptors. *Am. J. Physiol. Gastrointest. Liver Physiol.* **284,** G734–738.

Huang, Y., He, Q., Hillman, M. J., Rong, R., and Sheikh, M. S. (2001). Sulindac sulfide-induced apoptosis involves death receptor 5 and the caspase 8-dependent pathway in human colon and prostate cancer cells. *Cancer Res.* **61,** 6918–6924.

Jeon, K.-I., Rih, J.-K., Kim, H. J., Lee, Y. J., Cho, C.-H., Goldberg, I. D., Rosen, E. M., and Bae, I. (2003). Pretreatment of indole-3-carbinol augments TRAIL-inducing apoptosis in a prostate cancer cell line, LNCaP. *FEBS Lett.* **544,** 246–251.

Kischkal, F. C., Lawrence, D. A., Chuntharapai, A., Schow, P., Kim, K. J., and Ashkenazi, A. (2000). Apo2L/TRAIL-dependent recruitment of endogenous FADD and caspase-8 to death receptors 4 and 5. *Immunity* **12,** 611–620.

LaVallee, T. M., Zhan, X. H., Johnson, M. S., Herbstritt, C. J., Swartz, G., Williams, M. S., Hembrough, W. A., Green, S. J., and Pribluda, V. S. (2003). 2-Methoxyestradiol up-regulates death receptor 5 and induces apoptosis through activation of the extrinsic pathway. *Cancer Res.* **63,** 468–475.

Lawrence, D., Shahrokh, Z., Marsters, S., Achilles, K., Shih, D., Mounho, B., Hillan, K., Totpal, K., DeForge, L., Schow, P., Hooley, J., Sherwood, S., Pai, R., Leung, S., Khan, L., Gliniak, B., Bussiere, J., Smith, C. A., Strom, S. S., Kelley, S., Fox, J. A., Thomas, D., and

Ashkenazi, A. (2001). Differential hepatocyte toxicity of recombinant Apo2L/TRAIL versions. *Nature Med.* **7**, 383–385.

MacFarlane, M., Ahmad, M., Srinivasula, S. M., Fernandes-Alnemri, T., Cohen, G. M., and Alnemri, E. (1997). Identification and molecular cloning of two novel receptors for the cytotoxic ligand TRAIL. *J. Biol. Chem.* **272**, 25417–25420.

Marsters, S. A., Sheridan, J. P., Pitti, R. M., Huang, A., Skubatch, M., Baldwin, D., Yuan, J., Gurney, A., Goddard, A. D., Godowski, P., and Ashkenazi, A. (1997). A novel receptor for Apo2L/TRAIL contains a truncated death domain. *Curr. Biol.* **7**, 1003–1006.

Melton, D. W., McEwan, C., McKie, A. B., and Reid, A. M. (1986). Expression of the mouse HPRT gene; deletional analysis of the promoter region of an X-chromosome linked housekeeping gene. *Cell* **44**, 319–328.

Meng, R. D., and El-Deiry, W. S. (2001). p53-independent upregulation of *KILLER/DR5* TRAIL receptor expression by glucocorticoids and interferon-γ. *Exp. Cell Res.* **262**, 154–169.

Mitsiades, C. S., Treon, S. P., Mitsiades, N., Shima, Y., Richardson, P., Schlossman, R., Hideshima, T., and Anderson, K. C. (2001). TRAIL/Apo2L ligand selectively induces apoptosis and overcomes drug resistance in multiple myeloma: Therapeutic applications. *Blood* **98**, 795–804.

Mongkolsapaya, J., Cowper, A. E., Xu, X. N., Morris, G., McMichael, A. J., Bell, J. I., and Screaton, G. R. (1998). Lymphocyte inhibitor of TRAIL (TNF-related apoptosis-inducing ligand): A new receptor protecting lymphocytes from the death ligand TRAIL. *J. Immunol.* **160**, 3–6.

Nagane, M., Pan, G., Weddle, J. J., Dixit, V. M., Cavenee, W. K., and Huang, H.-J. S. (2000). Increased death receptor 5 expression by chemotherapeutic agents in human gliomas causes synergistic cytotoxicity with tumor necrosis factor-related apoptosis-inducing ligand *in vitro* and *in vivo*. *Cancer Res.* **60**, 847–853.

Naka, T., Sugamura, K., Hylander, B. L., Widmer, M. B., Rustum, Y. M., and Repasky, E. A. (2002). Effects of tumor necrosis factor-related apoptosis-inducing ligand alone and in combination with chemotherapeutic agents on patients' colon tumors growth in SCID mice. *Cancer Res.* **62**, 5800–5806.

Nimmanapalli, R., Perkins, C. L., Orlando, M., O'Bryan, E., Nguyen, D., and Bhalla, K. N. (2001). Pretreatment with paclitaxel enhances Apo-2 ligand/tumor necrosis factor-related apoptosis-inducing ligand-induced apoptosis of prostate cancer cells by inducing death receptors 4 and 5 protein levels. *Cancer Res.* **61**, 759–763.

Ohbayashi, T., Schmidt, E. E., Makino, Y., Kishimoto, T., Nabeshima, Y., Muramatsu, M., and Tamura, T. (1996). Promoter structure of the mouse TATA-binding protein (TBP) gene. *Biochem. Biophys. Res. Commun.* **225**, 275–280.

Pan, G., Ni, J., Wei, Y.-F., Yu, G.-L., Gentz, R., and Dixit, V. M. (1997a). An antagonist decoy receptor and a death domain-containing receptor for TRAIL. *Science* **277**, 815–818.

Pan, G., O'Rourki, K., Chinnaiyan, A. M., Gentz, R., Ebner, R., Ni, J., and Dixit, V. M. (1997b). The receptor for the cytotoxic ligand TRAIL. *Science* **276**, 111–113.

Pan, G. H., Ni, J., Yu, G. L., Wei, Y. F., and Dixit, V. M. (1998). TRUNDD, a new member of the TRAIL receptor family that antagonizes TRAIL signaling. *FEBS Lett.* **424**, 41–45.

Ravi, R., Bedi, G. C., Engstrom, L. W., Zeng, Q., Mokerjee, B., Gelinas, C., Fuchs, E. J., and Bedi, A. (2001). Regulation of death receptor expression and TRAIL/Apo2L-induced apoptosis by NF-κB. *Nature Cell Biol.* **3**, 409–416.

Sakai, T. (1996). Molecular cancer epidemiology—the present status and future possibilities. *Jpn. J. Hyg.* **50**, 1036–1046.

Screaton, G. R., Mongkolsapaya, J., Xu, X. N., Cowper, A. E., McMichael, A. J., and Bell, J. (1997). TRICK2, a new alternatively spliced receptor that transduces the cytotoxic signal from TRAIL. *Curr. Biol.* **7**, 693–696.

Sedger, L. M., Shows, D. M., Blanton, R. A., Peschon, J. J., Goodwin, R. G., Cosman, D., and Wiley, S. R. (1999). INF-γ mediates a novel antiviral activity through dynamic modulation of TRAIL and TRAIL receptor expression. *J. Immunol.* **163**, 920–926.

Sheikh, M. S., Burns, T. F., Huang, Y., Wu, G. S., Amundson, S., Brooks, K. S., Fornace, A. J., Jr., and El-Deiry, W. S. (1998). p53-dependent and -independent regulation of the death receptor *KILLER/DR5* gene expresion in response to genotoxic stress and tumor necrosis factor α. *Cancer Res.* **58**, 1593–1598.

Sheridan, J. P., Marsters, S. A., Pitti, R. M., Gurney, A., Skubatch, M., Baldwin, D., Ramakrishnan, L., Gray, C. L., Baker, K., Wood, W. I., Goddard, A. D., Godowski, P., and Ashkenazi, A. (1997). Control of TRAIL-induced apoptosis by a family of signaling and decoy receptors. *Science* **277**, 818–821.

Shigeno, M., Nakao, K., Ichikawa, T., Suzuki, K., Kawakami, A., Abiru, S., Miyazoe, S., Nakagawa, Y., Ishikawa, H., Hamasaki, K., Nakata, K., Ishii, N., and Eguchi, K. (2003). Interferon-α sensitizes human hepatoma cells to TRAIL-induced apoptosis through DR5 upregulation and NF-κB inactivation. *Oncogene* **22**, 1653–1662.

Sowa, Y., and Sakai, T. (2000). Butyrate as a model for "gene-regulating chemoprevention and chemotherapy." *BioFactors* **12**, 283–287.

Sprick, M. R., Weigand, M. A., Rieser, E., Rauch, C. T., Juo, P., Blenis, J., Krammer, P. H., and Walczac, H. (2000). FADD/MORT1 and caspase-8 are recruited to TRAIL receptors 1 and 2 and are essential for apoptosis mediated by TRAIL receptor 2. *Immunity* **12**, 599–609.

Sun, S.-Y., Yue, P., Wu, G. S., El-Deiry, W. S., Shroot, B., Hong, W. K., and Lotan, R. (1999a). Implication of p53 in growth arrest and apoptosis induced by the synthetic retinoid CD437 in human lung cancer cells. *Cancer Res.* **59**, 2829–2833.

Sun, S.-Y., Yue, P., Wu, G. S., El-Deiry, W. S., Shroot, B., Hong, W. K., and Lotan, R. (1999b). Mechanisms of apoptosis induced by the synthetic retinoid CD437 in human non-small cell lung carcinoma cells. *Oncogene* **18**, 2357–2365.

Sun, S.-Y., Yue, P., Hong, W. K., and Lotan, R. (2000a). Augmentation of tumor necrosis factor-related apoptosis-inducing ligand (TRAIL)-induced apoptosis by the synthetic retinoid 6-[3-(1-adamantyl)-4-hydroxyphenyl]-2-naphthalene carboxylic acid (CD437) through up-regulation of TRAIL receptors in human lung cancer cells. *Cancer Res.* **60**, 7149–7155.

Sun, S.-Y., Yue, P., and Lotan, R (2000b). Implication of multiple mechanisms in apoptosis induced by the synthetic retinoid CD437 in human prostate carcinoma cells. *Oncogene* **19**, 4513–4522.

Takimoto, R., and El-Deiry, W. S. (2000). Wild-type p53 transactivates the KILLER/DR5 gene through an intronic sequence-specific DNA-binding site. *Oncogene* **19**, 1735–1743.

van Noesel, M. M., van Bezouw, S., Salomons, G. S., Voute, P. A., Pieters, R., Baylin, S. B., Herman, J. G., and Versteeg, R. (2002). Tumor-specific down-regulation of the Tumor Necrosis Factor-related apoptosis inducing ligand decoy receptors DcR1 and DcR2 is associated with dense promoter hypermethylation. *Cancer Res.* **62**, 2157–2161.

Walczak, H., Degli-Eaposti, M. A., Johnson, R. S., Smolak, P. J., Waugh, J. Y., Boiani, N., Timour, M. S., Gerhart, M. J., Schooley, K. A., Smith, C. A., Goodwin, R. G., and Rauch, C. T. (1997). TRAIL-R2: A novel apoptosis-mediating receptor for TRAIL. *EMBO J.* **16**, 5386–5397.

Walczak, H., Miller, R. E., Ariail, K., Gliniak, B., Griffith, T. S., Kubin, M., Chin, W., Jones, J., Woodward, A., Le, T., Smith, C., Smolak, P., Goodwin, R. G., Rauch, C. T., Schuh, J. C. L., and Lynch, D. H. (1999). Tumoricidal activity of tumor necrosis factor-related apoptosis-inducing ligand *in vivo*. *Nature Med.* **5**, 157–163.

Wen, J., Ramadevi, N., Nguyen, D., Perkins, C., Worthington, E., and Bhalla, K. (2000). Antileukemic drugs increase death receptor 5 levels and enhance Apo-2L-iduced apoptosis of human acute leukemia cells. *Blood* **96**, 3900–3906.

Wu, G. S., Burns, T. F., McDonald, E. R., Jiang, W., Meng, R., Krantz, I. D., Kao, G., Gan, D. D., Zhou, J. Y., Muschel, R., Hamilton, S. R., Spinner, N. B., Markowitz, S., Wu, G., and El-Deiry, W. S. (1997). *KILLER/DR5* is a DNA damage-inducible p53-regulated death receptor gene. *Nat. Genet.* **17,** 141–143.

Wu, G. S., Burns, T. F., Zhan, Y., Alnemri, E. S., and El-Deiry, W. S. (1999a). Molecular cloning and functional analysis of the mouse homologue of the *KILLER/DR5* tumor necrosis factor related apoptosis-inducing ligand (TRAIL) death receptor. *Cancer Res.* **59,** 2770–2775.

Wu, G. S., Burns, T. F., McDonald, E. R., Meng, R. D., Kao, G., Muschel, R., Yen, T., and El-Deiry, W. S. (1999b). Induction of the TRAIL receptor *KILLER/DR5* in p53-dependent apoptosis but not growth arrest. *Oncogene* **18,** 6411–6418.

Yamaguchi, H., Bhalla, K., and Wang, H.-G. (2003). Bax plays a pivotal role in Thapsigargin-induced apoptosis of human colon cancer HCT116 cells by controlling Smac/Diablo and Omi/HtrA2 release from mitochondria. *Cancer Res.* **63,** 1483–1489.

Yoshida, T., Maeda, A., Tani, N., and Sakai, T. (2001). Promoter structure and transcription initiation sites of the human death receptor 5/TRAIL-R2 gene. *FEBS Lett.* **507,** 381–385.

4

Transcriptional Regulation of the TRAIL-R3 Gene

Carmen Ruiz de Almodóvar,* Abelardo López-Rivas,* Juan Miguel Redondo,[†] and Antonio Rodríguez[†]

*Department of Cellular Biology and Immunology
Instituto de Parasitología y Biomedicina
Consejo Superior de Investigaciones Científicas
Granada E-18001, Spain
[†]Centro de Biología Molecular 'Severo Ochoa,' Universidad Autónoma de Madrid
Madrid E-28049, Spain, and
Centro Nacional de Investigaciones Cardiovasculares, Tres Cantos E-28760, Spain

I. Introduction: TRAIL and its Receptors
II. Decoy Receptors
III. TRAIL-R3
   A. Introduction
   B. TRAIL-R3 as an Inhibitor of TRAIL-Induced Apoptosis
   C. Promoter Structure of TRAIL Receptors
   D. Characterization of the Human TRAIL-R3 Promoter
   E. TRAIL-R3: Transcriptional Regulation
   F. Conclusions
IV. Future Directions
   References

TRAIL-R3 is a decoy receptor for TRAIL (tumor necrosis factor–related apoptosis-inducing ligand), a member of the tumor necrosis factor (TNF) ligand family. TRAIL induces apoptosis in a broad range of cancer cell lines, but not in many normal cells—a finding that generated extraordinary excitement about its potential as a specific antitumor agent. In several cell types, decoy receptors inhibit TRAIL-induced apoptosis by binding to it and preventing its binding to TRAIL proapoptotic or death receptors. However, recently published data regarding the role of these receptors in TRAIL-induced cellular death are contradictory. The key to resolving this controversy may lie in the regulation and cellular localization of TRAIL receptors. In this regard, cloning and analysis of the TRAIL-R3 promoter will help to identify the cellular factors that regulate its transcriptional expression. This chapter summarizes current knowledge in this field and outlines directions for future research. © 2004 Elsevier Inc.

I. INTRODUCTION: TRAIL AND ITS RECEPTORS

TRAIL (tumor necrosis factor–related apoptosis-inducing ligand) is a type II transmembrane protein (that is, it has an N-terminal cytoplasmic tail and a C-terminal extracellular domain) belonging to the TNF family (Wiley *et al.*, 1995). Its extracellular domain can be proteolytically cleaved from the cell surface. Unlike other members of this family, TRAIL mRNA is expressed constitutively in most tissues of the human body (Wiley *et al.*, 1995). TRAIL can selectively induce apoptosis in tumor cells but not in nontransformed cells. TRAIL has also been found to be systemically nontoxic in preclinical studies with mice and nonhuman primates (Walczak *et al.*, 1999). These characteristics make TRAIL a potential candidate for use in cancer treatment.

Like most other TNF family members, TRAIL forms a homotrimer that can bind to any one of four specific trimeric receptors of the TNF receptor superfamily (Fig. 1). It also binds to a soluble receptor called osteoprotegerin (OPG) with low affinity at physiologic temperatures (Truneh *et al.*, 2000). The transmembrane receptors are classical type I proteins, homologous to other members of the TNF receptor superfamily. Two of them, TRAIL-R1 (DR4) (Pan *et al.*, 1997b) and TRAIL-R2 (DR5, KILLER or TRICK2) (Schneider *et al.*, 1997; Walczak *et al.*, 1997), are called proapoptotic receptors because they contain the intracellular death domain essential for transmitting apoptotic stimuli after binding their ligand. In contrast, TRAIL-R3 (TRID, LIT or DcR1) (Degli-Esposti *et al.*,

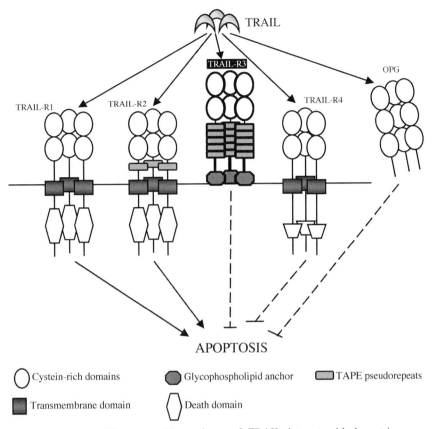

FIGURE 1. TRAIL system. Homotrimers of TRAIL interact with homotrimers or heterotrimers of TRAIL-R1 and -R2, inducing apoptosis through their cytoplasmic death domains. Trimers of TRAIL-R3 and TRAIL-R4 are also capable of binding TRAIL but do not trigger the apoptotic signal. OPG is a soluble receptor that binds TRAIL and inhibits its activity.

1997b; MacFarlane *et al.*, 1997) and TRAIL-R4 (TRUNDD or DcR2) (Degli-Esposti *et al.*, 1997a) are known as antiapoptotic or decoy receptors. Both bind TRAIL specifically, but do not signal for apoptosis. TRAIL-R3 lacks the cytoplasmic and transmembrane domains, and is bound to the cell surface through a glycosyl phosphatidyl inositol lipid anchor. TRAIL-R4 has a truncated cytoplasmic death domain, which is unable to signal for cell death. The genes for all of these four receptors are located on chromosome 8p22-21 (Fig. 2A), suggesting that they evolved from a common precursor (Degli-Esposti, 1999). TRAIL receptors, like TRAIL, are detected in many human tissues.

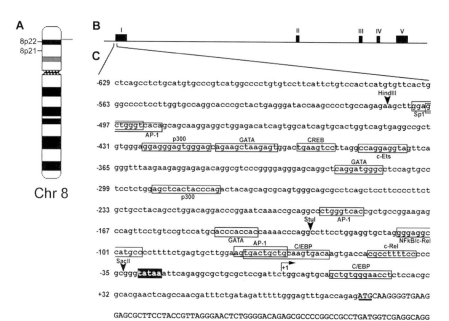

FIGURE 2. TRAIL-R3 genomic structure. A, TRAIL receptors are chromosomally linked on chromosome 8p22–21. B, The TRAIL-R3 gene contains five exons (numbered black boxes) separated by four introns. C, The promoter sequence of the gene. Predicted transcription factor binding sites identified by the TFSEARCH ver. 1.3. program are indicated as boxes. The TATA box is highlighted in black. Position +1 (bent arrow) indicates the transcription initiation site, determined by S1 nuclease protection and primer extension assays. Uppercase denotes the coding region, and the first methionine codon is underlined.

II. DECOY RECEPTORS

TRAIL and CD95L, both members of the TNF superfamily, can bind to antiapoptotic receptors, also called decoy receptors. Two decoy receptors have been described for CD95L. DcR3 is a secreted receptor with four cysteine-rich subdomains and binds CD95L in competition with CD95 (Pitti et al., 1998). FDR (Fas/CD95 decoy receptor) is a membrane-bound receptor composed of the extracellular part of Fas, but lacking its intracellular domain (Jenkins et al., 2000).

TRAIL binds to two specific decoy receptors. As previously mentioned, many human tumor cell lines are sensitive to apoptosis induced by TRAIL, while most normal cells are not. Initially it was postulated that resistance to TRAIL was due to the expression of TRAIL decoy receptors, and this was supported by early reports. Ectopic expression of TRAIL-R3 and TRAIL-R4 in TRAIL-sensitive cells made them resistant to TRAIL-induced apoptosis by sequestering the ligand and impeding its binding to proapoptotic receptors

(Degli-Esposti et al., 1997b; Pan et al., 1997a, 1998; Sheridan et al., 1997). Decoy receptors can also protect against TRAIL by forming heterotrimers with TRAIL-R2 (Munoz-Pinedo et al., 2002). Furthermore, it was found that TRAIL-R4 can activate the NF-κB pathway, thereby activating survival signals and protecting cells from apoptosis (Degli-Esposti et al., 1997a). This last result is controversial because other groups reported no NF-κB activation by TRAIL-R4 in a variety of cell types (Harper et al., 2001; Marsters et al., 1997). Moreover, recent reports suggest that there is no correlation between the expression of decoy receptors and sensitivity to TRAIL (Griffith et al., 1999; Kim et al., 2000; Walczak and Krammer, 2000).

Most of the studies published to date are based on mRNA or total protein expression of decoy receptors. A more complete elucidation of the role of TRAIL decoy receptors in TRAIL-induced apoptosis requires study of their surface expression and regulation. In this regard, Lacour et al. were able to detect TRAIL decoy receptors in colon cancer cells only after they had been permeabilized (Lacour et al., 2001). Zhang et al. have shown by confocal microscopy and flow cytometry that TRAIL-sensitive melanoma cells express decoy receptors, but that they are located in the nucleus and only relocate to the cytoplasm and cell membranes after exposure to TRAIL (Zhang et al., 2000a). These same authors also showed that the intracellular location of decoy receptors varies between cells from different tissues. They suggested that posttranslational mechanisms may play a critical role in determining whether the decoy receptors can be deployed on the cell surface and thus protect cells from TRAIL-induced apoptosis (Zhang et al., 2000b).

The regulation of the expression of TRAIL decoy receptors is poorly understood. Whereas in normal human tissues the TRAIL death and decoy receptors are coexpressed, many cancer cell lines preferentially express TRAIL-R1 and TRAIL-R2, suggesting a differential regulation of the death and decoy receptors. In this regard, an analysis of the gene promoter regions of these receptors is needed to provide an understanding of the transcriptional regulation of these genes. Beyond this, their function may also be regulated depending on their subcellular localization, suggesting the existence of more complex levels of regulation (Baetu and Hiscott, 2002; Zhang et al., 2000a).

The physiologic roles of TRAIL decoy receptors are as yet not well characterized, and it is not fully understood how these receptors modulate TRAIL signaling. At present, it is thought that they may regulate TRAIL-induced apoptosis in concert with other intracellular apoptosis inhibitors. Indeed, NF-κB, p53 and various protein kinase signaling pathways may also modulate sensitivity to TRAIL (Bortul et al., 2003; Chen et al., 2001; Meng et al., 2000; Nesterov et al., 2001; Oya et al., 2001; Shigeno et al., 2003).

III. TRAIL-R3

A. INTRODUCTION

The TRAIL-R3 sequence was identified in 1997 by searching an expressed sequence tag (EST) database sequences related to the extracellular, cysteine-rich, ligand-binding domain of TRAIL-R1 (Degli-Esposti et al., 1997b; Pan et al., 1997a). Like TRAIL-R1 and TRAIL-R2, it contains only two of the four cysteine-rich domains characteristic of most members of the TNF receptor family (Fig. 1). It shares 58% identity with TRAIL-R1, 54% with TRAIL-R2, and is stably associated to the cell surface by covalent linkage to glycosyl phosphatidyl inositol (Degli-Esposti et al., 1997b). A few groups reported that TRAIL-R3 is a 299 amino-acid protein with a predicted signal cleavage site after amino acid 69 (Degli-Esposti et al., 1997b; MacFarlane et al., 1997). Others described it as a 259-amino-acid protein with a 63-amino-acid-long signal sequence (Pan et al., 1997a; Sheridan et al., 1997). It contains an extracellular linker sequence of 88 amino acids with five copies of a 15-amino-acid pseudorepeat rich in Thr, Ala, Pro, and Glu (TAPE repeats) (see Fig. 1). This linker sequence is absent in TRAIL-R1 and appears as a single copy in TRAIL-R2. Two methionines are present within the first 60 amino acids of the ORF (Degli-Esposti et al., 1997b). Four different sizes of TRAIL-R3 transcript, resulting from alterations in the 3'-untranslated region, have been detected in different tissues (Pan et al., 1997a).

The genomic sequence of human TRAIL-R3 was published last year, and is contained within a chromosome 8-reference genomic contig (accession number NT_023666, from position 1334747 to 1349261). The TRAIL-R3 gene contains five exons separated by four introns of 8540, 2832, 738, and 1025 bp, respectively (Fig. 2B). The first exon encodes the 5' untranslated region (UTR) and the first 60 amino-acid residues of the open reading frame (ORF). Exon II codes for the next 35 amino acids, from position 61 to 95, and includes the predicted signal peptide cleavage site (Degli-Esposti et al., 1997b). Exons III and IV encode for two of the cysteine-rich domains characteristic of the TNF receptor family. The rest of the protein is encoded by exon V. This region contains the extreme carboxy-terminal hydrophobic region, the 88 amino-acid linker sequence, and the 3' UTR. The linker contains the five copies of the TAPE (see Fig. 1) 15 amino-acid pseudo-repeat and separates the cysteine-rich extracellular domain from the hydrophobic C-terminus.

The tissue and cellular distribution of TRAIL-R3 transcripts seems to be more restricted compared with the other receptors (Degli-Esposti et al., 1997b; Pan et al., 1997a; Schneider et al., 1997). Its transcripts have been detected in several normal human tissues but exist at substantially lower amounts in most transformed cells lines (Pan et al., 1997a; Sheridan et al., 1997).

B. TRAIL-R3 AS AN INHIBITOR OF TRAIL-INDUCED APOPTOSIS

Transient high-level expression of TRAIL-R3 in TRAIL-sensitive cells confers resistance against TRAIL-induced apoptosis (Pan et al., 1997a; Sheridan et al., 1997). Because TRAIL-R3 is anchored to the cell surface by a phospatidylinositol tail, it can be specifically removed by treating cells with phosphatidylinositol-specific phospholipase C (PI-PLC) (Kothny-Wilkes et al., 1998; Sheridan et al., 1997). This specific method has been used by some researchers to demonstrate that TRAIL-R3 protects different cell types from TRAIL-induced apoptosis. Bernard et al. showed that high expression levels of c-Rel or activation of endogenous Rel/NF-κB factors in HeLa cells, a human cervical carcinoma cell line, made them resistant to apoptosis induced by TRAIL. TRAIL-R3 was upregulated under these conditions, and this upregulation was responsible for the acquired resistance because removal of TRAIL-R3 from the cell surface with PI-PLC rendered these cells sensitive to TRAIL once again (Bernard et al., 2001). TRAIL-R3 can be detected at the surface of human umbilical vein endothelial cells (HUVEC) by flow cytometry, and pretreatment of these cells with recombinant PI-PLC depletes TRAIL-R3 from the cell surface and sensitizes them to TRAIL-induced apoptosis (Zhang et al., 2000b). Similar findings have been reported in granulosa cells, where removing TRAIL-R3 from healthy follicles with PI-PLC made them more sensitive to TRAIL-induced apoptosis, demonstrating an inhibitory role for TRAIL-R3 in apoptosis in this cell type (Wada et al., 2002).

C. PROMOTER STRUCTURE OF TRAIL RECEPTORS

The promoter regions for TRAIL-R1, TRAIL-R2, and TRAIL-R3, but not TRAIL-R4, have been characterized by several groups. The TRAIL-R1 promoter lacks a typical TATA box but contains Sp1 and MyoD binding sites between nucleotide positions -117 and -19, which suggests that these transcription factors may contribute to the basic promoter activity of this gene. Regulation of TRAIL-R1 expression via the AP-1 binding site located at -350/-344 has been conformed by luciferase reporter and electrophoretic mobility shift assays (Guan et al., 2001).

The TRAIL-R2 promoter also lacks a consensus TATA box sequence but instead has two Sp1 sites responsible for the basal transcriptional activity of the gene (Yoshida et al., 2001). Transcriptional regulation of TRAIL-R2 by p53 has been described, and although there are no functional binding sites for p53 within TRAIL-R2 promoter, there is a p53 response element in the first intron, which is responsible for the observed p53-dependent transactivation (Takimoto and El-Deiry, 2000). This feature has been also found in the CD95 gene, another receptor of the TNF superfamily (Muller et al., 1998; Ruiz-Ruiz et al., 2003).

D. CHARACTERIZATION OF THE HUMAN TRAIL-R3 PROMOTER

The TRAIL-R3 promoter region was described last year by our group (Ruiz de Almodóvar et al., 2002). Despite the publication of most of the human genome sequence, at that time the genomic organization of human TRAIL-R3 was not available. We therefore used the Human Promoter Finder DNA Walking Kit (Clontech) to find the genomic sequence upstream of human TRAIL-R3 cDNA. The cloning of the TRAIL-R3 5′-upstream region was based on nested PCR, using two gene-specific primers spanning regions near the translation start site. The amplified TRAIL-R3 upstream sequence was analyzed for regulatory regions with the TFSEARCH ver. 1.3. program (Fig. 2B). The presence of a consensus TATA box sequence as well as potential binding sites for known transcription factors suggested that this genomic region contained the human TRAIL-R3 promoter. This was confirmed by the identification of the transcription start site by S1 nuclease protection and primer extension assays. To confirm that the 5′ upstream region of TRAIL-R3 contained the basic elements for basal promoter activity, luciferase reporter assays were performed with serial deletions of the TRAIL-R3 promoter. The results of these transfection assays revealed that the basal transcriptional activity progressively increased with the length of the 5′ upstream region of the TRAIL-R3 promoter, indicating that all these regions contain elements that contribute to the overall promoter activity, and that the region spanning nucleotides −33 to +42 contains the minimal promoter element of the human TRAIL-R3 gene.

E. TRAIL-R3: TRANSCRIPTIONAL REGULATION

Little is known about TRAIL-R3 regulation. As mentioned previously, overexpression of c-Rel or activation of endogenous Rel/NF-κB factors by TNF-α in HeLa cells upregulates TRAIL-R3 and makes cells resistant to TRAIL-induced apoptosis. An analysis of potential transcription binding sites within the TRAIL-R3 promoter reveals that cRel/NF-κB–dependent regulation of TRAIL-R3 could occur through direct binding to any of the predicted binding sites for NF-κB within the promoter sequence (Fig. 2B). As is the case for CD95 (Muller et al., 1998; Ruiz-Ruiz et al., 2003) and other p53-regulated TRAIL receptors (Guan et al., 2001; Meng et al., 2000; Sheikh et al., 1998), TRAIL-R3 is induced by ionizing radiation and the genotoxic agent methyl methasulfonate in a p53-dependent manner (Sheikh et al., 1999). Although TRAIL-R3 appears to be p53-regulated, there are no p53 binding sites within the described 5′-upstream sequence of the gene (Ruiz de Almodóvar et al., 2002). We observed a p53-independent upregulation of TRAIL-R3 mRNA upon doxorubicin treatment of the breast tumor cell line MCF-7. Cells stably transfected with the human

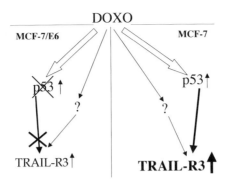

FIGURE 3. Doxorubicin-induced upregulation of TRAIL-R3 mRNA. Although TRAIL-R3 upregulation mediated by doxorubicin (DOXO) treatment is predominantly p53-dependent in MCF-7 cells *(right)*, a weaker induction of TRAIL-R3 mRNA can be detected in the absence of p53 *(left)*, as observed in MCF-7/E6 cell line (see text for details).

papilloma virus type 16 E6 protein, which induces p53 degradation and thus prevents p53 accumulation in response to genotoxic damage (Ruiz-Ruiz and López-Rivas, 1999), showed a weaker induction of TRAIL-R3 mRNA than wild-type MCF-7 cells, suggesting that p53 mediates additional signals that contribute to TRAIL-R3 gene expression (Fig. 3) (Ruiz de Almodóvar *et al.*, 2002). We have not excluded the existence of p53 responsive sites in other regions of the gene and it may prove to be that such a site underlies the responsiveness of TRAIL-R3 to p53, as occurs in TRAIL-R2 and CD95 (Muller *et al.*, 1998; Takimoto and El-Deiry, 2000). In fact, we have recently characterized a p53 consensus element located within the first intron of the human TRAIL-R3 gene (Ruiz de Almodóvar *et al.*, 2003). The significance of the upregulation of TRAIL decoy receptors by p53 remains unclear. It may be that just as p53 activates transcription of the TRAIL-R2 gene to mediate apoptosis, it activates TRAIL decoy receptors as a way of controlling its own apoptotic effects by blunting TRAIL-R1/R2–dependent apoptosis. More studies are needed to gain a full understanding of TRAIL-R3's functions as an antiapoptotic receptor and in p53-mediated apoptosis. Finally, as a new level of transcriptional regulation, it has been suggested that hypermethylation of the promoter of TRAIL decoy receptors is involved in the downregulation of the expression of these genes in neuroblastoma and other tumor types, such as brain, colon, and skin cancers (van Noesel *et al.*, 2002).

F. CONCLUSIONS

The data accumulated so far could be the starting point for a deeper analysis of the human TRAIL-R3 gene at the transcriptional level. As mentioned previously, and based on bioinformatic analysis, several different transcription factors could bind to the human TRAIL-R3 promoter region and regulate its expression. Further studies dissecting the cis- and trans-acting elements involved in the transcriptional regulation of TRAIL-R3 will facilitate the study of the mechanisms that control the expression of this antiapoptotic decoy receptor, a protein with potent tumor resistance properties.

IV. FUTURE DIRECTIONS

TRAIL and its complex receptor system, formed by widely expressed death and decoy receptors, present a complex picture that is far from being fully understood. Death and decoy receptors are often coexpressed within the same cell, giving rise to poor correlations between TRAIL receptor expression and TRAIL sensitivity. To explain the differential susceptibility to TRAIL-induced apoptosis, other factors must be taken into consideration. One such factor is the differential regulation of the death and decoy receptors by key regulators of apoptosis, such as NF-κB and p53, as well as their cellular localization after TRAIL treatment. The recent molecular characterization of the promoter region of several TRAIL receptors is helping to elucidate how these transcription factors and others may affect the expression of these proteins. So far, we are still a long way from knowing the key regulators of their expression and subcellular localization. Further knowledge of the regulation and physiologic role of TRAIL and its receptors may enable the future exploitation of the TRAIL signaling pathway to selectively eliminate tumor cells.

REFERENCES

Baetu, T. M., and Hiscott, J. (2002). On the TRAIL to apoptosis. *Cytokine Growth Factor Rev.* **13,** 199–207.

Bernard, D., Quatannens, B., Vandenbunder, B., and Abbadie, C. (2001). Rel/NF-kappaB transcription factors protect against tumor necrosis factor (TNF)-related apoptosis-inducing ligand (TRAIL)-induced apoptosis by up-regulating the TRAIL decoy receptor DcR1. *J. Biol. Chem.* **276,** 27322–27328.

Bortul, R., Tazzari, P. L., Cappellini, A., Tabellini, G., Billi, A. M., Bareggi, R., Manzoli, L., Cocco, L., and Martelli, A. M. (2003). Constitutively active Akt1 protects HL60 leukemia cells from TRAIL-induced apoptosis through a mechanism involving NF-kappaB activation and cFLIP(L) up-regulation. *Leukemia* **17,** 379–389.

Chen, X., Thakkar, H., Tyan, F., Gim, S., Robinson, H., Lee, C., Pandey, S. K., Nwokorie, C., Onwudiwe, N., and Srivastava, R. K. (2001). Constitutively active Akt is an important regulator of TRAIL sensitivity in prostate cancer. *Oncogene* **20,** 6073–6083.

Degli-Esposti, M. (1999). To die or not to die—the quest of the TRAIL receptors. *J. Leukocyte Biol.* **65,** 535–542.

Degli-Esposti, M. A., Dougall, W. C., Smolak, P. J., Waugh, J. Y., Smith, C. A., and Goodwin, R. G. (1997a). The novel receptor TRAIL-R4 induces NF-kappaB and protects against TRAIL-mediated apoptosis, yet retains an incomplete death domain. *Immunity* **7,** 813–820.

Degli-Esposti, M. A., Smolak, P. J., Walczak, H., Waugh, J., Huang, C. P., DuBose, R. F., Goodwin, R. G., and Smith, C. A. (1997b). Cloning and characterization of TRAIL-R3, a novel member of the emerging TRAIL receptor family. *J. Exp. Med.* **186,** 1165–1170.

Griffith, T. S., Rauch, C. T., Smolak, P. J., Waugh, J. Y., Boiani, N., Lynch, D. H., Smith, C. A., Goodwin, R. G., and Kubin, M. Z. (1999). Functional analysis of TRAIL receptors using monoclonal antibodies. *J. Immunol.* **162,** 2597–2605.

Guan, B., Yue, P., Clayman, G. L., and Sun, S. Y. (2001). Evidence that the death receptor DR4 is a DNA damage-inducible, p53-regulated gene. *J. Cell Physiol.* **188,** 98–105.

Harper, N., Farrow, S. N., Kaptein, A., Cohen, G. M., and MacFarlane, M. (2001). Modulation of tumor necrosis factor apoptosis-inducing ligand-induced NF-kappa B activation by inhibition of apical caspases. *J. Biol. Chem.* **276,** 34743–34752.

Jenkins, M., Keir, M., and McCune, J. M. (2000). A membrane-bound Fas decoy receptor expressed by human thymocytes. *J. Biol. Chem.* **275,** 7988–7993.

Kim, K., Fisher, M. J., Xu, S. Q., and El-Deiry, W. S. (2000). Molecular determinants of response to TRAIL in killing of normal and cancer cells. *Clin. Cancer Res.* **6,** 335–346.

Kothny-Wilkes, G., Kulms, D., Poppelmann, B., Luger, T. A., Kubin, M., and Schwarz, T. (1998). Interleukin-1 protects transformed keratinocytes from tumor necrosis factor-related apoptosis-inducing ligand. *J. Biol. Chem.* **273,** 29247–29253.

Lacour, S., Hammann, A., Wotawa, A., Corcos, L., Solary, E., and Dimanche-Boitrel, M. T. (2001). Anticancer agents sensitize tumor cells to tumor necrosis factor-related apoptosis-inducing ligand-mediated caspase-8 activation and apoptosis. *Cancer Res.* **61,** 1645–1651.

MacFarlane, M., Ahmad, M., Srinivasula, S. M., Fernandes-Alnemri, T., Cohen, G. M., and Alnemri, E. S. (1997). Identification and molecular cloning of two novel receptors for the cytotoxic ligand TRAIL. *J. Biol. Chem.* **272,** 25417–25420.

Marsters, S. A., Sheridan, J. P., Pitti, R. M., Huang, A., Skubatch, M., Baldwin, D., Yuan, J., Gurney, A., Goddard, A. D., Godowski, P., and Ashkenazi, A. (1997). A novel receptor for Apo2L/TRAIL contains a truncated death domain. *Curr. Biol.* **7,** 1003–1006.

Meng, R. D., McDonald, E. R., 3rd, Sheikh, M. S., Fornace, A. J., Jr., and El-Deiry, W. S. (2000). The TRAIL decoy receptor TRUNDD (DcR2, TRAIL-R4) is induced by adenovirus-p53 overexpression and can delay TRAIL-, p53-, and KILLER/DR5-dependent colon cancer apoptosis. *Mol. Ther.* **1,** 130–144.

Muller, M., Wilder, S., Bannasch, D., Israeli, D., Lehlbach, K., Li-Weber, M., Friedman, S. L., Galle, P. R., Stremmel, W., Oren, M., and Krammer, P. H. (1998). p53 activates the CD95 (APO-1/Fas) gene in response to DNA damage by anticancer drugs. *J. Exp. Med.* **188,** 2033–2045.

Munoz-Pinedo, C., Ruiz de Almodóvar, C., and Ruiz-Ruiz, C. (2002). Death on the beach: A rosy forecast for the 21st century. *Cell Death Differ.* **9,** 1026–1029.

Nesterov, A., Lu, X., Johnson, M., Miller, G. J., Ivashchenko, Y., and Kraft, A. S. (2001). Elevated AKT activity protects the prostate cancer cell line LNCaP from TRAIL-induced apoptosis. *J. Biol. Chem.* **276,** 10767–10774.

Oya, M., Ohtsubo, M., Takayanagi, A., Tachibana, M., Shimizu, N., and Murai, M. (2001). Constitutive activation of nuclear factor-kappaB prevents TRAIL-induced apoptosis in renal cancer cells. *Oncogene* **20,** 3888–3896.

Pan, G., Ni, J., Wei, Y. F., Yu, G., Gentz, R., and Dixit, V. M. (1997a). An antagonist decoy receptor and a death domain-containing receptor for TRAIL. *Science* **277,** 815–818.

Pan, G., Ni, J., Yu, G., Wei, Y. F., and Dixit, V. M. (1998). TRUNDD, a new member of the TRAIL receptor family that antagonizes TRAIL signalling. *FEBS Lett.* **424,** 41–45.

Pan, G., O'Rourke, K., Chinnaiyan, A. M., Gentz, R., Ebner, R., Ni, J., and Dixit, V. M. (1997b). The receptor for the cytotoxic ligand TRAIL. *Science* **276,** 111–113.

Pitti, R. M., Marsters, S. A., Lawrence, D. A., Roy, M., Kischkel, F. C., Dowd, P., Huang, A., Donahue, C. J., Sherwood, S. W., Baldwin, D. T., Godowski, P. J., Wood, W. I., Gurney, A. L., Hillan, K. J., Cohen, R. L., Goddard, A. D., Botstein, D., and Ashkenazi, A. (1998). Genomic amplification of a decoy receptor for Fas ligand in lung and colon cancer. *Nature* **396,** 699–703.

Ruiz de Almodóvar, C., López-Rivas, A., Redondo, J. M., and Rodriguez, A. (2002). Transcription initiation sites and promoter structure of the human TRAIL-R3 gene. *FEBS Lett.* **531,** 304–308.

Ruiz de Almodóvar, C., Ruiz-Ruiz, C., Rodríguez, A., Ortiz-Ferrón, G., Redondo, J. M., and López-Rivas, A. (2003). TRAIL decoy receptor TRAIL-R3 is upregulated by p53 in breast tumor cells through a mechanism involving an intrinsic p53 binding site. *J. Biol. Chem.* Nov. 17 [Epub ahead of print].

Ruiz-Ruiz, C., Robledo, G., Cano, E., Redondo, J. M., and López-Rivas, A. (2003). Characterization of p53-mediated up-regulation of CD95 gene expression upon genotoxic treatment in human breast tumor cells. *J. Biol. Chem.* **4,** 4.

Ruiz-Ruiz, M. C., and López-Rivas, A. (1999). p53-mediated up-regulation of CD95 is not involved in genotoxic drug-induced apoptosis of human breast tumor cells. *Cell Death Differ.* **6,** 271–280.

Schneider, P., Bodmer, J. L., Thome, M., Hofmann, K., Holler, N., and Tschopp, J. (1997). Characterization of two receptors for TRAIL. *FEBS Lett.* **416,** 329–334.

Sheikh, M. S., Burns, T. F., Huang, Y., Wu, G. S., Amundson, S., Brooks, K. S., Fornace, A. J., Jr., and El-Deiry, W. S. (1998). p53-dependent and -independent regulation of the death receptor KILLER/DR5 gene expression in response to genotoxic stress and tumor necrosis factor alpha. *Cancer Res.* **58,** 1593–1598.

Sheikh, M. S., Huang, Y., Fernandez-Salas, E. A., El-Deiry, W. S., Friess, H., Amundson, S., Yin, J., Meltzer, S. J., Holbrook, N. J., and Fornace, A. J., Jr. (1999). The antiapoptotic decoy receptor TRID/TRAIL-R3 is a p53-regulated DNA damage-inducible gene that is overexpressed in primary tumors of the gastrointestinal tract. *Oncogene* **18,** 4153–4159.

Sheridan, J. P., Marsters, S. A., Pitti, R. M., Gurney, A., Skubatch, M., Baldwin, D., Ramakrishnan, L., Gray, C. L., Baker, K., Wood, W. I., Goddard, A. D., Godowski, P., and Ashkenazi, A. (1997). Control of TRAIL-induced apoptosis by a family of signaling and decoy receptors. *Science* **277,** 818–821.

Shigeno, M., Nakao, K., Ichikawa, T., Suzuki, K., Kawakami, A., Abiru, S., Miyazoe, S., Nakagawa, Y., Ishikawa, H., Hamasaki, K., Nakata, K., Ishii, N., and Eguchi, K. (2003). Interferon-alpha sensitizes human hepatoma cells to TRAIL-induced apoptosis through DR5 upregulation and NF-kappa B inactivation. *Oncogene* **22,** 1653–1662.

Takimoto, R., and El-Deiry, W. S. (2000). Wild-type p53 transactivates the KILLER/DR5 gene through an intronic sequence-specific DNA-binding site. *Oncogene* **19,** 1735–1743.

Truneh, A., Sharma, S., Silverman, C., Khandekar, S., Reddy, M. P., Deen, K. C., McLaughlin, M. M., Srinivasula, S. M., Livi, G. P., Marshall, L. A., Alnemri, E. S., Williams, W. V., and Doyle, M. L. (2000). Temperature-sensitive differential affinity of TRAIL for its receptors. DR5 is the highest affinity receptor. *J. Biol. Chem.* **275,** 23319–23325.

van Noesel, M. M., van Bezouw, S., Salomons, G. S., Voute, P. A., Pieters, R., Baylin, S. B., Herman, J. G., and Versteeg, R. (2002). Tumor-specific down-regulation of the tumor necrosis factor-related apoptosis-inducing ligand decoy receptors DcR1 and DcR2 is associated with dense promoter hypermethylation. *Cancer Res.* **62,** 2157–2161.

Wada, S., Manabe, N., Nakayama, M., Inou, N., Matsui, T., and Miyamoto, H. (2002). TRAIL-decoy receptor 1 plays inhibitory role in apoptosis of granulosa cells from pig ovarian follicles. *J. Vet. Med. Sci.* **64,** 435–439.

Walczak, H., Degli-Esposti, M. A., Johnson, R. S., Smolak, P. J., Waugh, J. Y., Boiani, N., Timour, M. S., Gerhart, M. J., Schooley, K. A., Smith, C. A., Goodwin, R. G., and

Rauch, C. T. (1997). TRAIL-R2: A novel apoptosis-mediating receptor for TRAIL. *EMBO J.* **16,** 5386–5397.

Walczak, H., and Krammer, P. H. (2000). The CD95 (APO-1/Fas) and the TRAIL (APO-2L) apoptosis systems. *Exp. Cell Res.* **256,** 58–66.

Walczak, H., Miller, R. E., Ariail, K., Gliniak, B., Griffith, T. S., Kubin, M., Chin, W., Jones, J., Woodward, A., Le, T., Smith, C., Smolak, P., Goodwin, R. G., Rauch, C. T., Schuh, J. C., and Lynch, D. H. (1999). Tumoricidal activity of tumor necrosis factor-related apoptosis-inducing ligand in vivo. *Nat. Med.* **5,** 157–163.

Wiley, S. R., Schooley, K., Smolak, P. J., Din, W. S., Huang, C. P., Nicholl, J. K., Sutherland, G. R., Smith, T. D., Rauch, C., Smith, C. A. *et al.* (1995). Identification and characterization of a new member of the TNF family that induces apoptosis. *Immunity* **3,** 673–682.

Yoshida, T., Maeda, A., Tani, N., and Sakai, T. (2001). Promoter structure and transcription initiation sites of the human death receptor 5/TRAIL-R2 gene. *FEBS Lett.* **507,** 381–385.

Zhang, X. D., Franco, A. V., Nguyen, T., Gray, C. P., and Hersey, P. (2000a). Differential localization and regulation of death and decoy receptors for TNF-related apoptosis-inducing ligand (TRAIL) in human melanoma cells. *J. Immunol.* **164,** 3961–3970.

Zhang, X. D., Nguyen, T., Thomas, W. D., Sanders, J. E., and Hersey, P. (2000b). Mechanisms of resistance of normal cells to TRAIL induced apoptosis vary between different cell types. *FEBS Lett.* **482,** 193–199.

5

Monoclonal Antibodies Against TRAIL

Nina-Beate Liabakk and Terje Espevik

Institute of Cancer Research and Molecular Medicine
Norwegian University of Science and Technology, N-7489 Trondheim, Norway

    I. Introduction
   II. Overview of Monoclonal Antibodies Against TRAIL
  III. Applications of Monoclonal Antibodies Against TRAIL
      A. *Immunoassays for the Detection of Soluble TRAIL*
      B. *Binding of TRAIL Monoclonal Antibodies to Cells*
  IV. Physiologic Significance of TRAIL
   V. Future Aspects
      References

TRAIL (tumor necrosis factor related apoptosis inducing ligand) is a cytokine proposed to be used in cancer therapy, since it kills cancer cells but not normal cells. Also, recent studies report that TRAIL inhibits the development of arthritis. In order to investigate the role of TRAIL in health and disease, monoclonal antibodies against TRAIL have been developed. This chapter gives an overview of different monoclonal antibodies against TRAIL which are published or commercially available. Monoclonal antibodies against TRAIL are useful in different immunological techniques, and this chapter presents an overview of the applications of these antibodies with a focus on immunoassays for detection of soluble TRAIL. In addition, the physiological significance of some results obtained by using monoclonal antibodies against TRAIL are discussed. © 2004 Elsevier Inc.

I. INTRODUCTION

Tumor necrosis factor (TNF)–related apoptosis-inducing ligand (TRAIL), also called apoptosis 2 ligand (APO2L) because of its similarity in sequence, structure, and function to Fas ligand/Apo1L and TNF, is a member of the TNF superfamily (TNFSF), designated TNFSF10. TRAIL is constitutively expressed in many tissues such as liver, lung, placenta, kidney, spleen, peripheral lymphocytes, and immune-privileged sites (Pitti et al., 1996; Wiley et al., 1995). TRAIL exists as a membrane bound form and a soluble form, and both forms can induce apoptosis in various transformed cell lines (Sheridan et al., 1997), while no cytotoxicity is observed in normal human cell lines (Atkins et al., 2002). Despite the fact that TRAIL is expressed in normal nontransformed tissue, its role in this tissue is unknown. The selective killing of the cancer cells involves binding to two cell surface receptors, death receptor (DR) 4 and DR5. Two additional TRAIL decoy receptors (DcR1 and DcR2) and a soluble receptor called osteoprotegerin (OPG) do not trigger an apoptotic signal and have been proposed to protect against TRAIL-induced apoptosis (Holen et al., 2002).

In addition to its antitumor potential, TRAIL is supposed to have antiviral (Mundt et al., 2003; Sedger et al., 1999; Vidalain et al., 2000), anti-inflammatory (Hilliard et al., 2001; Lunemann et al., 2002; Song et al., 2000), and antiautoimmune (Hilliard et al., 2001; Ichikawa et al., 2003; Kayagaki et al., 2002; Song et al., 2000; Wandinger et al., 2003) properties. The different properties of TRAIL make it promising as a therapeutic agent in different disease conditions.

TRAIL also exists as a soluble form, representing the extracellular domain of membrane-bound TRAIL. Soluble TRAIL (sTRAIL) can be proteolytically cleaved from the cell surface. Bioactive sTRAIL has a homotrimeric structure and contains a zinc atom bound to the cystein residue at position 230, which is crucial for trimer stability and biologic activity (Bodmer et al., 2000; Lawrence et al., 2001). However, the physiologic significance of membrane-bound TRAIL and sTRAIL are still not completely understood. Monoclonal antibodies against TRAIL may therefore be a useful tool to investigate the regulation of TRAIL expression and the release of sTRAIL into the circulation during different disease states.

In this chapter we will give an overview of different monoclonal antibodies against TRAIL that are published or commercially available. Thereafter, we will describe the application of some monoclonal antibodies against TRAIL in different immunologic methods with focus on immunoassays for detection of sTRAIL. Finally, we will discuss the physiologic significance of some results obtained by using monoclonal antibodies against TRAIL.

II. OVERVIEW OF MONOCLONAL ANTIBODIES AGAINST TRAIL

Monoclonal antibodies against TRAIL can be made by using the hybridoma technology (Köhler and Milstein, 1975). Hybridoma cells are made by fusing spleen cells from mice immunized with the immunogen TRAIL and NSO plasmacytoma cells. This technique gives rise to many hybridoma clones, each secreting a unique monoclonal antibody against TRAIL. Antibodies in the growth media can further be purified and used in different immunological techniques. The hybridoma technology gives rise to many antibodies, which differ with respect to binding sites against TRAIL, affinity for TRAIL, and isotype. The diversity of monoclonal antibodies make them suitable in different immunologic techniques such as enzyme-linked immunosorbent assays (ELISAs), flow cytometric analysis, Western blot analysis, immunoprecipitation, immunohistochemistry, and biologic analysis. Each technique needs antibodies that are sensitive and specific against TRAIL. Antibodies with defined binding characterisics are important, especially in ELISA and flow cytometric analysis. In addition, antibodies that give the best signal and lowest background are preferred, especially in immunohistochemistry. In other immunologic techniques, like *in vivo* experiments and cytotoxicity assays, antibodies that neutralize bioactivity of TRAIL are needed. TRAIL is able to induce apoptosis in a large number of tumor cell lines, and a neutralizing antibody against TRAIL is able to inhibit this apoptotic effect.

During past years, the availability of monoclonal antibodies against TRAIL has increased considerably. Many monoclonal antibodies against TRAIL are now commercially available. Table I gives an overview of some of the available monoclonal antibodies against TRAIL. Table I also includes information about applications, references, and commercial availability. References are based on use of the actual antibody and the origin, if possible. The antibodies are listed randomly.

All the monoclonal antibodies against TRAIL listed in Table I are murine monoclonal antibodies against human TRAIL, except one. Much TRAIL reseach is done in mice, needing monoclonal antibodies against murine TRAIL. In most of the reported studies the rat antimurine TRAIL,

TABLE I. Overview of Monoclonal Antibodies Against TRAIL

Clone	Isotype	Immunogen	Application	References	Commercial available from
2E5	Mouse IgG1	rhuTRAIL, soluble aa 95-281	NB	Plasilova et al., 2002	Novus Biological, Inc., Littleton, CO
2E5	Mouse IgG1	rhuTRAIL, soluble aa 95-281	FC, NB	Plasilova et al., 2002	Alexis Biochemicals, San Diego, CA
2E5	Mouse IgG1	rhuTRAIL, soluble aa 95-281	NB	Plasilova et al., 2002	Serotec, Oxford, UK
RIK-2	Mouse IgG1	hTRAIL/2PK-3 cells	FC, NB, IFS	Kayagaki et al., 1999; Oshima et al., 2001; Miura et al., 2001	Bioscience, San Diego, CA
M181	Mouse IgG1	ns	FC	Sedger et al., 1999; Griffith et al., 1999; Dorothee et al., 2002	Immunex Corp., Seattle, WA
M15	ns	ns	FC	Zhang et al., 2000	Immunex Corp., Seattle, WA
5C2	ns	aa114-281	NB	Martinez-Lorenzo et al., 1998	Genentech, San Francisco, CA
VI10E	Mouse IgG2b	NSO-derived rhTRAIL	ELISA, FC, IP	Liabakk et al., 2002	Alexis Biochemicals, San Diego, CA
III6F	Mouse IgG2b	NSO-derived rhTRAIL	ELISA, WB, FC, IP, IH	Liabakk et al., 2002	Alexis Biochemicals, San Diego, CA
MAB375	Mouse IgG1	NSO-derived rhTRAIL	ELISA, WB, IH, NB	Liabakk et al., 2002	R&D Systems, Abingdon, UK
MAB687	Mouse IgG1	NSO-derived rhTRAIL	ELISA, WB, FC, IH		R&D Systems, Abingdon, UK
HS501	Mouse IgG1	r hu sTRAIL	WB	Washburn et al., 2003	Alexis Biochemicals, San Diego, CA
Ab against TRAIL	ns	ns		Baetu et al., 2001	Immunex Corp., Seattle, WA
Ab against TRAIL	ns	ns		Herbeuval et al., 2003	
clone 75411.11	Mouse IgG1	rhuTRAIL	ELISA, WB, NB		Sigma-Aldrich, Inc., St. Louis, MO

Clone	Isotype	Immunogen	Application	References	Source
T8175-50	Mouse IgG	KLH-conjugated huTRAIL, aa17-35	ELISA, WB		United States Biological, Swampscott, MA
2B2.108	Mouse IgG1	rTRAIL	FC		United States Biological, Swampscott, MA
B-T24	Mouse IgG1	rTRAIL	FC, NB		BioSource International, Inc., Camarillo, CA
55B709.3	Mouse IgG1	aa17-35	WB		IMGENEX, San Diego, CA
D3	Mouse IgG1	aa25-281	WB, IH		Santa Cruz Biotechnology, Inc., Santa Cruz, CA
C19	Goat IgG	C-terminus	WB, IH		Santa Cruz Biotechnology, Inc., Santa Cruz, CA
H257	Rabbit IgG	aa25-281	WB, IP, IH		Santa Cruz Biotechnology, Inc., Santa Cruz, CA
500-M49	Mouse IgG	rhuTRAIL	ELISA		PeproTech Inc., Cytokines for the Americas, Rocky Hill, NJ
05-607	Mouse IgG	KLH-conjugated huTRAIL, aa17-35	WB		Upstate, Lake Placid, NY
B-T24	Mouse IgG1	rhuTRAIL	ELISA, FC, NB		Diaclone, Besancon, France
B-S23*	Mouse IgG1	rhuTRAIL	ELISA, FC, enhances TRAIL induced apoptosis		Diaclone, Besancon, France
N2B2[a]	Rat IgG	ns	FC, NB	Kayagaki et al., 1999; Seki et al., 2003; Cretney et al., 2002; Takeda et al., 2002; Smyth et al., 2001; Lee et al., 2002	Bioscience, San Diego, CA
2E11[a]	ns	ns	ELISA, FC	Ashkenazi et al., 1999; Halaas et al., 2000; Johnsen et al., 1999	Genentech, San Francisco, CA
2G9[a]	ns	ns	ELISA	Ashkenazi et al., 1999	Genentech, San Francisco, CA

ns: not specified; [a]: antimurine; FC: flow cytometric analysis; WB: Western blot; IP: immunoprecipitation; IH: immunohistochemistry; NB: neutralization of bioactivity; and IFS: immunofluorescence staining.
*Enhances TRAIL-induced apoptosis.

named N2B2 is used (Cretney *et al.*, 2002; Kayagaki *et al.*, 1999; Lee *et al.*, 2002; Seki *et al.*, 2003; Takeda *et al.*, 2002).

III. APPLICATIONS OF MONOCLONAL ANTIBODIES AGAINST TRAIL

A. IMMUNOASSAYS FOR DETECTION OF SOLUBLE TRAIL

So far, the physiologic significance of soluble TRAIL (sTRAIL) in health and disease is poorly understood. To investigate the role of sTRAIL, immunoassays for detection of sTRAIL may be useful. Monoclonal antibodies against TRAIL have been used to establish ELISAs for detection of sTRAIL.

Liabakk *et al.* 2002 describe the development of an ELISA for the detection of sTRAIL. The principle behind the ELISA involves coating of a monoclonal antibody (VI10E or III6F) onto a microplate. Standards and samples are added to the wells and any TRAIL present is bound to the immobilized antibody. After washing away any unbound substances, a digitoxin (DIG) labeled monoclonal antibody specific for TRAIL (MAB375, R&D Systems) is added to the wells. Following a wash to remove any unbound DIG labeled antibody, enzyme labeled anti DIG Fab fragments are added. Following a wash to remove any unbound enzyme labeled anti-DIG Fab fragments, a substrate solution is added to the wells and color develops in proportion to the amount of TRAIL bound in the initial step. The color development is stopped and the intensity of the color is measured.

Recently, several ELISA kits for detection of sTRAIL have become commercially available. Table II gives an overview of available ELISAs for detection of sTRAIL. Table II also includes commercial availability, detection limit, assay range, and sample material. In addition to the ELISA kits listed in Table II, several monoclonal antibodies against TRAIL are reported to be useful in ELISAs for detection of sTRAIL (see Table I).

Monoclonal antibodies are essential in developing ELISAs. The binding properties of the antibodies, like the affinity to TRAIL, are of great importance. Usually, high-affinity antibodies are most suitable because the reaction between the antibody and TRAIL becomes both quicker and more specific. To establish a TRAIL ELISA, it is necessary to have two specific antibodies against TRAIL that recognize different epitopes on the TRAIL molecule. The quality of the antibodies used in an ELISA may influence both the specificity, i.e. if the antibodies cross react with other similar molecules and the sensitivity, i.e. the detection limit of the assay. The specificity of the antibodies is important if TRAIL is measured in a biologic

TABLE II. Overview of ELISAs for Detection of sTRAIL

Reference/Commercial availability	Detection limit (pg/mL)	Assay range (pg/mL)	Serum	Plasma	CCS	CL	Saliva	FS	BS
Ashkenazi et al., 1999[a]	ns	ns	x		x				
Liabakk et al., 2002	150	150–40.000	x	x	x	x	nt		
BioSource Internation Inc., USA	20	46.8–3000	x	x	x	x			x
R&D Systems	2.9	15.6–1000	x	x	x	x	x		
BioSource Internation Inc., USA	20	46.8–3000	x						
Peprotech	0.2[b]	ns							
ImgenexTRAIL ActivELISA kit, BIOCARTA US	ns	ns				x			
Active Motif Europe, Belgium	ns	30–2000				x		x	
Pharmacia, Peapack, NJ; Robertson et al., 2002									
Wandinger et al., 2003; Trinova Biochem, Giesen, Germany	64	93.75–3000	x						

[a] ELISA for detection of murine TRAIL.
[b] ng/well.
nt: not tested; ns: not specified; CCS: cell culture supernatant; CL: cell lysates; FS: fluid samples; BS: buffered solutions.

fluid, like a serum sample, which contains other factors with sequence homology that may cross react with the antibody. Liabakk et al. 2002 describe an ELISA based on two monoclonal antibodies. However, also polyclonal antibodies, i.e., antibodies that recognize several epitopes on the TRAIL molecule, are useful as secondary antibodies in TRAIL ELISAs (Active Motif Europe, Belgium). A combination between a monoclonal and a polyclonal antibody may even increase the specificity and the sensitivity of an ELISA. In conclusion, the ELISA needs a pair of antibodies that binds TRAIL in a specific and sensitive manner.

Little is known about the physiologic significance of sTRAIL in the circulation of the body. Because sTRAIL induces apoptosis in cancer cell lines, it may have a role in the natural immunologic defense mechanism against initiation and metastasis of cancer cells. The availability of ELISAs for detection of sTRAIL may help us to evaluate the significance of sTRAIL in different immunologic processes. Recent studies have shown that sTRAIL is released into the circulation during different disease states such as cancer, viral infections, autoimmune diseases and inflammation. Table III summarizes the major findings in different disease states using TRAIL ELISAs.

Another application of ELISAs for detection of sTRAIL is to study the mechanisms behind release of sTRAIL, i.e., identify what cell types release sTRAIL and furthermore what stimuli are needed to induce release of

TABLE III. Detection of Soluble TRAIL in Different Disease States

Disease state	Major findings	Reference
Cancer		
Anticancer treatment	Measurement of sTRAIL in serum from monkeys injected with rTRAIL	Ashkenazi et al., 1999
Lymphoma	Detection of sTRAIL in plasma from a lymphoma patient	Liabakk et al., 2002
Liver cancer	Increased serum levels of sTRAIL in patients with liver cancer	Han et al., 2002
Virus		
HIV	Increased serum levels of sTRAIL in HIV-infected patients	Liabakk et al., 2002
Reovirus	sTRAIL is released from cells infected with reovirus	Clarke et al., 2000
Hepatitis B	Detection of sTRAIL in HBV patients	Han et al., 2002
Autoimmune disease		
MS	Detection of sTRAIL in serum from MS patients treated with IFNβ	Wandinger et al., 2003
Inflammation		
Asthma	Increased serum levels of sTRAIL in patients with allergic asthma	Robertson et al., 2002

sTRAIL. Table IV gives an overview of results obtained by using TRAIL ELISAs to study release of sTRAIL from different cell types. As illustrated in Table IV, different immune cells release sTRAIL in response to different stimulus. These results must be seen in connection with results obtained by other techniques, and are further discussed in section IV.

B. BINDING OF TRAIL MONOCLONAL ANTIBODIES TO CELLS

Monoclonal antibodies against TRAIL are important tools to investigate intracellular and surface cell distributions. This can be done with flow cytometric analysis, which involves binding between membrane-bound TRAIL on a specific cell type and a fluorescence/fluorochrome labeled monoclonal antibody against TRAIL. Following a wash to remove any unbound fluorescence labeled antibody, the cell sample is further analyzed with a flow cytometer to determine the amount of binding between antibody and cell.

Several studies have mapped the distribution of TRAIL on different cell types. The results from these studies are summarized in Table V. As can be seen from Table V, TRAIL is expressed by several cell types in the immune system. The antibody N2B2 against murine TRAIL and several other monoclonal antibodies against human TRAIL (Table I) have been used to determine which cells express TRAIL. Studies in both mice and humans have shown that freshly isolated T-cells, B-cells, dendritic cells, monocytes, and NK cells do not express detectable levels of TRAIL on their surface. However, there are some exceptions. A subset of mouse liver NK cells

TABLE IV. Detection of sTRAIL in Cell Culture Supernatant and in Cell Lysates from Different Stimulated Cell Types Using TRAIL ELISAs

Cell type	Stimuli	Reference
PBMC	PHA	Liabakk et al., 2002
NK-cells	IL-2	Liabakk et al., 2002
NK-cells	IFNα, IFNβ	Johnsen et al., unpublished results
Monocytes	IFNα, IFNβ, IFNγ	Halaas et al., unpublished results
Monocytes	COH-1[b]	Halaas et al., unpublished results
Macrophages	Tumor cells	Herbeuval et al., 2003
REH[a]	Unstimulated	Liabakk et al., 2002
Daudi[a]	Unstimulated	Liabakk et al., 2002
Jurkat[a]	TNFα	Active Motif Europe, Belgium
N1186 T-cells[a]	Unstimulated	R&D Systems, Abingdon, UK

[a] cell lysate.
[b] COH: heat-killed group B streptococci.

TABLE V. The Use of Monoclonal Antibodies Against TRAIL in Flow Cytometric Analysis

Cell type	Reference
B-cells	Mariani and Krammer, 1998a,b
T-cells	Kayagaki et al., 1999
NK-cells	Johnsen et al., 1999
Dendritic cells	Fanger et al., 1999
Monocytes	Griffith et al., 1999
Microglia	Genc et al., 2003
Endothelial cells	Li et al., 2003
Plasma cells	Ursini-Siegel, 2002
Human osteoblasts-like cells (NHBC)	Atkins et al., 2002

express TRAIL constitutively (Smyth et al., 2001). In addition, recent publications have shown that both unstimulated endothelial cells (Li et al., 2003) and plasma cells (Ursini-Siegel et al., 2002) express TRAIL. One may therefore ask what physiologic role TRAIL has in these cells. On the other side, TRAIL is expressed on activated immune cells, including NK cells, T-cells, B-cells, DC, and monocytes, as listed in Table V.

IV. PHYSIOLOGIC SIGNIFICANCE OF TRAIL

All the results obtained by using monoclonal antibodies against TRAIL, as described previously (Tables III–V), may help us to understand the physiologic significance of TRAIL. TRAIL is expressed by activated cells (Table V); activated cells release sTRAIL (Table IV), and sTRAIL is detected in different disease states (Table III). Because of these observations, TRAIL may be involved in different conditions, including immune defense mechanisms, immune surveillance, and different disease states such as cancer, viral infections, autoimmune diseases, and inflammation.

TRAIL may be a part of the natural immune defense mechanism against cancer. IFNs activate immune cells (T-cells, NK cells, monocytes, dendritic cells) to express membrane-bound TRAIL (see Table V). Furthermore, the TRAIL-expressing cells kill the tumor cells by a mechanism involving binding of DR5 on the tumor cells to membrane-bound TRAIL on the immune cell. The killing was due to TRAIL because it could be inhibited by a neutralizing monoclonal antibody against TRAIL/soluble TRAIL receptor. These observations indicate that TRAIL may have a physiologic role as a tumor suppressor. In addition, NK-cells and monocytes release sTRAIL after stimulation with IFNs (see Table IV). These observations indicate that TRAIL plays a central role in innate immune responses involving IFNs, NK-cells, and monocytes.

TRAIL is also supposed to have a function in immune surveillance against tumor development. Takeda *et al.* (2002) illustrated that TRAIL eliminates developing tumors, in a mechanism involving NK cells and IFNs.

ELISAs for detection of sTRAIL are important tools to investigate the role of sTRAIL, because sTRAIL are released during different disease conditions (Table III). Regarding viral infection, Liabakk *et al.* (2002) found higher serum levels of sTRAIL in HIV-infected patients compared to healthy controls and the increase was restricted to those with the most advanced clinical and immunologic disease, i.e. AIDS patients. Han *et al.* (2003) detected increased levels of sTRAIL in serum from patients with hepatitis B virus compared to levels in healthy controls. These results indicate that viral infections may lead to systemic release of sTRAIL. Advanced HIV-related disease is characterized by markedly enhanced T-cell apoptosis, and it has been suggested that TRAIL-related mechanisms may be involved in this process (Jeremias *et al.*, 1998; Katsikis *et al.*, 1997). Another study supporting this hypothesis showed that monocytes treated with HIV Tat protein induced release of sTRAIL, and this sTRAIL was cytotoxic against T-cells (Yang *et al.*, 2003). In addition, measles virus induces expression of TRAIL in dendritic cells (Vidalain *et al.*, 2000). In conclusion, viral infections lead to TRAIL expression and release of sTRAIL. The physiologic significance of TRAIL in viral infections may be apoptosis of infected cells.

Recent data suggest that TRAIL is important in regulating natural defense mechanisms because it inhibits autoimmune inflammation (Song *et al.*, 2000) and autoimmune encephalomyelitis (animal model of multiple sclerosis [MS]) (Hilliard *et al.*, 2001). Wandinger *et al.* (2003) used an ELISA for detection of sTRAIL in serum samples from patients with MS treated with IFN-β to evaluate the mechanism behind IFN-β action in these patients. This study suggested TRAIL as a potential response marker for IFN-β treatment in MS patients.

Robertson *et al.* (2002) investigated the role of TRAIL in the chronic inflammatory process in asthma. They found increased levels of sTRAIL in patients with allergic asthma by using an ELISA. In addition, TRAIL concentrations increased in allergic asthmatic subjects after antigen challenge, and these values correlated with an increased eosinophil presence in the airway.

This chapter gives an overview of monoclonal antibodies against TRAIL. We have described some of the published and commercially available antibodies. In addition, we have shown the application of these antibodies and results obtained by using them. In conclusion, monoclonal antibodies against TRAIL are useful tools to investigate the mechanisms behind the physiological significance of TRAIL.

V. FUTURE ASPECTS

TRAIL was originally proposed as a cancer specific molecule, because it killed cancer cells but not normal cells. Administration of sTRAIL in mice (Walczak et al., 1999) and primates (Ashkenazi et al., 1999) induced tumor regression without systemic toxicity, which is the disadvantage of other members of the TNF family, such as TNF and FasL.

Because the binding between TRAIL and TRAIL-R2/DR5 induces apoptosis in tumors, TRAIL may have a potential in cancer therapy. Human Genome Sciences and Cambridge Antibody Technology have developed an agonistic human monoclonal antibody to TRAIL-R2 that mimics the activity of native TRAIL, but have higher specificity and longer serum half-life than TRAIL. Therefore, TRAIL specifically binds TRAIL-R2 and induces apoptosis in a broad range of tumor types, both alone and in combination with chemotherapy. TRAIL-R2 monoclonal antibody is now going to be tested in phase 1 clinical trial in cancer patients.

REFERENCES

Ashkenazi, A., Pai, R. C., Fong, S., Leung, S., Lawrence, D. A., Marsters, S. A., Blackie, C., Chang, L., Mcmurtrey, A. E., Hebert, A., DeForge, L., Koumenis, I. L., Lewis, D., Harris, L., Bussiere, J., Koeppen, H., Shahrokh, Z., and Schwall, R. H. (1999). Safety and antitumor activity of recombinant soluble Apo2 ligand. *J. Clin. Invest.* **104**, 155–162.

Atkins, G. J., Evdokiou, S., Labrinidis, A., Zannettini, A. C. W., Haynes, D. R., and Findlay, D. M. (2002). Human osteoblasts are resistant to Apo2L/TRAIL-mediated apoptosis. *Bone* **31**, 448–456.

Baetu, T. M., Kwon, H., Sharma, S., Grandvaux, N., and Hiscott, J. (2001). Disruption of NF-kappaB signaling reveals a novel role for NF-kappaB in the regulation of TNF-related apoptosis-inducing ligand expression. *J. Immunol.* **167**, 3164–3173.

Bodmer, J. L., Meier, P., Tschopp, J., and Schneider, P. (2000). Cysteine 230 is essential for the structure and activity of the cytotoxic ligand TRAIL. *J. Biol. Chem.* **275**, 20632–20637.

Clarke, P., Meintzer, S. M., Gibson, S., Widmann, C., Garrington, T. P., Johnson, G. L., and Tyler, K. L. (2000). Reovirus-induced apoptosis is mediated by TRAIL. *J. Virol.* **74**, 8135–8139.

Cretney, E., Takeda, K., Yagita, H., Glaccum, M., Peschon, J. J., and Smyth, M. J. (2002). Increased susceptibility to tumor initiation and metastasis in TNF-related apoptosis-inducing ligand-deficient mice. *J. Immunol.* **168**, 1356–1361.

Dorothee, G., Vergnon, I., Menez, J., Echchakir, H., Grunenwald, D., Kubin, M., Chouaib, S., and Mami-Chouaib, F. (2002). Tumor-infiltrating CD4+ T lymphocytes express APO2 ligand (APO2L)/TRAIL upon specific stimulation with autologous lung carcinoma cells: Role of IFN-alpha on APO2L/TRAIL expression and -mediated cytotoxicity. *J. Immunol.* **169**, 809–817.

Fanger, N. A., Maliszewski, C. R., Schooley, K., and Griffith, T. S. (1999). Human dendritic cells mediate cellular apoptosis via tumor necrosis factor-related apoptosis-inducing ligand (TRAIL). *J. Exp. Med.* **190**, 1155–1164.

Genc, S., Kizyldag, S., Genc, K., Ates, H., and Atabey, N. (2003). Interferon gamma and lipopolysaccharide upregulate TNF-related apoptosis-inducing ligand expression in murine microglia. *Immunol. Lett.* **85**, 271–274.

Griffith, T. S., Wiley, S. R., Kubin, M. Z., Sedger, L. M., Maliszewski, C. R., and Fanger, N. A. (1999). Monocyte-mediated tumoricidal activity via the tumor necrosis factor-related cytokine, TRAIL. *J. Exp. Med.* **189,** 1343–1354.

Halaas, Ø., Vik, R., Ashkenazi, A., and Espevik, T. (2000). Lipopolysaccharide induces expression of APO2 Ligand/TRAIL in human monocytes and macrophages. *Scand. J. Immunol.* **51,** 244–250.

Han, L. H., Sun, W. S., Ma, C. H., Zhang, L. N., Liu, S. X., Zhang, Q., Gao, L. F., and Chen, Y. H. (2002). Detection of soluble TRAIL in HBV infected patients and its clinical implications. *World J. Gastroenterol.* **8,** 1077–1080.

Herbeuval, J. P., Lambert, C., Sabido, O., Cottier, M., Fournel, P., Dy, M., and Genin, C. (2003). Macrophages from cancer patients: Analysis of TRAIL, TRAIL receptors and colon tumor cell apoptosis. *J. Natl. Cancer. Inst.* **95,** 611–621.

Hilliard, B., Wilmen, A., Seidel, C., Liu, T-S., Göke, R., and Chen, Y. (2001). Roles of TNF-related apoptosis-inducing ligand in experimental autoimmune encephalomyelitis. *J. Immunol.* **166,** 1314–1319.

Holen, I., Croucher, P. I., Hamdy, F. C., and Eaton, C. L. (2002). Osteoprotegerin (OPG) is a survival factor for human prostate cancer cells. *Cancer Res.* **62,** 1619–1623.

Ichikawa, K., Liu, W., Fleck, M., Zhang, H., Zhao, L., Ohtsuka, T., Wang, Z., Liu, D., Mountz, J. D., Ohtsuki, M., Koopman, W. J., Kimberly, R., and Zhou, T. (2003). TRAIL-R2 (DR5) mediates apoptosis of synovial fibroblasts in rheumatoid arthritis. *J. Immunol.* **171,** 1061–1069.

Jeremias, I., Herr, I., Boehler, T., and Debatin, K. M. (1998). TRAIL/Apo-2-ligand-induced apoptosis in human T cells. *Eur. J. Immunol.* **28,** 143–152.

Johnsen, A. C., Haux, J., Steinkjer, B., Nonstad, U., Egeberg, K., Sundan, A., Ashkenazi, A., and Espevik, T. (1999). Regulation of Apo-2 ligand/TRAIL expression in NK cells—involvement in NK cell-mediated cytotoxicity. *Cytokine* **11,** 664–672.

Katsikis, P. D., Garcia-Ojeda, M. E., Torres-Roca, J. F., Tijoe, I. M., Smith, C. A., Herzenberg, L. A., and Herzenberg, L. A. (1997). Interleukin-1 beta converting enzyme-like protease involvement in Fas-induced and activation-induced peripheral blood T cell apoptosis in HIV infection. TNF-related apoptosis-inducing ligand can mediate activation-induced T cell death in HIV infection. *J. Exp. Med.* **186,** 1365–1372.

Kayagaki, N., Yamaguchi, N., Nakayama, M., Okumura, K., and Yagita, H. (1999). Type I interferons (IFNs) regulate tumor necrosis factor-related apoptosis-inducing ligand (TRAIL) expression on human T cells: A novel mechanism for the antitumor effects of type I IFNs. *J. Exp. Med.* **189,** 1451–1460.

Kayagaki, N., Yamaguchi, N., Abe, M., Hirose, S., Shirai, T., Okumura, K., and Yagita, H. (2002). Suppression of antibody production by TNF-related apoptosis-inducing ligand (TRAIL). *Cell. Immunol.* **219,** 82–91.

Köhler, G., and Milstein, C. (1975). Continous cultures of fused cells secreting antibody of predefined specificity. *Nature* **256,** 495–497.

Lawrence, D., Shahrokh, Z., Marsters, S., Achilles, K., Shih, D., Mounho, B., Hillan, K., Totpal, K., Deforge, L., Schow, P., Hooley, J., Sherwood, S., Pai, R., Leung, S., Khan, L., Gliniak, B., Bussiere, J., Smith, C. A., Strom, S. S., Kelley, S., Fox, J. A., Thomas, D., and Ashkenazi, A. (2001). Differential hepatocyte toxicity of recombinant Apo2L/TRAIL versions. *Nat. Med.* **7,** 383–385.

Lee, H., Herndon, J. M., Barreiro, R., Griffith, T. S., and Ferguson, T. A. (2002). TRAIL: A mechanism of tumor surveillance in an immune privileged site. *J. Immunol.* **169,** 4739–4744.

Li, J. H., Kirkiles-Smith, N. C., McNiff, J. M., and Pober, J. S. (2003). TRAIL induces apoptosis and inflammatory gene expression in human endothelial cells. *J. Immunol.* **171,** 1526–1533.

Liabakk, N. B., Sundan, A., Torp, S., Aukrust, P., Frøland, S. S., and Espevik, T. (2002). Development, characterization and use of monoclonal antibodies against sTRAIL: Measurement of sTRAIL by ELISA. *J. Immunol. Meth.* **259,** 119–128.

Lunemann, J. D., Waiczies, S., Ehrlich, S., Wendling, U., Seeger, B., Kamradt, T., and Zipp, F. (2002). Death ligand TRAIL induces no apoptosis but inhibits activation of human (auto)antigen-specific T cells. *J. Immunol.* **168,** 4881–4888.

Mariani, S. M., and Krammer, P. H. (1998a). Surface expression of TRAIL/Apo-2 ligand in activated mouse T and B cells. *Eur. J. Immunol.* **28,** 1492–1498.

Mariani, S. M., and Krammer, P. H. (1998b). Differential regulation of TRAIL and CD95 ligand in transformed cells of the T and B lymphocyte lineage. *Eur. J. Immunol.* **28,** 973–982.

Martinez-Lorenzo, M. J., Alava, M. A., Gamen, S., Kim, K. J., Chuntharapai, A., Pineiro, A., Naval, J., and Anel, A. (1998). Involvement of APO2 ligand/TRAIL in activation-induced death of Jurkat and human peripheral blood T cells. *Eur. J. Immunol.* **28,** 2714–2725.

Miura, Y., Misawa, N., Maeda, N., Inagaki, Y., Tanaka, Y., Ito, M., Kayagaki, N., Yamamoto, N., Yagita, H., Mizusawa, H., and Koyanagi, Y. (2001). Critical contribution of tumor necrosis factor-related apoptosis-inducing ligand (TRAIL) to apoptosis of human CD4+ T cells in HIV-1-infected hu-PBL-NOD-SCID mice. *J. Exp. Med.* **193,** 651–659.

Mundt, B., Kuhnel, F., Zender, L., Paul, Y., Tillmann, H., Trautwein, C., Manns, M. P., and Kubicka, S. (2003). Involvement of TRAIL and its receptors in viral hepatitis. *FASEB J.* **17,** 94–96.

Oshima, K., Yanase, N., Ibukiyama, C., Yamashina, A., Kayagaki, N., Yagita, H., and Mizuguchi, J. (2001). Involvement of TRAIL/TRAIL-R interaction in IFN-alpha-induced apoptosis of Daudi B lymphoma cells. *Cytokine* **14,** 193–201.

Pitti, R. M., Marsters, S. A., Ruppert, S., Donahue, C. J., Moore, A., and Ashkenazi, A. (1996). Induction of apoptosis by Apo-2 ligand, a new member of the tumor necrosis factor cytokine family. *J. Biol. Chem.* **271,** 12687–12690.

Plasilova, M., Zivny, J., Jelinek, J., Neuwirtova, R., Cermak, J., Necas, E., Andera, L., and Stopka, T. (2002). TRAIL (Apo2L) suppresses growth of primary human leukemia and myelodysplasia progenitors. *Leukemia* **16,** 67–73.

Robertson, N. M., Zangrilli, J. G., Steplewski, A., Hastie, A., Lindemeyer, R. G., Planeta, M. A., Smith, M. K., Innocent, N., Musani, A. M., Pascual, R., Peters, S., and Litwack, G. (2002). Differential expression of TRAIL and TRAIL receptors in allergic asthmatics following segmental antigen challenge: Evidence for a role of TRAIL in eosinophil survival. *J. Immunol.* **169,** 5986–5996.

Sedger, L. M., Shows, D. M., Blanton, R. A., Peschon, J. J., Goodwin, R. G., Cosman, D., and Wiley, S. R. (1999). IFN-γ mediates a novel antiviral activity through dynamic modulation of TRAIL and TRAIL receptor expression. *J. Immunol.* **163,** 920–926.

Seki, N., Hayakawi, Y., Brooks, A. D., Wine, J., Wiltrout, R. H., Yagita, H., Tanner, J. E., Smyth, M. J., and Sayers, T. J. (2003). Tumor necrosis factor-related apoptosis-inducing ligand-mediated apoptosis is an important endogenous mechanism for resistance to liver metastasis in murine renal cancer. *Cancer Res.* **63,** 207–213.

Sheridan, J. P., Marsters, S. A., Pitti, R. M., Gurney, A., Skubatch, M., Baldwin, D., Ramakrishnan, L., Gray, C. L., Baker, K., Wood, W. I., Goddard, A. D., Godowski, P., and Ashkenazi, A. (1997). Control of TRAIL-induced apoptosis by a family of signaling and decoy receptors. *Science* **277,** 818–821.

Smyth, M. J., Cretney, E., Takeda, K., Wiltrout, R. H., Sedger, L. M., Kayagaki, N., Yagita, H., and Okumura, K. (2001). Tumor necrosis factor-related apoptosis-inducing ligand (TRAIL) contributes to interferon gamma-dependent natural killer cell protection from tumor metastasis. *J. Exp. Med.* **193,** 661–670.

Song, K., Chen, Y., Goke, R., Wilmen, A., Seidel, C., Goke, A., Hillard, B., and Chen, Y. (2000). Tumor necrosis factor-related apoptosis-inducing ligand (TRAIL) is an inhibitor of autoimmune inflammation and cell cycle progression. *J. Exp. Med.* **191,** 1095–1103.

Takeda, K., Smyth, M. J., Cretney, E., Hayakawa, Y., Kayagaki, N., Yagita, H., and Okumura, K. (2002). Critical role for tumor necrosis factor-related apoptosis-inducing ligand in immune surveillance against tumor development. *J. Exp. Med.* **195,** 161–169.

Ursini-Siegel, J., Zhang, W., Altmeyer, A., Hatada, E. N., Do, R. K. G., Yagita, H., and Chen-Kiang, S. (2002). TRAIL/Apo-2 ligand induces primary plasma cell apoptosis. *J. Immunol.* **169,** 5505–5513.

Vidalain, P. O., Azocar, O., Lamouille, B., Astier, A., Rabourdin-Combe, C., and Servet-Delprat, C. (2000). Measles virus induces functional TRAIL production by human dendritic cells. *J. Virol.* **74,** 556–559.

Walczak, H., Miller, R. E., Ariail, K., Gliniak, K., Griffith, T. S., Kubin, M., Chin, W., Jones, J., Woodward, A., Le, T., Smith, C., Schuh, J. C. L., and Lynch, D. H. (1999). Tumoricidal activity of tumor necrosis factor-related apoptosis-inducing ligand in vivo. *Nat. Med.* **5,** 157–163.

Wandinger, K. P., Lunemann, J. D., Wengert, O., Bellmann-Strobl, J., Aktas, O., Weber, A., Grundstrøm, E., Ehrlich, S., Wernecke, K. D., Volk, H. D., and Zipp, F. (2003). TNF-related apoptosis inducing ligand (TRAIL) as a potential response marker for interferon-beta treatment in multiple sclerosis. *Lancet* **361,** 2036–2043.

Washburn, B., Weigand, M. A., Grosse-Wilde, A., Janke, M., Stahl, H., Rieser, E., Sprick, M. R., Schirrmacher, V., and Walczak, H. (2003). TNF-related apoptosis-inducing ligand mediates tumoricidal activity of human monocytes stimulated by Newcastle disease virus. *J. Immunol.* **179,** 1814–1821.

Wiley, S. R., Schooley, K., Smolak, P. J., Din, W. S., Huang, C-P., Nicholl, J. K., Sutherland, G. R., Smoth, T. D., Rauch, C., Smith, C. A., and Goodwin, R. G. (1995). Identification and characterization of a new member of the TNF family that induces apoptosis. *Immunity* **3,** 673–682.

Yang, Y., Tikhonov, I., Ruckwardt, T. J., Djavani, M., Zapata, J. C., Pauza, C. D., and Salvato, M. S. (2003). Monocytes treated with human immunodeficiency virus Tat kill uninfected CD4(+) cells by a tumor necrosis factor-related apoptosis-induced ligand-mediated mechanism. *J. Virol.* **77,** 6700–6708.

Zhang, X. D., Franco, A. V., Nguyen, T., Gray, C. P., and Hersey, P. (2000). Differential localization and regulation of death and decoy receptors for TNF-related apoptosis-inducing ligand (TRAIL) in human melanoma cells. *J. Immunol.* **164,** 3961–3970.

6

Modulation of TRAIL Signaling Complex

Chunhai Hao,* Jin H. Song,* Urosh Vilimanovich,* and Norman M. Kneteman[†]

Department of Laboratory Medicine & Pathology
[†]*Department of Surgery, University of Alberta*
Edmonton, Alberta T6G 2B7, Canada

    I. TRAIL: A New Member of TNF Family
   II. TRAIL Death Receptors: DR4 and DR5
  III. TRAIL Decoy Receptors: DcR1, DcR2, and OPG
  IV. TRAIL-Induced DISC: Cleavage of Apoptosis Initiating Protease Caspase-8
   V. TRAIL-Induced Bid Cleavage: Activation of Mitochondrial Pathways
  VI. TRAIL-Induced Cell Proliferation: RIP-mediated NF-κB Activation
 VII. TRAIL-Induced c-FLIP Recruitment: DISC Modulation
VIII. TRAIL-Induced PED Recruitment: Differential Modulation of the DISC
  IX. TRAIL Physiologic Functions: Tumor Surveillance
   X. Recombinant TRAIL: A Novel Cancer Therapeutic Agent
      References

I. TRAIL: A NEW MEMBER OF TNF FAMILY

Programmed cell death (apoptosis) is a genetically controlled mechanism that is essential for the maintenance of cellular homeostasis within tissues involving the development and elimination of unwanted cells. This phenomenon is found under physiologic and pathologic conditions. There are two major signaling pathways that control apoptosis initiation: the extrinsic pathway through receptor-mediated death-signaling pathways (Ashkenazi and Dixit, 1998) and the intrinsic pathway via mitochondria (Green and Reed, 1998). A number of ligand-receptor families are involved in apoptosis. Death ligands and death receptors of tumor necrosis factor (TNF) family play crucial roles in instructive apoptosis and the ligand-receptor interactions within this family induce diverse cellular responses including apoptosis as well as cell proliferation and differentiation. The particular response depends on the receptor that is signaling, the cell type, and the concurrent signals received by the cells.

The death ligands of the TNF family are expressed as type II transmembrane proteins, and their carboxy-terminal extracellular domains are processed proteolytically to form a soluble molecule (Ashkenazi and Dixit, 1998). In contrast, the death receptors of TNF family are type I transmembrane proteins. In both death ligands and receptors, the sequence homology is mainly in the extracellular regions, which mediate ligand-receptor binding. The best characterized death ligands and death receptors of TNF family include TNFα/TNFR1, FasL (Fas ligand, CD95L, Apo1L)/ Fas (CD95, Apo1), Apo3L/Apo3 (DR3), and TRAIL (TNF-related apoptosis-inducing ligand, Apo2L)/DR4 and DR5.

TRAIL is a recently identified member of the TNF family (Pitti et al., 1996; Wiley et al., 1995). TRAIL consists of 281 (human TRAIL) or 291 (murine TRAIL) amino acids. The human TRAIL gene is located on chromosome 3 at position 3q26 and the human TRAIL protein has a calculated molecular mass of 32.5 kDa and isoelectric point of 7.63. TRAIL is primarily expressed as a type II membrane protein with a very short intracellular amino-terminal tail of 17 amino acids and an extracellular carboxy-terminal domain (amino acids 114–281). The extracellular carboxy-terminal domain forms a bell-shaped homotrimeric structure of the receptor-binding site (Cha et al., 1999) and can be cleaved at the cell surface to yield a soluble, biologically active form (114–281 amino acids, 19–20 kDa) (Mariani and Krammer, 1998). The extracellular carboxy-terminal domain of TRAIL shows homology to FasL (23.2%) and TNF (19.0%), but in contrast to FasL and TNF, TRAIL carries a zinc ion at the trimer interface coordinated by a free cysteine residue (Cys^{230}) of each monomer, which is an essential moiety of TRAIL structural integrity and its capacity to induce apoptosis (Bodmer et al., 2000b; Hymowitz et al., 2000).

II. TRAIL DEATH RECEPTORS: DR4 AND DR5

TRAIL interacts with five distinct receptors: two death receptors (DR), and two decoy receptors (DcR) and osteoprotegerin (OPG), all belonging to the TNF receptor superfamily. DR4 (TRAIL-R1) (Pan *et al.*, 1997a,b; Schneider *et al.*, 1997a) and DR5 (TRAIL-R2, TRICK2, KILLER) (Chaudhary *et al.*, 1997; Screaton *et al.*, 1997; Sheridan *et al.*, 1997; Walczak *et al.*, 1997; Wu *et al.*, 1997) are type I transmembrane proteins of 468 (DR4) or 411 (DR5) amino acids. The DR4 and DR5 genes map to chromosome 8p21. Both receptors contain an extracellular amino-terminal signal peptide, two extracellular cysteine-rich domains, a transmembrane domain, and a cytoplasmic carboxy-terminal region. DR4 and DR5 exhibit homology in their two extracellular cysteine-rich domains of ligand binding sites. The intracellular carboxy-terminal regions of DR4 and DR5 contain a 70-amino acid stretch with significant similarity to the death domains of TNFR1 and Fas that, upon ligand binding, initiate an intracellular apoptosis-signaling cascade.

TRAIL interacts via two cysteine-rich domains with DR4 or DR5 to form homotrimers (Cha *et al.*, 2000; Hymowitz *et al.*, 1999; Mongkolsapaya *et al.*, 1999). This 3:3 ratio of ligand to receptor complex was initially reported in TNFα/TNFR1 complexes (Banner *et al.*, 1993) and thus a ligand-induced trimerization model was suggested that three death ligands bind to three monomeric receptors, forming a complex that juxtaposes the cytoplasmic death domains (Smith *et al.*, 1994). According to the ligand-induced trimerization model, however, addition of TRAIL to cells that express both DR4 and DR5 should lead to the formation of DR4 and DR5 heterotrimers, which would fail to transduce signals because the recruitment of downstream signaling proteins requires the formation of homotrimeric death domain complexes. In reality, many cells express both DR4 and DR5, but are still sensitive to TRAIL (Hao *et al.*, 2001), suggesting that an alternative mechanism may exist.

Recent studies of TNFR1 and Fas report that each of the receptors pre-associates as a homotrimer before ligand binding and this preassociated receptor homotrimer is necessary for ligand binding (Chan *et al.*, 2000; Siegel *et al.*, 2000); the results suggest an alternative preligand binding assembly model. The first carboxy-terminal cysteine-rich domain of TNFR1 and Fas appears to be the preligand–binding assembly domain that mediates specific ligand-independent assembly of receptor homotrimers (Chan *et al.*, 2000; Siegel *et al.*, 2000). The signal peptide of DR4 and DR5 shares part sequence similarity with the first carboxy-terminal cysteine-rich domain of TNFR1 and Fas (Cha *et al.*, 1999; Hymowitz *et al.*, 1999), but whether the signal peptide of the TRAIL death receptors function as a preligand-binding assembly domain and mediate the preassociation of DR4 and DR5 into homotrimers remains to be examined.

III. TRAIL DECOY RECEPTORS: DcR1, DcR2, AND OPG

TRAIL also interacts with two decoy receptors (DcR). DcR1 (TRAIL-R3, LIT, TRID) is a glycosyl phosphatidylinositol (GPI)-anchored membrane protein (Degli-Esposti *et al.*, 1997b; MacFarlane *et al.*, 1997; Pan *et al.*, 1997a; Sheridan *et al.*, 1997). The extracellular domain of DcR1 has two cysteine-rich pseudorepeats, however, unlike DR4 and DR5, DcR1 lacks a cytoplasmic death domain. DcR2 (TRAIL-R4, TRUNDD) is a type I transmembrane protein of 386 amino acids with an extracellular domain similar to DR4 and DR5; however, the cytoplasmic death domain is substantially truncated and does not mediate apoptosis upon TRAIL binding (Degli-Esposti *et al.*, 1997a; Marsters *et al.*, 1997; Pan *et al.*, 1998). Although the truncated cytoplasmic region was reported to be functional in activation of nuclear factor κB (NF-κB) (Degli-Esposti *et al.*, 1997a), others failed to show this function of DcR2 (Marsters *et al.*, 1997).

DcR1 and DcR2 bind to TRAIL with high affinity, but they are unable to transduce an apoptotic signal. Overexpression of DcR2 results in cell resistance to TRAIL-induced apoptosis and deletion of the DcR2 cytoplasmic region does not abrogate the inhibitory activity of DcR2 (Marsters *et al.*, 1997; Pan *et al.*, 1998). These studies suggest that DcR1 and DcR2 may inhibit TRAIL-induced apoptosis by competing with DR4 and DR5 for binding to TRAIL. However, it remains possible that DcR1 and DcR2 may also act as dominant-negatives and form heteromeric complexes with DR4 and DR5, thus preventing TRAIL signal transduction through DR4 and DR5. DcR1 and DcR2 are mainly expressed in normal tissues, but their expression in the cells does not correlate with cell sensitivity to TRAIL. Therefore, the physiologic functions of DcR1 and DcR2 and their signal events remain to be established.

Osteoprotegerin (OPG) is a soluble member of the TNF receptor superfamily that regulates osteoclastogenesis (Simonet *et al.*, 1997). OPG competes with RANK (receptor activator of nuclear factor κB [NF-κB]) for binding of RANK ligand (RANKL, OPG1) that is also a TNF family member. OPG is a secreted glycoprotein of 401 amino acids (55 kDa monomer or 110 kDa disulfide-linked dimer). The amino-terminal half of the OPG protein has four cysteine-rich domains, but its carboxy-terminal half shows no homologies to any other known proteins. OPG interacts with TRAIL and thus inhibits TRAIL-induced apoptosis in osteoblasts and some tumor cell lines (Emery *et al.*, 1998; Holen *et al.*, 2002). However, whether OPG can efficiently act as a decoy receptor for the TRAIL under physiologic conditions is rather unclear, because its affinity to TRAIL is rather weak compared to the other TRAIL receptors (Truneh *et al.*, 2000).

IV. TRAIL-INDUCED DISC: CLEAVAGE OF APOPTOSIS INITIATING PROTEASE CASPASE-8

TRAIL-induced apoptosis occurs through its binding of DR4 and DR5 that in turn activate, via their death domains, the apoptosis-initiating protease caspase-8 (Pan et al., 1997a,b; Schneider et al., 1997a; Screaton et al., 1997; Sheridan et al., 1997; Walczak et al., 1997; Wu et al., 1997). However, initially there were contradicting results in studies of the intracellular signal transduction that links death domains to caspase-8. Some groups reported that DR4 and DR5 interact with two intracellular adaptors, TNF receptor 1–associated death domain (TRADD) and Fas-associated death domain (FADD) to activate caspase-8 (Schneider et al., 1997a; Walczak et al., 1997), whereas others reported no such interaction (Pan et al., 1997a,b). Recent analysis of nontransfected cells has revealed that DR4 and DR5 recruit FADD, but not TRADD, to activate caspase-8, which is similar to the signaling pathway in Fas-mediated apoptosis (Bodmer et al., 2000a; Kischkel et al., 2000; Kuang et al., 2000; Xiao et al., 2002).

FADD is a 245-amino acid intracellular protein that has a carboxy-terminal death domain and an amino-terminal death effector domain (Boldin et al., 1995; Chinnaiyan et al., 1995). The death effector domain is a specific example of a more global homophilic interaction domain termed the caspase recruitment domain, which can be found in apoptosis-initiating proteases caspase-8 (Muzio et al., 1996). FADD binds to death receptors via homophilic interaction of death domains and then recruits caspase-8 to the receptors via a homophilic interaction of death effector domains, leading to the assembly of a death-inducing signaling complex (DISC) (Kischkel et al., 1995). Within the DISC, caspase-8 is activated through transcatalytic and autocatalytic cleavage due to a close proximity of the molecules within the DISC (Muzio et al., 1998) (Fig. 1).

Caspase-8 cleavage in the DISC occurs in two consecutive steps: the first step cleavage generates large p43 and p41 and small p12 subunits from p55 and p53 proenzymes and the second step cleavage produces a prodomain and active p18 and p10 subunits (Medema et al., 1997). The active caspase-8 subunits are released from the DISC into the cytosol where they cleave downstream effector caspases such as caspase-3 (Boldin et al., 1996; Muzio et al., 1996). The caspase-8–mediated cleavage of caspase-3 p32 proenzyme generates large p20 and p17 and small p10 subunits (Samali et al., 1999), which in turn cleave their substrates such as DNA fragmentation factor (Liu et al., 1997), leading to programmed cell death (apoptosis).

Caspase-10 shares a homologous structure with caspase-8 and can process caspase-3, suggesting that it may be involved in initiation of DR4

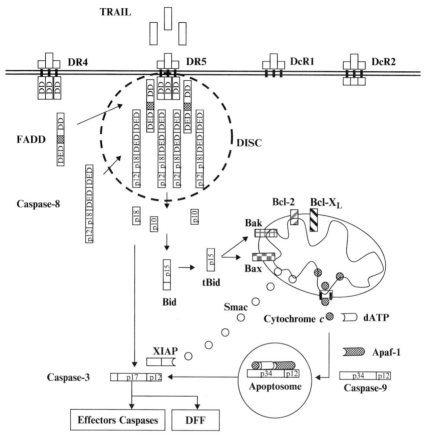

FIGURE 1. TRAIL-induced apoptosis through receptor-mediated extrinsic and mitochondria-involved intrinsic pathways.

and DR5-induced apoptosis (Fernandes-Alnemri et al., 1996). Initial studies in transfectants generated a controversy concerning DR4 and DR5 recruitment of caspase-10 (Bodmer et al., 2000a; Kischkel et al., 2000; MacFarlane et al., 1997; Pan et al., 1997b); however, several groups have recently shown that DR4 and DR5 recruit caspase-10 through FADD to the DISC (Kischkel et al., 2001; Sprick et al., 2002; Wang et al., 2001; Xiao et al., 2002). Once recruited, caspase-10 is cleaved and initiates apoptosis through cleavage of caspase-3 in caspase-8–deficit cells, suggesting that caspase-10 is an apoptosis initiator independent of caspase-8 (Xiao et al., 2002). However, caspase-10 appears to be unable to fully substitute for a loss of caspase-8, which suggests that it may play a role different from that of caspase-8 (Sprick et al., 2002; Xiao et al., 2002).

V. TRAIL-INDUCED BID CLEAVAGE: ACTIVATION OF MITOCHONDRIAL PATHWAYS

There is increasing evidence for cross talk between extrinsic and intrinsic pathways in TRAIL-induced apoptosis (Fig. 1). Once caspase-8 is activated in the DISC, it can either directly cleave caspase-3 to activate extrinsic pathways or cleave Bcl-2 inhibitory BH3-domain–containing protein (Bid) to mediate mitochondrial damage (Li *et al.*, 1998). Truncated Bid (tBid) translocates to the mitochondrial membrane and induces loss of mitochondrial transmembrane potential and release of apoptotic factors such as cytochrome *c* (Luo *et al.*, 1998) and Smac/DIABLO (second mitochondria-derived activator of caspase/direct inhibitor of apoptosis binding protein [IAP] with low pI) (Du *et al.*, 2000; Verhagen *et al.*, 2000). In the cytosol, cytochrome *c* binds to Apaf1 to recruit dATP and caspase-9 to form an apoptosome where caspase-9 is activated and in turn cleaves caspase-3 in execution of apoptosis (Li *et al.*, 1997). Once Smac/DIABLO is released from the mitochondria, it interacts with X-linked inhibitor of apoptosis proteins (XIAP), to release XIAP inhibition of caspase-3 (Deng *et al.*, 2002) and caspase-9 (Srinivasula *et al.*, 2001), thus promoting apoptosis.

Bcl-2 family members control the mitochondrial membrane potential. Antiapoptotic members, Bcl-2 and Bcl-X_L, inhibit the release of mitochondrial apoptotic factors whereas proapoptotic members, Bax and Bak, trigger their release (Kroemer and Reed, 2000). Bak resides on the mitochondrial membrane while Bax translocates from cytosol to mitochondria through interaction with tBid. tBid induces Bax and Bak oligomerization and thus loss of mitochondrial membrane potential (Gross *et al.*, 1998; Wei *et al.*, 2000). Bcl-2 and Bcl-X_L are bound to the outer mitochondrial membrane and interact with Bax and Bak to maintain the mitochondrial membrane potential (Kroemer and Reed, 2000). The overexpression of Bcl-2 or Bcl-X_L blocks TRAIL-induced apoptosis (Hinz *et al.*, 2000), whereas deletion of Bax from TRAIL-sensitive cells results in the cell resistance to TRAIL (LeBlanc *et al.*, 2002).

VI. TRAIL-INDUCED CELL PROLIFERATION: RIP-MEDIATED NF-κB ACTIVATION

Early studies reported that DR4 and DR5 recruit, via their death domains, receptor-interacting protein (RIP) to activate the nuclear factor κB (NF-κB) (Chaudhary *et al.*, 1997; Schneider *et al.*, 1997b). RIP is a cytoplasmic adaptor that has three domains: a carboxy-terminal death domain, an intermediate domain, and an amino-terminal serine/threonine protein kinase (Stanger *et al.*, 1995). The functions of RIP serine/threonine protein kinase remain unclear, although one report suggests that it mediates caspase-8–independent cell necrosis (Holler et al., 2000). In the TNFα/TNFR system, RIP interacts via its death domain with death receptors and via its intermediate domain with inhibitors of κB kinase (IKK) complex (Inohara et al., 2000). The IKK complex is comprised of IKKα, IKKβ, and IKKγ, and RIP interacts with IKKγ to recruit IKKα and IKKβ, which in turn phosphorylate inhibitors of κB (IκB), leading to IκB degradation and NF-κB activation (Mercurio et al., 1997; Zandi et al., 1997).

NF-κB is a collective term referring to dimeric transcription factors that share a 300-amino-acid Rel homology domain, which mediates dimerization, nuclear translocation, DNA binding, and interaction with a family of IκB (Karin and Lin, 2002). IκB proteins bind to NF-κB dimers in such a way as to mask their nuclear translocation sequence, thus retaining the entire complex in the cytoplasm. Phosphorylation of IκB results in its polyubiquitination and rapid targeting to the 26S proteasome for degradation, releasing NF-κB dimers from IκB inhibition and allows for their translocation to the nucleus where they bind target cell cycle genes for cell proliferation and antiapoptotic genes to inhibit apoptosis (Karin and Lin, 2002).

RIP-mediated NF-κB activation appears to be responsible for TRAIL-induced cell proliferation (Ehrhardt et al., 2003) (Fig. 2). RIP has been detected in TRAIL-induced DISC (Harper et al., 2001) and mediates TRAIL-induced IKK activation (Lin et al., 2000). TRAIL-induced NF-κB activation requires recruitment of FADD and caspase-8 to the TRAIL-DISC (Harper et al., 2001) and is blocked by RIP dominant-negative and in IKKγ-deficit cells (Ehrhardt et al., 2003). Collectively, the results suggest that RIP mediates TRAIL-induced NF-κB activation through TRAIL-DISC where RIP mediates IKK activation, leading to IκB phosphorylation and NF-κB activation. TRAIL has also been shown to activate c-Jun-kinase, mitogen-activated protein kinase kinase, and protein kinase Akt pathways (Muhlenbeck et al., 1998; Secchiero et al., 2003), but the molecular mechanisms that link the receptors to the kinases remain to be established.

VII. TRAIL-INDUCED c-FLIP RECRUITMENT: DISC MODULATION

TRAIL-DISC assembly is regulated at three protein-protein interaction levels: ligand-receptor binding on cell surface, intracellular interaction of death receptors and cytoplasmic adaptors via their death domains, and adaptor recruitment of caspase-8 through their death effector domains (Figs. 1 and 2). RIP-mediated NF-κB activation occurs through interaction of death domains whereas caspase-8 cleavage in the DISC is regulated at downstream interaction of death effector domains. Indeed, death effector domain-containing proteins have recently been detected in the TRAIL-DISC where they modulate caspase-8 cleavage and activation (Xiao et al.,

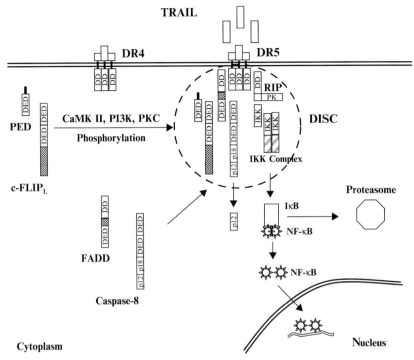

FIGURE 2. TRAIL-induced DISC and its modulation through the DD-DD and DED-DED protein interaction in inhibition of caspase-8 and NF-κB activation.

2002). One such death effector domain-containing protein is referred to as "cellular FADD-like, IL-1β–converting enzyme-inhibitory protein" (c-FLIP) (Irmler et al., 1997), which is also termed CASH (Goltsev et al., 1997), Casper (Shu et al., 1997), Harakiri (Inohara et al., 1997), FLAME-1 (Srinivasula et al., 1997), I-FLICE (Hu et al., 1997), MRIT (Han et al., 1997), and Usurpin (Rasper et al., 1998).

The *c-FLIP* gene is composed of 13 exons that are clustered within approximately 200 kb within the caspase-8 and caspase-10 genes on human chromosome 2q33 to 2q34 (Kischkel et al., 1998; Rasper et al., 1998). *c-FLIP* is expressed as four mRNA splice variants, but only two forms of protein are found in cells (Irmler et al., 1997; Scaffidi et al., 1999; Shu et al., 1997). The short form protein (c-FLIP$_S$, $M_r \sim 28$ kDa) contains two death effector domains, whereas the long form (c-FLIP$_L$, $M_r \sim 55$ kDa) has two death effector domains and a caspase-like domain that lacks catalytic activity because the active-center tyrosine is replaced with cysteine. c-FLIP$_L$ is expressed in many tissues, but c-FLIP$_S$ is mainly found in lymphatic tissue (Irmler et al., 1997). The c-FLIP$_L$ proteins are expressed in three isoforms, one of which appears to be phosphorylated by calcium/calmodulin-dependent protein kinases II (CaMKII) (Yang et al., 2003). Expression of c-FLIP is upregulated by CaMKII (Yang et al., 2003), mitogen-activated protein kinases kinase (Yeh et al., 1998), and phosphatidylinositol 3-kinase (Panka et al., 2001).

TRAIL triggers recruitment of c-FLIP$_L$ and c-FLIP$_S$ to the DISC, where c-FLIP$_L$ is cleaved to produce intermediate p43 fragments that, together with c-FLIP$_S$, remain bound in the DISC to prevent the second step cleavage of caspase-8, thus preventing TRAIL-induced apoptosis (Xiao et al., 2002) (Fig. 2). It appears that only the phosphorylated form of c-FLIP$_L$ is recruited to the DISC. Overexpression of c-FLIP$_L$ activates NF-κB whereas c-FLIP$_S$ transfection upregulates the proto-oncogene c-Fos in a FADD- and caspase-8–dependent manner (Hu et al., 2000; Siegmund et al., 2001). These results suggest the dual functions of c-FLIP proteins that inhibit caspase-8 cleavage to prevent TRAIL-induced apoptosis and activate NF-γB-mediated pathways for cell growth. Further studies are required to define the c-FLIP–mediated signal events that link these proteins from the DISC to downstream NF-κB pathways.

VIII. TRAIL-INDUCED PED RECRUITMENT: DIFFERENTIAL MODULATION OF THE DISC

TRAIL-DISC analysis has revealed another death effector domain–containing protein termed "phosphoprotein enriched in diabetes" (PED) (Condorelli et al., 1998) or "phosphoprotein enriched in astrocytes-15 kDa" (PEA-15) (Danziger et al., 1995). PED has a smaller molecular weight (15 kDa) than c-FLIP (25–55 kDa) and it possesses an amino-terminal death effector domain but not a caspase-like domain. It is phosphorylated on two residues: Ser^{104} by protein kinase C (PKC) (Araujo et al., 1993) and Ser^{116} by CaMKII (Kubes et al., 1998). CaMKII phosphorylation of Ser^{116} facilitates subsequent phosphorylation of Ser^{104} by PKC (Kubes et al., 1998). Thus, PED may exist in one of three forms: unphosphorylated, phosphorylated at Ser^{116}, and doubly phosphorylated at Ser^{116} and Ser^{104}.

TRAIL triggers PED recruitment to the DISC and PED *sense* cDNA transfection results in cell resistance to TRAIL whereas PED *antisense* cDNA transfection converts cells from a TRAIL-resistant to a TRAIL-sensitive phenotype (Hao et al., 2001). Of the three PED isoforms, only the doubly phosphorylated PED is expressed and recruited to the DISC in TRAIL-resistant cells, indicating that full PED phosphorylation is required for its recruitment (Xiao et al., 2002). These results would imply that PED is recruited to the TRAIL-induced DISC where it inhibits caspase-8 cleavage (Fig. 2). However, a recent report suggests that PED mediates activation of extracellular regulated kinase (ERK) MAP kinase (Formstecher et al.,

2001), raising possibilities that PED, like c-FLIP, may have dual functions. PED is prominently expressed in astrocytes (Danziger et al., 1995). PED recruitment to TRAIL-induced DISC in glioma, but not melanomas (Xiao et al., 2002), suggests that PED may regulate TRAIL-induced DISC in neural cells.

IX. TRAIL PHYSIOLOGIC FUNCTIONS: TUMOR SURVEILLANCE

TRAIL mRNA has been detected in a variety of tissues (Wiley et al., 1995), but the cell surface expression of TRAIL protein has been reported only in a subset of natural killer (NK) cells in mouse livers (Smyth et al., 2001a; Takeda et al., 2001). The finding that TRAIL contributes to NK cell protection of livers from tumor metastasis provides the first evidence for the physiologic function of TRAIL as a tumor suppressor (Takeda et al., 2001). This was further supported by the finding that TRAIL-deficit mice were more susceptible to experimental and spontaneous tumor metastasis (Cretney et al., 2002). Moreover, neutralization of TRAIL promoted tumor development induced by a chemical carcinogen methylchoranthrene in mice and this protective effect of TRAIL was partly mediated by NK cells (Takeda et al., 2002). These studies further implicate TRAIL in immune surveillance against tumor development.

Immune surveillance against tumors is mediated by both innate and adaptive components of the cellular immunity (Smyth et al., 2001b). NK cells in the lung, liver, and spleen normally do not constitutively express TRAIL, but they can be induced to express cell surface TRAIL and thereby kill tumor cells *in vivo* (Smyth et al., 2001a). These studies suggest the roles for TRAIL in NK cell–mediated innate immunity in immune surveillance against tumors. Moreover, human blood T cells, dendritic cells, and monocytes can also be induced to express TRAIL (Fanger et al., 1999; Griffith et al., 1999; Kayagaki et al., 1999). Activated human CD4+V alpha 24NKT cells from patients with acute myeloid leukemia (AML) have been reported to induce TRAIL-dependent apoptosis in autologous AML cells (Nieda et al., 2001). Collectively, the studies implicate a broad role of TRAIL in adaptive cellular immunity against tumors.

X. RECOMBINANT TRAIL: A NOVEL CANCER THERAPEUTIC AGENT

There is a considerable amount of interest in the application of apoptotic agents in cancer therapy. The role of death ligands of TNF family in apoptosis of cancerous cells has placed them in the limelight of chemotherapeutic development. TNFα and FasL can trigger apoptosis in solid cancers, but their potential clinical usage is limited due to their severe toxicity to normal tissues (Ogasawara et al., 1993). In contrast, TRAIL and its receptors are currently of particular interest, as their interaction has been shown to result in the selective death of cancer cells, while sparing normal cells (Ashkenazi et al., 1999; Hao et al., 2001; Leverkus et al., 2000). Preclinical experiments in mice and nonhuman primates have shown that systemic administration of TRAIL inhibit tumor growth with little toxicity in animals (Ashkenazi et al., 1999; Walczak et al., 1999).

However, several epitope-tagged recombinant soluble forms of human TRAIL have been reported to be toxic to normal tissues, raising the concern of TRAIL safety in humans. The extracellular carboxy-terminal of approximately 150 amino acids of TRAIL contains the receptor-binding sites (Cha et al., 1999) and thus recombinant soluble forms of human TRAIL are commonly epitope-tagged at the amino-terminus. A recombinant soluble human TRAIL (amino acids 95–281) amino-terminally fused to a trimerizing leucine zipper motif was shown to be toxic to isolated human astrocytes (Walczak et al., 1999) and keratinocytes (Qin et al., 2001). Antibody cross-linked Flag-tagged human TRAIL (amino acids 95–281) induced apoptosis in astrocytes and neurons (Nitsch et al., 2000) and a polyhistidine-tagged human TRAIL (amino acids 114–281) resulted in hepatotoxicity (Jo et al., 2000). In contrast, recombinant nontagged human native sequence of TRAIL (amino acids 114–281) are not associated with apoptotic death of human astrocytes (Hao et al., 2001) or human hepatocytes (Lawrence et al., 2001). Injection of nontagged soluble human TRAIL (amino acids 114–281) in mice with chimeric human livers (Mercer et al., 2001) has a profound effect on tumor growth without causing hepatotoxicity (our unpublished data). Therefore, this nontagged native sequence soluble human TRAIL may prove to be a safe and effective candidate for cancer therapy in future human clinical trials.

Another obstacle in TRAIL clinical therapy for cancers is the fact that TRAIL-induced apoptosis is limited in cancers because many tumor cell lines and primary tumor cultures are resistant to TRAIL (Song et al., 2003). DR5 mutations are responsible for TRAIL resistance in some tumors (Shin et al., 2001), but the majority of resistant tumor cells express DR5 and are resistant to TRAIL due to intracellular inhibition of the DISC and mitochondria (Hersey and Zhang, 2001; Song et al., 2003). Recently, conventional chemotherapeutic drugs such as cisplatin, etoposide, and camptothecin have been shown to downregulate c-FLIP, but upregulation of Bak to enhance TRAIL-induced apoptosis in resistant cell lines and primary tumor cultures (LeBlanc et al., 2002; Song et al., 2003), providing combined therapeutic approaches that target the mechanisms of TRAIL resistance. However, further investigations into the signal events involved in the modulation of TRAIL signaling by chemotherapeutic drugs in normal and tumor cells are warranted to generate novel therapeutic strategies in combined treatment of TRAIL and conventional chemotherapeutic drugs that specifically target tumor cells.

REFERENCES

Araujo, H., Danziger, N., Cordier, J., Glowinski, J., and Chneiweiss, H. (1993). Characterization of PEA-15, a major substrate for protein kinase C in astrocytes. *J. Biol. Chem.* **268,** 5911–5920.

Ashkenazi, A., and Dixit, V. M. (1998). Death receptors: Signaling and modulation. *Science* **281,** 1305–1308.

Ashkenazi, A., Pai, R. C., Fong, S., Leung, S., Lawrence, D. A., Marsters, S. A., Blackie, C., Chang, L., McMurtrey, A. E., Hebert, A. *et al.* (1999). Safety and antitumor activity of recombinant soluble Apo2 ligand. *J. Clin. Invest.* **104,** 155–162.

Banner, D. W., D'Arcy, A., Janes, W., Gentz, R., Schoenfeld, H. J., Broger, C., Loetscher, H., and Lesslauer, W. (1993). Crystal structure of the soluble human 55 kd TNF receptor-human TNF beta complex: Implications for TNF receptor activation. *Cell* **73,** 431–445.

Bodmer, J. L., Holler, N., Reynard, S., Vinciguerra, P., Schneider, P., Juo, P., Blenis, J., and Tschopp, J. (2000a). TRAIL receptor-2 signals apoptosis through FADD and caspase-8. *Nat. Cell. Biol.* **2,** 241–243.

Bodmer, J. L., Meier, P., Tschopp, J., and Schneider, P. (2000b). Cysteine 230 is essential for the structure and activity of the cytotoxic ligand TRAIL. *J. Biol. Chem.* **275,** 20632–20637.

Boldin, M. P., Goncharov, T. M., Goltsev, Y. V., and Wallach, D. (1996). Involvement of MACH, a novel MORT1/FADD-interacting protease, in Fas/APO-1- and TNF receptor-induced cell death. *Cell* **85,** 803–815.

Boldin, M. P., Varfolomeev, E. E., Pancer, Z., Mett, I. L., Camonis, J. H., and Wallach, D. (1995). A novel protein that interacts with the death domain of Fas/APO1 contains a sequence motif related to the death domain. *J. Biol. Chem.* **270,** 7795–7798.

Cha, S. S., Kim, M. S., Choi, Y. H., Sung, B. J., Shin, N. K., Shin, H. C., Sung, Y. C., and Oh, B. H. (1999). 2.8 A resolution crystal structure of human TRAIL, a cytokine with selective antitumor activity. *Immunity* **11,** 253–261.

Cha, S. S., Sung, B. J., Kim, Y. A., Song, Y. L., Kim, H. J., Kim, S., Lee, M. S., and Oh, B. H. (2000). Crystal structure of TRAIL-DR5 complex identifies a critical role of the unique frame insertion in conferring recognition specificity. *J. Biol. Chem.* **275,** 31171–31177.

Chan, F. K., Chun, H. J., Zheng, L., Siegel, R. M., Bui, K. L., and Lenardo, M. J. (2000). A domain in TNF receptors that mediates ligand-independent receptor assembly and signaling. *Science* **288,** 2351–2354.

Chaudhary, P. M., Eby, M., Jasmin, A., Bookwalter, A., Murray, J., and Hood, L. (1997). Death receptor 5, a new member of the TNFR family, and DR4 induce FADD-dependent apoptosis and activate the NF-kappaB pathway. *Immunity* **7,** 821–830.

Chinnaiyan, A. M., O'Rourke, K., Tewari, M., and Dixit, V. M. (1995). FADD, a novel death domain-containing protein, interacts with the death domain of Fas and initiates apoptosis. *Cell* **81,** 505–512.

Condorelli, G., Vigliotta, G., Iavarone, C., Caruso, M., Tocchetti, C. G., Andreozzi, F., Cafieri, A., Tecce, M. F., Formisano, P., Beguinot, L., and Beguinot, F. (1998). PED/PEA-15 gene controls glucose transport and is overexpressed in type 2 diabetes mellitus. *EMBO J.* **17,** 3858–3866.

Cretney, E., Takeda, K., Yagita, H., Glaccum, M., Peschon, J. J., and Smyth, M. J. (2002). Increased susceptibility to tumor initiation and metastasis in TNF-related apoptosis-inducing ligand-deficient mice. *J. Immunol.* **168,** 1356–1361.

Danziger, N., Yokoyama, M., Jay, T., Cordier, J., Glowinski, J., and Chneiweiss, H. (1995). Cellular expression, developmental regulation, and phylogenic conservation of PEA-15, the astrocytic major phosphoprotein and protein kinase C substrate. *J. Neurochem.* **64**, 1016–1025.

Degli-Esposti, M. A., Dougall, W. C., Smolak, P. J., Waugh, J. Y., Smith, C. A., and Goodwin, R. G. (1997a). The novel receptor TRAIL-R4 induces NF-kappaB and protects against TRAIL-mediated apoptosis, yet retains an incomplete death domain. *Immunity* **7**, 813–820.

Degli-Esposti, M. A., Smolak, P. J., Walczak, H., Waugh, J., Huang, C. P., DuBose, R. F., Goodwin, R. G., and Smith, C. A. (1997b). Cloning and characterization of TRAIL-R3, a novel member of the emerging TRAIL receptor family. *J. Exp. Med.* **186**, 1165–1170.

Deng, Y., Lin, Y., and Wu, X. (2002). TRAIL-induced apoptosis requires Bax-dependent mitochondrial release of Smac/DIABLO. *Genes Dev.* **16**, 33–45.

Du, C., Fang, M., Li, Y., Li, L., and Wang, X. (2000). Smac, a mitochondrial protein that promotes cytochrome c-dependent caspase activation by eliminating IAP inhibition. *Cell* **102**, 33–42.

Ehrhardt, H., Fulda, S., Schmid, I., Hiscott, J., Debatin, K. M., and Jeremias, I. (2003). TRAIL induced survival and proliferation in cancer cells resistant towards TRAIL-induced apoptosis mediated by NF-kappaB. *Oncogene* **22**, 3842–3852.

Emery, J. G., McDonnell, P., Burke, M. B., Deen, K. C., Lyn, S., Silverman, C., Dul, E., Appelbaum, E. R., Eichman, C., DiPrinzio, R. *et al.* (1998). Osteoprotegerin is a receptor for the cytotoxic ligand TRAIL. *J. Biol. Chem.* **273**, 14363–14367.

Fanger, N. A., Maliszewski, C. R., Schooley, K., and Griffith, T. S. (1999). Human dendritic cells mediate cellular apoptosis via tumor necrosis factor-related apoptosis-inducing ligand (TRAIL). *J. Exp. Med.* **190**, 1155–1164.

Fernandes-Alnemri, T., Armstrong, R. C., Krebs, J., Srinivasula, S. M., Wang, L., Bullrich, F., Fritz, L. C., Trapani, J. A., Tomaselli, K. J., Litwack, G., and Alnemri, E. S. (1996). In vitro activation of CPP32 and Mch3 by Mch4, a novel human apoptotic cysteine protease containing two FADD-like domains. *Proc. Natl. Acad. Sci. USA* **93**, 7464–7469.

Formstecher, E., Ramos, J. W., Fauquet, M., Calderwood, D. A., Hsieh, J. C., Canton, B., Nguyen, X. T., Barnier, J. V., Camonis, J., Ginsberg, M. H., and Chneiweiss, H. (2001). PEA-15 mediates cytoplasmic sequestration of ERK MAP kinase. *Dev. Cell.* **1**, 239–250.

Goltsev, Y. V., Kovalenko, A. V., Arnold, E., Varfolomeev, E. E., Brodianskii, V. M., and Wallach, D. (1997). CASH, a novel caspase homologue with death effector domains. *J. Biol. Chem.* **272**, 19641–19644.

Green, D. R., and Reed, J. C. (1998). Mitochondria and apoptosis. *Science* **281**, 1309–1312.

Griffith, T. S., Wiley, S. R., Kubin, M. Z., Sedger, L. M., Maliszewski, C. R., and Fanger, N. A. (1999). Monocyte-mediated tumoricidal activity via the tumor necrosis factor-related cytokine, TRAIL. *J. Exp. Med.* **189**, 1343–1354.

Gross, A., Jockel, J., Wei, M. C., and Korsmeyer, S. J. (1998). Enforced dimerization of BAX results in its translocation, mitochondrial dysfunction and apoptosis. *EMBO J.* **17**, 3878–3885.

Han, D. K., Chaudhary, P. M., Wright, M. E., Friedman, C., Trask, B. J., Riedel, R. T., Baskin, D. G., Schwartz, S. M., and Hood, L. (1997). MRIT, a novel death-effector domain-containing protein, interacts with caspases and BclXL and initiates cell death. *Proc. Natl. Acad. Sci. USA* **94**, 11333–11338.

Hao, C., Beguinot, F., Condorelli, G., Trencia, A., Van Meir, E. G., Yong, V. W., Parney, I. F., Roa, W. H., and Petruk, K. C. (2001). Induction and intracellular regulation of tumor necrosis factor-related apoptosis-inducing ligand (TRAIL) mediated apoptosis in human malignant glioma cells. *Cancer Res.* **61**, 1162–1170.

Harper, N., Farrow, S. N., Kaptein, A., Cohen, G. M., and MacFarlane, M. (2001). Modulation of tumor necrosis factor apoptosis-inducing ligand-induced NF-kappaB activation by inhibition of apical caspases. *J. Biol. Chem.* **276**, 34743–34752.

Hersey, P., and Zhang, X. D. (2001). How melanoma cells evade trail-induced apoptosis. *Natl. Rev. Cancer* **1**, 142–150.

Hinz, S., Trauzold, A., Boenicke, L., Sandberg, C., Beckmann, S., Bayer, E., Walczak, H., Kalthoff, H., and Ungefroren, H. (2000). Bcl-XL protects pancreatic adenocarcinoma cells against CD95- and TRAIL-receptor-mediated apoptosis. *Oncogene* **19**, 5477–5486.

Holen, I., Croucher, P. I., Hamdy, F. C., and Eaton, C. L. (2002). Osteoprotegerin (OPG) is a survival factor for human prostate cancer cells. *Cancer Res.* **62**, 1619–1623.

Holler, N., Zaru, R., Micheau, O., Thome, M., Attinger, A., Valitutti, S., Bodmer, J. L., Schneider, P., Seed, B., and Tschopp, J. (2000). Fas triggers an alternative, caspase-8-independent cell death pathway using the kinase RIP as effector molecule. *Nat. Immunol.* **1**, 489–495.

Hu, S., Vincenz, C., Ni, J., Gentz, R., and Dixit, V. M. (1997). I-FLICE, a novel inhibitor of tumor necrosis factor receptor-1- and CD-95-induced apoptosis. *J. Biol. Chem.* **272**, 17255–17257.

Hu, W. H., Johnson, H., and Shu, H. B. (2000). Activation of NF-kappaB by FADD, Casper, and caspase-8. *J. Biol. Chem.* **275**, 10838–10844.

Hymowitz, S. G., Christinger, H. W., Fuh, G., Ultsch, M., O'Connell, M., Kelley, R. F., Ashkenazi, A., and de Vos, A. M. (1999). Triggering cell death: The crystal structure of Apo2L/TRAIL in a complex with death receptor 5. *Mol. Cell.* **4**, 563–571.

Hymowitz, S. G., O'Connell, M. P., Ultsch, M. H., Hurst, A., Totpal, K., Ashkenazi, A., de Vos, A. M., and Kelley, R. F. (2000). A unique zinc-binding site revealed by a high-resolution X-ray structure of homotrimeric Apo2L/TRAIL. *Biochemistry* **39**, 633–640.

Inohara, N., Ding, L., Chen, S., and Nunez, G. (1997). Harakiri, a novel regulator of cell death, encodes a protein that activates apoptosis and interacts selectively with survival-promoting proteins Bcl-2 and Bcl-X(L). *EMBO J.* **16**, 1686–1694.

Inohara, N., Koseki, T., Lin, J., del Peso, L., Lucas, P. C., Chen, F. F., Ogura, Y., and Nunez, G. (2000). An induced proximity model for NF-kappaB activation in the Nod1/RICK and RIP signaling pathways. *J. Biol. Chem.* **275**, 27823–27831.

Irmler, M., Thome, M., Hahne, M., Schneider, P., Hofmann, K., Steiner, V., Bodmer, J. L., Schroter, M., Burns, K., Mattmann, C. *et al.* (1997). Inhibition of death receptor signals by cellular FLIP. *Nature* **388**, 190–195.

Jo, M., Kim, T. H., Seol, D. W., Esplen, J. E., Dorko, K., Billiar, T. R., and Strom, S. C. (2000). Apoptosis induced in normal human hepatocytes by tumor necrosis factor-related apoptosis-inducing ligand. *Nat. Med.* **6**, 564–567.

Karin, M., and Lin, A. (2002). NF-kappaB at the crossroads of life and death. *Nat. Immunol.* **3**, 221–227.

Kayagaki, N., Yamaguchi, N., Nakayama, M., Kawasaki, A., Akiba, H., Okumura, K., and Yagita, H. (1999). Involvement of TNF-related apoptosis-inducing ligand in human CD4+ T cell-mediated cytotoxicity. *J. Immunol.* **162**, 2639–2647.

Kischkel, F. C., Hellbardt, S., Behrmann, I., Germer, M., Pawlita, M., Krammer, P. H., and Peter, M. E. (1995). Cytotoxicity-dependent APO-1 (Fas/CD95)-associated proteins form a death-inducing signaling complex (DISC) with the receptor. *EMBO J.* **14**, 5579–5588.

Kischkel, F. C., Kioschis, P., Weitz, S., Poustka, A., Lichter, P., and Krammer, P. H. (1998). Assignment of CASP8 to human chromosome band 2q33→q34 and Casp8 to the murine syntenic region on chromosome 1B-proximal C by in situ hybridization. *Cytogenet. Cell Genet.* **82**, 95–96.

Kischkel, F. C., Lawrence, D. A., Chuntharapai, A., Schow, P., Kim, K. J., and Ashkenazi, A. (2000). Apo2L/TRAIL-dependent recruitment of endogenous FADD and caspase-8 to death receptors 4 and 5. *Immunity* **12**, 611–620.

Kischkel, F. C., Lawrence, D. A., Tinel, A., LeBlanc, H., Virmani, A., Schow, P., Gazdar, A., Blenis, J., Arnott, D., and Ashkenazi, A. (2001). Death receptor recruitment of endogenous caspase-10 and apoptosis initiation in the absence of caspase-8. *J. Biol. Chem.* **276**, 46639–46646.

Kroemer, G., and Reed, J. C. (2000). Mitochondrial control of cell death. *Nat. Med.* **6**, 513–519.

Kuang, A. A., Diehl, G. E., Zhang, J., and Winoto, A. (2000). FADD is required for DR4- and DR5-mediated apoptosis: Lack of trail-induced apoptosis in FADD-deficient mouse embryonic fibroblasts. *J. Biol. Chem.* **275**, 25065–25068.

Kubes, M., Cordier, J., Glowinski, J., Girault, J. A., and Chneiweiss, H. (1998). Endothelin induces a calcium-dependent phosphorylation of PEA-15 in intact astrocytes: Identification of Ser104 and Ser116 phosphorylated, respectively, by protein kinase C and calcium/calmodulin kinase II in vitro. *J. Neurochem.* **71**, 1307–1314.

Lawrence, D., Shahrokh, Z., Marsters, S., Achilles, K., Shih, D., Mounho, B., Hillan, K., Totpal, K., DeForge, L., Schow, P. *et al.* (2001). Differential hepatocyte toxicity of recombinant Apo2L/TRAIL versions. *Nat. Med.* **7**, 383–385.

LeBlanc, H., Lawrence, D., Varfolomeev, E., Totpal, K., Morlan, J., Schow, P., Fong, S., Schwall, R., Sinicropi, D., and Ashkenazi, A. (2002). Tumor-cell resistance to death receptor–induced apoptosis through mutational inactivation of the proapoptotic Bcl-2 homolog Bax. *Nat. Med.* **8**, 274–281.

Leverkus, M., Neumann, M., Mengling, T., Rauch, C. T., Brocker, E. B., Krammer, P. H., and Walczak, H. (2000). Regulation of tumor necrosis factor-related apoptosis-inducing ligand sensitivity in primary and transformed human keratinocytes. *Cancer Res.* **60**, 553–559.

Li, H., Zhu, H., Xu, C. J., and Yuan, J. (1998). Cleavage of BID by caspase 8 mediates the mitochondrial damage in the Fas pathway of apoptosis. *Cell* **94**, 491–501.

Li, P., Nijhawan, D., Budihardjo, I., Srinivasula, S. M., Ahmad, M., Alnemri, E. S., and Wang, X. (1997). Cytochrome c and dATP-dependent formation of Apaf-1/caspase-9 complex initiates an apoptotic protease cascade. *Cell* **91**, 479–489.

Lin, Y., Devin, A., Cook, A., Keane, M. M., Kelliher, M., Lipkowitz, S., and Liu, Z. G. (2000). The death domain kinase RIP is essential for TRAIL (Apo2L)-induced activation of IkappaB kinase and c-Jun N-terminal kinase. *Mol. Cell. Biol.* **20**, 6638–6645.

Liu, X., Zou, H., Slaughter, C., and Wang, X. (1997). DFF, a heterodimeric protein that functions downstream of caspase-3 to trigger DNA fragmentation during apoptosis. *Cell* **89**, 175–184.

Luo, X., Budihardjo, I., Zou, H., Slaughter, C., and Wang, X. (1998). Bid, a Bcl2 interacting protein, mediates cytochrome c release from mitochondria in response to activation of cell surface death receptors. *Cell* **94**, 481–490.

MacFarlane, M., Ahmad, M., Srinivasula, S. M., Fernandes-Alnemri, T., Cohen, G. M., and Alnemri, E. S. (1997). Identification and molecular cloning of two novel receptors for the cytotoxic ligand TRAIL. *J. Biol. Chem.* **272**, 25417–25420.

Mariani, S. M., and Krammer, P. H. (1998). Differential regulation of TRAIL and CD95 ligand in transformed cells of the T and B lymphocyte lineage. *Eur. J. Immunol.* **28**, 973–982.

Marsters, S. A., Sheridan, J. P., Pitti, R. M., Huang, A., Skubatch, M., Baldwin, D., Yuan, J., Gurney, A., Goddard, A. D., Godowski, P., and Ashkenazi, A. (1997). A novel receptor for Apo2L/TRAIL contains a truncated death domain. *Curr. Biol.* **7**, 1003–1006.

Medema, J. P., Scaffidi, C., Kischkel, F. C., Shevchenko, A., Mann, M., Krammer, P. H., and Peter, M. E. (1997). FLICE is activated by association with the CD95 death-inducing signaling complex (DISC). *EMBO J.* **16**, 2794–2804.

Mercer, D. F., Schiller, D. E., Elliott, J. F., Douglas, D. N., Hao, C., Rinfret, A., Addison, W. R., Fischer, K. P., Churchill, T. A., Lakey, J. R. *et al.* (2001). Hepatitis C virus replication in mice with chimeric human livers. *Nat. Med.* **7**, 927–933.

Mercurio, F., Zhu, H., Murray, B. W., Shevchenko, A., Bennett, B. L., Li, J., Young, D. B., Barbosa, M., Mann, M., Manning, A., and Rao, A. (1997). IKK-1 and IKK-2: Cytokine-activated IkappaB kinases essential for NF-kappaB activation. *Science* **278**, 860–866.

Mongkolsapaya, J., Grimes, J. M., Chen, N., Xu, X. N., Stuart, D. I., Jones, E. Y., and Screaton, G. R. (1999). Structure of the TRAIL-DR5 complex reveals mechanisms conferring specificity in apoptotic initiation. *Nat. Struct. Biol.* **6**, 1048–1053.

Muhlenbeck, F., Haas, E., Schwenzer, R., Schubert, G., Grell, M., Smith, C., Scheurich, P., and Wajant, H. (1998). TRAIL/Apo2L activates c-Jun NH2-terminal kinase (JNK) via caspase-dependent and caspase-independent pathways. *J. Biol. Chem.* **273,** 33091–33098.

Muzio, M., Chinnaiyan, A. M., Kischkel, F. C., O'Rourke, K., Shevchenko, A., Ni, J., Scaffidi, C., Bretz, J. D., Zhang, M., Gentz, R. *et al.* (1996). FLICE, a novel FADD-homologous ICE/CED-3-like protease, is recruited to the CD95 (Fas/APO-1) death-inducing signaling complex. *Cell* **85,** 817–827.

Muzio, M., Stockwell, B. R., Stennicke, H. R., Salvesen, G. S., and Dixit, V. M. (1998). An induced proximity model for caspase-8 activation. *J. Biol. Chem.* **273,** 2926–2930.

Nieda, M., Nicol, A., Koezuka, Y., Kikuchi, A., Lapteva, N., Tanaka, Y., Tokunaga, K., Suzuki, K., Kayagaki, N., Yagita, H. *et al.* (2001). TRAIL expression by activated human CD4(+)V alpha 24NKT cells induces in vitro and in vivo apoptosis of human acute myeloid leukemia cells. *Blood* **97,** 2067–2074.

Nitsch, R., Bechmann, I., Deisz, R. A., Haas, D., Lehmann, T. N., Wendling, U., and Zipp, F. (2000). Human brain-cell death induced by tumour-necrosis-factor-related apoptosis-inducing ligand (TRAIL). *Lancet* **356,** 827–828.

Ogasawara, J., Watanabe-Fukunaga, R., Adachi, M., Matsuzawa, A., Kasugai, T., Kitamura, Y., Itoh, N., Suda, T., and Nagata, S. (1993). Lethal effect of the anti-Fas antibody in mice. *Nature* **364,** 806–809.

Pan, G., Ni, J., Wei, Y. F., Yu, G., Gentz, R., and Dixit, V. M. (1997a). An antagonist decoy receptor and a death domain-containing receptor for TRAIL. *Science* **277,** 815–818.

Pan, G., O'Rourke, K., Chinnaiyan, A. M., Gentz, R., Ebner, R., Ni, J., and Dixit, V. M. (1997b). The receptor for the cytotoxic ligand TRAIL. *Science* **276,** 111–113.

Pan, G., Ni, J., Yu, G., Wei, Y. F., and Dixit, V. M. (1998). TRUNDD, a new member of the TRAIL receptor family that antagonizes TRAIL signalling. *FEBS Lett.* **424,** 41–45.

Panka, D. J., Mano, T., Suhara, T., Walsh, K., and Mier, J. W. (2001). Phosphatidylinositol 3-kinase/Akt activity regulates c-FLIP expression in tumor cells. *J. Biol. Chem.* **276,** 6893–6896.

Pitti, R. M., Marsters, S. A., Ruppert, S., Donahue, C. J., Moore, A., and Ashkenazi, A. (1996). Induction of apoptosis by Apo-2 ligand, a new member of the tumor necrosis factor cytokine family. *J. Biol. Chem.* **271,** 12687–12690.

Qin, J., Chaturvedi, V., Bonish, B., and Nickoloff, B. J. (2001). Avoiding premature apoptosis of normal epidermal cells. *Nat. Med.* **7,** 385–386.

Rasper, D. M., Vaillancourt, J. P., Hadano, S., Houtzager, V. M., Seiden, I., Keen, S. L., Tawa, P., Xanthoudakis, S., Nasir, J., Martindale, D. *et al.* (1998). Cell death attenuation by "Usurpin," a mammalian DED-caspase homologue that precludes caspase-8 recruitment and activation by the CD-95 (Fas, APO-1) receptor complex *Cell Death Differ.* **5,** 271–288.

Samali, A., Cai, J., Zhivotovsky, B., Jones, D. P., and Orrenius, S. (1999). Presence of a pre-apoptotic complex of pro-caspase-3, Hsp60 and Hsp10 in the mitochondrial fraction of jurkat cells. *EMBO J.* **18,** 2040–2048.

Scaffidi, C., Schmitz, I., Krammer, P. H., and Peter, M. E. (1999). The role of c-FLIP in modulation of CD95-induced apoptosis. *J. Biol. Chem.* **274,** 1541–1548.

Schneider, P., Bodmer, J. L., Thome, M., Hofmann, K., Holler, N., and Tschopp, J. (1997a). Characterization of two receptors for TRAIL. *FEBS Lett.* **416,** 329–334.

Schneider, P., Thome, M., Burns, K., Bodmer, J. L., Hofmann, K., Kataoka, T., Holler, N., and Tschopp, J. (1997b). TRAIL receptors 1 (DR4) and 2 (DR5) signal FADD-dependent apoptosis and activate NF-kappaB. *Immunity* **7,** 831–836.

Screaton, G. R., Mongkolsapaya, J., Xu, X. N., Cowper, A. E., McMichael, A. J., and Bell, J. I. (1997). TRICK2, a new alternatively spliced receptor that transduces the cytotoxic signal from TRAIL. *Curr. Biol.* **7,** 693–696.

Secchiero, P., Gonelli, A., Carnevale, E., Milani, D., Pandolfi, A., Zella, D., and Zauli, G. (2003). TRAIL promotes the survival and proliferation of primary human vascular endothelial cells by activating the Akt and ERK pathways. *Circulation* **107,** 2250–2256.

Sheridan, J. P., Marsters, S. A., Pitti, R. M., Gurney, A., Skubatch, M., Baldwin, D., Ramakrishnan, L., Gray, C. L., Baker, K., Wood, W. I. *et al.* (1997). Control of TRAIL-induced apoptosis by a family of signaling and decoy receptors. *Science* **277**, 818–821.

Shin, M. S., Kim, H. S., Lee, S. H., Park, W. S., Kim, S. Y., Park, J. Y., Lee, J. H., Lee, S. K., Lee, S. N., Jung, S. S. *et al.* (2001). Mutations of tumor necrosis factor-related apoptosis-inducing ligand receptor 1 (TRAIL-R1) and receptor 2 (TRAIL-R2) genes in metastatic breast cancers. *Cancer Res.* **61**, 4942–4946.

Shu, H. B., Halpin, D. R., and Goeddel, D. V. (1997). Casper is a FADD-and caspase-related inducer of apoptosis. *Immunity* **6**, 751–763.

Siegel, R. M., Frederiksen, J. K., Zacharias, D. A., Chan, F. K., Johnson, M., Lynch, D., Tsien, R. Y., and Lenardo, M. J. (2000). Fas preassociation required for apoptosis signaling and dominant inhibition by pathogenic mutations. *Science* **288**, 2354–2357.

Siegmund, D., Mauri, D., Peters, N., Juo, P., Thome, M., Reichwein, M., Blenis, J., Scheurich, P., Tschopp, J., and Wajant, H. (2001). Fas-associated death domain protein (FADD) and caspase-8 mediate up-regulation of c-Fos by Fas ligand and tumor necrosis factor-related apoptosis-inducing ligand (TRAIL) via a FLICE inhibitory protein (FLIP)-regulated pathway. *J. Biol. Chem.* **276**, 32585–32590.

Simonet, W. S., Lacey, D. L., Dunstan, C. R., Kelley, M., Chang, M. S., Luthy, R., Nguyen, H. Q., Wooden, S., Bennett, L., Boone, T. *et al.* (1997). Osteoprotegerin: A novel secreted protein involved in the regulation of bone density. *Cell* **89**, 309–319.

Smith, C. A., Farrah, T., and Goodwin, R. G. (1994). The TNF receptor superfamily of cellular and viral proteins: Activation, costimulation, and death. *Cell* **76**, 959–962.

Smith, C. A., Farrah, T., and Goodwin, R. G. (1994). The TNF receptor superfamily of cellular and viral proteins: Activation, costimulation, and death. *Cell* **76**, 959–962.

Smyth, M. J., Godfrey, D. I., and Trapani, J. A. (2001b). A fresh look at tumor immunosurveillance and immunotherapy. *Nat. Immunol.* **2**, 293–299.

Song, J. H., Song, D. K., Pyrzynska, P., Petruk, K. C., Van Meir, E. G., and Hao, C. (2003). TRAIL induces apoptosis in human glioma cells through extrinsic and intrinsic pathways. *Brain Pathol.* **13**, 539–553.

Sprick, M. R., Rieser, E., Stahl, H., Grosse-Wilde, A., Weigand, M. A., and Walczak, H. (2002). Caspase-10 is recruited to and activated at the native TRAIL and CD95 death-inducing signalling complexes in a FADD-dependent manner but can not functionally substitute caspase-8. *EMBO J.* **21**, 4520–4530.

Srinivasula, S. M., Ahmad, M., Ottilie, S., Bullrich, F., Banks, S., Wang, Y., Fernandes-Alnemri, T., Croce, C. M., Litwack, G., Tomaselli, K. J. *et al.* (1997). FLAME-1, a novel FADD-like anti-apoptotic molecule that regulates Fas/TNFR1-induced apoptosis. *J. Biol. Chem.* **272**, 18542–18545.

Srinivasula, S. M., Hegde, R., Saleh, A., Datta, P., Shiozaki, E., Chai, J., Lee, R. A., Robbins, P. D., Fernandes-Alnemri, T., Shi, Y., and Alnemri, E. S. (2001). A conserved XIAP-interaction motif in caspase-9 and Smac/DIABLO regulates caspase activity and apoptosis. *Nature* **410**, 112–116.

Stanger, B. Z., Leder, P., Lee, T. H., Kim, E., and Seed, B. (1995). RIP: A novel protein containing a death domain that interacts with Fas/APO-1 (CD95) in yeast and causes cell death. *Cell* **81**, 513–523.

Takeda, K., Hayakawa, Y., Smyth, M. J., Kayagaki, N., Yamaguchi, N., Kakuta, S., Iwakura, Y., Yagita, H., and Okumura, K. (2001). Involvement of tumor necrosis factor-related apoptosis-inducing ligand in surveillance of tumor metastasis by liver natural killer cells. *Nat. Med.* **7**, 94–100.

Takeda, K., Smyth, M. J., Cretney, E., Hayakawa, Y., Kayagaki, N., Yagita, H., and Okumura, K. (2002). Critical role for tumor necrosis factor-related apoptosis-inducing ligand in immune surveillance against tumor development. *J. Exp. Med.* **195**, 161–169.

Truneh, A., Sharma, S., Silverman, C., Khandekar, S., Reddy, M. P., Deen, K. C., McLaughlin, M. M., Srinivasula, S. M., Livi, G. P., Marshall, L. A. *et al.* (2000). Temperature-sensitive differential affinity of TRAIL for its receptors. DR5 is the highest affinity receptor. *J. Biol. Chem.* **275**, 23319–23325.

Verhagen, A. M., Ekert, P. G., Pakusch, M., Silke, J., Connolly, L. M., Reid, G. E., Moritz, R. L., Simpson, R. J., and Vaux, D. L. (2000). Identification of DIABLO, a mammalian protein that promotes apoptosis by binding to and antagonizing IAP proteins. *Cell* **102**, 43–53.

Walczak, H., Degli-Esposti, M. A., Johnson, R. S., Smolak, P. J., Waugh, J. Y., Boiani, N., Timour, M. S., Gerhart, M. J., Schooley, K. A., Smith, C. A. *et al.* (1997). TRAIL-R2: A novel apoptosis-mediating receptor for TRAIL. *EMBO J.* **16**, 5386–5397.

Walczak, H., Miller, R. E., Ariail, K., Gliniak, B., Griffith, T. S., Kubin, M., Chin, W., Jones, J., Woodward, A., Le, T. *et al.* (1999). Tumoricidal activity of tumor necrosis factor-related apoptosis-inducing ligand *in vivo*. *Nat. Med.* **5**, 157–163.

Wang, J., Chun, H. J., Wong, W., Spencer, D. M., and Lenardo, M. J. (2001). Caspase-10 is an initiator caspase in death receptor signaling. *Proc. Natl. Acad. Sci. USA* **98**, 13884–13888.

Wei, M. C., Lindsten, T., Mootha, V. K., Weiler, S., Gross, A., Ashiya, M., Thompson, C. B., and Korsmeyer, S. J. (2000). tBID, a membrane-targeted death ligand, oligomerizes BAK to release cytochrome c. *Genes Dev.* **14**, 2060–2071.

Wiley, S. R., Schooley, K., Smolak, P. J., Din, W. S., Huang, C. P., Nicholl, J. K., Sutherland, G. R., Smith, T. D., Rauch, C., Smith, C. A. *et al.* (1995). Identification and characterization of a new member of the TNF family that induces apoptosis. *Immunity* **3**, 673–682.

Wu, G. S., Burns, T. F., McDonald, E. R. 3rd, Jiang, W., Meng, R., Krantz, I. D., Kao, G., Gan, D. D., Zhou, J. Y., Muschel, R. *et al.* (1997). KILLER/DR5 is a DNA damage-inducible p53-regulated death receptor gene. *Nat. Genet.* **17**, 141–143.

Xiao, C., Yang, B. F., Asadi, N., Beguinot, F., and Hao, C. (2002). Tumor necrosis factor-related apoptosis-inducing ligand-induced death-inducing signaling complex and its modulation by c-FLIP and PED/PEA-15 in glioma cells. *J. Biol. Chem.* **277**, 25020–25025.

Yang, B. F., Xiao, C., Roa, W. H., Krammer, P. H., and Hao, C. (2003). Calcium/calmodulin-dependent protein kinase II regulation of c-FLIP expression and phosphorylation in modulation of Fas-mediated signaling in malignant glioma cells. *J. Biol. Chem.* **278**, 7043–7050.

Yeh, J. H., Hsu, S. C., Han, S. H., and Lai, M. Z. (1998). Mitogen-activated protein kinase kinase antagonized fas-associated death domain protein-mediated apoptosis by induced FLICE-inhibitory protein expression. *J. Exp. Med.* **188**, 1795–1802.

Zandi, E., Rothwarf, D. M., Delhase, M., Hayakawa, M., and Karin, M. (1997). The IkappaB kinase complex (IKK) contains two kinase subunits, IKKalpha and IKKbeta, necessary for IkappaB phosphorylation and NF-kappaB activation. *Cell* **91**, 243–252.

7

TRAIL AND NFκB SIGNALING—A COMPLEX RELATIONSHIP

Harald Wajant

*Department of Molecular Internal Medicine
Medical Polyclinic, University of Würzburg, D-97070 Würzburg, Germany*

I. Introduction
   A. *The NFκB Signaling Pathway(s)*
   B. *TRAIL*
II. Regulation of TRAIL-Induced Apoptosis by NFκB
   A. *NFκB can Globally Regulate Cellular Apoptosis-Sensitivity*
   B. *Regulation of Death Receptor-Induced Apoptosis by NFκB*
   C. *Selective Regulation of TRAIL-Induced Apoptosis by NFκB*
III. TRAIL Activates NFκB
   References

Tumor necrosis factor (TNF)-related apoptosis-inducing ligand (TRAIL) or Apo2L is a ligand of the TNF family interacting with five different receptors of the TNF receptor superfamily, including two death receptors. It has attracted wide interest as a potential anticancer therapy because some recombinant soluble forms of TRAIL induce cell death predominantly in transformed cells. The nuclear factor-kappaB (NFκB)/Rel family of proteins are composed of a group of dimeric transcription factors that have an outstanding role in the regulation of inflammation and immunity. Control of transcription by NFκB proteins can be of relevance to the function of TRAIL in three ways. First, induction of antiapoptotic NFκB dependent genes critically determines cellular susceptibility toward apoptosis induction by TRAIL-R1, TRAIL-R2, and other death receptors. Each of the multiple of known NFκB inducers therefore has the potential to interfere with TRAIL-induced cell death. Second, TRAIL and some of its receptors are inducible by NFκB, disclosing the possibility of autoamplifying TRAIL signaling loops. Third, the TRAIL death receptors can activate the NFκB pathway. This chapter summarizes basic knowledge regarding the understanding of the NFκB pathway and focuses on its multiple roles in TRAIL signaling. © 2004 Elsevier Inc.

I. INTRODUCTION

A. THE NFκB SIGNALING PATHWAY(S)

1. NFκB Proteins

The phylogenetically conserved family of nuclear factor-kappaB (NFκB) transcription factors in mammals are composed of more than 10 defined homo- and heterodimers of the cRel, RelA (p65), RelB, p50 (NFκB1), and p52 (NFκB2) proteins (Fig. 1). In contrast to the Rel NFκB proteins, which are directly produced as mature proteins, NFκB1 and NFκB2 result from proteolytic processing from their p105 and p100 precursor proteins (Karin and Lin, 2002; Li and Verma, 2002). The NFκB transcription factors become activated by a variety of extracellular stimuli and physical stresses, and they regulate a large number of genes involved in inflammation, differentiation, and apoptosis control. The structural hallmark of the NFκB transcription factors is the highly conserved Rel homology region (RHR), a module of two immunoglobulin-like domains of about 300 amino acids. The RHR has been implicated in dimerization, recognition of DNA binding sites, and interaction with members of the IκB family of NFκB regulatory proteins (Karin and Lin, 2002; Li and Verma, 2002). The RHR also buries a nuclear localization sequence (NLS). While RelA, cRel, and RelB have a carboxy-terminal transactivation domain, p50 and p52 have none and homodimers act as transcriptional repressors. However, in complex with RelA, cRel, and RelB, both proteins can also be a part of transcriptionally active heterodimers. Although certain NFκB dimers interact differentially with some κB DNA binding sites, in general, NFκB dimers interact with high affinity with most κB DNA binding sites. The decision of which type of NFκB dimer will finally bind to a given promoter is therefore not only dependent on the expression levels of NFκB proteins, differences in the affinity of each NFκB dimer to this promoter, but also from the presence of interacting proteins of the NFκB dimer and their concentration in the nucleus. Moreover, each NFκB dimer elicits transcription with different efficiency and can be the subject of posttranslational modifications that affect its activity.

2. IκB Proteins

NFκB dimers are ubiquitously expressed and their transcriptional activation capacity is tightly regulated through high affinity interactions with IκB family proteins (Karin and Lin, 2002; Li and Verma, 2002; Rothwarf and Karin, 1999). These proteins contain six or seven copies of a small, helical domain called ankyrin domain (ankyrin repeat domain, ARD), which is also present in a variety of functionally diverse proteins not related to NFκB signaling. The mammalian IκB family is composed of IκBα, two splice variants of IκBβ1 and IκBβ2, IκBγ, IκBε, Bcl3, and the p100 and p105 proteins, which are also precursors of p52 and p50 (Fig. 1). Thus, these precursor proteins contain in addition to their amino-terminal NFκB domain an ankyrin repeat domain, which is removed during maturation (Karin and Lin, 2002; Li and Verma, 2002; Rothwarf and Karin, 1999). A single IκB protein binds asymmetrically to the dimerization domain of NFκB homo- and heterodimers and masks the nuclear localization sequences of the NFκB subunits to a different extent. For example, IκBα makes excessive contacts with the NLS of RelA of a p50-RelA dimer, but interacts only moderately with the NLS of the p50 subunit (Huxford et al., 1998; Jacobs and Harrison, 1998). A differential interaction between an IκB protein and the nuclear localization sequences of the NFκB subunits is also evident with homodimeric NFκBs [e.g., in the IκBα-(RelA)$_2$ complex]. The asymmetric mode of IκB-NFκB association suggests that the stability and specificity of this interaction is mainly determined by the interaction of an IκB protein and one subunit of an NFκB dimer. In accordance with this concept, it has been found that IκBα preferentially interacts with RelA containing homo- and heterodimers, whereas IκBγ interacts with p50 containing dimers (Moorthy and Ghosh, 2003).

The extensively studied complexes of p50-RelA and IκBα or IκBβ are both predominately located in the cytoplasm of nonstimulated cells. However, the mechanisms by which this distribution is achieved is mechanistically different between IκBα- and IκBβ-containing complexes. IκBα-NFκB complexes shuttle in and out of the nucleus, thus their predominant cytoplasmic localization reflects an equilibrium in which the nuclear export rate exceeds the nuclear import rate (Harhaj and Sun, 1999; Huang and Miyamoto, 2001; Johnson et al., 1999). Despite their transient presence in the nucleus, IκBα-NFκB complexes do not drive transcription as IκBα prevents DNA binding of the NFκB dimer. In contrast, IκBβ-NFκB complexes are constitutively located in the cytoplasm of nonstimulated cells, suggesting that an as yet unknown cellular factor antagonizes the function of the IκB-free NLS within the NFκB dimer (Malek et al., 2001; Tam and Sen, 2001).

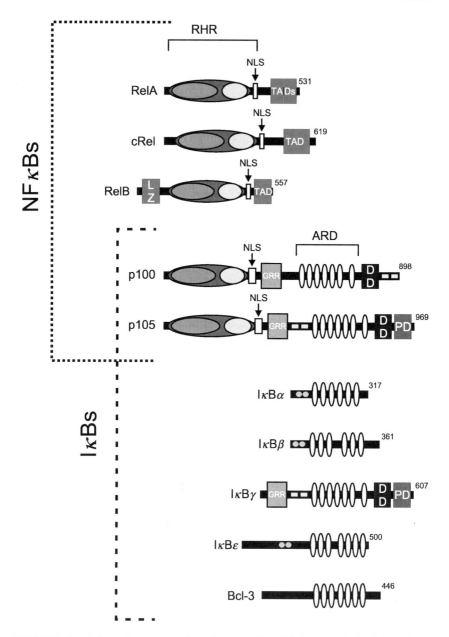

FIGURE 1. Schematic representation of mammalian NFκB and IκB family members. RHR, Rel homology domain (dark gray ellipse) consisting of a DNA binding domain (medium gray ellipse), a dimerization domain (light gray ellipse) and a nuclear localization sequence (NLS); LZ, leucine zipper; TAD, transactivation domain; GRR, glycine-rich region; ARD, ankyrin repeat domain; DD, death domain (note that the DDs of p100 and p105 are not involved in apoptosis signaling); PD, PEST domain. The serines of the signal response domain

Three of the IκB proteins, namely IκBα, IκBβ, and IκBε, have an amino-terminal signal response domain containing a pair of conserved serine residues. NFκB inducing signals (e.g., cytokines or microbial products) trigger dual phosphorylation of this domain (Brockman et al., 1995; Brown et al., 1995; DiDonato et al., 1996), allowing recognition by the Skp1-Cullin-F-box protein beta-transducin repeat-containing protein (SCF$^{\beta TRCP}$) ubiquitin ligase complex (Yaron et al., 1998) which catalyzes multi-ubiquitination of two adjacent lysine residues. This leads subsequently to degradation by the 26S proteasome (Baldi et al., 1996; Scherer et al., 1995). The ARD containing NFκB precursor proteins p105 and p100 can also form inactive complexes with a variety of NFκB proteins (Mercurio et al., 1993; Naumann et al., 1993; p105, Rice et al., 1992) or specifically with RelB (p100, Dobrzanski et al., 1995). The precursor-containing complexes become activated by signal-induced processing and/or degradation of the precursor proteins by the 26S proteasome. Release of mature p50 and 52 by limited proteolysis by the proteasome depends on a glycine-rich region in the carboxy-terminal part of the mature NFκB proteins. This glycine-rich domain inhibits the entry of the amino-terminal NFκB fragment of p100 and p105 into the proteasome and thereby prevents complete degradation (Heusch et al., 1999; Lin and Ghosh, 1996; Orian et al., 1999). The processing of p100 to p52 is inducible, for example, by some members of the TNF ligand family. As explained in the following, the mechanism of p100 processing differs from the classical pathway that triggers degradation of the other IκB proteins. p105 processing occurs constitutively with poor efficiency and is accompanied by its complete degradation. The latter can be stimulated in a mode related to signal-induced IκB degradation. Thus, extracellular cues [e.g., lipopolysaccharides (LPS), tumor necrosis factor (TNF), or interleukin-1 (IL1)] trigger phosphorylation of two serines in the carboxy-terminal domain rich in proline, glutamic acid, serine, and threonine (PEST domain). Again, this leads to recognition and ubiquitination by the SCF$^{\beta TRCP}$ ubiquitin ligase complex and accelerated proteasomal degradation (Heissmeyer et al., 2001; Lang et al., 2003; Salmeron et al., 2001). Taken together, proteolysis of IκB proteins releases NFκB dimers and admits their translocation in the nucleus and binding to DNA enhancer elements.

3. The IKK Complex

Inducible dual phosphorylation of degradation-relevant serines in IκB proteins and p105 is mediated by the IκB kinase (IKK) complex, which contains two related serine kinases, IKK1 and IKK2 assembled with a

---

of IκBα, IκBβ, and IκBε are indicated by small, light gray ellipses. Regulatory phosphorylation sites in p105 and IκBγ are shown as small, light gray boxes. IκBγ is identical to the C-terminal part of p105.

regulatory subunit, NEMO (IKKγ, IKKAP1, FIP3; Karin and Delhase, 2000; Rothwarf and Karin, 1999). A recent report has also identified the chaperones Hsp90 and Cdc37 as part of the IKK complex (Chen et al., 2002a). IKK1 and 2 have an overall identity of about 50% and consist of an amino-terminal kinase domain followed by a central leucin zipper mediating homo- and heterodimerization and a carboxy-terminal helix-loop–helix motif. The IKK complex is the integrating bottleneck of a variety of upstream signaling pathways. This is in particular due to its capability to interact with a pleothora of kinases either directly and or via IKK-interacting adaptor proteins (e.g., members of the TNF receptor associated factor family) (Karin and Delhase, 2000; Wajant et al., 2001). The IKK interacting kinases regulate two critical aspects of NFκB activation. First, they can phosphorylate the activation loop of the IKKs leading to their activation. Second, they can regulate the transcriptional activation ability of the NFκB subunits (Karin and Delhase, 2000). While mitogen-activated protein kinase kinase kinases (MAP3Ks or MEKKs), in particular MEKK1-3 (Lee et al., 1998; Nakano et al., 1998; Zhao and Lee, 1999), NIK (Nemoto et al., 1998; Woronicz et al., 1997) and TAK1 (Ninomiya-Tsuji et al., 1999; Sakurai et al., 1999b), have a central role in activation of the IKKs, phosphorylation-dependent regulation of the transcriptional activity of NFκBs is mediated by a diverse group of kinases, including protein kinase B/Akt (Madrid et al., 2000; Sizemore et al., 1999), atypical protein kinase Cs (PKCs; Duran et al., 2003; Leitges et al., 2001), the catalytic subunit of protein kinase A (PKA; Zhong et al., 1997), casein kinase II (Wang et al., 2000) and IKK2 (Sakurai et al., 1999b) itself. For example, phosphorylation of serine 276 in RelA by PKA and serine 311 by PKCζ promotes interaction with the transcriptional coactivator CREB-binding protein (CBP/p300; Duran et al., 2003). In addition, CKII and IKK2 have been implicated in phosphorylation of serine 529 and 536 of RelA, causing stimulation of the transcriptional function of NFκB (Sakurai et al., 1999a; Wang et al., 2000). As the contribution and exact roles of the different kinases involved in IKK phosphorylation and transcriptional activation of NFκBs are at least partly redundant and subject of secondary regulation by associated factors, signal-induced NFκB activation is very complex and depends also on the cell type regarded.

With respect to the IKKs, genetic studies clearly show that these kinases have overlapping and nonoverlapping functions. Signal-induced degradation of IκBs and p105 is completely blocked in NEMO deficient cells emphasizing the essential role of NEMO and an intact IKK complex in this process (Makris et al., 2000; Rudolph et al., 2000; Salmeron et al., 2001; Schmidt-Supprian et al., 2000). However, in either IKK1 or IKK2 deficient mouse embryonic fibroblasts (MEFs) TNF-induced p105 degradation is not affected, suggesting a redundant role of the IKKs in this response (Salmeron et al., 2001). In contrast, signal-induced degradation of IκBα is almost completely absent in IKK2 deficient cells but only marginally affected in IKK1 deficient cells (Hu *et al.*, 1999; Li *et al.*, 1999a; Takeda *et al.*, 1999; Tanaka *et al.*, 1999). However, cytokine-induced transcription of NFκB responsive genes is severely reduced in both cases (Hu *et al.*, 1999; Li *et al.*, 1999a,b; Takeda *et al.*, 1999; Tanaka *et al.*, 1999). Thus, both IKKs have an critical role in cytokine-mediated induction of NFκB target genes by largely distinct mechanisms. Indeed, it has been recently shown that IKK1, but not IKK2 or NEMO, associates with NFκB regulated promoters and enhances transcription by phosphorylation of histone H3, facilitating its acetylation by the histone acetyltransferase activity of CBP (Anest *et al.*, 2003; Yamamoto *et al.*, 2003). The prototypic IKK complex described previously is the central part of the so called canonical or classical NFκB pathway which is based on the signal-induced degradation of IκBs and p105 (Fig. 2A). A second, nonclassical pathway of NFκB activation has been recently revealed. This pathway regulates the signal-induced processing of the inhibitory p100 precursor (Fig. 2B) and is of special relevance for B-cell development. p100 processing, thus NFκB2 activation, can be triggered by a subgroup of ligands of the TNF ligand family (Claudio *et al.*, 2002; Coope *et al.*, 2002; Dejardin *et al.*, 2002; Kayagaki *et al.*, 2002; Saitoh *et al.*, 2003). The noncanonical pathway triggers phosphorylation of the p100 precursor protein by IKK1 activation via the MAP3 kinase NFκB-inducing kinase (NIK), (Senftleben *et al.*, 2001; Xiao *et al.*, 2001a,b). Remarkably, IKK1 acts in this pathway again independently from IKK2 and NEMO. Phosphorylation finally stimulates p100 ubiquitination by the $SCF^{\beta TRCP}$ ubiquitin ligase complex and subsequent proteasomal degradation (Fong and Sun, 2002).

4. Termination of NFκB Signaling

NFκB transcription factors induce a plethora of antiapoptotic and proinflammatory proteins, giving them a central indispensable role in host inflammatory and immune responses. However, the aberrant expression of these molecules can have severe consequences for health-promoting oncogenesis, cancer therapy resistance, and diverse inflammatory diseases, including rheumatoid arthritis, Crohn's disease, multiple sclerosis, and psoriasis. It is therefore no surprise that termination of NFκB signaling is regulated by a redundant network of negative feedback mechanisms acting at different stages of NFκB activation. According to where these mechanisms act in the NFκB signaling pathway, they can globally down regulate NFκB signaling or affect a group or even only a single activator of NFκB. Two powerful global mechanisms that cooperate in downregulation of NFκB are inactivation of the IKK complex and transcriptional upregulation of IκBα. IKK activation, in particular upon cytokine stimulation, is typically a transient event that reaches a peak within 5 to 15 min. There is evidence that the carboxy-terminal helix–loop–helix

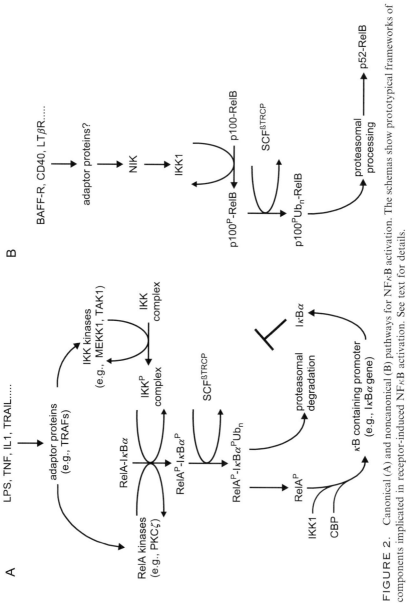

FIGURE 2. Canonical (A) and noncanonical (B) pathways for NFκB activation. The schemas show prototypical frameworks of components implicated in receptor-induced NFκB activation. See text for details.

domain and a neighboring serine cluster of IKK interact with the kinase domain of the molecule and facilitate its activation upon phosphorylation of the activation loop (Delhase *et al.*, 1999). Autophosphorylation of the serine cluster might then weaken this intramolecular interaction, resulting in downregulation of IKK activity. In accordance with this fast kinetic of IKK activation, IκBα is normally completely degraded within 10 to 30 min. However, IκBα is a bona fide NFκB target gene and is rapidly resynthesized (Arenzana-Seisdedos *et al.*, 1995). It translocates to the nucleus, binds transcriptionally active p50-RelA complexes, and dissociates them from the DNA, thus ensuring transient activation (Arenzana-Seisdedos *et al.*, 1995, 1997; Sachdev *et al.*, 1998, 2000). However, persistent NFκB activation can also occur and is regulated by IκBβ. In contrast to its phosphorylated form, unphosphorylated IκBβ does not interfere with DNA-binding of NFκB, but forms a ternary complex with DNA-NFκB complexes which is insensitive against newly synthesized IκBα (DeLuca *et al.*, 1999; Suyang *et al.*, 1996). Aside from IκBα, there are a couple of other NFκB target genes, including A20, IL1 receptor-associated kinase-M (IRAK-M), and the short splice form of the myeloid differentiation primary response gene 88 (MyD88-s), that also interfere with NFκB signaling. In contrast to IκBα, these regulators block only a subset or a single activator of NFκB. A20 selectively blocks TNF-induced NFκB activation by the inhibition of TNF-R1 signaling complex formation (He and Ting, 2002). IRAK-M and MyD88-s interfere with NFκB signaling by Toll-like receptors and the interleukin-1 receptor (Janssens *et al.*, 2002; Kobayashi *et al.*, 2002).

B. TRAIL

1. TRAIL and its Receptors

TNF-related apoptosis-inducing ligand (TRAIL), also designated as Apo2L, was identified in expressed sequence tag database screens for novel members of the TNF ligand family (Pitti *et al.*, 1996; Wiley *et al.*, 1995). TRAIL has five receptors, all of which are members of the TNF receptor superfamily (Fig. 3; Almasan and Ashkenazi, 2003; LeBlanc and Ashkenazi, 2003). TRAIL-R1 and TRAIL-R2 belong to the death receptor subgroup that triggers apoptosis (programmed cell death) by virtue of a conserved cytoplasmic signaling domain, the death domain. TRAIL-R3 lacks a cytoplasmic domain and is attached to the plasma membrane via a glycosylphosphatidylinositol anchor and seems to act primarily as a decoy receptor. TRAIL-R4 contains a truncated death domain devoid of apoptosis–inducing potential and can act as a decoy receptor, too. As it has a cytoplasmic domain of significant size, TRAIL-R4 might have additional, yet unknown, signaling capabilities. While TRAIL-R1 to TRAIL-R4 have closely related extracellular domains, osteoprotegerin (OPG), the fifth TRAIL interacting receptor, is only distantly related to these receptors and can also interact with

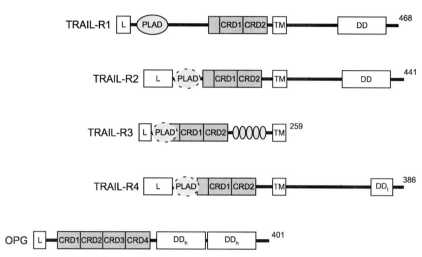

FIGURE 3. Domain organization of TRAIL receptors. L, leader sequence; PLAD, pre-ligand assembly domain (note that the PLAD has only been experimentally proven in TRAIL-R1); CRD, cysteine-rich domain; TM, transmembrane domain; DD, death domain; DD_t, truncated nonfunctional death domain; DD_h, death domain homology region. Copies of a short 15 aa repeat in TRAIL-R3 are indicated by light gray ellipses.

receptor activator of NFκB ligand (RANK-L), an additional ligand of the TNF family (Almasan and Ashkenazi, 2003; LeBlanc and Ashkenazi, 2003). OPG is a soluble decoy receptor that was originally identified as an inhibitor of RANK-L induced osteoclast differentiation and bone resorption (Bengtsson and Ryan, 2002). At physiological temperature, OPG has a comparable low affinity for TRAIL (KD = 400 nM versus KD = 2 nM for TRAIL-TRAIL-R2 interaction), (Truneh et al., 2000). Nevertheless, endogenously produced OPG can interfere with TRAIL-induced apoptosis in some cellular systems (Holen et al., 2002; Shipman and Croucher, 2003). Future studies must show whether the TRAIL inhibitory effect of OPG is only due to TRAIL depletion or an unknown mechanism works in addition.

2. Mechanisms of TRAIL-Induced Apoptosis

TRAIL death receptors stimulate the apoptotic machinery in a way similar to Fas, the best studied prototypic death receptor. For some members of the TNF receptor superfamily, including TNF-R1, TNF-R2, Fas, CD40, and TRAIL-R1, self-association into signaling-incompetent preassembled receptor complexes has been shown in nonstimulated cells (Chan et al., 2000; Siegel et al., 2000). These preformed receptor complexes hold together due to interaction of receptor monomers via their amino-terminal preligand assembly domain (PLAD; Chan et al., 2000; Siegel et al., 2000). Due to their TRAIL-R1-related primary structure, this should also apply accordingly to TRAIL-R2 and TRAIL-R4. Ligand binding to such preassembled receptor complexes induces their structural reorganization to signaling competent complexes and allows engagement of intracellular signaling pathways. In sensitive cells, Fas and the TRAIL death receptors induce and get part of a so-called death inducing signaling complex (DISC; Bodmer *et al.*, 2000; Kischkel *et al.*, 1995, 2000; Sprick *et al.*, 2000). As a first step in DISC formation, death receptors, including TRAIL-R1 and TRAIL-R2, recruit the death domain-containing adaptor protein FADD (Fas-associated death domain protein) by a death domain–death domain interaction between receptor and adaptor protein (Bodmer *et al.*, 2000; Kischkel *et al.*, 2000; Sprick *et al.*, 2000). In contrast to "free" cytosolic FADD, receptor-bound FADD is able to interact with the proform of caspase-8 and conducts this molecule into the DISC, too. The FADD–caspase-8 interaction is again mediated by homophilic association of a protein–protein-interaction domain, this time by death effector domains that are structurally related to the death domain and are present in the amino-termini of caspase-8 and FADD (Boldin *et al.*, 1996; Muzio *et al.*, 1996). The stoichiometry of death receptors, FADD and caspase-8 in the DISC of TRAIL-R1, TRAIL2, and Fas is still unknown. However, the fact that secondary clustering of trimeric Fas-FasL complexes is necessary for apoptosis-induction suggests that several procaspase-8 molecules are brought together in the context of DISC formation (Mundle and Raza, 2002). DISC-induced proximity of procaspase-8 molecules might then facilitate formation of enzymatically active DISC-bound procaspase-8 dimers (Boatright *et al.*, 2003; Chang *et al.*, 2003; Chen *et al.*, 2002b; Donepudi *et al.*, 2003) that convert to mature DISC-independent active caspase-8 $(p20/p10)_2$ heterotetramers. This conversion requires a two-step autoproteolytical processing mechanism. First, procaspase-8 (p55/53) is cleaved between the p20 and p10 subunits of the carboxy-terminal caspase homology domain, releasing the p10 subunit, which still remains in the DISC by interaction with the FADD-bound p43/41 intermediate. Then mature enzymatically active caspase-8 is freed by cleavage of the p43/41 intermediate between the prodomain and the p20 subunit. The still FADD/DISC-bound death effector domain-containing caspase-8 prodomain can be finally replaced by a new procaspase-8 molecule starting a new cycle of procaspase-8 activation (Medema *et al.*, 1997). Dependent on the cell-type, generation of active cytosolic caspase-8 $(p20/p10)_2$ heterotetramers can lead to apoptosis by two different ways (Fig. 4). In so-called type I cells, active caspase-8 robustly stimulates effector caspases, especially caspase-3, ensuring apoptotic cell death. In contrast, in type II cells, caspase-8 activation is less efficient and/or effector caspases are held in check by antiapoptotic inhibitor of apoptosis (IAP) proteins. In these cells TRAIL-mediated apoptosis therefore requires the support from proapoptotic mitochondrial proteins (Fig. 4).

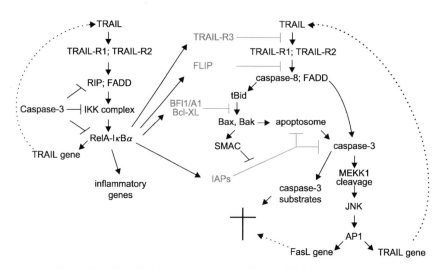

FIGURE 4. The TRAIL signaling network. The TRAIL death receptors can signal apoptosis (right) but also NFκB activation (left). In type I cells TRAIL strongly induces caspase activation directly leading to apoptosis. Thus, the tBid-mediated signaling branch is here dispensable. In type II cells only a suboptimal amount of active caspase-8 is generated by the DISC and/or effector caspase action is antagonized by IAP proteins. Apoptosis is therefore dependent on a tBid-induced mitochondrial amplification loop. Under nonapoptotic conditions, TRAIL-induced NFκB activation might amplify the initial trigger by transcriptional activation of the TRAIL gene (dotted arrow left). Under apoptotic conditions, TRAIL-induced caspase activation can trigger JNK activation and transcription of TRAIL and other death ligands thereby creating a positive apoptotic feedback loop (dotted arrow right). NFκB induced factors interfering with apoptosis are shown in gray. See text for details.

Caspase-8 triggers this apoptotic "activation" of mitochondria in type II cells by cleavage of the B-cell lymphoma gene-2 (Bcl-2) family member Bid resulting in a proapoptotic truncated Bid peptide, tBid, which targets mitochondria (Li et al., 1998; Luo et al., 1998) and induces oligomerization of Bax and Bak in the outer mitochondrial membrane (Korsmeyer et al., 2000). Pores formed by oligomerized Bax/Bak then allow the exit of apoptogenic proteins, including cytochrome c, second mitochondria-derived activator of caspase/direct IAP-binding protein (SMAC/DIABLO), and the serine protease HtrA2/Omi, into the cytosol (Wang, 2001). While cytochrome c assembles with Apaf1, ATP, and procaspase-9 to the caspase-9 activating apoptosome, which like mature caspase-8 induces caspase-3 activity (Shi, 2002; Wang, 2001), SMAC/DIABLO (Chai et al., 2000; Verhagen et al., 2000) and HtrA2/Omi (Suzuki et al., 2001a) block the caspase-inhibitory action of the before-mentioned IAP proteins and thereby enhance caspase-3 activation. It is worthy to note that active caspase-3 itself can process and activate its upstream initiator caspases (caspase-8 and caspase-9), leading to self-amplification of an initial apoptotic signal of sufficient strength (Engels *et al.*, 2000; Fujita *et al.*, 2001; Tang *et al.*, 2000). Experimentally, type I and type II cells can be distinguished by ectopic overexpression of Bcl-2, an inhibitor of Bax/Bak-mediated apoptosis (Scaffidi *et al.*, 1998). Thus, Bcl-2 expression has no protective effect on TRAIL-induced apoptosis in type I cells, but confers protection in type II cells by blocking the release of the proapoptotic mitochondrial factors.

II. REGULATION OF TRAIL-INDUCED APOPTOSIS BY NFκB

The first evidence for an antiapoptotic role of NFκB came from the analysis of ReIA deficient mice, which are embryonal lethal due to TNF-induced liver cell apoptosis (Beg and Baltimore, 1996; Beg *et al.*, 1995). Subsequent studies also implicated NFκB activity in resistance against ionizing radiation, chemotherapeutic drugs, and p53-induced apoptosis (Baldwin, 2001; Van Antwerp *et al.*, 1996). It is obvious that some of the NFκB target genes have to encode antiapoptotic proteins. In fact, currently, antiapoptotic features have been shown for almost a dozen of NFκB target genes, including those for cellular inhibitor of apoptosis-1 (cIAP1; Wang *et al.*, 1998), cIAP2 (Chu *et al.*, 1997; Wang *et al.*, 1998), x-linked IAP (xIAP; Stehlik *et al.*, 1998), TNF-receptor associated factor-1 (TRAF1; Schwenzer *et al.*, 1999; Wang *et al.*, 1998), cellular FLICE-inhibitory protein (cFLIP; Kreuz *et al.*, 2001; Micheau *et al.*, 2001), SCC-22 (You *et al.*, 2001), Bcl2-related protein A1 (Bfl1/A1; Lee *et al.*, 1999; Wang *et al.*, 1999; Zong *et al.*, 1999), Bcl-xL (Chen *et al.*, 2000; Lee *et al.*, 1999), and TRAIL-R3 (Bernard *et al.*, 2001). Remarkably, apoptosis-regulation by NFκB activation is no one-way street because NFκB signaling is reversionary inhibited during apoptosis (Fig. 4). Thus, while NFκB activation prevents apoptosis by induction of antiapoptotic proteins, active caspases in turn blocks the NFκB signaling pathway by cleavage of components involved in this pathway, in particular, RIP (Lin *et al.*, 1999; Martinon *et al.*, 2000), TRAF1 (Irmler *et al.*, 2000), IκBα (Barkett *et al.*, 1997), IKK2 (Tang *et al.*, 2001b), Akt (Bachelder *et al.*, 1999), NIK (Hu *et al.*, 2000), HPK (Arnold *et al.*, 2001), and p50 (Ravi *et al.*, 1998) and ReIA (Levkau *et al.*, 1999; Ravi *et al.*, 1998) itself. Caspase-mediated cleavage of these signaling intermediates does not simply inactivate these proteins but converts them into dominant-negative acting variants of their noncleaved counterparts.

A. NFκB CAN GLOBALLY REGULATE CELLULAR APOPTOSIS-SENSITIVITY

Dependent on the targets of the antiapoptotic NFκB regulated proteins, these factors can be specific for a single apoptosis-inducer (e.g., TRAIL) common to all death receptors or globally affect core components of the apoptotic machinery of the cell. Well-recognized NFκB target genes that regulate overall cellular apoptosis sensitivity are the IAP family members cIAP1, cIAP2, and xIAP and the Bcl-2 family members Bfl1/A1 and Bcl-XL.

The first members of the phylogenetically conserved family of inhibitor of apoptosis protein family have been discovered in the genomes of baculoviruses as factors preventing apoptosis of infected host cells (Salvesen and Duckett, 2002). Later, cellular orthologoues of the viral IAP proteins also were discovered in various metazoans, including nematodes, flies, and vertebrates. IAP proteins are characterized by having up to three copies of a zinc-binding protein domain of about 70 amino acids called the baculoviral IAP repeat (BIR), (Salvesen and Duckett, 2002). Deletion mutagenesis using various IAPs showed uniformly a dependency for at least one BIR domain for the inhibition of apoptosis. In addition to their amino-terminally located BIR domains, most IAP proteins also contain a carboxy-terminal really interesting new gene (RING) domain (Salvesen and Duckett, 2002). The NFκB regulated IAP proteins cIAP1, cIAP2, and xIAP have three BIR domains and one RING domain. In addition, cIAP1 and cIAP2 have a caspase recruitment domain (CARD) located between the BIR domains and the RING domain (Rothe et al., 1995). xIAP associates with and blocks processed caspase-3, -7, and -9 with affinities in the low nanomolar range (Roy et al., 1997). With a roughly 100-fold lower affinity, cIAP1 and cIAP2 inhibit caspase-3 and -7 (Roy et al., 1997). Remarkably, caspase-3/-7 and caspase-9 are blocked by xIAP by distinct mechanisms. While xIAP interact with its BIR3 domain with the amino-terminal end of the small subunit released during caspase-9 processing (Srinivasula et al., 2001; Sun et al., 1999), a segment preceeding the BIR2 domain interacts with high affinity with caspase-3/-7 and masks their active site (Chai et al., 2001; Huang et al., 2001; Riedl et al., 2001). The RING domain of cIAP1 (Li et al., 2002; Yang et al., 2000), cIAP2 (Huang et al., 2000) and xIAP (Suzuki et al., 2001b; Yang et al., 2000) have ubiquitin ligase (E3) activity and can therefore have a role in targeted protein degradation of proteins by the ubiquitin proteasome system. In fact, xIAP triggers its own proteasomal degradation, but also ubiquitination and degradation of caspase-3 and SMAC/DIABLO (MacFarlane et al., 2002; Suzuki et al., 2001b; Yang et al., 2000). Degradation of the latter is also triggered by cIAP1/2-mediated ubiquitination (Hu and Yang, 2003). SMAC/DIABLO is encoded by a nuclear gene and resides in its mature homodimeric form in the mitochondrial intermembrane space. It is released from mitochondria of type II cells in TRAIL-induced apoptosis and binds to the similar site of IAP proteins as caspase-9, thereby disrupting IAP-caspase interaction (Vaux and Silke, 2003). Thus, in addition to their well characterized caspase-inhibitory functions, cIAP1, cIAP2, and xIAP may also counteract apoptosis by targeting the proapoptotic SMAC/DIABLO molecule for proteasomal degradation. Proteasome inhibitors suppress IAP-mediated degradation of SMAC/DIABLO and caspases, and this might explain why cotreatment with proteasome inhibitors sensitizes cells for TRAIL-induced apoptosis. It is worthy of note that cell-permeable SMAC/DIABLO peptides, recognizing IAP proteins, overcome the antiapoptotic effect of Bcl-2/Bcl-XL expression and sensitize tumor cells for TRAIL-induced apoptosis (Fulda *et al.*, 2002). These data suggest that SMAC/DIABLO has a critical role in antagonizing inhibition of effector caspases by IAPs in apoptosis induction in type II cells by TRAIL-R1 and TRAIL-R2 and other death receptors (Fulda *et al.*, 2002; Leverkus *et al.*, 2003; Srinivasula *et al.*, 2000). SMAC/DIABLO is therefore a possible target to enhance the antitumoral effect of TRAIL. However, downregulation of SMAC/DIABLO degradation with proteasome inhibitors, which might be used in cancer therapy, leads in primary keratinocytes to sensitization for TRAIL-induced apoptosis (Leverkus *et al.*, 2003), thus challenging the clinical applicability of a combination therapy with TRAIL and proteasome inhibitors. The apoptosis decisive balance of processed effector caspases and IAP proteins can be further fine-tuned by additional mechanisms, including those listed in the following paragraphs:

- HtrA2/Omi is a serine protease, and like SMAC/DIABLO, is targeted to the mitochondrial intermembrane space, where its targeting peptide is removed, yielding an IAP-binding amino-terminus related to that of the small unit of caspase-9 and mature SMAC/DIABLO (Martins *et al.*, 2002; Suzuki *et al.*, 2001a). Thus, mature HtrA2/Omi inhibits xIAP function by direct binding in a similar way to SMAC/DIABLO. Moreover, it can inactivate cIAP1 by cleavage reducing its E3 ligase activity on caspase substrates (Srinivasula *et al.*, 2003). In addition, HtrA2/Omi might enforce also caspase-independent cell death by its serine protease activity (Verhagen *et al.*, 2002).
- xIAP has been identified as a signaling intermediate in TGFβ receptor-induced activation of NFκB and JNK (Birkey Reffey *et al.*, 2001). Further, its overexpression alone is sufficient to trigger activation of JNK1 by a transforming growth factor-beta (TGF-β) activated kinase-1 (TAK1)-dependent pathway (Sanna *et al.*, 2002). The effects of JNK activation are highly dependent on the circumstances and cell type and promote survival effects or apoptosis induction. Moreover, one study claimed xIAP as a factor mediating NFκB-induced inhibition of JNK in a situation of proapoptotic JNK activation, again giving xIAP antiapoptotic properties (Tang *et al.*, 2001a).

NFκB also induces antiapoptotic proteins of the Bcl-2 family, such as Bfl1/A1 and Bcl-xL (Chen *et al.*, 2000; Lee *et al.*, 1999; Wang *et al.*, 1999; Zong *et al.*, 1999). The promoters of both genes contain functional NFκB sites. The Bcl-2 family members are characterized by containing at least one of the so-called Bcl-2 homology (BH) domains 1 to 4 that have been originally defined in the prototypic Bcl-2 protein (Gross *et al.*, 1999). Dependent on their BH domain architecture, Bcl-2 family members exert pro- or antiapoptotic properties. Bcl-XL as well as Bfl-1/A1 exert their antiapoptotic function by regulating the mitochondrial release of proapoptotic factors such as cytochrome c, SMAC/DIABLO, HtrA2/Omi, and AIF (apoptosis-inducing factor) into the cytoplasm (Gross *et al.*, 1999). This is caused by oligomerization of monomeric Bax (and most likely Bak) in the outer mitochondrial membrane leading to the formation of pores that allow the translocation of proapoptotic mitochondrial proteins during apoptosis (Wei *et al.*, 2001). The antiapoptotic Bcl-XL protein inhibits apoptosis by sequestering proforms of death-inducing caspases and, more importantly, by preventing the formation of the apoptotic mitochondrial Bax/Bak pore (Gross *et al.*, 1999). As discussed previously, tBid transmits the apoptotic signal from death receptors to mitochondria in type II cells. Bfl-1/A1 has been found to interact with both full-length Bid and truncated tBid, via the Bid BH3 domain (Werner *et al.*, 2002). Bfl-1/A1 expression can prevent TRAIL-induced cytochrome c release, but does not interfere with Bid processing by caspase-8, or with its translocation to mitochondria. Thus, Bfl-1/A1 blocks the interaction of tBid with Bax and/or Bak and thereby prevents apoptotic pore formation by the latter (Werner *et al.*, 2002). Future studies will show whether Bfl-1/A1 sequesters other BH3 domain-only proteins in a similar fashion. Bfl-1/A1 and Bcl-XL are often induced via NFκB in the same type of cells by the same inducers, but it is not clear yet to which extent, if any, both proteins are coregulated. In B-cells of mice lacking cRel and RelA, Bcl-2 expression is impaired (Grossmann *et al.*, 2000) suggesting that this antiapoptotic protein can also be controlled by NFκB, but possibly by an indirect mechanism.

B. REGULATION OF DEATH RECEPTOR-INDUCED APOPTOSIS BY NFκB

Among the antiapoptotic NFκB target genes, cellular FLICE-inhibitory protein (cFLIP) is certainly central with respect to the inhibition of death receptor-induced apoptosis. Originally cFLIP has been identified as the cellular homologoue of a family of antiapoptotic viral proteins containing a death effector domain (Krueger *et al.*, 2001a; Thome and Tschopp, 2001). It is now clear that cFLIP is a major regulator of apoptosis induction by controlling processing and DISC release of caspase-8. More than 10 splice forms of FLIP have been defined at the RNA level (Djerbi *et al.*, 2001), but only two of them, called FLIP-short (FLIP-S) and FLIP-long (FLIP-L), have been studied at the protein level in detail (Krueger *et al.*, 2001a; Thome and Tschopp, 2001). The genes for cFLIP, caspase-8, and caspase-10 form a multigene family. Similar to caspase-8, FLIP-L consists of two amino-terminal death effector domains and a carboxy-terminal caspase homology domain (Krueger *et al.*, 2001a; Thome and Tschopp, 2001). However, the latter lacks enzymatic activity due to the substitution of active side residues. FLIP-S consists solely of the amino-terminal death effector domains and a short carboxy-terminal stretch of amino acids, which is unique for this FLIP isoform. Both FLIP isoforms are recruited into the DISC of Fas, TRAIL-R1, and TRAIL-R2. Here they interfere with different steps in the generation of mature caspase-8 (Krueger *et al.*, 2001b). While FLIP-S inhibits the first caspase-8 cleavage event between the p20 and the p10 subunit of the caspase homology domain, FLIP-L supports this cleavage step but prevents the second cleavage between the prodomain and the p20 subunit of the DISC-bound caspase-8 p43/41 intermediate, and thereby blocks the release of mature heterotetrameric caspase-8 into the cytosol. The latter might be of special relevance, as there is evidence that DISC-bound heterodimers of caspase-8 and FLIP-L are enzymatically active (Chang *et al.*, 2003; Micheau *et al.*, 2002). Thus, as FLIP-L arrests the DISC in this state, a local nonapoptotic activation of caspase-8 is feasible and is a possible explanation for reports describing caspase-dependent death-receptor signaling in the absence of apoptosis. Remarkably, the FLIP gene is not only targeted by the NFκB pathway but also by other major antiapoptotic pathways, including the Akt/PKB and MAPK pathways (Krueger et al., 2001a; Thome and Tschopp, 2001). The central role of FLIP for resistance against death receptor-induced apoptosis has been demonstrated in studies showing that selective downregulation of FLIP expression by antisense strategies (Hyer *et al.*, 2002; Mitsiades *et al.*, 2002; Pedersen *et al.*, 2002; Xiao *et al.*, 2003), RNA interference (Siegmund *et al.*, 2002), or transgenic mouse technology (Yeh *et al.*, 2000) is sufficient to sensitize cells for the apoptotic effects of death receptors.

C. SELECTIVE REGULATION OF TRAIL-INDUCED APOPTOSIS BY NFκB

Basically, a powerful mean by which receptor signaling can be controlled is regulation of ligand and/or receptor expression. In fact, this is of relevance for TRAIL biology. While most cells express at least one of the TRAIL death receptors, the TRAIL decoy receptors and TRAIL itself are constitutively expressed only on a few cell types but can be induced in most cells under defined circumstances. Remarkably, the predominantly antiapoptotic NFκB pathway is a major regulator of inducible expression of TRAIL and its two death receptors as well as of the TRAIL-R3 decoy.

1. Regulation of TRAIL Receptor Expression by NFκB

First evidence for a NFκB dependent expression of TRAIL receptors came from the analysis of RelA deficient and cRel deficient MEFs. The investigation of expression of proapoptotic proteins in these cells was prompted by *in vivo* observations pointing to a possible proapoptotic role of NFκB. It was found that wild-type and RelA deficient cells are TRAIL sensitive whereas cRel deficient cells are protected (Ravi *et al.*, 2001). This effect was specific for TRAIL as TNF-induced apoptosis was normal in cRel −/− cells. In accordance with an antiapoptotic function of RelA, MEFs deficient for this protein become sensitive for TNF and are also somewhat more sensitive against TRAIL-induced apoptosis. The TRAIL resistant phenotype of cRel correlates with a strong reduction in basal and TNF-inducible TRAIL-R2 expression and could be reverted by ectopic expression of TRAIL-R2 (Ravi *et al.*, 2001). Although in this system the pro- and antiapoptotic effects of NFκB can be attributed to two distinct NFκB proteins, this is no general principle. RelA can also induce proapoptotic proteins [e.g., Fas (Ouaaz *et al.*, 1999) or FasL (Kavurma and Khachigian, 2003)], and cRel also upregulates antiapoptotic factors (Wang *et al.*, 1999; Zong *et al.*, 1999). Indeed, NFκB dependent upregulation of proapoptotic proteins could be a common theme that is not apparent as long as the concomitantly induced antiapoptotic proteins hold in check their action. In accordance with this idea, the TRAIL decoy receptor TRAIL-R3 has been identified in a DNA-Mikroarray screen for cRel regulated genes in Hela cells (Bernard *et al.*, 2001). Moreover the removal of cell surface TRAIL-R3 by phosphatidylinositol-specific phospholipase C reverted the protective effect against TRAIL-induced apoptosis acquired by cRel expression (Bernard *et al.*, 2001).

2. Regulation of TRAIL Expression by NFκB

TRAIL is upregulated upon activation of T-cells or T-cell lines with T-cell receptor activation-mimicking agents (PMA/ionomycin; phytohemagglutinin, anti-CD3 + anti-CD28 treatment), which, among others things, activate the NFκB pathway. Studies with T-cell lines deficient in essential components of the NFκB pathway or cells overexpressing a super-repressor variant of IκBα completely blocked TRAIL induction indicating an essential role of NFκB in TRAIL upregulation in activated T-cells (Baetu *et al.*, 2001; Rivera-Walsh *et al.*, 2001; Siegmund *et al.*, 2001). NFκB-dependent upregulation has also been shown for other ligands of death receptors (Kavurma and Khachigian, 2003; Weingartner *et al.*, 2002). Indeed, NFκB-induced production of TRAIL, FasL, and TNF may explain the rare occasions when NFκB activation has a proapoptotic role. TRAIL-R1 and 2, but also other death receptors, can stimulate the NFκB pathway. It seems feasible therefore that a self-maintaining circuit of NFκB-dependent TRAIL induction and TRAIL-receptor-mediated NFκB activation can be constituted (Shetty *et al.*, 2002). However, such a NFκB dependent positive feedback loop should not contribute to death receptor-induced apoptosis, as the latter blocks NFκB driven gene transcription. Nevertheless, death receptor-induced transcriptional upregulation of death receptor ligands can contribute to death receptor-induced apoptosis. In this case upregulation of TRAIL, or FasL and TNF, occurs via the JNK pathway and the activation protein 1 (AP1) and ATF-2 transcription factors (Herr *et al.*, 2000). Indeed, JNK activity is strongly induced during apoptosis, for example, by cleavage of MEKK1 (Cardone *et al.*, 1997).

III. TRAIL ACTIVATES NFκB

Although early studies came up with controversial results concerning the importance of FADD for TRAIL-induced apoptosis (Almasan and Ashkenazi, 2003; LeBlanc and Ashkenazi, 2003), a pivotal role of this death domain-containing adaptor protein in TRAIL-R1 and TRAIL-R2-induced apoptosis is now generally accepted due to the study of FADD deficient mice (Kuang *et al.*, 2000). In contrast, a potential function of FADD in TRAIL-R1/2-mediated NFκB activation has only been poorly and preliminarily addressed. So, it has been shown that FADD overexpression is sufficient to stimulate NFκB driven reporter genes and upregulates endogenous target genes of the NFκB pathway (Hsu *et al.*, 1996b; Hu *et al.*, 2000; Wajant *et al.*, 2000). In addition, reduced NFκB activation in response to TRAIL has been observed in FADD deficient Jurkat clones (Wajant *et al.*, 2000). As overexpression of the death domain of FADD is sufficient to induce NFκB, it seems possible that this domain is enough to transduce the TRAIL receptor-dependent NFκB response. This could be related to FADD's capability to facilitate TRAIL-induced recruitment of the serine threonine kinase RIP into the TRAIL-R1 signaling complex (Lin *et al.*, 2000). RIP has been originally identified as a Fas-interacting protein (Stanger *et al.*, 1995), and an essential role in TRAIL-induced NFκB activation has been shown in RIP deficient MEFs (Lin *et al.*, 2000). RIP interacts with the death domain of FADD by virtue of its own carboxy-terminal death domain (Varfolomeev *et al.*, 1996). An essential role of RIP was also shown earlier for TNF-R1-induced NFκB activation (Kelliher *et al.*, 1998). In this case, RIP recruitment is facilitated by the death domain-containing adaptor protein TNF receptor-1 associated death domain protein (TRADD), whereas FADD is dispensable (Hsu *et al.*, 1996a). In context of TNF-R1 signaling, RIP activates by an unknown mechanism the kinases of the receptor-recruited IKK complex (Devin *et al.*, 2000). Although RIP is able to interact with NEMO (Devin *et al.*, 2001; Li *et al.*, 1999c; Zhang *et al.*, 2000), the structural and regulatory subunit of the IKK complex, recruitment of the IKK complex seems to be dependent on the TNF receptor associated factor 2 (TRAF2) adaptor protein (Devin et al., 2000). Eventually TRAF5 can substitute to some extent for TRAF2 as TRAF2 deficient MEFs are only slightly reduced in TNF-induced NFκB activation, whereas MEFs derived from TRAF2-TRAF5 double knockout mice are completely blocked in this respect (Tada et al., 2001). TRAF2 deficient MEFs respond normally to TRAIL, but a possible redundant action as observed with respect to TNF signaling has not been tested yet (Devin et al., 2000). Indeed, dominant-negative forms of TRAF2 interfere with NFκB activation by TRAIL in reporter gene assays, pointing to a possible role of TRAF2 or TRAF2-related proteins (Wajant et al., 2000). Remarkably, RIP is one of the rare caspase-8 substrates (Lin et al., 1999; Martinon et al., 2000). Thus, in the context of the DISC of TRAIL-R1/2, caspase-8 might cleave RIP generating a fragment containing the RIP death domain, which acts in a dominant-negative manner in NFκB signaling by competition with complete RIP for receptor recruitment.

In accordance with the idea that TRAIL death receptors utilize a RIP-dependent caspase-sensitive pathway resembling that of TNF-R1 to trigger NFκB activation, most studies have shown that inhibition of apoptosis, for example, by caspase inhibitors, enhances the ability of TRAIL to stimulate the NFκB pathway (Harper et al., 2001; Kumar-Sinha et al., 2002; Li et al., 2003; Trauzold et al., 2001; Wajant et al., 2000). However, there is growing evidence that TRAIL can also engage an alternative caspase-dependent pathway to activate NFκB (Secchiero et al., 2003). Maturation of monocytic HL60 cells can be induced via TNF-R1, and TRAIL-R1 by activation of NFκB. While TNF-induced NFκB activation and maturation is not affected by the addition of caspase-inhibitors, the TRAIL-R1-mediated response is blocked by caspase-inhibitors in HL60 cells (Secchiero et al., 2003). The TRAIL-R1-induced caspase-dependent activation of NFκB in monocyte maturation could be mediated by FADD, caspase-8, and/or FLIP, as all these molecules are involved in caspase-regulation by TRAIL receptors and induce NFκB activity upon overexpression (Chaudhary et al., 2000; Hu et al., 2000). Remarkably, NFκB activation by overexpression of FADD or FLIP is blocked by caspase-inhibitory peptides and the caspase-inhibitory cytokine response modifier A protein from cowpox, whereas NFκB activation by caspase-8 needs caspase inhibition and is also achieved by an active site mutant (Chaudhary et al., 2000; Hu et al., 2000). The differential caspase-dependencies of FADD/FLIP and caspase-8 mediated NFκB activation could be related to two basically different scenarios. One possibility is that a common FADD-FLIP-caspase-8 dependent pathway, leading to NFκB activation, exists and only the interaction of FADD and FLIP with caspase-8 is dependent on a functional active site of the latter. Alternatively, FADD, FLIP, and caspase-8 might stimulate NFκB activation by two or more at least partly independent pathways that are not necessarily connected to death receptor signaling. This possibility may also explain the differential inhibitory effects of the dominant-negative RIP mutant mentioned previously on FADD/caspase-8 and FLIP-dependent NFκB activation (Hu et al., 2000). Death receptor-independent NFκB activation by these molecules might also explain the puzzling finding that FLIP-L and FLIP-S activate to a considerable extent NFκB but at the same time significantly block the TRAIL-induced NFκB response, which reaches a significantly higher strength (Wajant et al., 2000). An inhibitory effect of FLIP-L on TRAIL-R1/2- and Fas-induced NFκB signaling is also in accordance with the fact that RIP cleavage, yielding an NFκB inhibitory fragment, occurs in the FLIP-L arrested DISC of Fas (Micheau et al., 2002). The NFκB activating effects of FLIP-L might then be related to the death receptor-independent activation of the IKK complex as it was observed for some viral FLIP proteins (Field et al., 2003; Liu et al., 2002). Future studies must show whether and how FADD, FLIP, and caspase-8 contributes to TRAIL-R1-induced caspase-dependent NFκB activation during monocyte maturation.

REFERENCES

Almasan, A., and Ashkenazi, A. (2003). Apo2L/TRAIL: Apoptosis signaling, biology, and potential for cancer therapy. *Cytokine Growth Factor Rev.* **14,** 337–348.

Anest, V., Hanson, J. L., Cogswell, P. C., Steinbrecher, K. A., Strahl, B. D., and Baldwin, A. S. (2003). A nucleosomal function for IkappaB kinase-alpha in NF-kappaB-dependent gene expression. *Nature* **423,** 659–663.

Arenzana-Seisdedos, F., Thompson, J., Rodriguez, M. S., Bachelerie, F., Thomas, D., and Hay, R. T. (1995). Inducible nuclear expression of newly synthesized I kappa B alpha negatively regulates DNA-binding and transcriptional activities of NF-kappa B. *Mol. Cell. Biol.* **15,** 2689–2696.

Arenzana-Seisdedos, F., Turpin, P., Rodriguez, M., Thomas, D., Hay, R. T., Virelizier, J. L., and Dargemont, C. (1997). Nuclear localization of I kappa B alpha promotes active transport of NF-kappa B from the nucleus to the cytoplasm. *J. Cell. Sci.* **110,** 369–378.

Arnold, R., Liou, J., Drexler, H. C., Weiss, A., and Kiefer, F. (2001). Caspase-mediated cleavage of hematopoietic progenitor kinase 1 (HPK1) converts an activator of NFkappaB into an inhibitor of NFkappaB. *J. Biol. Chem.* **276,** 14675–14684.

Bachelder, R. E., Ribick, M. J., Marchetti, A., Falcioni, R., Soddu, S., Davis, K. R., and Mercurio, A. M. (1999). p53 inhibits alpha 6 beta 4 integrin survival signaling by promoting the caspase 3-dependent cleavage of AKT/PKB. *J. Cell Biol.* **147,** 1063–1072.

Baetu, T. M., Kwon, H., Sharma, S., Grandvaux, N., and Hiscott, J. (2001). Disruption of NF-kappaB signaling reveals a novel role for NF-kappaB in the regulation of TNF-related apoptosis-inducing ligand expression. *J. Immunol.* **167,** 3164–3173.

Baldi, L., Brown, K., Franzoso, G., and Siebenlist, U. (1996). Critical role for lysines 21 and 22 in signal-induced, ubiquitin-mediated proteolysis of I kappa B-alpha. *J. Biol. Chem.* **271,** 376–379.

Baldwin, A. S. (2001). Control of oncogenesis and cancer therapy resistance by the transcription factor NF-kappaB. *J. Clin. Invest.* **107,** 241–246.

Barkett, M., Xue, D., Horvitz, H. R., and Gilmore, T. D. (1997). Phosphorylation of IkappaB-alpha inhibits its cleavage by caspase CPP32 *in vitro. J. Biol. Chem.* **272,** 29419–29422.

Beg, A. A., and Baltimore, D. (1996). An essential role for NF-kappaB in preventing TNF-alpha-induced cell death. *Science* **274**, 782–784.

Beg, A. A., Sha, W. C., Bronson, R. T., Ghosh, S., and Baltimore, D. (1995). Embryonic lethality and liver degeneration in mice lacking the RelA component of NF-kappa B. *Nature* **376**, 167–170.

Bengtsson, A. K., and Ryan, E. J. (2002). Immune function of the decoy receptor osteoprotegerin. *Crit. Rev. Immunol.* **22**, 201–215.

Bernard, D., Quatannens, B., Vandenbunder, B., and Abbadie, C. (2001). Rel/NF-kappaB transcription factors protect against tumor necrosis factor (TNF)-related apoptosis-inducing ligand (TRAIL)-induced apoptosis by up-regulating the TRAIL decoy receptor DcR1. *J. Biol. Chem.* **276**, 27322–27328.

Birkey Reffey, S., Wurthner, J. U., Parks, W. T., Roberts, A. B., and Duckett, C. S. (2001). X-linked inhibitor of apoptosis protein functions as a cofactor in transforming growth factor-beta signaling. *J. Biol. Chem.* **276**, 26542–26549.

Boatright, K. M., Renatus, M., Scott, F. L., Sperandio, S., Shin, H., Pedersen, I. M., Ricci, J. E., Edris, W. A., Sutherlin, D. P., Green, D. R., and Salvesen, G. S. (2003). A unified model for apical caspase activation. *Mol. Cell* **11**, 529–541.

Bodmer, J. L., Holler, N., Reynard, S., Vinciguerra, P., Schneider, P., Juo, P., Blenis, J., and Tschopp, J. (2000). TRAIL receptor-2 signals apoptosis through FADD and caspase-8. *Nat. Cell. Biol.* **2**, 241–243.

Boldin, M. P., Goncharov, T. M., Goltsev, Y. V., and Wallach, D. (1996). Involvement of MACH, a novel MORT1/FADD-interacting protease, in Fas/APO-1- and TNF receptor-induced cell death. *Cell* **85**, 803–815.

Brockman, J. A., Scherer, D. C., McKinsey, T. A., Hall, S. M., Qi, X., Lee, W. Y., and Ballard, D. W. (1995). Coupling of a signal response domain in I kappa B alpha to multiple pathways for NF-kappa B activation. *Mol. Cell. Biol.* **15**, 2809–2818.

Brown, K., Gerstberger, S., Carlson, L., Franzoso, G., and Siebenlist, U. (1995). Control of I kappa B-alpha proteolysis by site-specific, signal-induced phosphorylation. *Science* **267**, 1485–1488.

Cardone, M. H., Salvesen, G. S., Widmann, C., Johnson, G., and Frisch, S. M. (1997). The regulation of anoikis: MEKK-1 activation requires cleavage by caspases. *Cell* **90**, 315–323.

Chai, J., Du, C., Wu, J. W., Kyin, S., Wang, X., and Shi, Y. (2000). Structural and biochemical basis of apoptotic activation by Smac/DIABLO. *Nature* **406**, 855–862.

Chai, J., Shiozaki, E., Srinivasula, S. M., Wu, Q., Datta, P., Alnemri, E. S., Shi, Y., and Dataa, P. (2001). Structural basis of caspase-7 inhibition by XIAP. *Cell* **104**, 769–780.

Chan, F. K., Chun, H. J., Zheng, L., Siegel, R. M., Bui, K. L., and Lenardo, M. J. (2000). A domain in TNF receptors that mediates ligand-independent receptor assembly and signaling. *Science* **288**, 2351–2354.

Chang, D. W., Xing, Z., Capacio, V. L., Peter, M. E., and Yang, X. (2003). Interdimer processing mechanism of procaspase-8 activation. *EMBO J.* **22**, 4132–4142.

Chaudhary, P. M., Eby, M. T., Jasmin, A., Kumar, A., Liu, L., and Hood, L. (2000). Activation of the NF-kappaB pathway by caspase 8 and its homologs. *Oncogene* **19**, 4451–4460.

Chen, C., Edelstein, L. C., and Gelinas, C. (2000). The Rel/NF-kappaB family directly activates expression of the apoptosis inhibitor Bcl-x(L). *Mol. Cell. Biol.* **20**, 2687–2695.

Chen, G., Cao, P., and Goeddel, D. V. (2002a). TNF-induced recruitment and activation of the IKK complex require Cdc37 and Hsp90. *Mol. Cell* **9**, 401–410.

Chen, M., Orozco, A., Spencer, D. M., and Wang, J. (2002b). Activation of initiator caspases through a stable dimeric intermediate. *J. Biol. Chem.* **277**, 50761–50767.

Chu, Z. L., McKinsey, T. A., Liu, L., Gentry, J. J., Malim, M. H., and Ballard, D. W. (1997). Suppression of tumor necrosis factor-induced cell death by inhibitor of apoptosis c-IAP2 is under NF-kappaB control. *Proc. Natl. Acad. Sci. USA* **94**, 10057–10062.

Claudio, E., Brown, K., Park, S., Wang, H., and Siebenlist, U. (2002). BAFF-induced NEMO-independent processing of NF-kappa B2 in maturing B cells. *Nat. Immunol.* **3**, 958–965.

Coope, H. J., Atkinson, P. G., Huhse, B., Belich, M., Janzen, J., Holman, M. J., Klaus, G. G., Johnston, L. H., and Ley, S. C. (2002). CD40 regulates the processing of NF-kappaB2 p100 to p52. *EMBO J.* **21**, 5375–5385.

Dejardin, E., Droin, N. M., Delhase, M., Haas, E., Cao, Y., Makris, C., Li, Z. W., Karin, M., Ware, C. F., and Green, D. R. (2002). The lymphotoxin-beta receptor induces different patterns of gene expression via two NF-kappaB pathways. *Immunity* **17**, 525–535.

Delhase, M., Hayakawa, M., Chen, Y., and Karin, M. (1999). Positive and negative regulation of IkappaB kinase activity through IKKbeta subunit phosphorylation. *Science* **284**, 309–313.

DeLuca, C., Petropoulos, L., Zmeureanu, D., and Hiscott, J. (1999). Nuclear IkappaBbeta maintains persistent NF-kappaB activation in HIV-1-infected myeloid cells. *J. Biol. Chem.* **274**, 13010–13016.

Devin, A., Cook, A., Lin, Y., Rodriguez, Y., Kelliher, M., and Liu, Z. (2000). The distinct roles of TRAF2 and RIP in IKK activation by TNF-R1: TRAF2 recruits IKK to TNF-R1 while RIP mediates IKK activation. *Immunity* **12**, 419–429.

Devin, A., Lin, Y., Yamaoka, S., Li, Z., Karin, M., and Liu, Z. (2001). The alpha and beta subunits of IkappaB kinase (IKK) mediate TRAF2-dependent IKK recruitment to tumor necrosis factor (TNF) receptor 1 in response to TNF. *Mol. Cell. Biol.* **21**, 3986–3994.

DiDonato, J., Mercurio, F., Rosette, C., Wu-Li, J., Suyang, H., Ghosh, S., and Karin, M. (1996). Mapping of the inducible IkappaB phosphorylation sites that signal its ubiquitination and degradation. *Mol. Cell. Biol.* **16**, 1295–1304.

Djerbi, M., Darreh-Shori, T., Zhivotovsky, B., and Grandien, A. (2001). Characterization of the human FLICE-inhibitory protein locus and comparison of the anti-apoptotic activity of four different flip isoforms. *Scand. J. Immunol.* **54**, 180–189.

Dobrzanski, P., Ryseck, R. P., and Bravo, R. (1995). Specific inhibition of RelB/p52 transcriptional activity by the C-terminal domain of p100. *Oncogene* **10**, 1003–1007.

Donepudi, M., Mac Sweeney, A., Briand, C., and Grutter, M. G. (2003). Insights into the regulatory mechanism for caspase-8 activation. *Mol. Cell* **11**, 543–549.

Duran, A., Diaz-Meco, M. T., and Moscat, J. (2003). Essential role of RelA Ser311 phosphorylation by zetaPKC in NF-kappaB transcriptional activation. *EMBO J.* **22**, 3910–3918.

Engels, I. H., Stepczynska, A., Stroh, C., Lauber, K., Berg, C., Schwenzer, R., Wajant, H., Janicke, R. U., Porter, A. G., Belka, C., Gregor, M., Schulze-Osthoff, K., and Wesselborg, S. (2000). Caspase-8/FLICE functions as an executioner caspase in anticancer drug-induced apoptosis. *Oncogene* **19**, 4563–4573.

Field, N., Low, W., Daniels, M., Howell, S., Daviet, L., Boshoff, C., and Collins, M. (2003). KSHV vFLIP binds to IKK-(gamma) to activate IKK. *J. Cell. Sci.* **116**, 3721–3728.

Fong, A., and Sun, S. C. (2002). Genetic evidence for the essential role of beta-transducin repeat-containing protein in the inducible processing of NF-kappa B2/p100. *J. Biol. Chem.* **277**, 22111–22114.

Fujita, E., Egashira, J., Urase, K., Kuida, K., and Momoi, T. (2001). Caspase-9 processing by caspase-3 via a feedback amplification loop *in vivo*. *Cell Death Differ.* **8**, 335–344.

Fulda, S., Wick, W., Weller, M., and Debatin, K. M. (2002). Smac agonists sensitize for Apo2L/TRAIL- or anticancer drug-induced apoptosis and induce regression of malignant glioma *in vivo*. *Nat. Med.* **8**, 808–815.

Gross, A., McDonnell, J. M., and Korsmeyer, S. J. (1999). BCL-2 family members and the mitochondria in apoptosis. *Genes Dev.* **13**, 1899–1911.

Grossmann, M., O'Reilly, L. A., Gugasyan, R., Strasser, A., Adams, J. M., and Gerondakis, S. (2000). The anti-apoptotic activities of Rel and RelA required during B-cell maturation involve the regulation of Bcl-2 expression. *EMBO J.* **19**, 6351–6360.

Harhaj, E. W., and Sun, S. C. (1999). Regulation of RelA subcellular localization by a putative nuclear export signal and p50. *Mol. Cell. Biol.* **19,** 7088–7095.

Harper, N., Farrow, S. N., Kaptein, A., Cohen, G. M., and MacFarlane, M. (2001). Modulation of tumor necrosis factor apoptosis-inducing ligand-induced NF-kappa B activation by inhibition of apical caspases. *J. Biol. Chem.* **276,** 34743–34752.

He, K. L., and Ting, A. T. (2002). A20 inhibits tumor necrosis factor (TNF) alpha-induced apoptosis by disrupting recruitment of TRADD and RIP to the TNF receptor 1 complex in Jurkat T cells. *Mol. Cell. Biol.* **22,** 6034–6045.

Heissmeyer, V., Krappmann, D., Hatada, E. N., and Scheidereit, C. (2001). Shared pathways of IkappaB kinase-induced SCF(betaTrCP)-mediated ubiquitination and degradation for the NF-kappaB precursor p105 and IkappaBalpha. *Mol. Cell. Biol.* **21,** 1024–1035.

Herr, I., Posovszky, C., Di Marzio, L. D., Cifone, M. G., Boehler, T., and Debatin, K. M. (2000). Autoamplification of apoptosis following ligation of CD95-L, TRAIL and TNF-alpha. *Oncogene* **19,** 4255–4262.

Heusch, M., Lin, L., Geleziunas, R., and Greene, W. C. (1999). The generation of nfkb2 p52: Mechanism and efficiency. *Oncogene* **18,** 6201–6208.

Holen, I., Croucher, P. I., Hamdy, F. C., and Eaton, C. L. (2002). Osteoprotegerin (OPG) is a survival factor for human prostate cancer cells. *Cancer Res.* **62,** 1619–1623.

Hsu, H., Huang, J., Shu, H. B., Baichwal, V., and Goeddel, D. V. (1996a). TNF-dependent recruitment of the protein kinase RIP to the TNF receptor-1 signaling complex. *Immunity* **4,** 387–396.

Hsu, H., Shu, H. B., Pan, M. G., and Goeddel, D. V. (1996b). TRADD-TRAF2 and TRADD-FADD interactions define two distinct TNF receptor 1 signal transduction pathways. *Cell* **84,** 299–308.

Hu, S., and Yang, X. (2003). Cellular inhibitor of apoptosis 1 and 2 are ubiquitin ligases for the apoptosis inducer Smac/DIABLO. *J. Biol. Chem.* **278,** 10055–10060.

Hu, W. H., Johnson, H., and Shu, H. B. (2000). Activation of NF-kappaB by FADD, Casper, and caspase-8. *J. Biol. Chem.* **275,** 10838–10844.

Hu, Y., Baud, V., Delhase, M., Zhang, P., Deerinck, T., Ellisman, M., Johnson, R., and Karin, M. (1999). Abnormal morphogenesis but intact IKK activation in mice lacking the IKKalpha subunit of IkappaB kinase. *Science* **284,** 316–320.

Huang, H., Joazeiro, C. A., Bonfoco, E., Kamada, S., Leverson, J. D., and Hunter, T. (2000). The inhibitor of apoptosis, cIAP2, functions as a ubiquitin-protein ligase and promotes *in vitro* monoubiquitination of caspases 3 and 7. *J. Biol. Chem.* **275,** 26661–26664.

Huang, T. T., and Miyamoto, S. (2001). Postrepression activation of NF-kappaB requires the amino-terminal nuclear export signal specific to IkappaBalpha. *Mol. Cell. Biol.* **21,** 4737–4747.

Huang, Y., Park, Y. C., Rich, R. L., Segal, D., Myszka, D. G., and Wu, H. (2001). Structural basis of caspase inhibition by XIAP: Differential roles of the linker versus the BIR domain. *Cell* **104,** 781–790.

Huxford, T., Huang, D. B., Malek, S., and Ghosh, G. (1998). The crystal structure of the IkappaBalpha/NF-kappaB complex reveals mechanisms of NF-kappaB inactivation. *Cell* **95,** 759–770.

Hyer, M. L., Sudarshan, S., Kim, Y., Reed, J. C., Dong, J. Y., Schwartz, D. A., and Norris, J. S. (2002). Downregulation of c-FLIP sensitizes DU145 prostate cancer cells to Fas-mediated apoptosis. *Cancer Biol. Ther.* **1,** 401–406.

Irmler, M., Steiner, V., Ruegg, C., Wajant, H., and Tschopp, J. (2000). Caspase-induced inactivation of the anti-apoptotic TRAF1 during Fas ligand-mediated apoptosis. *FEBS Lett.* **468,** 129–133.

Jacobs, M. D., and Harrison, S. C. (1998). Structure of an IkappaBalpha/NF-kappaB complex. *Cell* **95,** 749–758.

Janssens, S., Burns, K., Tschopp, J., and Beyaert, R. (2002). Regulation of interleukin-1- and lipopolysaccharide-induced NF-kappaB activation by alternative splicing of MyD88. *Curr. Biol.* **12,** 467–471.

Johnson, C., Van Antwerp, D., and Hope, T. J. (1999). An N-terminal nuclear export signal is required for the nucleocytoplasmic shuttling of IkappaBalpha. *EMBO J.* **18,** 6682–6693.

Karin, M., and Delhase, M. (2000). The I kappa B kinase (IKK) and NF-kappa B: Key elements of proinflammatory signalling. *Semin. Immunol.* **12,** 85–98.

Karin, M., and Lin, A. (2002). NF-kappaB at the crossroads of life and death. *Nat. Immunol.* **3,** 221–227.

Kavurma, M. M., and Khachigian, L. M. (2003). Signaling and transcriptional control of Fas ligand gene expression. *Cell Death Differ.* **10,** 36–44.

Kayagaki, N., Yan, M., Seshasayee, D., Wang, H., Lee, W., French, D. M., Grewal, I. S., Cochran, A. G., Gordon, N. C., Yin, J., Starovasnik, M. A., and Dixit, V. M. (2002). BAFF/BLyS receptor 3 binds the B cell survival factor BAFF ligand through a discrete surface loop and promotes processing of NF-kappaB2. *Immunity* **17,** 515–524.

Kelliher, M. A., Grimm, S., Ishida, Y., Kuo, F., Stanger, B. Z., and Leder, P. (1998). The death domain kinase RIP mediates the TNF-induced NF-kappaB signal. *Immunity* **8,** 297–303.

Kischkel, F. C., Hellbardt, S., Behrmann, I., Germer, M., Pawlita, M., Krammer, P. H., and Peter, M. E. (1995). Cytotoxicity-dependent APO-1 (Fas/CD95)-associated proteins form a death-inducing signaling complex (DISC) with the receptor. *EMBO J.* **14,** 5579–5588.

Kischkel, F. C., Lawrence, D. A., Chuntharapai, A., Schow, P., Kim, K. J., and Ashkenazi, A. (2000). Apo2L/TRAIL-dependent recruitment of endogenous FADD and caspase-8 to death receptors 4 and 5. *Immunity* **12,** 611–620.

Kobayashi, K., Hernandez, L. D., Galan, J. E., Janeway, C. A., Jr., Medzhitov, R., and Flavell, R. A. (2002). IRAK-M is a negative regulator of Toll-like receptor signaling. *Cell* **110,** 191–202.

Korsmeyer, S. J., Wei, M. C., Saito, M., Weiler, S., Oh, K. J., and Schlesinger, P. H. (2000). Pro-apoptotic cascade activates BID, which oligomerizes BAK or BAX into pores that result in the release of cytochrome c. *Cell Death Differ.* **7,** 1166–1173.

Kreuz, S., Siegmund, D., Scheurich, P., and Wajant, H. (2001). NF-kappaB inducers upregulate cFLIP, a cycloheximide-sensitive inhibitor of death receptor signaling. *Mol. Cell. Biol.* **21,** 3964–3973.

Krueger, A., Baumann, S., Krammer, P. H., and Kirchhoff, S. (2001a). FLICE-inhibitory proteins: Regulators of death receptor-mediated apoptosis. *Mol. Cell. Biol.* **21,** 8247–8254.

Krueger, A., Schmitz, I., Baumann, S., Krammer, P. H., and Kirchhoff, S. (2001b). Cellular FLICE-inhibitory protein splice variants inhibit different steps of caspase-8 activation at the CD95 death-inducing signaling complex. *J. Biol. Chem.* **276,** 20633–20640.

Kuang, A. A., Diehl, G. E., Zhang, J., and Winoto, A. (2000). FADD is required for DR4- and DR5-mediated apoptosis: Lack of trail-induced apoptosis in FADD-deficient mouse embryonic fibroblasts. *J. Biol. Chem.* **275,** 25065–25068.

Kumar-Sinha, C., Varambally, S., Sreekumar, A., and Chinnaiyan, A. M. (2002). Molecular cross-talk between the TRAIL and interferon signaling pathways. *J. Biol. Chem.* **277,** 575–585.

Lang, V., Janzen, J., Fischer, G. Z., Soneji, Y., Beinke, S., Salmeron, A., Allen, H., Hay, R. T., Ben-Neriah, Y., and Ley, S. C. (2003). betaTrCP-mediated proteolysis of NF-kappaB1 p105 requires phosphorylation of p105 serines 927 and 932. *Mol. Cell. Biol.* **23,** 402–413.

LeBlanc, H. N., and Ashkenazi, A. (2003). Apo2L/TRAIL and its death and decoy receptors. *Cell Death Differ.* **10,** 66–75.

Lee, F. S., Peters, R. T., Dang, L. C., and Maniatis, T. (1998). MEKK1 activates both IkappaB kinase alpha and IkappaB kinase beta. *Proc. Natl. Acad. Sci. USA* **95,** 9319–9324.

Lee, H. H., Dadgostar, H., Cheng, Q., Shu, J., and Cheng, G. (1999). NF-kappaB-mediated up-regulation of Bcl-x and Bfl-1/A1 is required for CD40 survival signaling in B lymphocytes. *Proc. Natl. Acad. Sci. USA* **96,** 9136–9141.

Leitges, M., Sanz, L., Martin, P., Duran, A., Braun, U., Garcia, J. F., Camacho, F., Diaz-Meco, M. T., Rennert, P. D., and Moscat, J. (2001). Targeted disruption of the zetaPKC gene results in the impairment of the NF-kappaB pathway. *Mol. Cell* **8**, 771–780.

Leverkus, M., Sprick, M. R., Wachter, T., Mengling, T., Baumann, B., Serfling, E., Brocker, E. B., Goebeler, M., Neumann, M., and Walczak, H. (2003). Proteasome inhibition results in TRAIL sensitization of primary keratinocytes by removing the resistance-mediating block of effector caspase maturation. *Mol. Cell. Biol.* **23**, 777–790.

Levkau, B., Scatena, M., Giachelli, C. M., Ross, R., and Raines, E. W. (1999). Apoptosis overrides survival signals through a caspase-mediated dominant-negative NF-kappa B loop. *Nat. Cell. Biol.* **1**, 227–233.

Li, H., Zhu, H., Xu, C. J., and Yuan, J. (1998). Cleavage of BID by caspase-8 mediates the mitochondrial damage in the Fas pathway of apoptosis. *Cell* **94**, 491–501.

Li, J. H., Kirkiles-Smith, N. C., McNiff, J. M., and Pober, J. S. (2003). TRAIL induces apoptosis and inflammatory gene expression in human endothelial cells. *J. Immunol.* **171**, 1526–1533.

Li, Q., Lu, Q., Hwang, J. Y., Buscher, D., Lee, K. F., Izpisua-Belmonte, J. C., and Verma, I. M. (1999a). IKK1-deficient mice exhibit abnormal development of skin and skeleton. *Genes Dev.* **13**, 1322–1328.

Li, Q., Van Antwerp, D., Mercurio, F., Lee, K. F., and Verma, I. M. (1999b). Severe liver degeneration in mice lacking the IkappaB kinase 2 gene. *Science* **284**, 321–325.

Li, Y., Kang, J., Friedman, J., Tarassishin, L., Ye, J., Kovalenko, A., Wallach, D., and Horwitz, M. S. (1999c). Identification of a cell protein (FIP-3) as a modulator of NF-kappaB activity and as a target of an adenovirus inhibitor of tumor necrosis factor alpha-induced apoptosis. *Proc. Natl. Acad. Sci. USA* **96**, 1042–1047.

Li, Q., and Verma, I. M. (2002). NF-kappaB regulation in the immune system. *Nat. Rev. Immunol.* **2**, 725–734.

Li, X., Yang, Y., and Ashwell, J. D. (2002). TNF-RII and c-IAP1 mediate ubiquitination and degradation of TRAF2. *Nature* **416**, 345–347.

Lin, L., and Ghosh, S. (1996). A glycine-rich region in NF-kappaB p105 functions as a processing signal for the generation of the p50 subunit. *Mol. Cell. Biol.* **16**, 2248–2254.

Lin, Y., Devin, A., Cook, A., Keane, M. M., Kelliher, M., Lipkowitz, S., and Liu, Z. G. (2000). The death domain kinase RIP is essential for TRAIL (Apo2L)-induced activation of IkappaB kinase and c-Jun N-terminal kinase. *Mol. Cell. Biol.* **20**, 6638–6645.

Lin, Y., Devin, A., Rodriguez, Y., and Liu, Z. G. (1999). Cleavage of the death domain kinase RIP by caspase-8 prompts TNF-induced apoptosis. *Genes Dev.* **13**, 2514–2526.

Liu, L., Eby, M. T., Rathore, N., Sinha, S. K., Kumar, A., and Chaudhary, P. M. (2002). The human herpes virus 8-encoded viral FLICE inhibitory protein physically associates with and persistently activates the Ikappa B kinase complex. *J. Biol. Chem.* **277**, 13745–13751.

Luo, X., Budihardjo, I., Zou, H., Slaughter, C., and Wang, X. (1998). Bid, a Bcl2 interacting protein, mediates cytochrome c release from mitochondria in response to activation of cell surface death receptors. *Cell* **94**, 481–490.

MacFarlane, M., Merrison, W., Bratton, S. B., and Cohen, G. M. (2002). Proteasome-mediated degradation of Smac during apoptosis: XIAP promotes Smac ubiquitination *in vitro*. *J. Biol. Chem.* **277**, 36611–36616.

Madrid, L. V., Wang, C. Y., Guttridge, D. C., Schottelius, A. J., Baldwin, A. S. Jr., and Mayo, M. W. (2000). Akt suppresses apoptosis by stimulating the transactivation potential of the RelA/p65 subunit of NF-kappaB. *Mol. Cell. Biol.* **20**, 1626–1638.

Makris, C., Godfrey, V. L., Krahn-Senftleben, G., Takahashi, T., Roberts, J. L., Schwarz, T., Feng, L., Johnson, R. S., and Karin, M. (2000). Female mice heterozygous for IKK gamma/NEMO deficiencies develop a dermatopathy similar to the human X-linked disorder incontinentia pigmenti. *Mol. Cell* **5**, 969–979.

Malek, S., Chen, Y., Huxford, T., and Ghosh, G. (2001). IkappaBbeta, but not IkappaB-alpha, functions as a classical cytoplasmic inhibitor of NF-kappaB dimers by masking both NF-kappaB nuclear localization sequences in resting cells. *J. Biol. Chem.* **276**, 45225–45235.

Martinon, F., Holler, N., Richard, C., and Tschopp, J. (2000). Activation of a pro-apoptotic amplification loop through inhibition of NF-kappaB-dependent survival signals by caspase-mediated inactivation of RIP. *FEBS Lett.* **468**, 134–136.

Martins, L. M., Iaccarino, I., Tenev, T., Gschmeissner, S., Totty, N. F., Lemoine, N. R., Savopoulos, J., Gray, C. W., Creasy, C. L., Dingwall, C., and Downward, J. (2002). The serine protease Omi/HtrA2 regulates apoptosis by binding XIAP through a reaper-like motif. *J. Biol. Chem.* **277**, 439–444.

Medema, J. P., Scaffidi, C., Kischkel, F. C., Shevchenko, A., Mann, M., Krammer, P. H., and Peter, M. E. (1997). FLICE is activated by association with the CD95 death-inducing signaling complex (DISC). *EMBO J.* **16**, 2794–2804.

Mercurio, F., DiDonato, J. A., Rosette, C., and Karin, M. (1993). p105 and p98 precursor proteins play an active role in NF-kappa B-mediated signal transduction. *Genes Dev.* **7**, 705–718.

Micheau, O., Lens, S., Gaide, O., Alevizopoulos, K., and Tschopp, J. (2001). NF-kappaB signals induce the expression of c-FLIP. *Mol. Cell. Biol.* **21**, 5299–5305.

Micheau, O., Thome, M., Schneider, P., Holler, N., Tschopp, J., Nicholson, D. W., Briand, C., and Grutter, M. G. (2002). The long form of FLIP is an activator of caspase-8 at the Fas death-inducing signaling complex. *J. Biol. Chem.* **277**, 45162–45171.

Mitsiades, N., Mitsiades, C. S., Poulaki, V., Anderson, K. C., and Treon, S. P. (2002). Intracellular regulation of tumor necrosis factor-related apoptosis-inducing ligand-induced apoptosis in human multiple myeloma cells. *Blood* **99**, 2162–2171.

Moorthy, A. K., and Ghosh, G. (2003). p105.Ikappa Bgamma and prototypical Ikappa Bs use a similar mechanism to bind but a different mechanism to regulate the subcellular localization of NF-kappa B. *J. Biol. Chem.* **278**, 556–566.

Mundle, S. D., and Raza, A. (2002). Defining the dynamics of self-assembled Fas-receptor activation. *Trends Immunol.* **23**, 187–194.

Muzio, M., Chinnaiyan, A. M., Kischkel, F. C., O'Rourke, K., Shevchenko, A., Ni, J., Scaffidi, C., Bretz, J. D., Zhang, M., Gentz, R., Mann, M., Krammer, P. H., Peter, M. E., and Dixit, V. M. (1996). FLICE, a novel FADD-homologous ICE/CED-3-like protease, is recruited to the CD95 (Fas/APO-1) death—inducing signaling complex. *Cell* **85**, 817–827.

Nakano, H., Shindo, M., Sakon, S., Nishinaka, S., Mihara, M., Yagita, H., and Okumura, K. (1998). Differential regulation of IkappaB kinase alpha and beta by two upstream kinases, NF-kappaB-inducing kinase and mitogen-activated protein kinase/ERK kinase kinase-1. *Proc. Natl. Acad. Sci. USA* **95**, 3537–3542.

Naumann, M., Wulczyn, F. G., and Scheidereit, C. (1993). The NF-kappa B precursor p105 and the proto-oncogene product Bcl-3 are I kappa B molecules and control nuclear translocation of NF-kappa B. *EMBO J.* **12**, 213–222.

Nemoto, S., DiDonato, J. A., and Lin, A. (1998). Coordinate regulation of IkappaB kinases by mitogen-activated protein kinase kinase kinase 1 and NF-kappaB-inducing kinase. *Mol. Cell. Biol.* **18**, 7336–7343.

Ninomiya-Tsuji, J., Kishimoto, K., Hiyama, A., Inoue, J., Cao, Z., and Matsumoto, K. (1999). The kinase TAK1 can activate the NIK-I kappaB as well as the MAP kinase cascade in the IL-1 signalling pathway. *Nature* **398**, 252–256.

Orian, A., Schwartz, A. L., Israel, A., Whiteside, S., Kahana, C., and Ciechanover, A. (1999). Structural motifs involved in ubiquitin-mediated processing of the NF-kappaB precursor p105: Roles of the glycine-rich region and a downstream ubiquitination domain. *Mol. Cell. Biol.* **19**, 3664–3673.

Ouaaz, F., Li, M., and Beg, A. A. (1999). A critical role for the RelA subunit of nuclear factor kappaB in regulation of multiple immune-response genes and in Fas-induced cell death. *J. Exp. Med.* **189,** 999–1004.

Pedersen, I. M., Kitada, S., Schimmer, A., Kim, Y., Zapata, J. M., Charboneau, L., Rassenti, L., Andreeff, M., Bennett, F., Sporn, M. B., Liotta, L. D., Kipps, T. J., and Reed, J. C. (2002). The triterpenoid CDDO induces apoptosis in refractory CLL B cells. *Blood* **100,** 2965–2972.

Pitti, R. M., Marsters, S. A., Ruppert, S., Donahue, C. J., Moore, A., and Ashkenazi, A. (1996). Induction of apoptosis by Apo-2 ligand, a new member of the tumor necrosis factor cytokine family. *J. Biol. Chem.* **271,** 12687–12690.

Ravi, R., Bedi, A., and Fuchs, E. J. (1998). CD95 (Fas)-induced caspase-mediated proteolysis of NF-kappaB. *Cancer Res.* **58,** 882–886.

Ravi, R., Bedi, G. C., Engstrom, L. W., Zeng, Q., Mookerjee, B., Gelinas, C., Fuchs, E. J., and Bedi, A. (2001). Regulation of death receptor expression and TRAIL/Apo2L-induced apoptosis by NF-kappaB. *Nat. Cell. Biol.* **3,** 409–416.

Rice, N. R., MacKichan, M. L., and Israel, A. (1992). The precursor of NF-kappa B p50 has I kappa B-like functions. *Cell* **71,** 243–253.

Riedl, S. J., Renatus, M., Schwarzenbacher, R., Zhou, Q., Sun, C., Fesik, S. W., Liddington, R. C., and Salvesen, G. S. (2001). Structural basis for the inhibition of caspase-3 by XIAP. *Cell* **104,** 791–800.

Rivera-Walsh, I., Waterfield, M., Xiao, G., Fong, A., and Sun, S. C. (2001). NF-kappaB signaling pathway governs TRAIL gene expression and human T-cell leukemia virus-I Tax-induced T-cell death. *J. Biol. Chem.* **276,** 40385–40388.

Rothe, M., Pan, M. G., Henzel, W. J., Ayres, T. M., and Goeddel, D. V. (1995). The TNFR2-TRAF signaling complex contains two novel proteins related to baculoviral inhibitor of apoptosis proteins. *Cell* **83,** 1243–1252.

Rothwarf, D. M., and Karin, M. (1999). The NF-kappa B activation pathway: A paradigm in information transfer from membrane to nucleus. *Sci. STKE* **1999,** RE1.

Roy, N., Deveraux, Q. L., Takahashi, R., Salvesen, G. S., and Reed, J. C. (1997). The c IAP-1 and c-IAP-2 proteins are direct inhibitors of specific caspases. *EMBO J.* **16,** 6914–6925.

Rudolph, D., Yeh, W. C., Wakeham, A., Rudolph, B., Nallainathan, D., Potter, J., Elia, A. J., and Mak, T. W. (2000). Severe liver degeneration and lack of NF-kappaB activation in NEMO/IKKgamma-deficient mice. *Genes Dev.* **14,** 854–862.

Sachdev, S., Bagchi, S., Zhang, D. D., Mings, A. C., and Hannink, M. (2000). Nuclear import of IkappaBalpha is accomplished by a ran-independent transport pathway. *Mol. Cell. Biol.* **20,** 1571–1582.

Sachdev, S., Hoffmann, A., and Hannink, M. (1998). Nuclear localization of IkappaB alpha is mediated by the second ankyrin repeat: The IkappaB alpha ankyrin repeats define a novel class of cis-acting nuclear import sequences. *Mol. Cell. Biol.* **18,** 2524–2534.

Saitoh, T., Nakayama, M., Nakano, H., Yagita, H., Yamamoto, N., and Yamaoka, S. (2003). TWEAK induces NF-kB2 p100 processing and long-lasting NF-kB activation. *J. Biol. Chem.* **278,** 36005–36012.

Sakurai, H., Chiba, H., Miyoshi, H., Sugita, T., and Toriumi, W. (1999a). IkappaB kinases phosphorylate NF-kappaB p65 subunit on serine 536 in the transactivation domain. *J. Biol. Chem.* **274,** 30353–30356.

Sakurai, H., Miyoshi, H., Toriumi, W., and Sugita, T. (1999b). Functional interactions of transforming growth factor beta-activated kinase 1 with IkappaB kinases to stimulate NF-kappaB activation. *J. Biol. Chem.* **274,** 10641–10648.

Salmeron, A., Janzen, J., Soneji, Y., Bump, N., Kamens, J., Allen, H., and Ley, S. C. (2001). Direct phosphorylation of NF-kappaB1 p105 by the IkappaB kinase complex on serine 927 is essential for signal-induced p105 proteolysis. *J. Biol. Chem.* **276,** 22215–22222.

Salvesen, G. S., and Duckett, C. S. (2002). IAP proteins: Blocking the road to death's door. *Nat. Rev. Mol. Cell. Biol.* **3,** 401–410.

Sanna, M. G., da Silva Correia, J., Ducrey, O., Lee, J., Nomoto, K., Schrantz, N., Deveraux, Q. L., and Ulevitch, R. J. (2002). IAP suppression of apoptosis involves distinct mechanisms: The TAK1/JNK1 signaling cascade and caspase inhibition. *Mol. Cell. Biol.* **22,** 1754–1766.

Scaffidi, C., Fulda, S., Srinivasan, A., Friesen, C., Li, F., Tomaselli, K. J., Debatin, K. M., Krammer, P. H., and Peter, M. E. (1998). Two CD95 (APO-1/Fas) signaling pathways. *EMBO J.* **17,** 1675–1687.

Scherer, D. C., Brockman, J. A., Chen, Z., Maniatis, T., and Ballard, D. W. (1995). Signal-induced degradation of I kappa B alpha requires site-specific ubiquitination. *Proc. Natl. Acad. Sci. USA* **92,** 11259–11263.

Schmidt-Supprian, M., Bloch, W., Courtois, G., Addicks, K., Israel, A., Rajewsky, K., and Pasparakis, M. (2000). NEMO/IKK gamma-deficient mice model incontinentia pigmenti. *Mol. Cell* **5,** 981–992.

Schwenzer, R., Siemienski, K., Liptay, S., Schubert, G., Peters, N., Scheurich, P., Schmid, R. M., and Wajant, H. (1999). The human tumor necrosis factor (TNF) receptor-associated factor 1 gene (TRAF1) is up-regulated by cytokines of the TNF ligand family and modulates TNF-induced activation of NF-kappaB and c-Jun N-terminal kinase. *J. Biol. Chem.* **274,** 19368–19374.

Secchiero, P., Milani, D., Gonelli, A., Melloni, E., Campioni, D., Gibellini, D., Capitani, S., and Zauli, G. (2003). Tumor necrosis factor (TNF)-related apoptosis-inducing ligand (TRAIL) and TNF-alpha promote the NF-kappaB-dependent maturation of normal and leukemic myeloid cells. *J. Leukoc. Biol.* **74,** 223–232.

Senftleben, U., Cao, Y., Xiao, G., Greten, F. R., Krahn, G., Bonizzi, G., Chen, Y., Hu, Y., Fong, A., Sun, S. C., and Karin, M. (2001). Activation by IKKalpha of a second, evolutionary conserved, NF-kappa B signaling pathway. *Science* **293,** 1495–1499.

Shetty, S., Gladden, J. B., Henson, E. S., Hu, X., Villanueva, J., Haney, N., and Gibson, S. B. (2002). Tumor necrosis factor-related apoptosis inducing ligand (TRAIL) up-regulates death receptor 5 (DR5) mediated by NFkappaB activation in epithelial derived cell lines. *Apoptosis* **7,** 413–420.

Shi, Y. (2002). Apoptosome: The cellular engine for the activation of caspase-9. *Structure (Camb.)* **10,** 285–288.

Shipman, C. M., and Croucher, P. I. (2003). Osteoprotegerin is a soluble decoy receptor for tumor necrosis factor-related apoptosis-inducing ligand/Apo2 ligand and can function as a paracrine survival factor for human myeloma cells. *Cancer Res.* **63,** 912–916.

Siegel, R. M., Frederiksen, J. K., Zacharias, D. A., Chan, F. K., Johnson, M., Lynch, D., Tsien, R. Y., and Lenardo, M. J. (2000). Fas preassociation required for apoptosis signaling and dominant inhibition by pathogenic mutations. *Science* **288,** 2354–2357.

Siegmund, D., Hadwiger, P., Pfizenmaier, K., Vornlocher, H. P., and Wajant, H. (2002). Selective inhibition of FLICE-like inhibitory protein expression with small interfering RNA oligonucleotides is sufficient to sensitize tumor cells for TRAIL-induced apoptosis. *Mol. Med.* **8,** 725–732.

Siegmund, D., Hausser, A., Peters, N., Scheurich, P., and Wajant, H. (2001). Tumor necrosis factor (TNF) and phorbol ester induce TNF-related apoptosis-inducing ligand (TRAIL) under critical involvement of NF-kappa B essential modulator (NEMO)/IKKgamma. *J. Biol. Chem.* **276,** 43708–43712.

Sizemore, N., Leung, S., and Stark, G. R. (1999). Activation of phosphatidylinositol 3-kinase in response to interleukin-1 leads to phosphorylation and activation of the NF-kappaB p65/RelA subunit. *Mol. Cell. Biol.* **19,** 4798–4805.

Sprick, M. R., Weigand, M. A., Rieser, E., Rauch, C. T., Juo, P., Blenis, J., Krammer, P. H., and Walczak, H. (2000). FADD/MORT1 and caspase-8 are recruited to TRAIL receptors 1 and 2 and are essential for apoptosis mediated by TRAIL receptor 2. *Immunity* **12,** 599–609.

Srinivasula, S. M., Datta, P., Fan, X. J., Fernandes-Alnemri, T., Huang, Z., and Alnemri, E. S. (2000). Molecular determinants of the caspase-promoting activity of Smac/DIABLO and its role in the death receptor pathway. *J. Biol. Chem.* **275,** 36152–36157.

Srinivasula, S. M., Gupta, S., Datta, P., Zhang, Z., Hegde, R., Cheong, N., Fernandes-Alnemri, T., and Alnemri, E. S. (2003). Inhibitor of apoptosis proteins are substrates for the mitochondrial serine protease Omi/HtrA2. *J. Biol. Chem.* **278,** 31469–31472.

Srinivasula, S. M., Hegde, R., Saleh, A., Datta, P., Shiozaki, E., Chai, J., Lee, R. A., Robbins, P. D., Fernandes-Alnemri, T., Shi, Y., and Alnemri, E. S. (2001). A conserved XIAP-interaction motif in caspase-9 and Smac/DIABLO regulates caspase activity and apoptosis. *Nature* **410,** 112–116.

Stanger, B. Z., Leder, P., Lee, T. H., Kim, E., and Seed, B. (1995). RIP: A novel protein containing a death domain that interacts with Fas/APO-1 (CD95) in yeast and causes cell death. *Cell* **81,** 513–523.

Stehlik, C., de Martin, R., Kumabashiri, I., Schmid, J. A., Binder, B. R., and Lipp, J. (1998). Nuclear factor (NF)-kappaB-regulated X-chromosome-linked iap gene expression protects endothelial cells from tumor necrosis factor alpha-induced apoptosis. *J. Exp. Med.* **188,** 211–216.

Sun, C., Cai, M., Gunasekera, A. H., Meadows, R. P., Wang, H., Chen, J., Zhang, H., Wu, W., Xu, N., Ng, S. C., and Fesik, S. W. (1999). NMR structure and mutagenesis of the inhibitor-of-apoptosis protein XIAP. *Nature* **401,** 818–822.

Suyang, H., Phillips, R., Douglas, I., and Ghosh, S. (1996). Role of unphosphorylated, newly synthesized I kappa B beta in persistent activation of NF-kappa B. *Mol. Cell. Biol.* **16,** 5444–5449.

Suzuki, Y., Imai, Y., Nakayama, H., Takahashi, K., Takio, K., and Takahashi, R. (2001a). A serine protease, HtrA2, is released from the mitochondria and interacts with XIAP, inducing cell death. *Mol. Cell* **8,** 613–621.

Suzuki, Y., Nakabayashi, Y., and Takahashi, R. (2001b). Ubiquitin-protein ligase activity of X-linked inhibitor of apoptosis protein promotes proteasomal degradation of caspase-3 and enhances its anti-apoptotic effect in Fas-induced cell death. *Proc. Natl. Acad. Sci. USA* **98,** 8662–8667.

Tada, K., Okazaki, T., Sakon, S., Kobarai, T., Kurosawa, K., Yamaoka, S., Hashimoto, H., Mak, T. W., Yagita, H., Okumura, K., Yeh, W. C., and Nakano, H. (2001). Critical roles of TRAF2 and TRAF5 in tumor necrosis factor-induced NF-kappa B activation and protection from cell death. *J. Biol. Chem.* **276,** 36530–36534.

Takeda, K., Takeuchi, O., Tsujimura, T., Itami, S., Adachi, O., Kawai, T., Sanjo, H., Yoshikawa, K., Terada, N., and Akira, S. (1999). Limb and skin abnormalities in mice lacking IKKalpha. *Science* **284,** 313–316.

Tam, W. F., and Sen, R. (2001). IkappaB family members function by different mechanisms. *J. Biol. Chem.* **276,** 7701–7704.

Tanaka, M., Fuentes, M. E., Yamaguchi, K., Durnin, M. H., Dalrymple, S. A., Hardy, K. L., and Goeddel, D. V. (1999). Embryonic lethality, liver degeneration, and impaired NF-kappa B activation in IKK-beta-deficient mice. *Immunity* **10,** 421–429.

Tang, D., Lahti, J. M., and Kidd, V. J. (2000). Caspase-8 activation and bid cleavage contribute to MCF7 cellular execution in a caspase-3-dependent manner during staurosporine-mediated apoptosis. *J. Biol. Chem.* **275,** 9303–9307.

Tang, G., Minemoto, Y., Dibling, B., Purcell, N. H., Li, Z., Karin, M., and Lin, A. (2001a). Inhibition of JNK activation through NF-kappaB target genes. *Nature* **414,** 313–317.

Tang, G., Yang, J., Minemoto, Y., and Lin, A. (2001b). Blocking caspase-3-mediated proteolysis of IKKbeta suppresses TNF-alpha-induced apoptosis. *Mol. Cell* **8,** 1005–1016.

Thome, M., and Tschopp, J. (2001). Regulation of lymphocyte proliferation and death by FLIP. *Nat. Rev. Immunol.* **1,** 50–58.

Trauzold, A., Wermann, H., Arlt, A., Schutze, S., Schafer, H., Oestern, S., Roder, C., Ungefroren, H., Lampe, E., Heinrich, M., Walczak, H., and Kalthoff, H. (2001). CD95 and TRAIL receptor-mediated activation of protein kinase C and NF-kappaB contributes to apoptosis resistance in ductal pancreatic adenocarcinoma cells. *Oncogene* **20**, 4258–4269.

Truneh, A., Sharma, S., Silverman, C., Khandekar, S., Reddy, M. P., Deen, K. C., McLaughlin, M. M., Srinivasula, S. M., Livi, G. P., Marshall, L. A., Alnemri, E. S., Williams, W. V., and Doyle, M. L. (2000). Temperature-sensitive differential affinity of TRAIL for its receptors. DR5 is the highest affinity receptor. *J. Biol. Chem.* **275**, 23319–23325.

Van Antwerp, D. J., Martin, S. J., Kafri, T., Green, D. R., and Verma, I. M. (1996). Suppression of TNF-alpha-induced apoptosis by NF-kappaB. *Science* **274**, 787–789.

Varfolomeev, E. E., Boldin, M. P., Goncharov, T. M., and Wallach, D. (1996). A potential mechanism of "cross-talk" between the p55 tumor necrosis factor receptor and Fas/APO1: Proteins binding to the death domains of the two receptors also bind to each other *J. Exp. Med.* **183**, 1271–1275.

Vaux, D. L., and Silke, J. (2003). Mammalian mitochondrial IAP binding proteins. *Biochem. Biophys. Res. Commun.* **304**, 499–504.

Verhagen, A. M., Ekert, P. G., Pakusch, M., Silke, J., Connolly, L. M., Reid, G. E., Moritz, R. L., Simpson, R. J., and Vaux, D. L. (2000). Identification of DIABLO, a mammalian protein that promotes apoptosis by binding to and antagonizing IAP proteins. *Cell* **102**, 43–53.

Verhagen, A. M., Silke, J., Ekert, P. G., Pakusch, M., Kaufmann, H., Connolly, L. M., Day, C. L., Tikoo, A., Burke, R., Wrobel, C., Moritz, R. L., Simpson, R. J., and Vaux, D. L. (2002). HtrA2 promotes cell death through its serine protease activity and its ability to antagonize inhibitor of apoptosis proteins. *J. Biol. Chem.* **277**, 445–454.

Wajant, H., Haas, E., Schwenzer, R., Muhlenbeck, F., Kreuz, S., Schubert, G., Grell, M., Smith, C., and Scheurich, P. (2000). Inhibition of death receptor-mediated gene induction by a cycloheximide-sensitive factor occurs at the level of or upstream of Fas-associated death domain protein (FADD). *J. Biol. Chem.* **275**, 24357–24366.

Wajant, H., Henkler, F., and Scheurich, P. (2001). The TNF-receptor-associated factor family: Scaffold molecules for cytokine receptors, kinases and their regulators. *Cell Signal* **13**, 389–400.

Wang, C. Y., Guttridge, D. C., Mayo, M. W., and Baldwin, A. S. Jr. (1999). NF-kappaB induces expression of the Bcl-2 homologue A1/Bfl-1 to preferentially suppress chemotherapy-induced apoptosis. *Mol. Cell. Biol.* **19**, 5923–5929.

Wang, C. Y., Mayo, M. W., Korneluk, R. G., Goeddel, D. V., and Baldwin, A. S. Jr. (1998). NF-kappaB antiapoptosis: Induction of TRAF1 and TRAF2 and c-IAP1 and c-IAP2 to suppress caspase-8 activation. *Science* **281**, 1680–1683.

Wang, D., Westerheide, S. D., Hanson, J. L., and Baldwin, A. S. Jr. (2000). Tumor necrosis factor alpha-induced phosphorylation of RelA/p65 on Ser529 is controlled by casein kinase II. *J. Biol. Chem.* **275**, 32592–32597.

Wang, X. (2001). The expanding role of mitochondria in apoptosis. *Genes Dev.* **15**, 2922–2933.

Wei, M. C., Zong, W. X., Cheng, E. H., Lindsten, T., Panoutsakopoulou, V., Ross, A. J., Roth, K. A., MacGregor, G. R., Thompson, C. B., and Korsmeyer, S. J. (2001). Proapoptotic BAX and BAK: A requisite gateway to mitochondrial dysfunction and death. *Science* **292**, 727–730.

Weingartner, M., Siegmund, D., Schlecht, U., Fotin-Mleczek, M., Scheurich, P., and Wajant, H. (2002). Endogenous membrane tumor necrosis factor (TNF) is a potent amplifier of TNF receptor 1-mediated apoptosis. *J. Biol. Chem.* **277**, 34853–34859.

Werner, A. B., de Vries, E., Tait, S. W., Bontjer, I., and Borst, J. (2002). Bcl-2 family member Bfl-1/A1 sequesters truncated bid to inhibit is collaboration with pro-apoptotic Bak or Bax. *J. Biol. Chem.* **277**, 22781–22788.

Wiley, S. R., Schooley, K., Smolak, P. J., Din, W. S., Huang, C. P., Nicholl, J. K., Sutherland, G. R., Smith, T. D., Rauch, C., Smith, C. A. *et al.* (1995). Identification and characterization of a new member of the TNF family that induces apoptosis. *Immunity* **3**, 673–682.

Woronicz, J. D., Gao, X., Cao, Z., Rothe, M., and Goeddel, D. V. (1997). IkappaB kinase-beta: NF-kappaB activation and complex formation with IkappaB kinase-alpha and NIK. *Science* **278**, 866–869.

Xiao, C. W., Yan, X., Li, Y., Reddy, S. A., and Tsang, B. K. (2003). Resistance of human ovarian cancer cells to tumor necrosis factor alpha is a consequence of nuclear factor kappaB-mediated induction of Fas-associated death domain-like interleukin-1beta-converting enzyme-like inhibitory protein. *Endocrinology* **144**, 623–630.

Xiao, G., Cvijic, M. E., Fong, A., Harhaj, E. W., Uhlik, M. T., Waterfield, M., and Sun, S. C. (2001a). Retroviral oncoprotein Tax induces processing of NF-kappaB2/p100 in T cells: Evidence for the involvement of IKKalpha. *EMBO J.* **20**, 6805–6815.

Xiao, G., Harhaj, E. W., and Sun, S. C. (2001b). NF-kappaB-inducing kinase regulates the processing of NF-kappaB2 p100. *Mol. Cell* **7**, 401–409.

Yamamoto, Y., Verma, U. N., Prajapati, S., Kwak, Y. T., and Gaynor, R. B. (2003). Histone H3 phosphorylation by IKK-alpha is critical for cytokine-induced gene expression. *Nature* **423**, 655–659.

Yang, Y., Fang, S., Jensen, J. P., Weissman, A. M., and Ashwell, J. D. (2000). Ubiquitin protein ligase activity of IAPs and their degradation in proteasomes in response to apoptotic stimuli. *Science* **288**, 874–877.

Yaron, A., Hatzubai, A., Davis, M., Lavon, I., Amit, S., Manning, A. M., Andersen, J. S., Mann, M., Mercurio, F., and Ben-Neriah, Y. (1998). Identification of the receptor component of the IkappaBalpha-ubiquitin ligase. *Nature* **396**, 590–594.

Yeh, W. C., Itie, A., Elia, A. J., Ng, M., Shu, H. B., Wakeham, A., Mirtsos, C., Suzuki, N., Bonnard, M., Goeddel, D. V., and Mak, T. W. (2000). Requirement for Casper (c-FLIP) in regulation of death receptor-induced apoptosis and embryonic development. *Immunity* **12**, 633–642.

You, Z., Ouyang, H., Lopatin, D., Polver, P. J., and Wang, C. Y. (2001). Nuclear factor-kappa B-inducible death effector domain-containing protein suppresses tumor necrosis factor-mediated apoptosis by inhibiting caspase-8 activity. *J. Biol. Chem.* **276**, 26398–26404.

Zhang, S. Q., Kovalenko, A., Cantarella, G., and Wallach, D. (2000). Recruitment of the IKK signalosome to the p55 TNF receptor: RIP and A20 bind to NEMO (IKKgamma) upon receptor stimulation. *Immunity* **12**, 301–311.

Zhao, Q., and Lee, F. S. (1999). Mitogen-activated protein kinase/ERK kinase kinases 2 and 3 activate nuclear factor-kappaB through IkappaB kinase-alpha and IkappaB kinase-beta. *J. Biol. Chem.* **274**, 8355–8358.

Zhong, H., SuYang, H., Erdjument-Bromage, H., Tempst, P., and Ghosh, S. (1997). The transcriptional activity of NF-kappaB is regulated by the IkappaB-associated PKAc subunit through a cyclic AMP-independent mechanism. *Cell* **89**, 413–424.

Zong, W. X., Edelstein, L. C., Chen, C., Bash, J., and Gelinas, C. (1999). The prosurvival Bcl-2 homolog Bfl-1/A1 is a direct transcriptional target of NF-kappaB that blocks TNFalpha-induced apoptosis. *Genes Dev.* **13**, 382–387.

8

CARDINAL ROLES IN APOPTOSIS AND NFκB ACTIVATION

LISA BOUCHIER-HAYES* AND SEAMUS J. MARTIN[†]

*Division of Cellular Immunology, La Jolla Institute for Allergy and Immunology
San Diego, California 92121
[†]Molecular Cell Biology Laboratory, Department of Genetics, The Smurfit Institute
Trinity College, Dublin 2, Ireland

    I. Introduction
   II. Regulation of NFκB Activation
  III. CARD Proteins
  IV. CARD Proteins and NFκB Activation
   V. Functions of CARDINAL
  VI. CARDINAL as a Negative Regulator of Diverse NFκB Activation Pathways
 VII. CARDINAL and Apoptosis
VIII. CARDINAL and Caspase-1 Activation
  IX. The Inflammasome
   X. CARDINAL Contradictions
  XI. Conclusions
     References

Numerous proteins containing the caspase recruitment domain (CARD) have now been identified. While certain CARD-containing proteins are involved in caspase activation in the context of apoptosis, many others participate in NFκB signaling pathways associated with innate or adaptive immune responses. Here, we discuss the CARD-containing proteins that have been implicated as participants in immune response signaling pathways that culminate in NFκB activation. © 2004 Elsevier Inc.

I. INTRODUCTION

NFκB is a transcription factor that controls the expression of an array of genes that regulate the innate and adaptive immune responses (reviewed in Ghosh and Karin, 2002). A number of infectious agents trigger NFκB activation, which in turn induces expression of effector molecules, such as interleukin (IL)-1β and tumor necrosis factor (TNF), that are secreted from the cell and can act to amplify the immune response. The activation of the immune response leads to the migration of neutrophils, monocytes, and lymphocytes to the site of injury to remove the infectious agent. However, inflammatory mediators, such as lipopolysaccharides (LPS) and IL-1β, which can trigger NFκB activation, can also result in endotoxic shock when present in high doses and can lead to multiple organ failure and sometimes death. Thus it is essential that this process be strictly controlled, both positively and negatively, to ensure that infectious agents are efficiently recognized while simultaneously preventing minor proinflammatory insults from escalating into conditions of persistent inflammation.

II. REGULATION OF NFκB ACTIVATION

Much of the control of the inflammatory response occurs at the level of activation of the NFκB transcription factor. NFκB is activated by a regulatory complex that consists of three proteins: IKKα, IKKβ, and IKKγ (reviewed in Ghosh and Karin, 2002). Activation of the IκB kinase (IKK) complex leads to the phosphorylation of the alpha and beta subunits of neighboring IKK complexes—thus amplifying the signal—and results in the phosphorylation of the NFκB inhibitor, IκB. In quiescent cells, NFκB is sequestered in the cytoplasm through its interaction with IκB and the complex formed between these two proteins results in masking of the two nuclear localization signals on the surface of the NFκB protein. Upon phosphorylation of IκB, this protein becomes targeted to the proteasome for degradation leaving NFκB free to translocate to the nucleus where it can activate the transcription of a number of target genes. Because one of these target genes is IκB, this protein rapidly reaccumulates in the cell leading to the inactivation of NFκB. The latter feedback loop serves as an important homeostatic control mechanism for the regulation of NFκB activity.

Upstream of the IKK complex, a number of proteins participate in the many and varied pathways that lead to NFκB activation. A recently discovered subset of proteins that can regulate NFκB activation all possess a motif called the caspase recruitment domain (CARD) that was originally identified in molecules that participate in the regulation of apoptosis.

III. CARD PROTEINS

Apoptosis, or programmed cell death, is a process that allows the removal of defective, infected, transformed, or senescent cells from a cell population. The signaling pathways that control this process converge on a family of proteases known as caspases. Caspases are normally present in cells as inactive precursors that undergo maturation at the onset of apoptosis. Upon activation, caspases cleave a number of specific substrates, including many structural and regulatory proteins, that ultimately results in the death of the cell (Martin and Green, 1995). The CARD domain is a protein-protein interaction domain that was originally identified in certain caspases and their interaction partners. Thus, CARD-containing proteins were originally classified as proapoptotic molecules that act as docking platforms for caspases (Hofmann *et al.*, 1997). In this context, the CARD motif facilitates interactions between inactive procaspase molecules, which promotes their activation (Salvesen and Dixit, 1999).

Recently it has also become evident that the functions of proteins containing CARD motifs are not restricted to caspase recruitment and the regulation of apoptosis (Fig. 1). Indeed, a number of proteins with CARD motifs have been found to be involved in the regulation of NFκB activation (for a recent review see Bouchier-Hayes and Martin, 2002). Many of the CARD-containing proteins implicated in NFκB-activation pathways can either directly or indirectly lead to activation of the IKK complex and are thus positive regulators of NFκB. Interestingly, among this large group of proteins, only one protein has been characterized to date that can act as a negative regulator of NFκB. This protein, called CARDINAL (*CARD inhibitor of NFκB activating ligands*), may play an important role in the negative regulation of stress responses and inflammation and is the focus of this review (Bouchier-Hayes *et al.*, 2001).

IV. CARD PROTEINS AND NFκB ACTIVATION

CARD proteins that have been implicated in NFκB activation pathways include Nod1/CARD4 and Nod2 (Bertin *et al.*, 1999; Inohara *et al.*, 1999; Ogura *et al.*, 2001). Both of these proteins are structurally similar to the proapoptotic molecule Apaf-1 that plays an important role in regulating the activation of caspase-9 in an important pathway to apoptosis (see Adrain and Martin, 2001 for a review of this topic). Apaf-1, Nod1, and Nod2 each

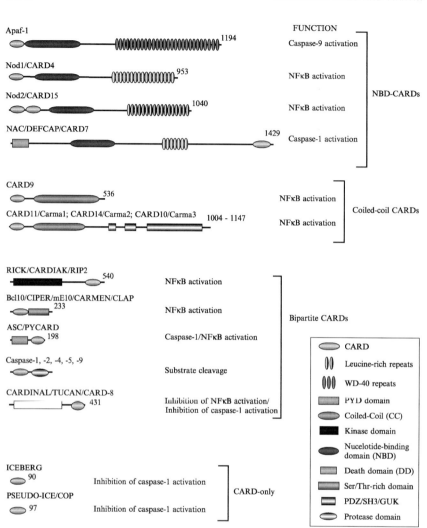

FIGURE 1. Schematic representation of the domain structures of a number of the known CARD-containing proteins that function in caspase-activation and NFκB-activation pathways.

contain a nucleotide-binding domain (NBD) in addition to the CARD motif, but the Nod proteins contain C-terminal leucine-rich repeats (LRRs) in place of the WD40 repeats found in Apaf-1 (Fig. 1).

Both Nod1 and Nod2 have been implicated as intracellular receptors for bacterial pathogens. Current evidence suggests that both Nod proteins recognize components of peptidoglycans found in bacterial cell walls. Nod1 acts as a sensor for peptidoglycans containing diaminopimelic acid and Nod2 recognizes peptidoglycans containing muramyl dipeptide (Chamaillard *et al.*, 2003; Girardin *et al.*, 2003a,b). The C-terminal LRRs are considered to be the sensory domains in this context (Inohara *et al.*, 2001). Although the precise details have yet to be resolved, upon activation by pathogen components, Nod1 or Nod2 molecules are likely to oligomerize via their NBD domains to permit recruitment and oligomerization of the CARD-containing kinase RICK (Inohara *et al.*, 2000). RICK in turn recruits IKKγ, the regulatory subunit of the IKK complex, leading to activation of the complex and thus NFκB downstream (Poyet *et al.*, 2000).

The Nod proteins are also part of a larger family of intracellular LRR-containing proteins defined as the CATERPILLER protein family [CARD, transcription enhancer, R (purine)-binding, pyrin, lots of LRRs; Harton *et al.*, 2002]. Many of these proteins have been implicated in NFκB activation and inflammatory pathways. A subfamily of the CATERPILLAR proteins contains the NALP subfamily, which consists of 14 proteins, all with similar structure to the Nod proteins with the exception that these proteins all contain a pyrin domain (PYD) in place of the CARD domain found at the N-terminal end of the Nod proteins (Tschopp *et al.*, 2003). The PYD motif is similar in structure to the CARD motif and derives its name from its discovery in a protein called Pyrin (Bertin and DiStefano, 2000; Martinon *et al.*, 2001; Masumoto *et al.*, 2001; Stehlik *et al.*, 2003). Interestingly, Pyrin is found in mutant form in familial Mediterranean fever, an inherited disorder characterized by frequent occurrences of fever and inflammation (The French FMF Consortium, 1997). Although the functions of most of the NALP proteins have yet to be elucidated, current evidence suggests that these proteins may also act as intracellular sensors for pathogen components (Tschopp *et al.*, 2003).

Another group of CARD proteins implicated in NFκB activation are broadly similar in structure to the NBD-CARDs but the central NBD is replaced by a coiled-coil (CC) motif and thus, similar to the Nods, are likely to be scaffold proteins for protein complex formation (Fig. 1). This group includes CARD-9, -10, -11 and -14 (Bertin *et al.*, 2000, 2001; Wang *et al.*, 2001). Each of these proteins is capable of recruiting Bcl-10, which is an important activator of NFκB in response to antigen-driven T-cell and B-cell receptor activation (Willis *et al.*, 1999).

Thus a general picture emerges for a role for CARD-containing proteins in NFκB activation pathways. It remains unclear how these molecules are activated by antigenic stimulation but it is possible that some of these proteins act as intracellular receptors for pathogen components just as the Toll receptor family act as sensors for extracellular pathogen components. Oligomerization of such proteins may allow for recruitment and clustering of downstream adaptors, such as Bcl-10 or RICK, which ultimately results in activation of the IKK complex. These pathways, which rely on induced proximity of activator proteins, are reminiscent of two well-characterized pathways that lead to caspase activation and apoptosis, the Apaf-1 and Fas pathways (Cain *et al.*, 2000; Kischkel *et al.*, 1995). In the latter cell death pathways, negative regulators or decoy molecules exist that inhibit signaling for apoptosis. An example is FLIP which acts as a decoy in the Fas signaling pathway by preventing recruitment of caspase-8 to the Fas receptor complex (Tschopp *et al.*, 1998). Given the similarities between the molecules that participate in pathways leading to apoptosis or NFκB activation, it is likely that similar decoy molecules exist in the NFκB activating pathways described previously. Indeed, the CARD-containing protein CARDINAL has been proposed to play such a role.

V. FUNCTIONS OF CARDINAL

The CARD motif is composed of a stretch of approximately 100 amino acids, arranged into six or seven antiparallel alpha-helices tightly packed around a hydrophobic core (Chou *et al.*, 1998). The CARD motif can be recognized by the presence of a number of hydrophobic residues in conserved regions throughout the motif (Hofmann *et al.*, 1997). The identification of CARDINAL (also called TUCAN/CARD8/NDPP1) was facilitated due to the presence of a CARD motif within its C-terminus (Bouchier-Hayes *et al.*, 2001; Pathan *et al.*, 2001; Razmara *et al.*, 2002; Zhang and Fu, 2002). However, CARDINAL is quite a small protein and does not contain any other well-defined motifs or protein domains. This sequence structure suggests that CARDINAL may act either as an adaptor protein or as a decoy molecule. CARDINAL was independently identified by a number of different groups and, as detailed subsequently, has been proposed to play a variety of roles in caspase as well as NFκB activation.

VI. CARDINAL AS A NEGATIVE REGULATOR OF DIVERSE NFκB ACTIVATION PATHWAYS

We and others have observed that CARDINAL has the ability to dramatically suppress the level of NFκB activation induced by a number of stimuli (RIP, RICK, Bcl-10, and TRADD) that can induce activation of this transcription factor (Bouchier-Hayes *et al.*, 2001; Razmara *et al.*, 2002; Stilo *et al.*, 2002; Zhang and Fu, 2002). Moreover, stimulation of CARDINAL-expressing cells with the proinflammatory cytokines TNF and IL-1 resulted in a marked suppression of their ability to induce activation of NFκB (Bouchier-Hayes *et al.*, 2001).

CARDINAL seems to be capable of inhibiting across a range of NFκB activation pathways. For example, Bcl-10 is known to be essential for

NFκB activation in response to T and B-cell receptor stimulation (Ruland *et al.*, 2001), while RICK is a component of the Nod1 pathway, which can be activated by bacterial components (Chamaillard *et al.*, 2003; Girardin *et al.*, 2003a; Inohara *et al.*, 2000). CARDINAL can inhibit NFκB activation associated with RICK or Bcl-10 in a comparable fashion (Bouchier-Hayes *et al.*, 2001). This suggests that CARDINAL operates downstream of proteins such as Bcl-10, RICK, RIP, and others. Thus CARDINAL is likely to act at the point of convergence of these diverse routes to NFκB activation, which is the IKK complex.

Both RICK and RIP have been shown to bind to IKKγ, the regulatory subunit of the IKK complex. It is thought that this induces clustering of a number of subunits of the complex and it is predicted that other downstream activators of NFκB might also act through the recruitment of IKKγ (Inohara *et al.*, 2000; Poyet *et al.*, 2000). CARDINAL, similar to RIP and RICK, has been found to interact with the IKKγ subunit of the IKK complex upon overexpression (Bouchier-Hayes *et al.*, 2001). Thus CARDINAL may act to antagonize the ability of proteins such as RIP and RICK to promote NFκB activation by sequestering IKKγ. In the case of RIP and RICK, these proteins bind to IKKγ not through their death or CARD domains (respectively) or through their kinase domains, but through an intermediate domain that has not yet been defined as possessing a particular motif (Inohara *et al.*, 2000; Poyet *et al.*, 2000). Similarly, in the case of CARDINAL, it is not the CARD domain that is responsible for the inhibitory function, but the N-terminal domain. These results suggest that CARDINAL functions to antagonize the function of NFκB activators, such as RIP and RICK, by competing with such proteins for recruitment of IKKγ.

However, one problem with this model is that all of these observations are based on overexpression experiments in transformed cell lines. Moreover, it is not yet known how CARDINAL may become activated in this context. Tissue expression analysis has revealed that CARDINAL is constitutively expressed in a number of cell lines and tissues suggesting that the activity of this protein is not controlled at a transcriptional level (Bouchier-Hayes *et al.*, 2001; Pathan *et al.*, 2001). Thus it seems likely that the activity of CARDINAL is controlled through protein-protein interactions with other CARD-containing proteins. Interestingly, a protein named DRAL has been identified as an interaction partner for CARDINAL (Stilo *et al.*, 2002). DRAL is a p53-responsive protein and, conversely to CARDINAL, has the effect of augmenting NFκB activation in certain contexts. However co-expression of CARDINAL with DRAL has been shown to abrogate the stimulatory effects of the latter on NFκB activation (Stilo *et al.*, 2002). This suggests that DRAL operates upstream of CARDINAL, but the functional significance of the CARDINAL-DRAL interaction remains obscure.

VII. CARDINAL AND APOPTOSIS

Proteins that contain the CARD motif have been found in three different contexts; in caspase activation pathways leading to apoptosis, in caspase activation pathways leading to inflammation, and as participants in NFκB activation pathways (Bouchier-Hayes and Martin, 2002). Although the results described previously suggest that CARDINAL acts as an inhibitor of NFκB activation, other reports have suggested alternative roles for CARDINAL in other pathways. CARDINAL has been proposed to act as an inhibitor of apoptosis through inhibiting assembly of the Apaf-1 apoptosome (Pathan *et al.*, 2001). These authors suggest that CARDINAL binds to caspase-9 and thus competes with Apaf-1 for recruitment of the latter. Consistent with this, CARDINAL has been found to dramatically inhibit Bax-mediated cell death (Pathan *et al.*, 2001). However, other groups were unable to confirm a role for CARDINAL as an inhibitor of apoptosis (Bouchier-Hayes *et al.*, 2002; Razmara *et al.*, 2002). To further complicate matters, CARDINAL has also been reported to sensitize cells to proapoptotic stimuli (Razmara *et al.*, 2002).

However, there exists a certain degree of interplay between pathways that lead to NFκB activation and apoptosis. In particular, it is well established that suppression of NFκB activation sensitizes cells to apoptosis through TNF receptor stimulation or exposure to chemotherapeutic drugs (Beg and Baltimore, 1996; Van Antwerp *et al.*, 1996). One could thus extrapolate from this that CARDINAL, as an NFκB inhibitor, may be capable of sensitizing cells to certain proapoptotic stimuli. The cytotoxic effects of CARDINAL observed by Alnemri and colleagues may thus be a side effect of NFκB inhibition (Razmara *et al.*, 2002).

VIII. CARDINAL AND CASPASE-1 ACTIVATION

An interesting feature of the sequence of the CARDINAL protein is the high degree of homology between the CARD of the latter and the prodomain of caspase-1. Given the recent discovery of a number of CARD proteins that may function as regulators of caspase-1 activation, it is plausible that CARDINAL may also act to regulate activation of this caspase. Indeed, observations by Alnemri and colleagues suggest that CARDINAL may act as an antagonist of caspase-1 activation (Razmara *et al.*, 2002). These studies suggest that CARDINAL has similar properties to the caspase-1 inhibitory proteins ICEBERG and pseudoICE. Razmara *et al.* have reported that CARDINAL interacts with caspase-1 and also has the ability to bind to ICEBERG and pseudoICE (Razmara *et al.*, 2002).

Caspase-1 was initially characterized as an interleukin-1β processing enzyme (Thornberry *et al.*, 1992), and shares sequence homology with the

C. elegans ced-3 gene, which is essential for cell death in the nematode. The sequence similarity between CED-3 and caspase-1 suggested that caspase-1 might promote cell death (Yuan *et al.*, 1993). However, unlike other caspases that are exclusively involved in the initiation or execution of apoptosis, caspase-1 is one of a number of caspases (including caspase-4, -5, and -11) that have a primary role in inflammation rather than apoptosis (Lin *et al.*, 2000). Caspase-1 is a highly selective protease and only two *bona fide* caspase-1 substrates are known; IL-1β and IL-18 (Gu *et al.*, 1997; Thornberry *et al.*, 1992). IL-1β plays an important role as an inflammatory response mediator and is involved in the pathogenesis of many conditions including septic shock, wound healing, bone resorption, and myelogenous leukemia (Dinarello, 1996). Targeted disruption of the *CASP-1* gene in the mouse results in a failure of IL-1β export and resistance to lipopolysaccharide-induced septic shock (Li *et al.*, 1995).

A number of proteins have been reported to bind to and potentially regulate caspase-1 activation. RICK, the CARD containing kinase, has been shown to bind caspase-1 and to weakly promote caspase-1 activation upon overexpression (Thome *et al.*, 1998). It is thought that RICK induces oligomerization of caspase-1 molecules leading to activation of the latter. ICEBERG and pseudoICE/COP are two small CARD proteins that are very similar to the CARD of caspase-1 (Fig. 1, Druilhe *et al.*, 2001; Humke *et al.*, 2000; Lee *et al.*, 2001). Both bind to caspase-1 and have been suggested to disrupt RICK-mediated oligomerization of this caspase. CARDINAL has been suggested to behave in a similar manner to ICEBERG and pseudoICE. Overexpression of CARDINAL in combination with RICK and caspase-1 resulted in an inhibition of RICK-induced processing of caspase-1 (Razmara *et al.*, 2002). However the ability of RICK to activate caspase-1 has recently been called into question with the phenotype of the *RICK* null mouse (Chin *et al.*, 2002; Kobayashi *et al.*, 2002). If RICK played a crucial role in the regulation of caspase-1 function, a phenotype with similarities to the caspase-1 knockout mouse would be expected. However, while targeted inactivation of *RICK* produced defects in toll receptor-initiated NFκB activation pathways, IL-1β production was normal in these animals (Chin *et al.*, 2002; Kobayashi *et al.*, 2002).

IX. THE INFLAMMASOME

The phenotype of the *RICK* knockout mouse suggests that RICK is not a major player in the pathway that regulates caspase-1 activation (Chin *et al.*, 2002; Kobayashi *et al.*, 2002). An alternative model was recently described for the activation of caspase-1, which has been termed the inflammasome (Martinon *et al.*, 2002). The inflammasome has been proposed to consist of four proteins: caspase-1, caspase-5, NAC (NALP1/DEFCAP), and a further CARD-containing protein called ASC (PYCARD).

Caspase-5 is a member of the sub-group of caspases that have been implicated in inflammation and is considered to be the human orthologue of the murine caspase, caspase-11. Caspase-11 has much in common with caspase-1 (Wang *et al.*, 1996). The phenotype of the *CASP-11* knockout mouse is similar to that of the *CASP-1* null mouse, also displaying resistance to LPS-induced septic shock and defects in IL-1β processing (Wang *et al.*, 1998). One notable difference between the two caspases is that while caspase-1 is constitutively expressed in cells, caspase-11 is inducible by LPS (Wang *et al.*, 1998). Caspase-5 is also LPS-inducible suggesting that this caspase is the human orthologue of caspase-11 and may cooperate with caspase-1 to regulate the processing of IL-1β (Lin *et al.*, 2000).

NAC/NALP1 is an NBD-CARD protein, similar in structure to the Nod proteins, but the CARD domain is uniquely located at its C-terminus and this molecule has a PYD domain at the N-terminus (Fig. 1, Hlaing *et al.*, 2001). It is through this PYD domain that NAC can bind to a second PYD-containing protein called ASC/PYCARD (Masumoto *et al.*, 1999). In addition to the PYD motif, ASC also contains a CARD motif, which in this case recruits caspase-1 to the NAC/ASC complex (Fig. 1, Srinivasula *et al.*, 2002). NAC has been proposed to simultaneously interact with caspase-5 through its C-terminal CARD motif (Martinon *et al.*, 2002). NAC also contains a centrally located nucleotide-binding domain, which likely acts as an oligomerization platform to enable a number of these molecules to be brought together to increase the local concentration of caspase-1 and caspase-5. Once in proximity, caspase-1 and caspase-5 may activate each other resulting in robust caspase-1 activation and subsequent IL-1β processing.

Interestingly, the N-terminus of CARDINAL and the C-terminus of NAC share significant homology but the significance of this similarity remains unclear (Bouchier-Hayes *et al.*, 2001). The strong homology between NAC and CARDINAL may suggest that these two proteins are able to interact, although this has yet to be explored. If these proteins do indeed interact, this would provide a possible means to enable CARDINAL to inhibit caspase-1 activation. This would suggest that, through binding to NAC, CARDINAL might act to antagonize signals routed through the inflammasome by competing with the inflammasome for caspase-1 (or possibly caspase-5) recruitment.

X. CARDINAL CONTRADICTIONS

Thus there is evidence that CARDINAL functions as an inhibitor of inflammatory pathways either through suppression of NFκB activation or inhibition of caspase-1 activation and indirect evidence suggests that the latter function may be routed through interruption of inflammasome complex formation (Bouchier-Hayes et al., 2001; Razmara et al., 2002). So is it possible that CARDINAL can function in both these respects?

Interestingly many of the proteins that lead to NFκB activation have been also implicated in caspase-1 regulation. As discussed previously, RICK has been suggested to activate caspase-1 as well as NFκB although the latter function is more likely. Similarly, Nod1 has also been shown to activate caspase-1 (Yoo et al., 2002), but it has not been reported whether the Nod1 knockout displays any defects in IL-1β processing (Chamaillard et al., 2003). ASC, a component of the inflammasome, has also been shown to activate NFκB (Masumoto et al., 2003). As well as binding to NAC, ASC also has the ability to bind to many other members of the NALP family and activate NFκB via this route. It has yet to be explored whether the NAC/ASC complex can recruit components of the IKK complex, which would allow the inflammasome to regulate NFκB activation. A bifunctional inflammasome would provide an explanation for the observations that CARDINAL appears capable of inhibiting both NFκB activation as well as caspase-1 activation.

XI. CONCLUSIONS

CARD-containing proteins have previously been implicated in pathways leading to caspase activation or NFκB activation. CARDINAL is the first example of a CARD-containing protein reported to inhibit NFκB activation. Clearly it will be important to further examine the role of CARDINAL by analyzing its function at endogenous levels and through targeted inactivation in the mouse.

ACKNOWLEDGMENTS

We thank Science Foundation Ireland (PI1/B038) for their generous support of ongoing work in the Martin laboratory. We also thank The Wellcome Trust for provision of a Prize studentship to LBH.

REFERENCES

Adrain, C., and Martin, S. J. (2001). The mitochondrial apoptosome: A killer unleashed by the cytochrome seas. *Trends Biochem. Sci.* **26,** 390–397.

Beg, A. A., and Baltimore, D. (1996). An essential role for NF-kappaB in preventing TNF-alpha–induced cell death. *Science* **274,** 782–784.

Bertin, J., and DiStefano, P. S. (2000). The PYRIN domain: A novel motif found in apoptosis and inflammation proteins. *Cell Death Differ.* **7,** 1273–1274.

Bertin, J., Guo, Y., Wang, L., Srinivasula, S. M., Jacobson, M. D., Poyet, J. L., Merriam, S., Du, M. Q., Dyer, M. J., Robison, K. E., DiStefano, P. S., and Alnemri, E. S. (2000). CARD9 is a novel caspase recruitment domain-containing protein that interacts with BCL10/CLAP and activates NF-kappaB. *J. Biol. Chem.* **275,** 41082–41086.

Bertin, J., Nir, W. J., Fischer, C. M., Tayber, O. V., Errada, P. R., Grant, J. R., Keilty, J. J., Gosselin, M. L., Robison, K. E., Wong, G. H., Glucksmann, M. A., and DiStefano, P. S. (1999). Human CARD4 protein is a novel CED-4/Apaf-1 cell death family member that activates NF-kappaB. *J. Biol. Chem.* **274,** 12955–12958.

Bertin, J., Wang, L., Guo, Y., Jacobson, M. D., Poyet, J. L., Srinivasula, S. M., Merriam, S., DiStefano, P. S., and Alnemri, E. S. (2001). CARD11 and CARD14 are novel caspase recruitment domain (CARD)/membrane-associated guanylate kinase (MAGUK) family members that interact with BCL10 and activate NF-kappaB. *J. Biol. Chem.* **276,** 11877–11882.

Bouchier-Hayes, L., Conroy, H., Egan, H., Adrain, C., Creagh, E. M., MacFarlane, M., and Martin, S. J. (2001). CARDINAL, a novel caspase recruitment domain protein, is an inhibitor of multiple NF-kappaB activation pathways. *J. Biol. Chem.* **276,** 44069–44077.

Bouchier-Hayes, L., and Martin, S. J. (2002). CARD games in apoptosis and immunity. *EMBO Rep.* **3,** 616–621.

Cain, K., Bratton, S. B., Langlais, C., Walker, G., Brown, D. G., Sun, X. M., and Cohen, G. M. (2000). Apaf-1 oligomerizes into biologically active approximately 700-kDa and inactive approximately 1.4-MDa apoptosome complexes. *J. Biol. Chem.* **275,** 6067–6070.

Chamaillard, M., Hashimoto, M., Horie, Y., Masumoto, J., Qiu, S., Saab, L., Ogura, Y., Kawasaki, A., Fukase, K., Kusumoto, S., Valvano, M. A., Foster, S. J., Mak, T. W., Nunez, G., and Inohara, N. (2003). An essential role for NOD1 in host recognition of bacterial peptidoglycan containing diaminopimelic acid. *Nat. Immunol.* **4,** 702–707.

Chin, A. I., Dempsey, P. W., Bruhn, K., Miller, J. F., Xu, Y., and Cheng, G. (2002). Involvement of receptor-interacting protein 2 in innate and adaptive immune responses. *Nature* **416,** 190–194.

Chou, J. J., Matsuo, H., Duan, H., and Wagner, G. (1998). Solution structure of the RAIDD CARD and model for CARD/CARD interaction in caspase-2 and caspase-9 recruitment. *Cell* **94,** 171–180.

Dinarello, C. A. (1996). Biologic basis for interleukin-1 in disease. *Blood* **87,** 2095–2147.

Druilhe, A., Srinivasula, S. M., Razmara, M., Ahmad, M., and Alnemri, E. S. (2001). Regulation of IL-1 beta generation by Pseudo-ICE and ICEBERG, two dominant negative caspase recruitment domain proteins. *Cell Death Differ.* **8,** 649–657.

Ghosh, S., and Karin, M. (2002). Missing pieces in the NF-kappaB puzzle. *Cell* **109**(Suppl), S81–S96.

Girardin, S. E., Boneca, I. G., Carneiro, L. A., Antignac, A., Jehanno, M., Viala, J., Tedin, K., Taha, M. K., Labigne, A., Zathringer, U., Coyle, A. J., DiStefano, P. S., Bertin, J., Sansonetti, P. J., and Philpott, D. J. (2003a). Nod1 detects a unique muropeptide from gram-negative bacterial peptidoglycan. *Science* **300,** 1584–1587.

Girardin, S. E., Boneca, I. G., Viala, J., Chamaillard, M., Labigne, A., Thomas, G., Philpott, D. J., and Sansonetti, P. J. (2003b). Nod2 is a general sensor of peptidoglycan through muramyl dipeptide (MDP) detection. *J. Biol. Chem.* **278,** 8869–8872.

Gu, Y., Kuida, K., Tsutsui, H., Ku, G., Hsiao, K., Fleming, M. A., Hayashi, N., Higashino, K., Okamura, H., Nakanishi, K., Kurimoto, M., Tanimoto, T., Flavell, R. A., Sato, V., Harding, M. W., Livingston, D. J., and Su, M. S. (1997). Activation of interferon-gamma inducing factor mediated by interleukin-1beta converting enzyme. *Science* **275,** 206–209.

Harton, J. A., Linhoff, M. W., Zhang, J., and Ting, J. P. (2002). Cutting edge: CATERPILLER: A large family of mammalian genes containing CARD, pyrin, nucleotide-binding, and leucine-rich repeat domains. *J. Immunol.* **169,** 4088–4093.

Hlaing, T., Guo, R. F., Dilley, K. A., Loussia, J. M., Morrish, T. A., Shi, M. M., Vincenz, C., and Ward, P. A. (2001). Molecular cloning and characterization of DEFCAP-L and -S, two isoforms of a novel member of the mammalian Ced-4 family of apoptosis proteins. *J. Biol. Chem.* **276,** 9230–9238.

Hofmann, K., Bucher, P., and Tschopp, J. (1997). The CARD domain: A new apoptotic signaling motif. *Trends Biochem. Sci.* **22,** 155–156.

Humke, E. W., Shriver, S. K., Starovasnik, M. A., Fairbrother, W. J., and Dixit, V. M. (2000). ICEBERG: A novel inhibitor of interleukin-1 beta generation. *Cell* **103,** 99–111.

Inohara, N., Koseki, T., del Peso, L., Hu, Y., Yee, C., Chen, S., Carrio, R., Merino, J., Liu, D., Ni, J., and Nunez, G. (1999). Nod1, an Apaf-1-like activator of caspase-9 and nuclear factor-kappaB. *J. Biol. Chem.* **274,** 14560–14567.

Inohara, N., Koseki, T., Lin, J., del Peso, L., Lucas, P. C., Chen, F. F., Ogura, Y., and Nunez, G. (2000). An induced proximity model for NF-kappaB activation in the Nod1/RICK and RIP signaling pathways. *J. Biol. Chem.* **275,** 27823–27831.

Inohara, N., Ogura, Y., Chen, F. F., Muto, A., and Nunez, G. (2001). Human Nod1 confers responsiveness to bacterial lipopolysaccharides. *J. Biol. Chem.* **276,** 2551–2554.

Kischkel, F. C., Hellbardt, S., Behrmann, I., Germer, M., Pawlita, M., Krammer, P. H., and Peter, M. E. (1995). Cytotoxicity-dependent APO-1 (Fas/CD95)-associated proteins form a death-inducing signaling complex (DISC) with the receptor. *EMBO J.* **14,** 5579–5588.

Kobayashi, K., Inohara, N., Hernandez, L. D., Galan, J. E., Nunez, G., Janeway, C. A., Medzhitov, R., and Flavell, R. A. (2002). RICK/Rip2/CARDIAK mediates signaling for receptors of the innate and adaptive immune systems. *Nature* **416,** 194–199.

Lee, S. H., Stehlik, C., and Reed, J. C. (2001). Cop, a caspase recruitment domain-containing protein and inhibitor of caspase-1 activation processing. *J. Biol. Chem.* **276,** 34495–34500.

Li, P., Allen, H., Banerjee, S., Franklin, S., Herzog, L., Johnston, C., McDowell, J., Paskind, M., Rodman, L., Salfeld, J. *et al.* (1995). Mice deficient in IL-1 beta-converting enzyme are defective in production of mature IL-1 beta and resistant to endotoxic shock. *Cell* **80,** 401–411.

Lin, X. Y., Choi, M. S., and Porter, A. G. (2000). Expression analysis of the human caspase-1 subfamily reveals specific regulation of the CASP5 gene by lipopolysaccharide and interferon-gamma. *J. Biol. Chem.* **275,** 39920–39926.

Martin, S. J., and Green, D. R. (1995). Protease activation during apoptosis: Death by a thousand cuts? *Cell* **82,** 349–352.

Martinon, F., Burns, K., and Tschopp, J. (2002). The inflammasome: A molecular platform triggering activation of inflammatory caspases and processing of proIL-beta. *Mol. Cell* **10,** 417–426.

Martinon, F., Hofmann, K., and Tschopp, J. (2001). The pyrin domain: A possible member of the death domain-fold family implicated in apoptosis and inflammation. *Curr. Biol.* **11,** R118–R120.

Masumoto, J., Dowds, T. A., Schaner, P., Chen, F. F., Ogura, Y., Li, M., Zhu, L., Katsuyama, T., Sagara, J., Taniguchi, S., Gumucio, D. L., Nunez, G., and Inohara, N. (2003). ASC is an activating adaptor for NF-kappaB and caspase-8-dependent apoptosis. *Biochem. Biophys. Res. Commun.* **303,** 69–73.

Masumoto, J., Taniguchi, S., Ayukawa, K., Sarvotham, H., Kishino, T., Niikawa, N., Hidaka, E., Katsuyama, T., Higuchi, T., and Sagara, J. (1999). ASC, a novel 22-kDa protein, aggregates during apoptosis of human promyelocytic leukemia HL-60 cells. *J. Biol. Chem.* **274,** 33835–33838.

Masumoto, J., Taniguchi, S., and Sagara, J. (2001). Pyrin N-terminal homology domain- and caspase recruitment domain-dependent oligomerization of ASC. *Biochem. Biophys. Res. Commun.* **280,** 652–655.

Ogura, Y., Inohara, N., Benito, A., Chen, F. F., Yamaoka, S., and Nunez, G. (2001). Nod2, a Nod1/Apaf-1 family member that is restricted to monocytes and activates NF-kappaB. *J. Biol. Chem.* **276,** 4812–4818.

Pathan, N., Marusawa, H., Krajewski, M., Matsuzawa, S., Kim, H., Okada, K., Torii, S., Kitada, S., Krajewski, S., Welsh, K., Pio, F., Godzik, A., and Reed, J. C. (2001). TUCAN, an antiapoptotic caspase-associated recruitment domain family protein overexpressed in cancer. *J. Biol. Chem.* **276,** 32220–32229.

Poyet, J. L., Srinivasula, S. M., Lin, J. H., Fernandes-Alnemri, T., Yamaoka, S., Tsichlis, P. N., and Alnemri, E. S. (2000). Activation of the Ikappa B kinases by RIP via IKKgamma/NEMO-mediated oligomerization. *J. Biol. Chem.* **275,** 37966–37977.

Razmara, M., Srinivasula, S. M., Wang, L., Poyet, J. L., Geddes, B. J., DiStefano, P. S., Bertin, J., and Alnemri, E. S. (2002). CARD-8 protein, a new CARD family member that regulates caspase-1 activation and apoptosis. *J. Biol. Chem.* **277,** 13952–13958.

Ruland, J., Duncan, G. S., Elia, A., del Barco Barrantes, I., Nguyen, L., Plyte, S., Millar, D. G., Bouchard, D., Wakeham, A., Ohashi, P. S., and Mak, T. W. (2001). Bcl10 is a positive regulator of antigen receptor-induced activation of NF-kappaB and neural tube closure. *Cell* **104,** 33–42.

Salvesen, G. S., and Dixit, V. M. (1999). Caspase activation: The induced-proximity model. *Proc. Natl. Acad. Sci. USA* **96,** 10964–10967.

Srinivasula, S. M., Poyet, J. L., Razmara, M., Datta, P., Zhang, Z., and Alnemri, E. S. (2002). The PYRIN-CARD protein ASC is an activating adaptor for caspase-1. *J. Biol. Chem.* **277,** 21119–21122.

Stehlik, C., Krajewska, M., Welsh, K., Krajewski, S., Godzik, A., and Reed, J. C. (2003). The PAAD/PYRIN-only protein POP1/ASC2 is a modulator of ASC-mediated nuclear-factor-kappaB and pro-caspase-1 regulation. *Biochem. J.* **373,** 101–113.

Stilo, R., Leonardi, A., Formisano, L., Di Jeso, B., Vito, P., and Liguoro, D. (2002). TUCAN/CARDINAL and DRAL participate in a common pathway for modulation of NF-kappaB activation. *FEBS Lett.* **521,** 165–169.

The French FMF Consortium. (1997). A candidate gene for familial Mediterranean fever. *Nat. Genet.* **17,** 25–31.

Thome, M., Hofmann, K., Burns, K., Martinon, F., Bodmer, J. L., Mattmann, C., and Tschopp, J. (1998). Identification of CARDIAK, a RIP-like kinase that associates with caspase-1. *Curr. Biol.* **8,** 885–888.

Thornberry, N. A., Bull, H. G., Calaycay, J. R., Chapman, K. T., Howard, A. D., Kostura, M. J., Miller, D. K., Molineaux, S. M., Weidner, J. R., Aunins, J. *et al.* (1992). A novel heterodimeric cysteine protease is required for interleukin-1 beta processing in monocytes. *Nature* **356,** 768–774.

Tschopp, J., Irmler, M., and Thome, M. (1998). Inhibition of fas death signals by FLIPs. *Curr. Opin. Immunol.* **10,** 552–558.

Tschopp, J., Martinon, F., and Burns, K. (2003). NALPs: A novel protein family involved in inflammation. *Nat. Rev. Mol. Cell. Biol.* **4,** 95–104.

Van Antwerp, D. J., Martin, S. J., Kafri, T., Green, D. R., and Verma, I. M. (1996). Suppression of TNF-alpha-induced apoptosis by NF-kappaB. *Science* **274,** 787–789.

Wang, L., Guo, Y., Huang, W. J., Ke, X., Poyet, J. L., Manji, G. A., Merriam, S., Glucksmann, M. A., DiStefano, P. S., Alnemri, E. S., and Bertin, J. (2001). Card10 is a novel caspase recruitment domain/membrane-associated guanylate kinase family member that interacts with BCL10 and activates NF-kappaB. *J. Biol. Chem.* **276,** 21405–21409.

Wang, S., Miura, M., Jung, Y., Zhu, H., Gagliardini, V., Shi, L., Greenberg, A. H., and Yuan, J. (1996). Identification and characterization of Ich-3, a member of the interleukin-1beta converting enzyme (ICE)/Ced-3 family and an upstream regulator of ICE. *J. Biol. Chem.* **271,** 20580–20587.

Wang, S., Miura, M., Jung, Y. K., Zhu, H., Li, E., and Yuan, J. (1998). Murine caspase-11, an ICE-interacting protease, is essential for the activation of ICE. *Cell* **92,** 501–509.

Willis, T. G., Jadayel, D. M., Du, M. Q., Peng, H., Perry, A. R., Abdul-Rauf, M., Price, H., Karran, L., Majekodunmi, O., Wlodarska, I., Pan, L., Crook, T., Hamoudi, R., Isaacson, P. G., and Dyer, M. J. (1999). Bcl10 is involved in t(1;14)(p22;q32) of MALT B cell lymphoma and mutated in multiple tumor types. *Cell* **96,** 35–45.

Yoo, N. J., Park, W. S., Kim, S. Y., Reed, J. C., Son, S. G., Lee, J. Y., and Lee, S. H. (2002). Nod1, a CARD protein, enhances pro-interleukin-1 beta processing through the interaction with procaspase-1. *Biochem. Biophys. Res. Commun.* **299,** 652–658.

Yuan, J., Shaham, S., Ledoux, S., Ellis, H. M., and Horvitz, H. R. (1993). The *C. elegans* cell death gene *ced-3* encodes a protein similar to mammalian interleukin-1 beta-converting enzyme. *Cell* **75,** 641–652.

Zhang, H., and Fu, W. (2002). NDPP1 is a novel CARD domain containing protein which can inhibit apoptosis and suppress NF-kappaB activation. *Int. J. Oncol.* **20,** 1035–1040.

9

TRAIL IN THE AIRWAYS

NOREEN M. ROBERTSON,* MARY ROSEMILLER,*
ROCHELLE G. LINDEMEYER,[†]
ANDRZEJ STEPLEWSKI,* JAMES G. ZANGRILLI,[‡]
AND GERALD LITWACK*,**

*Department of Biochemistry and Molecular Pharmacology
Jefferson Medical College, Thomas Jefferson University
Philadelphia, Pennsylvania 19107
**and Toluca Lake, California 91602
[†]Department of Pediatric Dentistry, School of Dental Medicine
University of Pennsylvania, Philadelphia, Pennsylvania 19104
[‡]Division of Critical Care, Pulmonary, Allergic, and Immunologic Diseases
Thomas Jefferson University, Philadelphia, Pennsylvania 19107

    I. Introduction
   II. TRAIL: Synthetic and Signaling Pathways
      A. TRAIL Expression
      B. Mechanisms of TRAIL Activation
  III. Effects of TRAIL in the Airways
      A. Physiologic Effects of TRAIL
      B. Effects of TRAIL in Asthma
  IV. Conclusions
     References

Tumor necrosis factor–related apoptosis inducing ligand (TRAIL) is an important immunomodulatory factor that may play a role in the structural changes observed in the asthmatic airways. *In vitro* as well as *in vivo* studies have evidenced a dual role for TRAIL: it can either function as a pro- or anti-inflammatory cytokine on inflammatory cells, participating in the initiation and resolution of inflammatory and immune responses. TRAIL is expressed in the airways by inflammatory cells infiltrated in the bronchial mucosa, as well as by structural cells of the airway wall including fibroblasts, epithelial, endothelial, and smooth muscle cells. By releasing TRAIL, these different cell types may then participate in the increased levels of TRAIL observed in bronchoalveolar lavage fluid from asthmatic patients. Taken together, this suggests that TRAIL may play a role in inflammation in asthma. However, concerning its role is dual in the modulation of inflammation, further studies are needed to elucidate the precise role of TRAIL in the airways. © 2004 Elsevier Inc.

I. INTRODUCTION

Tumor necrosis factor–related apoptosis inducing ligand (TRAIL) is a member of the TNF superfamily that consists of 18 ligands and 28 receptors (Wiley *et al.*, 1995). TRAIL is a type II membrane protein that induces apoptosis in a variety of target cells. In some cell types, TRAIL binding to DR4, DR5, and/or DcR2 may also activate NFκB, leading to transcription of genes known to antagonize the death signaling pathway and/or to promote inflammation (Jeremias and Debatin, 1998; Schneider *et al.*, 1997). TRAIL may regulate cellular processes related to pulmonary physiology and pathology. This immunomodulatory factor may be involved in the structural changes that occur in asthmatic airways, including subepithelial fibrosis and the presence of inflammatory infiltrates. This review examines the signaling pathways of TRAIL especially in the airways, and suggests that it plays a role in inflammatory diseases, especially in asthma.

II. TRAIL: SYNTHETIC AND SIGNALING PATHWAYS

A. TRAIL EXPRESSION

TRAIL mRNA has been found in a variety of tissues and a wide range of cells from the immune system, including T cells, dendritic cells, NK cells, macrophages and monocytes (Wiley *et al.*, 1995). In studies with mice and humans, freshly isolated T cells, NK cells, B cells, dendritic cells, or monocytes do not express a detectable level of TRAIL on their surface (Fanger *et al.*, 1999; Griffith *et al.*, 1999; Kayagaki *et al.*, 1999b; Smyth *et al.*, 2001; Takeda *et al.*, 2001).

As shown *in vivo* with immunohistochemistry, TRAIL is expressed in the airways by various cell types: fibroblasts, epithelial, endothelial, and smooth muscle cells and is also expressed by infiltrated inflammatory cells, such as eosinophils and alveolar macrophages (Robertson et al., 2002). Its primary location in normal human lungs appears to be the bronchial epithelium and airway smooth muscle cell. Thus, TRAIL may be produced by numerous cell types in the airways; both structural and infiltrated inflammatory cells. Therefore, the TRAIL that is released can have a paracrine or autocrine influence on nearby cells.

1. TRAIL Promoter

The TRAIL gene is located on chromosome 3 at position 3q26, which is not close to other TNF family members (Wang et al., 2000b). The genomic structure of the TRAIL gene spans approximately 20 kb and is composed of five exonic segments of 222, 138, 42, 106, and 1245 nucleotides and four introns of approximately 8.2, 3.2, 2.3, and 2.3 kb. In contrast to FASL and tumor necrosis factor (TNF)-α, TRAIL lacks TATA and CAAT boxes. The promoter region contains putative responsive elements for GATA, AP-1, C/EBP, SP-1, OCT-1, AP3, PEA3, CF-1, and ISRE (Wang et al., 2000b). However, the role of these transcription regulatory factor binding sites for promoter activity awaits further investigation.

2. Regulation of TRAIL Expression

Cytokines affect TRAIL expression. For example, inflammatory cells such as human peripheral blood T lymphocytes, CD11c$^+$ DC, and monocytes express TRAIL, after activation by type 1 interferons (Almasan and Ashkenazi, 2003; Fanger et al., 1999; Santini et al., 2000; Yu et al., 2002). TRAIL is highly expressed on most NK cells following stimulation with IL-2, interferons, or IL-15 (Kayagaki et al., 1999a,b; Sato et al., 2001; Zamai et al., 1998). Moreover, stimulation of human lung fibroblasts with TRAIL induces the release of transforming growth factor (TGF)-β (Yurovsky, 2003). The TGF-β–induced release of proinflammatory cytokines in the airways by inflammatory cells may thus enable nearby structural cells such as lung fibroblasts and bronchial, epithelial, and smooth muscle cells, to produce additional TGF-β. The airways contain several potential sources of TRAIL and its expression can be modulated by different mediators, including pro- and anti-inflammatory compounds.

3. TRAIL Expression in Asthma

The recent description of increased TRAIL expression in asthma suggests that this cytokine plays a role in asthmatic airways (Robertson et al., 2002). We observed the overexpression of TRAIL in bronchial biopsy specimens from subjects with moderate asthma, compared to controls without asthma. This increased expression was particularly elevated in airway smooth muscle and epithelial cells in bronchial biopsy samples and infililtrated inflammatory cells from subjects with asthma especially in activated eosinophils (Robertson et al., 2002).

We reported increased TRAIL levels in the bronchoalveolar lavage (BAL) fluid of stable atopic asthmatics than in that of control subjects (Robertson et al., 2002). Moreover, we examined TRAIL levels 24 h after a segmental antigen bronchoprovocation and found that they are higher in the BAL of patients with mild asthma after allergen exposure than before. We therefore proposed that TRAIL is involved in the late phase response to allergen. TRAIL may thus play a role in the events preceding an asthma attack, as the airway prepares for bronchoconstriction. TNF-α has been implicated in airway remodeling and fibrosis (Broide et al., 1992; Hakonarson et al., 2001; Sato et al., 1997). Moreover, a polymorphism in the TNF-α promoter resulting in its increased production has also been linked to asthma in genotypic studies (Winchester et al., 2000).

B. MECHANISMS OF TRAIL ACTIVATION

TRAIL is composed of 281 amino acids, and has the characteristics of a type II membrane protein (i.e., no leader sequence and an internal transmembrane domain) (Pitti et al., 1996; Wiley et al., 1995). TRAIL has an N-terminal cytoplasmic domain, which is not conserved across family members and extracellular C-terminal domains that are highly conserved, and can be proteolytically cleaved from the cell surface. TRAIL forms a homotrimer that binds three receptor molecules.

TRAIL action is mediated through binding to four specific transmembrane receptors. DR4 (TRAIL-RI) and DR5 (TRAIL-RII) contain a functional death domain and can therefore signal for apoptosis (Ashkenazi and Dixit, 1998; Pan et al., 1997; Pitti et al., 1996; Sheridan et al., 1997; Walczak et al., 1997). Two receptors, DcR1 (TRAIL-R3), DcR2 (TRAIL-R4), act as TRAIL-neutralizing decoy receptors. DcR1 lacks an intracytoplasmic domain, thereby sequestering TRAIL, abolishing its ability to transmit apoptotic signaling through its death receptors (Degli-Esposito et al., 1997a,b). DcR2 exhibits high sequence homology to the extracellular domains of DR4, DR5, and DcR1; however, its cytoplasmic domain contains a truncated death domain (Degli-Esposito et al., 1997a,b). Unlike DcR1, DcR2 has a functional intracellular signaling domain, and therefore protects cells from apoptosis by either acting as a decoy receptor or transmitting antiapoptotic signals. Stimulation of DcR2 by TRAIL activates the transcription factor NF-κB and thereby prevents apoptosis. It has been proposed that the expression of TRAIL receptors may regulate a cell's sensitivity to TRAIL. Specifically, the presence or absence of TRAIL decoy receptors may determine whether a cell is resistant or sensitive to TRAIL stimulation (Ashkenazi and Dixit, 1999; Degli-Esposti, 1999).

TRAIL induces its cellular responses through the formation of receptor complexes. DR4 and DR5 induce apoptosis through the binding of the adaptor protein FADD to their death domains. Caspase-8 and/or -10 is activated by FADD recruitment to DRs. This activation causes the cleavage of specific intracellular proteins leading to the induction of apoptosis. The BCL2 family member BID is cleaved by caspase-8. On cleavage, truncated BID translocates to the mitochondria contributing to the release of cytochrome c, which leads to caspase-3 cleavage and apoptosis (Almasan and Ashkenazi, 2003).

III. EFFECTS OF TRAIL IN THE AIRWAYS

A. PHYSIOLOGIC EFFECTS OF TRAIL

1. TRAIL and Fibroblasts

Stimulation of lung fibroblasts by TRAIL may play a role in the regulation of the extracellular matrix. At high concentrations, TRAIL induces apoptosis. In contrast, lower concentrations stimulate lung fibroblast proliferation (Yurovsky, 2003). In addition, TRAIL stimulates the production of TGF-β to induce the production of matrix components, such as collagen (Coker et al., 1997; Fine and Goldstein, 1987; Krupsky et al., 1996; Yurovsky, 2003). TGF-β also enhances matrix growth factors, such as connective tissue growth factor (Kucich et al., 2001; Ricupero et al., 1999). It depresses the synthesis of matrix-degrading proteases and upregulates the production of protease inhibitors, in particular tissue inhibitor of metalloproteinase-1 (TIMP-1) and plasminogen activator inhibitor-1 (PAI-1) (Edwards et al., 1987; Eickelberg et al., 1999; Laiho et al., 1986).

DNA microarray hybridization was used to evaluate changes in mRNA levels in lung fibroblasts treated for 24 h with TRAIL (Yurovsky, 2003). The expression of genes involved in signal transduction, DNA transcription, and tissue remodeling were significantly increased. Upregulated genes included JNK1, transcription factors Sp3 and c-fos, collagen types I through VI and XV, fibronectin, laminin, and elastin (Yurovsky, 2003). Moreover, TRAIL increased the transcription of regulators of extracellular matrix synthesis: TGF-β1, TGF-β2, connective tissue growth factor, TGF-β binding protein 1 and 2, insulin-like growth factor 2 receptor, thrombospondins 1 and 2, and the accessory TGF-β receptors (endoglin and betaglycan) (Yurovsky, 2003). These results suggest that TRAIL plays a physiological role in the regulation of lung fibroblast in the airways.

2. TRAIL and the Epithelium

The direct effects of TRAIL on airway epithelium are presently unknown. The ability of TRAIL to stimulate production of TGF-β in fibroblasts suggests that TRAIL can enhance extracellular matrix synthesis in fibroblasts by triggering TGF-β production that acts in an autocrine manner and may potentially act in a paracrine manner in epithelial cells. TGF-β is involved in controlling the growth and differentiation of epithelial cells. TGF-β induces squamous differentiation in normal human bronchial epithelial cells, while it regulates cell growth through inhibiting cell proliferation and may also induce apoptosis of normal human lung epithelial cells (Bogdanowicz and Pujol, 2000; Masui et al., 1986; Yanagisawa et al., 1998). Structural cells stimulated with TGF-β may participate in the inflammation observed in asthmatic airways (Ariazi et al., 1999). TGF-β treatment enhances the in vitro release of proinflammatory mediators from airway structural cells, such as IL-1 from epithelial cells and eotaxin from fibroblasts (Kumar et al., 1996; Wenzel et al., 2002). Moreover, RANTES and GM-CSF, which are involved in the chemotaxis of eosinophils, lymphocytes or monocytes, and their expression is also enhanced after TGF-β treatment of airway epithelial cells (Adachi et al., 1996; Minshall et al., 1997).

3. TRAIL and Vessels

TRAIL is expressed by the medial smooth cell layer of aorta and pulmonary artery (Secchiero et al., 2003). Moreover, endothelial cells express the mRNA for all TRAIL receptors. Secchiero et al. reported that in vascular endothelial cells, TRAIL stimulates the phosphorylation of the serine/threonine kinase, Akt in a PI3K-dependent manner. Activation of Akt-protected cells from apoptosis induced by trophic withdrawal. TRAIL also stimulated the ERK 1/2 but not the p38 or the JNK pathways. Moreover, TRAIL increased endothelial cell proliferation in an ERK-dependent manner. TRAIL did not activate NF-κB or affect the surface expression of E-selectin, intercellular adhesion molecule-1, and vascular cell adhesion molecule-1 (Secchiero et al., 2003). In contrast, Li et al. observed that TRAIL treatment of endothelial cells induced apoptosis (30%) and NF-κB dependent activation and inflammatory gene expression that results in adhesion of leukocytes (Li et al., 1998). In vivo, TRAIL induced endothelial cell injury. The basis of these differences may be attributed to the state of the endothelial cell cultures or the potency of the TRAIL preparations.

4. TRAIL and Airway Smooth Muscle

Airway wall inflammation is a central feature to the pathophysiology of asthma, and the induction and perpetuation of inflammation involves a complex and coordinated response of multiple inflammatory cells, mediators, and cytokines (Hallsworth et al., 2001). Evidence from our studies and others suggests that human airway smooth muscle (HASM) cells, in addition to their traditional role as regulators of airway tone, can contribute to the pathogenesis of asthma by expressing and secreting a host of glucocorticoid-sensitive genes that contribute to the altered structure and function of the airways in the asthmatic state (Ammit *et al.*, 2002; Chung, 2000; Hakonarson *et al.*, 2001; Hallsworth *et al.*, 1998; Robertson *et al.*, 2002).

TNF-α, IL-11, IL-6, RANTES, GM-CSF, IL-4, IL-5, IL-13, and other cytokines have been found in increased amounts in the bronchoalveolar lavage of asthmatic patients and have been implicated in allergic inflammatory processes (Broide *et al.*, 1992; Lukacs *et al.*, 1999). TNF-α, a pleiotropic cytokine, has been implicated in the pathophysiology of a host of proinflammatory disorders, including bronchial asthma. Recently, TNF-α 308 allele2 was found to be significantly associated with self-reported childhood asthma and with bronchial hyperreactivity in atopic and nonatopic asthmatic patients (Winchester *et al.*, 2000). Increased release of TNF-α from inflammatory cells is thought to cause proliferation of HASM and upregulation of ICAM-1 and VCAM adhesion molecules, favoring interaction with T cells (Muller, 2002).

One of the hallmarks of airway inflammation is the increased survival of infiltrating eosinophils (Colavita *et al.*, 2000; Shaver *et al.*, 1997; Zangrilli *et al.*, 1995). Physiologically, the role of eosinophils is host defense, particularly in responding to parasitic infection (Giembycz and Lindsay, 1999). Eosinophils are equipped with a range of cytotoxic proteins and enzymes able to generate reactive oxygen metabolites with the intent of killing parasites. In asthma, the inappropriate recruitment of eosinophils results in persistent tissue damage (Barnes, 1994; Gleich, 2000). We reported that TRAIL significantly enhances the survival of cultured eosinophils from allergic/asthmatic patients. Increased levels of TRAIL have been seen in bronchial biopsy and bronchoalveolar lauage fluid (BAL) from asthmatic patients (Robertson *et al.*, 2002).

a. Modulation of TRAIL Expression in Airway Smooth Muscle Cells

We investigated whether HASM cells are a source of TRAIL production. TRAIL expression levels were analyzed with reverse transcriptase–polymerase chain reaction (RT-PCR) after treatment with cytokines (Fig. 1). The RT-PCR data indicate that clear differences exist in the relative abundance of TRAIL mRNA in HASM cells following treatment with cytokines and glucocorticoid (GC). Thus, cytokines and GC either transcriptionally regulate the expression of TRAIL in HASM cells or alter the stability or half-life of the mRNA.

Flow cytometry was used to determine whether these treatments induced changes in the levels of TRAIL protein expression. TRAIL expression levels were examined by flow cytometry on permeabilized HASM cells as shown in Figure 2. Cytokine and glucocorticoid (triamcinolone acetonide) treatments produced modest changes in total TRAIL protein that parallel the changes seen in mRNA expression. TNF-α, IFN-β, and IFN-γ treatments all induced TRAIL (10–30%), while TA treatment decreased total TRAIL levels (10%). The modulation of TRAIL protein correlates with the pattern of mRNA expression seen with cytokine and GC treatment. Both TRAIL mRNA and protein levels in HASM cells increase following stimulation

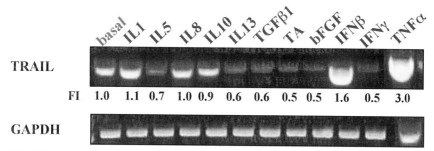

FIGURE 1. RT-PCR analysis of TRAIL mRNA in HASM cells following cytokine or GC stimulation. After growth arrest for 24 h, confluent HASM cells were treated with cytokines (10 ng/ml) or glucocorticoid (TA 100 nM) for 18 h. Total RNA was extracted and TRAIL mRNA expression was examined by RT-PCR. Data are presented as fold-induction (FI) in reference to basal levels which were set at 1.0 after normalization of expression using GAPDH mRNA.

with the proinflammatory cytokines, TNF-α, IFN-β, and IFN-γ, and decrease with GC.

b. Glucocorticoid Inhibits TNF-α–Induced TRAIL mRNA and Protein

To assess whether TNF-α–induced changes in TRAIL mRNA and protein are glucocorticoid sensitive, time-course experiments were performed with TNF-α–treated and untreated HASM cells in the presence or absence of TA. As shown in Figure 3, all treatments affected basal levels of TRAIL mRNA after 4 h. TA decreased TRAIL mRNA 20% below basal levels. TNF-α caused an 80% increase while the combination of TNF-α and TA reduced the induction seen with TNF alone to just 10% above basal levels. An enzyme-linked immunosorbent assay performed with HASM cell lysates confirmed that TNF-α plus TA reduced TNF-α–induced TRAIL protein to just below basal levels as shown in Figure 4. It has been shown that supernatants of anti-CD59 mAb-induced activated T cells are cytotoxic against Jurkat cells and that this cytotoxicity is due to TRAIL release. PHA activation resulted in the secretion of internal microvesicles loaded with functional TRAIL molecules able to retain their multimerization ability. This ability to release bioactive death ligands must be considered to understand the precise role these death messengers play in diverse physiologic and pathologic situations. Their potential apoptotic ability would then depend not only on the balance between cell-associated and proteolyzed forms but also on the capacity of a given cell to synthesize and secrete TRAIL-containing microvesicles. More studies are needed to characterize both the sorting and signal transduction pathways that trigger the differential release of TRAIL.

c. Transcriptional Regulation of the TRAIL Promoter

Transcriptional regulation of the TRAIL promoter was investigated in HASM cells transiently transfected with a reporter gene construct containing the full-length TRAIL promoter upstream of the luciferase gene

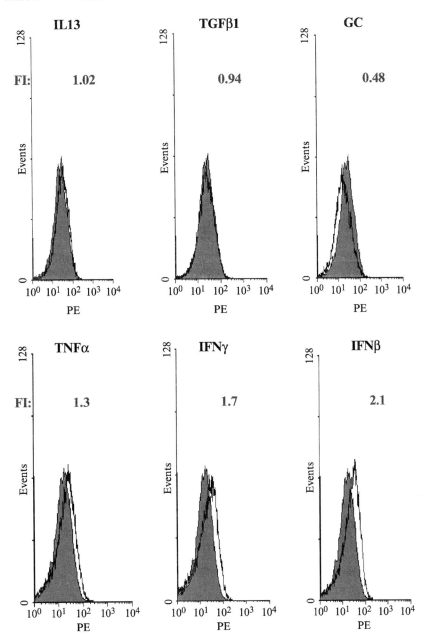

FIGURE 2. Flow cytometry analysis of TRAIL in permeabilized cells after cytokine and GC treatment. Confluent HASM cells were growth arrested, treated with indicated cytokines or glucocorticoid (GC) for 18 h, harvested, permeabilized with saponin and stained with anti-TRAIL monoclonal antibody. Solid histograms represent basal levels of total TRAIL staining. Open histograms of TRAIL staining in cytokine/GC treated HASM are overlaid. Data are represented as fold induction (FI) of TRAIL staining in treated cells compared to basal levels in untreated HASM. The results shown are representative of at least three studies performed on HASM cultures obtained from different patients.

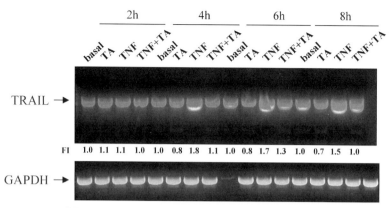

FIGURE 3. RT-PCR analysis of TRAIL mRNA in a TNF-α and glucocorticoid timecourse experiment. After growth arrest for 24 h, cells were treated with TNF-α (10 ng/ml), TA (100 nM), or TNF-α + TA for 2, 4, 6, and 8 h. Total RNA was extracted and TRAIL mRNA expression was examined by RT-PCR. Data are presented as fold-induction (FI) in reference to basal levels that were set at 1.0. mRNA expression was normalized using GAPDH. An extra well (*) was included in GAPDH as a control for genomic DNA contamination in the RT-PCR reaction.

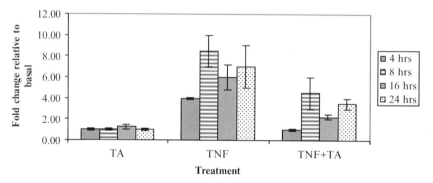

FIGURE 4. Time-course of transcriptional regulation of TRAIL by TNF-α and glucocorticoid. The TRAIL-luciferase reporter construct and β-galactosidase vector were transiently co-transfected into HASM cells and treated with TNF-α (10 ng/ml), TA (100 nM) or TNF-α + TA for the indicated times. Cell lysates were tested for luciferase activity. Results are expressed as fold change of luciferase activity in treated lysates relative to untreated basal levels. They represent average values of luciferase activity of triplicate samples normalized for transfection efficiency. The values correspond to the mean of three independent transfections and error bars indicate the standard error.

in the pGL3 vector. Promoter activity was induced by TNF-α, IFN-β, and IL1 above basal levels (data not shown). A time-course study (4–24 h) showed that TNF-α–induced TRAIL promoter activity peaked at 8 h and remained elevated as late as 24 h after addition of the cytokine as shown in

Figure 4. TA had no direct effect on TRAIL promoter activity, but decreased TNF-α–induced activation at every time-point.

This analysis of TRAIL promoter activity in HASM cells demonstrates that TNF-α stimulates an increase in TRAIL expression. While GC had no direct effect on the TRAIL promoter, it was able to minimize the effects of TNF-α in a dose-dependent manner. GCs exert their effects by binding and activating the cytosolic GC receptor (GR), which dimerizes and translocates to the nucleus. Activated GR can alter gene transcription directly, (cis) by binding to a glucocorticoid response element (GRE) on a promoter or indirectly, (trans) by binding to other transcription factors (Wang et al., 2000a). Recently it has been shown that the activated GR acts in part by recruiting corepressor proteins containing histone deacetylases (HDACs) to the site of active inflammatory gene transcription. HDACs then deacetylate histones that were previously acetylated by NFκB activation, thereby suppressing inflammatory gene expression (Ito et al., 2002). Because the TRAIL promoter contains no GRE, GR's effect is expected to be an indirect one (Gong and Almasan, 2000). AP-1 binding in the lung has been shown to be significantly repressed by GC treatment (Adcock et al., 1994). It has been suggested that GCs exert an inhibitory effect on TNF-α–induced RANTES expression by transrepression of AP-1 (Ammit et al., 2000). The TRAIL promoter also contains an AP-1 site, which may account for the observed induction of TRAIL by TNF-α and its transrepression by GC.

We observed that TRAIL treatment had no effect on HASM cell viability or proliferation (data not shown). Therefore, TRAIL seems to have no autocrine effect on HASM cells in terms of inducing apoptosis or affecting survival. We and others have reported previously that TRAIL is able to increase survival of eosinophils *in vitro* (Daigle and Simon, 2001; Robertson et al., 2002). HASM cells enhance eosinophil survival after IL1β treatment, an effect that is blocked by GC. Proinflammatory cytokine-induced TRAIL secretion by HASM cells may therefore prolong eosinophil survival and contribute to the chronic inflammation that is typical of asthma. There is evidence that TRAIL plays a role in the early elimination of virus-infected epithelial cells in the normal gut (Wang et al., 2002a,b). The increased levels of TRAIL found in the asthmatic airway could therefore play a role in the destruction of the lung epithelial layer as it has long been suspected that there is an infectious component to asthma (Bucchieri et al., 2002).

B. EFFECTS OF TRAIL IN ASTHMA

Asthma is a disease characterized by bronchial inflammation, reversible airflow obstruction, and tissue remodeling of the airway wall. Inflammatory and structural cells are important because they release proinflammatory mediators and thus induce an acute inflammatory response that includes increased mucus secretion, epithelial shedding and airway narrowing (McKay and Sharma, 2002; Wardlaw et al., 2002). Cytokines play a key role in the coordination and persistence of the chronic inflammatory process in asthma. The airway inflammation that occurs after repeated exposure to allergens or during the late-phase reaction results from a complex network of interactions between inflammatory cells (mast cells, eosinophils, macrophages, and activated T and B cells) and cells comprising the lung structure (endothelial cells, fibroblasts, bronchial epithelial cells, and smooth muscle cells) (Dunill, 1960; Poston et al., 1995). The proinflammatory role of eosinophils has been revealed in bronchial biopsy specimens, BAL fluid, and peripheral blood of patients with mild-to-moderate asthma, and the degree of blood eosinophilia has been correlated with disease severity (Colavita et al., 2000; Motojima et al., 1996; Poston et al., 1995; Zangrilli et al., 1995).

Eosinophils mediate some of their unique cytotoxic and inflammatory functions by the production, storage, and regulated release of their granule proteins contributing to tissue damage on bronchial epithelium and secretion of various cytokines (Adachi et al., 1995; Bousquet et al., 1994; Seminarino and Gleich, 1994). Thus, there is great interest in defining the mechanisms by which the selective accumulation, activation, and maintenance of eosinophils at the site of inflammation in allergic type responses are regulated. Known mechanisms of this complex series of events include selective adhesion pathways, specific chemotactic factors, and enhanced survival by certain cytokines, principally the T-cell derived eosinopoietins, IL-5, IL-3, and GM-CSF (Lopez et al., 1986; Rothenberg et al., 1988; Rothenberg et al., 1989). In addition, eosinophils secrete cytokines and chemokines including TGF-α and -β, TNF-α and -β, macrophage inflammatory protein-1α (MIP-1α), IL-1, IL-3, IL-5, IL-6, IL-8, and GM-CSF. The secretion of these factors may continue eosinophil participation in the inflammatory response, in part, by promoting their survival.

Simultaneously, the release of chemokines and growth factors causes airway inflammatory infiltrate to persist in asthmatic airways and thus induces structural changes within the airway wall, including subepithelial fibrosis accompanied by thickening of the basement membrane, increased vascularity, edema, increased collagen and matrix deposition, and smooth muscle hypertrophy and hyperplasia (Carroll et al., 1997; Wilson and Bamford, 2001). The combination of chronic airway inflammation and remodeling increases the thickness of the airway wall and exaggerates airway narrowing.

TRAIL can function either as a proinflammatory or an anti-inflammatory cytokine, as its effects on inflammatory cells show. Its proinflammatory effect is exerted, for example, by prolonging the survival of eosinophils. Its anti-inflammatory effects on inflammatory cells have also been demonstrated. TRAIL expression was characterized on IFN-α–stimulated peripheral blood T cells and IFN-stimulated human monocytes and dendritic cells (Almasan and Ashkenazi, 2003; Fanger et al., 1999; Santini et al., 2000; Yu et al., 2002). The monocytes become resistant to TRAIL. IL-2 stimulation induces TRAIL in human NK cells (Kayagaki et al., 1999b). Moreover, CD4+ T cells that expressed TRAIL were capable of mediating apoptosis via the TRAIL signaling pathways.

In a recent study, we demonstrated that in the allergic asthmatic subjects there was an increased level of TRAIL in BAL fluid after SAC, with levels dramatically increasing at 24 h, and for the most part returning to baseline at two weeks (Robertson et al., 2002). In allergic asthmatic subjects, increases in total cells and eosinophils were marked on day 2 and persisted through day 9. TRAIL concentrations were higher in BAL from the allergen challenged asthmatic subjects and strongly correlated with eosinophil influx, particularly on day 2 after challenge. Bronchial biopsy specimens obtained throughout the SAC protocol further revealed that both airway smooth muscle and epithelial cells of asthmatics express increased levels of TRAIL. Both BAL eosinophils and alveolar macrophages from asthmatic patients express higher levels of TRAIL than BAL cells from normal subjects. This suggests that the sources of TRAIL are likely to be local production by both airway smooth muscle and epithelial cells with additional contribution from infiltrating inflammatory cells.

Presently, the *in vivo* role of TRAIL remains unknown. We demonstrated that peripheral blood eosinophils cultured in the presence of exogenous TRAIL exhibit prolonged viability *ex vivo*. Thus, TRAIL appears to inhibit the spontaneous apoptosis of peripheral blood eosinophils induced by growth factor withdrawal. Similarly, we found that BAL fluid concentrates supported prolonged peripheral blood eosinophil viability, and that this enhanced viability was reduced by anti-TRAIL antibody (Robertson et al., 2002).

Once an eosinophil has exited the circulation, its continued presence in the tissue space as a viable effector depends upon the balance between the cell's natural tendency to undergo apoptosis and the local eosinophil-viability enhancing activity. This accumulation of eosinophils is regulated by the generation of survival and activation factors (i.e., the type I hematopoietic cytokines: IL-3, IL-5, and GM-CSF and several members of the IL-2 family of cytokines, IL-9, IL-13, and IL-15. Using microarray technology, Temple et al. identified candidate genes involved in eosinophil survival and apoptosis in peripheral blood eosinophils and IL-5 dependent TF1.8 cells. Interestingly, following IL-5 withdrawal, TRAIL expression was downregulated more than twofold in TF1.8 cells, supporting a role for TRAIL in survival (Temple et al., 2001).

BAL cells migrating into the airway of allergic asthmatic subjects post challenge were examined by immunocytochemistry for the expression of TRAIL and TRAIL receptors and compared with control nonasthmatic subjects (Robertson *et al.*, 2002). In the asthmatic subjects, at baseline both TRAIL and DcR2 were expressed by alveolar macrophages/monocytes, the predominant cell type present. Twenty-four hours after antigen challenge, eosinophils migrate into the asthmatic airway and the expression levels of both TRAIL and DcR2 on eosinophils and alveolar macrophages were significantly increased. We also observed that BAL cells from non-asthmatic subjects express TRAIL and DcR2 in alveolar macrophages, which suggests a role in the innate immune response. These data suggested that the differential expression of TRAIL and TRAIL receptor(s) and their interactions in the asthmatic airway may play a role in modulating eosinophil survival, thereby prolonging the inflammatory response.

IV. CONCLUSIONS

Airway wall inflammation is a central feature to the pathophysiology of asthma, and the induction and perpetuation of inflammation involves a complex and coordinated response of multiple inflammatory cells, mediators and cytokines. We have demonstrated that TRAIL is directly inducible in primary human airway smooth muscle cells by TNF-α, IFN-β, and IFN-γ, cytokines that play a role in asthma. The differential regulation of TRAIL expression in asthma in response to increased levels of proinflammatory cytokines may represent a pathologic response to a normal immune mechanism.

The maintenance of hematopoietic cell homeostasis by apoptosis is a critical regulatory mechanism in the normal immune response. Because inflammatory cells infiltrate the mucosa of asthmatic airways, it is likely that the increased levels of TRAIL reported in inflamed airways in asthma are involved in their survival and activation. It may thus participate in the inflammation occurring in asthmatic airways, by increasing the number of inflammatory cells and the time they spend there. Increased generation of TRAIL in the airway could be a key factor in promoting eosinophil survival, and prolonging injury in allergic asthmatic subjects after antigen exposure.

REFERENCES

Adachi, T., Motojima, S., Hirata, A., Fukuda, T., Kihara, N., and Makino, S. (1996). Detection of transforming growth factor-beta in sputum from patients with bronchial asthma by eosinophil survival assay and enzyme-linked immunosorbent assay. *Clin. Exp. Allergy* **26**, 557–562.

Adachi, T., Motojima, S., Hirata, A., Fukuda, T., and Makino, S. (1995). Eosinophil-viability enhancing activity in sputum from patients with bronchial asthma. Contributions of interleukin-5 and granulocyte/macrophage colony-stimulating factor. *Am. J. Respir. Crit. Care. Med.* **151**, 618–623.

Adcock, I. M., Brown, C. R., Shirasaki, H., and Barnes, P. J. (1994). Effects of dexamethasone on cytokine and phorbol ester stimulated c-Fos and c-Jun DNA binding and gene expression in human lung. *Eur. Respir. J.* **7,** 2117–2123.

Almasan, A., and Ashkenazi, A. (2003). Apo2L/TRAIL: Apoptosis signaling, biology, and potential for cancer therapy. *Cytokine Growth Factor Rev.* **14,** 337–348.

Ammit, A. J., Hoffman, R. K., Amrani, Y., Lazaar, A. L., Hay, D. W., Torphy, T. J., Penn, R. B., and Panettieri, R. A., Jr. (2000). Tumor necrosis factor-alpha-induced secretion of RANTES and interleukin-6 from human airway smooth-muscle cells. Modulation by cyclic adenosine monophosphate. *Am. J. Respir. Cell. Mol. Biol.* **23,** 794–802.

Ammit, A. J., Lazaar, A. L., Irani, C., O'Neill, G. M., Gordon, N. D., Amrani, Y., Penn, R. B., and Panettieri, R. A., Jr. (2002). Tumor necrosis factor-alpha-induced secretion of RANTES and interleukin-6 from human airway smooth muscle cells: Modulation by glucocorticoids and beta-agonists. *Am. J. Respir. Cell. Mol. Biol.* **26,** 465–474.

Ariazi, E. A., Satomi, Y., Ellis, M. J., Haag, J. D., Shi, W., Sattler, C. A., and Gould, M. N. (1999). Activation of the transforming growth factor beta signaling pathway and induction of cytostasis and apoptosis in mammary carcinomas treated with the anticancer agent perillyl alcohol [in process citation]. *Cancer Res.* **59,** 1917–1928.

Ashkenazi, A., and Dixit, V. M. (1998). Death receptors: Signaling and modulation. *Science* **281,** 1305–1308.

Ashkenazi, A., and Dixit, V. M. (1999). Apoptosis control by death and decoy receptors. *Curr. Opin. Cell. Biol.* **11,** 255–260.

Barnes, P. (1994). Cytokines as mediators of chronic asthma. *Am. J. Respir. Crit. Care Med.* **150,** S42–S49.

Bogdanowicz, P., and Pujol, J. P. (2000). Glycosylphosphatidylinositol (GPI) hydrolysis by transforming growth factor-beta1 (TGF-beta1) as a potential early step in the inhibition of epithelial cell proliferation. *Mol. Cell. Biochem.* **208,** 143–150.

Bousquet, J., Chanez, P., Vignola, A. M., Lacoste, J.-Y., and Michel, F. B. (1994). Eosinophil inflammation in asthma. *Am. Rev. Crit. Care Med.* **150,** S33–S38.

Broide, D. H., Lotz, M., Cuomo, A. J., Coburn, D. A., Federman, E. C., and Wasserman, S. I. (1992). Cytokines in symptomatic asthma airways. *J. Allergy Clin. Immunol.* **89,** 958–967.

Bucchieri, F., Puddicombe, S. M., Lordan, J. L., Richter, A., Buchanan, D., Wilson, S. J., Ward, J., Zummo, G., Howarth, P. H., Djukanovic, R., Holgate, S. T., and Davies, D. E. (2002). Asthmatic bronchial epithelium is more susceptible to oxidant-induced apoptosis. *Am. J. Respir. Cell. Mol. Biol.* **27,** 179–185.

Carroll, N. G., Cooke, C., and James, A. L. (1997). Bronchial blood vessel dimensions in asthma. *Am. J. Respir. Crit. Care Med.* **155,** 689–695.

Chung, K. F. (2000). Airway smooth muscle cells: Contributing to and regulating airway mucosal inflammation? *Eur. Respir. J.* **15,** 961–968.

Coker, R. K., Laurent, G. J., Shahzeidi, S., Lympany, P. A., du Bois, R. M., Jeffery, P. K., and McAnulty, R. J. (1997). Transforming growth factors-beta 1, -beta 2, and -beta 3 stimulate fibroblast procollagen production in vitro but are differentially expressed during bleomycin-induced lung fibrosis. *Am. J. Pathol.* **150,** 981–991.

Colavita, A. M., Hastie, A. T., Musani, A. J., Pascual, R. M., Reinach, A. J., Lustine, H. T., Galati, S. A., Zangrilli, J. G., Fish, J. E., and Peters, S. P. (2000). Kinetics of IL-10 production after segmental antigen challenge of atopic asthmatic subjects. *J. Allergy Clin. Immunol.* **106,** 880–886.

Daigle, I., and Simon, H. U. (2001). Alternative functions for TRAIL receptors in eosinophils and neutrophils. *Swiss Med. Wkly.* **131,** 231–237.

Degli-Esposito, M. A., Dougall, W. C., Smolak, P. J., Waugh, J. Y., Smith, C. A., and Goodwin, R. G. (1997a). The novel receptor TRAIL-R4 induces NF-κB and protects against TRAIL-mediated apoptosis, yet remains an incomplete death domain. *Immunity* **7,** 813–820.

Degli-Esposito, M. A., Smolak, P. J., Walczak, H., Waugh, J. Y., Huang, C. P., DuBose, R. F., Goodwin, R. G., and Smith, C. A. (1997b). Cloning and characterization of TRAIL-R3, a novel member of the emerging TRAIL receptor family. *J. Exp. Med.* **186**, 1165–1170.

Degli-Esposti, M. (1999). To die or not to die—the quest of the TRAIL receptors. *J. Leukoc. Biol.* **65**, 535–542.

Dunill, M. S. (1960). The pathology of asthma, with special reference to changes in the bronchial mucosa. *J. Clin. Pathol.* **13**, 27–31.

Edwards, D. R., Murphy, G., Reynolds, J. J., Whitham, S. E., Docherty, A. J., Angel, P., and Heath, J. K. (1987). Transforming growth factor beta modulates the expression of collagenase and metalloproteinase inhibitor. *EMBO J.* **6**, 899–904.

Eickelberg, O., Kohler, E., Reichenberger, F., Bertschin, S., Woodtli, T., Erne, P., Perruchoud, A. P., and Roth, M. (1999). Extracellular matrix deposition by primary human lung fibroblasts in response to TGF-beta1 and TGF-beta3. *Am. J. Physiol.* **276**, L814–L824.

Fanger, N. A., Maliszewski, C. R., Schooley, K., and Griffith, T. S. (1999). Human dendritic cells mediate cellular apoptosis via tumor necrosis factor-related apoptosis-inducing ligand (TRAIL). *J. Exp. Med.* **190**, 1155–1164.

Fine, A., and Goldstein, R. H. (1987). The effect of transforming growth factor-beta on cell proliferation and collagen formation by lung fibroblasts. *J. Biol. Chem.* **262**, 3897–3902.

Giembycz, M. A., and Lindsay, M. A. (1999). Pharmacology of the eosinophil. *Pharmacol. Rev.* **51**, 213–340.

Gleich, G. J. (2000). Mechanisms of eosinophil-associated inflammation. *J. Allergy Clin. Immunol.* **105**, 651–663.

Gong, B., and Almasan, A. (2000). Genomic organization and transcriptional regulation of human Apo2/TRAIL gene. *Biochem. Biophys. Res. Commun.* **278**, 747–752.

Griffith, T. S., Wiley, S. R., Kubin, M. Z., Sedger, L. M., Maliszewski, C. R., and Fanger, N. A. (1999). Monocyte-mediated tumoricidal activity via the tumor necrosis factor-related cytokine, TRAIL. *J. Exp. Med.* **189**, 1343–1354.

Hakonarson, H., Halapi, E., Whelan, R., Gulcher, J., Stefansson, K., and Grunstein, M. M. (2001). Association between IL-1beta/TNF-alpha-induced glucocorticoid-sensitive changes in multiple gene expression and altered responsiveness in airway smooth muscle. *Am. J. Respir. Cell. Mol. Biol.* **25**, 761–771.

Hallsworth, M. P., Moir, L. M., Lai, D., and Hirst, S. J. (2001). Inhibitors of mitogen-activated protein kinases differentially regulate eosinophil-activating cytokine release from human airway smooth muscle. *Am. J. Respir. Crit. Care Med.* **164**, 688–697.

Hallsworth, M. P., Soh, C. P., Twort, C. H., Lee, T. H., and Hirst, S. J. (1998). Cultured human airway smooth muscle cells stimulated by interleukin-1beta enhance eosinophil survival. *Am. J. Respir. Cell. Mol. Biol.* **19**, 910–919.

Ito, K., Lim, S., Caramori, G., Cosio, B., Chung, K. F., Adcock, I. M., and Barnes, P. J. (2002). A molecular mechanism of action of theophylline: Induction of histone deacetylase activity to decrease inflammatory gene expression. *Proc. Natl. Acad. Sci. USA* **99**, 8921–8926.

Jeremias, I., and Debatin, K. M. (1998). TRAIL induces apoptosis and activation of NFκB. *Eur. Cytokine Netw.* **9**, 687–688.

Kayagaki, N., Yamaguchi, N., Nakayama, M., Eto, H., Okumura, K., and Yagita, H. (1999a). Type I interferons (IFNs) regulate tumor necrosis factor-related apoptosis-inducing ligand (TRAIL) expression on human T cells: A novel mechanism for the antitumor effects of type I IFNs. *J. Exp. Med.* **189**, 1451–1460.

Kayagaki, N., Yamaguchi, N., Nakayama, M., Kawasaki, A., Akiba, H., Okumura, K., and Yagita, H. (1999b). Involvement of TNF-related apoptosis-inducing ligand in human CD4+ T cell-mediated cytotoxicity. *J. Immunol.* **162**, 2639–2647.

Krupsky, M., Fine, A., Kuang, P. P., Berk, J. L., and Goldstein, R. H. (1996). Regulation of type I collagen production by insulin and transforming growth factor-beta in human lung fibroblasts. *Connect Tissue Res.* **34**, 53–62.

Kucich, U., Rosenbloom, J. C., Herrick, D. J., Abrams, W. R., Hamilton, A. D., Sebti, S. M., and Rosenbloom, J. (2001). Signaling events required for transforming growth factor-beta stimulation of connective tissue growth factor expression by cultured human lung fibroblasts. *Arch. Biochem. Biophys.* **395**, 103–112.

Kumar, N. M., Rabadi, N. H., Sigurdson, L. S., Schunemann, H. J., and Lwebuga-Mukasa, J. S. (1996). Induction of interleukin-1 and interleukin-8 mRNAs and proteins by TGF beta 1 in rat lung alveolar epithelial cells. *J. Cell. Physiol.* **169**, 186–199.

Laiho, M., Saksela, O., Andreasen, P. A., and Keski-Oja, J. (1986). Enhanced production and extracellular deposition of the endothelial-type plasminogen activator inhibitor in cultured human lung fibroblasts by transforming growth factor-beta. *J. Cell. Biol.* **103**, 2403–2410.

Li, F., Ambrosini, G., Chu, E. Y., Plescia, J., Tognin, S., Marchisio, P. C., and Altieri, D. C. (1998). Control of apoptosis and mitotic spindle checkpoint by survivin. *Nature* **396**, 580–584.

Lopez, F. F., Williamson, D. J., Gamble, J. R., Begley, C. J., Harlan, J. M., Klebanoff, S. J., Waltersdorf, A., Wong, S., Clark, S. C., and Vadas, M. A. (1986). Recombinant granulocyte-macrophage colony-stimulating factor stimulates in vitro mature human neutrophil and eosinophil function, surface receptor expression, and survival. *J. Clin. Invest.* **78**, 1220–1228.

Lukacs, N. W., Oliveira, S. H., and Hogaboam, C. M. (1999). Chemokines and asthma: Redundancy of function or a coordinated effort? *J. Clin. Invest.* **104**, 995–999.

Masui, T., Wakefield, L. M., Lechner, J. F., LaVeck, M. A., Sporn, M. B., and Harris, C. C. (1986). Type beta transforming growth factor is the primary differentiation-inducing serum factor for normal human bronchial epithelial cells. *Proc. Natl. Acad. Sci. USA* **83**, 2438–2442.

McKay, S., and Sharma, H. S. (2002). Autocrine regulation of asthmatic airway inflammation: Role of airway smooth muscle. *Respir. Res.* **3**, 11.

Minshall, E. M., Leung, D. Y., Martin, R. J., Song, Y. L., Cameron, I., Ernst, P., and Hamid, Q. (1997). Eosinophil-associated TGF-beta1 mRNA expression and airways fibrosis in bronchial asthma. *Am. J. Resp. Cell. Mol. Biol.* **17**, 326–333.

Motojima, S., Adachi, T., Manaka, K., Arima, M., Fukuda, T., and Makino, S. (1996). Eosinophil peroxidase stimulates the release of granulocyte-macrophage colony-stimulating factor from bronchial epithelial cells. *J. Allergy Clin. Immunol.* **98**, S216–S223.

Muller, B. (2002). Cytokine imbalance in non-immunological chronic disease. *Cytokine* **18**, 334–339.

Pan, G., O'Rouke, K., Chinnaiyan, A. M., Gentz, R., Ebner, R., Ni, J., and Dixit, V. M. (1997). The receptor for the cytotoxic ligand TRAIL. *Nature* **276**, 111–113.

Pitti, R. M., Marsters, S. A., Ruppert, S., Donahue, C. J., Moore, A., and Ashkenazi, A. (1996). Induction of apoptosis by Apo-2 ligand, a new member of the tumor necrosis factor cytokine family. *J. Biol. Chem.* **271**, 12687–12690.

Poston, R. N., Chanez, P., Lacoste, J. Y., Litchfield, T., Lee, T. H., and Bousquet, J. (1995). Immunohistochemical characterization of the cellular infiltration in asthmatic bronchi. *Am. Rev. Resp. Dis.* **145**, 918–921.

Ricupero, D. A., Rishikof, D. C., Kuang, P. P., Poliks, C. F., and Goldstein, R. H. (1999). Regulation of connective tissue growth factor expression by prostaglandin E(2). *Am. J. Physiol.* **277**, L1165–1171.

Robertson, N. M., Zangrilli, J. G., Steplewski, A., Hastie, A., Lindemeyer, R. G., Planeta, M. A., Smith, M. K., Innocent, N., Musani, A., Pascual, R., Peters, S., and Litwack, G. (2002). Differential expression of TRAIL and TRAIL receptors in allergic asthmatics following segmental antigen challenge: Evidence for a role of TRAIL in eosinophil survival. *J. Immunol.* **169**, 5986–5996.

Rothenberg, M. E., Owen, W. F., Silberstein, D. S., Woods, J., Soberman, R. J., Austen, K. F., and Stevens, R. L. (1988). Human eosinophils have prolonged survival, enhanced functional properties and become hypodense when exposed to GM-CSF. *J. Clin. Invest.* **78**, 1220–1228.

Rothenberg, M. E., Peterson, J., Stevens, R. L., Silberstein, D. S., McKenzie, D. T., Austen, K. F., and Owen, W. F. (1989). IL-5 dependent conversion of normodense human eosinophils to the hypodense phenotype uses 3T3 fibroblasts for enhanced vibility, accelerated hypodense and sustained antibody-dependent cytotoxicity. *J. of Immunol.* **143**, 3211–3216.

Santini, S. M., Lapenta, C., Logozzi, M., Parlato, S., Spada, M., Di Pucchio, T., and Belardelli, F. (2000). Type I interferon as a powerful adjuvant for monocyte-derived dendritic cell development and activity in vitro and in Hu-PBL-SCID mice. *J. Exp. Med.* **191**, 1777–1788.

Sato, K., Hida, S., Takayanagi, H., Yokochi, T., Kayagaki, N., Takeda, K., Yagita, H., Okumura, K., Tanaka, N., Taniguchi, T., and Ogasawara, K. (2001). Antiviral response by natural killer cells through TRAIL gene induction by IFN-alpha/beta. *Eur. J. Immunol.* **31**, 3138–3146.

Sato, M., Takizawa, H., Kohyama, T., Ohtoshi, T., Takafuji, S., Kawasaki, S., Tohma, S., Ishii, A., Shoji, S., and Ito, K. (1997). Eosinophil adhesion to human bronchial epithelial cells: Regulation by cytokines. *Int. Arch. Allergy Immunol.* **113**, 203–205.

Schneider, P., Thome, M., Burns, K., Bodmer, J. L., Hofmann, K., Kataoka, T., Holler, N., and Tschopp, J. (1997). TRAIL receptors 1 (DR4) and 2 (DR5) signal FADD-dependent apoptosis and activate NF-κB. *Immunity* **7**, 831–836.

Secchiero, P., Gonelli, A., Carnevale, E., Milani, D., Pandolfi, A., Zella, D., and Zauli, G. (2003). TRAIL promotes the survival and proliferation of primary human vascular endothelial cells by activating the Akt and ERK pathways. *Circulation* **107**, 2250–2256.

Seminarino, M., and Gleich, G. J. (1994). The role of eosinophils in the pathogenesis of asthma. *Curr. Opin. Immunol.* **6**, 860–864.

Shaver, J. R., Zangrilli, J., Cho, S. K., Cirelli, R. A., Pollice, M., Hastie, A. T., Fish, J. E., and Peters, S. P. (1997). Kinetics of the development and recovery of the lung from IgE-mediated inflammation: Dissociation of pulmonary eosinophilia, lung injury, and eosinophil-active cytokines. *Am. J. Respir. Crit. Care Med.* **155**, 442–448.

Sheridan, J. P., Marsters, S. A., Pitti, R. M., Gurney, A., Skubatch, M., Baldwin, D., Ramakrishnan, L., Gray, C. L., Baker, K., Wood, W. I., Goddard, A. D., Godowski, P., and Ashkenazi, A. (1997). Control of TRAIL-induced apoptosis by a family of signaling and decoy receptors [see comments]. *Science* **277**, 18–21.

Smyth, M. J., Cretney, E., Takeda, K., Wiltrout, R. H., Sedger, L. M., Kayagaki, N., Yagita, H., and Okumura, K. (2001). Tumor necrosis factor-related apoptosis-inducing ligand (TRAIL) contributes to interferon gamma-dependent natural killer cell protection from tumor metastasis. *J. Exp. Med.* **193**, 661–670.

Takeda, K., Hayakawa, Y., Smyth, M. I., Kayagaki, N., Yamaguchi, N., Kakuta, S., Iwakura, Y., Yagita, H., and Okumura, K. (2001). Involvement of tumor necrosis factor-related apoptosis-inducing ligand in surveillance of tumor metastasis by liver natural killer cells. *Nat. Med.* **7**, 94–100.

Temple, R., Allen, E., Fordham, J., Phipps, S., Schneider, H. C., Lindauer, K., Hayes, I., Lockey, J., Pollock, K., and Jupp, R. (2001). Microarray analysis of eosinophils reveals a number of candidate survival and apoptosis genes. *Am. J. Respir. Cell. Mol. Biol.* **25**, 425–433.

Walczak, H., Degliesposti, M. A., Johnson, R. S., Smolak, P. J., Waugh, J. Y., Boiani, N., Timour, M. S., Gerhart, M. J., Schooley, K. A., Smith, C. A., Goodwin, R. G., and Rauch, C. T. (1997). Trail-R2–a novel apoptosis-mediating receptor for TRAIL. *EMBO J.* **16**, 5386–5397.

Wang, Q., Ji, Y., Wang, X., and Evers, B. M. (2000a). Isolation and molecular characterization of the 5′-upstream region of the human TRAIL gene. *Biochem. Biophys. Res. Commun.* **276**, 466–471.

Wang, Q., Yanshan, J., Wang, X., and Evers, B. M. (2000b). Isolation and molecular characterization of the 5′-upstream region of the human TRAIL gene. *Biochem. Biophys. Res. Commun.* **264**, 813–819.

Wang, Q., Wang, X., Hernandez, A., Hellmich, M. R., Gatalica, Z., and Evers, B. M. (2002a). Regulation of trail expression by the PI3-KINASE/Akt/GSK-3 pathway in human colon cancer cells. *J. Biol. Chem.* **24,** 24.

Wang, Y., Yi, S., Tay, Y. C., Feng, X., Kairaitis, L., and Harris, D. C. (2002b). Transfection of tubule cells with Fas ligand causes leukocyte apoptosis. *Kidney Int.* **61,** 1303–1311.

Wardlaw, A. J., Brightling, C. E., Green, R., Woltmann, G., Bradding, P., and Pavord, I. D. (2002). New insights into the relationship between airway inflammation and asthma. *Clin. Sci. (Lond.)* **103,** 201–211.

Wenzel, S. E., Trudeau, J. B., Barnes, S., Zhou, X., Cundall, M., Westcott, J. Y., McCord, K., and Chu, H. W. (2002). TGF-beta and IL-13 synergistically increase eotaxin-1 production in human airway fibroblasts. *J. Immunol.* **169,** 4613–4619.

Wiley, S. R., Schooley, K., Smolak, P. J., Din, W. S., Huang, C. P., Nicholl, J. K., Sutherland, G. R., Smith, T. D., Rauch, C., Smith, C. A. *et al.* (1995). Identification and characterization of a new member of the TNF family that induces apoptosis. *Immunity* **3,** 673–682.

Wilson, J. W., and Bamford, T. L. (2001). Assessing the evidence for remodelling of the airway in asthma. *Pulm. Pharmacol. Ther.* **14,** 229–247.

Winchester, E. C., Millwood, I. Y., Rand, L., Penny, M. A., and Kessling, A. M. (2000). Association of the TNF-alpha-308 (G→A) polymorphism with self-reported history of childhood asthma. *Hum. Genet.* **107,** 591–596.

Yanagisawa, K., Osada, H., Masuda, A., Kondo, M., Saito, T., Yatabe, Y., Takagi, K., and Takahashi, T. (1998). Induction of apoptosis by Smad3 and down-regulation of Smad3 expression in response to TGF-beta in human normal lung epithelial cells. *Oncogene* **17,** 1743–1747.

Yu, Y., Liu, S., Wang, W., Song, W., Zhang, M., Zhang, W., Qin, Z., and Cao, X. (2002). Involvement of tumour necrosis factor-alpha-related apoptosis-inducing ligand in enhanced cytotoxicity of lipopolysaccharide-stimulated dendritic cells to activated T cells. *Immunology* **106,** 308–315.

Yurovsky, V. V. (2003). Tumor necrosis factor-related apoptosis-inducing ligand enhances collagen production by human lung fibroblasts. *Am. J. Respir. Cell. Mol. Biol.* **28,** 225–231.

Zamai, L., Ahmad, M., Bennett, I. M., Azzoni, L., Alnemri, E. S., and Perussia, B. (1998). Natural killer (NK) cell-mediated cytotoxicity: Differential use of TRAIL and Fas ligand by immature and mature primary human NK cells. *J. Exp. Med.* **188,** 2375–2380.

Zangrilli, J. G., Shaver, J. R., Cirelli, R. A., Cho, S. K., Garlisi, C. G., Falcone, A., Cuss, F. M., Fish, J. E., and Peters, S. P. (1995). sVCAM-1 levels after segmental antigen challenge correlate with eosinophil influx, IL-4 and IL-5 production, and the late phase response. *Am. J. Resp. Crit. Care Med.* **151,** 1346–1353.

10

TRAIL Death Receptors, Bcl-2 Protein Family, and Endoplasmic Reticulum Calcium Pool

M. Saeed Sheikh and Ying Huang

Department of Pharmacology
State University of New York Upstate Medical University
Syracuse, New York 13210

I. Introduction
II. The Endoplasmic Reticulum
   A. Endoplasmic Reticulum Ca^{2+} and Signal Transduction
III. Endoplasmic Reticulum Ca^{2+} Pools and Regulation of Apoptosis
   A. Bcl-2 Protein Family
   B. Bcl-2 Protein Family and Control of Ca^{2+} Homeostasis
   C. TRAIL and TRAIL Receptors
   D. TRAIL Receptor Regulation and Control of Ca^{2+} Homeostasis
IV. Concluding Remarks
References

Calcium (Ca^{2+}) is one of the highly versatile second messengers critical in cellular pathophysiology. Alterations in Ca^{2+} homeostasis affect many cellular processes, including apoptosis. Recent studies have started to unravel the molecular mechanisms of apoptosis regulation in context to intracellular Ca^{2+} pools. In this regard, Bcl-2 has been reported to mediate its anti-apoptotic effects, partly, by lowering the endoplasmic reticulum (ER) Ca^{2+} load and by inhibiting the mitochondrial uptake of Ca^{2+}. However, the opposite is true for Bax and Bak that promote apoptosis, in part, by increasing the ER Ca^{2+} load and Ca^{2+} transfer from the ER to mitochondria. Massive ER Ca^{2+} depletion coupled with upregulation of DR5 has also been reported to induce apoptosis. The mechanistic details of how some of these molecules affect intracellular Ca^{2+} contents and sense perturbations in Ca^{2+} homeostasis remain to be elucidated. The recent explosion of information in the fields of cell signaling and apoptosis is likely to facilitate the future investigations aiming to explore these issues. © 2004 Elsevier Inc.

I. INTRODUCTION

The importance of calcium (Ca^{2+}) in cell physiology is known since 1883. Ca^{2+} functions as a second messenger and alterations in Ca^{2+} homeostasis affect many cellular processes, including apoptosis. The molecular details of the apoptotic events modulated by alterations in Ca^{2+} homeostasis are still emerging. Available evidence implicates Bcl-2 family members, certain endoplasmic reticulum (ER)–associated proteins, tumor necrosis factor (TNF)–related apoptosis-inducing ligand (TRAIL) receptor-2 (TRAIL-R2 also known as DR5), and TRAIL. This chapter presents the current state of knowledge on the role of these molecules in relation to alterations in Ca^{2+} homeostasis and apoptosis.

II. THE ENDOPLASMIC RETICULUM

The ER is a cellular organelle with an extensive web of interconnected, closed membrane tubules and cisternae. In general, the ER is subdivided into the rough ER (RER) and the smooth ER (SER). The RER is continuous with the nuclear membrane and due to the physical connection between these membranes some proteins present in the nuclear envelope also reside in reticular part of the RER. Proteins bound for plasma membrane or extracellular secretion and those entering different cellular compartments are synthesized in the RER. Unlike the ribosome-coupled RER, the SER devoid of ribosomes facilitates the transport of newly synthesized proteins to the Golgi apparatus. The ER performs several important cellular functions including protein and lipid biosynthesis. Approximately one third of all cellular proteins is post-translationally modified, folded, and oligomerized in the ER lumen (Kaufman, 1999). The transmembrane proteins destined for most of the cellular organelles generally originate at the ER membrane. Soluble secretory proteins or proteins bound to other cellular organelles enter completely into the ER lumen before transport to their final destination. The transmembrane proteins destined for the ER itself or other cellular membranes are not transferred into the ER lumen; instead, they are embedded into the lipid bilayer via their hydrophobic membrane-spanning alpha-helices. A stretch of amino acids serving as a signal sequence directs transmembrane proteins to anchor in the ER membrane. Proteins predestined for initial transferring to the ER often contain a signal sequence at their N-terminal end, which typically includes a stretch of approximately 5–10 hydrophobic residues. In addition to the N-terminal signal sequence, many ER-resident proteins contain a four amino acid (KDEL) ER-retrieval signal at the C-terminal ends that serves to direct their return back to the ER. Some ER-resident proteins, although not carrying the characteristic N-terminal signal sequence, harbor a signal patch that is believed to facilitate their transport toward the ER.

A. ENDOPLASMIC RETICULUM Ca^{2+} AND SIGNAL TRANSDUCTION

The ER (or the sarcoplasmic reticulum in muscle) is the major intracellular Ca^{2+} storage site. Consequently, regulation of Ca^{2+}-dependent signaling events is one of the key functions of the ER. Several ER-associated proteins have been identified that modulate intracellular Ca^{2+} signaling. Some of these proteins include calreticulin, glucose-regulated proteins 78 and 94 (Grp78/Bip and Grp94), calsequstrin, protein disulfide isomerase (PDI) and PDI-like proteins, such as ERp72, ERp57, and ERp29 (Jethmalani and Henle, 1998; Liu *et al.*, 1997; Papp *et al.*, 2003). Calcium channels constitute yet another class of proteins that affect intracellular Ca^{2+} signaling; proteins in this group include the Ca^{2+} uptake channels, such as sarcoplasmic/endoplasmic reticulum Ca^{2+}-transporting ATPases (SERCAs) and the Ca^{2+} releasing channels, such as the inositol-1,4,5-triphosphate receptor (IP_3R) and ryanodine receptors (RYRs) (Carafoli, 2002; Hajnoczky *et al.*, 2003; Papp *et al.*, 2003) (Fig. 1).

Ca^{2+} is one of the key regulators of cell death and proliferation. A large body of evidence indicates that Ca^{2+} can modulate several signaling pathways that control cell proliferation including the (i) mitogen-activated protein kinase (MAPK), (ii) phosphotidylinositol-3 kinase (PI3-K), and (iii) Ca^{2+}/calmodulin-dependent protein kinase pathways (Berridge *et al.*, 2000). For example, binding of extracellular ligands to the tyrosine kinase–linked or G protein–coupled receptors present in the cell membrane activates phospholipase C (PLC) to generate inositol-1,4,5-triphosphate (IP_3) and diacylglycerol (DAG). IP_3 in turn binds to IP_3R present in the ER membrane and promotes Ca^{2+} release from the ER into cytosol. An increase

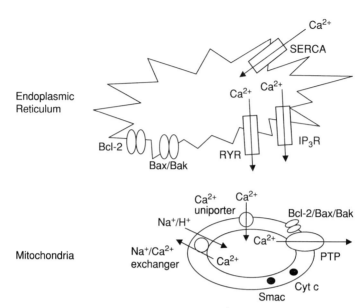

FIGURE 1. Schematic illustration of calcium (Ca^{2+}) mobilization in the endoplasmic reticulum and mitochondria. Cyt c, cytochrome c; IP_3R, inositol-1,4,5-triphosphate receptor; PTP, permeability transition pore; RYR, ryanodine receptor; SERCA, sarco(endo)plasmic reticulum Ca^{2+}-ATPase.

in the cytosolic Ca^{2+} levels can activate the MAPK signaling pathway via proline-rich tyrosine kinase 2 (PYK2), which then acts through the Ras/Raf/MEK/ERK cascade to further activate the downstream target genes that control cell proliferation (Berridge et al., 2000). Ca^{2+} also activates protein kinase C (PKC), which in turn induces cell proliferation or survival via activation of MAPK pathway or transcription factor NF kappa-B (Berridge et al., 2000; Swarthout et al., 2001). Intracellular Ca^{2+} also plays an important role in modulating the signaling pathways regulated by Ca^{2+}/calmodulin (CaM)/CaM-dependent kinases. Calcium-activated CaM-dependent kinases phosphorylate CREB, a transcription factor that binds to the cyclic AMP response element (CRE), and activate the transcription of cell proliferative genes (Berridge et al., 2000).

III. ENDOPLASMIC RETICULUM CA^{2+} POOLS AND REGULATION OF APOPTOSIS

A role for Ca^{2+} in the control of apoptosis has long been known. A connection between Ca^{2+} and apoptosis was established in the earlier studies when it was found that DNA fragmentation noted during glucocorticoid-induced apoptosis in thymocytes was dependent on Ca^{2+} (Cohen and Duke, 1984; McConkey et al., 1989; Wyllie, 1980). Such induction of DNA fragmentation was attributed to endonucleases that require Ca^{2+} and Mg^{2+} for their activation (Cohen and Duke, 1984; McConkey et al., 1989; Wyllie, 1980). There has now been a renewed interest in studying the molecular mechanisms by which Ca^{2+} controls apoptosis. In particular, the regulation of apoptosis in relation to ER Ca^{2+} pools has recently become the focus of intense research (Demaurex and Distelhorst, 2003; Hajnoczky et al., 2003; Orrenius et al., 2003). Several studies indicate that the ER Ca^{2+} contents determine a cell's sensitivity to apoptotic stimuli and that Ca^{2+} movement between the ER and mitochondria has a profound effect in regulating apoptosis (Foyouzi-Youssefi et al., 2000; Nakamura et al., 2000; Pinton et al., 2001; Rapizzi et al., 2002; Scorrano et al., 2003). Under normal conditions, Ca^{2+} release from the ER is transient and partial and it continuously circulates between the ER and mitochondria. Ca^{2+} is pumped into ER by Ca^{2+} uptake ATPase channels, such as SERCAs, and released by Ca^{2+} releasing channels such as the IP_3R and RYR (Demaurex and Distelhorst, 2003; Hajnoczky et al., 2003; Orrenius et al., 2003) (Fig. 1). Massive ER Ca^{2+} overload or ER Ca^{2+} pool depletion is believed to induce a type of stress known as "ER stress" that triggers apoptosis. For example, it has been shown that an increase in the ER Ca^{2+} concentration via overexpression of SERCAs or calreticulin (the ER-luminal Ca^{2+} binding protein) sensitizes cells to apoptotic stimuli (Nakamura et al., 2000). Similarly, thapsigargin (TG), a sesqueterpene lactone, promotes a massive ER Ca^{2+} pool depletion by irreversibly inhibiting SERCAs (Sagara and Inesi, 1991; Thastrup et al., 1990) and thereby induces apoptosis (He et al., 2002a; Wertz and Dixit, 2000). Whether an increase in the cytosolic Ca^{2+} due to ER Ca^{2+} pool depletion activates apoptotic signals or the "ER stress" due to massive exit of Ca^{2+} from the ER is the trigger of apoptotic signals is unclear. Based on several lines of evidence indicating that just an increase in the cytosolic Ca^{2+} levels is not enough to induce apoptosis, it could be argued that the "ER stress" due to depletion of ER Ca^{2+} pools appears to be the trigger (Hajnoczky et al., 2003; Orrenius et al., 2003). However, further in-depth studies are needed to resolve this issue. Mitochondria also modulate and synchronize Ca^{2+} signal along with the ER. Ca^{2+} released from the ER is rapidly taken up by mitochondria via a Ca^{2+} uniporter (Demaurex and Distelhorst, 2003; Orrenius et al., 2003; Rizzuto et al., 1992) (Fig. 1). Likewise, Ca^{2+} is released from mitochondria by a Na^+/Ca^{2+} exchanger (Demaurex and Distelhorst 2003; Orrenius et al., 2003; Rizzuto et al., 1992) (Fig. 1). This ER-mitochondria connection is believed to play a key role in modulating the ability of mitochondria to transduce apoptotic signals (Pacher and Hajnoczky, 2001; Szalai et al., 1999).

A. BCL-2 PROTEIN FAMILY

Bcl-2 protein family members include both the anti- and proapoptotic molecules that are key regulators of the intrinsic (mitochondrial) pathway of apoptosis. Some members of Bcl-2 family also bridge a cross-talk between the extrinsic (death receptor) and intrinsic pathways. To date, several members of Bcl-2 family have been identified and are grouped into three subfamilies including (1) the antiapoptotic subfamily, (2) the multidomain proapoptotic subfamily, and (3) the BH-3-only protein subfamily (Scorrano and Korsmeyer, 2003; Tsujimoto, 2003) (Fig. 2).

1. The Antiapoptotic Subfamily

The proteins in this subfamily are grouped based on their structure and function in that they contain regions of homology dubbed Bcl-2 homology (BH) regions (Fig. 2) and function to inhibit apoptosis. Bcl-2, Bcl-X_L, Mcl-1, Bcl-w, and A1(Bfl-1) are the members that contain four BH domains including BH1 through BH4 whereas NR-13 lacking the BH3 domain contains only three. BH3 domain is also absent in Boo/Diva whose role has been implicated in both anti- and proapoptosis (Scorrano and Korsmeyer, 2003; Tsujimoto, 2003). With the exception of A1(Bfl-1), most of these proteins also contain a transmembrane domain that serves to anchor them in the membranes of mitochondria and the ER (Fig. 2).

2. The Multidomain Proapoptotic Subfamily

As the name indicates, the proteins in this subfamily are proapoptotic molecules that contain three BH domains including BH1 through BH3. To date several members have been grouped into this subfamily including Bax, Bak, Mtd (Bok), and Bcl-rambo (Scorrano and Korsmeyer, 2003; Tsujimoto, 2003). In general, these proteins also harbor a membrane anchoring region (Fig. 2).

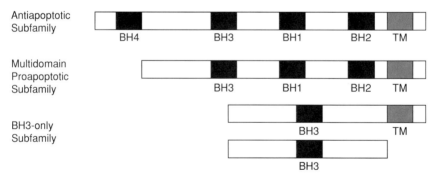

FIGURE 2. Structural illustration of Bcl-2 family members. Prototype structures representative of each subfamily are illustrated. BH, Bcl-2 homology domain; TM, transmembrane.

3. The BH3-only Protein Subfamily

As the name implies, the members in this subfamily harbor sequence conservation only in the BH3 domain (Fig. 2). In general, the proteins in this subfamily are implicated in proapoptotic function and include Bad, Bcl-G, Bid, Bik (Nbk), Bim (Bod), Blk, Bmf, Bnip3, Bnip3L, Hrk (DP5), Nix, Noxa, p193, and Puma (Scorrano and Korsmeyer, 2003; Tsujimoto, 2003). Some BH3-only proteins also contain transmembrane domain while some do not (Fig. 2).

B. BCL-2 PROTEIN FAMILY AND CONTROL OF Ca^{2+} HOMEOSTASIS

Recent years have witnessed significant progress in our understanding of how the Bcl-2 family members modulate apoptosis. The current state of knowledge indicates that members in this family modulate apoptotic events predominantly via the intrinsic (mitochondrial) pathway. For example, the mitochondrial membrane permeability is altered in response to a variety of apoptotic insults that engage the intrinsic pathway resulting in the release of apoptogenic factors from mitochondria into the cytosol. Several such factors have been identified and include cytochrome c, Smac (also known as Diablo), HtrA2/Omi, AIF (apoptosis-inducing factor), and endonuclease G (Du *et al.*, 2000; Liu *et al.*, 1996; Li *et al.*, 2001; Susin *et al.*, 1999; Suzuki *et al.*, 2001; Tsujimoto, 2003; Verhagen *et al.*, 2000). Of these apoptogenic factors, cytochrome c is the most extensively investigated molecule that in association with Apaf-1 and ATP activates caspase 9. Smac is known to function by inhibiting the IAPs (inhibitors of apoptosis) that block the activation of caspases (Tsujimoto, 2003). HtrA2/Omi is a serine protease that is also believed to mediate its effect by inhibiting the IAPs whereas endonuclease G and AIF modulate the nuclear events of apoptosis (Tsujimoto, 2003). In general, members of the antiapoptotic subfamily block the release of these factors while the proapoptotic members including the BH3-only molecules promote their release from mitochondria into cytosol (Scorrano and Korsmeyer, 2003; Tsujimoto, 2003).

Several Bcl-2 family proteins, such as Bcl-2, Bax, and Bak, which are key regulators of apoptosis, localize in the membranes of both mitochondria and the ER (Demaurex and Distelhorst, 2003; Tsujimoto, 2003). In addition to their well-known function in controlling the release of apoptogenic factors from mitochondria into the cytosol, the Bcl-2 family members regulate ER Ca^{2+}. For example, by decreasing the ER Ca^{2+} load and mitochondrial Ca^{2+} uptake, Bcl-2 appears to protect cells from apoptosis (Demaurex and Distelhorst, 2003; Foyouzi-Youssefi *et al.*, 2000; Pinton *et al.*, 2000). In contrast, Bax and Bak overexpression by increasing ER Ca^{2+} load promotes Ca^{2+} transfer from the ER to mitochondria and thus favors apoptosis (Nutt et al., 2002a,b). Consistent with these findings, murine cells deficient in Bax, Bak, or both proteins have been found to exhibit lower ER Ca^{2+} load and resistance to apoptosis induced by several agents (Scorrano et al., 2003). Bax and Bak double knockout (DKO) cells do not display defects in their ability to transfer Ca^{2+} to mitochondria but their ER Ca^{2+} contents appear to be too low to transfer enough Ca^{2+} to mitochondria (Scorrano et al., 2003). In this context it is of note that whereas Bcl-2 tends to protect from apoptosis by decreasing the ER Ca^{2+} load and Bax/Bak promote apoptosis by increasing the ER Ca^{2+} load, massive ER Ca^{2+} pool depletion induced by agents such as TG also proves detrimental. Thus, although it remains unclear as to how Bcl-2, Bax, and Bak modulate the ER Ca^{2+} pools, it is nonetheless clear that Ca^{2+} homeostasis is critical for normal cellular physiology and alteration in the intracellular Ca^{2+} concentration is likely to have untoward effects on normal cellular function.

C. TRAIL AND TRAIL RECEPTORS

TRAIL (tumor necrosis factor–related apoptosis-inducing ligand; also known as Apo2L) has garnered significant attention because of its potential as a novel anticancer agent. TRAIL belongs to the TNF family of ligands that are expressed as type 2 membrane proteins (LeBlanc and Ashkenazi, 2003). Due to proteolytic cleavage of TRAIL at its extracellular domain it is also detected in soluble form. As is the case for other ligands in this family, TRAIL binds to its membrane receptors as a homotrimer (LeBlanc and Ashkenazi, 2003). TRAIL-R1 and TRAIL-R2 (hereafter referred to as DR4 and DR5, respectively) are the death domain–containing receptors that are activated by TRAIL and transduce apoptotic signals (Askkenazi and Dixit, 1999; LeBlanc and Ashkenazi, 2003; Schulze-Osthoff et al., 1998; Singh et al., 1998) (Fig. 3). In addition to DR4 and DR5, two other TRAIL receptors named TRAIL-R3 (TRID) (Degli-Esposti et al., 1997a; Pan et al., 1997; Sheridan et al., 1997) and TRAIL-R4 (TRUNDD) (Degli-Esposti et al., 1997b; Marsters et al., 1997; Pan et al., 1998) have been identified (Fig. 3). TRAIL-R3 is a GPI-linked molecule without death domain whereas TRAIL-R4 contains a deleted version of death domain and thus, both are known as antiapoptotic decoy receptors incapable of transducing apoptotic signals (Degli-Esposti et al., 1997a,b; Marsters et al., 1997; Pan et al., 1997, 1998; Sheridan et al., 1997). Osteoprotegerin (OPG) is yet another TRAIL-interacting molecule that exists in soluble form and functions as a decoy molecule (Simonet et al., 1997) (Fig. 3). TRAIL-R3, TRAIL-R4, and OPG compete with DR4 and DR5 for TRAIL-binding and upon overexpression have been shown to inhibit TRAIL-induced apoptosis.

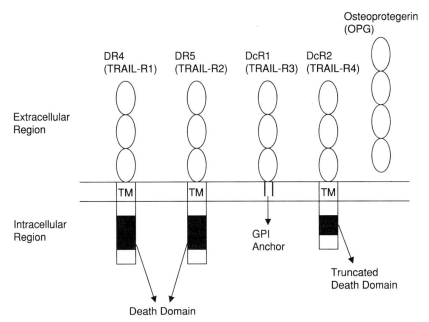

FIGURE 3. Structural illustration of TRAIL death and decoy receptors. TRAIL-R1 and -R2 (DR4 and DR5) are death receptors whereas TRAIL-R3 and -R4 (DcR1 and DcR2) are decoy receptors. TRAIL-R1 is a GPI-linked receptor devoid of death domain whereas TRAIL-R4 harbors a nonfunctional truncated death domain. Osteoprotegrin (OPG) is a soluble protein that also binds TRAIL.

Undoubtedly, the discoveries of TRAIL and its death-inducing receptors have greatly advanced the field of apoptosis. A wealth of information now demonstrates that TRAIL mediates apoptosis in a wide variety of different tumor cell lines but not in various normal cells of different lineages (LeBlanc and Ashkenazi, 2003). Several lines of evidence indicate that exogenously administered TRAIL is not only tolerated by animals but also inhibits the growth of implanted tumors (Ashkenazi et al., 1999; Walczak et al., 1999). These properties distinguish TRAIL from its other family members, such as TNF and FasL, both of which have proven extremely toxic and thus have limited use as anticancer agents. Recent evidence also indicates that p53 and several DNA damage-inducing agents upregulate DR5 (Nimmanapalli et al., 2001; Sheikh et al., 1998; Wu et al., 1997). Accordingly, TRAIL can augment the effects of other anticancer agents that also upregulate DR5. Thus, targeting DR5 with TRAIL in combination with other anticancer agents could prove superior than using TRAIL alone in some malignancies. Clearly, TRAIL shows significant promise as an anticancer agent and opportunities now exist to exploit its therapeutic potential. Unfortunately, not all tumor types are sensitive to TRAIL (Nesterov et al., 2001). The underlying mechanisms of TRAIL resistance noted in some tumors remain unclear. The apparent lack of a correlation between TRAIL resistance and the levels of its decoy receptors in some tumors points to mechanisms of TRAIL resistance other than the decoy receptor expression (Griffith and Lynch, 1998). Thus, an improved understanding of the regulation of TRAIL receptors and that of the signaling events originating from them could prove beneficial in improving the therapeutic value of targeting TRAIL death receptors in cancer therapy. Such information will also help to elucidate why normal cells seem to better tolerate this novel cytokine.

D. TRAIL RECEPTOR REGULATION AND CONTROL OF Ca^{2+} HOMEOSTASIS

Work performed in our laboratories has shown that TRAIL and its receptor DR5 (TRAIL-R2) are regulated during ER Ca^{2+} pool depletion–induced apoptosis. As noted previously, thapsigargin (TG), a sesqueterpene lactone, inhibits the ER Ca^{2+} ATPases (Sagara and Inesi, 1991; Thastrup *et al.*, 1990). Inhibition of ER Ca^{2+} ATPases by TG abrogates the re-uptake of cytosolic Ca^{2+} into the ER, which in turn increases the cytoplasmic Ca^{2+} levels and depletion of ER Ca^{2+} pools (Sagara and Inesi, 1991; Thastrup *et al.*, 1990).

In keeping with the notion that perturbations in cellular Ca^{2+} homeostasis induce apoptosis, TG was found to induce apoptosis in a variety of cell lines representing different malignancies including cancers of the prostate, breast, colon and lung. Figure 4 shows morphologic features of TG-induced apoptosis in human colon cancer cells. As is shown, TG induced characteristic features of apoptosis including chromatin condensation and nuclear fragmentation. TG-induced apoptosis was coupled with upregulation of DR5 at both mRNA and protein levels in various cancer cell lines. TG

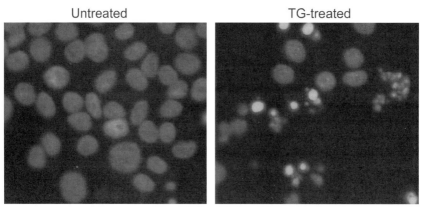

FIGURE 4. Representative photomicrographs show the morphologic features of thapsigargin-induced apoptosis in HT29 human colon cancer cells. Reproduced from He *et al.* (2002a) with permission from the publisher, Nature Publishing Group.

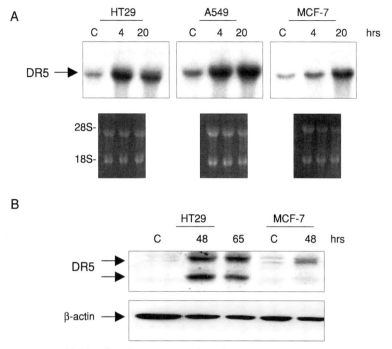

FIGURE 5. (A) Thapsigargin (TG) regulation of death receptor 5 (DR5) mRNA in three different cancer cell lines. Northern blots show that TG increases DR5 mRNA levels in human colon (HT29), lung (A549) and breast (MCF-7) cancer cells. Cells were treated with vehicle (DMSO) alone as control (C) or TG for the periods indicated in hours. Cells were exposed to 1 μM TG except for 4 h treatment in the case of HT29 cells. A human DR5 cDNA fragment was used as a probe. Ethidium bromide staining of the same gels shows RNA integrity and comparable loading. (B) TG effect on death receptor 5 (DR5) protein levels in human cancer cell lines. Cells were treated with vehicle (DMSO) alone as control (C) or TG (2 μM) for the periods shown in hours and Western blot analyses were performed using anti-DR5 and anti-β-actin antibodies. Reproduced from He *et al.* (2002a) with permission from the publisher, Nature Publishing Group.

modulating the DR5 expression at both the transcriptional and post-transcriptional levels specifically regulated DR5 but not DR4, the other TRAIL death receptor. Figure 5 shows TG-induced upregulation of DR5 mRNA and protein levels in human colon, lung, and breast cancer cells. TG-induced upregulation of DR5 expression was noted in a number of cancer cell lines tested whereas TRAIL was found to be upregulated in some cell lines (He *et al.*, 2002a).

These findings would suggest that Ca^{2+} pool depletion–induced apoptosis engages both the ligand and receptor but in some cells only receptor regulation appears to be sufficient to promote apoptosis. Figure 6 shows that TG-induced apoptosis was associated with activation of caspases 3, 8, and 9 as well as PARP cleavage. The fact that TG induced activation of

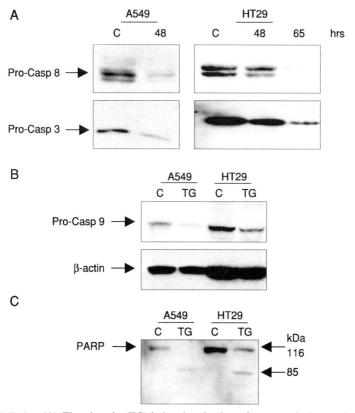

FIGURE 6. (A) Thapsigargin (TG)-induced activation of caspases in human lung and colon cancer cells. Lung (A549) and colon (HT29) cancer cells were treated with vehicle (DMSO) alone as control (C) or TG (2 μM) for the indicated time periods. Western blot analyses were performed using antipro-caspase 8 and 3 antibodies. A decrease in the levels of the pro-caspases indicates their activation. TG-induced activation of caspase 9 (B) and PARP-cleavage (C) in A549 and HT29 cells. Adapted from He *et al.* (2002a) with permission from the publisher, Nature Publishing Group.

caspase 8, the apical caspase that is recruited at the active DR5 site, would further support the notion that ER-Ca^{2+} pool depletion–induced apoptosis involves DR5-dependent pathways. Previous evidence suggested that ER-Ca^{2+} pool depletion–induced apoptosis predominantly engaged the intrinsic pathway as was evidenced by activation of caspase 9 and the release of cytochrome c from mitochondria into the cytosols of TG-treated cells (Wertz and Dixit, 2000). Based on the fact that TG also induces Bid cleavage (He *et al.*, 2002a), which is known to bridge the cross talk between extrinsic and intrinsic pathways, it is highly likely that the DR5 pathway talks with the intrinsic pathway via truncated Bid (tBid) during ER Ca^{2+} pool depletion–induced apoptosis.

As noted previously, the Bcl-2 family members play an important role in maintaining cellular Ca^{2+} homeostasis and inducing apoptosis when Ca^{2+} homeostasis is perturbed. Burns and El-Deiry (2001) have recently identified Bcl-X_L, the antiapoptotic member of Bcl-2 family, as a bona fide molecule that was responsible for conferring TRAIL resistance. Furthermore, Bax-deficient human colon cancer cells were found to exhibit significant resistance to TRAIL (Burns and El-Deiry, 2001; He *et al.*, 2002b). Together these findings support the notion that Bcl-2 family members play an important role in modulating the cellular response to TRAIL. These findings would also highlight the significance of a cross talk between the extrinsic and intrinsic pathways during TRAIL-induced apoptosis.

Because ER Ca^{2+} pool depletion–induced apoptosis engages the TRAIL pathway (extrinsic pathway) and the mitochondrial pathway (intrinsic pathway) and the fact that Bax-deficient cells become resistant to TRAIL, He *et al.* (2003) investigated the effect of Bax deficiency on the TRAIL and mitochondrial pathways during TG-induced apoptosis. The authors used isogenic Bax-proficient and Bax-deficient human colon cancer cells and found that Bax-deficient cells were clearly more resistant to TG-induced apoptosis than the Bax-proficient counterparts (Fig. 7). Similar results were reported by Yamaguchi *et al.* (2003), who independently studied the same

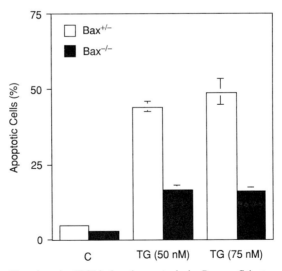

FIGURE 7. Thapsigargin (TG)-induced apoptosis in Bax-proficient and Bax-deficient cells. Bax-proficient ($Bax^{+/-}$) or Bax-deficient ($Bax^{-/-}$) HCT116 cells were exposed to indicated concentrations of TG for approximately 24 h and then processed for apoptosis detection by counting floating and adherent cells using a phase contrast microscope. The values represent mean ± s.e.m. of two independent experiments. Reproduced from He *et al.* (2003) with permission from the publisher, Nature Publishing Group.

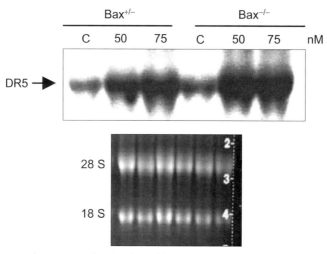

FIGURE 8. A representative Northern blot shows the effect of thapsigargin (TG) on death receptor 5 (DR5) mRNA levels in Bax-proficient and Bax-deficient HCT116 cells. Cells were either left untreated (C) or treated with the indicated concentrations of TG for approximately 24 h. A human DR5 cDNA fragment was used as a probe; ethidium bromide staining of the gel shows RNA integrity. Reproduced from He *et al.* (2003) with permission from the publisher, Nature Publishing Group.

set of Bax-deficient and Bax-proficient cells. Bax deficiency however, did not affect TG regulation of DR5, and caspase 8 activation and Bid cleavage were only minimally affected in these cells (Figs. 8 and 9). TG-induced caspase 3 activation, on the other hand, was greatly reduced and caspase 9 activation was completely abrogated due to Bax deficiency (Fig. 9). TG also did not promote the release of cytochrome c from mitochondria into cytosol in Bax-deficient cells although it did so in Bax-proficient cells (He *et al.*, 2003). The absence of TG-induced caspase 9 activation in Bax-deficient cells was suggested to be due to the lack of cytochrome c release from mitochondria into cytosol. Yamaguchi *et al.* (2003) also reported similar findings and additionally found that TG-induced release of Smac and HtrA2/Omi from mitochondria into cytosol was also abrogated in Bax-deficient cells. Together these findings suggest that the DR5-dependent pathway cross talks with the intrinsic pathway during ER Ca^{2+} pool depletion–induced apoptosis. To investigate whether TRAIL and TG would cooperate to overcome Bax-deficiency, He *et al.* (2003) used TRAIL and TG in combination and found that the combination of both was more effective at inducing apoptosis in both Bax-proficient and Bax-deficient cells (Fig. 10). Although both agents also displayed enhanced cooperation at activating caspase 9 and promoting cytochrome c release into the cytosol in Bax-proficient cells, they did not do so in Bax-deficient cells (He *et al.*, 2003).

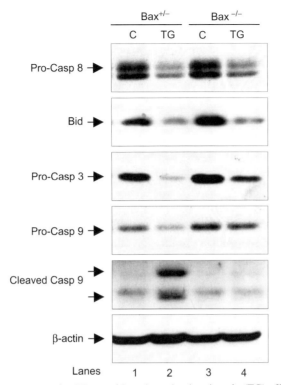

FIGURE 9. Representative Western blots show the thapsigargin (TG) effects on various intracellular apoptotic molecules in Bax-proficient and Bax-deficient HCT116 cells. Bax-proficient ($Bax^{+/-}$) or Bax-deficient ($Bax^{-/-}$) cells were either left untreated or treated with TG (50 nM) for approximately 24 h. Cells were harvested and processed for Western blotting using the indicated antibodies; same blot was sequentially probed with the indicated antibodies. Reproduced from He et al. (2003) with permission from the publisher, Nature Publishing Group.

Together these findings indicate that (i) ER Ca^{2+} pool depletion–induced apoptosis involves extrinsic and intrinsic pathways of apoptosis, and (ii) although the activation of mitochondrial events is compromised in Bax-deficient cells, cooperation between TRAIL and TG by engaging DR5-dependent signals could bypass these defects (He et al., 2003).

IV. CONCLUDING REMARKS

Ca^{2+} has proven to be a versatile second messenger and alterations in Ca^{2+} homeostasis profoundly affect major cellular processes including apoptosis. Recent studies have started to unravel the molecular mechanisms

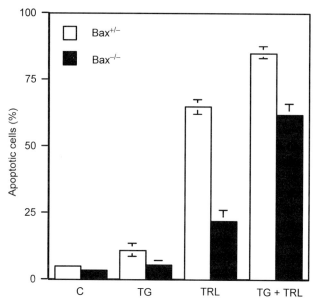

FIGURE 10. TRAIL and thapsigargin (TG) interact to mediate apoptosis in Bax-proficient and Bax-deficient HCT116 cells. Bax-proficient (Bax$^{+/-}$) or Bax-deficient (Bax$^{-/-}$) cells were either left untreated or treated with TG (50 nM), TRAIL (50 ng/ml) or the combination of TG and TRAIL for approximately 7 h. Cells were processed for apoptosis detection by counting floating and adherent cells using a phase contrast microscope as reported previously (He et al., 2002a,b). Values are the means of two independent experiments; bars, s.e.m. Reproduced from He et al. (2003) with permission from the publisher, Nature Publishing Group.

of apoptosis regulation in context to intracellular Ca^{2+} pools. However, several key issues still remain unclear and will be the focus of future investigations. For example, it is believed that massive ER Ca^{2+} overload or the ER Ca^{2+} pool depletion due to massive exit of Ca^{2+} from the ER triggers an "ER stress" response that induces apoptosis, but the molecular details as to how the "ER stress" activates apoptotic signals remain to be fully elucidated. In this context, the role of Bcl-2 family members such as Bcl-2, Bax, and Bak appear critical in controlling the ER Ca^{2+} load and Ca^{2+} movement from the ER to mitochondria but the mechanistic details of how Bcl-2 family members mediate these effects remain to be explored. Similarly, DR5 is transcriptionally regulated in response to ER stress induced by ER Ca^{2+} pool depletion but how the stress sensing signals originating in the ER are transduced into the nucleus to activate the expression of this death receptor gene is another important issue that remains to be investigated. The recent explosion of information in the fields of cell signaling and apoptosis is likely to facilitate the future studies aiming to explore this fertile area of investigation.

ACKNOWLEDGMENTS

Work in authors' laboratories was supported in part by the NIH grants CA86945, CA89043, DK062136 and a Department of Defense grant DAMD 170010722.

REFERENCES

Askkenazi, A., and Dixit, V. M. (1999). Apoptosis control by death and decoy receptors. *Curr. Opin. Cell Biol.* **11,** 255–260.
Ashkenazi, A., Pai, R. C., Fong, S., Leung, S., Lawrence, D. A., Marsters, S. A., Blackie, C., Chang, L., McMurtrey, A. E., Hebert, A., DeForge, L., Koumenis, I. L., Lewis, D., Harris, L., Bussiere, J., Koeppen, H., Shahrokh, Z., and Schwall, R. H. (1999). Safety and antitumor activity of recombinant soluble Apo2 ligand. *J. Clin. Invest.* **104,** 155–162.
Berridge, M. J., Lipp, P., and Bootman, M. D. (2000). The versatility and universality of calcium signaling. *Nature Rev. Mol. Cell. Biol.* **1,** 11–20.
Burns, T. F., and El-Deiry, W. S. (2001). Identification of inhibitors of TRAIL-induced death (ITIDs) in the TRAIL-sensitive colon carcinoma cell line SW480 using a genetic approach. *J. Biol. Chem.* **276,** 37879–37886.
Carafoli, E. (2002). Calcium signaling: A tale for seasons. *Proc. Natl. Acad. Sci. USA* **99,** 1115–1122.
Cohen, J. J., and Duke, R. C. (1984). Glucocorticoid activation of a calcium-dependent endonuclease in thymocyte nuclei leads to cell death. *J. Immunol.* **132,** 38–42.
Degli-Esposti, M. A., Smolak, P. J., Walczak, H., Waugh, J. Y., Huang, C.-P., DuBose, R. F., Goodwin, R. G., and Smith, C. A. (1997a). Cloning and characterization of TRAIL-R3, a novel member of the emerging TRAIL receptor family. *J. Exp. Med.* **186,** 1165–1170.
Degli-Esposti, M. A., Dougall, W. C., Smolak, P. J., Waugh, J. Y., Smith, C. A., and Goodwin, R. G. (1997b). The novel receptor TRAIL-R4 induces NFκB and protects against TRAIL-mediated apoptosis, yet retains an incomplete death domain. *Immunity* **7,** 813–820.
Demaurex, N., and Distelhorst, C. (2003). Apoptosis—the calcium connection. *Science* **300,** 65–67.
Du, C., Fang, M., Li, Y., Li, L., and Wang, X. (2000). Smac, a mitochondrial protein that promotes cytochrome c-dependent caspase activation by eliminating IAP inhibition. *Cell* **102,** 33–42.
Foyouzi-Youssefi, R., Arnaudeau, S., Borner, C., Kelley, W. L., Tschopp, J., Lew, D. P., Demaurex, N., and Krause, K. H. (2000). Bcl-2 decreases the free Ca^{2+} concentration within the endoplasmic reticulum. *Proc. Natl. Acad. Sci. USA* **97,** 5723–5728.
Griffith, T. S., and Lynch, D. H. (1998). TRAIL: A molecule with multiple receptors and control mechanisms. *Curr. Opin. Immunol.* **10,** 559–563.
Hajnoczky, G., Davies, E., and Madesh, M. (2003). Calcium signaling and apoptosis. *Biochem. Biophys. Res. Commun.* **304,** 445–454.
He, Q., Lee, D. I., Rong, R., Yu, M., Luo, X., Klein, M., El-Deiry, W. S., Huang, Y., Hussain, A., and Sheikh, M. S. (2002a). Endoplasmic reticulum calcium pool depletion-induced apoptosis is coupled with activation of the death receptor 5 pathway. *Oncogene* **21,** 2623–2633.
He, Q., Luo, X., Huang, Y., and Sheikh, M. S. (2002b). Apo2L/TRAIL differentially modulates the apoptotic effects of sulindac and a COX-2 selective non-steroidal anti-inflammatory agent in Bax-deficient cells. *Oncogene* **21,** 6032–6040.
He, Q., Montalbano, J., Corcoran, C., Jin, W., Huang, Y., and Sheikh, M. S. (2003). Effect of Bax deficiency on death receptor 5 and mitochondrial pathways during endoplasmic reticulum calcium pool depletion-induced apoptosis. *Oncogene* **22,** 2674–2679.

Jethmalani, S. M., and Henle, K. J. (1998). Calreticulin associates with stress proteins: Implications for chaperone function during heat stress. *J. Cell. Biochem.* **69,** 30–43.

Kaufman, R. J. (1999). Stress signaling from the lumen of the endoplasmic reticulum: Coordination of gene transcriptional and translational controls. *Genes Dev.* **13,** 1211–1233.

LeBlanc, H. N., and Ashkenazi, A. (2003). Apo2L/TRAIL and its death and decoy. *Cell Death Differ.* **10,** 66–75.

Li, L. Y., Luo, X., and Wang, X. (2001). Endonuclease G is an apoptotic DNase when released from mitochondria. *Nature* **412,** 95–99.

Liu, X., Kim, C. N., Yang, J., Jemmerson, R., and Wang, X. (1996). Induction of apoptotic program in cell-free extracts: Requirement for dATP and cytochrome c. *Cell* **86,** 147–157.

Liu, H., Bowes, R. C., 3rd, van de Water, B., Sillence, C., Nagelkerke, J. F., and Stevens, J. L. (1997). Endoplasmic reticulum chaperones GRP78 and calreticulin prevent oxidative stress, Ca^{2+} disturbances, and cell death in renal epithelial cells. *J. Biol. Chem.* **272,** 21751–21759.

Marsters, S. A., Sheridan, J. P., Pitti, R. M., Huang, A., Skubatach, M., Baldwin, D., Yuan, J., Gurney, A., Goddard, A. D., Godowski, P., and Ashkenazi, A. (1997). A novel receptor for Apo2L/TRAIL contains a truncated death domain. *Curr. Biol.* **7,** 1003–1006.

McConkey, D. J., Nicotera, P., Hartzell, P., Bellomo, G., Wyllie, A. H., and Orrenius, S. (1989). Glucocorticoids activate a suicide process in thymocytes through an elevation of cytosolic Ca2+ concentration. *Arch. Biochem. Biophys.* **269,** 365–370.

Nakamura, K., Bossy-Wetzel, E., Burns, K., Fadel, M. P., Lozyk, M., Goping, I. S., Opas, M., Bleackley, R. C., Green, D. R., and Michalak, M. (2000). Changes in endoplasmic reticulum luminal environment affect cell sensitivity to apoptosis. *J. Cell Biol.* **150,** 731–740.

Nesterov, A., Lu, X., Johnson, M., Miller, G. J., Ivashchenko, Y., and Kraft, A. S. (2001). Elevated Akt activity protects the prostate cancer cell line LNCap from TRAIL-induced apoptosis. *J. Biol. Chem.* **276,** 10767–10774.

Nimmanapalli, R., Perkins, C. L., Orlando, M., O'Bryan, E., Nguyen, D., and Bhalla, K. N. (2001). Pretreatment with paclitaxel enhances apo-2 ligand/tumor necrosis factor-related apoptosis-inducing ligand-induced apoptosis of prostate cancer cells by inducing death receptors 4 and 5 protein levels. *Cancer Res.* **61,** 759–763.

Nutt, L. K., Chandra, J., Pataer, A., Fang, B., Roth, J. A., Swisher, S. G., O'Neil, R. G., and McConkey, D J. (2002a). Bax-mediated Ca^{2+} mobilization promotes cytochrome c release during apoptosis. *J. Biol. Chem.* **277,** 20301–20308.

Nutt, L. K., Pataer, A., Pahler, J., Fang, B., Roth, J., McConkey, D. J., and Swisher, S. G. (2002b). Bax and Bak promote apoptosis by modulating endoplasmic reticular and mitochondrial Ca^{2+} stores. *J. Biol. Chem.* **277,** 9219–9225.

Orrenius, S., Zhivotovsky, B., and Nicotera, P. (2003). Regulation of cell death: The calcium-apoptosis link. *Nat. Rev. Mol. Cell. Biol.* **4,** 552–565.

Pacher, P., and Hajnoczky, G. (2001). Propagation of the apoptotic signal by mitochondrial waves. *EMBO J.* **20,** 4107–4121.

Pan, G., Ni, J., Wei, Y.-F., Yu, G.-L., Gentz, R., and Dixit, V. M. (1997). An antagonist decoy receptor and a death domain-containing receptor for TRAIL. *Science* **277,** 815–818.

Pan, G., Ni, J., Yu, G., Wei, Y.-F., and Dixit, V. M. (1998). TRUNDD, a new member of the TRAIL receptor family that antagonizes TRAIL signaling. *FEBS Lett.* **424,** 41–45.

Papp, S., Dziak, E., Michalak, M., and Opas, M. (2003). Is all of the endoplasmic reticulum created equal? The effects of the heterogeneous distribution of endoplasmic reticulum Ca^{2+}-handling proteins *J. Cell. Biol.* **160,** 475–479.

Pinton, P., Ferrari, D., Magalhaes, P., Schulze-Osthoff, K., Di Virgilio, F., Pozzan, T., and Rizzuto, R. (2000). Reduced loading of intracellular Ca^{2+} stores and downregulation of capacitative Ca^{2+} influx in Bcl-2-overexpressing cells. *J. Cell Biol.* **148,** 857–862.

Pinton, P., Ferrari, D., Rapizzi, E., Di Virgilio, F. D., Pozzan, T., and Rizzuto, R. (2001). The Ca^{2+} concentration of the endoplasmic reticulum is a key determinant of ceramide-induced apoptosis: Significance for the molecular mechanism of Bcl-2 action. *EMBO J.* **20**, 2690–2701.

Rapizzi, E., Pinton, P., Szabadkai, G., Wieckowski, M. R., Vandecasteele, G., Baird, G., Tuft, R. A., Fogarty, K. E., and Rizzuto, R. (2002). Recombinant expression of the voltage-dependent anion channel enhances the transfer of Ca^{2+} microdomains to mitochondria. *J. Cell Biol.* **159**, 613–624.

Rizzuto, R., Simpson, A. W., Brini, M., and Pozzan, T. (1992). Rapid changes of mitochondrial Ca^{2+} revealed by specifically targeted recombinant aequorin. *Nature* **358**, 325–327.

Sagara, Y., and Inesi, G. (1991). Inhibition of the sarcoplasmic reticulum Ca^{2+} transport ATPase by thapsigargin at subnanomolar concentrations. *J. Biol. Chem.* **266**, 13503–13506.

Schulze-Osthoff, K., Ferrari, D., Los, M., Wesselborg, S., and Peter, M. E. (1998). Apoptosis signaling by death receptors. *Eur. J. Biochem.* **254**, 439–459.

Scorrano, L., Oakes, S. A., Opferman, J. T., Cheng, E. H., Sorcinelli, M. D., Pozzan, T., and Korsmeyer, S. J. (2003). BAX and BAK regulation of endoplasmic reticulum Ca^{2+}: A control point for apoptosis. *Science* **300**, 135–139.

Scorrano, L., and Korsmeyer, S. J. (2003). Mechanisms of cytochrome c release by proapoptotic BCL-2 family members. *Biochem. Biophys. Res. Commun.* **304**, 437–444.

Sheikh, M. S., Burns, T. F., Huang, Y., Wu, G. S., Amundson, S., Brooks, K. S., Fornace, A. J., Jr., and El-Deiry, W. S. (1998). p53-dependent and independent regulation of the death receptor KILLER/DR5 gene expression in response to genotoxic stress and TNFα. *Cancer Res.* **58**, 1593–1598.

Sheridan, J. P., Marsters, S. A., Pitti, R. M., Gurney, A., Skubatch, M., Baldwin, D., Ramakrishnan, L., Gray, C. L., Baker, K., Wood, W. I., Goddard, A. D., Godowski, P., and Ashkenazi, A. (1997). Control of TRAIL-induced apoptosis by a family of signaling and decoy receptrors. *Science* **277**, 818–821.

Simonet, W. S., Lacey, D. L., Dunstan, C. R., Kelley, M., Chang, M. S., Luthy, R., Nguyen, H. Q., Wooden, S., Bennett, L., Boone, T., Shimamoto, G., DeRose, M., Elliott, R., Colombero, A., Tan, H. L., Trail, G., Sullivan, J., Davy, E., Bucay, N., Renshaw-Gegg, L., Hughes, T. M., Hill, D., Pattison, W., Campbell, P., Boyle, W. J. *et al.* (1997). Osteoprotegerin: A novel secreted protein involved in the regulation of bone density. *Cell* **89**, 309–319.

Singh, A., Ni, J., and Aggarwal, B. B. (1998). Death domain receptors and their role in cell demise. *J. Interferon Cytol. Res.* **18**, 439–450.

Susin, S. A., Lorenzo, H. K., Zamzami, N., Marzo, I., Snow, B. E., Brothers, G. M., Mangion, J., Jacotot, E., Costantini, P., Loeffler, M., Larochette, N., Goodlett, D. R., Aebersold, R., Siderovski, D. P., Penninger, J. M., and Kroemer, G. (1999). Molecular characterization of mitochondrial apoptosis-inducing factor. *Nature* **397**, 441–446.

Suzuki, Y., Imai, Y., Nakayama, H., Takahashi, K., Takio, K., and Takahashi, R. (2001). A serine protease, HtrA2, is released from the mitochondria and interacts with XIAP, inducing cell death. *Mol. Cell* **8**, 613–621.

Swarthout, J. T., Doggett, T. A., Lemker, J. L., and Partridge, N. C. (2001). Stimulation of extracellular signal-regulated kinases and proliferation in rat osteoblastic cells by parathyroid hormone is protein kinase C-dependent. *J. Biol. Chem.* **276**, 7586–7592.

Szalai, G., Krishnamurthy, R., and Hajnoczky, G. (1999). Apoptosis driven by IP(3)-linked mitochondrial calcium signals. *EMBO J.* **18**, 6349–6361.

Thastrup, O., Cullen, P. J., Drobak, B. K., Hanley, M. R., and Dawson, A. P. (1990). Thapsigargin, a tumor promoter, discharges intracellular Ca2+ stores by specific inhibition of the endoplasmic reticulum Ca^{2+}-ATPase. *Proc. Natl. Acad. Sci. USA* **87**, 2466–2470.

Tsujimoto, Y. (2003). Cell death regulation by the Bcl-2 protein family in the mitochondria. *J. Cell. Physiol.* **195**, 158–167.

Verhagen, A. M., Ekert, P. G., Pakusch, M., Silke, J., Connolly, L. M., Reid, G. E., Moritz, R. L., Simpson, R. J., and Vaux, D. L. (2000). Identification of DIABLO, a mammalian protein that promotes apoptosis by binding to and antagonizing IAP proteins. *Cell* **102,** 43–53.

Walczak, H., Miller, R. E., Ariail, K., Gliniak, B., Griffith, T. S., Kubin, M., Chin, W., Jones, J., Woodward, A., Le, T., Smith, C., Smolak, P., Goodwin, R. G., Rauch, C. T., Schuh, J. C., and Lynch, D. H. (1999). Tumoricidal activity of tumor necrosis factor–related apoptosis-inducing ligand in vivo. *Nat. Med.* **5,** 157–163.

Wertz, I. E., and Dixit, V. M. (2000). Characterization of calcium release-activated apoptosis of LNCaP prostate cancer cells. *J. Biol. Chem.* **275,** 11470–11477.

Wu, G. S., Burns, T. F., McDonald, E. R., III, Jiang, W., Meng, R., Krantz, I. D., Kao, G., Gan, D. D., Zhou, J-Y, Muschel, R., Hamilton, S. R., Spinner, N. B., Markowitz, S., Wu, G., and El-Deiry, W. S. (1997). KILLER/DR5 is DNA damage-inducible p53-regulated death receptor gene. *Nature Gen.* **17,** 141–143.

Wyllie, A. H. (1980). Glucocorticoid-induced thymocyte apoptosis is associated with endogenous endonuclease activation. *Nature* **284,** 555–556.

Yamaguchi, H., Bhalla, K., and Wang, H. G. (2003). Bax plays a pivotal role in thapsigargin-induced apoptosis of human colon cancer HCT116 cells by controlling Smac/Diablo and Omi/HtrA2 release from mitochondria. *Cancer Res.* **63,** 1483–1489.

11

FLIP PROTEIN AND TRAIL-INDUCED APOPTOSIS

WILFRIED ROTH AND JOHN C. REED

The Burnham Institute, La Jolla, California 92037

I. Introduction
II. FLIP: Structure and Mechanisms of Action
    A. Domain Structure of the FLIP Protein
    B. Mechanisms of Action
    C. Alternative Mechanisms
    D. Dual Functionality
    E. Expression of FLIP
III. FLIP Function in Normal Physiology and Disease
    A. Cardiovascular System
    B. Immune System
    C. Cancer
IV. TRAIL and FLIP
    A. Role of TRAIL for Immunosurveillance of Cancer
    B. Therapeutic Possibilities of TRAIL
    C. FLIP and Resistance to TRAIL
    D. Sensitizing Tumor Cells by Antagonizing FLIP
References

Death ligands (such as Fas/CD95 ligand and TRAIL/Apo2L) and death receptors (such as Fas/CD95, TRAIL-R1/DR4, and TRAIL-R2/DR5) are involved in immune-mediated neutralization of activated or autoreactive lymphocytes, virus-infected cells, and tumor cells. Consequently, dysregulation of death receptor–dependent apoptotic signaling pathways has been implicated in the development of autoimmune diseases, immunodeficiency, and cancer. Moreover, the death ligand TRAIL has gained considerable interest as a potential anticancer agent, given its ability to induce apoptosis of tumor cells without affecting most types of untransformed cells. The FLICE-inhibitory protein (FLIP) potently blocks TRAIL-mediated cell death by interfering with caspase-8 activation. Pharmacologic down-regulation of FLIP might serve as a therapeutic means to sensitize tumor cells to apoptosis induction by TRAIL. © 2004 Elsevier Inc.

I. INTRODUCTION

Apoptosis is a tightly regulated process that is fundamental for the maintainance of cellular homeostasis in multicellular organisms. Because virtually all cells possess armaments to commit cell suicide, powerful mechanisms must exist to prevent accidental triggering of cell death. Therefore, cells are well equipped with a wide range of antiapoptotic proteins, such as Bcl-2 and IAP family members. In many cell types, apoptosis can be inhibited by the FLICE-inhibitory protein (FLIP), which is also known by the synonyms CFLAR, CASH, CLARP, Casper, FLAME1, I-FLICE, MRIT, or Usurpin (Goltsev *et al.*, 1997; Han *et al.*, 1997; Hu *et al.*, 1997; Inohara *et al.*, 1997; Irmler *et al.*, 1997; Rasper *et al.*, 1998; Shu *et al.*, 1997; Srinivasula *et al.*, 1997). FLIP is best known for its inhibitory effects on the death receptor–mediated apoptotic signaling cascade. Several studies suggest that FLIP expression might be dysregulated in cancer, autoimmune disorders, and cardiovascular diseases (reviewed in French and Tschopp, 2002; Igney and Krammer, 2002; Thome and Tschopp, 2001).

II. FLIP: STRUCTURE AND MECHANISMS OF ACTION

A. DOMAIN STRUCTURE OF THE FLIP PROTEIN

The first FLIP orthologs discovered were of viral origin (Thome *et al.*, 1997). By inhibiting death receptor–induced apoptosis, viruses might delay cytotoxic T lymphocyte–mediated eradication of the infected host cells, thereby prolonging viral replication time (Bertin *et al.*, 1997; Sturzl *et al.*, 1999; Tschopp *et al.*, 1998; Wang *et al.*, 1997). Mammalian FLIPs were discovered shortly thereafter (Goltsev *et al.*, 1997; Han *et al.*, 1997; Hu *et al.*, 1997; Inohara *et al.*, 1997; Irmler *et al.*, 1997; Rasper *et al.*, 1998; Shu *et al.*, 1997; Srinivasula *et al.*, 1997). The domain structures of viral and

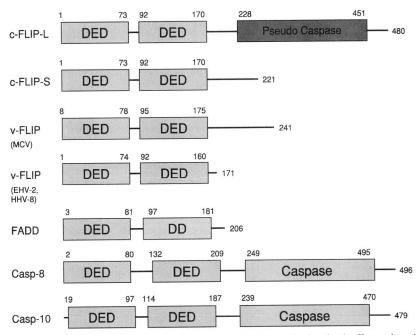

FIGURE 1. Domain structure of cellular and viral FLIPs and other death effector domain (DED)-containing proteins. Numbers represent amino acid residue positions. DD, death domain; MCV, molluscipoxvirus; EHV, equine herpesvirus; HHV, human herpesvirus; Casp-8, Caspase-8; Casp-10, Caspase-10.

cellular FLIPs are quite similar (Fig. 1). The human *FLIP* gene, located on chromosome 2q33-q34, shares 65% amino acid sequence identity with the mouse ortholog, which is located on chromosome 1 (Reed *et al.*, 2003). Mammalian "cellular" FLIP (c-FLIP) shares extensive amino acid sequence similarity with procaspase-8 and -10 and features two death effector domains (DEDs) at the N-terminus and one caspase-like domain at the C-terminus. The caspase-like domain lacks enzymatic activity because it does not carry the critical residues required for protease activity, including the catalytic cysteine (Cohen, 1997). However, similar to caspase-8, the FLIP protein can be cleaved at a conserved aspartic-acid cleavage site between the p20 and p12 domains (Srinivasula *et al.*, 1997) (Fig. 2). Several splicing isoforms of FLIPs have been reported, with the two most predominant translating into FLIP-L (55 kD) and FLIP-S (26 kD) proteins.

B. MECHANISMS OF ACTION

Following binding of death ligands of the tumor necrosis factor (TNF) family (such as Fas/CD95 ligand and TRAIL/Apo2L) to their corresponding death receptors (Fas/CD95, TRAIL-R1/DR4, and TRAIL-R2/DR5,

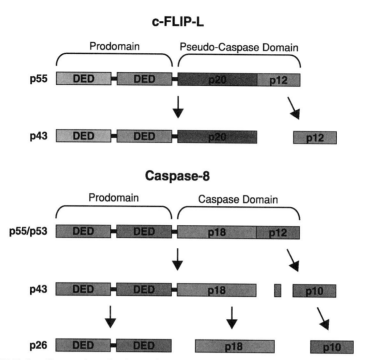

FIGURE 2. Proposed mechanisms of FLIP-L and caspase-8 cleavage. Cleavage sites and the size of cleavage products are indicated.

respectively), the "death-inducing signalling complex" (DISC) is formed (Ashkenazi and Dixit, 1998; Kischkel et al., 1995). In this protein complex, the receptors recruit the adapter protein FADD and caspase-8 to their cytosolic domains, which activates caspase-8 by an "induced proximity" mechanism involving clustering of the pro-caspase-8 zymogen around aggregated receptor complexes (Salvesen and Dixit, 1999; Sprick et al., 2000). This caspase-8 clustering results in its cleavage to p43 and p10 fragments and, after further processing of p43, the active p18 subunit is released (Medema et al., 1997) (Fig. 2). If FLIP is present, however, it binds via homotypic DED interactions to FADD, resulting in recruitment and proteolytic cleavage of FLIP in the DISC (Goltsev et al., 1997; Han et al., 1997; Irmler et al., 1997; Srinivasula et al., 1997) (Fig. 3). The partially processed FLIP (p43) retains in the DISC because it binds to the FADD/death receptor complexes (Krueger et al., 2001). In addition to the p43 fragment of FLIP-L, full length FLIP-L and FLIP-S have also been reported to inhibit activation of caspase-8 in the DISC (Krueger et al., 2001; Scaffidi et al., 1999). Thus, FLIP proteins act as dominant-negative inhibitors of caspase-8 by preventing the further recruitment and processing of pro-caspase-8, as well as the release of active caspase-8 from the receptor complex.

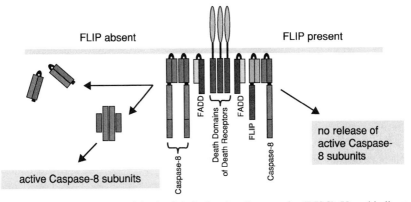

FIGURE 3. Components of the death-inducing signaling complex (DISC). Upon binding to the DISC, FLIP (p43) prevents the processing of caspase-8 and the release of its active subunits.

C. ALTERNATIVE MECHANISMS

Apart from the well-characterized inhibition of death receptor–induced signaling by interfering with DISC assembly or function, FLIP might also exercise antiapoptotic functions by other mechanisms of action.

1. Interaction with DEDD Proteins

DEDD1 and DEDD2 are two related proteins that contain an N-terminal DED and two nuclear localization sequences (NLS) (Roth et al., 2002; Stegh et al., 1998). The DEDD proteins induce apoptosis in several cell types, and structure-function studies indicate that nuclear localization of DEDD1 and DEDD2 is required for their proapoptotic activity. Interestingly, DEDD2 binds to FLIP, but not to other DED-containing proteins (Roth et al., 2002). FLIP can be sequestered to the nucleus by over-expressed DEDD2, which could explain why DEDD2 promotes predominantly death receptor–induced cell death (Roth et al., 2002; Zhan et al., 2002). Further studies are needed to elucidate the extent to which FLIP is involved in apoptosis induced by DEDD proteins.

2. Activation of NF-κB

Overexpression of FLIP is known to activate the transcription factor nuclear factor κB (NF-κB) (Chaudhary et al., 1999, 2000; Hu et al., 2000; Kataoka et al., 2000). Therefore, FLIP might have a role in the regulation of NF-κB–dependent gene expression, which could affect cellular proliferation in response to stimulation of death receptors. Consistent with its role in the NF-κB pathway, FLIP can interact with other NF-κB–related signaling proteins, such as TRAF1, TRAF2, and IKK2 (Chaudhary et al., 1999; Shu et al., 1997).

D. DUAL FUNCTIONALITY

Several reports indicated that when overexpressed, FLIP induces rather than inhibits apoptosis (Goltsev *et al.*, 1997; Han *et al.*, 1997; Inohara *et al.*, 1997; Irmler *et al.*, 1997; Shu *et al.*, 1997). This finding was explained by the non-physiologic aggregation of large amounts of FLIP and caspase-8, leading to processing and activation of caspase-8. However, at lower and nearer to physiologic expression levels, FLIP protects against death receptor–induced apoptosis (Goltsev *et al.*, 1997; Hu *et al.*, 1997; Irmler *et al.*, 1997; Srinivasula *et al.*, 1997). The antiapoptotic properties of FLIP *in vivo* were also supported by the finding that embryonic fibroblasts derived from FLIP knockout mice exhibit increased susceptibility to death receptor–induced cell death, whereas they show normal responses to etoposide, sorbitol, staurosporine and anisomycin (Yeh *et al.*, 2000). In contrast, recent studies have suggested that even at low near-physiologic concentrations, FLIP can be both an inhibitor and enhancer of death receptor–mediated apoptosis, indicating that a rheostat effect between the levels of caspase-8 and FLIP might determine the ultimate sensitivity of cells to apoptosis induced by death receptors (Chang *et al.*, 2002; Micheau *et al.*, 2002). This hypothesis is based on the findings that FLIP can induce caspase-8 activation in an inducible dimerization system and that antisense-mediated reductions in FLIP decrease rather than increase the sensitivity of HeLa cells to Fas/CD95-involved apoptosis (Chang *et al.*, 2002). Thus, subtle changes in intracellular FLIP levels could determine the outcome of death receptor–mediated signaling resulting either in cell death or cell survival. If this scenario proves true, strict control of FLIP synthesis and degradation would be of utmost importance for the regulation of cell death and survival.

E. EXPRESSION OF FLIP

FLIP mRNA is widely expressed, with high FLIP-L protein levels occurring in heart, skeletal muscle, lymphoid tissues and kidney (Rasper *et al.*, 1998). Little or no FLIP-L expression was detected in mammary glands, lung, pancreas, colon, placenta, and testis. FLIP-S expression is more restricted and is characterized by high protein levels in lymphoid tissues (thymus, lymph nodes, and spleen) (Rasper *et al.*, 1998).

III. FLIP FUNCTION IN NORMAL PHYSIOLOGY AND DISEASE

A. CARDIOVASCULAR SYSTEM

The generation of FLIP knockout mice has provided interesting insights into the physiologic roles of the FLIP protein. FLIP knockout mice die in utero (E10.5–11.5) from cardiac failure associated with severely impaired heart development (Yeh *et al.*, 2000). A potential role for FLIP in cardiovascular disease is supported by the finding of decreased FLIP expression in myocardial tissue in the setting of ischemia-reperfusion injury (Rasper *et al.*, 1998). Thus, downregulation of FLIP might be associated with increased sensitivity to apoptosis in ischemic heart tissue. FLIP is highly expressed not only in the adult human heart but also in endothelial and smooth muscle cells in arteries (Imanishi *et al.*, 2000b; Rasper *et al.*, 1998). Normal endothelial cells are resistant to death ligand–mediated apoptosis despite expression of death receptors. However, FLIP expression is reduced in smooth muscle cells in atherosclerotic plaques, which is accompanied by sensitization to CD95-mediated cell death (Imanishi *et al.*, 2000a; Sata and Walsh, 1998). In light of these findings, a role for FLIP in pathologic processes in the cardiovascular system, such as atherosclerosis, can be presumed.

B. IMMUNE SYSTEM

By means of their antiapoptotic activity, viral FLIP proteins can interfere with the immune system of the host by delaying the apoptotic death of infected cells. On the other hand, c-FLIP is expressed by immune cells themselves and might play an essential role in immunologic processes such as avoidance of activation-induced cell death (AICD) of T lymphocytes (Algeciras-Schimnich *et al.*, 1999; Inaba *et al.*, 1999; Irmler *et al.*, 1997; Kirchhoff *et al.*, 2000; Refaeli *et al.*, 1998; Van Parijs *et al.*, 1999; Yeh *et al.*, 1998) or the neutralization of peripheral autoreactive T cells (Algeciras-Schimnich *et al.*, 1999; Holler *et al.*, 2000; Irmler *et al.*, 1997; Refaeli *et al.*, 1998; Semra *et al.*, 2001). Defects in either of these two mechanisms can result in the accumulation of autoreactive immune effector cells and consequently in autoimmune diseases. In multiple sclerosis, a disease in which autoreactive T cells attack antigens in the myelin of nerve sheaths in the central nervous system, FLIP was found to be overexpressed in immune effector cells in the cerebrospinal fluid of patients (Conlon *et al.*, 1999; Semra *et al.*, 2001; Sharief, 2000). In patients with Graves disease, a form of autoimmune hyperthyroidism characterized by autoantibody-mediated thyrotropin receptor stimulation, FLIP expression was increased in thyrocytes, which was associated with resistance to Fas/CD95-mediated cell death (Stassi *et al.*, 2000).

C. CANCER

1. Immune Evasion

Several tumor types have been reported to have inappropriately elevated levels of FLIP, e.g. melanoma, colon carcinoma, and Hodgkin lymphoma (Bullani *et al.*, 2001; Ryu *et al.*, 2001; Thomas *et al.*, 2002a). High expression levels of FLIP in tumor cells could lead to resistance to apoptosis induction by death ligand–expressing cytotoxic lymphocytes (French and Tschopp, 1999), which is exemplified by the finding that resistance to Fas/CD95-mediated apoptosis of B cell lymphoma cell lines increases with high FLIP expression (Irisarri et al., 2000; Mueller and Scott, 2000; Tepper and Seldin, 1999). In an immunocompetent mouse model, FLIP overexpressing B lymphoma tumor cells developed into more aggressive tumors than the respective control cells, both in syngenic and semiallogeneic tumor-host systems (Djerbi et al., 1999; Medema et al., 1999), indicating that FLIP expression was associated with a selective advantage for tumor growth. In another study, FLIP overexpression in tumor cells was capable of preventing tumor rejection by perforin-deficient NK cells (Taylor et al., 2001). In conclusion, de-regulated FLIP expression in tumor cells could act as a tumor-progression factor and promote the evasion of cancer cells from immune surveillance mechanisms.

2. Counter-attack Hypothesis

Although controversial, FLIP-mediated resistance to Fas/CD95-mediated apoptosis may allow tumor cells to tolerate expressing Fas/CD95 ligand and to use this death ligand as a weapon against neighboring normal cells or to trigger apoptosis of cytotoxic immune effector cells (Bennett et al., 1998; O'Connell et al., 1996, 1998).

3. Drug Resistance

The role of the death receptor pathway for chemotherapeutic drug-mediated cell death of tumor cells remains unclear, but at least in some cancer cell lines the expression of death receptors can be induced by treating the cells with cytotoxic drugs (Fulda et al., 1997; Herr and Debatin, 2001). In leukemia T cell lines, a direct involvement of the death ligand/death receptor system in doxorubicin-mediated apoptosis has been suggested (Friesen et al., 1996). However, FADD knockout and caspase-8 knockout cells were as sensitive to chemotherapeutic drugs as their respective control counterparts, which is inconsistent with a general involvement of death receptor-mediated signaling in drug-induced cell death (reviewed in Los et al., 1999). Similarly, drug-induced cytotoxicity of glioblastoma and T-acute lymphatic leukemia cells was independent of death receptors (Glaser et al., 1999; Villunger et al., 1997). Provided that the contribution of death receptor–mediated pathways to drug-induced cell death depends on the specific cell type, it may not be surprising that conflicting data exist about the capability of FLIP to block chemotherapy-induced apoptosis (Engels et al., 2000; Kataoka et al., 1998; Matta et al., 2002).

IV. TRAIL AND FLIP

A. ROLE OF TRAIL FOR IMMUNOSURVEILLANCE OF CANCER

In addition to the Perforin/Granzyme pathway, death ligand/death receptor–mediated apoptosis plays an important role in cytotoxic T cell–mediated lysis of tumor cells. Recent findings suggest that the death ligand TRAIL might be crucial for immunosurveillance of cancer. TRAIL is a transmembrane type-II protein and its extracellular domain can be proteolytically cleaved from the cell surface. The apoptosis-inducing receptors for TRAIL include TRAIL-R1 (DR4) and TRAIL-R2 (DR5), which are transmembrane type-I proteins expressed on the surface of many cell types. However, TRAIL also binds to non-apoptosis–inducing decoy receptors, which compete with death receptors for ligand and suppress apoptosis, including TRAIL-R3 (DcR1), TRAIL-R4 (DcR2), and osteoprotegerin (OPG). TRAIL is constitutively expressed on murine liver natural killer (NK) cells, and neutralizing anti-TRAIL antibodies increased metastasis of liver carcinoma in a mouse model, suggesting a role for TRAIL in NK cytotoxicity toward cancer cells (Takeda *et al.*, 2001). Moreover, primary human immature NK cells can kill Jurkat cells by TRAIL-dependent pathways, whereas the main mechanism of tumor cell lysis by mature NK cells may be Perforin/Granzyme or Fas/CD95 pathways (Zamai *et al.*, 1998). Killing of target cells by cytotoxic T cells (Kayagaki *et al.*, 1999; Thomas and Hersey, 1998) as well as monocytes (Griffith *et al.*, 1999) could also depend on TRAIL.

B. THERAPEUTIC POSSIBILITIES OF TRAIL

Considerable interest in the possibility of exploiting the apoptotic effects of TRAIL for the treatment of cancer has emerged. Analysis of the effects of TRAIL on normal and malignant cells has provided compelling evidence that recombinant TRAIL protein preferentially induces apoptosis of cancer cells without harming most types of untransformed cells (Walczak *et al.*, 1999). Recombinant TRAIL also lacks significant toxicity in primate species that possess receptors capable of binding human TRAIL, suggesting that most normal cells are protected from TRAIL-induced apoptosis *in vivo* (Ashkenazi *et al.*, 1999). Preclinical studies of recombinant TRAIL in mice have demonstrated impressive antitumor activity and synergy with cytotoxic anticancer drugs in several tumor models (Ashkenazi *et al.*, 1999; Nagane *et al.*, 2000). Promising preclinical data have also been reported by Sankyo, Inc. for an agonistic antibody that binds TRAIL-R2 (Ichikawa *et al.*, 2001). Encouraged by the impressive preclinical data, recombinant nontagged, trimeric TRAIL is scheduled to enter human clinical trials soon, through a joint venture by

Genentech, Inc. and Amgen, Inc. Moreover, agonistic monoclonal antibodies that bind and activate TRAIL-R1, have recently completed phase 1 clinical trials, by Human Genome Sciences, Inc. in collaboration with Cambridge Antibody Technologies, Inc. (www.hgsi.com/). Human Genome Sciences, Inc. has also initiated a clinical phase 1 study to evaluate the safety and pharmacology of TRAIL-R2 monoclonal antibodies in patients with advanced tumors.

C. FLIP AND RESISTANCE TO TRAIL

On the basis of their potential to directly induce apoptosis of cancer cells, death receptor-targeting therapeutic approaches are expected to circumvent some of the resistance mechanisms that often hamper chemotherapy and radiation therapy (Johnstone *et al.*, 2002; Reed, 2002; Yount *et al.*, 1999). However, not all tumor cells respond to death ligands. Many cancer cell lines are partially or even completely resistant to TRAIL. In some cases, the increased ratio of decoy receptors over death receptors on the surface of tumor cells may account for resistance, but more experimental evidence supports the importance of intracellular blocks to apoptosis for cancer cell resistance to TRAIL (Griffith *et al.*, 1998; Leverkus *et al.*, 2000). With respect to intracellular resistance mechanisms, antiapoptotic proteins such as FLIP have been implicated in resistance of some cancer cell lines. This is supported by the finding that TRAIL-resistant tumor cells can be re-sensitized by specifically downregulating FLIP using antisense oligonucleotides (Kim *et al.*, 2002; Mitsiades *et al.*, 2002; Siegmund *et al.*, 2002). Moreover, endogenous FLIP expression might correlate with resistance to Fas/CD95 or TRAIL in several cancer cell lines, but obviously not in all (Griffith *et al.*, 1998; Hao *et al.*, 2001; Irmler *et al.*, 1997; Mitsiades *et al.*, 2001; Raisova *et al.*, 2000; Ugurel *et al.*, 1999; Zhang *et al.*, 1999). Similarly, inhibition of the mitochondrial apoptosis pathway by Bcl-2 or Bcl-XL inhibits TRAIL-induced apoptosis in some cancer cell lines but not in others (reviewed in LeBlanc and Ashkenazi, 2003). Loss of *bax* gene expression might be an important mechanism for TRAIL resistance in colon cancer, because TRAIL-mediated apoptosis was completely blocked in Bax-deficient colon carcinoma cell lines (Deng *et al.*, 2002; LeBlanc *et al.*, 2002). Given this differential dependence on mitochondrial pathways, an urgent need exists to uncover the causes of TRAIL resistance in specific tumor types, because knowledge of such mechanisms is a prerequisite to developing improved therapeutic strategies for cancer.

D. SENSITIZING TUMOR CELLS BY ANTAGONIZING FLIP

During evolution of tumors *in vivo*, selection occurs for malignant clones that are capable of withstanding the immune attack of the host. Given the importance of death ligands/death receptors for the anti-tumor defense of the immune system, a successful biologic therapy will substantially benefit by restoring competency of the death receptor–mediated apoptotic signaling pathway. Therefore, pharmacologic agents that sensitize tumor cells to apoptosis-inducing proteins such as TRAIL or anti-TRAIL receptor antibodies would be a welcome addition to the battery of weapons available to medical oncologists. A possible mechanism for accomplishing this goal has recently been revealed by studies using synthetic triterpenoids, such as CDDO, that trigger a poorly defined pathway for ubiquitination and degradation of FLIP (Kim *et al.*, 2002; Pedersen *et al.*, 2002). At least *in vitro*, these compounds sensitize solid tumor cell lines to TRAIL, and induce apoptosis as single agents in leukemia cells, through a caspase-8–dependent mechanism that remains operative even in chemorefractory cells (Ito *et al.*, 2000; Kim *et al.*, 2002; Pedersen *et al.*, 2002). It remains to be determined whether these promising observations will hold true *in vivo*.

Other substances have also been reported to sensitize tumor cells to TRAIL, presumably by downregulating FLIP, including doxorubicin (Kelly *et al.*, 2002), cycloheximide (Kreuz *et al.*, 2001), actinomycin (Olsson *et al.*, 2001), sodium butyrate (Hernandez *et al.*, 2001), bisindolylmaleimide (Poulaki *et al.*, 2002), the proteasome inhibitor PS-341 (Sayers *et al.*, 2003), the permeable NF-κB inhibitor SN50 (Bortul *et al.*, 2003), NEMO-binding domain (NBD) peptide (Thomas *et al.*, 2002b), and 9-nitrocamptothecin (Chatterjee *et al.*, 2001). Modulation of TRAIL sensitivity of cancer cells also occurs after altering FLIP expression levels by affecting the activity of transcription factors, e.g. NF-κB (Kreuz *et al.*, 2001) and p53 (Bartke *et al.*, 2001; Fukazawa *et al.*, 2001), by regulating the activity of kinases, e.g. phosphatidylinositol 3-kinase/Akt (Asakuma *et al.*, 2003; Panka *et al.*, 2001) and MEK1 (Panka *et al.*, 2001), or by other stimuli, e.g. infection of cells with FasL-expressing adenovirus (Hyer *et al.*, 2002) or altering glucose levels (Munoz-Pinedo *et al.*, 2003). However, in most of the previously mentioned studies, further evidence is desirable to corroborate that downregulation of FLIP is mainly responsible for sensitization of cells to TRAIL-mediated apoptosis. Because FLIP is short-lived, protein levels will rapidly decrease if inhibition of transcription or protein synthesis occurs as a secondary effect of treatment.

To what extent these novel approaches can be exploited for possible apoptosis-based therapies for cancer remains to be determined. Continued efforts to understand the regulation of cell death and the *in vivo* functions of FLIP and other antiapoptotic proteins should reveal ways of restoring competency of the death receptor–mediated apoptotic signaling pathway in cancer cells.

REFERENCES

Algeciras-Schimnich, A., Griffith, T. S., Lynch, D. H., and Paya, C. V. (1999). Cell cycle-dependent regulation of FLIP levels and susceptibility to Fas-mediated apoptosis. *J. Immunol.* **162**, 5205–5211.

Asakuma, J., Sumitomo, M., Asano, T., and Hayakawa, M. (2003). Selective Akt inactivation and tumor necrosis factor-related apoptosis-inducing ligand sensitization of renal cancer cells by low concentrations of paclitaxel. *Cancer Res.* **63**, 1365–1370.

Ashkenazi, A., and Dixit, V. M. (1998). Death receptors: Signaling and modulation. *Science* **281**, 1305–1308.

Ashkenazi, A., Pai, R. C., Fong, S., Leung, S., Lawrence, D. A., Marsters, S. A., Blackie, C., Chang, L., McMurtrey, A. E., Hebert, A., DeForge, L., Koumenis, I. L., Lewis, D., Harris, L., Bussiere, J., Koeppen, H., Shahrokh, Z., and Schwall, R. H. (1999). Safety and antitumor activity of recombinant soluble Apo2 ligand. *J. Clin. Invest.* **104**, 155–162.

Bartke, T., Siegmund, D., Peters, N., Reichwein, M., Henkler, F., Scheurich, P., and Wajant, H. (2001). p53 upregulates cFLIP, inhibits transcription of NF-kappaB-regulated genes and induces caspase-8-independent cell death in DLD-1 cells. *Oncogene* **20**, 571–580.

Bennett, M. W., O'Connell, J., O'Sullivan, G. C., Brady, C., Roche, D., Collins, J. K., and Shanahan, F. (1998). The Fas counterattack in vivo: Apoptotic depletion of tumor-infiltrating lymphocytes associated with Fas ligand expression by human esophageal carcinoma. *J. Immunol.* **160**, 5669–5675.

Bertin, J., Armstrong, R. C., Ottilie, S., Martin, D. A., Wang, Y., Banks, S., Wang, G. H., Senkevich, T. G., Alnemri, E. S., Moss, B., Lenardo, M. J., Tomaselli, K. J., and Cohen, J. I. (1997). Death effector domain-containing herpesvirus and poxvirus proteins inhibit both Fas- and TNFR1-induced apoptosis. *Proc. Natl. Acad. Sci. USA* **94**, 1172–1176.

Bortul, R., Tazzari, P. L., Cappellini, A., Tabellini, G., Billi, A. M., Bareggi, R., Manzoli, L., Cocco, L., and Martelli, A. M. (2003). Constitutively active Akt1 protects HL60 leukemia cells from TRAIL-induced apoptosis through a mechanism involving NF-kappaB activation and cFLIP(L) up-regulation. *Leukemia* **17**, 379–389.

Bullani, R. R., Huard, B., Viard-Leveugle, I., Byers, H. R., Irmler, M., Saurat, J. H., Tschopp, J., and French, L. E. (2001). Selective expression of FLIP in malignant melanocytic skin lesions. *J. Invest. Dermatol.* **117**, 360–364.

Chang, D. W., Xing, Z., Pan, Y., Algeciras-Schimnich, A., Barnhart, B. C., Yaish-Ohad, S., Peter, M. E., and Yang, X. (2002). c-FLIP(L) is a dual function regulator for caspase-8 activation and CD95-mediated apoptosis. *EMBO J.* **21**, 3704–3714.

Chatterjee, D., Schmitz, I., Krueger, A., Yeung, K., Kirchhoff, S., Krammer, P. H., Peter, M. E., Wyche, J. H., and Pantazis, P. (2001). Induction of apoptosis in 9-nitrocamptothecin-treated DU145 human prostate carcinoma cells correlates with de novo synthesis of CD95 and CD95 ligand and down-regulation of c-FLIP(short). *Cancer Res.* **61**, 7148–7154.

Chaudhary, P. M., Jasmin, A., Eby, M. T., and Hood, L. (1999). Modulation of the NF-kappaB pathway by virally encoded death effector domains-containing proteins. *Oncogene* **18**, 5738–5746.

Chaudhary, P. M., Eby, M. T., Jasmin, A., Kumar, A., Liu, L., and Hood, L. (2000). Activation of the NF-kappaB pathway by caspase 8 and its homologs. *Oncogene* **19**, 4451–4460.

Cohen, G. M. (1997). Caspases: The executioners of apoptosis. *Biochem. J.* **326**(Pt 1), 1–16.

Conlon, P., Oksenberg, J. R., Zhang, J., and Steinman, L. (1999). The immunobiology of multiple sclerosis: An autoimmune disease of the central nervous system. *Neurobiol. Dis.* **6**, 149–166.

Deng, Y., Lin, Y., and Wu, X. (2002). TRAIL-induced apoptosis requires Bax-dependent mitochondrial release of Smac/DIABLO. *Genes Dev.* **16**, 33–45.

Djerbi, M., Screpanti, V., Catrina, A. I., Bogen, B., Biberfeld, P., and Grandien, A. (1999). The inhibitor of death receptor signaling, FLICE-inhibitory protein defines a new class of tumor progression factors. *J. Exp. Med.* **190**, 1025–1032.

Engels, I. H., Stepczynska, A., Stroh, C., Lauber, K., Berg, C., Schwenzer, R., Wajant, H., Janicke, R. U., Porter, A. G., Belka, C., Gregor, M., Schulze-Osthoff, K., and Wesselborg, S. (2000). Caspase-8/FLICE functions as an executioner caspase in anticancer drug-induced apoptosis. *Oncogene* **19**, 4563–4573.

French, L. E., and Tschopp, J. (1999). Inhibition of death receptor signaling by FLICE-inhibitory protein as a mechanism for immune escape of tumors. *J. Exp. Med.* **190**, 891–894.

French, L. E., and Tschopp, J. (2002). Defective death receptor signaling as a cause of tumor immune escape. *Semin. Cancer Biol.* **12**, 51–55.

Friesen, C., Herr, I., Krammer, P. H., and Debatin, K. M. (1996). Involvement of the CD95 (APO-1/FAS) receptor/ligand system in drug-induced apoptosis in leukemia cells. *Nat. Med.* **2**, 574–577.

Fukazawa, T., Fujiwara, T., Uno, F., Teraishi, F., Kadowaki, Y., Itoshima, T., Takata, Y., Kagawa, S., Roth, J. A., Tschopp, J., and Tanaka, N. (2001). Accelerated degradation of cellular FLIP protein through the ubiquitin-proteasome pathway in p53-mediated apoptosis of human cancer cells. *Oncogene* **20**, 5225–5231.

Fulda, S., Sieverts, H., Friesen, C., Herr, I., and Debatin, K. M. (1997). The CD95 (APO-1/Fas) system mediates drug-induced apoptosis in neuroblastoma cells. *Cancer Res.* **57**, 3823–3829.

Glaser, T., Wagenknecht, B., Groscurth, P., Krammer, P. H., and Weller, M. (1999). Death ligand/receptor-independent caspase activation mediates drug-induced cytotoxic cell death in human malignant glioma cells. *Oncogene* **18**, 5044–5053.

Goltsev, Y. V., Kovalenko, A. V., Arnold, E., Varfolomeev, E. E., Brodianskii, V. M., and Wallach, D. (1997). CASH, a novel caspase homologue with death effector domains. *J. Biol. Chem.* **272**, 19641–19644.

Griffith, T. S., Chin, W. A., Jackson, G. C., Lynch, D. H., and Kubin, M. Z. (1998). Intracellular regulation of TRAIL-induced apoptosis in human melanoma cells. *J. Immunol.* **161**, 2833–2840.

Griffith, T. S., Wiley, S. R., Kubin, M. Z., Sedger, L. M., Maliszewski, C. R., and Fanger, N. A. (1999). Monocyte-mediated tumoricidal activity via the tumor necrosis factor-related cytokine, TRAIL. *J. Exp. Med.* **189**, 1343–1354.

Han, D. K., Chaudhary, P. M., Wright, M. E., Friedman, C., Trask, B. J., Riedel, R. T., Baskin, D. G., Schwartz, S. M., and Hood, L. (1997). MRIT, a novel death-effector domain-containing protein, interacts with caspases and BclXL and initiates cell death. *Proc. Natl. Acad. Sci. USA* **94**, 11333–11338.

Hao, C., Beguinot, F., Condorelli, G., Trencia, A., Van Meir, E. G., Yong, V. W., Parney, I. F., Roa, W. H., and Petruk, K. C. (2001). Induction and intracellular regulation of tumor necrosis factor-related apoptosis-inducing ligand (TRAIL) mediated apotosis in human malignant glioma cells. *Cancer Res.* **61**, 1162–1170.

Hernandez, A., Thomas, R., Smith, F., Sandberg, J., Kim, S., Chung, D. H., and Evers, B. M. (2001). Butyrate sensitizes human colon cancer cells to TRAIL-mediated apoptosis. *Surgery* **130**, 265–272.

Herr, I., and Debatin, K. M. (2001). Cellular stress response and apoptosis in cancer therapy. *Blood* **98**, 2603–2614.

Holler, N., Zaru, R., Micheau, O., Thome, M., Attinger, A., Valitutti, S., Bodmer, J. L., Schneider, P., Seed, B., and Tschopp, J. (2000). Fas triggers an alternative, caspase-8-independent cell death pathway using the kinase RIP as effector molecule. *Nat. Immunol.* **1**, 489–495.

Hu, S., Vincenz, C., Ni, J., Gentz, R., and Dixit, V. M. (1997). I-FLICE, a novel inhibitor of tumor necrosis factor receptor-1- and CD-95-induced apoptosis. *J. Biol. Chem.* **272**, 17255–17257.

Hu, W. H., Johnson, H., and Shu, H. B. (2000). Activation of NF-kappaB by FADD, Casper, and caspase-8. *J. Biol. Chem.* **275**, 10838–10844.

Hyer, M. L., Sudarshan, S., Kim, Y., Reed, J. C., Dang, J. Y., Schwartz, D. A., and Norris, J. S. (2002). Downregulation of c-FLIP sensitizes DU145 prostate cancer cells to Fas-mediated apoptosis. *Cancer Biol. Ther.* **1**, 401–406.

Ichikawa, K., Liu, W., Zhao, L., Wang, Z., Liu, D., Ohtsuka, T., Zhang, H., Mountz, J. D., Koopman, W. J., Kimberly, R. P., and Zhou, T. (2001). Tumoricidal activity of a novel anti-human DR5 monoclonal antibody without hepatocyte cytotoxicity. *Nat. Med.* **7,** 954–960.

Igney, F. H., and Krammer, P. H. (2002). Death and anti-death: Tumour resistance to apoptosis. *Nat. Rev. Cancer* **2,** 277–288.

Imanishi, T., McBride, J., Ho, Q., O'Brien, K. D., Schwartz, S. M., and Han, D. K. (2000a). Expression of cellular FLICE-inhibitory protein in human coronary arteries and in a rat vascular injury model. *Am. J. Pathol.* **156,** 125–137.

Imanishi, T., Murry, C. E., Reinecke, H., Hano, T., Nishio, I., Liles, W. C., Hofsta, L., Kim, K., O'Brien, K. D., Schwartz, S. M., and Han, D. K. (2000b). Cellular FLIP is expressed in cardiomyocytes and down-regulated in TUNEL-positive grafted cardiac tissues. *Cardiovasc. Res.* **48,** 101–110.

Inaba, M., Kurasawa, K., Mamura, M., Kumano, K., Saito, Y., and Iwamoto, I. (1999). Primed T cells are more resistant to Fas-mediated activation-induced cell death than naive T cells. *J. Immunol.* **163,** 1315–1320.

Inohara, N., Koseki, T., Hu, Y., Chen, S., and Nunez, G. (1997). CLARP, a death effector domain-containing protein interacts with caspase-8 and regulates apoptosis. *Proc. Natl. Acad. Sci. USA* **94,** 10717–10722.

Irisarri, M., Plumas, J., Bonnefoix, T., Jacob, M. C., Roucard, C., Pasquier, M. A., Sotto, J. J., and Lajmanovich, A. (2000). Resistance to CD95-mediated apoptosis through constitutive c-FLIP expression in a non-Hodgkin's lymphoma B cell line. *Leukemia* **14,** 2149–2158.

Irmler, M., Thome, M., Hahne, M., Schneider, P., Hofmann, K., Steiner, V., Bodmer, J. L., Schroter, M., Burns, K., Mattmann, C., Rimoldi, D., French, L. E., and Tschopp, J. (1997). Inhibition of death receptor signals by cellular FLIP. *Nature* **388,** 190–195.

Ito, Y., Pandey, P., Place, A., Sporn, M. B., Gribble, G. W., Honda, T., Kharbanda, S., and Kufe, D. (2000). The novel triterpenoid 2-cyano-3,12-dioxoolean-1,9-dien-28-oic acid induces apoptosis of human myeloid leukemia cells by a caspase-8-dependent mechanism. *Cell Growth Differ.* **11,** 261–267.

Johnstone, R. W., Ruefli, A. A., and Lowe, S. W. (2002). Apoptosis. A link between cancer genetics and chemotherapy. *Cell* **108,** 153–164.

Kataoka, T., Schroter, M., Hahne, M., Schneider, P., Irmler, M., Thome, M., Froelich, C. J., and Tschopp, J. (1998). FLIP prevents apoptosis induced by death receptors but not by perforin/granzyme B, chemotherapeutic drugs, and gamma irradiation. *J. Immunol.* **161,** 3936–3942.

Kataoka, T., Budd, R. C., Holler, N., Thome, M., Martinon, F., Irmler, M., Burns, K., Hahne, M., Kennedy, N., Kovacsovics, M., and Tschopp, J. (2000). The caspase-8 inhibitor FLIP promotes activation of NF-kappaB and Erk signaling pathways. *Curr. Biol.* **10,** 640–648.

Kayagaki, N., Yamaguchi, N., Nakayama, M., Eto, H., Okumura, K., and Yagita, H. (1999). Type I interferons (IFNs) regulate tumor necrosis factor-related apoptosis-inducing ligand (TRAIL) expression on human T cells: A novel mechanism for the antitumor effects of type I IFNs. *J. Exp. Med.* **189,** 1451–1460.

Kelly, M. M., Hoel, B. D., and Voelkel-Johnson, C. (2002). Doxorubicin pretreatment sensitizes prostate cancer cell lines to TRAIL induced apoptosis which correlates with the loss of c-FLIP expression. *Cancer Biol. Ther.* **1,** 520–527.

Kim, Y., Suh, N., Sporn, M., and Reed, J. C. (2002). An inducible pathway for degradation of FLIP protein sensitizes tumor cells to TRAIL-induced apoptosis. *J. Biol. Chem.* **277,** 22320–22329.

Kirchhoff, S., Muller, W. W., Krueger, A., Schmitz, I., and Krammer, P. H. (2000). TCR-mediated up-regulation of c-FLIPshort correlates with resistance toward CD95-mediated apoptosis by blocking death-inducing signaling complex activity. *J. Immunol.* **165,** 6293–6300.

Kischkel, F. C., Hellbardt, S., Behrmann, I., Germer, M., Pawlita, M., Krammer, P. H., and Peter, M. E. (1995). Cytotoxicity-dependent APO-1 (Fas/CD95)-associated proteins form a death-inducing signaling complex (DISC) with the receptor. *EMBO J.* **14,** 5579–5588.

Kreuz, S., Siegmund, D., Scheurich, P., and Wajant, H. (2001). NF-kappaB inducers upregulate cFLIP, a cycloheximide-sensitive inhibitor of death receptor signaling. *Mol. Cell. Biol.* **21**, 3964–3973.

Krueger, A., Schmitz, I., Baumann, S., Krammer, P. H., and Kirchhoff, S. (2001). Cellular FLICE-inhibitory protein splice variants inhibit different steps of caspase-8 activation at the CD95 death-inducing signaling complex. *J. Biol. Chem.* **276**, 20633–20640.

LeBlanc, H., Lawrence, D., Varfolomeev, E., Totpal, K., Morlan, J., Schow, P., Fong, S., Schwall, R., Sinicropi, D., and Ashkenazi, A. (2002). Tumor-cell resistance to death receptor-induced apoptosis through mutational inactivation of the proapoptotic Bcl-2 homolog Bax. *Nat. Med.* **8**, 274–281.

LeBlanc, H. N., and Ashkenazi, A. (2003). Apo2L/TRAIL and its death and decoy receptors. *Cell Death Differ.* **10**, 66–75.

Leverkus, M., Neumann, M., Mengling, T., Rauch, C. T., Brocker, E. B., Krammer, P. H., and Walczak, H. (2000). Regulation of tumor necrosis factor-related apoptosis-inducing ligand sensitivity in primary and transformed human keratinocytes. *Cancer Res.* **60**, 553–559.

Los, M., Wesselborg, S., and Schulze-Osthoff, K. (1999). The role of caspases in development, immunity, and apoptotic signal transduction: Lessons from knockout mice. *Immunity* **10**, 629–639.

Matta, H., Eby, M. T., Gazdar, A. F., and Chaudhary, P. M. (2002). Role of MRIT/cFLIP in protection against chemotherapy-induced apoptosis. *Cancer Biol. Ther.* **1**, 652–660.

Medema, J. P., de Jong, J., van Hall, T., Melief, C. J., and Offringa, R. (1999). Immune escape of tumors in vivo by expression of cellular FLICE-inhibitory protein. *J. Exp. Med.* **190**, 1033–1038.

Medema, J. P., Scaffidi, C., Kischkel, F. C., Shevchenko, A., Mann, M., Krammer, P. H., and Peter, M. E. (1997). FLICE is activated by association with the CD95 death-inducing signaling complex (DISC). *EMBO J.* **16**, 2794–2804.

Micheau, O., Thome, M., Schneider, P., Holler, N., Tschopp, J., Nicholson, D. W., Briand, C., and Grutter, M. G. (2002). The long form of FLIP is an activator of caspase-8 at the Fas death-inducing signaling complex. *J. Biol. Chem.* **277**, 45162–45171.

Mitsiades, N., Poulaki, V., Mitsiades, C., and Tsokos, M. (2001). Ewing's sarcoma family tumors are sensitive to tumor necrosis factor-related apoptosis-inducing ligand and express death receptor 4 and death receptor 5. *Cancer Res.* **61**, 2704–2712.

Mitsiades, N., Mitsiades, C. S., Poulaki, V., Anderson, K. C., and Treon, S. P. (2002). Intracellular regulation of tumor necrosis factor-related apoptosis-inducing ligand-induced apoptosis in human multiple myeloma cells. *Blood* **99**, 2162–2171.

Mueller, C. M., and Scott, D. W. (2000). Distinct molecular mechanisms of Fas resistance in murine B lymphoma cells. *J. Immunol.* **165**, 1854–1862.

Munoz-Pinedo, C., Ruiz-Ruiz, C., Ruiz De Almodovar, C., Palacios, C., and Lopez-Rivas, A. (2003). Inhibition of glucose metabolism sensitizes tumor cells to death receptor-triggered apoptosis through enhancement of death-inducing signaling complex formation and apical procaspase-8 processing. *J. Biol. Chem.* **278**, 12759–12768.

Nagane, M., Pan, G., Weddle, J. J., Dixit, V. M., Cavenee, W. K., and Huang, H. J. (2000). Increased death receptor 5 expression by chemotherapeutic agents in human gliomas causes synergistic cytotoxicity with tumor necrosis factor-related apoptosis-inducing ligand in vitro and in vivo. *Cancer Res.* **60**, 847–853.

O'Connell, J., O'Sullivan, G. C., Collins, J. K., and Shanahan, F. (1996). The Fas counterattack: Fas-mediated T cell killing by colon cancer cells expressing Fas ligand. *J. Exp. Med.* **184**, 1075–1082.

O'Connell, J., Bennett, M. W., O'Sullivan, G. C., Roche, D., Kelly, J., Collins, J. K., and Shanahan, F. (1998). Fas ligand expression in primary colon adenocarcinomas: Evidence that the Fas counterattack is a prevalent mechanism of immune evasion in human colon cancer. *J. Pathol.* **186**, 240–246.

Olsson, A., Diaz, T., Aguilar-Santelises, M., Osterborg, A., Celsing, F., Jondal, M., and Osorio, L. M. (2001). Sensitization to TRAIL-induced apoptosis and modulation of FLICE-inhibitory protein in B chronic lymphocytic leukemia by actinomycin D. *Leukemia* **15**, 1868–1877.

Panka, D. J., Mano, T., Suhara, T., Walsh, K., and Mier, J. W. (2001). Phosphatidylinositol 3-kinase/Akt activity regulates c-FLIP expression in tumor cells. *J. Biol. Chem.* **276**, 6893–6896.

Pedersen, I. M., Kitada, S., Schimmer, A., Kim, Y., Zapata, J. M., Charboneau, L., Rassenti, L., Andreeff, M., Bennett, F., Sporn, M. B., Liotta, L. D., Kipps, T. J., and Reed, J. C. (2002). The triterpenoid CDDO induces apoptosis in refractory CLL B cells. *Blood* **100**, 2965–2972.

Poulaki, V., Mitsiades, C. S., Kotoula, V., Tseleni-Balafouta, S., Ashkenazi, A., Koutras, D. A., and Mitsiades, N. (2002). Regulation of Apo2L/tumor necrosis factor-related apoptosis-inducing ligand-induced apoptosis in thyroid carcinoma cells. *Am. J. Pathol.* **161**, 643–654.

Raisova, M., Bektas, M., Wieder, T., Daniel, P., Eberle, J., Orfanos, C. E., and Geilen, C. C. (2000). Resistance to CD95/Fas-induced and ceramide-mediated apoptosis of human melanoma cells is caused by a defective mitochondrial cytochrome c release. *FEBS Lett.* **473**, 27–32.

Rasper, D. M., Vaillancourt, J. P., Hadano, S., Houtzager, V. M., Seiden, I., Keen, S. L., Tawa, P., Xanthoudakis, S., Nasir, J., Martindale, D., Koop, B. F., Peterson, E. P., Thornberry, N. A., Huang, J., MacPherson, D. P., Black, S. C., Hornung, F., Lenardo, M. J., Hayden, M. R., Roy, S., and Nicholson, D. W. (1998). Cell death attenuation by 'Usurpin,' a mammalian DED-caspase homologue that precludes caspase-8 recruitment and activation by the CD-95 (Fas, APO-1) receptor complex. *Cell Death Differ.* **5**, 271–288.

Reed, J. C. (2002). Apoptosis-based therapies. *Nat. Rev. Drug Discov.* **1**, 111–121.

Reed, J. C., Doctor, K., Rojas, A., Zapata, J. M., Stehlik, C., Fiorentino, L., Damiano, J., Roth, W., Matsuzawa, S., Newman, R., Takayama, S., Marusawa, H., Xu, F., Salvesen, G., and Godzik, A. (2003). Comparative analysis of apoptosis and inflammation genes of mice and humans. *Genome Res.* **13**, 1376–1388.

Refaeli, Y., Van Parijs, L., London, C. A., Tschopp, J., and Abbas, A. K. (1998). Biochemical mechanisms of IL-2-regulated Fas-mediated T cell apoptosis. *Immunity* **8**, 615–623.

Roth, W., Stenner-Liewen, F., Pawlowski, K., Godzik, A., and Reed, J. C. (2002). Identification and characterization of DEDD2, a death effector domain-containing protein. *J. Biol. Chem.* **277**, 7501–7508.

Ryu, B. K., Lee, M. G., Chi, S. G., Kim, Y. W., and Park, J. H. (2001). Increased expression of cFLIP(L) in colonic adenocarcinoma. *J. Pathol.* **194**, 15–19.

Salvesen, G. S., and Dixit, V. M. (1999). Caspase activation: The induced-proximity model. *Proc. Natl. Acad. Sci. USA* **96**, 10964–10967.

Sata, M., and Walsh, K. (1998). Endothelial cell apoptosis induced by oxidized LDL is associated with the down-regulation of the cellular caspase inhibitor FLIP. *J. Biol. Chem.* **273**, 33103–33106.

Sayers, T. J., Brooks, A. D., Koh, C. Y., Ma, W., Seki, N., Raziuddin, A., Blazar, B. R., Zhang, X., Elliott, P. J., and Murphy, W. J. (2003). The proteasome inhibitor PS-341 sensitizes neoplastic cells to TRAIL-mediated apoptosis by reducing levels of c-FLIP. *Blood* **102**, 303–310.

Scaffidi, C., Schmitz, I., Krammer, P. H., and Peter, M. E. (1999). The role of c-FLIP in modulation of CD95-induced apoptosis. *J. Biol. Chem.* **274**, 1541–1548.

Semra, Y. K., Seidi, O. A., and Sharief, M. K. (2001). Overexpression of the apoptosis inhibitor FLIP in T cells correlates with disease activity in multiple sclerosis. *J. Neuroimmunol.* **113**, 268–274.

Sharief, M. K. (2000). Increased cellular expression of the caspase inhibitor FLIP in intrathecal lymphocytes from patients with multiple sclerosis. *J. Neuroimmunol.* **111**, 203–209.

Shu, H. B., Halpin, D. R., and Goeddel, D. V. (1997). Casper is a FADD- and caspase-related inducer of apoptosis. *Immunity* **6,** 751–763.

Siegmund, D., Hadwiger, P., Pfizenmaier, K., Vornlocher, H. P., and Wajant, H. (2002). Selective inhibition of FLICE-like inhibitory protein expression with small interfering RNA oligonucleotides is sufficient to sensitize tumor cells for TRAIL-induced apoptosis. *Mol. Med.* **8,** 725–732.

Sprick, M. R., Weigand, M. A., Rieser, E., Rauch, C. T., Juo, P., Blenis, J., Krammer, P. H., and Walczak, H. (2000). FADD/MORT1 and caspase-8 are recruited to TRAIL receptors 1 and 2 and are essential for apoptosis mediated by TRAIL receptor 2. *Immunity* **12,** 599–609.

Srinivasula, S. M., Ahmad, M., Ottilie, S., Bullrich, F., Banks, S., Wang, Y., Fernandes-Alnemri, T., Croce, C. M., Litwack, G., Tomaselli, K. J., Armstrong, R. C., and Alnemri, E. S. (1997). FLAME-1, a novel FADD-like anti-apoptotic molecule that regulates Fas/TNFR1-induced apoptosis. *J. Biol. Chem.* **272,** 18542–18545.

Stassi, G., Di Liberto, D., Todaro, M., Zeuner, A., Ricci-Vitiani, L., Stoppacciaro, A., Ruco, L., Farina, F., Zummo, G., and De Maria, R. (2000). Control of target cell survival in thyroid autoimmunity by T helper cytokines via regulation of apoptotic proteins. *Nat. Immunol.* **1,** 483–488.

Stegh, A. H., Schickling, O., Ehret, A., Scaffidi, C., Peterhansel, C., Hofmann, T. G., Grummt, I., Krammer, P. H., and Peter, M. E. (1998). DEDD, a novel death effector domain-containing protein, targeted to the nucleolus. *EMBO J.* **17,** 5974–5986.

Sturzl, M., Hohenadl, C., Zietz, C., Castanos-Velez, E., Wunderlich, A., Ascherl, G., Biberfeld, P., Monini, P., Browning, P. J., and Ensoli, B. (1999). Expression of K13/v-FLIP gene of human herpesvirus 8 and apoptosis in Kaposi's sarcoma spindle cells. *J. Natl. Cancer Inst.* **91,** 1725–1733.

Takeda, K., Hayakawa, Y., Smyth, M. J., Kayagaki, N., Yamaguchi, N., Kakuta, S., Iwakura, Y., Yagita, H., and Okumura, K. (2001). Involvement of tumor necrosis factor-related apoptosis-inducing ligand in surveillance of tumor metastasis by liver natural killer cells. *Nat. Med.* **7,** 94–100.

Taylor, M. A., Chaudhary, P. M., Klem, J., Kumar, V., Schatzle, J. D., and Bennett, M. (2001). Inhibition of the death receptor pathway by cFLIP confers partial engraftment of MHC class I-deficient stem cells and reduces tumor clearance in perforin-deficient mice. *J. Immunol.* **167,** 4230–4237.

Tepper, C. G., and Seldin, M. F. (1999). Modulation of caspase-8 and FLICE-inhibitory protein expression as a potential mechanism of Epstein-Barr virus tumorigenesis in Burkitt's lymphoma. *Blood* **94,** 1727–1737.

Thomas, W. D., and Hersey, P. (1998). TNF-related apoptosis-inducing ligand (TRAIL) induces apoptosis in Fas ligand-resistant melanoma cells and mediates CD4 T cell killing of target cells. *J. Immunol.* **161,** 2195–2200.

Thomas, R. K., Kallenborn, A., Wickenhauser, C., Schultze, J. L., Draube, A., Vockerodt, M., Re, D., Diehl, V., and Wolf, J. (2002a). Constitutive expression of c-FLIP in Hodgkin and Reed-Sternberg cells. *Am. J. Pathol.* **160,** 1521–1528.

Thomas, R. P., Farrow, B. J., Kim, S., May, M. J., Hellmich, M. R., and Evers, B. M. (2002b). Selective targeting of the nuclear factor-kappaB pathway enhances tumor necrosis factor-related apoptosis-inducing ligand-mediated pancreatic cancer cell death. *Surgery* **132,** 127–134.

Thome, M., Schneider, P., Hofmann, K., Fickenscher, H., Meinl, E., Neipel, F., Mattmann, C., Burns, K., Bodmer, J. L., Schroter, M., Scaffidi, C., Krammer, P. H., Peter, M. E., and Tschopp, J. (1997). Viral FLICE-inhibitory proteins (FLIPs) prevent apoptosis induced by death receptors. *Nature* **386,** 517–521.

Thome, M., and Tschopp, J. (2001). Regulation of lymphocyte proliferation and death by FLIP. *Nat. Rev. Immunol.* **1,** 50–58.

Tschopp, J., Thome, M., Hofmann, K., and Meinl, E. (1998). The fight of viruses against apoptosis. *Curr. Opin. Genet. Dev.* **8,** 82–87.

Ugurel, S., Seiter, S., Rappl, G., Stark, A., Tilgen, W., and Reinhold, U. (1999). Heterogenous susceptibility to CD95-induced apoptosis in melanoma cells correlates with bcl-2 and bcl-x expression and is sensitive to modulation by interferon-gamma. *Int. J. Cancer* **82,** 727–736.

Van Parijs, L., Refaeli, Y., Abbas, A. K., and Baltimore, D. (1999). Autoimmunity as a consequence of retrovirus-mediated expression of C-FLIP in lymphocytes. *Immunity* **11,** 763–770.

Villunger, A., Egle, A., Kos, M., Hartmann, B. L., Geley, S., Kofler, R., and Greil, R. (1997). Drug-induced apoptosis is associated with enhanced Fas (Apo-1/CD95) ligand expression but occurs independently of Fas (Apo-1/CD95) signaling in human T-acute lymphatic leukemia cells. *Cancer Res.* **57,** 3331–3334.

Walczak, H., Miller, R. E., Ariail, K., Gliniak, B., Griffith, T. S., Kubin, M., Chin, W., Jones, J., Woodward, A., Le, T., Smith, C., Smolak, P., Goodwin, R. G., Rauch, C. T., Schuh, J. C., and Lynch, D. H. (1999). Tumoricidal activity of tumor necrosis factor-related apoptosis-inducing ligand in vivo. *Nat. Med.* **5,** 157–163.

Wang, G. H., Bertin, J., Wang, Y., Martin, D. A., Wang, J., Tomaselli, K. J., Armstrong, R. C., and Cohen, J. I. (1997). Bovine herpesvirus 4 BORFE2 protein inhibits Fas- and tumor necrosis factor receptor 1-induced apoptosis and contains death effector domains shared with other gamma-2 herpesviruses. *J. Virol.* **71,** 8928–8932.

Yeh, J. H., Hsu, S. C., Han, S. H., and Lai, M. Z. (1998). Mitogen-activated protein kinase kinase antagonized fas-associated death domain protein-mediated apoptosis by induced FLICE-inhibitory protein expression. *J. Exp. Med.* **188,** 1795–1802.

Yeh, W. C., Itie, A., Elia, A. J., Ng, M., Shu, H. B., Wakeham, A., Mirtsos, C., Suzuki, N., Bonnard, M., Goeddel, D. V., and Mak, T. W. (2000). Requirement for Casper (c-FLIP) in regulation of death receptor-induced apoptosis and embryonic development. *Immunity* **12,** 633–642.

Yount, G. L., Levine, K. S., Kuriyama, H., Haas-Kogan, D. A., and Israel, M. A. (1999). Fas (APO-1/CD95) signaling pathway is intact in radioresistant human glioma cells. *Cancer Res.* **59,** 1362–1365.

Zamai, L., Ahmad, M., Bennett, I. M., Azzoni, L., Alnemri, E. S., and Perussia, B. (1998). Natural killer (NK) cell-mediated cytotoxicity: Differential use of TRAIL and Fas ligand by immature and mature primary human NK cells. *J. Exp. Med.* **188,** 2375–2380.

Zhan, Y., Hegde, R., Srinivasula, S. M., Fernandes-Alnemri, T., and Alnemri, E. S. (2002). Death effector domain-containing proteins DEDD and FLAME-3 form nuclear complexes with the TFIIIC102 subunit of human transcription factor IIIC. *Cell Death Differ.* **9,** 439–447.

Zhang, X. D., Franco, A., Myers, K., Gray, C., Nguyen, T., and Hersey, P. (1999). Relation of TNF-related apoptosis-inducing ligand (TRAIL) receptor and FLICE-inhibitory protein expression to TRAIL-induced apoptosis of melanoma. *Cancer Res.* **59,** 2747–2753.

12

Epidermal Growth Factor and TRAIL Interactions in Epithelial-Derived Cells

Spencer Bruce Gibson

*Department of Biochemistry and Medical Genetics
Manitoba Institute of Cell Biology, University of Manitoba
Winnipeg, Manitoba R3E 0V9, Canada*

I. Epidermal Growth Factor Receptor–Mediated Cell Survival
II. Epidermal Growth Factor Signaling Pathways
III. Transcriptional Activation
IV. Antiapoptotic Proteins
V. Targeting Epidermal Growth Factor Receptor Signaling for Treatment of Cancer
VI. TRAIL Death Receptors
VII. TRAIL as a Cancer Treatment
VIII. Convergence of the Epidermal Growth Factor and TRAIL Signaling Pathways
References

In healthy tissues, there is a balance between cell survival and death. This balance ensures epithelial cells survive in the right milieu, but undergo programmed cell death (apoptosis) when the environment is no longer supportive. Cells sense these changes primarily through receptors on the cell surface that bind to specific ligands present in the extracellular environment. These receptors, through signal transduction pathways, lead to promotion of cell survival or induction of cell death. One of the most important types of receptors regulating cell survival is the epidermal growth factor (EGF) receptors, while one of the most important types of receptors regulating apoptosis is the tumor necrosis factor–related apoptosis-inducing ligand (TRAIL) death receptors. EGF receptors activate survival signaling pathways including PI3K/AKT, Ras/MAPK, and JAK/STAT signaling pathways leading to cell survival. TRAIL activates apoptotic signaling pathways leading to caspase activation and mitochondrial dysfunction. The balance between these two signaling pathways determine whether a cell survives or dies. In disease states, this balance is altered. For example, epithelial-derived cancer cells often have increased expression of EGF receptors and are resistant to apoptosis. Understanding the interactions between survival and apoptotic signaling pathways mediated by EGF receptors and TRAIL death receptors will be essential to explain the role these pathways play in healthy and diseased cells. © 2004 Elsevier Inc.

I. EPIDERMAL GROWTH FACTOR RECEPTOR–MEDIATED CELL SURVIVAL

Ligation of receptors expressed on the cell surface can protect cells against apoptosis. These receptors could be tyrosine kinase receptors, non–tyrosine kinase receptors, G-protein–coupled receptors or adhesion molecules (Danielsen and Maihle, 2002; Niiro and Clark, 2002; Parise et al., 2000; Swarthout and Walling, 2000). Their ability to protect cells from cell death is determined by the presence of their specific ligand found in the extracellular matrix. This leads to changes in cell signaling promoting cell survival. Therefore, changes in receptor ligation on the cell surface could alter the ability of a cell to survive. This ensures that cells perform their functions in the context of the correct environment and when an inappropriate environment exists undergo cell death.

Growth factors bind to tyrosine kinase receptors (Talapatra and Thompson, 2001). These receptors are some of the most important receptors regulating cell survival. One growth factor capable of inducing survival responses in epithelial-derived cells is EGF (Danielsen and Maihle, 2002). This growth factor binds to a family of tyrosine kinase receptors called the EGF receptor family composed of ErbB1 (EGFR, HER-1), ErbB2 (Neu, HER-2), ErbB3 (HER-3), and ErbB4 (HER-4) (Yarden, 2001). These receptors form homodimers and heterodimers with each other. EGF binds to ErbB1 homodimer and in combination with the other three EGF receptor members. We demonstrated that ligation of EGF to its receptors is sufficient to protect against death receptor–induced apoptosis. Besides EGF, other EGF receptor ligands have shown protective properties.

Heregulin binds to ErbB2–4 homodimerized or heterodimerized receptors (Yarden, 2001). Similar to EGF, heregulin treatment prevents apoptosis (Mandal *et al.*, 2002; Venkateswarlu *et al.*, 2002). ErbB2 receptor, unlike other EGF family members, does not bind to a ligand but forms heterodimers with other EGF receptor members. When overexpressed, it becomes constitutively active by forming homodimers and provides a survival response against genotoxins, serum starvation, or death receptor–induced apoptosis (Menard *et al.*, 2001; Navolanic *et al.*, 2003).

Non–tyrosine receptors, G-coupled receptors, and adhesion receptors are other sensors besides tyrosine kinase receptors used by cells to provide a survival response. Cytokines bind to non–tyrosine kinase receptors such as interleukin 6 (IL6) receptor and activated cytoplasmic tyrosine kinases that promote cell survival (Badache and Hynes, 2001; Qiu *et al.*, 1998). Lysophosphatidic acid (LPA) is a natural glycerophospholipid that possesses a range of biologic activities including cell survival in epithelial-derived cells. LPA interacts with specific G-protein receptors, including Edg-2 and Edg-7, leading to signaling pathways that promote cell survival (Swarthout and Walling, 2000). Integrins are a family of adhesion receptors that also activate cytoplasmic tyrosine kinases leading to cell survival. For example, binding of integrin such as $\alpha_5\beta_1$ that bind to fibronectin can protect intestinal epithelial cells against apoptosis caused by growth factor deprivation (Juliano, 2002). These receptors also activate the EGF receptor independent of EGF receptor ligands contributing to their cell survival responses (Badache and Hynes, 2001; Daub *et al.*, 1997; Reginato *et al.*, 2003). Thus, survival responses through activation of a variety of cell surface receptors may require engaging the EGF receptor signaling pathways independent of EGF in epithelial-derived cells.

II. EPIDERMAL GROWTH FACTOR SIGNALING PATHWAYS

EGF receptor family members transmit survival signals to the cytoplasm of cells. There are three major growth factor signaling pathways that contribute to cell survival. These include the PI3K/AKT pathway, Ras/MAPK pathway, and the JAK/STAT pathway (Fig. 1) (Danielsen and Maihle, 2002). The PI3K/AKT pathway is activated by EGF receptors through the recruitment of phosphatidylinositol 3-kinase (PI3K) to the plasma membrane. This recruitment could be through direct interaction of PI3K to the receptor or through binding to an adaptor protein which in turn associates with a receptor. In either case, the association is governed by the tyrosine phosphorylation of the receptor or adaptor molecules causing the binding of the SH2 domain of PI3K to these phospho-tyrosine residues. This activates PI3K, leading to the transfer of a phosphate group from

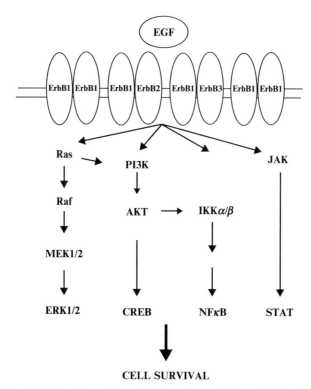

FIGURE 1. EGF signaling pathways. Epidermal growth factor binds to ErbB1 in combination with ErbB1–4. This causes the activation of the Ras/MAPK pathway, PI3K/AKT pathway and JAK/STAT pathway. These signaling pathways cause activation of transcription factors, leading to survival responses in epithelial-derived cells.

adenosine triphosphate to the D3 position of phosphatidylinositol (PI). This generates phosphatidylinositol 3,4 bisphosphate (PI-3,4P), and phosphatidylinositol 3,4,5 trisphosphate (PI-3,4,5P). By inhibiting PI3K activation either through addition of chemical inhibitors such as wortmannin or by expression of dominant negative forms of PI3K, the protective effects of growth factors such as EGF are eliminated (Navolanic et al., 2003). PI3K activation can be negatively regulated by a lipid/protein phosphatase PTEN by its ability to dephosphorylate 3′ phospholipid products generated by PI3K. Mutations in PTEN are found in many cancers, including breast cancer, and inactivation of PTEN leads to increased cell survival (Kandasamy and Srivastava, 2002; Stambolic et al., 1998). In mice lacking PTEN expression, induction of death receptor–induced apoptosis and anoikis is inhibited (Stambolic et al., 1998). This indicates that EGF protection against apoptosis engages the PI3K signaling pathway.

PI3K phospholipids bind to plecksrin homology (PH) domains, which are found on a wide variety of proteins. AKT is a serine/threonine kinase containing a PH domain and when overexpressed protects cells against apoptosis. PI3K-generated phospholipids bind to the PH domain of AKT causing its translocation to the plasma membrane (Lawlor and Alessi, 2001; Nicholson and Anderson, 2002; Vivanco and Sawyers, 2002). Once at the plasma membrane, AKT is phosphorylated on two critical residues—serine 473 and threonine 308. Phosphorylation of serine 473 is by serine/threonine kinase PDK while phosphorylation of threonine 308 is by an unknown kinase although cyclic adenosine monophosphate–dependent protein kinase has been implicated (Brazil *et al.*, 2002; Vivanco and Sawyers, 2002). Upon phosphorylation of these sites, AKT kinase activity is activated leading to the phosphorylation of a variety of protein. Kinase inactive AKT fails to protect cells against apoptosis and might act as dominant negative protein blocking protective responses (Vivanco and Sawyers, 2002). We and others have shown that overexpression of kinase inactive AKT effectively blocks EGF protection against apoptosis (Danielsen and Maihle, 2002; Gibson *et al.*, 1999, 2002; Lawlor and Alessi, 2001). One mechanism that dominant negative AKT could interfere with AKT survival function is to block AKT phosphorylation of its substrates.

AKT phosphorylates many proteins but only a subset of these proteins has been shown to have antiapoptotic effects. These proteins include Bcl-2 family member BAD, caspase 9, Forkhead, NFκB, and CREB (Fang *et al.*, 1999; Kops and Burgering, 2000; Lawlor and Alessi, 2001; Pianetti *et al.*, 2001). Phosphorylation of serine 136 in BAD inactivates it proapoptotic function. Unphosphorylated BAD binds to the antiapoptotic protein Bcl-2 in the mitochondria inhibiting its protection against apoptosis. Upon AKT phosphorylation of BAD, it associates with 14-3-3 (Datta *et al.*, 1999). This sequesters it away from the mitochondria and from binding to the antiapoptotic protein Bcl-2. This allows Bcl-2 to protect cells against apoptosis (Coultas and Strasser, 2003; Tsujimoto, 2003). The level of BAD phosphorylation might not be sufficient to protect cells from a significant apoptotic stimulus. Thus other AKT substrates in combination with BAD phosphorylation lead to cell survival. AKT also phosphorylates caspase-9 on serine 196 and thereby prevents its activation (Datta *et al.*, 1999). The importance of this phosphorylation remains controversial. In the mouse, this phosphorylation site for caspase-9 is not found. Furthermore, proteolytic cleavage of caspase-9, which phosphorylation of caspase-9 by AKT presumably inhibits, does not correlate with its proteolytic activity (Cardone *et al.*, 1998; Fujita *et al.*, 1999). Inhibition of caspase-9 however could still be a viable target for AKT action. AKT also phosphorylates transcription factor Forkhead. This phosphorylation initiates binding of 14-3-3 to Forkhead, preventing its translocation to the nucleus and thus inhibiting its transcriptional activity (Brunet *et al.*, 1999).

AKT also phosphorylates the transcription factor CREB. This phosphorylation, instead of being inhibitory, activates CREB transcription activity by increasing its DNA binding ability (Datta et al., 1999). CREB has been associated with upregulation of antiapoptotic proteins. CREB is primarily a nuclear protein while AKT is a cytosolic protein. The mechanism of how AKT phosphorylates CREB in a cell remains to be determined. Finally AKT activates the transcription factors NFκB (Nicholson and Anderson, 2002; Pianetti et al., 2001). It does this by phosphorylating IκB kinase α/β (IKK α/β). This leads to the phosphorylation of the N-terminus of IκB. Upon phosphorylation, IκB is targeted for degradation. Because IκB constitutively binds to NFκB in the cytoplasm, upon IκB degradation, NFκB translocates to the nucleus. This allows NFκB to activate transcription of antiapoptotic genes. The ability of AKT to activate NFκB could be cell-type and stimuli-dependent. Ligation of EGF receptors with EGF has shown NFκB activation mediated by both AKT-dependent and AKT-independent signaling pathways (Henson et al., 2003; Nicholson and Anderson, 2002; Pianetti et al., 2001). The exact mechanism of how AKT activation triggers NFκB activation remains unclear. We found that EGF activates AKT leading to NFκB activation and that this activation is essential for the ability of EGF to protect epithelial-derived cells from TRAIL-induced apoptosis (Henson et al., 2003). However, this does not eliminate other substrates for AKT from playing a role in EGF survival responses.

The Ras/MAPK signaling pathway is another survival pathway activated by EGF (Grant et al., 2002). By blocking MAPK (ERK) activation with PD098056 compound that specifically blocks MEK1/2 activity thereby blocking ERK activation, death receptor or serum starvation-induced apoptosis is increased (Widmann et al., 1999). The ERK signaling pathway is activated by a small G protein called Ras. Ras becomes activated when a Guanosine diphosphate is exchanged for a guanosine triphosphate mediated by a guanine nucleotide exchange factor such as son of sevenless (SOS). Expression of a dominant active form of SOS in transgenic mice results in skin papillomas but in cells lacking EGF receptor this transformation is eliminated (Sibilia et al., 2000). SOS is recruited to EGF receptors by binding to adaptor molecules such as Grb2. Grb2 binds to phospho-tyrosine residues on the cytoplasmic side of cell surface receptors. This leads to recruitment of SOS to the plasma membrane where it activates Ras causing a conformational change. This change allows Ras to bind to a MAPK kinase kinase called Raf1. Raf1 upon binding to Ras becomes active and phosphorylates the MAP kinase kinase MEK1 and MEK2. These MAP kinase kinases then phosphorylate and activate the MAPKs ERK1 and ERK2 (Widmann et al., 1999). This activation leads to a survival response. Dominant active Ras mutants have been shown to provide protective responses as well as overexpression of Raf1. Conversely, dominant negative mutant of Ras inhibited the protective effects of ERK. Several targets for ERK activation leading to cell survival have been proposed. One of these targets is the family of 90-kDa ribosomal S6 kinases (Rsk). In interleukin 3

(IL3)–dependent 32D cells, inhibition of Raf1 antagonizes IL3-modulated survival (von Gise et al., 2001). Expression of constitutively active Rsk1 is sufficient to rescue these cells. Rsk activity protects cells from death through phosphorylation of BAD on serine 136 similar to AKT (Lizcano et al., 2000). EGF-mediated ERK activation has also been shown to phosphorylate caspase-9 and inactivate its caspase activity that is also similar to AKT function (Allan et al., 2003). ERK-mediated transcriptional activation might also contribute to its survival response by upregulating antiapoptotic proteins such as Mcl-1. For example, EGF treatment of human esophageal carcinoma cells lead to induction of Mcl-1 expression via the Ras/MAPK pathways (Leu et al., 2000). Thus, the Ras/MAPK signaling pathway could contribute to EGF survival responses.

The Janus family of kinases (JAKs) plays a major role in signaling from cytokine receptors to the family of STAT transcription factors. The JAK/STAT signaling pathway is also activated by EGF, which leads to cell survival (David et al., 1996; Guren et al., 2003). There are at least four JAK isoforms and seven STAT molecules. These STAT transcription factors have specific DNA binding properties. Knockout studies on STAT3 indicate that it might have a role in cell survival. STAT3 was shown to be required for colony-stimulating factor–dependent cell survival (O'Shea et al., 2002). In addition, epithelial cells expressing a dominant negative form of STAT3 were more likely to undergo apoptosis (Song et al., 2003). Similar to MAPK activation, STAT3 has been implicated in increased expression of antiapoptotic proteins (Niu et al., 2002). Furthermore, blocking EGF-mediated STAT3 activation in human non–small-cell lung carcinoma results in induction of apoptosis (Song et al., 2003). Similar to PI3K/AKT and Ras/MAPK signaling pathways, the JAK/STAT signaling pathway could also contribute to EGF survival response.

III. TRANSCRIPTIONAL ACTIVATION

As mentioned previously, EGF activation of transcription factors contributes to cell survival. The regulation of transcription factors by EGF could be the key to determine if a cell undergoes cell death. The activation of Forkhead transcription factor has been implicated in the induction of FAS ligand expression (Kavurma and Khachigian, 2003). This ligand then binds to FAS death receptor and induces apoptosis. The ability of signals such as AKT to inhibit its activation is an important control on Forkhead activity (Brunet et al., 1999). For other transcription factors, activation leads to upregulation of antiapoptotic genes that inhibit apoptosis. It is important to note that activation of transcription factors involved in survival responses is also involved in other cellular functions. NFκB is an example of this in which its activation has been contributed to proliferation, differentiation, cell death, and cell survival (Nakshatri and Goulet, 2002; Santoro et al., 2003). NFκB upregulates many proteins involved in cellular survival including inhibitor of apoptosis (IAP) proteins, and antiapoptotic Bcl-2 family members (Wang et al., 1998). Indeed, we have shown that EGF through NFκB activation increases the antiapoptotic protein Mcl-1 (Henson et al., 2003). In contrast, NFκB has been implicated in upregulation of TRAIL and its death receptors, leading to apoptosis (Chen et al., 2003; Shetty et al., 2002). The context of NFκB stimulations will determine the cellular outcome. The mechanism that differentiates NFκB activation from a prosurvival to an apoptotic response is unknown but probably involves the different subunits of NFκB and/or different protein binding partners regulating the expression of prosurvival or apoptotic genes. It has been suggested that the p65 subunit of NFκB is involved in a protective response because mouse embryonic fibroblast cells lacking expression of p65 are more sensitive to apoptosis following treatment with the death receptor ligand TRAIL. MEF cells lacking c-Rel, another NFκB subunit, showed increased resistance to induction of apoptosis presumably through a lack of upregulation of TRAIL death receptors (Chen et al., 2003). This indicates that the subunits of NFκB could contribute to both cell survival and apoptosis. Activation of CREB also leads to cell survival (Bonni et al., 1999). CREB transcriptional regulation is mediated by phosphorylation of kinases such as AKT (Datta et al., 1999). This increases the expression of antiapoptotic proteins such as Mcl-1 (Datta et al., 1999). CREB activation is induced by both the PI3K/AKT pathway and the Ras/MAPK signaling pathway (Bonni et al., 1999; Datta et al., 1999). The role of CREB in EGF protection against death receptor–induced apoptosis remains unknown.

Besides activation of transcription factors mediated by signal transduction, there may be another more direct method for EGF receptors to activate gene expression. ErbB1 and ErbB4 receptors are localized to the nucleus following EGF stimulation (Marti and Wells, 2000; Ni et al., 2003). ErbB1 may act directly as a transcription factor because it binds to the promoter of cyclin D1 (Lin et al., 2001). The ability of these receptors to regulate expression of antiapoptotic genes is currently unknown. Overall, the coordinated activation of transcription factors by EGF provides another level of control mediating EGF survival responses.

IV. ANTIAPOPTOTIC PROTEINS

Apoptotic signaling pathways can be blocked in general in two ways; inhibition of caspases, and blockage of mitochondrial dysfunction such as protein release and opening of the PT pore. Inhibitors of caspases were first discovered in bacterioviruses. The family of caspase inhibitors is called inhibitors of apoptosis proteins (IAP) consisting of XIAP, IAP1, IAP2, and survivin (Salvesen and Duckett, 2002; Vaux and Silke, 2003). IAP family members inhibit the proteolytic activity of caspase-3, -6, -7, and -9. IAP binds to the cleaved active forms of the caspases with the exception of caspase-9. Because caspase-9 can be active without proteolytic cleavage, it is understandable that IAP would bind and inactivate full length caspase-9 (Salvesen and Duckett, 2002). The ratio of IAP and caspase association required to inactivate caspases is 1:1 or 2:1, indicating again that IAP need only to bind to one or two sites on caspases to inhibit their activity. Furthermore, the concentration of IAP required to inhibit caspases is 0.2–10 nM. This potent inhibitory function of IAP is also due to the ability of IAP family members to target caspases for proteolytic degradation (Yang and Yu, 2003). IAP's inhibitory properties can also be reversed unlike bacterial caspase inhibitors, such as p35, which bind covalently to caspases irreversibly inhibiting their function (Vaux and Silke, 2003). Thus, IAPs can be closely regulated to alter the ability of cells to undergo cell death. IAPs are also negatively regulated by binding of proteins that sequesters them away from caspases. One of these proteins is called Smac/Diablo, which is released by the mitochondria upon induction of apoptosis. This protein binds to XIAP and inhibits its ability to block caspase-3 activation (Verhagen and Vaux, 2002). In addition, XAF also binds to XIAP but instead of being released from the mitochondria, XAF binds and translocates XIAP to the nucleus (Liston *et al.*, 2001). This allows for caspase-3, which is found in the cytoplasm, to be activated. This sequestration of IAPs away from caspases effectively removes the inhibitors for caspase activation. Finally, the expression levels of IAPs are regulated by transcription factors such as NFκB so that a survival stimulus can increase the expression levels of these caspase inhibitors preventing apoptosis (Salvesen and Duckett, 2002; Wang *et al.*, 1998). Indeed, EGF has been shown to upregulate IAP1 and IAP2 expression (Herrera *et al.*, 2002). Thus caspase inhibitors are tightly regulated by their binding partners, localization and expression levels and could provide a mechanism for EGF protection against apoptosis.

The Bcl-2 family contains both proapoptotic and antiapoptotic members as mentioned previously. The antiapoptotic members include Bcl-2, Bclx, Bclw, A1, and Mcl-1. Generally antiapoptotic members have BH1–4 domains while proapoptotic Bcl-2 family members have a BH3 domain alone or in combination with BH1 or BH-2 domain (Coultas and Strasser, 2003; Deveraux *et al.*, 2001; Fleischer *et al.*, 2003). Bcl-2 family members are primarily associated with the mitochondria, endoplasmic reticulium, or nuclear membranes. The antiapoptotic Bcl-2 family members form heterodimers with proapoptotic family members thereby inhibiting their apoptotic potential (Tsujimoto, 2003). This association is governed by the BH1–3 domains of antiapoptotic members that form a hybrophobic cleft. This cleft binds to the BH3 domain of proapoptotic proteins such as Bax (Humlova, 2002). Association of antiapoptotic and proapoptotic Bcl-2 family members prevents cell death by blocking release of mitochondrial proteins, and by preventing changes in mitochondrial membrane potential (Scorrano and Korsmeyer, 2003). The relative expression of proapoptotic and antiapoptotic family members are important to determine if a cell will survive or die upon apoptotic stimulation. In addition, induction of expression of antiapoptotic Bcl-2 family members such as Bclx$_L$ and Mcl-1 might also control the survival response in cells (Humlova, 2002). Indeed, EGF has been implicated in increasing the expression of both Bclx$_L$ and Mcl-1 (Coultas and Strasser, 2003; Leu et al., 2000). Thus EGF regulation of antiapoptotic Bcl-2 family members inhibits apoptosis by protecting the mitochondria from proapoptotic members and inhibits caspase activation.

V. TARGETING EPIDERMAL GROWTH FACTOR RECEPTOR SIGNALING FOR TREATMENT OF CANCER

Inhibition of receptor tyrosine kinase activation could be an effective method of sensitizing epithelial-derived cancers to chemotherapy or radiation and, in some cases, this inhibition induces cell death directly (Brunelleschi et al., 2002). ErbB receptors are often overexpressed in epithelial-derived cancers such as breast, lung, and colon cancers (Coultas and Strasser, 2003). Blocking the activation of these receptors might eliminate the signaling pathways leading to cell survival. An antibody directed against ErbB2 receptor (Heceptin) is effective at sensitizing epithelial-derived cells to chemotherapeutic drugs. In clinical trials with metastatic breast cancer overexpressing ErbB2, Herceptin treatment in combination with taxol reduced tumor size and prolonged survival (Ranson and Sliwkowski, 2002). Herceptin is currently used generally as an adjunct therapy in the treatment of breast cancer in patients with high ErbB2 expression. We determined that Herceptin sensitizes epithelial-derived cells to apoptosis under hypoxic conditions (Kothari, 2003). ErbB2 overexpression is only found in 25% of breast cancer cells, which limits its clinical usefulness (Menard et al., 2001; Ranson and Sliwkowski, 2002). ErbB1 is overexpressed in a wide variety of cancers at a higher frequency. Two types of drugs that specifically inhibit ErbB1 are currently in clinical trials. The first type is antibodies directed against ErbB1 that down-regulate the receptor and blocks its activation by EGF (Herbst and Langer, 2002; Normanno et al., 2003a). One antibody in clinical trials, C225, is well tolerated by patients, blocks cell cycle progression, and induces cell death in epithelial-derived cells. The second type of drug is a kinase inhibitor directed specifically toward the adenosine triphosphate binding site in ErbB1 receptor. ZD1839 (Iressa) is one such drug that is also in clinical trails in a variety of epithelial-derived cancers (Normanno et al., 2003b). This targeting of EGF receptors by kinase inhibitors has also been applied to PDGF receptor (ST1571) and is in clinical trials treating cancers with overexpression of PDGF receptor (Roussidis and Karamanos, 2002). By blocking survival signals in cells, this allows conventional chemotherapy to be more effective. Besides directly inhibiting growth factor signaling, kinase inhibitors directed against signaling components of these pathways are being developed. In particular, AKT, ERK, and PI3K kinase inhibitors could provide reduced cell survival in cancer cells.

Proteosome inhibitors are also under development and are currently in clinical trials (Baldwin, 2001). These inhibitors block the degradation of proteins and thus inhibit specific signaling pathways. One pathway inhibited by proteosome inhibitors is the NFκB pathway. This blocks the ability of NFκB to upregulate antiapoptotic genes and sensitizes cells to chemotherapy (Darnell, 2002). Together this targeted approach to cancer treatment could provide better survival rates in patients with epithelial-derived cancers with EGF-mediated survival responses.

VI. TRAIL DEATH RECEPTORS

Death receptors belong to a subset of tumor necrosis factors receptor superfamily members. Death receptors in the TNF receptor family consist of Fas (CD95), tumor necrosis factor receptor (TNFR)-1, death receptor 3 (DR3/wsl-1), death receptor 4 (DR4), and death receptor 5 (DR5), as well as others (Gupta, 2001). These receptors bind to death ligands causing trimerization of the receptors and activation. The ligand for DR4 and DR5 is TRAIL. DR4 and DR5 contain cysteine-rich repeats in the extracellular domain and a death domain in their cytoplasm (LeBlanc and Ashkenazi, 2003). Point mutations or deletion of the death domain abrogate apoptosis. Death receptors activate caspases principally through two mechanisms. The first is direct activation of caspases and the second mechanism involves the mitochondria. Upon TRAIL ligation, the death-inducing signaling complex (DISC) is formed on DR4 or DR5 within seconds. The adaptor protein FADD binds to the death domain of DR4 or DR5 via its own death domain and recruits pro-caspase-8 to the DISC through it death effector domain. Upon recruitment to the death receptor, pro-caspase-8 is cleaved and activated. This initiates a caspase cascade by directly cleaving effector caspase-3 (Gupta, 2003). DR4 and DR5 also recruit TRAF-2 and RIP proteins through FADD, leading to NFκB activation (Schneider et al., 1997). This activation leads to upregulation of DR5 as a positive feedback mechanism promoting apoptosis (Shetty et al., 2002). The DISC formation also leads to mitochondria dysfunction. Activated caspase-8 cleaves pro-apoptotic Bcl-2 family member Bid into a truncated form (T-Bid), which translocates to the mitochondria and induces cytochrome c release. This results in the formation of the apoptosome and subsequent caspase-9 activation, leading to caspase-3 activation and apoptosis (LeBlanc and Ashkenazi, 2003). Besides cytochrome c release, Smac/Diablo is released from the mitochondria mediated by T-Bid. Smac/Diablo causes sequestration of IAP molecule away from caspases. This eventually leads to activation of caspase-3 and apoptosis (LeBlanc and Ashkenazi, 2003). This release of mitochondrial proteins also requires the presence of the pro-apoptotic Bcl2 family member Bax. In MEF cells lacking Bax, TRAIL-induced apoptosis is blocked (Ravi and Bedi, 2002). These two types of apoptotic signaling pathways could occur in the same cell at the same time. The strength of the death receptor ligation and thereby the magnitude of caspase-8 activation determine the type of apoptotic signaling pathway invoked. In either case, TRAIL death receptors represent a mechanism in which epithelial-derived cell directly activate caspases leading to apoptosis.

TRAIL death receptor activation is also negatively regulated. Decoy receptors, DcR1 and DcR2, are expressed on the cell surface thereby sequestering TRAIL away from its death receptors. Decoy receptors are expressed on healthy epithelial cells preventing the unwanted induction of apoptosis by TRAIL (LeBlanc and Ashkenazi, 2003). Another mechanism used by cells to block TRAIL-induced apoptosis is through FLICE-inhibitory proteins (FLIP). Two proteins, $FLIP_S$ and $FLIP_L$, can prevent apoptosis induced by TRAIL death receptors. $FLIP_S$ contains two DED domains but lacks the caspase recruitment domain, whereas $FLIP_L$ is identical to pro-caspase-8 but the critical cysteine residue is replaced by a tyrosine (Kim *et al.*, 2002; MacFarlane, 2003). Both proteins interact with FADD and pro-caspase-8, preventing apoptosis. Furthermore, overexpression of antiapoptotic Bcl-2 family members also blocks TRAIL-induced apoptosis by inhibiting mitochondrial dysfunction such as release of mitochondrial proteins (Ravi and Bedi, 2002; Suliman *et al.*, 2001). Finally, overexpression of IAP family members effectively blocks TRAIL-induced apoptosis (Spalding *et al.*, 2002; Zhang *et al.*, 2001). These antiapoptotic proteins are often overexpressed in epithelial-derived cancers effectively blocking death receptor–induced apoptosis. Thus changes in both activators and inhibitors of TRAIL apoptotic signaling will determine if TRAIL induces apoptosis in epithelial-derived cells.

VII. TRAIL AS A CANCER TREATMENT

Recombinant soluble TRAIL is effective at inducing apoptosis in a variety of cancer cell lines. It can also reduce epithelial-derived breast and colon cancer tumors in mice and monkeys without significant toxicity. TRAIL has been shown to be toxic to normal human hepatocytes but alternative forms of TRAIL, with reduced molecular weight, show no apparent toxicity to human hepatocytes (Almasan and Ashkenazi, 2003). Whether TRAIL will be nontoxic in humans is still unknown but clinical trials are currently underway to address these concerns. It is postulated that TRAIL could be an effective treatment against epithelial-derived cancers.

Combined treatment of TRAIL with the DNA damaging agents, etoposide or doxorubicin, leads to a synergistic apoptotic response in breast and lung cancer cells (Gibson et al., 2000; Keane et al., 1999). Combining TRAIL with chemotherapy can also eliminate human solid tumors grown in mice (Almasan and Ashkenazi, 2003). TRAIL and its receptors also play an important role in radiation- and genotoxin-induced apoptosis. Radiation induces increased expression of TRAIL and DR5 in Jurkat, MOLT-4, and CEM cell lines, while viral infection or etoposide treatment increases DR4 and DR5 expression in lung and breast cancer cell lines (Clarke et al., 2000; Gibson et al., 2000; Gong and Almasan, 2000). This etoposide upregulation of DR4/DR5 is mediated by the serine threonine kinase, MEKK1, and the transcription factor, NFκB, and is influenced by p53 activation (Clarke et al., 2000; Gibson et al., 2000; Gong and Almasan, 2000; Takimoto and El-Deiry, 2000). Etoposide-induced apoptosis in breast cancer cells is reduced by adding DR5:Fc fusion protein (binds to endogenous TRAIL preventing TRAIL binding to its receptors) indicating the involvement of TRAIL receptor activation for this activity (Gibson et al., 2000). Taken together, TRAIL treatment alone or in combination with conventional chemotherapy could be an effective treatment for epithelial-derived cancers.

VIII. CONVERGENCE OF THE EPIDERMAL GROWTH FACTOR AND TRAIL SIGNALING PATHWAYS

The multitude of signaling pathways and regulatory mechanisms controlling induction of apoptosis integrate into a network giving the cell the ability to either protect itself against apoptosis or induce programmed cell death. Understanding the integration of these pathways within the context of the cellular environment will allow for the prediction if a cell will live or die. The extracellular matrix contains the growth factors as well as other ligands that send a signal to cells that they should resist undergoing cell death. In epithelial cells, EGF and its receptors are critical sensors allowing cells to survive through activation of signaling pathways such as Ras/MAPK or PI3K/AKT pathways either through direct activation of EGF receptors or indirectly through activation of other receptors such as integrins (Liu et al., 2002; Qiu et al., 1998). These signaling pathways lead to either inactivation of proapoptotic proteins and/or the activation of antiapoptotic proteins.

The activation of antiapoptotic protein generally involves activation of transcription factors such as NFκB (Orlowski and Baldwin, 2002). This leads to the upregulation of antiapoptotic proteins including IAP1, FLIP, and Mcl-1. Increased expression of these proteins has been shown to block TRAIL-induced apoptosis. Indeed, we demonstrated that EGF upregulates Mcl-1, inhibiting TRAIL-induced apoptosis (Fig. 2) (Henson *et al.*, 2003). This blockage of TRAIL-induced apoptosis depends on both the concentration of TRAIL and EGF and the time cells are exposed to TRAIL or EGF. For example, if EGF is used to treat cells simultaneously with TRAIL treatment, no protective effect of EGF is observed (Gibson *et al.*, 2002). However, if EGF is exposed to cells at least 1 h before TRAIL treatment, EGF protects against apoptosis. This EGF protection also becomes less pronounced as time of cellular exposure of TRAIL increases. This dynamic regulation of apoptosis provides evidence that epithelial-derived cells can be

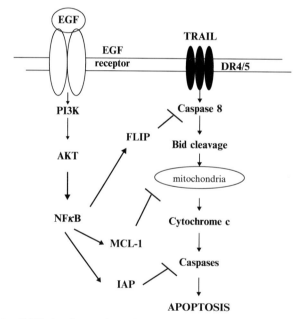

FIGURE 2. EGF signaling pathway interactions with the TRAIL apoptotic signaling pathway. Upon TRAIL binding to its death receptors DR4 or DR5, caspase-8 becomes activated and cleaves the proapoptotic Bcl-2 family member BID. Truncated BID translocates to the mitochondria and causes release of cytochrome *c*. This leads to further caspase activation and apoptosis. Upon EGF ligation to ErbB1 in combination with ErbB1 or ErbB2, PI3K becomes activated, leading to phosphorylation of the serine threonine kinase AKT. Upon phosphorylation, AKT kinase activity increases, leading to transcription factor NFκB activation. This causes increased expression of the antiapoptotic proteins such as FLIP, Mcl-1, and IAP. FLIP prevents caspase-8 activation and death receptor apoptosis. Mcl-1 localizes with the mitochondria and prevents TRAIL-mediated release of cytochrome *c* and apoptosis. IAP binds to caspases, inhibiting their activation and preventing apoptosis.

manipulated to undergo apoptosis depending on the balance between these survival and apoptotic signals.

Epithelial-derived cancers often overexpress EGF receptor family members and undergo apoptosis when exposed to TRAIL (Almasan and Ashkenazi, 2003; Gibson *et al.*, 2002; Yarden, 2001). With the emergence of antibodies directed against EGF receptors as legitimate therapies for these cancers, combination treatments including TRAIL and antibodies against EGF receptor family members could be an effective treatment in these cancers (Almasan and Ashkenazi, 2003; Normanno *et al.*, 2003). Indeed, Herceptin and ErbB1 kinase inhibitors have been shown to increase TRAIL sensitivity in epithelial-derived cells (Kari *et al.*, 2003; Normanno *et al.*, 2003a,b).

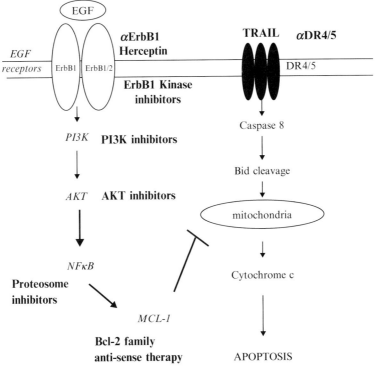

FIGURE 3. Molecular targeting the EGF and TRAIL signaling pathways for treatment of epithelial-derived cancers. The addition of TRAIL or activating antibodies against DR4 or DR5 initiates the death receptor apoptotic signaling pathway. In contrast, inhibiting antibodies against ErbB1/2 or ErbB1 kinase inhibitors block the EGF-mediated antiapoptotic signaling pathway. Furthermore, compounds that specifically inhibit PI3K and AKT also block the EGF survival signaling pathway. Proteasome inhibitors are a class of drugs currently in clinical trials that block transcription factor NFκB activation and survival signaling. Finally, antisense therapies directed against Bcl-2 family members such as Mcl-1 are also in clinical trials and effectively block antiapoptotic signals. Potential chemotherapeutic drugs are in bold type. Their molecular targeted proteins are in italic type.

In addition, conventional chemotherapy on epithelial-derived cells increased TRAIL death receptor expression and activation (Gibson et al., 2000). Blockage of TRAIL death receptor activation reduces chemotherapy-induced apoptosis. Not unexpected, combination treatment of TRAIL and chemotherapeutic drugs provides a synergistic apoptotic response in epithelial-derived cells possibly due to chemotherapy-mediated DR4 and DR5 upregulation (Gibson et al., 2000). It is thus likely that inhibiting survival signaling pathways (EGF receptor family member inhibitors) and activating apoptotic signaling pathways (TRAIL and/or conventional chemotherapy) will be an effective treatment for epithelial-derived cancers and may prevent development of drug resistance in these epithelial-derived cancers (Fig. 3).

In the future, antibodies directed specifically toward the extracellular domain of DR4 and DR5 will be developed giving another molecular targeted therapy towards epithelial-derived cancers. In addition, new potent inhibitors of EGF signaling pathways as well as to other growth factor signaling pathways will be more effective at eliminating survival signaling responses in cancer cells. This will reduce the probability that drug resistance will develop in epithelial-derived cancers. Conceivably, this manipulation of growth factor survival responses and death receptor–induced apoptotic responses could give effective treatments to diseases other than cancer where uncontrolled cell death occurs, such as in neurologic degenerative diseases and heart disease.

REFERENCES

Allan, L. A., Morrice, N., Brady, S., Magee, G., Pathak, S., and Clarke, P. R. (2003). Inhibition of caspase-9 through phosphorylation at Thr 125 by ERK MAPK. *Nat. Cell. Biol.* **5,** 647–654.

Almasan, A., and Ashkenazi, A. (2003). Apo2L/TRAIL: Apoptosis signaling, biology, and potential for cancer therapy. *Cytokine Growth Factor Rev.* **14,** 337–348.

Badache, A., and Hynes, N. E. (2001). Interleukin 6 inhibits proliferation and, in cooperation with an epidermal growth factor receptor autocrine loop, increases migration of T47D breast cancer cells. *Cancer Res.* **61,** 383–391.

Baldwin, A. S. (2001). Control of oncogenesis and cancer therapy resistance by the transcription factor NF-kappaB. *J. Clin. Invest.* **107,** 241–246.

Bonni, A., Brunet, A., West, A. E., Datta, S. R., Takasu, M. A., and Greenberg, M. E. (1999). Cell survival promoted by the Ras-MAPK signaling pathway by transcription-dependent and -independent mechanisms. *Science* **286,** 1358–1362.

Brazil, D. P., Park, J., and Hemmings, B. A. (2002). PKB binding proteins. Getting in on the Akt. *Cell* **111,** 293–303.

Brunelleschi, S., Penengo, L., Santoro, M. M., and Gaudino, G. (2002). Receptor tyrosine kinases as target for anti-cancer therapy. *Curr. Pharmacal. Res.* **8,** 1959–1972.

Brunet, A., Bonni, A., Zigmond, M. J., Lin, M. Z., Juo, P., Hu, L. S., Anderson, M. J., Arden, K. C., Blenis, J., and Greenberg, M. E. (1999). Akt promotes cell survival by phosphorylating and inhibiting a Forkhead transcription factor. *Cell* **96,** 857–868.

Cardone, M. H., Roy, N., Stennicke, H. R., Salvesen, G. S., Franke, T. F., Stanbridge, E., Frisch, S., and Reed, J. C. (1998). Regulation of cell death protease caspase-9 by phosphorylation. *Science* **282,** 1318–1321.

Chen, X., Kandasamy, K., and Srivastava, R. K. (2003). Differential roles of RelA (p65) and c-Rel subunits of nuclear factor kappa B in tumor necrosis factor-related apoptosis-inducing ligand signaling. *Cancer Res.* **63,** 1059–1066.

Clarke, P., Meintzer, S. M., Gibson, S., Widmann, C., Garrington, T. P., Johnson, G. L., and Tyler, K. L. (2000). Reovirus-induced apoptosis is mediated by TRAIL. *J. Virol.* **74,** 8135–8139.

Coultas, L., and Strasser, A. (2003). The role of the Bcl-2 protein family in cancer. *Semin. Cancer Biol.* **13,** 115–123.

Danielsen, A. J., and Maihle, N. J. (2002). The EGF/ErbB receptor family and apoptosis. *Growth Factors* **20,** 1–15.

Darnell, J. E., Jr. (2002). Transcription factors as targets for cancer therapy. *Nat. Rev. Cancer* **2,** 740–749.

Datta, S. R., Brunet, A., and Greenberg, M. E. (1999). Cellular survival: A play in three Akts. *Genes Dev.* **13,** 2905–2927.

Daub, H., Wallasch, C., Lankenau, A., Herrlich, A., and Ullrich, A. (1997). Signal characteristics of G protein-transactivated EGF receptor. *EMBO J.* **16,** 7032–7044.

David, M., Wong, L., Flavell, R., Thompson, S. A., Wells, A., Larner, A. C., and Johnson, G. R. (1996). STAT activation by epidermal growth factor (EGF) and amphiregulin. Requirement for the EGF receptor kinase but not for tyrosine phosphorylation sites or JAK1. *J. Biol. Chem.* **271,** 9185–9188.

Deveraux, Q. L., Schendel, S. L., and Reed, J. C. (2001). Antiapoptotic proteins. The bcl-2 and inhibitor of apoptosis protein families. *Cardiol. Clin.* **19,** 57–74.

Fang, X., Yu, S., Eder, A., Mao, M., Bast, R. C., Jr., Boyd, D., and Mills, G. B. (1999). Regulation of BAD phosphorylation at serine 112 by the Ras-mitogen-activated protein kinase pathway. *Oncogene* **18,** 6635–6640.

Fleischer, A., Rebollo, A., and Ayllon, V. (2003). BH3-only proteins: The lords of death. *Arch. Immunol. Ther. Exp. (Warsz)* **51,** 9–17.

Fujita, E., Jinbo, A., Matuzaki, H., Konishi, H., Kikkawa, U., and Momoi, T. (1999). Akt phosphorylation site found in human caspase-9 is absent in mouse caspase-9. *Biochem. Biophys. Res. Commun.* **264,** 550–555.

Gibson, E. M., Henson, E. S., Haney, N., Villanueva, J., and Gibson, S. B. (2002). Epidermal growth factor protects epithelial-derived cells from tumor necrosis factor-related apoptosis-inducing ligand-induced apoptosis by inhibiting cytochrome c release. *Cancer Res.* **62,** 488–496.

Gibson, S., Tu, S., Oyer, R., Anderson, S. M., and Johnson, G. L. (1999). Epidermal growth factor protects epithelial cells against Fas-induced apoptosis. Requirement for Akt activation. *J. Biol. Chem.* **274,** 17612–17618.

Gibson, S. B., Oyer, R., Spalding, A. C., Anderson, S. M., and Johnson, G. L. (2000). Increased expression of death receptors 4 and 5 synergizes the apoptosis response to combined treatment with etoposide and TRAIL. *Mol. Cell. Biol.* **20,** 205–212.

Gong, B., and Almasan, A. (2000). Apo2 ligand/TNF-related apoptosis-inducing ligand and death receptor 5 mediate the apoptotic signaling induced by ionizing radiation in leukemic cells. *Cancer Res.* **60,** 5754–5760.

Grant, S., Qiao, L., and Dent, P. (2002). Roles of ERBB family receptor tyrosine kinases, and downstream signaling pathways, in the control of cell growth and survival. *Front. Biosci.* **7,** d376–d389.

Gupta, S. (2001). Molecular steps of death receptor and mitochondrial pathways of apoptosis. *Life Sci.* **69,** 2957–2964.

Gupta, S. (2003). Molecular signaling in death receptor and mitochondrial pathways of apoptosis (review). *Int. J. Oncol.* **22,** 15–20.

Guren, T. K., Odegard, J., Abrahamsen, H., Thoresen, G. H., Susa, M., Andersson, Y., Ostby, E., and Christoffersen, T. (2003). EGF receptor-mediated, c-Src-dependent, activation of Stat5b is downregulated in mitogenically responsive hepatocytes. *J. Cell. Physiol.* **196,** 113–123.

Henson, E. S., Gibson, E. M., Bristow, N. A., Villanueva, J., Haney, N., and Gibson, S. B. (2003). Increased expression of Mcl-1 is responsible for the blockage of TRAIL-induced apoptosis mediated by EGF/ErbB1 signaling pathway. *J. Cell. Biochem.* **89,** 1177–1192.

Herbst, R. S., and Langer, C. J. (2002). Epidermal growth factor receptors as a target for cancer treatment: The emerging role of IMC-C225 in the treatment of lung and head and neck cancers. *Semin. Oncol.* **29,** 27–36.

Herrera, B., Fernandez, M., Benito, M., and Fabregat, I. (2002). cIAP-1, but not XIAP, is cleaved by caspases during the apoptosis induced by TGF-beta in fetal rat hepatocytes. *FEBS Lett.* **520,** 93–96.

Humlova, Z. (2002). Protooncogene bcl-2 in process of apoptosis. Review article. *Sb Lek.* **103,** 419–425.

Juliano, R. L. (2002). Signal transduction by cell adhesion receptors and the cytoskeleton: Functions of integrins, cadherins, selectins, and immunoglobulin-superfamily members. *Annu. Rev. Pharmacol. Toxicol.* **42,** 283–323.

Kandasamy, K., and Srivastava, R. K. (2002). Role of the phosphatidylinositol 3'-kinase/PTEN/Akt kinase pathway in tumor necrosis factor-related apoptosis-inducing ligand-induced apoptosis in non-small cell lung cancer cells. *Cancer Res.* **62,** 4929–4937.

Kari, C., Chan, T. O., Rocha de Quadros, M., and Rodeck, U. (2003). Targeting the epidermal growth factor receptor in cancer: Apoptosis takes center stage. *Cancer Res.* **63,** 1–5.

Kavurma, M. M., and Khachigian, L. M. (2003). Signaling and transcriptional control of Fas ligand gene expression. *Cell Death Differ.* **10,** 36–44.

Keane, M. M., Ettenberg, S. A., Nau, M. M., Russell, E. K., and Lipkowitz, S. (1999). Chemotherapy augments TRAIL-induced apoptosis in breast cell lines. *Cancer Res.* **59,** 734–741.

Kim, Y., Suh, N., Sporn, M., and Reed, J. C. (2002). An inducible pathway for degradation of FLIP protein sensitizes tumor cells to TRAIL-induced apoptosis. *J. Biol. Chem.* **277,** 22320–22329.

Kops, G. J., and Burgering, B. M. (2000). Forkhead transcription factors are targets of signalling by the proto-oncogene PKB (C-AKT). *J. Anat.* **197,** 571–574.

Kothari, S., Cizeau, J., McMillian-Ward, E., Isreals, S., Ens, K., Kirshenbaum, L., Bailes, M., and Gibson, S. B. (2003). BNIP3 plays a role in hypoxia induced cell death and is negatively regulated by growth factors in human epithelial-derived cells. *Oncogene* in press.

Lawlor, M. A., and Alessi, D. R. (2001). PKB/Akt: A key mediator of cell proliferation, survival and insulin responses? *J. Cell Sci.* **114,** 2903–2910.

LeBlanc, H. N., and Ashkenazi, A. (2003). Apo2L/TRAIL and its death and decoy receptors. *Cell Death Differ.* **10,** 66–75.

Leu, C. M., Chang, C., and Hu, C. (2000). Epidermal growth factor (EGF) suppresses staurosporine-induced apoptosis by inducing mcl-1 via the mitogen-activated protein kinase pathway. *Oncogene* **19,** 1665–1675.

Lin, S. Y., Makino, K., Xia, W., Matin, A., Wen, Y., Kwong, K. Y., Bourguignon, L., and Hung, M. C. (2001). Nuclear localization of EGF receptor and its potential new role as a transcription factor. *Nat. Cell Biol.* **3,** 802–880.

Liston, P., Fong, W. G., Kelly, N. L., Toji, S., Miyazaki, T., Conte, D., Tamai, K., Craig, C. G., McBurney, M. W., and Korneluk, R. G. (2001). Identification of XAF1 as an antagonist of XIAP anti-caspase activity. *Nat. Cell Biol.* **3,** 128–133.

Liu, D., Aguirre Ghiso, J., Estrada, Y., and Ossowski, L. (2002). EGFR is a transducer of the urokinase receptor initiated signal that is required for in vivo growth of a human carcinoma. *Cancer Cell* **1,** 445–457.

Lizcano, J. M., Morrice, N., and Cohen, P. (2000). Regulation of BAD by cAMP-dependent protein kinase is mediated via phosphorylation of a novel site, Ser155. *Biochem. J.* **349,** 547–557.

MacFarlane, M. (2003). TRAIL-induced signalling and apoptosis. *Toxicol. Lett.* **139,** 89–97.

Mandal, M., Li, F., and Kumar, R. (2002). Heregulin up-regulates heat shock protein-70 expression in breast cancer cells. *Anticancer Res.* **22,** 1965–1969.

Marti, U., and Wells, A. (2000). The nuclear accumulation of a variant epidermal growth factor receptor (EGFR) lacking the transmembrane domain requires coexpression of a full-length EGFR. *Mol. Cell Biol. Res. Commun.* **3,** 8–14.

Menard, S., Fortis, S., Castiglioni, F., Agresti, R., and Balsari, A. (2001). HER2 as a prognostic factor in breast cancer. *Oncology* **61,** 67–72.

Nakshatri, H., and Goulet, R. J., Jr. (2002). NF-kappaB and breast cancer. *Curr. Probl. Cancer* **26,** 282–309.

Navolanic, P. M., Steelman, L. S., and McCubrey, J. A. (2003). EGFR family signaling and its association with breast cancer development and resistance to chemotherapy (Review). *Int. J. Oncol.* **22,** 237–252.

Ni, C. Y., Yuan, H., and Carpenter, G. (2003). Role of the ErbB-4 carboxyl terminus in gamma-secretase cleavage. *J. Biol. Chem.* **278,** 4561–4565.

Nicholson, K. M., and Anderson, N. G. (2002). The protein kinase B/Akt signalling pathway in human malignancy. *Cell Signal* **14,** 381–395.

Niiro, H., and Clark, E. A. (2002). Regulation of B-cell fate by antigen-receptor signals. *Nat. Rev. Immunol.* **2,** 945–956.

Niu, G., Bowman, T., Huang, M., Shivers, S., Reintgen, D., Daud, A., Chang, A., Kraker, A., Jove, R., and Yu, H. (2002). Roles of activated Src and Stat3 signaling in melanoma tumor cell growth. *Oncogene* **21,** 7001–7010.

Normanno, N., Bianco, C., De Luca, A., Maiello, M. R., and Salomon, D. S. (2003a). Target-based agents against ErbB receptors and their ligands: A novel approach to cancer treatment. *Endocrinal. Relat. Cancer* **10,** 1–21.

Normanno, N., Maiello, M. R., and De Luca, A. (2003b). Epidermal growth factor receptor tyrosine kinase inhibitors (EGFR-TKIs): Simple drugs with a complex mechanism of action? *J. Cell. Physiol.* **194,** 13–19.

O'Shea, J. J., Gadina, M., and Schreiber, R. D. (2002). Cytokine signaling in 2002: New surprises in the Jak/Stat pathway. *Cell* **109,** S121–S131.

Orlowski, R. Z., and Baldwin, A. S., Jr. (2002). NF-kappaB as a therapeutic target in cancer. *Trends Mol. Med.* **8,** 385–389.

Parise, L. V., Lee, J., and Juliano, R. L. (2000). New aspects of integrin signaling in cancer. *Semin. Cancer Biol.* **10,** 407–414.

Pianetti, S., Arsura, M., Romieu-Mourez, R., Coffey, R. J., and Sonenshein, G. E. (2001). Her-2/neu overexpression induces NF-kappaB via a PI3-kinase/Akt pathway involving calpain-mediated degradation of IkappaB-alpha that can be inhibited by the tumor suppressor PTEN. *Oncogene* **20,** 1287–1299.

Qiu, Y., Ravi, L., and Kung, H. J. (1998). Requirement of ErbB2 for signalling by interleukin-6 in prostate carcinoma cells. *Nature* **393,** 83–85.

Ranson, M., and Sliwkowski, M. X. (2002). Perspectives on anti-HER monoclonal antibodies. *Oncology* **63,** 17–24.

Ravi, R., and Bedi, A. (2002). Requirement of BAX for TRAIL/Apo2L-induced apoptosis of colorectal cancers: Synergism with sulindac-mediated inhibition of Bcl-x(L). *Cancer Res.* **62,** 1583–1587.

Reginato, M. J., Mills, K. R., Paulus, J. K., Lynch, D. K., Sgroi, D. C., Debnath, J., Muthuswamy, S. K., and Brugge, J. S. (2003). Integrins and EGFR coordinately regulate the pro-apoptotic protein Bim to prevent anoikis. *Nat. Cell Biol.* **6**, 6.

Roussidis, A. E., and Karamanos, N. K. (2002). Inhibition of receptor tyrosine kinase-based signal transduction as specific target for cancer treatment. *In Vivo* **16**, 459–469.

Salvesen, G. S., and Duckett, C. S. (2002). IAP proteins: Blocking the road to death's door. *Nat. Rev. Mol. Cell. Biol.* **3**, 401–410.

Santoro, M. G., Rossi, A., and Amici, C. (2003). NF-kappaB and virus infection: Who controls whom. *EMBO J.* **22**, 2552–2560.

Schneider, P., Thome, M., Burns, K., Bodmer, J. L., Hofmann, K., Kataoka, T., Holler, N., and Tschopp, J. (1997). TRAIL receptors 1 (DR4) and 2 (DR5) signal FADD-dependent apoptosis and activate NF-kappaB. *Immunity* **7**, 831–836.

Scorrano, L., and Korsmeyer, S. J. (2003). Mechanisms of cytochrome c release by proapoptotic BCL-2 family members. *Biochem. Biophys. Res. Commun.* **304**, 437–444.

Shetty, S., Gladden, J. B., Henson, E. S., Hu, X., Villanueva, J., Haney, N., and Gibson, S. B. (2002). Tumor necrosis factor-related apoptosis inducing ligand (TRAIL) up-regulates death receptor 5 (DR5) mediated by NFkappaB activation in epithelial derived cell lines. *Apoptosis* **7**, 413–420.

Sibilia, M., Fleischmann, A., Behrens, A., Stingl, L., Carroll, J., Watt, F. M., Schlessinger, J., and Wagner, E. F. (2000). The EGF receptor provides an essential survival signal for SOS-dependent skin tumor development. *Cell* **102**, 211–220.

Song, L., Turkson, J., Karras, J. G., Jove, R., and Haura, E. B. (2003). Activation of Stat3 by receptor tyrosine kinases and cytokines regulates survival in human non-small cell carcinoma cells. *Oncogene* **22**, 4150–4165.

Spalding, A. C., Jotte, R. M., Scheinman, R. I., Geraci, M. W., Clarke, P., Tyler, K. L., and Johnson, G. L. (2002). TRAIL and inhibitors of apoptosis are opposing determinants for NF-kappaB-dependent, genotoxin-induced apoptosis of cancer cells. *Oncogene* **21**, 260–271.

Stambolic, V., Suzuki, A., de la Pompa, J. L., Brothers, G. M., Mirtsos, C., Sasaki, T., Ruland, J., Penninger, J. M., Siderovski, D. P., and Mak, T. W. (1998). Negative regulation of PKB/Akt-dependent cell survival by the tumor suppressor PTEN. *Cell* **95**, 29–39.

Suliman, A., Lam, A., Datta, R., and Srivastava, R. K. (2001). Intracellular mechanisms of TRAIL: Apoptosis through mitochondrial-dependent and -independent pathways. *Oncogene* **20**, 2122–2133.

Swarthout, J. T., and Walling, H. W. (2000). Lysophosphatidic acid: Receptors, signaling and survival. *Cell. Mol. Life Sci.* **57**, 1978–1985.

Takimoto, R., and El-Deiry, W. S. (2000). Wild-type p53 transactivates the KILLER/DR5 gene through an intronic sequence-specific DNA-binding site. *Oncogene* **19**, 1735–1743.

Talapatra, S., and Thompson, C. B. (2001). Growth factor signaling in cell survival: Implications for cancer treatment. *J. Pharmacol. Exp. Ther.* **298**, 873–878.

Tsujimoto, Y. (2003). Cell death regulation by the Bcl-2 protein family in the mitochondria. *J. Cell. Physiol.* **195**, 158–167.

Vaux, D. L., and Silke, J. (2003). Mammalian mitochondrial IAP binding proteins. *Biochem. Biophys. Res. Commun.* **304**, 499–504.

Venkateswarlu, S., Dawson, D. M., St Clair, P., Gupta, A., Willson, J. K., and Brattain, M. G. (2002). Autocrine heregulin generates growth factor independence and blocks apoptosis in colon cancer cells. *Oncogene* **21**, 78–86.

Verhagen, A. M., and Vaux, D. L. (2002). Cell death regulation by the mammalian IAP antagonist Diablo/Smac. *Apoptosis* **7**, 163–166.

Vivanco, I., and Sawyers, C. L. (2002). The phosphatidylinositol 3-kinase AKT pathway in human cancer. *Nat. Rev. Cancer* **2**, 489–501.

von Gise, A., Lorenz, P., Wellbrock, C., Hemmings, B., Berberich-Siebelt, F., Rapp, U. R., and Troppmair, J. (2001). Apoptosis suppression by Raf-1 and MEK1 requires MEK- and phosphatidylinositol 3-kinase-dependent signals. *Mol. Cell. Biol.* **21,** 2324–2336.

Wang, C. Y., Mayo, M. W., Korneluk, R. G., Goeddel, D. V., and Baldwin, A. S., Jr. (1998). NF-kappaB antiapoptosis: Induction of TRAF1 and TRAF2 and c-IAP1 and c-IAP2 to suppress caspase-8 activation. *Science* **281,** 1680–1683.

Widmann, C., Gibson, S., Jarpe, M. B., and Johnson, G. L. (1999). Mitogen-activated protein kinase: Conservation of a three-kinase module from yeast to human. *Physiol. Rev.* **79,** 143–180.

Yang, Y., and Yu, X. (2003). Regulation of apoptosis: The ubiquitous way. *FASEB J.* **17,** 790–799.

Yarden, Y. (2001). The EGFR family and its ligands in human cancer signalling mechanisms and therapeutic opportunities. *Eur. J. Cancer* **37,** S3–S8.

Zhang, X. D., Zhang, X. Y., Gray, C. P., Nguyen, T., and Hersey, P. (2001). Tumor necrosis factor-related apoptosis-inducing ligand-induced apoptosis of human melanoma is regulated by smac/DIABLO release from mitochondria. *Cancer Res.* **61,** 7339–7348.

13

TRAIL AND CERAMIDE

YONG J. LEE* AND ANDREW A. AMOSCATO[†]

*Department of Surgery and Pharmacology
[†]Department of Pathology
University of Pittsburgh, Pittsburgh, Pennsylvania 15213

I. Introduction
II. TRAIL
   A. Differential Cytotoxic Effect of TRAIL
   B. Factors Involved in the Differential Sensitivity to TRAIL
   C. TRAIL-Induced Death Signal
   D. Role of Death Inhibitors in TRAIL-Induced Cytotoxicity
   E. TRAIL and Smac/DIABLO
III. Sphingolipids
   A. The Generation of Ceramide
   B. The Use of Ceramide
IV. TRAIL and Ceramide
   References

Tumor necrosis factor–related apoptosis-inducing ligand (TRAIL) is a clinically useful cytokine. TRAIL induces apoptosis in a wide variety of transformed cells, but does not cause toxicity to most normal cells. Recent studies show that death receptors (DR4 and DR5), decoy receptors (DcR1 and DcR2), and death inhibitors (FLIP, FAP-1, and IAP) are responsible for the differential sensitivity to TRAIL of normal and tumor cells. Several researchers have also shown that genotoxic agents, such as chemotherapeutic agents and ionizing radiation, enhance TRAIL-induced cytotoxicity by increasing DR5 gene expression or decreasing the intracellular level of FLIP, an antiapoptotic protein. Previous studies have shown that ceramide helps to regulate a cell's response to various forms of stress. Stress-induced alterations in the intracellular concentration of ceramide occur through the activation of a variety of enzymes that synthesize or catabolize ceramide. Increases in intracellular ceramide levels modulate apoptosis by acting through key proteases, phosphatases, and kinases. This review discusses the interaction between TRAIL and ceramide signaling pathways in regulating apoptotic death. © 2004 Elsevier Inc.

I. INTRODUCTION

In recent years, it has become increasingly clear that ceramide has an important role in apoptosis. With the advent of newer methods of analysis and a greater variety of molecular tools to investigate ceramide metabolism, the challenge of addressing specific questions regarding ceramide topology, as well as defining roles for specific ceramide species now become possible.

Of critical importance is the role that ceramide plays in orchestrating a cell's response to various forms of stress. Alterations in the intracellular concentration of ceramide during a stress-induced response occur through the activation of a variety of enzymes that synthesize or catabolize ceramide. Indeed, early studies identified a variety of agents that induce the hydrolysis of sphingomyelin, generating ceramide (Mathisa et al., 1998; Pettus et al., 2002). The resulting biologic effects, including cytotoxicity, could be mimicked by addition of exogenous ceramide or overexpression of sphingomyelinases. These studies indicated that ceramide/sphingolipid metabolism is regulated in response to a variety of extracellular factors. Subsequent studies indicated that many of the typical inducers of apoptosis such as Fas ligand, tumor necrosis factor (TNF) alpha, etc., regulate one or more enzymes involved in ceramide metabolism, leading to the increase in intracellular levels. Increases in intracellular ceramide levels further modulate apoptosis by acting through key proteases, phosphatases, and kinases.

TRAIL is an apoptosis-inducing member of the TNF gene family, based on amino acid homology to TNF and Fas ligand. Human mononuclear phagocytes (MΦ), T cells, natural killer (NK) cells, and dendritic cells express TRAIL on the surface of their cellular membrane after activation with interferon (IFN)-α (type I) or IFN-γ (type II) (Fanger et al., 1999; Griffith et al., 1999; Kayagaki et al., 1999; Zamai et al., 1998). In addition to upregulation by IFNs, TRAIL induction occurs on measles virus–stimulated dendritic cells (Vidalain et al., 2000) and on cytomegalovirus (CMV)-infected fibroblasts (Sedger et al., 1999). These observations suggest that TRAIL may play a role in innate immune responses involving IFN. Although TRAIL protein expression *in vivo* has been relatively poorly studied, recent studies have shown that administration of interleukin (IL)-12 induces expression of TRAIL by stimulating IFN-γ production in natural killer (NK) cells (Smyth et al., 2001). These results reveal that TRAIL induction on NK cells plays a critical role in IFN-γ–mediated antimetastatic effects of IL-12. It is well known that TRAIL binds to death receptors and induces apoptotic death in a number of tumor cell lines, but not in most normal cells. TRAIL-mediated antitumor effects are enhanced by treatment with antitumor agents. For example, several researchers have shown that genotoxic agents such as chemotherapeutic agents and ionizing radiation enhance TRAIL-induced cytotoxicity by increasing DR5 gene expression or decreasing intracellular levels of FLIP, an antiapoptotic protein. Previous studies have shown an increase in endogenous ceramide levels in response to genotoxic stress (Dbaibo et al., 1998). Elevation of the intracellular level of ceramide promotes TRAIL-induced cytotoxicity (Nam et al., 2002).

II. TRAIL

TRAIL/APO-2L is a recently identified type II integral membrane protein belonging to the TNF family. TRAIL is a 281-amino acid protein, related most closely to the Fas/APO-1 ligand. Like Fas ligand (FasL) and TNF, the C-terminal extracellular region of TRAIL (amino acids 114–281) exhibits a homotrimeric subunit structure (Pitti et al., 1996). However, unlike FasL and TNF, it induces apoptosis in a variety of tumor cell lines more efficiently than in normal cells (Ashkenazi and Dixit, 1998). This differential cytotoxicity suggests that TRAIL is a good candidate for use as an effective anticancer agent.

A. DIFFERENTIAL CYTOTOXIC EFFECT OF TRAIL

Although several cytotoxin genes such as cytokines, apoptosis-related genes, and suicide genes are currently available for gene therapy, one important criterion in choosing such a gene is a greater toxicity in tumor cells relative to normal cells. Numerous studies reveal that TRAIL induces apoptosis in a wide variety of tumor cells, but does not cause toxicity to most normal cells. However, recent studies reveal that a polyhistidine-tagged TRAIL induces apoptosis in normal human hepatocytes in culture (Jo et al., 2000). This is probably due to an aberrant conformation and subunit structure of TRAIL in the presence of low zinc concentrations

(Lawrence et al., 2001). In contrast, native-sequence, nontagged recombinant TRAIL, when produced under optimized zinc concentrations, is markedly more active against tumor cells than the polyhistidine-tagged ligand, but has minimal toxicity toward human hepatocytes *in vitro* (Lawrence et al., 2001). Moreover, preclinical studies in mice and primates have shown that administration of TRAIL can induce apoptosis in human tumors, but no cytotoxicity to normal organs or tissue (Walczak et al., 1999). In addition, unlike TNF and FasL, TRAIL mRNA is expressed constitutively in many tissues (Pitti et al., 1996; Wiley et al., 1995). Recent studies also reveal that TRAIL, which is constitutively expressed on murine NK cells in the liver, plays an important role in surveillance of tumor metastasis (Hayakawa et al., 2001). Although TRAIL is an apoptosis-inducing member of the TNF gene family based on amino acid homology to the TNF and FasL (Pitti et al., 1996; Wiley et al., 1995), unlike TNF and FasL, TRAIL infusion does not cause a lethal inflammatory response.

B. FACTORS INVOLVED IN THE DIFFERENTIAL SENSITIVITY TO TRAIL

The apoptotic signal induced by TRAIL is transduced by its binding to the death receptors TRAIL-R1 (DR4) and TRAIL-R2 (DR5), which are members of the TNF receptor superfamily. Both DR4 and DR5 contain a cytoplasmic death domain that is required for TRAIL receptor-induced apoptosis. TRAIL also binds to TRAIL-R3 (DcR1) and TRAIL-R4 (DcR2), which act as decoy receptors by inhibiting TRAIL signaling (Degli-Esposti et al., 1997a,b; Marsters et al., 1997; Pan et al., 1997a,b; Sheridan et al., 1997; Walczak et al., 1997). Unlike DR4 and DR5, DcR1 does not have a cytoplasmic domain and DcR2 retains a cytoplasmic fragment containing a truncated form of the consensus death domain motif (Pan et al., 1997a). The relative resistance of normal cells to TRAIL has been explained by the presence of large numbers of the decoy receptors on normal cells (Ashkenazi and Dixit, 1999; Gura, 1997). Recently, this hypothesis has been challenged based on poor correlations between DR4, DR5, and DcR1 expression and sensitivity to TRAIL-induced apoptosis in normal and cancerous breast cell lines (Keane et al., 1999). This discrepancy indicates that other factors such as death inhibitors (FLIP, FAP-1, or IAP) are also involved in the differential sensitivity to TRAIL.

C. TRAIL-INDUCED DEATH SIGNAL

Recent studies demonstrated that the Fas-associated death domain (FADD) is required for TRAIL-induced apoptosis (Kischkel et al., 2000; Kuang et al., 2000; Sprick et al., 2000). FADD contains a C-terminal death domain (DD) as well as an N-terminal death effector domain (DED).

TRAIL triggers apoptosis by recruiting the apoptosis initiator procaspase-8 through the adaptor FADD (Bodmer *et al.*, 2000; Kischkel *et al.*, 2000) and forming a death-inducing signaling complex (DISC) (Medema *et al.*, 1997). Caspase-8 can directly activate downstream effector caspases including procaspase-3, -6, and -7 (Cohen, 1997). Caspase-8 also cleaves Bid and the 15 kDa form of truncated Bid binds to Bax and triggers a change in the confirmation of Bax. As a result, Bax oligomerizes and inserts into the outer mitochondrial membrane, which results in cytochrome *c* release from mitochondria (Eskes *et al.*, 2000). Cytochrome *c* then interacts with Apaf-1, dATP/ATP, and procaspase-9 to form a 700-kDa complex known as the apoptosome, which results in autocatalytic processing of caspase-9 (Hu *et al.*, 1999). Caspase-9 cleaves and activates procaspase-3 (Slee *et al.*, 1999). Previous studies have shown that induction of apoptosis in type I cells is mainly accompanied by activation of large amounts of caspase-8 and formation of DISC (mitochondrial-independent apoptosis pathway). In contrast, in type II cells, DISC formation is strongly reduced and activation of caspase-8 and caspase-3 occurred following the loss of mitochondrial transmembrane potential (mitochondrial-dependent apoptosis pathway). Recently, we observed that TRAIL-induced caspase-3 activation and apoptosis are mediated through two different apoptotic pathways, mitochondria-dependent and mitochondria-independent (Lee *et al.*, 2001). Interestingly, TRAIL-induced cytotoxicity is not significantly inhibited by Bcl-2 overexpression (Gazitt *et al.*, 1999; Lee *et al.*, 2001; Walczak *et al.*, 2000). These results suggest that TRAIL-induced apoptosis is primarily dependent on a mitochondria-independent pathway. However, TRAIL-induced apoptotic death can also be enhanced by promoting the mitochondria-dependent pathway (Lee *et al.*, 2001).

D. ROLE OF DEATH INHIBITORS IN TRAIL-INDUCED CYTOTOXICITY

Numerous proteins including cellular FLICE-inhibitory protein (FLIP) (Griffith *et al.*, 1998), Fas-associated protein (FAP-1) (Sato *et al.*, 1995), Bcl-2 (Wen *et al.*, 2000), Bcl-X_L (Wen *et al.*, 2000), Bruton's tyrosine kinase (BTK) (Vassilev *et al.*, 1999), silencer of death domain (SODD) (Tschopp *et al.*, 1999), toso (Hitoshi *et al.*, 1998), inhibitor of apoptosis (IAP) (Kothny-Wilkes *et al.*, 1999), X-linked inhibitor of apoptosis (XIAP) (Deveraux and Reed, 1999), and survivin (Tamm *et al.*, 1998) have been shown to inhibit apoptotic death. These proteins block either death receptor or mitochondrial pathways by inhibiting the activity of the effector caspase-3 and -7 and the initiator caspase-9 and -8. For example, c-FLIP is structurally similar to caspase-8, because it contains two death effector domains and a caspase-like domain. However, this domain lacks residues that are important for its catalytic activity, most notably the cysteine within the active site. c-FLIP is recruited to the DISC. Expressed at high levels in stable transfectants, c-FLIP completely blocks receptor-ligand apoptotic death through inhibition of caspase-8 processing at the DISC (Scaffidi et al., 1999).

E. TRAIL AND Smac/DIABLO

Second mitochondria-derived activator of caspase (Smac)/direct IAP binding protein with low pI (DIABLO) is a mitochondrial protein that is released together with cytochrome c from the mitochondria during apoptosis (Du et al., 2000). It promotes cytochrome c–dependent caspase activation by neutralizing IAPs. The N terminus of Smac/DIABLO is required for its ability to interact with the baculovirus IAP repeat (BIR3) of IAPs and to promote cytochrome c–dependent caspase activation (Srinivasula et al., 2000). Consistent with the ability of Smac/DIABLO to function at the level of the effector caspases, expression to a cytosolic Smac/DIABLO in type II cells allowed TRAIL to bypass Bcl-X_L inhibition of death receptor–induced apoptosis. Interestingly, recent studies have shown that Bax and Bak differentially regulate the release of cytochrome c and Smac/DIABLO from mitochondria (Kandasamy et al., 2003). TRAIL induces cytochrome c release and apoptosis in wild-type, $Bid^{-/-}$, $Bax^{-/-}$, or $Bak^{-/-}$ murine embryonic fibroblasts (MEFs), but not in $Bax^{-/-}$ $Bak^{-/-}$ double knockout MEFs. Unlike cytochrome c release, TRAIL-induced Smac/DIABLO release is blocked in $Bid^{-/-}$, $Bax^{-/-}$, $Bak^{-/-}$, or $Bax^{-/-}$ $Bak^{-/-}$ MEFs. Zhang et al. (2003) demonstrated that activation of ERK1/2 protects melanoma cells from TRAIL-induced apoptosis by inhibiting the relocalization of Bax from the cytosol to mitochondria and consequently inhibiting Smac/DIABLO release from mitochondria (Zhang et al., 2003).

III. SPHINGOLIPIDS

Sphingolipids comprise a family of lipid molecules that include the long chain bases sphingosine, sphinganine and their related phosphorylated and methylated derivatives; ceramides; gangliosides; neutral glycolipids and sulfatides; among others (Kanfer, 1983a; Hannun, 1996). Extensive studies on this family of lipid molecules clearly establish their role in cellular regulation. Several identified defects in a number of sphingolipid enzymes result in a variety of inherited sphingolipidoses (Kanfer, 1983b; Merrill and Wang, 1992). In addition, sphingolipids are involved in cell contact responses, participate as anchors/components for membrane proteins/receptors, are actively involved in cell differentiation, and are necessary for cell viability with regard to membrane integrity (Hannun and Bell, 1989).

The importance of sphingolipids in cellular regulation is further demonstrated by the structural identification of certain bacterial cytotoxins and hemolysins as sphingomyelinases (Merrill *et al.*, 1993). Toxic fungal metabolites also bear a striking structural resemblance to sphingosine, and target directly to enzymes of sphingolipid metabolism (Merrill *et al.*, 1993). These studies provide further support for the importance of this class of molecules in cell regulation.

A. THE GENERATION OF CERAMIDE

Ceramide levels are controlled by the coordinate actions of enzymes that are responsible for its formation and degradation. Their activities are sensitive to a variety of stress signals (Fig. 1).

1. The Hydrolysis of Sphingomyelin by Sphingomyelinases

Five forms of sphingomyelinase exist, distinguished from one another by their pH optima (acid, neutral, or alkaline), their subcellular localization, and their cation dependence. These enzymes hydrolyze the phosphodiester bond of sphingomyelin yielding ceramide and phosphocholine.

A neutral sphingomyelinase (N-SMase) activity has been found to be activated by a variety of stimuli, including TNF-alpha, interleukin-1, Fas ligand, chemotherapeutic agents, 1 alpha, 25-dihydroxyvitamin D_3, ischemia/reperfusion, heat stress, arachidonic acid and glutathione depletion

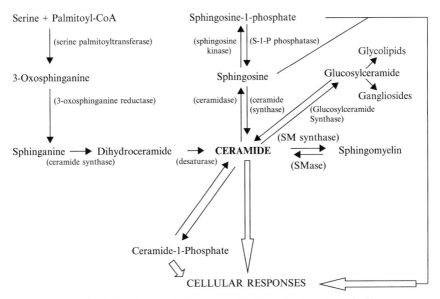

FIGURE 1. A flow chart for ceramide biosynthesis and metabolism.

(Horinouchi et al., 1995; Liu et al., 1998; Okazaki et al., 1994; Otterbach and Stoffel, 1995; Spence et al., 1982; Tepper et al., 1995). Several forms have been described and are products of a distinct gene, but have not been clearly characterized at the molecular level. FAN (factor associated with neutral sphingomyelinase activation) is an adapter protein that associates with the TNF alpha receptor and is necessary for neutral SMase activation. Magnesium-dependent and -independent isoforms of N-SMase appear to be associated with the plasma membrane, cytosolic, mitochondrial and nuclear compartments as well as with other organelles (Albi and Magni, 1997; Alessenko and Chatterjee, 1995; Gatt, 1976; Hofmann et al., 2000; Jaffrezou et al., 2001; Levade et al., 1995; Okazaki et al., 1994; Spence et al., 1982; Tamiya-Koizumi et al., 1989; Tomiuk et al., 1998).

Acid SMase (A-SMase) is a lysosomal/endosomal enzyme that is required for plasma membrane turnover and is defective in patients with Niemann-Pick disease (Schneider and Kennedy, 1967; Schuchman and Desnick, 1995). The enzymes in the human and murine systems have been cloned and several processing steps are required to activate the enzyme (Hurwitz et al., 1994). The acid form of SMase also exists as a zinc-dependent secreted form, stimulated by cytokines (Marathe et al., 1998). It is thought that the secreted form of acid SMase functions in hydrolyzing cell surface sphingomyelin, thereby initiating cell signaling (Schissel et al., 1996), and also in the formation of atherogenic lipoprotein particles (Tabas, 1999). Acid Smase-deficient Niemann-Pick fibroblasts are resistant to radiation-induced apoptosis, which suggests that this particular enzyme is necessary for the radiation-induced response in these cells.

Very little is known about alkaline SMases, but they appear to be involved in digestion and mucosal cell proliferation (Duan et al., 1996; Nilsson and Duan, 1999; Nyberg et al., 1996).

2. *De Novo* Synthesis of Ceramide

This pathway, which has its origins in the endoplasmic reticulum (ER), is initiated by the condensation of serine and palmitoyl-CoA via serine palmitoyltransferase, the rate-limiting enzyme, forming 3-oxosphinganine. A 3-oxosphinganine reductase then reduces the carbonyl group to a hydroxyl group forming sphinganine. This product is then acetylated by ceramide synthase to yield dihydroceramide. This acetylation step in ceramide biosynthesis can be inhibited by the mycotoxin fumonisin B1, and has been used in defining which ceramide pathway is involved in the stress-induced response. In the final step, a dihydroceramide desaturase adds the critical trans-4,5 double bond that is necessary for ceramide's biologic activity (Bielawska et al., 1993). This pathway of ceramide generation has been shown to be induced in response to TNF and several chemotherapeutic agents including daunorubicin, etoposide, CPT-11 and hexadecylphosphocholine (Bose and Kolesnick, 2000; Perry et al., 2000;

Wieder et al., 1998). We have also shown that the de novo pathway of ceramide biosynthesis is involved in tumor-induced dendritic cell apoptosis (Kanto et al., 2001) as well as apoptosis induced by androgen ablation in prostate cancer cells (Eto et al., 2003). Free palmitoyl CoA has also been shown to activate this pathway and the ceramide produced has been suggested to play a role in mediating complications seen in diabetes mellitus and obesity (Paumen et al., 1997).

B. THE USE OF CERAMIDE

The enzymes of the sphingomyelinase pathway or the de novo synthetic pathway all contribute toward increasing cellular ceramide levels as discussed previously. Once formed, ceramide can be used by a variety of biosynthetic pathways, which include the formation of sphingomyelin, the formation of glucosylceramide for eventual conversion to gangliosides and glycolipids, the formation of ceramide-1-phosphate, as well as the breakdown of ceramide by various ceramidases.

1. The Formation of Sphingomyelin

Ceramide can be used for the formation of sphingomyelin by the action of sphingomyelin synthase. Activities for this enzyme have been associated with the Golgi, plasma membrane, nuclear, and mitochondrial compartments (Luberto and Hannun, 1998; Luberto et al., 2000; Mathias et al., 1998; Pettus et al., 2002). This enzyme functions in transferring the phosphorylcholine group from phosphatidylcholine to ceramide.

2. The Formation of Glucosylceramide

The first step in the formation of gangliosides or glycolipids begins with the condensation of ceramide and glucose via a glucosylceramide synthase forming glucosylceramide. This mechanism allows for the "removal" of ceramide by generating glucosylceramides, glycolipids, or gangliosides and has been associated with multidrug resistance and resistance to TNF-mediated apoptosis (Liu et al., 1999b; Liu et al., 2001). Glucosylceramide synthase activity has been shown to be associated with the Golgi compartment.

3. Ceramide Catabolism by Ceramidases

Three forms of ceramidase, a neutral, alkaline, and acidic form, have been documented. They reside in the mitochondrial/endosomal, ER/Golgi, and lysosomal/caveolin-enriched light membrane compartments, respectively (Coroneos et al., 1995; El Bawab et al., 1999; El Bawab et al., 2000; Mao et al., 2001; Mitsutake et al., 2001; Okino et al., 1998; Strelow et al., 2000; Tani et al., 2000). These enzymes serve to cleave ceramide at the amide bond, forming sphingosine and a free fatty acid. The resulting sphingosine and its phosphorylated derivative have additional downstream biologic activities (Liu *et al.*, 1999a; Merrill and Wang, 1992). One study indicated that the overexpression of ceramidase resulted in protection from TNF-induced cell death (Strelow *et al.*, 2000). In addition, it has been shown that one form of ceramidase is inhibited by nitric oxide (NO), allowing ceramide to accumulate and apoptosis to occur (Huwiler *et al.*, 1999). These studies indicate that mechanisms of ceramide generation and depletion both contribute to ceramide's overall biologic effect.

4. Sites of Ceramide Formation

De novo synthesis of ceramide involves membrane-bound enzymes located on the cytoplasmic side of the ER as well as on the lumenal side of the Golgi (Merrill and Wang, 1992). Further conversion to either sphingomyelin or glucosylceramide also occurs in these subcellular compartments. Early studies indicated that pools of sphingomyelin and ceramide exist, each of which participates in distinct pathways of cell regulation (Zhang *et al.*, 1997a). Current studies indicate that the location of ceramide generation may be critical in determining its biologic effects. Thus, studies involving vitamin D3–induced sphingomyelin hydrolysis indicated that a cytosolic form and not a membrane bound form of SMase was responsible for ceramide generation (Okazaki *et al.*, 1994). Another study indicated that addition of exogenously added bacterial SMase caused sphingomyelin hydrolysis and ceramide generation in the outer leaflet of the plasma membrane but failed to cause cell death (Zhang *et al.*, 1997a). Other reports indicated that sphingomyelin hydrolysis could occur in plasma membrane microdomains, such as caveolae/lipid rafts (Bilderback *et al.*, 1997; Veldman *et al.*, 2001), which are necessary for CD95 clustering and capping (Cremesti *et al.*, 2001; Grassme *et al.*, 2001). Additional reports indicated that hydrolysis of SM in the inner leaflet of the plasma membrane is necessary for TNF-induced cell death (Andrieu *et al.*, 1996; Dobrowsky *et al.*, 1995; Linardic and Hannun, 1994). Taken together, these studies suggest that a certain topology exists with regard to sphingomyelin hydrolysis/ceramide generation and depends on the cell type and the stress-induced signal. Given the fact that several forms of sphingomyelinases and ceramidases exist in various subcellular compartments, this situation lends itself to a compartmentalized form of ceramide genesis and depletion.

5. Methods Used for Ceramide Analysis

There are a variety of methods that have been used to measure cellular ceramide content. These include the diacylglycerol kinase (DGK) assay, ceramide derivatization followed by high performance liquid chromatography, lipid charring, and radiolabeling techniques (Kolesnick, 1991; Schneider and Kennedy, 1967). The DGK assay is the most commonly used assay. In the past several years, a mass spectrometric approach has been used for the analysis of ceramide (Grullich *et al.*, 2000; Gu *et al.*, 1997; Liebish *et al.*, 1999; Merrill *et al.*, 1988; Sullards, 2000; Thomas *et al.*, 1999; Veldhoven *et al.*, 1989; Watts *et al.*, 1997). Advantages of this method of analysis include not only sensitivity but also the ability to identify and quantitate individual ceramide species. In most cases, results from mass spectrometric-based assays have confirmed results obtained by the DGK assay and have supported DGK findings, indicating that ceramide is generated during the initiation phase of apoptosis. Because of the resolving power of the mass spectrometric method, this type of analysis can begin to investigate the possibility that different ceramide species are associated with different modes of ceramide action.

6. Ceramide and Cell Regulation/Cell Fate

The list of extracellular/biologic inducers of sphingomyelin hydrolysis/ ceramide accumulation is long and varied. They include ionizing and ultraviolet irradiation, Fas ligand, tumor necrosis factor, oxidative stress, 1,25-dihydroxyvitamin D_3, interferon gamma, endotoxin, chemotherapeutic agents, nerve growth factor, heat, CD28, IL-1, progesterone, retinoic acid, serum withdrawal, as well as human immunodeficiency virus (Boland *et al.*, 1997; Bose *et al.*, 1995; Bradshaw *et al.*, 1996; Dbaibo *et al.*, 1995; Hannun, 1994; Herr *et al.*, 1997; Jaffrezou *et al.*, 1996; Jayadev *et al.*, 1995; Kolesnick and Golde, 1994; Michael *et al.*, 1997; Quintans *et al.*, 1994; Suzuki *et al.*, 1997; Verheij *et al.*, 1996; Whitman *et al.*, 1997; Zhang *et al.*, 1996). The role of ceramide in various stress-induced pathways and the kinetics regarding its generation/accumulation are complex and cell- and stress-specific. Some studies indicate increases in ceramide occur early in the stress-induced response (Fas, TNF, and radiation). Specifically with regard to radiation, this form of stress has been shown to directly activate SMase in isolated membranes (Haimovitz-Friedman *et al.*, 1994). In addition, it has recently been shown that Fas stimulation does indeed signal ceramide accumulation in the initiation phase of apoptosis, allowing Fas to cap and kill (El Bawab *et al.*, 1999; Grullich *et al.*, 2000). Thus, ceramide generation is obligate for induction of apoptosis in some stress response systems. Its generation is cell- and stress-specific and is dependent on the delicate balance of a variety of pathways leading to its formation and usage/breakdown.

Normal physiologic signals, such as those induced by cytokines and/or growth factors, can also increase cellular ceramide levels. Thus, depending on the type of stimulus and cell type, the ceramide signal may affect a variety of cellular functions, which include proliferation, differentiation, growth arrest, and apoptosis.

With regard to proliferation, mitogenic signals such as platelet-derived growth factor (PDGF) cause the rapid synthesis of sphingosine-1-phosphate, which is produced from the breakdown of ceramide by the action of ceramidase, followed by a phosphorylation step. It is believed that the critical balance between ceramide and this metabolite coordinately regulates a balance between an apoptotic state and a proliferative state, respectively (Cuvillier et al., 1996).

Other ceramide metabolites such as ceramide-1-phosphate (formed by the action of ceramide kinase) can also stimulate proliferation of 3T3 fibroblasts (Gomez-Munoz et al., 1997).

In some cases, treatment of cells with cytokines such as TNF, IL-1, or gamma interferon, also produces ceramide, only this time affecting differentiation. Specifically, ceramide generation in response to TNF treatment of HL-60 cells or in response to IL-1 treatment of EL-4 cells initiated differentiation in both cell types (Mathias et al., 1993; Okazaki et al., 1989).

Following cellular injury, cells appear to respond by undergoing cell cycle arrest, to allow for sufficient repair, or apoptosis, if the damage is too severe. Several studies support the role of ceramide in growth suppression and apoptosis. Conclusions have been based on several factors: a) most of the inducers of ceramide accumulation/generation are themselves inducers of apoptosis; b) exogenously added ceramides induce apoptosis, differentiation, cell senescence or cell cycle arrest and require the presence of the 4,5-trans double bond in the ceramide molecule; c) concentrations of exogenously added ceramide exert their biologic effect in the same concentration range as that measured in cells; d) addition of inhibitors of sphingolipid metabolism to actively growing cells results in either apoptosis or cell-cycle arrest; and e) ceramide activates proteases of the ICE family.

Induction of ceramide-specific cell-cycle arrest in many cell lines occurs via the inhibition of thymidine uptake and dephosphorylation of the retinoblastoma gene product (Bielawska, 1996; Dbaibo, 1995; Gomez-Munoz et al., 1995; Jarvis et al., 1994; Jayadev et al., 1995; Merrill, 1986; Rani, 1995; Riboni et al., 1992), although ceramide-independent pathways of regulating this particular gene product exist. Ceramide also suppresses expression of the c-myc proto-oncogene (Dbaibo, 1995) and interferes with other pathways of signal transduction by inhibiting phospholipase D activation (Chang et al., 1995; Dobrowsky and Hannun, 1992; Hayakawa et al., 1996; Latinis and Koretzky, 1996; Laulederkind et al., 1995; Pyne et al., 1996; Westwick et al., 1995).

In many instances, an increase in the ceramide level can signal apoptosis. Ceramide is generated at the initiation phase of apoptosis, subsequent to the assembly of the adaptor protein/receptor–death domain complex in some systems (Grullich et al., 2000). During ionizing or ultraviolet irradiation–induced apoptosis, the generation of ceramide appears to be independent of this signaling complex, but the exact mechanism has not been determined (Datta et al., 1996). Studies providing a critical link between ionizing radiation–induced ceramide generation and apoptosis include: a) acid SMase knockout experiments, which showed the lack of ceramide elevation in endothelial cells undergoing apoptosis in response to ionizing stresses or LPS/TNF treatment (Santana *et al.*, 1996); b) in studies involving Niemann-Pick cells (Santana *et al.*, 1996); and c) in experiments using acid sphingomyelinase knockout murine embryonic fibroblasts (MEFs) (Lozano *et al.*, 2001). Although apoptosis can occur in the absence of ceramide generation in some cases (perhaps in cells that display alternate apoptotic pathways), it does appear to be required for apoptosis. This supports the hypothesis that in addition to its generation in the initiation phase of apoptosis in some systems, ceramide could also provide an amplification signal in those types of cell death where mitochondrial-dependent and mitochondrial-independent pathways are operating.

7. Mitochondria, Ceramide, and Apoptosis

Mitochondria have gained much attention over the past several years because of the prominent role that they play in apoptosis through permeability changes, the generation of reactive oxygen species, and the release of cytochrome *c*. These events can be regulated by members of the Bcl-2 family of proteins. Interestingly, ceramide-induced apoptosis can be inhibited by Bcl-2 (Mathias *et al.*, 1998).

The mitochondrial events that appear to be "linked" to ceramide include the mitochondrial permeability transition (MPT) and the generation of reactive oxygen species (ROS). Ionizing radiation or TNF-alpha induces a MPT, which can be reproduced in isolated mitochondria using a cytoplasmic fraction from ceramide-treated cells (Castedo *et al.*, 1996; Pastorino *et al.*, 1996; Subham *et al.*, 1998). Ceramide has also been shown to generate ROS in isolated mitochondria (Garcia-Ruiz *et al.*, 1997) and can specifically exert its effect at site III of the electron transport system (Gudz *et al.*, 1997). Another study indicated that TNF-alpha was able to increase the level of ceramide in mitochondria (Garcia-Ruiz *et al.*, 1997). We have recently shown a specific increase in mitochondrial C16 ceramide during Fas or radiation-induced apoptosis (Matsko *et al.*, 2001). In addition, recent studies indicate that ceramide itself is capable of releasing cytochrome *c* from mitochondria (Zhang *et al.*, 1997b). Ceramide induces cytochrome *c* release from the mitochondria through cleavage of Bid followed by translocation of truncated Bid into the mitochondria (Ghafourifar *et al.*, 1999). The possibility that a neutral sphingomyelinase and a ceramidase exist at the mitochondrial level has been suggested (Albi and Magni, 1997; Alessenko and Chatterjee, 1995; Gatt, 1976; Hofmann *et al.*, 2000; Jaffrezou *et al.*, 2001; Levade *et al.*, 1995; Okazaki *et al.*, 1994; Ruvolo *et al.*, 1999; Spence *et al.*, 1982; Tamiya-Koizumi *et al.*, 1989; Tomiuk *et al.*, 1998). In addition, a CAPP has been shown to be involved in dephosphorylating mitochondrial PKC alpha and Bcl-2 (Ruvolo *et al.*, 1998). Taken together, all of these findings suggest that the ceramide-mitochondria interaction is important in programmed cell death.

8. Biologic Targets of Ceramide

Earlier studies indicated that ceramide increased the activity of the serine/threonine protein phosphatases PP1 and PP2A (ceramide activated protein phosphatases, CAPP) (Dobrowsky *et al.*, 1993; Wolff *et al.*, 1994). These ceramide activated protein phosphatases are not activated by dihydro-ceramide (Dobrowsky *et al.*, 1993; Fishbein *et al.*, 1993; Law and Rossie, 1995; Wolff *et al.*, 1994). Only the D- and L-erythro forms of ceramide with specific fatty acid chain lengths were capable of activating the enzymes. Cells deficient in specific subunits of these enzymes fail to respond to the effects of ceramide. Biologic substrates for these phosphatases include c-Jun, Rb, PKC alpha, Akt/PKB, SR and Bcl-2 proteins (Black *et al.*, 1991; Deng *et al.*, 1998; Sato *et al.*, 2000), all of which are involved in some way with cellular apoptotic pathways. CAPP may also participate in down regulation of *c-myc* expression (Mathias *et al.*, 1998).

A ceramide activated protein kinase (CAPK) has been reported in A-431 cells. This has subsequently been identified as the kinase suppressor of RAS (KSR). It has been suggested that activation of KSR by ceramide leads to activation of Raf (through direct interaction), ultimately activating stress kinase pathways (Basu and Kolesnick, 1998; Xing and Kolesnick, 2001). Ceramide synthesized by the de novo pathway has been shown to target the Raf-1/ERK pathway during astrocyte apoptosis (Blazquez *et al.*, 2000).

Ceramide also increases protein kinase C ζ activity through direct interaction, thereby modulating TNF-induced cellular responses (Stratford *et al.*, 2001). Ceramide has also been shown to be necessary for the complexation of the antiapoptotic protein phosphoinositide-3-kinase (PI-3 kinase) to caveolin-1, resulting in PI-3 kinase inactivation (Zundel and Giaccia, 1998; Zundel *et al.*, 2000).

In contrast to ceramide, sphingosine 1-phosphate, a sphingolipid metabolite generated from sphingosine through sphingosine kinase activation, has been implicated as a signaling molecule that promotes cell survival in response to apoptotic stimuli such as cell-permeable ceramide (Cuvillier and Levade, 2001). Sphingosine 1-phosphate inhibits executioner caspase activation by significantly counteracting the release of mitochondrial proteins, cytochrome *c*, and Smac/DIABLO, induced by exogenous ceramide.

Cathepsin D has also been shown to be a direct binding target of ceramide generated via acid sphingomyelinase in the lysosomal/endosomal compartment. Translocation of this enzyme from the lysosomal compartment has been shown to be important in cellular responses to oxidative stress (Kagedal *et al.*, 2001) and to a variety of other apoptosis inducing agents (Deiss *et al.*, 1996; Shibata *et al.*, 1998; Wu *et al.*, 1998).

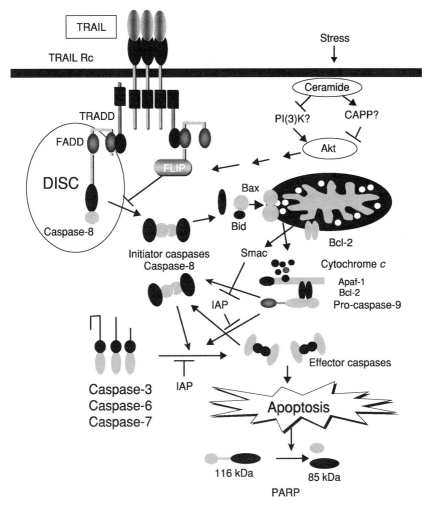

FIGURE 2. A model for the mechanism of TRAIL and ceramide-induced apoptotic pathways.

IV. TRAIL AND CERAMIDE

Recent studies show that low glucose or a low concentration of paclitaxel promotes ceramide formation and TRAIL-induced apoptosis by modulating Akt activity and subsequently cFLIP gene expression (Asakuma *et al.*, 2003; Nam *et al.*, 2002). The elevation of ceramide content may subsequently inactivate PI(3)K (Zundel and Giaccia, 1998) and/or activate CAPP, which is a member of the 2A class of Ser/Thr protein phosphatases (PP2A), as well as protein phosphatase-1 (PP1) (Chalfant *et al.*, 1999;

Dobrowsky and Hannun, 1992). The inactivated PI(3)K and/or CAPP probably dephosphorylates Akt and subsequently down-regulates FLIP gene expression (Panka et al., 2001; Fig. 2). Recent studies also reveal that ceramide inhibits Akt activity directly without affecting PI(3)K activity (Summers et al., 1998; Zhou et al., 1998). This possibility needs to be further examined.

PI(3)K consists of a regulatory subunit (p85) that binds to an activated growth factor/cytokine receptor and undergoes phosphorylation, which results in the activation of its catalytic subunit (p110) (Rodriguez-Viciana et al., 1996). PIP3 facilitates the recruitment of Akt to the plasma membrane through binding with the pleckstrin homology (PH) domain of Akt (Rameh and Cantley, 1999). Akt is activated by phosphoinositide-dependent kinase-1 (PDK1) through phosphorylation at threonine 308 and serine 473 (Alessi et al., 1997). A number of proapoptotic proteins have been identified as direct Akt substrates, including BAD, caspsase-9, and Forkhead transcription factors (Brunet et al., 1999; Cardone et al., 1998; Cross et al., 1995; Datta et al., 1997; del Peso et al., 1997; Hetman et al., 2000). The proapoptotic function of these molecules is suppressed or stimulated by phosphorylation or dephosphorylation of Akt, respectively. Recent studies also show that Akt induces the degradation of IκB by promoting IKKα activity and subsequently stimulating the nuclear translocation of NF-κB (Ozes et al., 1999). These results suggest that elevation of the intracellular level of ceramide promotes TRAIL-induced cytotoxicity through the ceramide-Akt-NF-κB-FLIP pathway.

ACKNOWLEDGMENTS

This work was supported by the following grants: NCI grants CA48000, CA95191, CA96989 (Y.J.L), and CA92389 (A.A.A), Competitive Medical Research Fund of The UPMC Health System (Y.J.L), Oral Cancer Center Grant Fund (Y.J.L), Elsa U. Pardee Foundation (Y.J.L), The Pittsburgh Foundation (Y.J.L) and DOD prostate program fund (PC020530) (Y.J.L).

REFERENCES

Albi, E., and Magni, M. P. (1997). Chromatin neutral sphingomyelinase and its role in hepatic regeneration. *Biochem. Biophys. Res. Commun.* **236**, 29–33.

Alessenko, A., and Chatterjee, S. (1995). Neutral sphingomyelinase: Localization in rat liver nuclei and involvement in regeneration/proliferation. *Mol. Cell. Biochem.* **143**, 169–174.

Alessi, D., James, S., Downes, C., Holmes, A., Gaffney, P., Reese, C., and Cohen, P. (1997). Characterization of a 3-phosphoinositide-dependent protein kinase which phosphorylates and activates protein kinase Balpha. *Curr. Biol.* **7**, 261–269.

Andrieu, N., Salvayre, R., and Levade, T. (1996). Comparative study of the metabolic pools of sphingomyelin and phosphatidylcholine sensitive to tumor necrosis factor. *Eur. J. Biochem.* **236**, 738–745.

Asakuma, J., Sumitomo, M., Asano, T., Asano, T., and Hayakawa, M. (2003). Selective Akt inactivation and tumor necrosis factor-related apoptosis-inducing ligand sensitization of renal cancer cells by low concentrations of paclitaxel. *Cancer Res.* **63,** 1365–1370.

Ashkenazi, A., and Dixit, V. M. (1998). Death receptors: Signaling and modulation. *Science* **281,** 1305–1308.

Ashkenazi, A., and Dixit, V. M. (1999). Apoptosis control by death and decoy receptors. *Curr. Opin. Cell Biol.* **11,** 255–260.

Basu, S., and Kolesnick, R. (1998). Stress signals for apoptosis: Ceramide and c-Jun kinase. *Oncogene* **17,** 3277–3285.

Bielawska, A. (1996). (1S,2R)-D-erythro-(N-myristoylamino)-1-phenyl-1-propanol as an inhibitor of ceramidase. *J. Biol. Chem.* **271,** 12646–12654.

Bielawska, A., Crane, H. M., Liotta, D., Obeid, L. M., and Hannun, Y. A. (1993). Selectivity of ceramide-mediated biology. Lack of activity of erythro-dihydroceramide. *J. Biol. Chem.* **268,** 26226–26232.

Bilderback, T. R., Grigsby, R. J., and Dobrowsky, R. T. (1997). Association of p75(NTR) with caveolin and localization of neurotrophin-induced sphingomyelin hydrolysis to caveolae. *J. Biol. Chem.* **272,** 10922–10927.

Black, E. J., Street, A. J., and Gillespie, D. A. (1991). Protein phosphatase 2A reverses phosphorylation of c-Jun specified by the delta domain in vitro: Correlation with oncogenic activation and deregulated transactivation activity of v-Jun. *Oncogene* **6,** 1949–1958.

Blazquez, C., Galve-Roperh, I., and Guzman, M. (2000). De novo-synthesized ceramide signals apoptosis in astrocytes via extracellular signal-regulated kinase. *FASEB J.* **14,** 2315–2322.

Bodmer, J. L., Holler, N., Reynard, S., Vinciguerra, P., Schneider, P., Juo, P., Blenis, J., and Tschopp, J. (2000). TRAIL receptor-2 signals apoptosis through FADD and caspase-8. *Nat. Cell Biol.* **2,** 241–243.

Boland, M., Foster, S., and O'Neill, L. A. (1997). Resistance to radiation-induced apoptosis in Burkitt's lymphoma cells is associated with defective ceramide signaling. *J. Biol. Chem.* **272,** 12952–12960.

Bose, R., and Kolesnick, R. (2000). Measurement of ceramide synthase activity. *Methods Enzymol.* **322,** 378–382.

Bose, R., Verheij, M., Haimovitz-Friedman, A., Scotto, K., Fuks, Z., and Kolesnick, R. (1995). Ceramide synthase mediates daunorubicin-induced apoptosis: An alternative mechanism for generating death signal. *Cell* **82,** 405–414.

Bradshaw, C., Ella, K., Thomas, A., Qi, A., and Meier, K. (1996). Effects of Ara-C on neutral sphingomyelinase and mitogen- and stress-activated protein kinases in T-lymphocyte cell lines. *Biochem. Mol. Biol. Int.* **40,** 709–719.

Brunet, A., Bonni, A., Zigmond, M. J., Lin, M. Z., Juo, P., Hu, L. S., Anderson, M. J., Arden, K. C., Blenis, J., and Greenberg, M. E. (1999). Akt promotes cell survival by phosphorylating and inhibiting a Forkhead transcription factor. *Cell* **96,** 857–868.

Cardone, M. H., Roy, N., Stennicke, H. R., Salvesen, G. S., Franke, T. F., Stanbridge, E., Frisch, S., and Reed, J. C. (1998). Regulation of cell death protease caspase-9 by phosphorylation. *Science* **282,** 1318–1321.

Castedo, M., Hirsch, T., Susin, S. A., Zamzami, N., Marchetti, P., Macho, A., and Kroemer, G. (1996). Sequential acquisition of mitochondrial and plasma membrane alterations during early lymphocyte apoptosis. *J. Immunol.* **157,** 512–521.

Chalfant, C. E., Kishikawa, K., Mumby, M. C., Kamibayashi, C., Bielawska, A., and Hannun, Y. A. (1999). Long chain ceramide activate protein phosphatase-1 and protein phosphatase-2A: Activation is stereospecific and regulated by phosphatidic acid. *J. Biol. Chem.* **274,** 20313–20317.

Chang, Y., Abe, A., and Shayman, J. A. (1995). Ceramide formation during heat shock: A potential mediator of alpha B-crystallin transcription. *Proc. Natl. Acad. Sci. USA* **92,** 12275–12279.

Cohen, G. M. (1997). Caspases: The executioners of apoptosis. *Biochem. J.* **326,** 1–16.
Coroneos, E., Martinez, M., McKenna, S., and Kester, M. (1995). Differential regulation of sphingomyelinase and ceramidase activities by growth factors and cytokines. Implications for cellular proliferation and differentiation. *J. Biol. Chem.* **270,** 23305–23309.
Cremesti, A., Paris, F., Grassme, H., Holler, N., Tschopp, J., Fuks, Z., Gulbins, E., and Kolesnick, R. (2001). Ceramide enables Fas to cap and kill. *J. Biol. Chem.* **276,** 23954–23961.
Cross, D. A., Alessi, D. R., Cohen, P., Andjelkovich, M., and Hemmings, B. A. (1995). Inhibition of glycogen synthase kinase-3 by insulin mediated by protein kinase B. *Nature* **378,** 785–789.
Cuvillier, O., and Levade, T. (2001). Sphigosine 1-phosphate antagonizes apoptosis of human leukemia cells by inhibiting release of cytochrome *c* and Smac/DIABLO from mitochondria. *Blood* **98,** 2828–2836.
Cuvillier, O., Pirianov, G., Kleuser, B., Vanek, P. G., Coso, O. A., Gutkind, J. S., and Spiegel, S. (1996). Suppression of ceramide-mediated programmed cell death by sphingosine-1-phosphate. *Nature (London)* **381,** 800–803.
Datta, S. R., Dudek, H., Tao, X., Masters, S., Fu, H., Gotoh, Y., and Greenberg, M. E. (1997). Akt phosphorylation of BAD couples survival signals to the cell-intrinsic death machinery. *Cell* **91,** 231–241.
Datta, R., Kojima, H., Banach, D., Bump, N. J., Talanian, R. V., Alnemri, E. S., Weichselbaum, R. R., Wong, W. W., and Kufe, D. W. (1996). Activation of a CrmA-insensitive, p35-sensitive pathway in ionizing radiation-induced apoptosis. *J. Biol. Chem.* **271,** 1965–1969.
Dbaibo, G. S. (1995). Retinoblastoma gene product as a downstream target for a ceramide dependent pathway of growth and arrest. *Proc. Natl. Acad. Sci. USA* **92,** 1347–1351.
Dbaibo, G. S., Pushkareva, M. Y., Jayadev, S., Schwarz, J. K., Horowitz, J. M., Obeid, L. M., and Hannun, Y. A. (1995). Retinoblastoma gene product as a downstream target for ceramide-dependent pathway of growth arrest. *Proc. Natl. Acad. Sci. USA* **92,** 1347–1351.
Dbaibo, G. S., Pushkareva, M. Y., Rachid, R. A., Alter, N., Smyth, M. J., Obeid, L. M., and Hannun, Y. A. (1998). p53-dependent ceramide responses to genotoxic stress. *J. Clin. Invest.* **102,** 329–339.
Degli-Esposti, M. A., Smolak, P. J., Walczak, H., Waugh, J., Huang, C. P., DuBose, R. F., Goodwin, R. G., and Smith, C. A. (1997a). Cloning and characterization of TRAIL-R3, a novel member of the emerging TRAIL receptor family. *J. Exp. Med.* **186,** 1165–1170.
Degli-Esposti, M. A., Dougall, W. C., Smolak, P. J., Waugh, J. Y., Smith, C. A., and Goodwin, R. G. (1997b). The novel receptor TRAIL-R4 induces NF-kappaB and protects against TRAIL-mediated apoptosis, yet retains an incomplete death domain. *Immunity* **7,** 813–820.
Deiss, L. P., Galinka, H., Berissi, H., Cohen, O., and Kimchi, A. (1996). Cathepsin D protease mediates programmed cell death induced by interferon-gamma, Fas/APO-1 and TNF-alpha. *EMBO J.* **15,** 3861–3870.
del Peso, L., Gonzalez-Garcia, M., Page, C., Herrera, R., and Nunez, G. (1997). Interleukin-3-induced phosphorylation of BAD through the protein kinase Akt. *Science* **278,** 687–689.
Deng, X., Ito, T., Carr, B., Mumby, M., and May, W. S., Jr. (1998). Reversible phosphorylation of Bcl2 following interleukin 3 or bryostatin 1 is mediated by direct interaction with protein phosphatase 2A. *J. Biol. Chem.* **273,** 34157–34163.
Deveraux, Q. L., and Reed, J. C. (1999). IAP family proteins: Suppressors of apoptosis. *Genes Dev.* **13,** 239–252.
Dobrowsky, R. T., and Hannun, Y. A. (1992). Ceramide stimulates a cytosolic protein phosphatase. *J. Biol. Chem.* **267,** 5048–5051.
Dobrowsky, R. T., Jenkins, G. M., and Hannun, Y. A. (1995). Neurotrophins induce sphingomyelin hydrolysis. Modulation by co-expression of p75NTR with Trk receptors. *J. Biol. Chem.* **270,** 22135–22142.

Dobrowsky, R. T., Kamibayashi, C., Mumby, M. C., and Hannun, Y. A. (1993). Ceramide activates heterotrimeric protein phosphatase 2A. *J. Biol. Chem.* **268**, 15523–15530.

Du, C., Fang, M., Li, Y., Li, L., and Wang, X. (2000). Smac, a mitochondrial protein that promotes cytochrome c-dependent caspase activation by eliminating IAP inhibition. *Cell* **102**, 33–42.

Duan, R. D., Hetervig, E., Nyberg, L., Hauge, T., Sternby, B., Lillieau, J., Farooqi, A., and Nilsson, A. (1996). Distribution of alkaline sphingomyelinase activity in human beings and animals. *Digest. Disc. Sci.* **41**, 1801–1806.

El Bawab, S., Bielawska, A., and Hannun, Y. A. (1999). Purification and characterization of a membrane-bound nonlysosomal ceramidase from rat brain. *J. Biol. Chem.* **274**, 27948–27955.

El Bawab, S., Roddy, P., Qian, T., Bielawska, A., Lemasters, J. J., and Hannun, Y. A. (2000). Molecular cloning and characterization of a human mitochondrial ceramidase. *J. Biol. Chem.* **275**, 21508–21513.

Eskes, R., Desagher, S., Antonsson, B., and Martinou, J. C. (2000). Bid induces the oligomerization and insertion of bax into the outer mitochondrial membrane. *Mol. Cell. Biol.* **20**, 929–935.

Eto, M., Bennouna, J., Hunter, O., Kanto, T., Hershberger, P., Johnson, C., Lotze, M. T., and Amoscato, A. A. (2003). Apoptosis associated C16 ceramide accumulates following androgen ablation in androgen-dependent prostate cancer cells. *Prostate*, in press.

Fanger, N. A., Maliszewski, C. R., Schooley, K., and Griffith, T. S. (1999). Human dendritic cells mediate cellular apoptosis via tumor necrosis factor-related apoptosis-inducing ligand (TRAIL) *J. Exp. Med.* **190**, 1155–1164.

Fishbein, J. D., Dobrowsky, R. T., Bielawska, A., Garrett, S., and Hannun, Y. A. (1993). Ceramide-mediated growth inhibition and CAPPase conserved in S. Cerevisiae. *J. Biol. Chem.* **268**, 9255–9261.

Garcia-Ruiz, C., Colell, A., Mari, M., Morales, A., and Fernandez-Checa, J. (1997). Direct effect of ceramide on the mitochondrial electron transport chain leads to generation of reactive oxygen species. *J. Biol. Chem.* **272**, 11369–11377.

Gatt, S. (1976). Magnesium-dependent sphingomyelinase. *Biochem. Biophys. Res. Commun.* **68**, 235–241.

Gazitt, Y., Shaughnessy, P., and Montgomery, W. (1999). Apoptosis-induced by TRAIL and TNF-α in human mutiple myeloma cells is not blocked by Bcl-2. *Cytokine* **11**, 1010–1019.

Ghafourifar, P., Klein, S. D., Schucht, O., Schenk, U., Pruschy, M., Rocha, S., and Richter, C. (1999). Ceramide induces cytochrome c release from isolated mitochondria: Important of mitochondrial redox state. *J. Biol. Chem.* **274**, 6080–6084.

Gomez-Munoz, A., Frago, L., Alvarez, L., and Varela-Nieto, I. (1997). Stimulation of DNA synthesis by natural ceramide-1-phosphate. *Biochem. J.* **325**, 435–440.

Gomez-Munoz, A., Waggoner, D. W., O'Brien, L., and Brindly, D. N. (1995). Interaction of ceramides, sphingosine and sphingosine-1-phosphate in regulating DNA synthesis and phospholipase D activity. *J. Biol. Chem.* **270**, 26318–26325.

Grassme, H., Jekle, A., Riehle, A., Schwarz, H., Berger, J., Sandhoff, K., Kolesnick, R., and Gulbins, E. (2001). CD95 signaling via ceramide-rich membrane rafts. *J. Biol. Chem.* **276**, 20589–20596.

Griffith, T. S., Chin, W. A., Jackson, G. C., Lynch, D. H., and Kubin, M. Z. (1998). Intracellular regulation of TRAIL-induced apoptosis in human melanoma cells. *J. Immunol.* **161**, 2833–2840.

Griffith, T. S., Wiley, S. R., Kubin, M. Z., Sedger, L. M., Maliszewski, C. R., and Fanger, N. A. (1999). *J. Exp. Med.* **189**, 1343–1353.

Grullich, C., Sullards, M. C., Fuks, Z., Merrill, A. H., Jr., and Kolesnick, R. (2000). CD95(Fas/APO-1) signals ceramide generation independent of the effector stage of apoptosis. *J. Biol. Chem.* **275**, 8650–8656.

Gu, M., Kerwin, J. L., Watts, J. D., and Aebersold, R. (1997). Ceramide profiling of complex lipid mixtures by electrospray ionization mass spectrometry. *Anal. Biochem.* **244**, 347–356.

Gudz, T., Tserng, K-Y., and Hoppel, C. (1997). Direct inhibiton of mitochondrial respiratory chain complex III by cell-permeable ceramide. *J. Biol. Chem.* **272**, 24154–24158.

Gura, T. (1997). How TRAIL kills cancer cells, but not normal cells. *Science* **277**, 768.

Haimovitz-Friedman, A., Kan, C-C., Ehleiter, D., Persaud, R. S., McLoughlin, M., Fuks, Z., and Kolesnick, R. N. (1994). Ionizing radiation acts on cellular membranes to generate ceramide and initiate apoptosis. *J. Exp. Med.* **180**, 525–535.

Hannun, Y. A. (1994). The sphingomyelin cycle and the second messenger function of ceramide. *J. Biol. Chem.* **269**, 3125–3128.

Hannun, Y. A. (1996). Functions of ceramide in coordinating cellular responses to stress. *Science* **275**, 1855–1859.

Hannun, Y. A., and Bell, R. M. (1989). Function of sphingolipids and sphingolipid breakdown products in cellular regulation. *Science* **243**, 500–506.

Hayakawa, M., Jayadev, S., Tsujimoto, M., Hannun, Y. A., and Ito, F. (1996). Role of ceramide in stimulation of the transcription of cytosolic phospholipase A2 and cyclooxygenase 2. *Biochem. Biophys. Res. Commun.* **220**, 681–686.

Hayakawa, Y., Smyth, M. J., Kayagaki, N., Yamaguchi, N., Kakuta, S., Iwakura, Y., Yagita, H., Okumura, K., and Takeda, K. (2001). Involvement of tumor necrosis factor-related apoptosis-inducing ligand in surveillance of tumor metastasis by liver natural killer cells. *Nat. Med.* **7**, 94–100.

Herr, I., Wilheim, D., Bohler, T., Angel, P., and Debatin, K. (1997). Activation of CD95 signaling by ceramide mediates cancer therapy-induced apoptosis. *EMBO J.* **16**, 6200–6208.

Hetman, M., Cavanaugh, J. E., Kimeiman, D., and Xia, Z. (2000). Role of glycogen synthase kinase-3beta in neuronal apoptosis induced by trophic withdrawal. *J. Neurosci.* **20**, 2567–2574.

Hitoshi, Y., Lorens, J., Kitada, S. I., Fisher, J., LaBarge, M., Ring, H. Z., Francke, U., Reed, J. C., Kinoshita, S., and Nolan, G. P. (1998). Toso, a cell surface, specific regulator of Fas-induced apoptosis in T cells. *Immunity* **8**, 461–471.

Hofmann, K., Tomiuk, S., Wolff, G., and Stoffel, W. (2000). Cloning and characterization of the mammalian brain-specific, Mg2+-dependent neutral sphingomyelinase. *Proc. Natl. Acad. Sci. USA* **97**, 5895–5900.

Horinouchi, K., Erlich, S., Perl, D. P., Ferlinz, K., Bisgaier, C. L., Sandhoff, K., Desnick, R. J., Stewart, C. L., and Schuchman, E. H. (1995). Acid sphingomyelinase deficient mice: A model for types A and B Nieman-Pick disease. *Nat. Genet.* **10**, 288–293.

Hu, Y., Benedict, M. A., Ding, L., and Nunez, G. (1999). Role of cytochrome c and dATP/ATP hydrolysis in Apaf-1-mediated caspase-9 activation and apoptosis. *EMBO J.* **18**, 3586–3595.

Hurwitz, R., Ferlinz, K., Vielhaber, G., Suzuki, K., and Sandhoff, K. (1994). Processing of human acid sphingomyelinase in normal and I-cell fibroblasts. *J. Biol. Chem.* **269**, 5440–5445.

Huwiler, A., Pfeilschifter, J., and van den Bosch, H. (1999). Nitric oxide donors induce stress signaling via ceramide formation in rat renal mesangial cells. *J. Biol. Chem.* **274**, 7190–7195.

Jaffrezou, J. P., Bruno, A. P., Moisand, A., Levade, T., and Laurent, G. (2001). Activation of a nuclear sphingomyelinase in radiation-induced apoptosis. *FASEB J.* **15**, 123–133.

Jaffrezou, J. P., Levade, T., Bettaieb, A., Andrieu, N., Bezombes, C., Maestre, N., Vermeersch, S., Rousse, A., and Laurent, G. (1996). Daunorubicin-induced apoptosis and nuclear factor kappa B. *EMBO J.* **15**, 2417–2424.

Jarvis, W. D., Turner, A. J., Povirk, L. F., Traylor, R. S., and Grant, S. (1994). Inhibition of apoptotic DNA fragmentation and cell death in HL-60 cells by pharmacological inhibitors of protein kinase C. *Cancer Res.* **54**, 1707–1714.

Jayadev, S., Liu, B., Bielawska, A. E., Lee, J. Y., Nazaire, F., Pushkareva, M., Obeid, L. M., and Hannun, Y. A. (1995). Role for ceramide in cell-cycle arrest. *J. Biol. Chem.* **270**, 2047–2052.

Jo, M., Kim, T. H., Seol, D. W., Esplen, J. E., Dorko, K., Billiar, T. R., and Strom, S. C. (2000). Apoptosis induced in normal human hepatocytes by tumor necrosis factor-related apoptosis-inducing ligand. *Nat. Med.* **6**, 564–567.

Kagedal, K., Johansson, U., and Ollinger, K. (2001). The lysosomal protease cathepsin D mediates apoptosis induced by oxidative stress. *FASEB J.* **15**, 1592–1594.

Kandasamy, K., Srinivasula, S. M., Alnemri, E. S., Thompson, C. B., Korsmeyer, S. J., Bryant, J. L., and Srivastava, R. K. (2003). Involvement of proapoptotic molecules Bax and Bak in tumor necrosis factor-related apoptosis-inducing ligand (TRAIL)-induced mitochondrial disruption and apoptosis: Differential regulation of cytochrome c and Smac/DIABLO release. *Cancer Res.* **63**, 1712–1721.

Kanfer, J. (1983a). Sphingolipid metabolism. *In* "Sphingolipid Biochemistry" (J. N. Kanfer and S. Hakomori, Eds.), vol. 3, pp. 167–240. Plenum Press, New York.

Kanfer, J. (1983b). The sphingolipidoses. *In* "Sphingolipid Biochemistry" (J. N. Kanfer and S. Hakomori, Eds.), vol. 3, pp. 252–326. Plenum Press, New York.

Kanto, T. M., Kalinski, P., Hunter, O., Lotze, M. T., and Amoscato, A. A. (2001). Ceramide mediates tumor-induced dendritic cell apoptosis. *J. Immunol.* **167**, 3773–3784.

Kayagaki, N., Yamaguchi, N., Nakayama, M., Eto, H., Okumura, K., and Yagita, H. (1999). Type I interferons (IFNs) regulate tumor necrosis factor-related apoptosis-inducing ligand (TRAIL) expression on human T cells: A novel mechanism for the antitumor effects of type I IFNs. *J. Exp. Med.* **189**, 1451–1460.

Keane, M. M., Ettenberg, S. A., Nau, M. M., Russell, E. K., and Lipkowitz, S. (1999). Chemotherapy augments TRAIL-induced apoptosis in breast cell lines. *Cancer Res.* **59**, 734–741.

Kischkel, F. C., Lawrence, D. A., Chuntharapai, A., Schow, P., Kim, K. J., and Ashkenazi, A. (2000). Apo2L/TRAIL-dependent recruitment of endogenous FADD and caspase-8 to death receptors 4 and 5. *Immunity* **12**, 611–620.

Kolesnick, R. N. (1991). Sphingomyelin and derivatives as cellular signals. *Prog. Lipid Res.* **30**, 1–38.

Kolesnick, R., and Golde, D. (1994). The sphingolipid pathway in TNF and Il-1 signaling. *Cell* **77**, 325–328.

Kothny-Wilkes, G., Kulms, D., Luger, T. A., Kubin, M., and Schwarz, T. (1999). Interleukin-1 protects transformed keratinocytes from tumor necrosis factor-related apoptosis-inducing ligand- and CD95-induced apoptosis but not from ultraviolet radiation-induced apoptosis. *J. Biol. Chem.* **274**, 28916–28921.

Kuang, A. A., Diehl, G., Zhang, J., and Winoto, A. (2000). FADD is required for DR4- and DR5-mediated apoptosis: Lack of TRAIL-induced apoptosis in FADD-deficient mouse embryonic fibroblasts. *J. Biol. Chem.* **275**, 25065–25068.

Latinis, K. M., and Koretzky, G. A. (1996). Fas ligation induces apoptosis and Jun kinase activation independently of CD45 and lck in human T cells. *Blood* **87**, 871–875.

Laulederkind, S. J. F., Bielawska, A., Raghow, R., Hannun, Y. A., and Ballou, L. R. (1995). Ceramide induces IL-6 gene expression in human fibroblasts. *J. Exp. Med.* **182**, 599–604.

Law, B., and Rossie, S. (1995). The dimeric and catalytic subunit forms of protein phosphatase 2A from rat brain are stimulated by C2-ceramide. *J. Biol. Chem.* **270**, 12808–12813.

Lawrence, D., Shahrokh, Z., Marsters, S., Achilles, K., Shih, D., Mounho, B., Hillan, K., Totpal, K., DeForge, L., Schow, P., Hooley, J., Sherwood, S., Pai, R., Leung, S., Khan, L., Gliniak, B., Bussiere, J., Smith, C. A., Strom, S. S., Kelley, S., Fox, J. A., Thomas, D., and Ashkenazi, A. (2001). Differential hepatocyte toxicity of recombinant Apo2L/TRAIL versions. *Nat. Med.* **7**, 383–385.

Lee, Y. J., Lee, K. H., Kim, H. R. C., Jessup, J. M., Seol, D. W., Kim, T. H., Billiar, T. R., and Song, Y. K. (2001). Sodium nitroprusside enhances TRAIL-induced apoptosis via a mitochondria-dependent pathway in human colorectal carcinoma CX-1 cells. *Oncogene* **20**, 1476–1485.

Levade, T., Vidal, F., Vermeesch, S., Andrieu, N., Gatt, S., and Salvayre, R. (1995). Degradation of fluorescent and radiolabelled sphingomyelins in intact cells by a non-lysosomal pathway. *Biochim. Biophys. Acta* **1258**, 277–287.

Liebish, G., Drobnik, W., Reil, M., Trunbach, B., Arnecke, R., Oligemuller, B., Roscher, A., and Schmitz, G. (1999). Quantitative measurement of different ceramide species from crude cellular extracts by electrospray ionization tandem mass spectrometry (ESI-MS/MS). *J. Lipid. Res.* **40**, 1539–1546.

Linardic, C., and Hannun, Y. A. (1994). Identification of a distinct pool of sphingomyelin involved in the sphingomyelin cycle. *J. Biol. Chem.* **269**, 23530–23537.

Liu, B., Andrieu-Abadie, N., Levade, T., Zhang, P., Obeid, L. M., and Hannun, Y. A. (1998). Glutathione regulation of neutral sphingomyelinase in tumor necrosis factor-alpha-induced cell death. *J. Biol. Chem.* **273**, 11313–11320.

Liu, G., Kleine, L., and Hebert, R. L. (1999a). Advances in the signal transduction of ceramide and related sphingolipids. *Crit. Rev. Clin. Lab. Sci.* **36**, 511–573.

Liu, Y. Y., Han, T. Y., Giuliano, A. E., Ichikawa, S., Hirabayashi, Y., and Cabot, M. C. (1999b). Glycosylation of ceramide potentiates cellular resistance to tumor necrosis factor-alpha-induced apoptosis. *Exp. Cell. Res.* **252**, 464–470.

Liu, Y. Y., Han, T. Y., Giuliano, A. E., and Cabot, M. C. (2001). Ceramide glycosylation potentiates cellular multidrug resistance. *FASEB J.* **15**, 719–730.

Lozano, J., Menendez, S., Morales, A., Ehleiter, D., Liao, W.-C., Wagman, R., Haimovitz-Friedman, A., Fuks, Z., and Kolesnick, R. (2001). Cell autonomous apoptosis defects in acid sphingomyelinase knockout fibroblasts. *J. Biol. Chem.* **276**, 442–448.

Luberto, C., and Hannun, Y. A. (1998). Sphingomyelin synthase, a potential regulator of intracellular levels of ceramide and diacylglycerol during SV40 transformation. Does sphingomyelin synthase account for the putative phosphatidylcholine-specific phospholipase C? *J. Biol. Chem.* **273**, 14550–14559.

Luberto, C., Yoo, D. S., Suidan, H. S., Bartoli, G. M., and Hannun, Y. A. (2000). Differential effects of sphingomyelin hydrolysis and resynthesis on the activation of NF-kappa B in normal and SV40-transformed human fibroblasts. *J. Biol. Chem.* **275**, 14760–14766.

Mao, C., Xu, R., Szulc, Z. M., Bielawska, A., Galadari, S. H., and Obeid, L. M. (2001). Cloning and characterization of a novel human alkaline ceramidase. A mammalian enzyme that hydrolyzes phytoceramide. *J. Biol. Chem.* **276**, 26577–26588.

Marathe, S., Schissel, S., Telin, M., Beatini, N., Mintzer, R., Williams, K., and Tabas, I. (1998). Human vascular endothelial cells are a rich and regulatable source of secretory sphingomyelinase. Implications for early atherogenesis and ceramide-mediated signaling. *J. Biol. Chem.* **273**, 4081–4088.

Marsters, S. A., Sheridan, J. P., Pitti, R. M., Huang, A., Skubatch, M., Baldwin, D., Yuan, J., Gurney, A., Goddard, A. D., Godowski, P., and Ashkenazi, A. (1997). A novel receptor for Apo2L/TRAIL contains a truncated death domain. *Curr. Biol.* **7**, 1003–1006.

Mathias, S., Pena, L. A., and Kolesnick, R. N. (1998). Signal transduction of stress via ceramide. *Biochem. J.* **335**, 465–480.

Mathias, S., Younes, A., Kan, C. C., Orlow, I., Joseph, C., and Kolesnick, R. N. (1993). Activation of the sphingomyelin pathway in intact EL-4 cells and in a cell free system by IL-1 beta. *Science* **259**, 519–522.

Matsko, C. M., Hunter, O. C., Rabinowich, H., Lotse, M. T., and Amoscato, A. A. (2001). Mitochondrial lipid alterations during Fas- and radiation-induced apoptosis. *Biochem. Biophys. Res. Commun.* **287**, 1112–1120.

Medema, J. P., Scaffidi, C., Kischkel, F. C., Shevchenko, A., Mann, M., Krammer, P. H., and Peter, M. E. (1997). FLICE is activated by association with the CD95 death-inducing signaling complex (DISC). *EMBO J.* **16**, 2794–2804.

Merrill, A. H., Jr. (1986). Inhibition of phorbol ester dependent differentiation of HL60 cells by sphinganine and other long chain bases. *J. Biol. Chem.* **261**, 12610–12615.

Merrill, A. H., Jr., and Wang, E. (1992). Sphingolipid biosynthesis enzymes. *In* "Methods in Enzymology" (E. A. Dennis and D. E. Vance, Eds.), vol. 209, pp. 427–436. Academic Press, New York.

Merrill, A. H., Wang, E., Gilchrist, D. G., and Riley, R. T. (1993). Fumonisins and other inhibitors of de novo sphingolipid biosynthesis. *In* "Advances in Lipid Research" (R. M. Bell, A. H. Merrill, and Y. A. Hannun, Eds.), vol. 26, pp. 215–234. Academic Press, New York.

Merrill, A. H., Jr., Wang, E., Mullins, R. E., Jamison, W. C., Nimkar, S., and Liotta, D. (1988). Quantitation of free sphingosine in liver by high-performance liquid chromatography. *Anal. Biochem.* **171**, 373–381.

Michael, J. M., Lavin, M. F., and Watters, D. J. (1997). Resistance to radiation-induced apoptosis in Burkitt's lymphoma cells is associated with defective ceramide signaling. *Cancer Res.* **57**, 3600–3605.

Mitsutake, S., Tani, M., Okino, N., Mori, K., Ichinose, S., Omori, A., Iida, H., Nakamura, T., and Ito, M. (2001). Purification, characterization, molecular cloning, and subcellular distribution of neutral ceramidase of rat kidney. *J. Biol. Chem.* **276**, 26249–26259.

Nam, S. Y., Amoscato, A. A., and Lee, Y. J. (2002). Low glucose-enhanced TRAIL cytotoxicity is mediated through the ceramide-Akt-FLIP pathway. *Oncogene* **21**, 337–346.

Nilsson, A., and Duan, R. D. (1999). Alkaline sphingomyelinases and ceramidases of the gastrointestinal tract. *Chem. Phys. Lipids* **102**, 97–105.

Nyberg, L., Duan, R. D., Axelson, J., and Nilsson, A. (1996). Identification of an alkaline sphingomyelinase activity in human bile. *Biochim. Biophys. Acta* **1300**, 42–48.

Okazaki, T., Bell, R. M., and Hannun, Y. (1989). Sphingomyelin turnover induced by vitamin D3 in HL-60 cells. Role in cell differentiation. *J. Biol. Chem.* **264**, 19076–19080.

Okazaki, T., Bielawska, A., Domae, N., Bell, R. M., and Hannun, Y. A. (1994). Characteristics and partial purification of a novel cytosolic, magnesium-independent, neutral sphingomyelinase activated in the early signal transduction of 1 α,25-dihydroxyvitamin D3-induced HL-60 cell differentiation. *J. Biol. Chem.* **269**, 4070–4077.

Okino, N., Tani, M., Imayama, S., and Ito, M. (1998). Purification and characterization of a novel ceramidase from Pseudomonas aeruginosa. *J. Biol. Chem.* **273**, 14368–14373.

Otterbach, B., and Stoffel, W. (1995). Acid sphingomyelinase-deficient mice mimic the neurovisceral form of human lysosomal storage disease. *Cell* **81**, 1053–1061.

Ozes, O. N., Mayo, L. D., Gustin, J. A., Pfeffer, S. R., Pfeffer, L. M., and Donner, D. B. (1999). NF-kappaB activation by tumour necrosis factor requires the Akt serine-threonine kinase. *Nature* **401**, 82–85.

Pan, G., O'Rourke, K., Chinnaiyan, A. M., Gentz, R., Ebner, R., Ni, J., and Dixit, V. M. (1997a). The receptor for the cytotoxic ligand TRAIL. *Science* **276**, 111–113.

Pan, G., Ni, J., Wei, Y. F., Yu, G., Gentz, R., and Dixit, V. M. (1997b). An antagonist decoy receptor and a death domain-containing receptor for TRAIL. *Science* **277**, 815–818.

Panka, D. J., Mano, T., Suhara, T., Walsh, K., and Mier, J. (2001). Phosphatidylinositol-3 kinase/Akt activity regulates c-FLIP expression in tumor cells. *J. Biol. Chem.* **276**, 6893–6896.

Pastorino, J., Simbula, G., Yamamoto, K., Glascott, P. J., Rothman, R., and Farber, J. (1996). The cytotoxicity of TNF depends on the induction of the mitochondrial permeability transition. *J. Biol. Chem.* **271**, 29792–29798.

Paumen, M. B., Ishida, Y., Muramatsu, M., Yamamoto, M., and Honjo, T. (1997). Inhibition of carnitine palmitoyltransferase I augments sphingolipid synthesis and palmitate-induced apoptosis. *J. Biol. Chem.* **272**, 3324–3329.

Perry, D. K., Carton, J., Shah, A. K., Meredith, F., Uhlinger, D. J., and Hannun, Y. A. (2000). Serine palmitoyltransferase regulates de novo ceramide generation during etoposide-induced apoptosis. *J. Biol. Chem.* **275,** 9078–9084.

Pettus, B. J., Chalfant, C. E., and Hannun, Y. A. (2002). Ceramide in apoptosis: An overview and current perspectives. *Biochim. Biophys. Acta* **1585,** 114–125.

Pitti, R. M., Marsters, S. A., Ruppert, S., Donahue, C. J., Moore, A., and Ashkenazi, A. (1996). Induction of apoptosis by Apo2 ligand, a new member of the tumor necrosis factor receptor family. *J. Biol. Chem.* **271,** 12687–12690.

Pyne, S., Chapman, J., Steele, L., and Pyne, N. J. (1996). Sphingomyelin derived lipids differentially regulate the extracellular signal-regulated kinase (ERK-2) and c-Jun and N-terminal kinase (JNK) signal cascades in airway smooth muscle. *Eur. J. Biochem.* **237,** 819–826.

Quintans, J., Kikus, J., McShan, C., Gottschalk, A., and Dawson, G. (1994). Ceramide mediates the apoptotic response of WEHI 231 cells to anti-immunoglobulin, corticosteroids and irradiation. *Biochem. Biophys. Res. Commun.* **202,** 710–714.

Rameh, L. E., and Cantley, L. C. (1999). The role of phosphoinositide 3-kinase lipid products in cell function. *J. Biol. Chem.* **274,** 8347–8350.

Rani, C. S. (1995). Cell-cycle arrest induced by an inhibitor of glucosylceramide synthase. Correlation with cyclin-dependent kinase. *J. Biol. Chem.* **270,** 2859–2867.

Riboni, L., Bassi, R., Sonnino, S., and Tettamanti, G. (1992). Formation of free sphingosine and ceramide form exogenous ganglioside GM1 by cerebellar granule cells in culture. *FEBS Lett.* **300,** 188–192.

Rodriguez-Viciana, P., Marte, B., Warne, P., and Downward, J. (1996). Phosphatidylinositol 3′ kinase: One of the effectors of Ras. *Philos. Trans. R. Soc. Lond. B Biol. Sci.* **351,** 225–231.

Ruvolo, P. P., Deng, X., Carr, B. K., and May, W. S. (1998). A functional role for mitochondrial protein kinase Calpha in Bcl2 phosphorylation and suppression of apoptosis. *J. Biol. Chem.* **273,** 25436–25442.

Ruvolo, P. P., Deng, X., Ito, T., Carr, B. K., and May, W. S. (1999). Ceramide induces Bcl2 dephosphorylation via a mechanism involving mitochondrial PP2A. *J. Biol. Chem.* **274,** 20296–20300.

Santana, P., Pena, L. A., Haimovitz-Friedman, A., Martin, S., Green, D., McLoughlin, M., Cordon-Cardo, C., Schuchman, E. H., Fuks, Z., and Kolesnick, R. N. (1996). Acid sphingomyelinase-deficient lymphoblasts are defective in radiation-induced apoptosis. *Cell* **86,** 189–199.

Sato, S., Fujita, N., and Tsuruo, T. (2000). Modulation of Akt kinase activity by binding to Hsp90. *Proc. Natl. Acad. Sci. USA* **97,** 10832–10837.

Sato, T., Irie, S., Kitada, S., and Reed, J. C. (1995). FAP-1: A protein tyrosine phosphatase that associates with Fas. *Science* **268,** 411–415.

Scaffidi, C., Schmitz, I., Krammer, P. H., and Peter, M. E. (1999). The role of c-FLIP in modulation of CD95-induced apoptosis. *J. Biol. Chem.* **274,** 1541–1548.

Schissel, S., Schuchman, E., Williams, K., and Tabas, I. (1996). The cellular trafficking and Zn-dependence of secretory and lysosomal sphingomyelinase, two products of the acid sphingomyelinase gene. *J. Biol. Chem.* **271,** 18431–18436.

Schneider, P., and Kennedy, E. (1967). Sphingomyelinase in normal human spleens and in spleens from subjects with Niemann-Pick disease. *J. Lipid Res.* **8,** 202–209.

Schuchman, E., and Desnick, R. (1995). Niemann-pick disease types A and B: Acid sphingomyelinase deficiencies. *In* "The Metabolic and Molecular Bases of Inherited Disease" (C. Scriver, A. Beaudet, W. Sly, and D. Valle, Eds.), pp. 2601–2624. McGraw Hill, New York.

Sedger, L. M., Shows, D. M., Blanton, R. A., Peschon, J. J., Goodwin, R. G., Cosman, D., and Wiley, S. R. (1999). IFN-γ mediates a novel antiviral activity through dynamic modulation of TRAIL and TRAIL receptor expression. *J. Immunol.* **163,** 920–926.

Sheridan, J. P., Marsters, S. A., Pitti, R. M., Gurney, A., Skubatch, M., Baldwin, D., Ramakrishnan, L., Gray, C. L., Baker, K., Wood, W. I., Goddard, A. D., Godowski, P., and Ashkenazi, A. (1997). Control of TRAIL-induced apoptosis by a family of signaling and decoy receptors. *Science* **277,** 818–821.

Shibata, M., Kanamori, S., Isahara, K., Ohsawa, Y., Konishi, A., Kametaka, S., Watanabe, T., Ebisu, S., Ishido, K., Kominami, E., and Uchiyama, Y. (1998). Participation of cathepsins B and D in apoptosis of PC12 cells following serum deprivation. *Biochem. Biophys. Res. Commun.* **251,** 199–203.

Slee, E. A., Harte, M. T., Kluck, R. M., Wolf, B. B., Casiano, C. A., Newmeyer, D. D., Wang, H. G., Reed, J. C., Nicholson, D. W., Alnemri, E. S., Green, D. R., and Martin, S. J. (1999). Ordering the cytochrome c-initiated caspase cascade: Hierarchical activation of caspases-2, -3, -6, -7, -8, and -10 in a caspase-9-dependent manner. *J. Cell Biol.* **144,** 281–292.

Smyth, M. J., Cretney, E., Takeda, K., Wiltrout, R. H., Sedger, L. M., Kayagaki, N., Yagita, H., and Okumura, K. (2001). Tumor necrosis factor-related apoptosis-inducing ligand (TRAIL) contributes to interferon γ-dependent natural killer cell protection from tumor metastasis. *J. Exp. Med.* **193,** 661–670.

Spence, M. W., Wakkary, J., Clarke, J. T., and Cooke, H. W. (1982). Localization of neutral Mg-stimulated sphingomyelinase in plasma membrane of cultured neuroblastoma cells. *Biochim. Biophys. Acta* **719,** 162–164.

Sprick, M. R., Weigand, M. A., Rieser, E., Rauch, C. T., Juo, P., Blenis, J., Krammer, P. H., and Walczak, H. (2000). FADD/MORT1 and caspase-8 are recruited to TRAIL receptor 1 and 2 are essential for apoptosis mediated by TRAIL receptor 2. *Immunity* **12,** 599–609.

Srinivasula, S. M., Datta, P., Fan, X. J., Fernandes-Alnemri, T., Huang, Z., and Alnemri, E. S. (2000). Molecular determinants of the caspase-promoting activity of Smac/DIABLO and its role in the death receptor pathway. *J. Biol. Chem.* **275,** 36152–36157.

Stratford, S., DeWald, D. B., and Summers, S. A. (2001). Ceramide dissociates 3′-phosphoinositide production from pleckstrin homology domain translocation. *Biochem. J.* **354,** 359–368.

Strelow, A., Bernardo, K., Adam-Klages, S., Linke, T., Sandhoff, K., Kronke, M., and Adam, D. (2000). Overexpression of acid ceramidase protects from tumor necrosis factor-induced cell death. *J. Exp. Med.* **192,** 601–612.

Subham, B., Bayoumy, S., Zhang, Y., Lozano, J., and Kolesnick, R. N. (1998). BAD enables ceramide to signal apoptosis via Ras and Raf-1. *J. Biol. Chem.* **273,** 30419–30426.

Sullards, M. C. (2000). Analysis of sphingomyelin, glucosylceramide, ceramide, sphingosine, and sphingosine 1-phosphate by tandem mass spectrometry. *Meth. Enzymol.* **312,** 32–45.

Summers, S. A., Garza, L. A., Zhou, H., and Birnbaum, M. J. (1998). Regulation of insulin-stimulated glucose transporter GLUT4 translocation and Akt kinase activity by ceramide. *Mol. Cell. Biol.* **18,** 5457–5464.

Suzuki, A., Iwasaki, M., Kato, M., and Wagai, N. (1997). Sequential operation of ceramide synthesis and ICE cascades in CPT-11 initiated apoptotic death signaling. *Exp. Cell. Res.* **233,** 41–47.

Tabas, I. (1999). Secretory sphingomyelinase. *Chem. Phys. Lipids* **102,** 123–130.

Tamiya-Koizumi, K., Umekawa, H., Yoshida, S., and Kojima, K. (1989). Existence of Mg^{2+}-dependent, neutral sphingomyelinase in nuclei of rat ascites hepatoma cells. *J. Biochem. (Tokyo)* **106,** 593–598.

Tamm, I., Wang, Y., Sausville, E., Scudiero, D. A., Vigna, N., Oltersdorf, T., and Reed, J. C. (1998). IAP-family protein survivin inhibits caspase activity and apoptosis induced by Fas (CD95), Bax, caspases, and anticancer drugs. *Cancer Res.* **58,** 5315–5320.

Tani, M., Okino, N., Mori, K., Tanigawa, T., Isu, H., and Ito, M. (2000). Molecular cloning of the full-length cDNA encoding mouse neutral ceramidase. A novel but highly conserved gene family of neutral/alkaline ceramidases. *J. Biol. Chem.* **275,** 11229–11234.

Tepper, C. G., Jayadev, S., Liu, B., Bielawska, A., Wolff, R., Hannun, S., Hannun, Y. A., and Seldin, M. F. (1995). Role for ceramide as an endogenous mediator of Fas-induced cytotoxicity. *Proc. Natl. Acad. Sci. USA* **92**, 8443–8447.

Thomas, R. L., Matsko, C. M., Lotze, M. T., and Amoscato, A. A. (1999). Mass spectrometric identification of increased C16 ceramide levels during apoptosis. *J. Biol. Chem.* **274**, 30580–30588.

Tomiuk, S., Hofmann, K., Nix, M., Zumbansen, M., and Stoffel, W. (1998). Cloned mammalian neutral sphingomyelinase: Functions in sphingolipid signaling? *Proc. Natl. Acad. Sci. USA* **95**, 3638–3643.

Tschopp, J., Martinon, F., and Hofmann, K. (1999). Apoptosis: Silencing the death receptors. *Curr. Biol.* **9**, R381–R384.

Vassilev, A., Ozer, Z., Navara, C., Mahajan, S., and Uckun, F. M. (1999). Bruton's tyrosine kinase as an inhibitor of the Fas/CD95 death-inducing signaling complex. *J. Biol. Chem.* **274**, 1646–1656.

Veldhoven, P. P. V., Bishop, W. R., and Bell, R. M. (1989). Enzymatic quantification of sphingosine in the picomole range in cultured cells. *Anal. Biochem.* **183**, 177–189.

Veldman, R. J., Maestre, N., Aduib, O. M., Medin, J. A., Salvayre, R., and Levade, T. (2001). A neutral sphingomyelinase resides in sphingolipid-enriched microdomains and is inhibited by the caveolin-scaffolding domain: Potential implications in tumour necrosis factor signalling. *Biochem. J.* **355**, 859–868.

Verheij, M., Bose, R., Lin, X. H., Yao, B., Jarvis, W. D., Grant, S., Birrer, M. J., Szabo, E., Zon, L. I., Kyriakis, J. M., Haimovitz-Friedman, A., Fuks, Z., and Kolesnick, R. N. (1996). Requirement for ceramide-initiated SAPK/JNK signaling in stress-induced apoptosis. *Nature* **380**, 75–79.

Vidalain, P. O., Azocar, O., Lamouille, B., Astier, A., Rabourdin-Combe, C., and Servet-Delprat, C. (2000). Measles virus induces functional TRAIL production by human dendritic cells. *J. Virol.* **74**, 556–559.

Walczak, H., Degli-Esposti, M. A., Johnson, R. S., Smolak, P. J., Waugh, J. Y., Boiani, N., Timour, M. S., Gerhart, M. J., Schooley, K. A., Smith, C. A., Goodwin, R. G., and Rauch, C. T. (1997). TRAIL-R2: A novel apoptosis-mediating receptor for TRAIL. *EMBO J.* **16**, 5386–5397.

Walczak, H., Miller, R. E., Ariail, K., Gliniak, B., Griffith, T. S., Kubin, M., Chin, W., Jones, J., Woodward, A., Le, T., Smith, C., Smolak, P., Goodwin, R. G., Rauch, C. T., Schuh, J. C. L., and Lynch, D. H. (1999). Tumoricidal activity of tumor necrosis factor-related apoptosis-inducing ligand *in vivo*. *Nat. Med.* **5**, 157–163.

Walczak, H., Bouchon, A., Stahl, H., and Krammer, P. H. (2000). Tumor necrosis factor-related apoptosis-inducing ligand retains its apoptosis-inducing capacity on Bcl-2- or Bcl-xL overexpressing chemotherapy-resistant tumor cells. *Cancer Res.* **60**, 3051–3057.

Watts, J. D., Gu, M., Polverino, A. J., Patterson, S. D., and Aebersold, R. (1997). Fas-induced apoptosis of T cells occurs independently of ceramide generation. *Proc. Natl. Acad. Sci. USA* **94**, 7292–7296.

Wen, J., Ramadevi, N., Nguyen, D., Perkins, C., Worthington, E., and Bhalla, K. (2000). Antileukemic drugs increase death receptor 5 levels and enhance Apo-2L-induced apoptosis of human acute leukemia cells. *Blood* **96**, 3900–3906.

Westwick, J. K., Bielawska, A. E., Dbaibo, G., Hannun, Y. A., and Brenner, D. A. (1995). Ceramide activates the stress-activated protein kinases. *J. Biol. Chem.* **270**, 22689–22692.

Whitman, S., Civoli, F., and Daniel, L. (1997). Protein kinase CbetaII activation by 1-beta-D-arabinofuranosylcytosine is antagonistic to stimulation of apoptosis and Bcl-2alpha down-regulation. *J. Biol. Chem.* **272**, 23481–23484.

Wieder, T., Orfanos, C. E., and Geilen, C. C. (1998). Induction of ceramide-mediated apoptosis by the anticancer phospholipid analog, hexadecylphosphocholine. *J. Biol. Chem.* **273**, 11025–11031.

Wiley, S. R., Schooley, K., Smolak, P. J., Din, W. S., Huang, C. P., Nicholl, J. K., Sutherland, G. R., Davis Smith, T., Rauch, C., Smith, C. A., and Goodwin, R. G. (1995). Identification and characterization of a new member of the TNF family that induces apoptosis. *Immunity* **3**, 673–682.

Wolff, R. A., Dobrowsky, R. T., Bielawska, A., Obeid, L. M., and Hannun, Y. A. (1994). Role of ceramide-activated protein phosphatase in ceramide-mediated signal transduction. *J. Biol. Chem.* **269**, 19605–19609.

Wu, G. S., Saftig, P., Peters, C., and El-Deiry, W. S. (1998). Potential role for cathepsin D in p53-dependent tumor suppression and chemosensitivity. *Oncogene* **16**, 2177–2183.

Xing, H. R., and Kolesnick, R. N. (2001). Kinase suppressor of Ras signals through Thr269 of c-Raf-1. *J. Biol. Chem.* **276**, 9733–9741.

Zamai, L., Ahmad, M., Bennett, I. M., Azzoni, L., Alnemri, E. S., and Perussia, B. (1998). Natural killer (NK) cell-mediated cytotoxicity: Differential use of TRAIL and Fas ligand by immature and mature primary human NK cells. *J. Exp. Med.* **188**, 2375–2380.

Zhang, J., Alter, N., Reed, J. C., Borner, C., Obeid, L. M., and Hannun, Y. A. (1996). Bcl-2 interrupts the ceramide mediated pathway of cell death. *Proc. Natl. Acad. Sci. USA* **93**, 5325–5328.

Zhang, X. D., Borrow, J. M., Zhang, X. Y., Nguyen, T., and Hersey, P. (2003). Activation of ERK1/2 protects melanoma cells from TRAIL-induced apoptosis by inhibiting Smac/DIABLO release from mitochondria. *Oncogene* **22**, 2869–2881.

Zhang, P., Liu, B., Jenkins, G. M., Hannun, Y. A., and Obeid, L. M. (1997a). Expression of neutral sphingomyelinase identifies a distinct pool of sphingomyelin involved in apoptosis. *J. Biol. Chem.* **272**, 9609–9612.

Zhang, P., Liu, B., Kang, S. W., Seo, M. S., Rhee, S. G., and Obeid, L. M. (1997b). Thioredoxin peroxidase is a novel inhibitor of apoptosis with a mechanism distinct from that of Bcl-2. *J. Biol. Chem.* **272**, 30615–30618.

Zhou, H., Summers, S. A., Birnbaum, M. J., and Pittman, R. N. (1998). Inhibition of Akt kinase by cell-permeable ceramide and its implications for ceramide-induced apoptosis. *J. Biol. Chem.* **273**, 16568–16575.

Zundel, W., and Giaccia, A. (1998). Inhibition of the anti-apoptotic PI(3)K/Akt/Bad pathway by stress. *Genes Dev.* **12**, 1941–1946.

Zundel, W., Swiersz, L. M., and Giaccia, A. (2000). Caveolin 1-mediated regulation of receptor tyrosine kinase-associated phosphatidylinositol 3-kinase activity by ceramide. *Mol. Cell. Biol.* **20**, 1507–1514.

14

TRAIL AND VIRAL INFECTION

JÖRN STRÄTER AND PETER MÖLLER

Department of Pathology, University Hospital of Ulm, D-89081 Ulm, Germany

I. Introduction
II. Apoptosis as a Process of Degradation of Host Cell and Viral Constituents
III. Death Receptor/Ligand Systems
   A. *TRAIL and its Receptors*
   B. *TRAIL Signaling*
IV. The Cytotoxic Activity of TRAIL Against Virus-Infected Cells
V. Viral Strategies to Circumvent TRAIL-Induced Apoptosis
VI. TRAIL in Virus-Induced Immunosuppression
VII. Viruses and TRAIL in Malignant Disease
VIII. Conclusions
    References

Tumor necrosis factor (TNF)–related apoptosis-inducing ligand (TRAIL) is a member of the TNF family that can induce apoptosis when binding to either of two receptors bearing an intracellular death domain. The physiologic function of the TRAIL system, which also comprises three receptors not mediating a death signal has just begun to be elucidated. Expression of TRAIL, mostly upon stimulation by interferons, in different cytotoxic immune cells suggested it has a role as an important effector molecule in immune surveillance. In addition to its ability to induce apoptosis in transformed tumor cells, TRAIL has attracted attention for its possibly critical role in the defense against viral infection. Viruses may induce TRAIL expression in host and/or immune cells and sensitize host cells toward TRAIL-mediated apoptosis. On the other hand, viruses have evolved a variety of strategies to prevent TRAIL-mediated host cell death early in infection, which may contribute to allowing their replication and the spread of viral progeny. The knowledge of the molecular mechanisms leading to modification of TRAIL sensitivity in virus-host cell interactions may also impact upon future (virus-based) strategies to increase TRAIL sensitivity of tumor cells. © 2004 Elsevier Inc.

I. INTRODUCTION

Apoptosis, programmed cell death, is a fundamental biologic process which, in the developing organism, plays an important role in the formation and shaping of organs and tissues. In the adult, it has tremendous implications for the maintenance of tissue homeostasis. When apoptotic death rates overcome proliferation, this results in atrophy, whereas hyperplasia and neoplasia are consequences of mitotic activity surpassing cell death. Finally, apoptosis plays a pivotal role in inflammatory conditions where programmed cell death is an important way of cutting down immune responses, but, on the other hand, may also contribute to tissue damage. This is also true for viral infections in which apoptosis of host cells is, in most cases, the only way to destroy their deleterious load and terminate the infection. Cytotoxic immune cells are the main effectors which force host cells into apoptosis. It is, however, known from a variety of *in situ* experiments that virus-infected cells can also enter apoptosis in a cell-autonomous way (Tyler *et al.*, 1995). Apoptosis of parenchymal cells, however, may contribute to infection-related organ failure, e.g., in fulminant viral hepatitis (Losser and Payen, 1996). On the other hand, viruses had to evolve strategies to delay or inhibit programmed death to enable their replication in host cells and the spread of viral progeny. These strategies may aim at avoiding recognition of infected cells by the immune system or activation of cytotoxic cells (for a recent review cf. Vossen *et al.*, 2002), but most often also involve interference with the apoptosis signaling cascade in host cells.

In recent years, some members of the tumor necrosis factor (TNF) family of proteins have attracted particular attention due to their ability to induce apoptosis in a very direct way by ligation of so-called death receptors on target cells. The TNF-related apoptosis-inducing ligand, TRAIL, also called APO-2 ligand (APO-2L), is such a member of the TNF family. Together with its five known receptors, it constitutes a quite complex death-inducing system that has recently been implicated in viral infection-associated apoptosis induction. In this review, we will discuss what is known to date about the potential role of TRAIL and its receptors in viral infections.

II. APOPTOSIS AS A PROCESS OF DEGRADATION OF HOST CELL AND VIRAL CONSTITUENTS

The central signaling cascade of apoptosis involves a hierarchical system of proteases known as caspases. Caspases are preformed in the cell as proenzymes that are activated either autocatalytically or by activated upstream caspases. Once activated, they cleave a variety of cellular proteins bearing specific peptide motifs with an aspartic acid residue. However, it is quite conceivable that viral proteins may also be degraded during host cell apoptosis. Actually, adenoviral E1A protein has only recently been shown to be cleaved by caspases (Grand *et al.*, 2002). Although only limited data are available to support this hypothesis at present, it will be interesting to see whether caspase-mediated degradation of viral antigens is a feature in host cell apoptosis.

An important cellular target protein for caspases is DFF45/ICAD. Cleavage of DFF45/ICAD during apoptosis activates DFF40, caspase-activated DNase (CAD), with which it forms a heterodimer as its inhibitor. Activation of CAD, in turn, gives rise to a genome-wide degradation of DNA into internucleosomal fragments, which is the biochemical hallmark of apoptosis. But degradation is not limited to DNA. Houge *et al.* provided evidence that ribosomal RNA (rRNA) is also cleaved in cells undergoing apoptosis (Houge *et al.*, 1995). Cleavage of rRNA is restricted to specific sites within the D domains that have shown exceptionally high divergence in evolution. This finding has led to the hypothesis that the apoptotic program may aim at the elimination of illegitimate (viral?) polynucleotides in general (Houge *et al.*, 1995).

To summarize, the biochemical processes involved in apoptosis may provide an efficient means of destroying viral constituents otherwise released from the dying cell. However, because viruses may also actively induce host cell apoptosis in lytic infections after a maximum of virus progeny has been produced, to allow their spread (Kawanishi, 1993), it may be critical that host cells are killed or enter apoptosis early during viral infection.

III. DEATH RECEPTOR/LIGAND SYSTEMS

There are two main pathways by which cytotoxic cells involved in the elimination of virally infected cells induce apoptosis: the perforin-granzyme pathway and the death receptor pathway. Death receptors possess the unique property of directly activating the intracellular death signaling machinery upon ligation by their natural ligands or cross-linking antibody. All death receptors known so far are members of the TNF receptor (TNFR) superfamily, a group of type I transmembrane proteins with one or several homologous cysteine-rich extracellular domains, and are characterized by a highly conserved intracytoplasmic death domain. Their natural ligands, on the other hand, belong to the TNF superfamily of homologous type II transmembrane proteins.

Although early studies suggested that death receptor-mediated apoptosis induction in virus infections is confined to CD95/CD95 ligand (CD95L) (Lowin et al., 1994), it has become clear that other death receptor/ligand pairs are also involved, namely TNFR-1/TNF-α (Elkon et al., 1997). There is now growing evidence that a third death-inducing ligand is an important effector in viral infections: TRAIL.

A. TRAIL AND ITS RECEPTORS

TRAIL, also known as APO-2L, is a more recently described apoptosis-inducing member of the TNF family (Pitti et al., 1996; Wiley et al., 1995). It exists in both a membrane-bound and, after proteolytic cleavage by cysteine proteases, a soluble form (Mariani and Krammer, 1998). So far, four non-promiscuous and one promiscuous receptor for TRAIL have been described, making this system the most complex of all death receptor-ligand systems. TRAIL receptor (TRAIL-R) 1/death receptor (DR)-4 and TRAIL-R2/DR-5 bear a cytoplasmic death domain and are able to transmit an apoptotic signal (Chaudhary et al., 1997; MacFarlane et al., 1997; Pan et al., 1997a,b; Schneider et al., 1997; Screaton et al., 1997; Sheridan et al., 1997; Walczak et al., 1997; Wu et al., 1997). TRAIL-R3 lacks an intracellular domain and is linked to the cell membrane by a phosphatidylinositol anchor. Consequently, TRAIL-R3, similar to TRAIL-R4, which has a truncated, non-functional death domain, is not able to induce cell death. It has been suggested that TRAIL-R3 and TRAIL-R4 compete with the apoptosis-inducing receptors, TRAIL-R1 and TRAIL-R2, for TRAIL binding and act as decoy receptors (hence their alternative names, decoy receptor [DcR]-1 and DcR-2, respectively) (Degli-Eposti et al., 1997a,b; Marsters et al., 1997; Pan et al., 1997b; Schneider et al., 1997; Sheridan et al., 1997). However, although TRAIL-R3 and TRAIL-R4 were shown to inhibit TRAIL-mediated apoptosis in over-expression experiments, their properties as decoy receptors under more physiologic conditions have been questioned when

TRAIL sensitivity of tumor cells was shown not to correlate with expression of decoy receptors (Lincz et al., 2001; Nimmanapalli et al., 2001; Petak et al., 2000). Osteoprotegerin (OPG) is a member of the TNFR superfamily which, in addition to osteoclast differentiation factor (ODF/RANKL), also binds to TRAIL, although with low affinity (Emery et al., 1998). OPG only exists in a soluble form, making it a further potential decoy receptor. Whether or not there is a physiologic role for OPG in the regulation of TRAIL activity *in vivo* is largely unknown.

B. TRAIL SIGNALING

Like most other TNF family members, TRAIL forms homotrimers that bind three receptors. Ligation of death receptors leads to the recruitment of intracellular proteins to the death domain of the receptor, where they form the so-called death-inducing signaling complex, DISC. The TRAIL DISC consists of the adaptor protein Fas-associated death domain (FADD) and caspase-8 which, once recruited, is activated by autoproteolysis (Kischkel et al., 2000; Sprick et al., 2000). Another initiator caspase with high homology to caspase-8, caspase-10, is able to replace caspase-8 in the DISC and to transmit the apoptotic signal (Kischkel et al., 2001).

Although initial studies suggested that TRAIL signaling is independent of the proapoptotic machinery of mitochondria (Walczak et al., 2000), there is now evidence that, in some cell types, TRAIL-induced apoptosis requires activation of the intrinsic (mitochondrial) signaling pathway influenced by the Bcl-2 family of proteins (Deng et al., 2002; LeBlanc et al., 2002). In this intrinsic apoptosis pathway, proapoptotic Bcl-2 family members such as Bax or Bak are counteracted by the antiapoptotic relatives Bcl-2 or Bcl-X_L (Bouillet and Strasser, 2002). As for CD95-mediated death, two cell types may be defined depending on the signaling pathway preferably used. In type I cells, robust DISC formation leads to a strong activation of caspase-8 which, in turn, directly activates downstream caspase-3 before loss of mitochondrial transmembrane potential. In type II cells, however, DISC formation and caspase-8 activity are low following death receptor engagement. Activation of caspase-3 and apoptosis occurs only following mitochondrial changes. There is evidence that this pathway is initiated by caspase-8–mediated cleavage of Bid, another Bcl-2 family member, which subsequently translocates to the mitochondria and activates Bax (Li et al., 1998; Luo et al., 1998). Bax, in turn, is essential in TRAIL-mediated apoptosis for mitochondrial depolarization and release of cytochrome *c* into the cytosol (LeBlanc et al., 2002). Cytochrome *c,* together with free Apaf-1, associates with caspase-9, which is cleaved and subsequently activates downstream effector caspases. While type II cells do not die upon death receptor ligation when antiapoptotic Bcl-2 family members such as Bcl-2 or Bcl-X_L are overexpressed, apoptosis in type I cells is independent of Bcl-2 family member expression. Importantly, cells that require the mitochondrial pathway to apoptose upon CD95 ligation and are type II cells with respect to CD95-mediated apoptosis may not do so following TRAIL receptor engagement and can be type I cells as to TRAIL receptor signaling (Walczak et al., 2000).

IV. THE CYTOTOXIC ACTIVITY OF TRAIL AGAINST VIRUS-INFECTED CELLS

In contrast to most other death-inducing TNF family members, TRAIL is broadly expressed in a variety of different tissues (Wiley et al., 1995). Although this is also true for the apoptosis-inducing TRAIL-R1 and 2, normal cells are largely resistant to TRAIL-induced apoptosis. TRAIL-induced apoptosis may, therefore, be a rare event in the healthy organism. Actually, injection of soluble TRAIL into mice is well tolerated and does not seem to provoke significant apoptosis (Ashkenazi et al., 1999; Walczak et al., 1999).

A first hint as to the physiologic function of TRAIL came from the observation that TRAIL is expressed and confers cytotoxic activity in a variety of cell types known to be involved in host defense against viral infections such as natural killer (NK) cells (Sato et al., 2001; Zamai et al., 1998) and cytotoxic T cells (Jeremias et al., 1998; Kayagaki et al., 1999a), monocytes/macrophages (Griffith et al., 1999), and dendritic cells (Fanger et al., 1999). Moreover, TRAIL expression in these cells is induced or enhanced by type I and II interferons (IFNs) known for their critical role in viral diseases (Fanger et al., 1999; Kayagaki et al., 1999b). Actually, the murine and human TRAIL promoters were recently shown to bear an IFN-stimulated response element (Gong and Almasan, 2000; Sato et al., 2001).

Direct proof of a role for the TRAIL system in viral infections, however, came from studies by Sedger and coworkers (1999). These authors impressively demonstrated that fibroblasts were sensitized toward TRAIL-induced apoptosis upon infection with human cytomegalovirus (HCMV). Apoptosis in infected cells was even more dramatic when IFN-γ was added. Increased TRAIL sensitivity of virally infected cells was later confirmed in different *in vitro* models involving adenovirus, HCMV, and reovirus (Clarke et al., 2000; Sträter et al., 2002). Sato et al. (2001) finally demonstrated that mice infected with encephalomyocarditis virus died earlier and had a much higher viral load in their hearts when treated with a neutralizing anti-TRAIL antibody compared to controls, for the first time providing evidence for the importance of TRAIL in an *in vivo* model of viral infection.

Some of these studies are also important in terms of which are the producers of TRAIL in viral infections. HCMV infection induces strong TRAIL expression on the surface of fibroblasts (Sedger et al., 1999) while HEK293 cells and a variety of tumor cell lines release soluble TRAIL into the supernatant following reovirus infection (Clarke et al., 2000, 2001). Thus, TRAIL can be expressed in cells other than immune cells providing a cell-autonomous means of eliminating infected cells by committing suicide or fratricide. Moreover, IFN-γ treatment of fibroblasts goes along with TRAIL expression, even on the surface of uninfected cells suggesting that non-infected bystanders may contribute to cytotoxicity toward infected neighbors in a juxtacrine or paracrine way (Sedger et al., 1999). In the mouse model by Sato and coworkers, however, protections against encephalomyocarditis virus appeared to be dependent on NK cells because injection of anti-asialoGM1 antibodies interfering with NK cell function resulted in high virus titers in the hearts of the animals similar to those observed after treatment with anti-TRAIL (Sato et al., 2001). It remains to be demonstrated whether NK cells are required for host cell sensitization to TRAIL, e.g. by secreting IFN-γ, or whether NK cells are the main source of TRAIL in this model.

But how are virus-infected cells sensitized to TRAIL? Interestingly, HCMV and reovirus infection may induce increased surface expression of the apoptosis-mediating receptors TRAIL-R1 and TRAIL-R2 on host cells while IFN-γ selectively downregulates death receptor expression in uninfected cells (Clarke et al., 2000; Sedger et al., 1999; Sträter et al., 2002). There is evidence that expression of both TRAIL receptors may be induced by the tumor suppressor protein p53 (Guan et al., 2001; Wu et al., 1997). Because p53 is stabilized by some viral factors such as the adenoviral E1A protein, due to their capacity to interfere with the host cell cycle, it will be interesting to see whether upregulation of TRAIL receptors following infection may be explained, at least in part, by activation of p53.

However, given the apparently low influence of TRAIL receptor expression on TRAIL sensitivity in a variety of systems (Griffith et al., 1998; Leverkus et al., 2000), it is unlikely that upregulation of death receptor expression is the main mechanism by which host cells become sensitized. In fact, TRAIL-sensitization of cells following reovirus infection is independent of altered receptor expression in different tumor cell lines (Clarke et al., 2001), suggesting that TRAIL sensitivity is regulated intracytoplasmatically. Interferon-γ has been shown to sensitize tumor cell lines with absent or low caspase-8 expression for death receptor–mediated apoptosis by increasing caspase-8 expression in a Stat1-dependent manner (Fulda and Debatin, 2002). Similarly, interferon-induced upregulation of caspase-8 could contribute to sensitize host cells also in viral infections.

V. VIRAL STRATEGIES TO CIRCUMVENT TRAIL-INDUCED APOPTOSIS

It is obvious that early death of host cells following viral infection would limit virus production and reduce or even eliminate the spread of progeny virus in the host. Viruses have therefore adapted a variety of different strategies to evade or delay apoptosis in an attempt to allow production of high yields of progeny virus. Receptor expression as well as different levels of the mitochondrion-dependent and -independent apoptosis signaling pathway may be affected. Due to homologies of the molecules involved and shared signaling pathways, it is not surprising that many viral inhibitors interfere with apoptosis induction by more than one apoptosis trigger.

A few years ago, Tollefson et al. revealed that the adenoviral E3-encoded receptor internalization and degradation (RID) integral membrane protein complex inhibited CD95-mediated apoptosis by internalization of cell surface CD95 and its destruction in endolysosomes (Tollefson et al., 1998). It was subsequently demonstrated by two independent groups that RID also suppresses TRAIL-induced cell death (Benedict et al., 2001; Tollefson et al., 2001) by a comparable mechanism involving TRAIL-R1 and TRAIL-R2. Mutation experiments suggest that a tyrosine residue near the C terminus of RIDβ may serve as a tyrosine-based sorting signal enabling receptor internalization (Lichtenstein et al., 2002). However, clearance of TRAIL-R2 from the cell surface, in addition to RID, requires another adenoviral protein, E3-6.7K, the function of which is not quite clear in this context (Benedict et al., 2001). Moreover, E3-6.7K alone was able to protect Jurkat T-cell lymphoma cells from apoptosis induced by CD95 antibodies and TRAIL (Moise et al., 2002).

CrmA is a cowpox virus–encoded protein which first attracted attention due to its ability to block apoptosis in response to granzyme B and cytotoxic T cells (Quan et al., 1995; Tewari et al., 1995). CrmA turned out to be a serine protease inhibitor (serpin) effectively inhibiting initiator caspase-8 while having little effect on downstream caspases-3, -6, and -7 (Zhou et al., 1997). Similarly, the baculovirus protein p35 can inhibit caspases but shows a much broader inhibitory spectrum. Thus, it is not surprising that both CrmA and p35 interfere with caspase-8 activation following death receptor engagement by several TNF family members, TRAIL included (Marsters et al., 1996; Suliman et al., 2001).

In 1997, two papers (Bertin et al., 1997; Thome et al., 1997) independently described a new family of viral inhibitors that were called viral FLICE-inhibitory proteins (v-FLIPs) by one group (Thome et al., 1997). Cells expressing these inhibitors encoded by several γ-herpesviruses (including HHV-8) and human molluscum contagiosum virus were protected against apoptosis induced by TRAIL and CD95L. DISC analysis revealed that FLIPs, through their death effector domain, interact with the adaptor protein

FADD and thus inhibit the recruitment and activation of caspase-8 (also known as FLICE) by the death receptor. Actually, v-FLIPs facilitate viral spread and persistence and seem to contribute to the transforming capacity of some herpesviruses (Thome et al., 1997). Shortly afterward, the discovery of a human homologue, c-FLIP, was reported, which may influence sensitivity of cells to death receptor engagement (Irmler et al., 1997).

Another strategy to inhibit death receptor-mediated apoptosis induction of host cells relies on interference with the intrinsic apoptosis signaling pathway. Many viruses encode Bcl-2 homologues such as BHRF1 in EBV, KSBcl-2 in HHV-8, or adenoviral E1B19K, which interfere with loss of mitochondrial membrane potential, release of cytochrome c and caspase activation (Cheng et al., 1997; Henderson et al., 1993). Interestingly, all herpesviruses encoding v-FLIPs also encode Bcl-2 homologues. Thus, it is conceivable that viruses developed this double strategy to ensure apoptosis inhibition in both type I and type II cells (Peter and Krammer, 1998).

VI. TRAIL IN VIRUS-INDUCED IMMUNOSUPPRESSION

Once host cells have been rendered resistant to TRAIL-mediated apoptosis, viruses may even force immune cells to die by turning their weapons upon themselves. Two viruses have been shown to induce apoptosis in immune cells by TRAIL receptor ligation: the human immunodeficiency virus (HIV) and measles virus.

It is well known that HIV infection induces an acquired immunodeficiency syndrome (AIDS) with marked lymphopenia as the main reason for opportunistic infections and mortality in this disease. Lymphopenia has been ascribed to an ongoing induction of apoptosis that mainly occurs in uninfected T cells in secondary lymphoid organs (Finkel et al., 1995). This view is further supported by *in vitro* experiments showing that peripheral blood T cells (PBL) from HIV patients are highly sensitive toward activation-induced cell death by CD3/T cell receptor (TCR)-engagement (Groux et al., 1992). This is also true for CD95-mediated apoptosis (Estaquier et al., 1995; Katsikis et al., 1995), although a functional link between increased activation-induced cell death and CD95 in HIV infection has been controversial (Bäumler et al., 1996; Katsikis et al., 1996). Instead, it turned out that TRAIL contributes considerably to anti-CD3–induced death in PBL from some HIV-infected patients (Katsikis et al., 1997). In contrast to healthy individuals, freshly isolated PBL from HIV-infected persons are spontaneously highly susceptible to TRAIL-induced apoptosis and even more so than to CD95 (Jeremias et al., 1998). Using an NOD-SCID-mouse model engrafted with human PBL, Miura and coworkers demonstrated that apoptosis in HIV-infected animals predominantly occurred in (uninfected) splenic CD4+

T cells in proximity to CD3+ CD4+ TRAIL-expressing human T cells. Moreover, neutralizing antibodies to TRAIL, but not CD95L, markedly inhibited T cell apoptosis in this model (Miura *et al.*, 2001). The finding that endogenous expression of the HIV *tat* gene in infected cells blocks TRAIL-mediated apoptosis to a much higher extent than does extracellular Tat protein may contribute to TRAIL-mediated apoptosis induction predominantly in non-infected bystander cells (Gibellini *et al.*, 2001).

To summarize, there is growing evidence indicating that, to a considerable extent, TRAIL may be responsible for apoptosis of T cells in HIV infection and could be a target for preventing the progression to AIDS. Nevertheless, there is a subpopulation of HIV patients in whom activation-induced cell death of T cells cannot be blocked by anti-TRAIL or caspase inhibitors (Katsikis *et al.*, 1997). Thus, mechanisms other than TRAIL-induced apoptosis appear also to contribute to T cell death in HIV infection.

Interestingly, a recently published study by Miura *et al.* (2003) using their mouse model suggests that TRAIL expressed by HIV-infected macrophages/microglia in the brain is the main cause for apoptosis of neuronal cells, leading to cognitive and motor dysfunction known as HIV encephalopathy.

Another viral disease accompanied by lymphopenia and immunosuppression, though transiently, is measles virus (MV) infection. As in HIV infection, the severity of lymphopenia is in striking contrast to the low number of virus-infected lymphocytes and it is mostly non-infected cells that are destroyed. In recent years, it has become clear that antigen-presenting dendritic cells (DC) are centered in the pathogenesis of MV-induced lymphopenia. In contrast to normal DC which, following pathogen-driven maturation, induce naive T cell activation, proliferation and maturation into effector T cells, MV-infected DC are able to kill activated T cells (Fugier-Vivier *et al.*, 1997). A recent study provides the first evidence that MV-infected monocyte-derived DC express TRAIL and show cytotoxic activity that can be inhibited by a TRAIL-R2:Fc fusion protein (Vidalain *et al.*, 2000). Thus, published data point to a model in which lymphopenia in the course of MV infection is caused by TRAIL-mediated apoptosis of T cells interacting with (TRAIL-expressing) MV-infected DC. Interestingly, a similar mechanism was reported recently to be active in human cytomegalovirus (HCMV) infection and may explain HCMV-triggered immunosuppression (Raftery *et al.*, 2001).

VII. VIRUSES AND TRAIL IN MALIGNANT DISEASE

During recent years, TRAIL has attracted most attention in cancer research as a promising anticancer therapeutic. This is due to the fact that, in contrast to normal cells, many tumor cell lines undergo apoptosis upon

TRAIL treatment *in vitro* (Griffith *et al.*, 1998; Pitti *et al.*, 1996; Wiley *et al.*, 1995) and, in contrast to FasL or TNF, its application in animal models does not show significant toxicity (Ashkenazi *et al.*, 1999; Walczak *et al.*, 1999). Moreover, tumor cells can be sensitized towards TRAIL by additional application of cytotoxic drugs such as 5-fluorouracil, doxorubicin, and CPT-11 (Ashkenazi *et al.*, 1999; Gliniak and Le, 1999; Keane *et al.*, 1999). However, in tumor xenograft models using tumor cell lines that are even highly sensitive to TRAIL *in vitro*, established tumors can be brought to only partial regression by repeated injection of human recombinant TRAIL. Also, due to rapid clearance of injected TRAIL from the blood, relatively large amounts of soluble TRAIL are required (Ashkenazi *et al.*, 1999; Walczak *et al.*, 1999). Finally, while about half of all established tumor cell lines are sensitive to TRAIL, there remains another 50% that are not. It is also likely that sporadic tumors are even more resistant to TRAIL-induced apoptosis than cell lines growing *in vitro* for many years (Nguyen *et al.*, 2001). Thus, the development of alternative methods to deliver TRAIL, i.e., high local TRAIL expression in tumors instead of exogenous administration, as well as to sensitize TRAIL-resistant tumor cells is highly desirable. In this respect too, viruses are becoming ever more interesting.

An interesting observation made during the *in vitro* studies on virus-infected cells was that the cytotoxic activity of host cells due to virus-induced upregulation of TRAIL could also be directed against (TRAIL-sensitive) tumor cells (Clarke *et al.*, 2000; Vidalain *et al.*, 2000). These findings point to a potential strategy in which viruses may be used to induce local TRAIL expression in tumors *in vivo*. A virus already being used beneficially in clinical studies as an antineoplastic and immunostimulatory agent is Newcastle disease virus (NDV). A postoperative tumor vaccination therapy with NDV-modified, live tumor cells is well tolerated, and no serious side effects have been observed to date. An important step toward understanding of the molecular basis of this therapy, however, was taken only very recently. It turned out that it is actually TRAIL that is upregulated in human monocytes upon stimulation with NDV *in vitro* and which mediates cytotoxic activity of macrophages against tumor cells (Washburn *et al.*, 2003).

A more direct way to provide TRAIL in tumors is the introduction of the TRAIL gene and its expression in tumor cells. This has been done by Griffith *et al.* who generated a replication-defective adenovirus encoding TRAIL. Infection of different tumor cell lines with adenovirus resulted in TRAIL expression and death in TRAIL-sensitive tumor cells *in vitro*. Importantly, infection also prompted normal prostatic epithelial cells to express TRAIL. While these normal cells remained resistant to TRAIL-induced apoptosis upon infection, they were able to kill TRAIL-sensitive PC-3 cells (Griffith *et al.*, 2000). This study underlines that it is not imperative specifically to target tumor cells, as infection in either the tumor itself or in the surrounding normal tissue would lead to local production of TRAIL and selective tumor cell death (Griffith et al., 2000).

It is not quite clear whether TRAIL sensitivity of tumor cells is also altered in these models. However, in both studies, tumor cell lines which were most effectively killed following infection were those which already spontaneously exhibited the highest sensitivity to exogenous TRAIL *in vitro* (Griffith et al., 2000; Washburn et al., 2003). This may be different when reovirus is used to infect tumor cells. Reovirus not only kills infected tumor cells by endogenous TRAIL induction but, most importantly, also synergistically sensitizes tumor cells to apoptosis induction by exogenous TRAIL so that even previously resistant cell lines die upon combined reovirus/TRAIL treatment (Clarke et al., 2001).

In conclusion, we have just entered the very promising TRAIL trail to tumor cell death. The more we know about how viruses influence TRAIL sensitivity of host cells, the better we will find ways to generate virus mutants for use in treatment of human malignancies. For instance, Routes *et al.* recently demonstrated that TRAIL-resistant tumor cells can be sensitized to undergo TRAIL-induced apoptosis when stably transfected with adenoviral E1A protein. However, the very same tumor cells remain TRAIL-resistant when infected with complete adenovirus since E1B and E3 gene products block TRAIL-induced killing (Routes et al., 2000). It is certainly conceivable that adenoviral mutants encoding the E1A gene but lacking both E1B and E3 coding regions effectively kill tumor cells in a TRAIL-dependent way (Routes et al., 2000). Finally, it is important to show for every new virus mutant that infected normal cells are not sensitized to TRAIL in a similar way to tumor cells. This would largely limit the use of a virus-based therapy due to unwanted tissue damage.

VIII. CONCLUSIONS

We just begun to gain awareness of the potential role of the TRAIL system in viral infections. However, it remains to be seen whether TRAIL actually has the dominant role in host defense against viruses that recent *in vitro* studies suggest or whether it is just one way that virus-infected cells can be forced to apoptosis or to commit suicide. Redundancies in the repertoire of cytotoxic cells to induce apoptosis may be important in light of the multiple strategies viruses have developed to escape host cell death. It is thus also likely that the dominant mechanism by which the host organism tries to get rid of infected cells varies with the type of virus it has to cope with. Also, different cells in different organs may have differential sensitivity towards different death-inducing TNF family members (Tay and Welsh, 1997). Finally, given the close similarity of means by which the immune system fights against virus-infected cells and malignant tumors, a better understanding of the molecular mechanisms by which viruses influence TRAIL sensitivity of host cells may also help to realize the exciting therapeutic potential of TRAIL in malignant tumors.

ACKNOWLEDGMENTS

This work was supported by grants from the Deutsche Krebshilfe to J. Sträter (10-1644-Str 2) and from the Deutsche Forschungsgemeinschaft (SFB518 TPA13) and the IZKF Ulm to P. Möller.

REFERENCES

Ashkenazi, A., Pai, R. C., Fong, S., Leung, S., Lawrence, D. A., Marsters, S. A., Blackie, C., Chang, L., McMurtrey, A. E., Hebert, A., DeForge, L., Koumenis, I. L., Lewis, D., Harris, L., Bussiere, J., Koeppen, H., Shahrokh, Z., and Schwall, R. H. (1999). Safety and antitumor activity of recombinant soluble Apo2 ligand. *J. Clin. Invest.* **104,** 155–162.

Bäumler, C. B., Böhler, T., Herr, I., Benner, A., Krammer, P. H., and Debatin, K.-M. (1996). Activation of the CD95 (APO-1/Fas) system in T cells from human immunodeficiency virus type-1-infected children. *Blood* **88,** 1741–1746.

Benedict, C. A., Norris, P. S., Prigozy, T. I., Bodmer, J.-L., Mahr, J. A., Garnett, C. T., Martinon, F., Tschopp, J., Gooding, L. R., and Ware, C. F. (2001). Three adenovirus E3 proteins cooperate to evade apoptosis by tumor necrosis factor-related apoptosis-inducing ligand receptor-1 and -2. *J. Biol. Chem.* **276,** 3270–3278.

Bertin, J., Armstrong, R. C., Ottilie, S., Martin, D. A., Wang, Y., Banks, S., Wang, G. H., Senkevich, T. G., Alnemri, E. S., Moss, B., Lenardo, M. J., Tomaselli, K. J., and Cohen, J. I. (1997). Death effector domain-containing herpesvirus and poxvirus proteins inhibit Fas- and TNFR1-induced apoptosis. *Proc. Natl. Acad. Sci. USA* **94,** 1172–1176.

Bouillet, P., and Strasser, A. (2002). BH3-only proteins – evolutionarily conserved proapoptotic Bcl-2 family members essential for initiating programmed cell death. *J. Cell Sci.* **115,** 1567–1574.

Chaudhary, P. M., Eby, M., Jasmin, A., Bookwalter, A., Murray, J., and Hood, L. (1997). Death receptor 5, a new member of the TNFR family, and DR4 induce FADD-dependent apoptosis and activate the NF-kB pathway. *Immunity* **7,** 821–830.

Cheng, E. H., Nicholas, J., Bellows, D. S., Hayward, G. S., Guo, H. G., Reitz, M. S., and Hardwick, J. M. (1997). A Bcl-2 homolog encoded by Kaposi sarcoma-associated virus, human herpesvirus 8, inhibits apoptosis but does not heterodimerize with Bax or Bak. *Proc. Natl. Acad. Sci. USA* **94,** 690–694.

Clarke, P., Meintzer, S. M., Gibson, S., Widmann, C., Garrington, T. P., Johnson, G. L., and Tyler, K. L. (2000). Reovirus-induced apoptosis is mediated by TRAIL. *J. Virol.* **74,** 8135–8139.

Clarke, P., Meintzer, S. M., Spalding, A. C., Johnson, G. L., and Tyler, K. L. (2001). Caspase 8–dependent sensitization of cancer cells to TRAIL-induced apoptosis following reovirus infection. *Oncogene* **20,** 6910–6919.

Degli-Eposti, M. A., Smolak, P. J., Walczak, H., Waugh, J., Huang, C.-P., DuBose, R. F., Goodwin, R. G., and Smith, C. A. (1997a). Cloning and characterization of TRAIL-R3, a novel member of the emerging TRAIL receptor family. *J. Exp. Med.* **186,** 1165–1170.

Degli-Eposti, M. A., Dougall, W. C., Smolak, P. J., Waugh, J. Y., Smith, C. A., and Goodwin, R. G. (1997b). The novel receptor TRAIL-R4 induces NF-kB and protects against TRAIL-mediated apoptosis, yet retains an incomplete death domain. *Immunity* **7,** 813–820.

Deng, Y., Lin, Y., and Wu, X. (2002). TRAIL-induced apoptosis requires Bax-dependent mitochondrial release of Smac/DIABLO. *Genes Dev.* **16,** 33–45.

Elkon, K. B., Liu, C. C., Gall, J. G., Trevejo, J., Marino, M. W., Abrahamsen, K. A., Song, X., Zhou, J. L., Old, L. J., Crystal, R. G., and Falck-Pedersen, E. (1997). Tumor necrosis factor alpha plays a central role in immune-mediated clearance of adenoviral vectors. *Proc. Natl. Acad. Sci. USA* **94,** 9814–9819.

Emery, J. G., McDonnell, P., Burke, M. B., Deen, K. C., Lyn, S., Silverman, C., Dul, E., Appelbaum, E. R., Eichman, C., DiPrinzio, R., Dodds, R. A., James, I. E., Rosenburg, M., Lee, J. C., and Young, P. R. (1998). Osteoprotegerin is a receptor for the cytotoxic ligand TRAIL. *J. Biol. Chem.* **273,** 14363–14367.

Estaquier, J., Idziorek, T., Zou, W., Emilie, D., Farber, C., Bourez, J., and Ameisen, J. C. (1995). T helper type 1/T helper type 2 cytokines and T cell death: Preventive effect of interleukin 12 on activation-induced and CD95 (Fas/APO-1)-mediated apoptosis of CD4+ T cells from human immunodeficiency virus-infected persons. *J. Exp. Med.* **182,** 1759–1767.

Fanger, N. A., Maliszewski, C. R., Schooley, K., and Griffith, T. S. (1999). Human dendritic cells mediate cellular apoptosis via tumor necrosis factor-related apoptosis-inducing ligand (TRAIL). *J. Exp. Med.* **190,** 1155–1164.

Finkel, T. H., Tudor-Williams, G., Banda, N. K., Cotton, M. F., Curiel, T., Monks, C., Baba, T. W., Ruprecht, R. M., and Kupfer, A. (1995). Apoptosis occurs predominantly in bystander cells and not in productively infected cells of HIV- and SIV-infected lymph nodes. *Nat. Med.* **1,** 129–134.

Fugier-Vivier, I., Servet-Delprat, C., Rivailler, P., Rissoan, M. C., Liu, Y. J., and Rabourdin-Combe, C. (1997). Measles virus suppresses cell-mediated immunity by interfering with the survival and functions of dendritic and T cells. *J. Exp. Med.* **186,** 813–823.

Fulda, S., and Debatin, K. M. (2002). IFNgamma sensitizes for apoptosis by upregulating caspase-8 expression through the Stat1 pathway. *Oncogene* **21,** 2295–2308.

Gibellini, D., Re, M. C., Ponti, C., Maldini, C., Celeghini, C., Cappelini, A., La Placa, M., and Zauli, G. (2001). HIV-1 Tat protects CD4+ Jurkat T lmyphoblastoid cells from apoptosis mediated by TNF-related apoptosis-inducing ligand. *Cell. Immunol.* **207,** 89–99.

Gliniak, B., and Le, T. (1999). Tumor necrosis factor-related apoptosis-inducing ligand's antitumor activity in vivo is enhanced by the chemotherapeutic agent CPT-11. *Cancer Res.* **59,** 6153–6158.

Gong, B., and Almasan, A. (2000). Genomic organization and transcriptional regulation of human APO2/TRAIL gene. *Biochem. Biophys. Res. Commun.* **278,** 747–752.

Grand, R. J., Schmeiser, K., Gordon, E. M., Zhang, X., Gallimore, P. H., and Turnell, A. S. (2002). Caspase-mediated cleavage of adenovirus early region 1A proteins. *Virology* **301,** 255–271.

Griffith, T. S., Chin, W. A., Jackson, G. C., Lynch, D. H., and Kubin, M. Z. (1998). Intracellular regulation of TRAIL-induced apoptosis in human melanoma cells. *J. Immunol.* **161,** 2833–2840.

Griffith, T. S., Wiley, S. R., Kubin, M. Z., Sedger, L. M., Maliszewski, C. R., and Fanger, N. A. (1999). Monocyte-mediated tumoricidal activity via the tumor necrosis factor-related cytokine, TRAIL. *J. Exp. Med.* **189,** 1343–1353.

Griffith, T. S., Anderson, R. D., Davidson, B. L., Williams, R. D., and Ratliff, T. L. (2000). Adenoviral-mediated transfer of the TNF-related apoptosis-inducing ligand/Apo-2 ligand gene induces tumor cell apoptosis. *J. Immunol.* **165,** 2886–2894.

Groux, H., Torpier, G., Monte, D., Mouton, Y., Capron, A., and Ameisen, J. C. (1992). Activation-induced death by apoptosis in CD4+ T cells from human immunodeficiency virus-infected asymptomatic individuals. *J. Exp. Med.* **175,** 331–340.

Guan, B., Yue, P., Clayman, G. L., and Sun, S.-Y. (2001). Evidence that the death receptor DR4 is a DNA damage-inducible, p53-regulated gene. *J. Cell. Physiol.* **188**, 98–105.

Henderson, S., Huen, D., Rowe, M., Dawson, C., Johnson, G., and Rickinson, A. (1993). Epstein-Barr virus-coded BHRF1 protein, a viral homologue of Bcl-2, protects human B cells from programmed cell death. *Proc. Natl. Acad. Sci. USA* **90**, 8479–8483.

Houge, G., Robaye, B., Eikhom, T. S., Golstein, J., Mellgren, G., Gjertsen, B. T., Lanotte, M., and Doskeland, S. O. (1995). Fine mapping of 28S rRNA sites specifically cleaved in cells undergoing apoptosis. *Mol. Cell. Biol.* **15**, 2051–2062.

Irmler, M., Thome, M., Hahne, M., Schneider, P., Hofmann, K., Steiner, V., Bodmer, J. L., Schröter, M., Burne, K., Mattmann, C., Rimoldi, D., French, L. E., and Tschopp, J. (1997). Inhibition of death receptor signals by cellular FLIP. *Nature* **388**, 190–195.

Jeremias, I., Herr, I., Boehler, T., and Debatin, K.-M. (1998). TRAIL/Apo-2-ligand-induced apoptosis in human T cells. *Eur. J. Immunol.* **28**, 143–152.

Katsikis, P. D., Wunderlich, E. S., Smith, C. A., Herzenberg, L. A., and Herzenberg, L. A. (1995). Fas antigen stimulation induces marked apoptosis of T lymphocytes in human immunodeficiency virus-infected individuals. *J. Exp. Med.* **181**, 2029–2036.

Katsikis, P. D., Garcia-Ojeda, M. E., Wunderlich, E. S., Smith, C. A., Yagita, H., Okumura, K., Kayagaki, N., Alderson, M., Herzenberg, L. A., and Herzenberg, L. A. (1996). Activation-induced peripheral blood T cell apoptosis is Fas independent in HIV-infected individuals. *Int. Immunol.* **8**, 1311–1317.

Katsikis, P. D., Garcia-Ojeda, M. E., Torres-Roca, J. F., Tijoe, I. M., Smith, C. A., Herzenberg, L. A., and Herzenberg, L. A. (1997). Interleukin-1b converting enzyme-like protease involvement in Fas-induced and activation-induced peripheral blood T cell apoptosis in HIV infection. TNF-related apoptosis-inducing ligand can mediate activation-induced T cell death in HIV infection. *J. Exp. Med.* **186**, 1365–1372.

Kawanishi, M. (1993). Epstein-Barr virus induces fragmentation of chromosomal DNA during lytic infection. *J. Virol.* **67**, 7654–7658.

Kayagaki, N., Yamaguchi, N., Nakayama, M., Kawasaki, A., Akiba, H., Okumura, K., and Yagita, H. (1999a). Involvement of TNF-related apoptosis-inducing ligand in human $CD4^+$ T cell-mediated cytotoxicity. *J. Immunol.* **162**, 2639–2647.

Kayagaki, N., Yamaguchi, N., Nakayama, M., Eto, H., Okumura, K., and Yagita, H. (1999b). Type I interferons (IFNs) regulate tumor necrosis factor-related apoptosis-inducing ligand (TRAIL) expression on human T cells: A novel mechanism for the antitumor effects of type I IFNs. *J. Exp. Med.* **189**, 1451–1460.

Keane, M. M., Ettenberg, S. A., Nau, M. M., Russell, E. K., and Lipkowitz, S. (1999). Chemotherapy augments TRAIL-induced apoptosis in breast cell lines. *Cancer Res.* **59**, 734–741.

Kischkel, F. C., Lawrence, D. A., Chuntharapai, A., Schow, P., Kim, K. J., and Ashkenazi, A. (2000). Apo2L/TRAIL-dependent recruitment of endogenous FADD and caspase-8 to death receptors 4 and 5. *Immunity* **12**, 611–620.

Kischkel, F. C., Lawrence, D. A., Tinel, A., LeBlanc, H., Virmani, A., Schow, P., Gazdar, A., Blenis, J., Arnott, D., and Ashkenazi, A. (2001). Death receptor recruitment of endogenous caspase-10 and apoptosis initiation in the absence of caspase-8. *J. Biol. Chem.* **276**, 46639–46646.

LeBlanc, H., Lawrence, D., Varfolomeev, E., Totpal, K., Morlan, J., Schow, P., Fong, S., Schwall, R., Sinicropi, D., and Ashkenazi, A. (2002). Tumor cell resistance to death receptor induced apoptosis through mutational inactivation of the proapoptotic Bcl-2 homolog Bax. *Nat. Med.* **8**, 274–281.

Leverkus, M., Neumann, M., Mengling, T., Rauch, C. T., Bröcker, E.-B., Krammer, P. H., and Walczak, H. (2000). Regulation of TRAIL sensitivity in primary and transformed human keratinocytes. *Cancer Res.* **60**, 553–559.

Li, H., Zhu, H., Xu, C.-J., and Yuan, J. (1998). Cleavage of BID by caspase 8 mediates the mitochondrial damage in the Fas pathway to apoptosis. *Cell* **94,** 491–501.

Lichtenstein, D. L., Krajcsi, P., Esteban, D. J., Tollefson, A. E., and Wold, W. S. M. (2002). Adenovirus RIDβ subunit contains a tyrosine residue that is critical for RID-mediated receptor internalization and inhibition of Fas- and TRAIL-induced apoptosis. *J. Virol.* **76,** 11329–11342.

Lincz, L. F., Yeh, T.-X., and Spencer, A. (2001). TRAIL-induced eradication of primary tumour cells from multiple myeloma patient bone marrows is not related to TRAIL receptor expression prior to chemotherapy. *Leukemia* **15,** 1650–1657.

Losser, M. R., and Payen, D. (1996). Mechanisms of liver damage. *Semin. Liver Dis.* **16,** 357–367.

Lowin, B., Hahne, M., Mattmann, C., and Tschopp, J. (1994). Cytolytic T-cell cytotoxicity is mediated through perforin and Fas lytic pathways. *Nature* **370,** 650–652.

Luo, X., Budihardjo, I., Zou, H., Slaughter, C., and Wang, X. (1998). Bid, a Bcl2 interacting protein, mediates cytochrome c release from mitochondria in response to activation of cell surface death receptors. *Cell* **94,** 481–490.

MacFarlane, M., Ahmad, M., Srinivasula, S. M., Fernandes-Alnemri, T., Cohen, G. M., and Alnemri, E. S. (1997). Identification and molecular cloning of two novel receptors for the cytotoxic ligand TRAIL. *J. Biol. Chem.* **272,** 25417–25420.

Mariani, S. M., and Krammer, P. H. (1998). Differential regulation of TRAIL and CD95 ligand in transformed cells of the T and B lymphocyte lineage. *Eur. J. Immunol.* **28,** 973–982.

Marsters, S. A., Pitti, R. M., Donahue, C. J., Ruppert, S., Bauer, K. D., and Ashkenazi, A. (1996). Activation of apoptosis by Apo-2 ligand is independent of FADD but blocked by CrmA. *Curr. Biol.* **6,** 750–752.

Marsters, S. A., Sheridan, J. P., Pitti, R. M., Huang, A., Skubatch, M., Baldwin, D., Yuan, J., Gurney, A., Goddard, A. D., Godowski, P., and Ashkenazi, A. (1997). A novel receptor for APO-2L/TRAIL contains a truncated death domain. *Curr. Biol.* **7,** 1003–1006.

Miura, Y., Misawa, N., Maeda, N., Inagaki, Y., Tanaka, Y., Ito, M., Kayagaki, N., Yamamoto, N., Yagita, H., Mizusawa, H., and Koyanagi, Y. (2001). Critical contribution of tumor necrosis factor-related apoptosis-inducing ligand (TRAIL) to apoptosis of human CD4+ T cells in HIV-1-infected hu-PBL-NOD-SCID mice. *J. Exp. Med.* **193,** 651–659.

Miura, Y., Misawa, N., Kawano, Y., Okada, H., Inagaki, Y., Yamamoto, N., Ito, M., Yagita, H., Okumura, K., Mizusawa, H., and Koyanagi, Y. (2003). Tumor necrosis factor-related apoptosis-inducing ligand induces neuronal death in a murine model of HIV central nervous system infection. *Proc. Natl. Acad. Sci. USA* **100,** 2777–2782.

Moise, A. R., Grant, J. R., Vitalis, T. Z., and Jefferies, W. A. (2002). Adenovirus E3-6.7K maintains calcium homeostasis and prevents apoptosis and arachidonic acid release. *J. Virol.* **76,** 1578–1587.

Nguyen, T., Zhang, X. D., and Hersey, P. (2001). Relative resistance of fresh isolates of melanoma to tumor necrosis factor-related apoptosis-inducing ligand (TRAIL)-induced apoptosis. *Clin. Cancer Res.* **7**(Suppl), 966s–973s.

Nimmanapalli, R., Perkins, C. L., Orlando, M., O'Bryan, E., Nguyen, D., and Bahlla, K. N. (2001). Pretreatment with paclitaxel enhances Apo-2 ligand/tumor necrosis factor-related apoptosis-inducing ligand-induced apoptosis of prostate cancer cells by inducing death receptors 4 and 5 protein levels. *Cancer Res.* **61,** 759–763.

Pan, G., O'Rourke, K., Chinnaiyan, A. M., Gentz, R., Ebner, R., Ni, J., and Dixit, V. M. (1997a). The receptor for the cytotoxic ligand TRAIL. *Science* **276,** 111–113.

Pan, G., Ni, J., Wei, Y.-F., Yu, G.-I., Gentz, R., and Dixit, V. M. (1997b). An antagonist decoy receptor and a death domain-containing receptor for TRAIL. *Science* **277,** 815–818.

Petak, I., Douglas, L., Tillman, D. M., Vernes, R., and Houghton, J. A. (2000). Pediatric rhabdomyosarcoma cell lines are resistant to Fas-induced apoptosis and highly sensitive to TRAIL-induced apoptosis. *Clin. Cancer Res.* **6,** 4119–4127.

Peter, M. E., and Krammer, P. H. (1998). Mechanisms of CD95 (APO-1/Fas)-mediated apopotsis. *Curr. Opin. Immunol.* **10,** 545–551.

Pitti, R. M., Marsters, S. A., Ruppert, S., Donahue, C. J., Moore, A., and Ashkenazi, A. (1996). Induction of apoptosis by APO-2 ligand, a new member of the tumor necrosis factor cytokine family. *J. Biol. Chem.* **271,** 12687–12690.

Quan, L. T., Caputo, A., Bleackley, R. C., Pickup, D. J., and Salvesen, G. S. (1995). Granzyme B is inhibited by the cowpox virus serpin cytokine response modifier A. *J. Biol. Chem.* **272,** 10377–10379.

Raftery, M. J., Schwab, M., Eibert, S. M., Samstag, Y., Walczak, H., and Schönrich, G. (2001). Targeting the function of mature dendritic cells by human cytomegalovirus: A multilayered viral defense strategy. *Immunity* **15,** 867–870.

Routes, J. M., Ryan, S., Clase, A., Miura, T., Kuhl, A., Potter, T. A., and Cook, J. L. (2000). Adenovirus E1A oncogene expression in tumor cells enhances killing by TNF-related apoptosis-inducing ligand (TRAIL). *J. Immunol.* **165,** 4522–4527.

Sato, K., Hida, S., Takayanagi, H., Yokochi, T., Kayagaki, N., Takeda, K., Yagita, H., Okumura, K., Tanaka, N., Taniguchi, T., and Ogasawara, K. (2001). Antiviral response by natural killer cells through TRAIL gene induction by IFN-α/β. *Eur. J. Immunol.* **31,** 3138–3146.

Schneider, P., Bodmer, J.-L., Thome, M., Hofmann, K., Holler, N., and Tschopp, J. (1997). Characterization of two receptors for TRAIL. *FEBS Lett.* **416,** 329–334.

Screaton, G. R., Mongkolsapaya, J., Xu, X.-N., Cowper, A. E., McMichael, A. J., and Bell, J. I. (1997). TRICK2, a new alternatively spliced receptor that transduces the cytotoxic signal from TRAIL. *Curr. Biol.* **7,** 693–696.

Sedger, L. M., Shows, D. M., Blanton, R. A., Peschon, J. J., Goodwin, R. G., Cosman, D., and Wiley, S. R. (1999). IFN-gamma mediates a novel antiviral activity through dynamic modulation of TRAIL and TRAIL receptor expression. *J. Immunol.* **163,** 920–926.

Sheridan, J. P., Marsters, S. A., Pitti, P. M., Gurney, A., Skubatch, M., Baldwin, D., Ramakrishnan, L., Gray, C. L., Baker, K., Wood, W. I., Goddard, A. D., Godowski, P., and Ashkenazi, A. (1997). Control of TRAIL-induced apoptosis by a family of signaling and decoy receptors. *Science* **277,** 818–821.

Sprick, M. R., Weigand, M. A., Rieser, E., Rausch, C. T., Juo, P., Blenis, J., and Krammer, P. H. (2000). FADD/MORT1 and caspase-8 are recruited to TRAIL receptors 1 and 2 and are essential for apoptosis mediated by TRAIL receptor 2. *Immunity* **12,** 599–609.

Sträter, J., Walczak, H., Pukrop, T., von Müller, L., Hasel, C., Kornmann, M., Mertens, T., and Möller, P. (2002). TRAIL and its receptors in the colonic epithelium: A putative role in the defence of viral infections. *Gastroenterol.* **122,** 659–666.

Suliman, A., Lam, A., Datta, R., and Srivastava, R. K. (2001). Intracellular mechanisms of TRAIL: Apoptosis through mitochondrial-dependent and -independent pathways. *Oncogene* **20,** 2122–2133.

Tay, C. H., and Welsh, R. M. (1997). Distinct organ-dependent mechanisms for the control of murine cytomegalovirus infection by natural killer cells. *J. Virol.* **71,** 267–275.

Tewari, M., Telford, W. G., Miller, R. A., and Dixit, V. M. (1995). CrmA, a poxvirus-encoded serpin, inhibits cytotoxic T-lymphocyte-mediated apoptosis. *J. Biol. Chem.* **270,** 22705–22708.

Thome, M., Schneider, P., Hofmann, K., Fickenscher, H., Meinl, E., Neipel, F., Mattmann, C., Burns, K., Bodmer, J.-L., Schröter, M., Scaffidi, C., Krammer, P. H., Peter, M. E., and Tschopp, J. (1997). Viral FLICE-inhibitory proteins (FLIPs) prevent apoptosis induced by death receptors. *Nature* **386,** 517–521.

Tollefson, A. E., Hermiston, T. W., Lichtenstein, D. L., Colle, C. F., Tripp, R. A., Dimitrov, T., Toth, K., Wells, C. E., Doherty, P. C., and Wold, W. S. M. (1998). Forced degradation of Fas inhibits apoptosis in adenovirus-infected cells. *Nature* **392,** 726–730.

Tollefson, A. E., Toth, K., Doronin, K., Kuppuswamy, M., Doronina, O. A., Lichtenstein, D. L., Hermiston, T. W., Smith, C. A., and Wold, W. S. M. (2001). Inhibition of TRAIL-induced apoptosis and forced internalization of TRAIL receptor 1 by adenovirus proteins. *J. Virol.* **75,** 8875–8887.

Tyler, K. L., Squier, M. K., Rodgers, S. E., Schneider, B. E., Oberhaus, S. M., Grdina, T. A., Cohen, J. J., and Dermody, T. S. (1995). Differences in the capacity of reovirus strains to induce apoptosis are determined by the viral attachment protein sigma 1. *J. Virol.* **69,** 6972–6997.

Vidalain, P.-O., Azocar, O., Lamouille, B., Astier, A., Rabourdin-Combe, C., and Servet-Delprat, C. (2000). Measles virus induces functional TRAIL production by human dendritic cells. *J. Virol.* **74,** 556–559.

Vossen, M. T. M., Westerhout, E. M., Söderberg-Nauclér, C., and Wiertz, E. J. H. J. (2002). Viral immune evasion: A masterpiece of evolution. *Immunogenetics* **54,** 527–542.

Walczak, H., Degli-Eposti, M. A., Johnson, R. S., Smolak, P. J., Waugh, J. Y., Boiani, N., Timour, M. S., Gerhart, M. J., Schooley, K. A., Smith, C. A., Goodwin, R. G., and Rauch, C. T. (1997). TRAIL-R2: A novel apoptosis-mediating receptor for TRAIL. *EMBO J.* **16,** 5386–5397.

Walczak, H., Miller, R. E., Ariail, K., Gliniak, B., Griffith, T. S., Kubin, M., Chin, W., Jones, J., Woodward, A., Le, T., Smith, C., Smolak, P., Goodwin, R. G., Rauch, C. T., Schuh, J. C., and Lynch, D. H. (1999). Tumoricidal activity of tumor necrosis factor-related apoptosis-inducing ligand in vivo. *Nat. Med.* **5,** 157–163.

Walczak, H., Bouchon, A., Stahl, H., and Krammer, P. H. (2000). Tumor necrosis factor-related apoptosis-inducing ligand retains its apoptosis-inducing capacity on Bcl-2 or Bcl-X_L-overexpressing chemotherapy-resistant tumor cells. *Cancer Res.* **60,** 3051–3057.

Washburn, B., Weigand, M. A., Grosse-Wilde, A., Janke, M., Stahl, H., Rieser, E., Sprick, M. R., Schirrmacher, V., and Walczak, H. (2003). TNF-related apoptosis-inducing ligand mediates tumoricidal activity of human monocytes stimulated by Newcastle disease virus. *J. Immunol.* **170,** 1814–1821.

Wiley, S. R., Schooley, K., Smolak, P. J., Din, W. S., Huang, C. P., Nicholl, J. K., Sutherland, G. R., Davis-Smith, T., Rauch, C. T., Smith, C. A., and Goodwin, R. G. (1995). Identification and characterization of a new member of the TNF family that induces apoptosis. *Immunity* **3,** 673–682.

Wu, G. S., Burns, T. F., McDonald III, E. R., Jiang, W., Meng, R., Krantz, I. D., Kao, G., Gan, D.-D., Zhou, J.-Y., Muschel, R., Hamilton, S. R., Spinner, N. B., Markowitz, S., Wu, G., and El-Deiry, W. S. (1997). KILLER/DR5 is a DNA damage-inducible p53-regulated death receptor gene. *Nat. Genet.* **17,** 141–143.

Zamai, L., Ahmad, M., Bennett, I. M., Azzoni, L., Alnemri, E. S., and Perussia, B. (1998). Natural killer (NK) cell-mediated cytotoxicity: Differential use of TRAIL and Fas ligand by immature and mature primary human NK cells. *J. Exp. Med.* **188,** 2375–2380.

Zhou, Q., Snipas, S., Orth, K., Muzio, M., Dixit, V. M., and Salvesen, G. S. (1997). Target protease specificity of the viral serpin CrmA. Analysis of five caspases. *J. Biol. Chem.* **272,** 7797–7800.

15

Modulation of TRAIL Signaling for Cancer Therapy

Simone Fulda and Klaus-Michael Debatin

University Children's Hospital, D-89075 Ulm, Germany

I. Introduction
II. The Core Apoptotic Machinery
III. TRAIL and its Receptors
IV. TRAIL Signaling
V. Defective TRAIL Signaling in Cancers
   A. TRAIL Receptors
   B. c-FLIP
   C. Bcl-2 Proteins
   D. Inhibitors of Apoptosis Proteins
   E. NFκB
VI. TRAIL and Cancer Therapy
   A. Safety of TRAIL Administration
   B. Antitumor Activity of TRAIL
   C. Combination Therapy with TRAIL
   D. Tumor Surveillance
VII. Conclusions
    References

Apoptosis, the cell's intrinsic death program, is a key regulator of tissue homeostasis, and an imbalance between cell death and proliferation may result in tumor formation. Also, killing of cancer cells by cytotoxic therapies such as chemotherapy, γ-irradiation or ligation of death receptors is predominantly mediated by triggering apoptosis in target cells. Tumor necrosis factor–related apoptosis-inducing ligans (TRAIL) is a member of the tumor necrosis factor (TNF) superfamily that induces apoptosis upon binding to its receptors. TRAIL is of special interest for cancer therapy, because TRAIL has been shown to predominantly kill cancer cells, while sparing normal cells. Importantly, combined treatment with TRAIL together with chemotherapy or γ-irradiation synergized to achieve antitumor activity in tumor cell lines and also in tumor models *in vivo*. However, failure to undergo apoptosis in response to TRAIL treatment may result in tumor resistance. Understanding the molecular events that regulate TRAIL-induced apoptosis and their deregulation in resistant forms of cancer may provide new opportunities for cancer therapy. Thus, novel strategies targeting tumor cell resistance will be based on further insights into the molecular mechanisms of cell death, e.g., triggered by TRAIL. © 2004 Elsevier Inc.

I. INTRODUCTION

Apoptosis, a distinct, intrinsic cell death program, occurs in various physiologic and pathologic situations and has an important regulatory function in tissue homeostasis and immune regulation (Hengartner, 2000; Thompson, 1995). Apoptosis is characterized by typical morphologic and biochemical hallmarks including cell shrinkage, nuclear DNA fragmentation, and membrane blebbing (Hengartner, 2000). Various stimuli can trigger an apoptosis response, e.g., withdrawal of growth factors or stimulation of cell surface receptors (Hengartner, 2000). Also, killing of tumor cells by diverse cytotoxic strategies such as anticancer drugs, γ-irradiation, or immunotherapy, has been shown to involve induction of apoptosis in target cells (Debatin *et al.*, 2002; Johnstone *et al.*, 2002; Lowe and Lin, 2000). In addition, T cells or NK cells may release cytotoxic compounds such as granzyme B, which can directly initiate apoptosis effector pathways inside the cell (Hengartner, 2000). Proteolytic enzymes called caspases are important effector molecules of different forms of cell death (Thornberry and Lazebnik, 1998). Apoptosis pathways are tightly controlled by a number of inhibitory and promoting factors (Igney and Krammer, 2002). The antiapoptotic mechanisms regulating apoptotic cell death have also been implicated in conferring drug resistance to tumor cells (Johnstone *et al.*, 2002). Importantly, combinations of anticancer agents together with death-inducing ligands have been shown to synergize in triggering apoptosis in cancer cells and may even overcome some forms of drug resistance (Fulda and Debatin, 2003). Further insights into the mechanisms controlling tumor cell death in response to death receptor ligation will provide a molecular basis for novel strategies targeting death pathways in apoptosis-resistant forms of cancer.

II. THE CORE APOPTOTIC MACHINERY

In most cases, anticancer therapies eventually result in activation of caspases, a family of cysteine proteases that act as common death effector molecules in various forms of cell death (Thornberry and Lazebnik, 1998). Caspases are synthesized as inactive proforms and upon activation, they cleave next to aspartate residues. The fact that caspases can activate each other by cleavage at identical sequences results in amplification of caspase activity through a protease cascade. Caspases cleave a number of different substrates in the cytoplasm or nucleus leading to many of the morphologic features of apoptotic cell death (Hengartner, 2000). For example, polynucleosomal DNA fragmentation is mediated by cleavage of ICAD (inhibitor of caspase-activated DNase), the inhibitor of the endonuclease CAD (caspase-activated DNase) that cleaves DNA into the characteristic oligomeric fragments (Nagata, 2000). Likewise, proteolysis of several cytoskeletal proteins such as actin or fodrin leads to loss of overall cell shape, while degradation of lamin results in nuclear shrinking (Hengartner, 2000).

Activation of caspases can be initiated from different angles, e.g., at the plasma membrane upon ligation of death receptor (receptor pathway) or at the mitochondria (mitochondrial pathway) (Fulda and Debatin, 2003). Stimulation of death receptors of the tumor necrosis factor (TNF) receptor superfamily such as CD95 (APO-1/Fas) or TRAIL receptors results in activation of the initiator caspase-8, which can propagate the apoptosis signal by direct cleavage of downstream effector caspases such as caspase-3 (Ashkenazi, 2002). The mitochondrial pathway is initiated by the release of apoptogenic factors such as cytochrome c, apoptosis inducing factor (AIF), Smac/Diablo, Omi/HtrA2, endonuclease G, caspase-2, or caspase-9 from the mitochondrial intermembrane space (van Loo *et al.*, 2002). The release of cytochrome c into the cytosol triggers caspase-3 activation through formation of the cytochrome c/Apaf-1/caspase-9-containing apoptosome complex (Salvesen and Renatus, 2002). Smac/Diablo and Omi/HtrA2 promote caspase activation through neutralizing the inhibitory effects of IAPs (van Loo *et al.*, 2002).

Links between the receptor and the mitochondrial pathway exist at different levels. Upon death receptor triggering, activation of caspase-8 may result in cleavage of Bid, a Bcl-2 family protein with a BH3 domain only, which in turn translocates to mitochondria to release cytochrome c thereby initiating a mitochondrial amplification loop (Cory and Adams, 2002). In addition, cleavage of caspase-6 downstream of mitochondria may feed back to the receptor pathway by cleaving caspase-8 (Cowling and Downward, 2002).

III. TRAIL AND ITS RECEPTORS

TNF-related apoptosis-inducing ligand (TRAIL)/Apo-2L was identified in 1995 based on its sequence homology to other members of the TNF superfamily (Marsters et al., 1996; Wiley et al., 1995). TRAIL is a type II transmembrane protein, the extracellular domain of which can be proteolytically cleaved from the cell surface. TRAIL is constitutively expressed in a wide range of tissues. Comprising five different receptors, the complexity of the TRAIL receptor system is unprecedented. TRAIL-R1 and TRAIL-R2, the two agonistic TRAIL receptors, contain a conserved cytoplasmic death domain motif, which enables them to engage the cell's apoptotic machinery upon ligand binding (Chaudhary et al., 1997; MacFarlane et al., 1997; Pan et al., 1997; Walczak et al., 1997; Wu et al., 1997). TRAIL-R3 to -R5 are antagonistic decoy receptors, which bind TRAIL, but do not transmit a death signal (Degli-Esposti et al., 1997a,b; Marsters et al., 1997; Pan et al., 1997, 1998). TRAIL-R3 is a glycosyl-phosphatidylinositol GPI-anchored cell surface protein, which lacks a cytoplasmic tail, while TRAIL-R4 harbors a substantially truncated cytoplasmic death domain. In addition to these four membrane-associated receptors, osteoprotegerin is a soluble decoy receptor that is involved in regulation of osteoclastogenesis (Emery et al., 1998).

Similar to CD95L, TRAIL rapidly triggers apoptosis in many tumor cells (LeBlanc and Ashkenazi, 2003; Wajant et al., 2002; Walczak and Krammer, 2000). The TRAIL ligand and its receptors are of special interest for cancer therapy because TRAIL has been shown to predominantly kill cancer cells, while sparing normal cells. The underlying mechanisms for the differential sensitivity of malignant versus non-malignant cells for TRAIL have not exactly been defined. One possible mechanism of protection of normal tissues is thought to be based on the set of antagonistic decoy receptors, which compete with TRAIL-R1 and TRAIL-R2 for binding to TRAIL (Ozoren and El-Deiry, 2003). However, screening of various different tumor cell types and normal cells did not reveal a consistent association between TRAIL sensitivity and TRAIL receptor expression. Therefore, susceptibility for TRAIL-induced cytotoxicity has been suggested to be regulated intracellularly by distinct patterns of pro- and antiapoptotic molecules.

IV. TRAIL SIGNALING

Similar to the CD95 system, ligation of the agonistic TRAIL receptors TRAIL-R1 and TRAIL-R2 by TRAIL or agonistic antibodies results in receptor trimerization and clustering of the receptors' death domains (Bodmer et al., 2000; Kischkel et al., 2000; LeBlanc and Ashkenazi, 2003;

Sprick *et al.*, 2000; Walczak and Krammer, 2000). This leads to recruitment of adaptor molecules such as FADD through homophilic interaction mediated by the death domain. FADD in turn recruits caspase-8 to the activated TRAIL receptor complex. Oligomerization of caspase-8 at the activated TRAIL receptor complex drives its activation through self-cleavage. Caspase-8 then activates downstream effector caspases such as caspase-3. Also, activated caspase-8 can cleave Bid, which then translocates to mitochondria to induce cytochrome *c* release. In addition to caspase-8, caspase-10 is recruited to the TRAIL DISC (Kischkel *et al.*, 2001). However, the importance of caspase-10 in the TRAIL DISC for apoptosis induction has been discussed (Kischkel *et al.*, 2001; Sprick *et al.*, 2002).

V. DEFECTIVE TRAIL SIGNALING IN CANCERS

Signaling by agonistic TRAIL receptors is regulated at various levels along the signaling pathway. Importantly, cancer cells have evolved numerous ways to evade induction of apoptosis triggered by the death ligand TRAIL resulting in TRAIL resistance as outlined subsequently.

A. TRAIL RECEPTORS

For example, loss of expression of the agonistic TRAIL receptors TRAIL-R1 and R2 may account for TRAIL resistance. Both receptors are located on chromosome 8p, a region of frequent loss of heterozygosity (LOH) in tumors (LeBlanc and Ashkenazi, 2003). In a small percentage of cancers, e.g., non-Hodgkin's lymphoma, colorectal, breast, head and neck or lung carcinoma, deletions or mutations were found, which resulted in loss of both copies of TRAIL-R1 or R2 (Arai *et al.*, 1998; Lee *et al.*, 1999, 2001; Pai *et al.*, 1998; Shin *et al.*, 2001). In addition, loss of TRAIL-R1 or R-2 expression may be caused by epigenetic alterations such as promotor hypermethylation, e.g., in neuroblastoma (van Noesel *et al.*, 2002).

B. c-FLIP

TRAIL signaling can also be negatively influenced by intracellular proteins that associate with the cytoplasmatic domain of TRAIL receptors such as c-FLIP (Krueger *et al.*, 2001). c-FLIP has homology to caspase-8 and caspase-10, but lacks protease activity. By binding to FADD, c-FLIP prevents the interaction between the adaptor molecule FADD and pro-caspase-8. c-FLIP exists as a long (c-FLIP$_L$) and a short isoform (c-FLIP$_s$), both of which can inhibit death receptor–induced apoptosis. Interestingly, c-FLIP$_s$ was identified in a screen for genes that could confer resistance to TRAIL-induced apoptosis (Burns and El-Deiry, 2001). High c-FLIP expression has been detected in many tumors and has been correlated with resistance to TRAIL-induced apoptosis, e.g., in melanoma. However, a consistent correlation between c-FLIP expression and TRAIL resistance has not always been found (Hersey and Zhang, 2001). Importantly, down-regulating of FLIP expression by inhibition of protein synthesis or RNA translation, cytotoxic drugs, proteasome inhibitors, PPAR$_\gamma$ ligand or siRNAI for c-FLIP restored sensitivity for TRAIL-induced apoptosis in several cell types (Fulda *et al.*, 2000; Kim *et al.*, 2002; Siegmund *et al.*, 2002).

C. BCL-2 PROTEINS

Bcl-2 family proteins play an important role in the regulation of the mitochondrial pathway, because these proteins localize to intracellular membranes such as the mitochondrial membrane (Cory and Adams, 2002). They comprise both antiapoptotic members, e.g., Bcl-2 or Bcl-X$_L$, as well as proapoptotic molecules such as Bax or Bid. Altered expression of Bcl-2 family proteins have been reported in various human cancers. Imbalances in the ratio of anti- and proapoptotic Bcl-2 proteins may favor tumor cell survival instead of cell death. Overexpression of Bcl-2 or Bcl-X$_L$ blocked TRAIL-triggered apoptosis in many tumor cell lines, e.g., prostate carcinoma, pancreatic carcinoma, or neuroblastoma or glioblastoma cells (Fulda and Debatin 2002; Hinz *et al.*, 2000; Munshi *et al.*, 2001). In addition, gene ablation studies showed that Bax was absolutely required for TRAIL-induced apoptosis in colon carcinoma cells (Burns and El-Deiry, 2001; Deng *et al.*, 2002). However, overexpression of Bcl-2 or Bcl-X$_L$ did not interfere with TRAIL-induced apoptosis in some cell types, e.g., Jurkat or CEM T cell leukemia cells (Walczak *et al.*, 2000). Thus, the contribution of the mitochondrial pathway to TRAIL-induced apoptosis may depend on the cell type.

D. INHIBITORS OF APOPTOSIS PROTEINS

The family of endogenous caspase inhibitors (inhibitor of apoptosis proteins [IAPs]) are highly conserved throughout evolution and comprise the human analogues XIAP, cIAP1, cIAP2, survivin, and livin (ML-IAP) (Salvesen and Duckett, 2002). IAPs have been reported to directly inhibit active caspase-3 and -7 and to block caspase-9 activation. In addition to regulation of apoptosis, IAP members such as survivin are involved in the regulation of mitosis (Altieri, 2003). The activity of IAPs are controlled at various levels, e.g., by the transcription factor NFκB that has been reported to stimulate expression of cIAP1, cIAP and XIAP (Salvesen and Duckett, 2002). In addition, Smac/Diablo and Omi, which are released from mitochondria upon apoptosis induction, neutralize IAPs through binding to

IAPs thereby displacing them from their caspase partners (van Loo et al., 2002). Increased IAPs expression is detected in many tumors and has been correlated with adverse prognosis (Hersey and Zhang, 2001; Salvesen and Duckett, 2002). Importantly, overexpression of XIAP blocked TRAIL-triggered apoptosis in several cellular systems (Fulda et al., 2002a; Ng and Bonavida, 2002).

E. NFκB

The transcription factor NFκB has been connected with multiple aspects of oncogenesis, including cell proliferation or inhibition of apoptosis (Karin et al., 2002). NFκB is composed of heterodimers or homodimers of NFκB/Rel family of proteins, which mediate protein dimerization, nuclear import, and specific DNA binding. In most cell types, NFκB is sequestered in the cytoplasm by its interaction with IκB proteins and therefore remains inactive. Upon stimulation, IκB becomes phosphorylated following activation of the IKK complex and is degraded via the proteasome thereby releasing NFκB to translocate into the nucleus for transcription of target genes. NFκB target genes include several antiapoptotic proteins, e.g., cIAP1, cIAP2, TRAF1, TRAF2, Bfl-1/A1, Bcl-X_L, and FLIP. Interestingly, promotor activation of certain proapoptotic molecules of the TRAIL system including TRAIL-R1, TRAIL-R2 or TRAIL is also controlled by NFκB consistent with reports that NFκB can promote apoptosis under certain circumstances (Baetu et al., 2001; Ravi et al., 2001). NFκB is constitutively active in certain tumor types such as Hodgkin's lymphoma or pancreatic carcinoma (Karin et al., 2002). Also, NFκB activity is induced in response to a variety of stimuli, e.g., in response to cellular stress or anticancer agents. In addition, TRAIL can activate NFκB, which is mediated by TRAIL-R1, TRAIL-R2, or TRAIL-R4 through a TRAF2-NIK-IKKα/β–dependent signaling cascade (Chaudhary et al., 1997; Schneider et al., 1997). Interestingly, TRAIL has been reported to activate NFκB especially under conditions when apoptosis is blocked (Erhardt et al., 2003; Harper et al., 2001). Inhibition of NFκB signaling, e.g., by proteasome inhibitors, which prevent IκBα degradation, or by overexpresison of non-degradable IκBα mutants, has been reported to sensitize tumor cells for TRAIL-induced apoptosis, at least in some contexts (Jeremias et al., 1998).

VI. TRAIL AND CANCER THERAPY

A. SAFETY OF TRAIL ADMINISTRATION

The idea to specifically target death receptors to trigger apoptosis in tumor cells is attractive for cancer therapy because death receptors have a direct link to the cell's death machinery (Ashkenazi, 2002). In addition, apoptosis upon death receptor ligation has been reported to occur independent of the p53 tumor suppressor gene, which is deleted or inactivated in more than half of human tumors (El-Deiry, 2001). However, the clinical application of CD95 ligand or TNFα is hampered by severe toxic side effects (Walczak and Krammer, 2000). Systemic administration of TNFα or CD95 ligand causes a severe inflammatory response syndrome or massive liver cell apoptosis. In contrast, TRAIL appears to be a relatively safe and promising candidate for clinical application, particularly in its nontagged, zinc-bound homotrimeric form (LeBlanc and Ashkenazi, 2003). Studies in nonhuman primates such as chimpanzees and cynomolgus monkeys showed no toxicity upon intravenous infusion, even at high doses (Ashkenazi et al., 1999). In addition, no cytotoxic activity of TRAIL was reported on a variety of normal human cells of different lineages including fibroblasts, endothelial cells, smooth muscle cells, epithelial cells, or astrocytes (Lawrence et al., 2001). However, some concerns about potential toxic side effects on human hepatocytes or brain tissue have also been raised (Jo et al., 2000; Nitsch et al., 2000). The loss of tumor selectivity may be related to the TRAIL preparations used in these studies. TRAIL preparations, which are antibody-crosslinked or not optimized for Zn content, have been reported to form multimeric aggregates thereby overpassing the threshold of sensitivity of normal cells (LeBlanc and Ashkenazi, 2003).

B. ANTITUMOR ACTIVITY OF TRAIL

Recombinant soluble TRAIL induced apoptosis in a broad spectrum of cancer cell lines, including colon carcinoma, breast carcinoma, lung carcinoma, pancreatic carcinoma, prostate carcinoma, renal carcinoma, thyroid carcinoma, malignant brain tumors, Ewing tumor, osteasarcoma, neuroblastoma, leukemia, and lymphoma (LeBlanc and Ashkenazi, 2003; Wajant et al., 2002). Also, TRAIL exhibited potent tumoricidal activity *in vivo* in several xenograft models of colon carcinoma, breast carcinoma, malignant glioma, or multiple myeloma. Furthermore, monoclonal antibodies that engage the TRAIL receptors DR4 or DR5 also demonstrated potent antitumor activity against tumor cell lines and in preclinical cancer models (Chuntharapai et al., 2001; Ichikawa et al., 2001).

C. COMBINATION THERAPY WITH TRAIL

Although these studies provided ample evidence of the potential of TRAIL for cancer therapy, many tumors remain refractory toward treatment with TRAIL, which has been related to the dominance of antiapoptotic signals. Importantly, numerous studies have shown that TRAIL synergized together with cytotoxic drugs or γ-irradiation to achieve antitumor activity in various cancers including malignant glioma, melanoma, leukemia, breast, colon, or prostate carcinoma (Belka et al., 2001; Chinnaiyan et al., 2000; Gliniak and Le, 1999; Keane et al., 2000; Nagane et al., 2000; Rohn et al., 2001). Remarkably, TRAIL and anticancer agents also cooperated to suppress tumor growth in different mouse models of human cancers. The molecular mechanisms, which account for this synergistic interaction, may include transcriptional upregulation of the agonistic TRAIL receptors TRAIL-R1 and -R2, which occurred in a p53-dependent or p53-independent manner (Meng and El-Deiry, 2001; Takimoto and El-Deiry, 2000). Of note, p53 has also been shown to transcriptionally activate the antagonistic TRAIL receptors TRAIL-R3 and -R4 (Meng et al., 2000). The synergistic interaction between cytotoxic drugs and TRAIL may also be mediated by downregulation of antiapoptotic proteins such as Bcl-2, Bcl-X_L, or FLIP upon drug treatment (Olsson et al., 2001). In addition, anticancer agents may sensitize tumor cells for TRAIL treatment by upregulating proapoptotic molecules incuding caspases or FADD (Micheau et al., 1999). Caspase-8 expression is frequently impaired by hypermethylation in several tumors including neuroblastoma, Ewing tumors, malignant brain tumors or lung carcinoma (Fulda et al., 2001; Teitz et al., 2000). Importantly, restoration of caspase-8 expression by gene transfer or by demethylation treatment sensitized resistant tumor cells for TRAIL-induced apoptosis (Fulda et al., 2001). Also, biological response modifiers such as IFNγ strongly enhanced the cytotoxic activity of TRAIL by upregulating caspase-8 expression in a STAT-1 dependent manner (Fulda and Debatin, 2002).

Because TRAIL demonstrated synergistic interaction with anticancer agents or irradiation, TRAIL may be most effective in combination with conventional cancer treatments. In addition, small molecules may serve as molecular therapeutics to specifically target tumor cell resistance toward TRAIL in resistant forms of cancer. To this end, Smac agonists have recently been reported to potentiate the efficacy of TRAIL treatment by antagonizing the inhibitory effect of IAPs, which are overexpressed in many tumors (Fulda et al., 2002b). Importantly, Smac peptides synergized with TRAIL to eradicate malignant glioma in an orthotopic mouse model without any detectable toxicities to the normal brain tissue (Fulda et al., 2002b).

Most studies investigating the TRAIL sensitivity of tumor cells have so far been performed *in vitro* (Wajant et al., 2002). Despite many limitations of cell culture experiments, *in vitro* studies may reflect some of the important features of tumor treatment *in vivo*. Importantly, experiments performed on primary cancer cells *ex vivo* in most cases yielded similar results. For example, combined treatment of primary cancer cells with TRAIL together with anticancer agents synergized in antitumor activity.

D. TUMOR SURVEILLANCE

There is also mounting evidence for an important role of TRAIL in tumor surveillance, e.g., from studies with TRAIL knockout mice (Cretney *et al.*, 2002; Smyth *et al.*, 2003). Although the biology of the TRAIL system may differ significantly between mice and humans, because there is only one TRAIL receptor in mice, which is homologous to both TRAIL-R1 and TRAIL-R2, the phenotype of these knockout mice are informative with respect to the physiologic function of TRAIL *in vivo*. Importantly, TRAIL-deficient mice were more susceptible to tumor metastasis than wild-type mice (Cretney *et al.*, 2002). These data are in accordance with studies showing an important role of NK cells, which constitutively express TRAIL, in the control of tumor metastasis (Smyth *et al.*, 2001; Takeda *et al.*, 2001). In addition, tumor formation induced by carcinogens was found to be enhanced in the presence of antagonistic TRAIL antibodies (Takeda *et al.*, 2001). Thus, TRAIL may play an essential role as innate effector molecule in immune surveillance during tumor formation and progression.

VII. CONCLUSIONS

Key elements of the TRAIL signaling pathway have been identified and numerous studies provided substantial insights into the molecular mechanisms regulating TRAIL-induced apoptosis. The death ligand TRAIL is of special interest for cancer therapy because of its differential toxicity towards transformed versus normal cells. Importantly, TRAIL strongly synergized together with chemotherapy or irradiation in antitumor activity, even against some resistant forms of cancer.

However, several points remain to be addressed in future studies. Because the tumor selective toxicity of TRAIL is presently not fully understood, further preclinical safety testing are necessary to assess the potential toxicity of TRAIL or agonistic TRAIL receptor antibodies on nonmalignant tissues. In addition, the possible toxicity of TRAIL on normal tissues under conditions of combined treatment with anticancer agents or irradiation remains to be determined. Also, the molecular mechanisms leading to activation of survival pathways such as NFκB or AKT upon treatment with TRAIL and their significance for cancer therapy are only partially understood. Moreover, the promise of TRAIL alone or in combination protocols remains to be tested in clinical settings. Future studies on the role of TRAIL in individual tumors both *in vitro* and *in vivo* in tumor cells of patients under chemotherapy, e.g., by DNA microarrays or proteomic studies, may provide the basis for "tailored" tumor therapy with TRAIL and may identify new targets for therapeutic interventions.

REFERENCES

Altieri, D. C. (2003). Validating survivin as a cancer therapeutic target. (Review) (126 refs). *Nature reviews. Cancer* **3**(1), 46–54.

Arai, T., Akiyama, Y., Okabe, S., Saito, K., Iwai, T., and Yuasa, Y. (1998). Genomic organization and mutation analyses of the DR5/TRAIL receptor 2 gene in colorectal carcinomas. *Cancer Lett.* **133**(2), 197–204.

Ashkenazi, A. (2002). Targeting death and decoy receptors of the tumour-necrosis factor superfamily. (Review) (141 refs). *Nature Reviews. Cancer* **2**(6), 420–430.

Ashkenazi, A., Pai, R. C., Fong, S., Leung, S., Lawrence, D. A., Marsters, S. A., Blackie, C., Chang, L., McMurtrey, A. E., Hebert, A., DeForge, L., Koumenis, I. L., Lewis, D., Harris, L., Bussiere, J., Koeppen, H., Shahrokh, Z., and Schwall, R. H. (1999). Safety and antitumor activity of recombinant soluble Apo2 ligand. *J. Clin. Invest.* **104**(2), 155–162.

Baetu, T. M., Kwon, H., Sharma, S., Grandvaux, N., and Hiscott, J. (2001). Disruption of NF-kappaB signaling reveals a novel role for NF-kappaB in the regulation of TNF-related apoptosis-inducing ligand expression. *J. Immunol.* **167**(6), 3164–3173.

Belka, C., Schmid, B., Marini, P., Durand, E., Rudner, J., Faltin, H., Bamberg, M., Schulze-Osthoff, K., and Budach, W. (2001). Sensitization of resistant lymphoma cells to irradiation-induced apoptosis by the death ligand TRAIL. *Oncogene* **20**(17), 2190–2196.

Bodmer, J. L., Holler, N., Reynard, S., Vinciguerra, P., Schneider, P., Juo, P., Blenis, J., and Tschopp, J. (2000). TRAIL receptor-2 signals apoptosis through FADD and caspase-8. *Nat. Cell Biol.* **2**(4), 241–243.

Burns, T. F., and El-Deiry, W. S. (2001). Identification of inhibitors of TRAIL-induced death (ITIDs) in the TRAIL sensitive colon carcinoma cell line SW480 using a genetic approach. *J. Biol. Chem.* **276**, 37879–37886.

Chaudhary, P. M., Eby, M., Jasmin, A., Bookwalter, A., Murray, J., and Hood, L. (1997). Death receptor 5, a new member of the TNFR family, and DR4 induce FADD-dependent apoptosis and activate the NF-kappaB pathway. *Immunity* **7**(6), 821–830.

Chinnaiyan, A. M., Prasad, U., Shankar, S., Hamstra, D. A., Shanaiah, M., Chenevert, T. L., Ross, B. D., and Rehemtulla, A. (2000). Combined effect of tumor necrosis factor-related apoptosis-inducing ligand and ionizing radiation in breast cancer therapy. *Proceed. Nat. Acad. Sci. USA* **97**(4), 1754–1759.

Chuntharapai, A., Dodge, K., Grimmer, K., Schroeder, K., Marsters, S. A., Koeppen, H., Ashkenazi, A., and Kim, K. J. (2001). Isotype-dependent inhibition of tumor growth in vivo by monoclonal antibodies to death receptor 4. *J. Immunol.* **166**(8), 4891–4898.

Cory, S., and Adams, J. M. (2002). The Bcl2 family: Regulators of the cellular life-or-death switch. (Review) (154 refs). *Nature Reviews. Cancer* **2**(9), 647–656.

Cowling, V., and Downward, J. (2002). Caspase-6 is the direct activator of caspase-8 in the cytochrome c-induced apoptosis pathway: Absolute requirement for removal of caspase-6 prodomain. *Cell Death Differ.* **9**(10), 1046–1056.

Cretney, E., Takeda, K., Yagita, H., Glaccum, M., Peschon, J. J., and Smyth, M. J. (2002). Increased susceptibility to tumor initiation and metastasis in TNF-related apoptosis-inducing ligand-deficient mice. *J. Immunol.* **168**(3), 1356–1361.

Debatin, K. M., Poncet, D., and Kroemer, G. (2002). Chemotherapy: Targeting the mitochondrial cell death pathway. (Review) (186 refs). *Oncogene* **21**(57), 8786–8803.

Degli-Esposti, M. A., Dougall, W. C., Smolak, P. J., Waugh, J. Y., Smith, C. A., and Goodwin, R. G. (1997a). The novel receptor TRAIL-R4 induces NF-kappaB and protects against TRAIL-mediated apoptosis, yet retains an incomplete death domain. *Immunity* **7**(6), 813–820.

Degli-Esposti, M. A., Smolak, P. J., Walczak, H., Waugh, J., Huang, C. P., DuBose, R. F., Goodwin, R. G., and Smith, C. A. (1997b). Cloning and characterization of TRAIL-R3, a novel member of the emerging TRAIL receptor family. *J. Exp. Med.* **186**(7), 1165–1170.

Deng, Y., Lin, Y., and Wu, X. (2002). TRAIL-induced apoptosis requires Bax-dependent mitochondrial release of Smac/DIABLO. *Genes Dev.* **16**, 33–45.

El-Deiry, W. S. (2001). Insights into cancer therapeutic design based on p53 and TRAIL receptor signaling. (Review) (105 refs). *Cell Death Differ.* **8**(11), 1066–1075.

Emery, J. G., McDonnell, P., Burke, M. B., Deen, K. C., Lyn, S., Silverman, C., Dul, E., Appelbaum, E. R., Eichman, C., DiPrinzio, R., Dodds, R. A., James, I. E., Rosenberg, M., Lee, J. C., and Young, P. R. (1998). Osteoprotegerin is a receptor for the cytotoxic ligand TRAIL. *J. Biol. Chem.* **273**(23), 14363–14367.

Erhardt, H., Fulda, S., Schmid, I., Hiscott, J., Debatin, K. M., and Jeremias, I. (2003). TRAIL induced survival and proliferation in cancer cells resistant towards TRAIL-induced apoptosis mediated by NF-kappaB. *Oncogene* **22**(25), 3842–3852.

Fulda, S., and Debatin, K. M. (2002). IFNgamma sensitizes for apoptosis by upregulating caspase-8 expression through the Stat1 pathway. *Oncogene* **21**(15), 2295–2308.

Fulda, S., and Debatin, K. M. (2003). Death receptor signaling in cancer therapy. *Curr. Med. Chem. Anti-Canc. Agents* **1**(4), 253–262.

Fulda, S., Kufer, M. U., Meyer, E., van Valen, F., Dockhorn-Dworniczak, B., and Debatin, K. M. (2001). Sensitization for death receptor- or drug-induced apoptosis by re-expression of caspase-8 through demethylation or gene transfer. *Oncogene* **20**(41), 5865–5877.

Fulda, S., Meyer, E., and Debatin, K. M. (2000). Metabolic inhibitors sensitize for CD95 (APO-1/Fas)-induced apoptosis by down-regulating Fas-associated death domain-like interleukin 1-converting enzyme inhibitory protein expression. *Cancer Res.* **60**(14), 3947–3956.

Fulda, S., Meyer, E., and Debatin, K. M. (2002a). Inhibition of TRAIL-induced apoptosis by Bcl-2 overexpression. *Oncogene* **21**(15), 2283–2294.

Fulda, S., Wick, W., Weller, M., and Debatin, K. M. (2002b). Smac agonists sensitize for Apo2L/TRAIL- or anticancer drug-induced apoptosis and induce regression of malignant glioma in vivo. *Nat. Med.* **8**(8), 808–815.

Gliniak, B., and Le, T. (1999). Tumor necrosis factor-related apoptosis-inducing ligand's antitumor activity in vivo is enhanced by the chemotherapeutic agent CPT-11. *Cancer Res.* **59**(24), 6153–6158.

Harper, N., Farrow, S. N., Kaptein, A., Cohen, G. M., and MacFarlane, M. (2001). Modulation of tumor necrosis factor apoptosis-inducing ligand-induced NF-kappaB activation by inhibition of apical caspases. *J. Biol. Chem.* **276**(37), 34743–34752.

Hengartner, M. O. (2000). The biochemistry of apoptosis. (Comment) (Review) (75 refs). *Nature* **407**(6805), 770–776.

Hersey, P., and Zhang, X. D. (2001). How melanoma cells evade trail-induced apoptosis. (Review) (101 refs). Nature reviews. *Cancer* **1**(2), 142–150.

Hinz, S., Trauzold, A., Boenicke, L., Sandberg, C., Beckmann, S., Bayer, E., Walczak, H., Kalthoff, H., and Ungefroren, H. (2000). Bcl-XL protects pancreatic adenocarcinoma cells against CD95- and TRAIL-receptor-mediated apoptosis. *Oncogene* **19**(48), 5477–5486.

Ichikawa, K., Liu, W., Zhao, L., Wang, Z., Liu, D., Ohtsuka, T., Zhang, H., Mountz, J. D., Koopman, W. J., Kimberly, R. P., and Zhou, T. (2001). Tumoricidal activity of a novel anti-human DR5 monoclonal antibody without hepatocyte cytotoxicity. *Nat. Med.* **7**(8), 954–960.

Igney, F. H., and Krammer, P. H. (2002). Immune escape of tumors: Apoptosis resistance and tumor counterattack. (Review) (260 refs). *J. Leukocyte Biol.* **71**(6), 907–920.

Jeremias, I., Kupatt, C., Baumann, B., Herr, I., Wirth, T., and Debatin, K. M. (1998). Inhibition of nuclear factor kappaB activation attenuates apoptosis resistance in lymphoid cells. *Blood* **91**(12), 4624–4631.

Jo, M., Kim, T. H., Seol, D. W., Esplen, J. E., Dorko, K., Billiar, T. R., and Strom, S. C. (2000). Apoptosis induced in normal human hepatocytes by tumor necrosis factor-related apoptosis-inducing ligand. (Comment). *Nat. Med.* **6**(5), 564–567.

Johnstone, R. W., Ruefli, A. A., and Lowe, S. W. (2002). Apoptosis: A link between cancer genetics and chemotherapy. (Review) (101 refs). *Cell* **108**(2), 153–164.

Karin, M., Cao, Y., Greten, F. R., and Li, Z. W. (2002). NF-kappaB in cancer: From innocent bystander to major culprit. (Review) (104 refs). Nature reviews. *Cancer* **2**(4), 301–310.

Keane, M. M., Rubinstein, Y., Cuello, M., Ettenberg, S. A., Banerjee, P., Nau, M. M., and Lipkowitz, S. (2000). Inhibition of NF-kappaB activity enhances TRAIL mediated apoptosis in breast cancer cell lines. *Breast Cancer Res. Treat.* **64**(2), 211–219.

Kim, Y., Suh, N., Sporn, M., and Reed, J. C. (2002). An inducible pathway for degradation of FLIP protein sensitizes tumor cells to TRAIL-induced apoptosis. *J. Biol. Chem.* **277**(25), 22320–22329.

Kischkel, F. C., Lawrence, D. A., Chuntharapai, A., Schow, P., Kim, K. J., and Ashkenazi, A. (2000). Apo2L/TRAIL-dependent recruitment of endogenous FADD and caspase-8 to death receptors 4 and 5. *Immunity* **12**(6), 611–620.

Kischkel, F. C., Lawrence, D. A., Tinel, A., LeBlanc, H., Virmani, A., Schow, P., Gazdar, A., Blenis, J., Arnott, D., and Ashkenazi, A. (2001). Death receptor recruitment of endogenous caspase-10 and apoptosis initiation in the absence of caspase-8. *J. Biol. Chem.* **276**(49), 46639–46646.

Krueger, A., Baumann, S., Krammer, P. H., and Kirchhoff, S. (2001). FLICE-inhibitory proteins: Regulators of death receptor-mediated apoptosis. (Review) (89 refs). *Mol. Cell. Biol.* **21**(24), 8247–82454.

Lawrence, D., Shahrokh, Z., Marsters, S., Achilles, K., Shih, D., Mounho, B., Hillan, K., Totpal, K., DeForge, L., Schow, P., Hooley, J., Sherwood, S., Pai, R., Leung, S., Khan, L., Gliniak, B., Bussiere, J., Smith, C. A., Strom, S. S., Kelley, S., Fox, J. A., Thomas, D., and Ashkenazi, A. (2001). Differential hepatocyte toxicity of recombinant Apo2L/TRAIL versions. *Nat. Med.* **7**(4), 383–385.

LeBlanc, H. N., and Ashkenazi, A. (2003). Apo-2L/TRAIL and its death and decoy rescepters. *Cell Death Differ.* **10**, 66–75.

Lee, S. H., Shin, M. S., Kim, H. S., Lee, H. K., Park, W. S., Kim, S. Y., Lee, J. H., Han, S. Y., Park, J. Y., Oh, R. R., Jang, J. J., Han, J. Y., Lee, J. Y., and Yoo, N. J. (1999). Alterations of the DR5/TRAIL receptor 2 gene in non-small cell lung cancers. *Cancer Res.* **59**(22), 5683–5686.

Lee, S. H., Shin, M. S., Kim, H. S., Lee, H. K., Park, W. S., Kim, S. Y., Lee, J. H., Han, S. Y., Park, J. Y., Oh, R. R., Kang, C. S., Kim, K. M., Jang, J. J., Nam, S. W., Lee, J. Y., and Yoo, N. J. (2001). Somatic mutations of TRAIL-receptor 1 and TRAIL-receptor 2 genes in non-Hodgkin's lymphoma. *Oncogene* **20**(3), 399–403.

Lowe, S. W., and Lin, A. W. (2000). Apoptosis in cancer. (Review) (169 refs). *Carcinogenesis* **21**(3), 485–495.

MacFarlane, M., Ahmad, M., Srinivasula, S. M., Fernandes-Alnemri, T., Cohen, G. M., and Alnemri, E. S. (1997). Identification and molecular cloning of two novel receptors for the cytotoxic ligand TRAIL. *J. Biol. Chem.* **272**(41), 25417–25420.

Marsters, S. A., Pitti, R. M., Donahue, C. J., Ruppert, S., Bauer, K. D., and Ashkenazi, A. (1996). Activation of apoptosis by Apo-2 ligand is independent of FADD but blocked by CrmA. death and decoy receptors. (Review) (66 refs). *Curr. Biol.* **6**, 750–752.

Marsters, S. A., Sheridan, J. P., Pitti, R. M., Huang, A., Skubatch, M., Baldwin, D., Yuan, J., Gurney, A., Goddard, A. D., Godowski, P., and Ashkenazi, A. (1997). A novel receptor for Apo2L/TRAIL contains a truncated death domain. *Curr. Biol.* **7**(12), 1003–1006.

Meng, R. D., and El-Deiry, W. S. (2001). p53-independent upregulation of KILLER/DR5 TRAIL receptor expression by glucocorticoids and interferon-gamma. *Exp. Cell Res.* **262**(2), 154–169.

Meng, R. D., McDonald, E. R. 3rd, Sheikh, M. S., Fornace, A. J., Jr., and El-Deiry, W. S. (2000). The TRAIL decoy receptor TRUNDD (DcR2, TRAIL-R4) is induced by adenovirus-p53 overexpression and can delay TRAIL-, p53-, and KILLER/DR5-dependent colon cancer apoptosis. *Mol. Ther. J. Am. Soc. Gene Ther.* **1**(2), 130–144.

Micheau, O., Hammann, A., Solary, E., and Dimanche-boitrel, M. T. (1999). Stat-1-independent upregulation of FADD and procaspase-3 and -8 in cancer cells treated with cytotoxic drugs. *Biochem. Biophys. Res. Commun.* **256**, 603–611.

Munshi, A., Pappas, G., Honda, T., McDonnell, T. J., Younes, A., Li, Y., and Meyn, R. E. (2001). TRAIL (APO-2L) induces apoptosis in human prostate cancer cells that is inhibitable by Bcl-2. *Oncogene* **20**(29), 3757–3765.

Nagane, M., Pan, G., Weddle, J. J., Dixit, V. M., Cavenee, W. K., and Huang, H. J. (2000). Increased death receptor 5 expression by chemotherapeutic agents in human gliomas causes synergistic cytotoxicity with tumor necrosis factor-related apoptosis-inducing ligand in vitro and in vivo. *Cancer Res.* **60**(4), 847–853.

Nagata, S. (2000). Apoptotic DNA fragmentation. (Review) (56 refs). *Exp. Cell Res.* **256**(1), 12–18.

Ng, C. P., and Bonavida, B. (2002). X-linked inhibitor of apoptosis (XIAP) blocks Apo2 ligand/tumor necrosis factor-related apoptosis-inducing ligand-mediated apoptosis of prostate cancer cells in the presence of mitochondrial activation: Sensitization by overexpression of second mitochondria-derived activator of caspase/direct IAP-binding protein with low pI (Smac/DIABLO). *Mol. Cancer Ther.* **1**(12), 1051–1058.

Nitsch, R., Bechmann, I., Deisz, R. A., Haas, D., Lehmann, T. N., Wendling, U., and Zipp, F. (2000). Human brain-cell death induced by tumour-necrosis-factor-related apoptosis-inducing ligand (TRAIL). (Comment). *Lancet* **356**(9232), 827–828.

Olsson, A., Diaz, T., and Aguilar-Santelises, M. (2001). Sensitization to TRAIL-induced apoptosis and modulation of FLICE-inhibitory protein in B chronic lymphocytic leukemia by actinomycin D. *Leukemia* **15**, 1868–1877.

Ozoren, N., and El-Deiry, W. S. (2003). Cell surface death receptor signaling in normal and cancer cells. *Semin. Cancer Biol.* **13**(2), 135–147.

Pai, S. I., Wu, G. S., Ozoren, N., Wu, L., Jen, J., Sidransky, D., and El-Deiry, W. S. (1998). Rare loss-of-function mutation of a death receptor gene in head and neck cancer. *Cancer Res.* **58**(16), 3513–3518.

Pan, G., Ni, J., Wei, Y. F., Yu, G., Gentz, R., and Dixit, V. M. (1997). An antagonist decoy receptor and a death domain-containing receptor for TRAIL. (Comment). *Science* **277**(5327), 815–818.

Pan, G., Ni, J., Yu, G., Wei, Y. F., and Dixit, V. M. (1998). TRUNDD, a new member of the TRAIL receptor family that antagonizes TRAIL signalling. *FEBS Lett.* **424**(1–2), 41–45.

Pan, G., O'Rourke, K., Chinnaiyan, A. M., Gentz, R., Ebner, R., Ni, J., and Dixit, V. M. (1997). The receptor for the cytotoxic ligand TRAIL. *Science* **276**(5309), 111–113.

Ravi, R., Bedi, G. C., Engstrom, L. W., Zeng, Q., Mookerjee, B., Gelinas, C., Fuchs, E. J., and Bedi, A. (2001). Regulation of death receptor expression and TRAIL/Apo2L-induced apoptosis by NF-kappaB. *Nat. Cell Biol.* **3**(4), 409–416.

Rohn, T. A., Wagenknecht, B., Roth, W., Naumann, U., Gulbins, E., Krammer, P. H., Walczak, H., and Weller, M. (2001). CCNU-dependent potentiation of TRAIL/Apo2L-induced apoptosis in human glioma cells is p53-independent but may involve enhanced cytochrome c release. *Oncogene* **20**(31), 4128–4137.

Salvesen, G. S., and Duckett, C. S. (2002). IAP proteins: Blocking the road to death's door. (Review) (154 refs). *Nat. Rev. Mol. Cell. Biol.* **3**(6), 401–410.

Salvesen, G. S., and Renatus, M. (2002). Apoptosome: The seven-spoked death machine. (Review) (9 refs). *Dev. Cell* **2**(3), 256–257.

Schneider, P., Thome, M., Burns, K., Bodmer, J. L., Hofmann, K., Kataoka, T., Holler, N., and Tschopp, J. (1997). TRAIL receptors 1 (DR4) and 2 (DR5) signal FADD-dependent apoptosis and activate NF-kappaB. *Immunity* **7**(6), 831–836.

Shin, M. S., Kim, H. S., Lee, S. H., Park, W. S., Kim, S. Y., Park, J. Y., Lee, J. H., Lee, S. K., Lee, S. N., Jung, S. S., Han, J. Y., Kim, H., Lee, J. Y., and Yoo, N. J. (2001). Mutations of tumor necrosis factor-related apoptosis-inducing ligand receptor 1 (TRAIL-R1) and receptor 2 (TRAIL-R2) genes in metastatic breast cancers. *Cancer Res.* **61**(13), 4942–4946.

Siegmund, D., Hadwiger, P., Pfizenmaier, K., Vornlocher, H. P., and Wajant, H. (2002). Selective inhibition of FLICE-like inhibitory protein expression with small interfering RNA oligonucleotides is sufficient to sensitize tumor cells for TRAIL-induced apoptosis. *Mol. Med.* **8**(11), 725–732.

Smyth, M. J., Cretney, E., Takeda, K., Wiltrout, R. H., Sedger, L. M., Kayagaki, N., Yagita, H., and Okumura, K. (2001). Tumor necrosis factor-related apoptosis-inducing ligand (TRAIL) contributes to interferon gamma-dependent natural killer cell protection from tumor metastasis. *J. Exp. Med.* **193**(6), 661–670.

Smyth, M. J., Takeda, K., Hayakawa, Y., Peschon, J. J., van den Brink, M. R., and Yagita, H. (2003). Nature's TRAIL—on a path to cancer immunotherapy. (Review) (37 refs). *Immunity* **18**(1), 1–6.

Sprick, M. R., Rieser, E., Stahl, H., Grosse-Wilde, A., Weigand, M. A., and Walczak, H. (2002). Caspase-10 is recruited to and activated at the native TRAIL and CD95 death-inducing signalling complexes in a FADD-dependent manner but can not functionally substitute caspase-8. *EMBO J.* **21**(17), 4520–4530.

Sprick, M. R., Weigand, M. A., Rieser, E., Rauch, C. T., Juo, P., Blenis, J., Krammer, P. H., and Walczak, H. (2000). FADD/MORT1 and caspase-8 are recruited to TRAIL receptors 1 and 2 and are essential for apoptosis mediated by TRAIL receptor 2. *Immunity* **12**(6), 599–609.

Takeda, K., Hayakawa, Y., Smyth, M. J., Kayagaki, N., Yamaguchi, N., Kakuta, S., Iwakura, Y., Yagita, H., and Okumura, K. (2001). Involvement of tumor necrosis factor-related apoptosis-inducing ligand in surveillance of tumor metastasis by liver natural killer cells. *Nat. Med.* **7**(1), 94–100.

Takimoto, R., and El-Deiry, W. S. (2000). Wild-type p53 transactivates the KILLER/DR5 gene through an intronic sequence-specific DNA-binding site. *Oncogene* **19**(14), 1735–1743.

Teitz, T., Wei, T., Valentine, M. B., Vanin, E. F., Grenet, J., Valentine, V. A., Behm, F. G., Look, A. T., Lahti, J. M., and Kidd, V. J. (2000). Caspase 8 is deleted or silenced preferentially in childhood neuroblastomas with amplification of MYCN. (Comment). *Nat. Med.* **6**(5), 529–535.

Thompson, C. B. (1995). Apoptosis in the pathogenesis and treatment of disease. (Review) (84 refs). *Science* **267**(5203), 1456–1462.

Thornberry, N. A., and Lazebnik, Y. (1998). Caspases: Enemies within. (Review) (50 refs). *Science* **281**(5381), 1312–1316.

van Loo, G., Saelens, X., van Gurp, M., MacFarlane, M., Martin, S. J., and Vandenabeele, P. (2002). The role of mitochondrial factors in apoptosis: A Russian roulette with more than one bullet. (Review) (146 refs). *Cell Death Differ.* **9**(10), 1031–1042.

van Noesel, M. M., van Bezouw, S., Salomons, G. S., Voute, P. A., Pieters, R., Baylin, S. B., Herman, J. G., and Versteeg, R. (2002). Tumor-specific down-regulation of the tumor necrosis factor-related apoptosis-inducing ligand decoy receptors DcR1 and DcR2 is associated with dense promoter hypermethylation. *Cancer Res.* **62**(7), 2157–2161.

Wajant, H., Pfizenmaier, K., and Scheurich, P. (2002). TNF-related apoptosis inducing ligand (TRAIL) and its receptors in tumor surveillance and cancer therapy. *Apoptosis* **7**, 449–459.

Walczak, H., Bouchon, A., Stahl, H., and Krammer, P. H. (2000). Tumor necrosis factor related apoptosis-inducing ligand retains its apoptosis-inducing capacity on Bcl-2 or Bcl-XL expressing chemotherapy-resistant tumor cells. *Cancer Res.* **60**, 3051–3057.

Walczak, H., Degli-Esposti, M. A., Johnson, R. S., Smolak, P. J., Waugh, J. Y., Boiani, N., Timour, M. S., Gerhart, M. J., Schooley, K. A., Smith, C. A., Goodwin, R. G., and Rauch, C. T. (1997). TRAIL-R2: A novel apoptosis-mediating receptor for TRAIL. *EMBO J.* **16**(17), 5386–5397.

Walczak, H., and Krammer, P. H. (2000). The CD95 (APO-1/Fas) and the TRAIL (APO-2L) apoptosis systems. (Review) (105 refs). *Exp. Cell Res.* **256**(1), 58–66.

Wiley, S. R., Schooley, K., Smolak, P. J., Din, W. S., Huang, C. P., Nicholl, J. K., Sutherland, G. R., Smith, T. D., Rauch, C., Smith, C. A. *et al.* (1995). Identification and characterization of a new member of the TNF family that induces apoptosis. *Immunity* **3**(6), 673–682.

Wu, G. S., Burns, T. F., McDonald, E. R. 3rd, Jiang, W., Meng, R., Krantz, I. D., Kao, G., Gan, D. D., Zhou, J. Y., Muschel, R., Hamilton, S. R., Spinner, N. B., Markowitz, S., Wu, G., and el-Deiry, W. S. (1997). KILLER/DR5 is a DNA damage-inducible p53-regulated death receptor gene. *Nat. Genet.* **17**(2), 141–143.

16

Interferon-Gamma and TRAIL in Human Breast Tumor Cells

Carmen Ruiz de Almodóvar,*
Abelardo López-Rivas,* and Carmen Ruiz-Ruiz†

*Department of Cellular Biology and Immunology
Instituto de Parasitología y Biomedicina
Consejo Superior de Investigaciones Científicas
Granada E-18001, Spain
†Facultad de Medicina, Universidad de Grenada
Granada E-18071, Spain

I. Introduction
II. Interferon-γ
  A. Antitumoral Effects of IFN-γ
  B. Interferon-γ and Breast Cancer Cells
III. TRAIL System
  A. TRAIL Signaling
  B. Mechanisms of Resistance to TRAIL-Induced Apoptosis
  C. Physiologic Roles of TRAIL
  D. Transcriptional Regulation of TRAIL
  E. TRAIL System in Breast Cancer Cells
IV. Regulation by IFN-γ of TRAIL-Induced Apoptosis in Breast Tumor Cells
  A. Sensitization by IFN-γ of Breast Tumor Cells to Death Receptor–Mediated Apoptosis
  B. Mechanism for IFN-γ–Induced Sensitization of Breast Tumor Cells to TRAIL-Mediated Apoptosis: Upregulation of Caspase-8

C. Role of Caspase-8 in Tumor Cells
D. Interplay Between IFN-γ and TRAIL in the
   Induction of Apoptosis in Breast Tumor Cells
V. Conclusions
   References

Induction of apoptosis in tumor cells by death receptor activation is a novel therapeutic strategy. However, in systemic antitumor treatments, severe toxic effects have been observed with tumor necrosis factor-α (TNF-α) and CD95 ligand. TNF-α causes a lethal inflammatory response and CD95L produces lethal liver damage. Preclinical studies in mice and nonhuman primates showed no systemic cytotoxicity upon injection of recombinant TNF-related apoptosis-inducing ligand (TRAIL) at doses that effectively suppressed solid tumors such as colon and mammary carcinomas. Although unwanted effects of some TRAIL preparations have been reported in normal cells, these data suggest that TRAIL could be a suitable approach in cancer therapy. However, several mechanisms of resistance to TRAIL-mediated apoptosis have been described in tumor cells such as lack of TRAIL apoptotic receptors, enhanced expression of TRAIL-decoy receptors, and expression of apoptosis inhibitors. In combination regimes, interferon-γ (IFN-γ) could provide a promising antitumor therapeutic approach as it has been described to enhance cellular susceptibility to apoptosis in a variety of tumor cells. The mechanism by which IFN-γ promotes cell death seems to be via the regulation of the expression of different proteins involved in apoptosis. Altogether, these data suggest a combination strategy to selectively kill tumor cells that need to be further explored. © 2004 Elsevier Inc.

I. INTRODUCTION

Apoptosis is a cell death process that is required for integrity and homeostasis of multicellular organisms and is also important in pathologic situations (Steller, 1995; Thompson, 1995). The past decade has witnessed an enormous progress in our knowledge of the pathway for the execution of apoptotic cell death and both mitochondria-dependent and -independent mechanisms have been involved in the induction of apoptosis by different treatments (Green and Reed, 1998). Apoptosis is a highly regulated process. Among the main regulators, members of the Bcl-2 family play a crucial role, mainly at the mitochondrial level (Gross et al., 1999). Deregulation of apoptosis can eventually lead to aberrant accumulation of cells and the development of cancer. Moreover, defects in the apoptosis machinery can result in drug resistance by tumor cells (Reed, 1999).

In contrast to many therapeutic drugs that induces apoptosis by acting at the mitochondrial level, the so-called extrinsic pathway of apoptosis uses membrane-bound "death receptors" to activate the caspase cascade and apoptosis upon ligand binding. These receptors are members of the tumor necrosis factor (TNF) receptor superfamily (Ashkenazi and Dixit, 1998). Among their ligands, TNF-related apoptosis-inducing ligand (TRAIL; APO-2 ligand), a recently discovered member of the TNF family, is a type II transmembrane protein that functions as an apoptosis inducer in activation-induced cell death, immune privilege, T cell-mediated cytotoxicity, and autoimmunity (Lamhamedi-Cherradi *et al.*, 2003; Martinez-Lorenzo *et al.*, 1998; Phillips *et al.*, 1999; Thomas and Hersey, 1998). In culture, TRAIL provokes apoptosis mainly in tumor cells (Ashkenazi *et al.*, 1999). However, in breast cancer cells, resistance to TRAIL-induced apoptosis is frequently observed (Keane *et al.*, 1999).

During the past few years several studies have demonstrated a critical role of interferon-γ in promoting protective host response to tumors. Thus, their antitumor activity has been reported against a variety of tumor cells such as lymphomas, melanomas, and multiple myeloma (Strander, 1986; Wadler and Schwartz, 1990). Moreover, clinically and experimentally, it has been demonstrated that IFN-γ can enhance the antitumor effects of antimetabolites on cancer cells (Koshiji *et al.*, 1997; Wadler and Schwartz, 1990). However, the mechanism of IFN-γ–mediated response against tumors has not been completely elucidated.

In the present chapter we review the current data on the sensitization by interferon-γ of breast tumor cells to apoptosis mediated by TRAIL receptor activation.

II. INTERFERON-γ

Interferons (IFNs) are a family of natural glycoproteins that play a key role in antiviral, antiproliferative, and immunomodulatory responses. According to the structure, function, and signaling pathways, IFNs can be divided into two different classes: type I and type II IFNs (Stark *et al.*, 1998). They also differ in the stimuli that induce their expression. Thus, type I IFNs (IFNα and IFNβ) are primarily induced in response to viral infection of cells (Domanski and Colamonici, 1996). In contrast, IFN-γ or type II IFN, also known as immune IFN, is produced mainly by T lymphocytes, natural killer (NK) T cells and NK cells in response to activation with immune and inflammatory stimuli, rather than viral infection (Boehm *et al.*, 1997; Farrar and Schreiber, 1993). It has also been reported that, in the presence of IL-12 and IL-18, activated dendritic cells and macrophages produce IFN-γ (Frucht *et al.*, 2001;

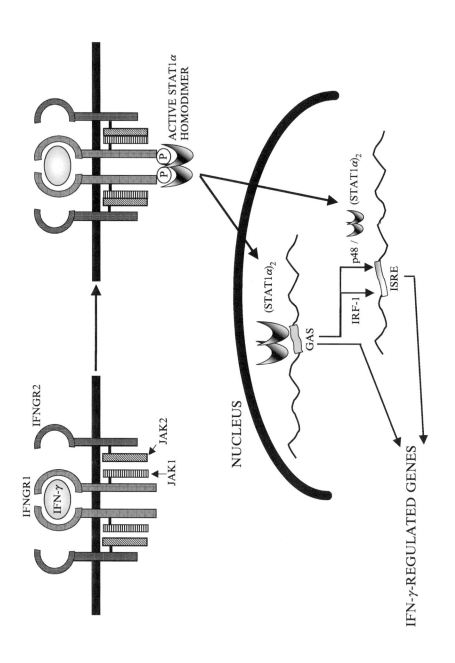

Gessani and Belardelli, 1998). The active form of IFN-γ is a homodimer that exerts its effect by interacting with its receptor expressed on the surface of almost all normal cells (Ealick *et al.*, 1991; Farrar and Schreiber, 1993). The IFN-γ receptor consists of two polypeptides: IFNGR1, that is responsible for ligand binding and IFNGR2 that plays a minor role in ligand binding but is required for IFN-γ signaling (Fig. 1). These subunits are associated via their intracellular domains with the Janus kinases JAK1 and JAK2, respectively (Pestka *et al.*, 1997). When a homodimeric IFN-γ molecule binds to its receptor JAKs become activated and phosphorylate key tyrosine residues within the IFNGR1 intracellular domain, which are recognized by the SH2 domain of STAT1 molecules. STAT1 proteins are then phosphorylated and form homodimers which translocate to the nucleus (Fig. 1) where they bind to specific gene promoter elements known as gamma-interferon activated sites or GAS elements (Darnell *et al.*, 1994; Pestka *et al.*, 1997). Phosphorylated STAT-1 factors can also bind to ISRE elements in the promoters of target genes and activate transcription in coordination with other proteins (Fig. 1) (Kimura *et al.*, 1996).

A. ANTITUMORAL EFFECTS OF IFN-γ

It seems that the antitumoral response promoted by IFN-γ is the result of different coordinated events that involve both immunologic and non-immunologic processes. IFN-γ activates cells of the innate immunity such as NK cells, NKT cells and macrophages to proliferate, produce cytokines and acquire lytic activity against tumor cells (Carnaud *et al.*, 1999; Nastala *et al.*, 1994; Schreiber *et al.*, 1986; Trinchieri, 1995). In addition IFN-γ, together with IL-12 produced by activated macrophages, promote the development of a Th1 antitumor response that involves cytotoxic CD8+ T cells (Fallarino and Gajewski, 1999; Trinchieri, 1995). On the other hand IFN-γ can exert direct anti-proliferative and proapoptotic effects on a wide variety of tumor cells by the induction of genes that encode inhibitors of cell cycle progression, such as p21 and p27, and different apoptosis-related proteins including death receptors and their respective ligands, caspases, and several members of the Bcl-2 family (Chin *et al.*, 1997; Mandal *et al.*, 1998; Ossina *et al.*, 1997; Shyu *et al.*, 2000). Several other proteins have been involved

---

FIGURE 1. Intracellular signaling induced by interferon-γ. Upon binding of homodimeric IFN-γ to its receptor, JAKs become activated and phosphorylate key tyrosine residues within the IFNGR1 intracellular domain. These residues are recognized by the SH2 domain of STAT1 molecules. STAT1 proteins are then phosphorylated and form homodimers that translocate to the nucleus where they bind to GAS elements in a number of gene promoters. In coordination with other factors, phosphorylated STAT-1 may also bind to ISRE elements in the promoters of target genes and activate transcription.

in the induction of apoptosis by IFN-γ. Transfection of Hela cells with an antisense cDNA library, followed by selection of transfectants that survived in the continuous presence of IFN-γ, led to the identification of seven different genes whose inactivation seemed to protect them from IFN-γ–induced cell death (Levy-Strumpf and Kimchi, 1998). Five of them were novel genes that coded for proteins DAP-1 to DAP-5 (death associated proteins), which display a diverse spectrum of biochemical activities. The other two were genes for thioredoxin and cathepsin-D. Furthermore the dsRNA-activated protein kinase, PKR, is an IFN-γ–inducible protein that has also been shown to function as an inducer of apoptosis and a tumor suppressor protein (Lee and Esteban, 1994; Meurs et al., 1993). Finally, it is worth mentioning that there exists evidence suggesting an antiangiogenic role for IFN-γ within the tumor that contributes to the inhibition of tumor growth. This angiostatic effect seems to be mediated by IP-10 and maybe other IFN-inducible chemokines such as Mig and I-TAC (Angiolillo et al., 1995; Sgadari et al., 1996, 1997).

B. INTERFERON-γ AND BREAST CANCER CELLS

Breast cancer is the most common neoplasia among women in the western world and therefore progress for its treatment is of prime interest. Genotoxic drugs as well as radiation therapy have been widely used to induce apoptosis in breast tumor cells (Fisher, 1994; McCloskey et al., 1996). Recently, different combined strategies have been investigated to improve the effects of chemotherapy and radiation therapy. In this respect, we and others have reported that DNA-damaging drugs and ionizing radiation can sensitize breast cancer cells to death receptor–induced apoptosis (Chinnaiyan et al., 2000; Ruiz-Ruiz and López-Rivas, 1999). The accumulation of the tumor suppressor protein p53 is generally required for the efficacy of those combined treatments involving drugs and irradiation. However, inactivating p53 mutations are frequently observed in breast cancer cells that abrogate p53-dependent gene transcription (Bartek et al., 1990; Forrester et al., 1995; Park et al., 1994).

On the other hand, it has been shown that breast cancer chemotherapy is immunosuppressive, leading to a severe impairment of host immune mechanisms (Zielinski et al., 1990). Immune reactivity to breast tumor has been suggested by the presence of tumor-associated and tumor-infiltrating lymphocytes (Aaltomaa et al., 1992) and the identification of several tumor-associated antigens, such as carcinoma embryonic antigen (CEA), HER-2/neu protein, p53 or the mucin MUC-1, which seem to be involved in cellular as well as humoral immune responses (Disis et al., 1994; Schlichtholz et al., 1992; Schlom et al., 1996; von Mensdorff-Pouilly et al., 1996). Therefore, an immunotherapeutic approach has been used in the treatment of breast cells and positive results have been obtained in early trials with natural interferons (IFNs) and interleukins, particularly in combination therapies (Borden et al., 1982; Gutterman et al., 1980).

Recent studies with several murine mammary adenocarcinomas have shown the efficacy of IFN-γ in the inhibition of tumor growth and the reduction of metastases (Coughlin et al., 1998; Nanni et al., 2001; Rakhmilevich et al., 2000; Wu et al., 2001). These results have further suggested that the antitumor activities of either different individual cytokines such as IL-12 and IL-10 or the combination of IL-12, IL-18 and IL-1 beta converting enzyme (ICE) are, at least in part dependent on the endogenous expression of IFN-γ and mainly mediated by CD8+ T cells and NK cells (Nanni et al., 2001; Oshikawa et al., 1999; Rakhmilevich et al., 2000; Sun et al., 2000). Moreover, in carcinogen-induced tumor models IFN-γ has been found to collaborate with lymphocytes in mediating host protection and preventing the development of breast adenocarcinomas (Shankaran et al., 2001). Phagocytic cells have also been shown to be responsible for the critical role of IFN-γ in the innate immune response against metastatic mammary carcinoma (Pulaski et al., 2002). Bearing in mind all these reports, it is not surprising that an increase in IFN-γ levels has been associated with a favorable prognosis in the response of breast cancer patients to the combination of interferons and hormonotherapy (Barak et al., 1998).

III. TRAIL SYSTEM

TRAIL is a type II transmembrane protein that belongs to the tumor necrosis factor (TNF) family of cytokines. It was identified by homology to the C-terminal extracellular domain of other TNF family members (Wiley et al., 1995). TRAIL can bind to four specific type I membrane receptors (Fig. 2); TRAIL-R1 (DR4) (Pan et al., 1997b) and TRAIL-R2 (DR5, KILLER or TRICK2) (Schneider et al., 1997; Walczak et al., 1997) are called proapoptotic receptors because they transmit apoptotic signals and contain the intracellular death domain (DD) essential for the induction of apoptosis. TRAIL-R3 (TRID, LIT or DcR1) (MacFarlane et al., 1997; Meurs et al., 1993; Pan et al., 1997a) and TRAIL-R4 (TRUNDD or DcR2) (Degli-Esposti et al., 1997) are known as antiapoptotic or decoy receptors because even though they can bind TRAIL, they are not capable of engaging the cell suicide apparatus. TRAIL-R3 does not have a cytoplasmic domain nor a transmembrane domain and is bound to the cell surface through a glycosyl phosphatidyl inositol linkage. TRAIL-R4 has a truncated cytoplasmic death domain that is also not able to signal cell death.

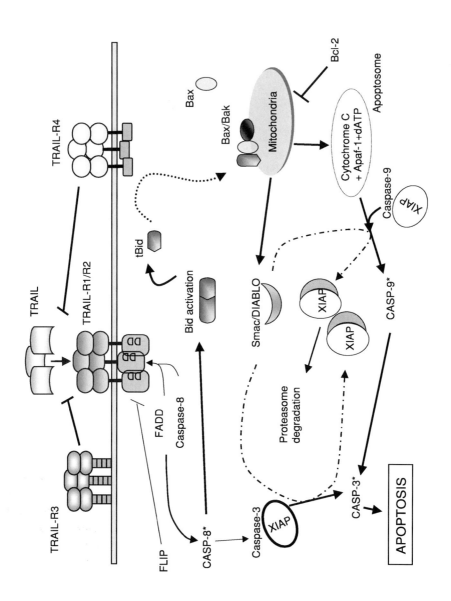

A. TRAIL SIGNALING

Once TRAIL binds to its proapoptotic receptors, a few changes take place to engage the caspase cascade and consequently induce cell death (Fig. 2). Proapoptotic receptors become clustered in trimers by crosslinking with their ligand. This clustering allows the formation of the DISC (death-inducing signaling complex) by recruitment of the dual adaptor molecule FADD (Fas-associated death domain) through its death domain (DD), which in turn recruits the initiator caspase-8 through its "death effector domain" (DED) (Suliman et al., 2001). Caspase-8 is then activated at the DISC by oligomerization and active caspase-8 can process Bid (a proapoptotic protein of the Bcl-2 family) to kill the cell by a mitochondrial-regulated apoptotic pathway (Sprick et al., 2000). Active caspase-8 can also directly activate caspase-3 to induce cell death by a mitochondria-independent pathway (Scaffidi et al., 1998). In the mitochondria-dependent apoptotic pathway, caspase-8–mediated cleavage of Bid generates a truncated form of Bid (tBid) that translocates to the mitochondria and promotes the release of cytochrome c (Li et al., 1998; Luo et al., 1998) and Smac/Diablo (Chai et al., 2000; Verhagen et al., 2000) in a Bax/Bak-dependent mechanism (Deng et al., 2002; Shimizu et al., 1999). Once cytochrome c is at the cytosol, it becomes an essential part of the apoptososme (a complex formed by Apaf-1, cytochrome c, dATP and caspase-9) (Saleh et al., 1999; Zou et al., 1999). Formation of the apoptosome leads to the activation of caspase-9, which in turn processes and activates the executioner caspases, caspase-3 and -7.

Great expectations have arisen regarding the future use of TRAIL as an antitumor agent because it has been extensively reported that TRAIL can selectively induce apoptosis in tumor cells but not in normal cells. Unlike CD95L, TRAIL is expressed constitutively and is widely distributed in normal organs and tissues (Wiley et al., 1995), suggesting that this ligand may be nontoxic to normal cells. TRAIL receptors, as TRAIL, appear to be ubiquitously expressed with transcripts detected in most human tissues as well as in tumor cell lines of different lineages. It has also been shown that TRAIL does not produce systemic toxicity in preclinical studies with mice and nonhuman primates (Walczak et al., 1999). However, a recombinant

FIGURE 2. Apoptotic pathways activated by TRAIL. Crosslinking of proapoptotic TRAIL receptors by trimeric TRAIL induces DISC formation in the tumor cells. Oligomerization of procaspase-8 at the DISC leads to its activation. Depending on the target cell, caspase-8 activation can result in a caspase cascade and apoptosis. In other cells, apoptosis by TRAIL may require a mitochondria-regulated pathway. In the later, Bid cleavage by caspase-8 induces the release of apoptotic factors from the mitochondria that promotes the activation of caspases and apoptosis. In some instances, expression of TRAIL decoy receptors at the cell surface and endogenous inhibitors (FLIP, Bcl-2, XIAP) may prevent TRAIL-induced apoptosis. *Signifies activated caspases.

form of human TRAIL has been shown to induce apoptosis in human hepatocytes (Jo et al., 2000). Nevertheless, more recent data have demonstrated that different recombinant versions of TRAIL vary considerably in toxicity toward normal human cells, but all of them maintain their antitumor properties (Lawrence et al., 2001). So, it is necessary to understand the biology and functions of TRAIL to progress to the use of TRAIL as a possible candidate for cancer treatment.

B. MECHANISMS OF RESISTANCE TO TRAIL-INDUCED APOPTOSIS

Not all types of tumor cells are sensitive to TRAIL-induced apoptosis. Cancer cells can express proapoptotic receptors (TRAIL-R1 and TRAIL-R2) and be resistant to TRAIL. Control of death receptor–mediated apoptosis is exerted at many stages; the first one is the level of expression of death receptors at the cell surface (Zhang et al., 2000). It was shown, by overexpressing decoy receptors, that expression of these receptors at the tumor cell surface could protect cells from TRAIL-induced apoptosis, either by binding to TRAIL and impeding its binding to the proapoptotic ones, or by forming heterotrimers with TRAIL-R2 thereby preventing this trimer from transmiting the apoptotic signal (Bernard et al., 2001; Munoz-Pinedo et al., 2002; Pan et al., 1997a). However, a correlation between survival of cancer cells and decoy receptors expression has not been conclusively demonstrated. So far, most of the studies of TRAIL receptors expression have been based in mRNA or total protein expression. In our opinion, further studies are needed to determine regulation of TRAIL decoy receptors expression at the cell surface and clarify their role in TRAIL-induced apoptosis. On the other hand, intracellular apoptotic molecules can block the apoptotic effect of TRAIL signaling pathway at different levels. The FLICE-like inhibitory protein (c-FLIP) blocks the apoptotic pathway right at the beginning because it competes with caspase-8 for binding to FADD (Griffith et al., 1998; MacFarlane et al., 2002a; Siegmund et al., 2002; Thome et al., 1997). Bcl-2 and Bcl-X_L impede or delay the activation of the mitochondrial pathway in nonlymphoid cancer cells, thereby blocking the release of cytochrome c from mitochondria to cytosol (de Almodovar et al., 2001; Fulda et al., 2002). XIAP, cIAP-1 and cIAP-2 act by inhibiting active caspases (Leverkus et al., 2003; MacFarlane et al., 2002b; Ng and Bonavida, 2002). The antiapoptotic signals induced by NF-κB (Oya et al., 2001; Ravi et al., 2001), protein kinase B/Akt (Nesterov et al., 2001; Yuan and Whang, 2002), PKCs (Guo and Xu, 2001; Sarker et al., 2001) or MAPK (Tran et al., 2001) have also been implicated in resistance to TRAIL-induced apoptosis.

C. PHYSIOLOGIC ROLES OF TRAIL

Little is known about the physiologic role of TRAIL *in vivo*. Several studies point to a role for TRAIL in immune functions. It has been shown that TRAIL-deficient mice have a severe defect in thymocyte apoptosis and are also hypersensitive to autoimmune diseases (Lamhamedi-Cherradi *et al.*, 2003). Dendritic cells are the most potent antigen presenting cells for initiation of immune responses including antitumor immune responses. Human immature dendritic cells have been shown to express TNF, FasL, and TRAIL on their cell surface (Lu *et al.*, 2002) and be capable of inducing apoptosis on cancer cells through them. Other authors have demonstrated that human dendritic cells express TRAIL on their surface after stimulation with IFN-α or IFN-γ and acquire the ability to kill TRAIL-sensitive tumor cells (Fanger *et al.*, 1999). Similar effects are seen when monocytes are treated with IFN-α or IFN-γ. TRAIL is upregulated and monocyte antitumor cytotoxicity seems to be dependent on TRAIL-induced apoptosis (Griffith *et al.*, 1999). TRAIL is also induced on the surface of human peripheral blood T cells upon stimulation of TCR receptor or treatment with type I IFNs (Fig. 3). The cytotoxic activity of these T cells was demonstrated to be TRAIL dependent (Kayagaki *et al.*, 1999). TRAIL is also expressed in mouse liver NK cells but not on other lymphocytes isolated from liver or spleen. By administration of neutralizing monoclonal antibody against TRAIL to mice subcutaneously inoculated with TRAIL-sensitive tumors, it was demonstrated that blocking of endogenous TRAIL activity promoted outgrowth of these tumors (Takeda *et al.*, 2001b). Also, in mouse inoculated with a chemical carcinogen, methylcolanthrene, Takeda *et al.* demonstrated that endogenous TRAIL plays a critical role, cooperatively with the Fas and perforin branches, in host surveillance against primary tumor development *in vivo* (Takeda *et al.*, 2002). It also makes a critical contribution to the NK cell–mediated suppression of liver metastasis of TRAIL-sensitive cells (Takeda *et al.*, 2001a). TRAIL expression in NK cells and the antitumor and antimetastatic potential of these cells were dependent on the presence of IFN-γ (Fig. 3).

Thus, TRAIL mediates an important part of the antitumor cytotoxicity of monocytes, NK cells, dendritic and T cells. In all these cell types, TRAIL is upregulated with IFN treatment, thereby augmenting their antitumor activity.

D. TRANSCRIPTIONAL REGULATION OF TRAIL

The human TRAIL gene locus spans approximately 20 kb and it is composed of five exons and four introns (Gong and Almasan, 2000). The 5'-upstream region of TRAIL gene has been described and analyzed (Gong and Almasan, 2000; Wang *et al.*, 2000) revealing that it lacks typical "TATA" or "CAAT" boxes but contains two potential Sp1 binding sites.

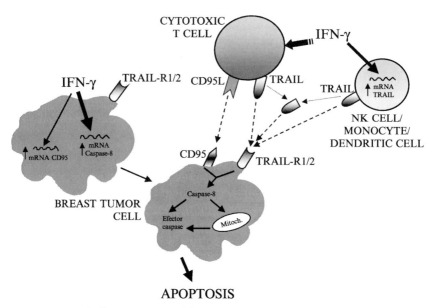

FIGURE 3. Antitumor response elicited by interferon-γ in breast tumor cells. IFN-γ can activate an antitumor response at two levels. One, acting on cells of the immune system, IFN-γ induces the expression of TRAIL in these cells and thus promotes a cytotoxic response against the tumor cells. A second mechanism involves the upregulation by IFN-γ of the expression in the tumor cells of proapoptotic proteins, i.e., procaspase-8 and CD95 receptors.

Using computer programs such as MatInspector V2.2 software and TRANSFAC putative transcription binding sites, which include NFAT, AP-1, GATA, C/EBP, GAS-like elements and ISRE, were identified. The last two transcription binding sites, Gas-like elements and ISRE, have been described to be responsible for the transcriptional upregulation of other genes by IFN treatment (Boehm et al., 1997). Analysis of promoter activity by luciferase reporter assays with different reporter gene constructs containing sequential deletions of the TRAIL gene 5′-flanking region revealed that a region between −165 and −35 bp upstream of the transcription start site contains an IFN-responsive element (Gong and Almasan, 2000; Wang et al., 2000). It seems that IFN-α and IFN-β induced TRAIL in leukemic, lymphoma, and myeloid cells (Chawla-Sarkar et al., 2001), whereas TRAIL is induced by all types of IFNs in other types of tumor cells (Shin et al., 2001a; Wang et al., 2000). So, further studies are needed to identify the site and the transcription factors involved in IFN-inducible transcriptional upregulation.

E. TRAIL SYSTEM IN BREAST CANCER CELLS

TRAIL and its receptors have been shown to be expressed in breast cancer cells (de Almodóvar *et al.*, 2001; Ruiz-Ruiz and López-Rivas, 2002). TRAIL ligand is located in chromosome 3 whereas all four TRAIL receptors are localized on human chromosome 8p21-22 (Golstein, 1997). Mutations in TRAIL-receptors genes in breast tumor cells have been detected. Specific mutations in TRAIL-R1 and TRAIL-R2 death domains that suppress apoptosis have been found only in breast cancer with metastasis (Shin *et al.*, 2001b). Allelic loss of chromosome 8p21-22 have been shown to be higher in metastatic breast cancers (Seitz *et al.*, 2002; Shin *et al.*, 2001b).

Recombinant TRAIL has been shown to have apoptotic activity against breast carcinoma cells but not normal epithelial cells (Chinnaiyan *et al.*, 2000). In mice challenged intraperitoneally with breast tumor cells or bearing subcutaneous implanted breast tumors, repeated treatment with recombinant TRAIL clearly resulted in tumor shrinkage and elimination of tumor cells, respectively (Walczak *et al.*, 1999). The combination of chemotherapeutic agents or radiation therapy with TRAIL have been used in a large number of studies; pretreatment of tumor cells with genotoxic agents sensitizes breast tumor cells to TRAIL-induced apoptosis (Keane *et al.*, 1999). Also, ionizing radiation can enhance the efficacy of TRAIL-induced apoptosis *in vitro* and *in vivo*; this combination induces the regression of established breast tumors in mice (Chinnaiyan *et al.*, 2000). Gene therapy is another approach for the treatment of cancer; several groups are working on proapoptotic gene therapy and its use for the treatment of cancers, including those resistant to conventional therapy. Because of its unique characteristics, TRAIL is a good candidate for use as a proapoptotic gene. With regard to this, it has been demonstrated that treatment of breast cancer cells with an adenoviral vector that express the GFP/TRAIL fusion gene driven by a tumor specific promoter (hTERT) elicited apoptosis in malignant cells both sensitive and resistant to doxorubicin and suppresses tumor growth, including complete regression *in vivo* with minimal toxicity to normal cells (Lin *et al.*, 2002).

IV. REGULATION BY IFN-γ OF TRAIL-INDUCED APOPTOSIS IN BREAST TUMOR CELLS

A. SENSITIZATION BY IFN-γ OF BREAST TUMOR CELLS TO DEATH RECEPTOR–MEDIATED APOPTOSIS

As mentioned previously, IFN-γ seems to be a promising antitumor therapeutic approach in combination regimes. It has been described to enhance cellular susceptibility to apoptosis in a variety of tumor cells, a property that could play an important role in its antitumor effects. The mechanism by which IFN-γ promotes cell death seems to involve the regulation of the expression of several pro- and antiapoptotic proteins. In different tumor models such as colorectal adenocarcinoma, renal cell carcinoma, cholangiocarcinoma, gastric cancer cells, multiple myeloma or myeloid leukemic cells, it has been shown that IFN-γ is able to enhance CD95-mediated apoptosis or even directly induce cell death by upregulating the levels of CD95 receptor, different caspases, Bax, or Bak, and downregulating the antiapoptotic proteins Bcl-2 and Bcl-x (Ahn et al., 2002; Koshiji et al., 1998; Shyu et al., 2000; Spets et al., 1998; Tomita et al., 2003; Varela et al., 2001). We, as well as other authors, have reported that IFN-γ also sensitizes breast cancer cells to CD95-induced apoptosis (Keane et al., 1996; Ruiz-Ruiz et al., 2000). However, activation of the CD95 pathway may not be an useful therapeutic strategy because normal human cells are killed by CD95 ligation and massive apoptosis is observed in the liver of experimental mice injected with CD95 antibody (Ogasawara et al., 1993).

Interferon-γ has also been shown to induce apoptosis in a model of human hepatoma cells through an upregulation of TRAIL expression and a decrease in the levels of the decoy receptor TRAIL-R4 (Shin et al., 2001a). Interestingly, cell lines of Ewing's sarcoma resistant to TRAIL-induced apoptosis were rendered sensitive by IFN-γ (Kontny et al., 2001). Moreover, pretreatment of Hela cells with IFN-γ was found to prevent TRAIL-induced inhibitor of apoptosis protein-2 (IAP-2) upregulation, thus enhancing TRAIL-mediated apoptosis in these cells (Park et al., 2002). Provided that IFN-γ induces the expression of TRAIL in several cells of the immune system such as NK cells, monocytes, lymphocytes, and dendritic cells, and that this death ligand can contribute to the antitumor activity of these cells *in vivo*, the possible modulation by IFN-γ of tumor sensitivity to TRAIL-mediated apoptosis seems to play an essential role in the development of new therapeutic strategies.

We have recently described the sensitization by IFN-γ of breast tumor cells to TRAIL-induced apoptosis (Ruiz-Ruiz and López-Rivas, 2002). In this study, IFN-γ has been shown to markedly enhance the activation by TRAIL of all the biochemical events that are known to take place following TRAIL receptor stimulation in MCF-7 and MDA-MB231 breast cancer cell lines (Sarker et al., 2001; Suliman et al., 2001). Activation of caspase-8, cleavage of Bid, translocation of cytosolic Bax to mitochondria, release of mitochondrial cytochrome *c*, activation of caspase-9, and cleavage of substrates for executioner caspases were all enhanced in the presence of IFN-γ. Together, those results support the hypothesis that in certain breast tumor cells, IFN-γ favors the activation by TRAIL of a mitochondria-operated apoptotic pathway.

B. MECHANISM FOR IFN-γ–INDUCED SENSITIZATION OF BREAST TUMOR CELLS TO TRAIL-MEDIATED APOPTOSIS: UPREGULATION OF CASPASE-8

Sensitization of cells to TRAIL-mediated apoptosis can take place at different stages in the apoptotic signaling pathway (Griffith et al., 1998; Zhang et al., 2000). We observed that MCF-7 and MDA-MB231 breast cancer cells expressed TRAIL-R1 and TRAIL-R2 death receptors as well as TRAIL-R3 and TRAIL-R4 decoy receptors for TRAIL, but IFN-γ treatment does not modify the mRNA level or protein levels at the cell surface of any of these receptors (Ruiz-Ruiz and López-Rivas, 2002, and unpublished results) in contrast to what has been reported for other tumoral cell lines (Meng and El-Deiry, 2001). We have also analyzed the expression, in both tumor cell lines, of several apoptosis-related proteins on treatment with a sensitizing concentration of IFN-γ. No changes in the levels of the Bcl-2 family proteins Bax, Bak, Bad, Bid, or Bcl-2 have been observed after incubation with IFN-γ. Likewise, the cellular levels of the adaptor molecule FADD, the apoptosis-inhibitors XIAP, IAP-1 and IAP-2 as well as the pro-apoptotic protein Smac/DIABLO are not modified in response to IFN-γ (Ruiz-Ruiz and López-Rivas, 2002; Ruiz-Ruiz et al., 2000).

Previous data have indicated that caspase-1/ICE is upregulated in some breast cancer cell lines on treatment with IFN-γ thus sensitizing these cells to CD95-mediated apoptosis (Keane et al., 1996). However, more recent results have demonstrated that this caspase does not play a role in the proteolytic cascade activated on CD95 cross-linking by CD95 ligand or CD95 antibody and there are no data involving it in the TRAIL signaling pathway. Interestingly, we have found that IFN-γ treatment of breast tumor cell lines upregulates caspase-8 mRNA and protein expression (Fig. 3) (Ruiz-Ruiz et al., 2000), the first caspase required in TRAIL-mediated apoptosis (Suliman et al., 2001). As already described, IFN-γ enhanced early steps in TRAIL-activated apoptotic pathway, accordingly the modulation of apical caspase-8 expression could be an important mechanism for IFN-γ–induced sensitization of breast tumor cells. The levels of expression of other caspases remained invariable upon IFN-γ treatment. In caspase-3–overexpressing MCF-7 cells, mitochondria seems not to be involved in TRAIL receptor–activated apoptotic pathway because activation of caspase-3 and PARP cleavage are observed before any mitochondria-regulated events can be detected. However, IFN-γ is also able to enhance this TRAIL-induced mitochondria-independent apoptotic pathway, further supporting the idea that early activation of caspase-8 in IFN-γ–treated cells may be sufficient to mediate the sensitization to TRAIL-induced cell death. Moreover, these data suggest that IFN-γ can be regarded as a general strategy to facilitate TRAIL-mediated apoptosis irrespective of the signaling pathway activated by TRAIL.

High levels of expression of the antiapoptotic protein Bcl-2 in breast carcinomas appear to be an important factor in the inhibition of cell death (Rochaix *et al.*, 1999). It has been demonstrated that Bcl-2-overexpressing MCF-7 cells are not sensitized by IFN-γ to TRAIL-induced apoptosis (Ruiz-Ruiz and López-Rivas, 2002). However, we have to keep in mind that this is a special cell line because of its deficiency in caspase-3 expression (Jänicke *et al.*, 1998), which appears to be required to activate the mitochondria-independent pathway of apoptosis (Scaffidi *et al.*, 1998). Thus it is possible that IFN-γ can sensitize different human breast carcinomas, without defects in the expression of caspase-3 and expressing high levels of Bcl-2, by facilitating the activation of apical caspase-8 and executioner caspases upon TRAIL receptors ligation.

C. ROLE OF CASPASE-8 IN TUMOR CELLS

Modulation of caspase-8 expression by IFN-γ has been described in other human tumor models such as colon adenocarcinoma cells, myeloid leukemic cells, neuroblastoma, medulloblastoma, Ewing tumor or small-cell lung carcinoma cell lines (Fulda and Debatin, 2002; Hopkins-Donaldson *et al.*, 2000; Langaas *et al.*, 2001; Varela *et al.*, 2001; Yang *et al.*, 2003). In these reports the authors also indicate that upregulation of caspase-8 expression is involved in the sensitization of tumor cells to TRAIL-mediated apoptosis by IFN-γ. Interestingly, loss of caspase-8 expression has been recently demonstrated in malignant neuroblastomas, medulloblastomas, Ewing tumor, rhabdomyosarcomas, retinoblastomas, primitive neuroectodermal brain tumors, and small-cell lung carcinomas, mainly by gene hypermethylation although other mechanisms of gene inactivation are present in some cases (Fulda and Debatin, 2002; Harada *et al.*, 2002; Hopkins-Donaldson *et al.*, 2000, 2003; Kontny *et al.*, 2001; Yang *et al.*, 2003; Zuzak *et al.*, 2002). These data suggest that caspase-8 may act as a tumor suppressor gene and that inhibition of the death receptor pathway to apoptosis may play an important role in the pathogenesis of many types of tumors. In this respect, it is worth remembering that regulation of the expression and/or activity of the DISC components seems to be a general strategy used by virally infected or tumor cells to escape from the host immune system, as other mechanisms such as the expression of FLIP or the modulation of death receptors expression, have been reported to mediate resistance to death ligand–induced apoptosis.

On the other hand, induction and activation of caspase-8 could facilitate not only death receptor–mediated apoptosis but also drug-induced cell death in tumor cells. Caspase-8 together with caspase-3 play a crucial role in apoptosis of colorectal adenocarcinoma induced by IFN-γ and/or 5-FU (Adachi *et al.*, 1999). Other drug-inducible apoptotic pathways in which caspase-8 mediates the activation of executioner caspases leading to cell death have also been described in different tumor cells (Ferreira *et al.*, 2000; Wesselborg *et al.*, 1999). Moreover, IFN-γ–induced caspase-8 upregulation in resistant tumor cell lines derived from neuroblastoma, medulloblastoma, or Ewing tumor seems to sensitize them to the cytotoxic action of chemotherapeutic drugs (Fulda and Debatin, 2002). Therefore, modulation of caspase-8 by IFN-γ might be envisaged as a more widespread mechanism for sensitization of different tumor cells to death receptor– and chemotherapy-induced apoptosis.

D. INTERPLAY BETWEEN IFN-γ AND TRAIL IN THE INDUCTION OF APOPTOSIS IN BREAST TUMOR CELLS

In view of the variety of proteins that can be regulated by IFN-γ and the multiple proposed mechanisms for the induction of cell death in response to this cytokine, it is possible that other molecules or signals, apart from the upregulation of caspase-8, might be involved in the sensitization by IFN-γ of breast tumor cells to TRAIL-mediated apoptosis. For instance, it has been reported that IFN-γ may inhibit NF-kappaB transcriptional activity (Sanceau *et al.*, 2002; Sedger *et al.*, 1999; Suk *et al.*, 2001). On the other hand, TRAIL has been shown to induce NF-kappaB activation, which may represent a mechanism of resistance to TRAIL-mediated apoptosis in tumor cells because NF-kappaB is known to upregulate the expression of antiapoptotic genes such as IAPs or the decoy receptor TRAIL-R3 (Bernard *et al.*, 2001; Harper *et al.*, 2001; Wang *et al.*, 1998). In fact, NF-kappaB seems to be involved in the resistance of some breast cancer cells to TRAIL-induced apoptosis (Keane *et al.*, 2000). These data suggest that IFN-γ might inhibit TRAIL-induced activation of NF-kappaB in breast tumor cells thus contributing to the sensitization of these cells to TRAIL-mediated cell death.

The IFN-γ–induced dsRNA-activated protein kinase, PKR, has been described to induce apoptosis through a mechanism involving FADD and activation of caspase-8 but independently of Fas/FasL and TNFR1-TNF2 interactions (Gil and Esteban, 2000). These authors have suggested that other death ligands, maybe TRAIL, could mediate PKR-induced caspase-8 activation and apoptosis (Gil and Esteban, 2000). Moreover, it has also been shown that activation of a mitochondria-regulated pathway, which can be inhibited by the overexpression of Bcl-2, is involved in PKR-induced apoptosis (Gil *et al.*, 2002; Lee *et al.*, 1997). These data further indicate that both TRAIL and PKR display similar apoptotic pathways. Thus one can speculate the possibility that IFN-γ upregulates the expression of PKR, which in turn may induce apoptosis through activation of the TRAIL system thereby enhancing the TRAIL signaling pathway and cell death in breast tumor cells. Alternatively, PKR, by an unknown mechanism independently of TRAIL, can induce the activation of FADD and caspase-8 and so synergize with TRAIL-mediated apoptosis. Deregulation of PKR expression and activity has been described in breast tumor cells (Savinova et al., 1999).

As has been previously discussed, IFN-γ induces the upregulation of TRAIL not only in several cells of the immune system but also in hepatoma cell lines thereby inducing apoptosis in those tumor cells. In breast cancer cells we have also observed IFN-γ–mediated regulation of TRAIL expression. Furthermore, IFN-γ is able to induce apoptosis in cultures of breast tumor cells after several days incubation (Ruiz de Almodóvar, Ruiz-Ruiz, and López-Rivas, unpublished data). At present we do not know the role of TRAIL, if any, in IFN-γ–induced cell death of breast tumor cells. However, in view of all the data presented, we can infer that there are different points of interaction between TRAIL and IFN-γ in the induction of apoptosis in tumor cells, and in particular, in breast cancer cells. In this respect, a recent report has shown the stimulation of several IFN pathway related genes, such as ISGF3γ, STAT1, PKR, or IFNGR2 in breast carcinoma cells by TRAIL, further indicating the existence of a cross-talk between TRAIL and IFN signaling pathways (Kumar-Sinha et al., 2002). This work also supports the synergistic apoptotic effect on MCF-7 cells of the combined treatment with IFN-γ and TRAIL that we have already described.

Additional evidence for the concept of an interplay between TRAIL and IFN-γ apoptotic pathways comes from studies with DAP3 (death-associated protein-3). DAP3 is a nucleotide binding protein initially described as a positive mediator of IFN-γ–induced cell death because antisense RNA-mediated inactivation of DAP3 gene allowed Hela cells to remain viable in the continuous presence of IFN-γ (Kissil et al., 1995). Moreover, it was found that antisense DAP3 RNA, as well as a dominant interfering form of DAP3, protected cells from apoptosis induced by Fas and TNFα (Kissil et al., 1999). Recently, DAP3 has been proposed as a requirement for TRAIL-induced apoptosis because it binds directly to TRAIL receptors and FADD to allow caspase-8 activation (Miyazaki and Reed, 2001). Although some discrepancies exist about the role of DAP3 in TRAIL-mediated cell death (Berger and Kretzler, 2002), DAP3 could represent another common factor in the signaling pathways stimulated by IFN-γ and TRAIL.

V. CONCLUSIONS

The development of new antitumor combined therapeutic strategies is a current subject of study as tumor cells often present different mechanisms of resistance to classical chemotherapy and radiation therapy. Moreover, in the case of breast tumor cells, chemotherapy seems to imply important immunosuppressive effects, which suggest the importance of finding alternative therapeutic approaches. IFN-γ has been shown to collaborate with the immune system in promoting protective host responses to primary and transplanted tumors. On the other hand, TRAIL seems to be a promising agent in cancer therapy based on its specificity against tumor cells and the safety of some recombinant versions of TRAIL upon systemic administration. It has been shown that TRAIL mediates, at least in part, the antitumor activity of NK cells, lymphocytes, monocytes, and dendritic cells, and what is more interesting, IFN-γ upregulates the expression of TRAIL in these cells. These data establish a link between both molecules and suggest that their combination could represent an effective antitumor strategy.

In breast tumor cells, we as well as other authors, have described the sensitization to TRAIL-induced apoptosis by IFN-γ. Apart from the upregulation of caspase-8, which has been suggested to play a certain role as a tumor suppressor gene, other molecules or signals activated by IFN-γ could be involved in the potentiation of TRAIL-mediated cell death, as already discussed. Therefore, breast cancer cells provide an important model to further study the interplay between TRAIL and IFN-γ and the potential use of both antitumor agents in cancer therapy.

ACKNOWLEDGMENTS

This work was supported by a grant from Ministerio de Ciencia y Tecnología (SAF2000-0118-C03-01). C. Ruiz-Ruiz and C. Ruiz de Almodóvar were recipients of a Ramón y Cajal contract from Ministerio de Ciencia y Tecnología and a fellowship from Fondo de Investigación Sanitaria (Exp. 00/9319), respectively.

REFERENCES

Aaltomaa, S., Lipponen, P., Eskelinen, M., Kosma, V. M., Marin, S., Alhava, E., and Syrjanen, K. (1992). Lymphocyte infiltrates as a prognostic variable in female breast cancer. *Eur. J. Cancer* **28A,** 859–864.

Adachi, Y., Taketani, S., Oyaizu, H., Ikebukuro, K., Tokunaga, R., and Ikehara, S. (1999). Apoptosis of colorectal adenocarcinoma induced by 5-FU and/or IFN-gamma through caspase 3 and caspase 8. *Int. J. Oncol.* **15,** 1191–1196.

Ahn, E. Y., Pan, G., Vickers, S. M., and McDonald, J. M. (2002). IFN-gamma upregulates apoptosis-related molecules and enhances Fas-mediated apoptosis in human cholangiocarcinoma. *Int. J. Cancer* **100,** 445–451.

Angiolillo, A. L., Sgadari, C., Taub, D. D., Liao, F., Farber, J. M., Maheshwari, S., Kleinman, H. K., Reaman, G. H., and Tosato, G. (1995). Human interferon-inducible protein 10 is a potent inhibitor of angiogenesis in vivo. *J. Exp. Med.* **182,** 155–162.

Ashkenazi, A., and Dixit, V. M. (1998). Death receptors: Signaling and modulation. *Science* **281,** 1305–1308.

Ashkenazi, A., Pai, R. C., Fong, S., Leung, S., Lawrence, D. A., Marsters, S. A., Blackie, C., Chang, L., McMurtrey, A. E., Hebert, A., DeForge, L., Koumenis, I. L., Lewis, D., Harris, L.,

Bussiere, J., Koeppen, H., Shahrokh, Z., and Schwall, R. H. (1999). Safety and antitumor activity of recombinant soluble Apo2 ligand. *J. Clin. Invest.* **104**, 155–162.

Barak, V., Kalickman, I., Nisman, B., Farbstein, H., Fridlender, Z. G., Baider, L., Kaplan, A., Stephanos, S., and Peretz, T. (1998). Changes in cytokine production of breast cancer patients treated with interferons. *Cytokine* **10**, 977–983.

Bartek, J., Iggo, R., Gannon, J., and Lane, D. P. (1990). Genetic and immunochemical analysis of mutant p53 in human breast cancer cell lines. *Oncogene* **5**, 893–899.

Berger, T., and Kretzler, M. (2002). TRAIL-induced apoptosis is independent of the mitochondrial apoptosis mediator DAP3. *Biochem. Biophys. Res. Commun.* **297**, 880–884.

Bernard, D., Quatannens, B., Vandenbunder, B., and Abbadie, C. (2001). Rel/NF-kappaB transcription factors protect against tumor necrosis factor (TNF)-related apoptosis-inducing ligand (TRAIL)-induced apoptosis by up-regulating the TRAIL decoy receptor DcR1. *J. Biol. Chem.* **276**, 27322–27328.

Boehm, U., Klamp, T., Groot, M., and Howard, J. C. (1997). Cellular responses to interferon-gamma. *Annu. Rev. Immunol.* **15**, 749–795.

Borden, E. C., Holland, J. F., Dao, T. L., Gutterman, J. U., Wiener, L., Chang, Y. C., and Patel, J. (1982). Leukocyte-derived interferon (alpha) in human breast carcinoma. The American Cancer Society phase II trial. *Ann. Intern. Med.* **97**, 1–6.

Carnaud, C., Lee, D., Donnars, O., Park, S. H., Beavis, A., Koezuka, Y., and Bendelac, A. (1999). Cutting edge: Cross-talk between cells of the innate immune system: NKT cells rapidly activate NK cells. *J. Immunol.* **163**, 4647–4650.

Chai, J., Du, C., Wu, J. W., Kyin, S., Wang, X., and Shi, Y. (2000). Structural and biochemical basis of apoptotic activation by Smac/DIABLO. *Nature* **406**, 855–862.

Chawla-Sarkar, M., Leaman, D. W., and Borden, E. C. (2001). Preferential induction of apoptosis by interferon (IFN)-beta compared with IFN-alpha2: Correlation with TRAIL/Apo2L induction in melanoma cell lines. *Clin. Cancer Res.* **7**, 1821–1831.

Chin, Y. E., Kitagawa, M., Kuida, K., Flavell, R. A., and Fu, X. Y. (1997). Activation of the STAT signaling pathway can cause expression of caspase 1 and apoptosis. *M. Cell Biol.* **17**, 5328–5337.

Chinnaiyan, A. M., Prasad, U., Shankar, S., Hamstra, D. A., Shanaiah, M., Chenevert, T. L., Ross, B. D., and Rehemtulla, A. (2000). Combined effect ot tumor necrosis factor-related apoptosis-inducing ligand and ionizing radiation in breast cancer therapy. *Proc. Natl. Acad. Sci. USA* **97**, 1754–1759.

Coughlin, C. M., Salhany, K. E., Gee, M. S., LaTemple, D. C., Kotenko, S., Ma, X., Gri, G., Wysocka, M., Kim, J. E., Liu, L., Liao, F., Farber, J. M., Pestka, S., Trinchieri, G., and Lee, W. M. (1998). Tumor cell responses to IFNgamma affect tumorigenicity and response to IL-12 therapy and antiangiogenesis. *Immunity* **9**, 25–34.

Darnell, J. E., Jr., Kerr, I. M., and Stark, G. R. (1994). Jak-STAT pathways and transcriptional activation in response to IFNs and other extracellular signaling proteins. *Science* **264**, 1415–1421.

de Almodóvar, C. R., Ruiz-Ruiz, C., Munoz-Pinedo, C., Robledo, G., and Lopez-Rivas, A. (2001). The differential sensitivity of Bcl-2-overexpressing human breast tumor cells to TRAIL or doxorubicin-induced apoptosis is dependent on Bcl-2 protein levels. *Oncogene* **20**, 7128–7133.

Degli-Esposti, M. A., Dougall, W. C., Smolak, P. J., Waugh, J. Y., Smith, C. A., and Goodwin, R. G. (1997). The novel receptor TRAIL-R4 induces NF-kappaB and protects against TRAIL-mediated apoptosis, yet retains an incomplete death domain. *Immunity* **7**, 813–820.

Deng, Y., Lin, Y., and Wu, X. (2002). TRAIL-induced apoptosis requires Bax-dependent mitochondrial release of Smac/DIABLO. *Genes Dev.* **16**, 33–45.

Disis, M. L., Calenoff, E., McLaughlin, G., Murphy, A. E., Chen, W., Groner, B., Jeschke, M., Lydon, N., McGlynn, E., and Livingston, R. B. (1994). Existent T-cell and antibody immunity to HER-2/neu protein in patients with breast cancer. *Cancer Res.* **54**, 16–20.

Domanski, P., and Colamonici, O. R. (1996). The type-I interferon receptor. The long and short of it. *Cytokine Growth Factor Rev.* **7,** 143–151.

Ealick, S. E., Cook, W. J., Vijay-Kumar, S., Carson, M., Nagabhushan, T. L., Trotta, P. P., and Bugg, C. E. (1991). Three-dimensional structure of recombinant human interferon-gamma. *Science* **252,** 698–702.

Fallarino, F., and Gajewski, T. F. (1999). Cutting edge: Differentiation of antitumor CTL in vivo requires host expression of Stat1. *J. Immunol.* **163,** 4109–4113.

Fanger, N. A., Maliszewski, C. R., Schooley, K., and Griffith, T. S. (1999). Human dendritic cells mediate cellular apoptosis via tumor necrosis factor-related apoptosis-inducing ligand (TRAIL). *J. Exp. Med.* **190,** 1155–1164.

Farrar, M. A., and Schreiber, R. D. (1993). The molecular cell biology of interferon-gamma and its receptor. *Annu. Rev. Immunol.* **11,** 571–611.

Ferreira, C. G., Span, S. W., Peters, G. J., Kruyt, F. A. E., and Giaccone, G. (2000). Chemotherapy triggers apoptosis in a caspase-8-dependent and mitochondria-controlled manner in the non-small cell lung cancer cell line NCI-H460. *Cancer Res.* **60,** 7133–7141.

Fisher, D. E. (1994). Apoptosis in cancer therapy: Crossing the threshold. *Cell* **78,** 539–542.

Forrester, K., Lupold, S. E., Ott, V. L., Chay, C. H., Band, V., Wang, X. W., and Harris, C. C. (1995). Effects of p53 mutants on wild-type p53-mediated transactivation are cell type dependent. *Oncogene* **10,** 2103–2111.

Frucht, D. M., Fukao, T., Bogdan, C., Schindler, H., O'Shea, J. J., and Koyasu, S. (2001). IFN-gamma production by antigen-presenting cells: Mechanisms emerge. *Trends Immunol.* **22,** 556–560.

Fulda, S., and Debatin, K. M. (2002). IFNgamma sensitizes for apoptosis by upregulating caspase-8 expression through the Stat1 pathway. *Oncogene* **21,** 2295–2308.

Fulda, S., Meyer, E., and Debatin, K. M. (2002). Inhibition of TRAIL-induced apoptosis by Bcl-2 overexpression. *Oncogene* **21,** 2283–2294.

Gessani, S., and Belardelli, F. (1998). IFN-gamma expression in macrophages and its possible biological significance. *Cytokine Growth Factor Rev.* **9,** 117–123.

Gil, J., and Esteban, M. (2000). The interferon-induced protein kinase (PKR), triggers apoptosis through FADD-mediated activation of caspase 8 in a manner independent of Fas and TNF-alpha receptors. *Oncogene* **19,** 3665–3674.

Gil, J., Garcia, M. A., and Esteban, M. (2002). Caspase 9 activation by the dsRNA-dependent protein kinase, PKR: Molecular mechanism and relevance. *FEBS Lett.* **529,** 249–255.

Golstein, P. (1997). Cell death: TRAIL and its receptors. *Curr. Biol.* **7,** R750–R753.

Gong, B., and Almasan, A. (2000). Genomic organization and transcriptional regulation of human Apo2/TRAIL gene. *Biochem. Biophys. Res. Commun.* **278,** 747–752.

Green, D. R., and Reed, J. C. (1998). Mitochondria and apoptosis. *Science* **281,** 1309–1312.

Griffith, T. S., Chin, W. A., Jackson, G. C., Lynch, D. H., and Kubin, M. Z. (1998). Intracellular regulation of TRAIL-induced apoptosis in human melanoma cells. *J. Immunol.* **161,** 2833–2840.

Griffith, T. S., Wiley, S. R., Kubin, M. Z., Sedger, L. M., Maliszewski, C. R., and Fanger, N. A. (1999). Monocyte-mediated tumoricidal activity via the tumor necrosis factor-related cytokine, TRAIL. *J. Exp. Med.* **189,** 1343–1354.

Gross, A., McDonnell, J. M., and Korsmeyer, S. J. (1999). BCL-2 family members and the mitochondria in apoptosis. *Genes Dev.* **13,** 1899–1911.

Guo, B. C., and Xu, Y. H. (2001). Bcl-2 over-expression and activation of protein kinase C suppress the trail-induced apoptosis in Jurkat T cells. *Cell Res.* **11,** 101–106.

Gutterman, J. U., Blumenschein, G. R., Alexanian, R., Yap, H. Y., Buzdar, A. U., Cabanillas, F., Hortobagyi, G. N., Hersh, E. M., Rasmussen, S. L., Harmon, M., Kramer, M., and Pestka, S. (1980). Leukocyte interferon-induced tumor regression in human metastatic breast cancer, multiple myeloma, and malignant lymphoma. *Ann. Intern. Med.* **93,** 399–406.

Harada, K., Toyooka, S., Shivapurkar, N., Maitra, A., Reddy, J. L., Matta, H., Miyajima, K., Timmons, C. F., Tomlinson, G. E., Mastrangelo, D., Hay, R. J., Chaudhary, P. M., and Gazdar, A. F. (2002). Deregulation of caspase 8 and 10 expression in pediatric tumors and cell lines. *Cancer Res.* **62,** 5897–5901.

Harper, N., Farrow, S. N., Kaptein, A., Cohen, G. M., and MacFarlane, M. (2001). Modulation of tumor necrosis factor apoptosis-inducing ligand-induced NF-kappa B activation by inhibition of apical caspases. *J. Biol. Chem.* **276,** 34743–34752.

Hopkins-Donaldson, S., Bodmer, J.-L., Bourloud, K. B., Brognara, C. B., Tschopp, J., and Gross, N. (2000). Loss of caspase-8 expression in highly malignant human neuroblastoma cells correlates with resistance to tumor necrosis factor-related apoptosis-inducing ligand-induced apoptosis. *Cancer Res.* **60,** 4315–4319.

Hopkins-Donaldson, S., Ziegler, A., Kurtz, S., Bigosch, C., Kandioler, D., Ludwig, C., Zangemeister-Wittke, U., and Stahel, R. (2003). Silencing of death receptor and caspase-8 expression in small cell lung carcinoma cell lines and tumors by DNA methylation. *Cell Death Differ.* **10,** 356–364.

Jänicke, R. U., Sprengart, M. L., Wati, M. R., and Porter, A. G. (1998). Caspase-3 is required for DNA fragmentation and morphological changes associated with apoptosis. *J. Biol. Chem.* **273,** 9357–9360.

Jo, M., Kim, T.-H., Seol, D.-W., Esplen, J. E., Dorko, K., Billiar, T. R., and Strom, S. C. (2000). Apoptosis induced in normal human hepatocytes by tumor necrosis factor-related apoptosis-inducing ligand. *Nat. Med.* **6,** 564–567.

Kayagaki, N., Yamaguchi, N., Nakayama, M., Eto, H., Okumura, K., and Yagita, H. (1999). Type I interferons (IFNs) regulate tumor necrosis factor-related apoptosis-inducing ligand (TRAIL) expression on human T cells: A novel mechanism for the antitumor effects of type I IFNs. *J. Exp. Med.* **189,** 1451–1460.

Keane, M. C., Ettenberg, S. A., Nau, M. M., Russell, E. K., and Lipkowitz, S. (1999). Chemotherapy augments TRAIL-induced apoptosis in breast cell lines. *Cancer Res.* **59,** 734–741.

Keane, M. M., Ettenberg, S. A., Lowrey, G. A., Russell, E. K., and Lipkowitz, S. (1996). Fas expression and function in normal and malignant breast cell lines. *Cancer Res.* **56,** 4791–4798.

Keane, M. M., Rubinstein, Y., Cuello, M., Ettenberg, S. A., Banerjee, P., Nau, M. M., and Lipkowitz, S. (2000). Inhibition of NF-kappaB activity enhances TRAIL mediated apoptosis in breast cancer cell lines. *Breast Cancer Res. Treat.* **64,** 211–219.

Kimura, T., Kadokawa, Y., Harada, H., Matsumoto, M., Sato, M., Kashiwazaki, Y., Tarutani, M., Tan, R. S., Takasugi, T., Matsuyama, T., Mak, T. W., Noguchi, S., and Taniguchi, T. (1996). Essential and non-redundant roles of p48 (ISGF3 gamma) and IRF-1 in both type I and type II interferon responses, as revealed by gene targeting studies. *Genes Cells* **1,** 115–124.

Kissil, J. L., Cohen, O., Raveh, T., and Kimchi, A. (1999). Structure-function analysis of an evolutionary conserved protein, DAP3, which mediates TNF-alpha- and Fas-induced cell death. *EMBO J.* **18,** 353–362.

Kissil, J. L., Deiss, L. P., Bayewitch, M., Raveh, T., Khaspekov, G., and Kimchi, A. (1995). Isolation of DAP3, a novel mediator of interferon-gamma-induced cell death. *J. Biol. Chem.* **270,** 27932–27936.

Kontny, H. U., Hammerle, K., Klein, R., Shayan, P., Mackall, C. L., and Niemeyer, C. M. (2001). Sensitivity of Ewing's sarcoma to TRAIL-induced apoptosis. *Cell Death Differ.* **8,** 506–514.

Koshiji, M., Adachi, Y., Sogo, S., Taketani, S., Oyaizu, N., Than, S., Inaba, M., Phawa, S., Hioki, K., and Ikehara, S. (1998). Apoptosis of colorectal adenocarcinoma (COLO 201) by tumour necrosis factor-alpha (TNF-alpha) and/or interferon-gamma (IFN-gamma), resulting from down-modulation of Bcl-2 expression. *Clin. Exp. Immunol.* **111,** 211–218.

Koshiji, M., Adachi, Y., Taketani, S., Takeuchi, K., Hioki, K., and Ikehara, S. (1997). Mechanisms underlying apoptosis induced by combination of 5-fluorouracil and interferon-gamma. *Biochem. Biophys. Res. Commun.* **240,** 376–381.

Kumar-Sinha, C., Varambally, S., Sreekumar, A., and Chinnaiyan, A. M. (2002). Molecular cross-talk between the TRAIL and interferon signaling pathways. *J. Biol. Chem.* **277,** 575–585.

Lamhamedi-Cherradi, S. E., Zheng, S. J., Maguschak, K. A., Peschon, J., and Chen, Y. H. (2003). Defective thymocyte apoptosis and accelerated autoimmune diseases in TRAIL-/- mice. *Nat. Immunol.* **4,** 255–260.

Langaas, V., Shahzidi, S., Johnsen, J. I., Smedsrod, B., and Sveinbjornsson, B. (2001). Interferon-gamma modulates TRAIL-mediated apoptosis in human colon carcinoma cells. *Anticancer Res.* **21,** 3733–3738.

Lawrence, D., Shahrokh, Z., Marsters, S., Achilles, K., Shih, D., Mounho, B., Hillan, K., Totpal, K., DeForge, L., Schow, P., Hooley, J., Sherwood, S., Pai, R., Leung, S., Khan, L., Gliniak, B., Bussiere, J., Smith, C. A., Strom, S. S., Kelley, S., Fox, J. A., Thomas, D., and Ashkenazi, A. (2001). Differential hepatocyte toxicity of recombinant Apo2L/TRAIL versions. *Nat. Med.* **7,** 383–385.

Lee, S. B., and Esteban, M. (1994). The interferon-induced double-stranded RNA-activated protein kinase induces apoptosis. *Virology* **199,** 491–496.

Lee, S. B., Rodriguez, D., Rodriguez, J. R., and Esteban, M. (1997). The apoptosis pathway triggered by the interferon-induced protein kinase PKR requires the third basic domain, initiates upstream of Bcl-2, and involves ICE-like proteases. *Virology* **231,** 81–88.

Leverkus, M., Sprick, M. R., Wachter, T., Mengling, T., Baumann, B., Serfling, E., Brocker, E. B., Goebeler, M., Neumann, M., and Walczak, H. (2003). Proteasome inhibition results in TRAIL sensitization of primary keratinocytes by removing the resistance-mediating block of effector caspase maturation. *Mol. Cell. Biol.* **23,** 777–790.

Levy-Strumpf, N., and Kimchi, A. (1998). Death associated proteins (DAPs): From gene identification to the analysis of their apoptotic and tumor suppressive functions. *Oncogene* **17,** 3331–3340.

Li, H., Zhu, H., Xu, C. J., and Yuan, J. (1998). Cleavage of BID by caspase 8 mediates the mitochondrial damage in the Fas pathway of apoptosis. *Cell* **94,** 491–501.

Lin, T., Huang, X., Gu, J., Zhang, L., Roth, J. A., Xiong, M., Curley, S. A., Yu, Y., Hunt, K. K., and Fang, B. (2002). Long-term tumor-free survival from treatment with the GFP-TRAIL fusion gene expressed from the hTERT promoter in breast cancer cells. *Oncogene* **21,** 8020–8028.

Lu, G., Janjic, B. M., Janjic, J., Whiteside, T. L., Storkus, W. J., and Vujanovic, N. L. (2002). Innate direct anticancer effector function of human immature dendritic cells. II. Role of TNF, lymphotoxin-alpha(1)beta(2), Fas ligand, and TNF-related apoptosis-inducing ligand. *J. Immunol.* **168,** 1831–1839.

Luo, X., Budihardjo, I., Zou, H., Slaughter, C., and Wang, X. (1998). Bid, a Bcl2 interacting protein, mediates cytochrome c release from mitochondria in response to activation of cell surface death receptors. *Cell* **94,** 481–490.

MacFarlane, M., Ahmad, M., Srinivasula, S. M., Fernandes-Alnemri, T., Cohen, G. M., and Alnemri, E. S. (1997). Identification and molecular cloning of two novel receptors for the cytotoxic ligand TRAIL. *J. Biol. Chem.* **272,** 25417–25420.

MacFarlane, M., Harper, N., Snowden, R. T., Dyer, M. J., Barnett, G. A., Pringle, J. H., and Cohen, G. M. (2002a). Mechanisms of resistance to TRAIL-induced apoptosis in primary B cell chronic lymphocytic leukaemia. *Oncogene* **21,** 6809–6818.

MacFarlane, M., Merrison, W., Bratton, S. B., and Cohen, G. M. (2002b). Proteasome-mediated degradation of Smac during apoptosis: XIAP promotes Smac ubiquitination in vitro. *J. Biol. Chem.* **277,** 36611–36616.

Mandal, M., Bandyopadhyay, D., Goepfert, T. M., and Kumar, R. (1998). Interferon-induces expression of cyclin-dependent kinase-inhibitors p21WAF1 and p27Kip1 that

prevent activation of cyclin-dependent kinase by CDK-activating kinase (CAK). *Oncogene* **16**, 217–225.
Martinez-Lorenzo, M. J., Alava, M. A., Gamen, S., Kim, K. J., Chuntharapai, A., Pineiro, A., Naval, J., and Anel, A. (1998). Involvement of APO2 ligand/TRAIL in activation-induced death of Jurkat and human peripheral blood T cells. *Eur. J. Immunol.* **28**, 2714–2725.
McCloskey, D. E., Armstrong, D. K., Jackisch, C., and Davidson, N. E. (1996). Programmed cell death in human breast cancer cells. *Recent Prog. Horm. Res.* **51**, 493–508.
Meng, R. D., and El-Deiry, W. S. (2001). p53-independent upregulation of KILLER/DR5 TRAIL receptor expression by glucocorticoids and interferon-gamma. *Exp. Cell Res.* **262**, 154–169.
Meurs, E. F., Galabru, J., Barber, G. N., Katze, M. G., and Hovanessian, A. G. (1993). Tumor suppressor function of the interferon-induced double-stranded RNA-activated protein kinase. *Proc. Natl. Acad. Sci. USA* **90**, 232–236.
Miyazaki, T., and Reed, J. C. (2001). A GTP-binding adapter protein couples TRAIL receptors to apoptosis-inducing proteins. *Nat. Immunol.* **2**, 493–500.
Munoz-Pinedo, C., Ruiz de Almodovar, C., and Ruiz-Ruiz, C. (2002). Death on the beach: A rosy forecast for the 21st century. *Cell Death Differ.* **9**, 1026–1029.
Nanni, P., Nicoletti, G., De Giovanni, C., Landuzzi, L., Di Carlo, E., Cavallo, F., Pupa, S. M., Rossi, I., Colombo, M. P., Ricci, C., Astolfi, A., Musiani, P., Forni, G., and Lollini, P. L. (2001). Combined allogeneic tumor cell vaccination and systemic interleukin 12 preventsmammary carcinogenesis in HER-2/neu transgenic mice. *J. Exp. Med.* **194**, 1195–1205.
Nastala, C. L., Edington, H. D., McKinney, T. G., Tahara, H., Nalesnik, M. A., Brunda, M. J., Gately, M. K., Wolf, S. F., Schreiber, R. D., Storkus, W. J. *et al.* (1994). Recombinant IL-12 administration induces tumor regression in association with IFN-gamma production. *J. Immunol.* **153**, 1697–1706.
Nesterov, A., Lu, X., Johnson, M., Miller, G. J., Ivashchenko, Y., and Kraft, A. S. (2001). Elevated AKT activity protects the prostate cancer cell line LNCaP from TRAIL-induced apoptosis. *J. Biol. Chem.* **276**, 10767–10774.
Ng, C. P., and Bonavida, B. (2002). X-linked inhibitor of apoptosis (XIAP) blocks Apo2 ligand/ tumor necrosis factor-related apoptosis-inducing ligand-mediated apoptosis of prostate cancer cells in the presence of mitochondrial activation: Sensitization by overexpression of second mitochondria-derived activator of caspase/direct IAP-binding protein with low pI (Smac/DIABLO). *Mol. Cancer Ther.* **1**, 1051–1058.
Ogasawara, J., Watanabe, F. R., Adachi, M., Matsuzawa, A., Kasugai, T., Kitamura, Y., Itoh, N., Suda, T., and Nagata, S. (1993). Lethal effect of the anti-Fas antibody in mice. *Nature* **364**, 806–809.
Oshikawa, K., Shi, F., Rakhmilevich, A. L., Sondel, P. M., Mahvi, D. M., and Yang, N. S. (1999). Synergistic inhibition of tumor growth in a murine mammary adenocarcinoma model by combinational gene therapy using IL-12, pro-IL-18, and IL-1beta converting enzyme cDNA. *Proc. Natl. Acad. Sci. USA* **96**, 13351–13356.
Ossina, N. K., Cannas, A., Powers, V. C., Fitzpatrick, P. A., Knight, J. D., Gilbert, J. R., Shekhtman, E. M., Tomei, L. D., Umansky, S. R., and Kiefer, M. C. (1997). Interferon-gamma modulates a p53-independent apoptotic pathway and apoptosis-related gene expression. *J. Biol. Chem.* **272**, 16351–16357.
Oya, M., Ohtsubo, M., Takayanagi, A., Tachibana, M., Shimizu, N., and Murai, M. (2001). Constitutive activation of nuclear factor-kappaB prevents TRAIL-induced apoptosis in renal cancer cells. *Oncogene* **20**, 3888–3896.
Pan, G., Ni, J., Wei, Y.-F., Yu, G.-L., Gentz, R., and Dixit, V. M. (1997a). An antagonist decoy receptor and a death domain-containing receptor for TRAIL. *Science* **277**, 815–818.
Pan, G., O'Rourke, K., Chinnaiyan, A. M., Gentz, R., Ebner, R., Ni, J., and Dixit, V. M. (1997b). The receptor for the cytotoxic ligand TRAIL. *Science* **276**, 111–113.

Park, D. J., Nakamura, H., Chumakov, A. M., Said, J. W., Miller, C. W., Chen, D. L., and Koeffler, H. P. (1994). Transactivational and DNA binding abilities of endogenous p53 in p53 mutant cell lines. *Oncogene* **9**, 1899–1906.

Park, S. Y., Billiar, T. R., and Seol, D. W. (2002). IFN-gamma inhibition of TRAIL-induced IAP-2 upregulation, a possible mechanism of IFN-gamma-enhanced TRAIL-induced apoptosis. *Biochem. Biophys. Res. Commun.* **291**, 233–236.

Pestka, S., Kotenko, S. V., Muthukumaran, G., Izotova, L. S., Cook, J. R., and Garotta, G. (1997). The interferon gamma (IFN-gamma) receptor: A paradigm for the multichain cytokine receptor. *Cytokine Growth Factor Rev.* **8**, 189–206.

Phillips, T. A., Ni, J., Pan, G., Ruben, S. M., Wei, Y. F., Pace, J. L., and Hunt, J. S. (1999). TRAIL (Apo-2L) and TRAIL receptors in human placentas: Implications for immune privilege. *J. Immunol.* **162**, 6053–6059.

Pulaski, B. A., Smyth, M. J., and Ostrand-Rosenberg, S. (2002). Interferon-gamma-dependent phagocytic cells are a critical component of innate immunity against metastatic mammary carcinoma. *Cancer Res.* **62**, 4406–4412.

Rakhmilevich, A. L., Janssen, K., Hao, Z., Sondel, P. M., and Yang, N. S. (2000). Interleukin-12 gene therapy of a weakly immunogenic mouse mammary carcinoma results in reduction of spontaneous lung metastases via a T-cell-independent mechanism. *Cancer Gene Ther.* **7**, 826–838.

Ravi, R., Bedi, G. C., Engstrom, L. W., Zeng, Q., Mookerjee, B., Gelinas, C., Fuchs, E. J., and Bedi, A. (2001). Regulation of death receptor expression and TRAIL/Apo2L-induced apoptosis by NF-kappaB. *Nat. Cell Biol.* **3**, 409–416.

Reed, J. C. (1999). Mechanisms of apoptosis avoidance in cancer. *Curr. Opin. Oncol.* **11**, 68–75.

Rochaix, P., Krajewski, S., Reed, J. C., Bonnet, F., Voigt, J. J., and Brousset, P. (1999). In vivo patterns of Bcl-2 family protein expression in breast carcinomas in relation to apoptosis. *J. Pathol.* **187**, 410–415.

Ruiz-Ruiz, C., and López-Rivas, A. (2002). Mitochondria-dependent and -independent mechanisms in tumour necrosis factor-related apoptosis-inducing ligand (TRAIL)-induced apoptosis are both regulated by interferon-gamma in human breast tumour cells. *Biochem. J.* **365**, 825–832.

Ruiz-Ruiz, C., Muñoz-Pinedo, C., and López-Rivas, A. (2000). Interferon-gamma treatment elevates caspase-8 expression and sensitizes human breast tumor cells to a death receptor-induced mitochondria-operated apoptotic program. *Cancer Res.* **60**, 5673–5680.

Ruiz-Ruiz, M. C., and López-Rivas, A. (1999). p53-mediated up-regulation of CD95 is not involved in genotoxic drug-induced apoptosis of human breast tumor cells. *Cell Death Differ.* **6**, 271–280.

Saleh, A., Srinivasula, S. M., Acharya, S., Fishel, R., and Alnemri, E. S. (1999). Cytochrome c and dATP-mediated oligomerization of Apaf-1 is a prerequisite for procaspase-9 activation. *J. Biol. Chem.* **274**, 17941–17945.

Sanceau, J., Boyd, D. D., Seiki, M., and Bauvois, B. (2002). Interferons inhibit tumor necrosis factor-alpha-mediated matrix metalloproteinase-9 activation via interferon regulatory factor-1 binding competition with NF-kappa B. *J. Biol. Chem.* **277**, 35766–35775.

Sarker, M., Ruiz-Ruiz, C., and López-Rivas, A. (2001). Activation of protein kinase C inhibits TRAIL-induced caspase activation, mitochondrial events and apoptosis in a human leukemic T cell line. *Cell Death Differ.* **8**, 172–181.

Savinova, O., Joshi, B., and Jagus, R. (1999). Abnormal levels and minimal activity of the dsRNA-activated protein kinase, PKR, in breast carcinoma cells. *Int. J. Biochem. Cell Biol.* **31**, 175–189.

Scaffidi, C., Fulda, S., Srinivasan, A., Friesen, C., Li, F., Tomaselli, K. J., Debatin, K.-M., Krammer, P. H., and Peter, M. E. (1998). Two CD95 (APO-1/Fas) signalling pathways. *EMBO J.* **17**, 1675–1687.

Schlichtholz, B., Legros, Y., Gillet, D., Gaillard, C., Marty, M., Lane, D., Calvo, F., and Soussi, T. (1992). The immune response to p53 in breast cancer patients is directed against immunodominant epitopes unrelated to the mutational hot spot. *Cancer Res.* **52,** 6380–6384.

Schlom, J., Kantor, J., Abrams, S., Tsang, K. Y., Panicali, D., and Hamilton, J. M. (1996). Strategies for the development of recombinant vaccines for the immunotherapy of breast cancer. *Breast Cancer Res. Treat.* **38,** 27–39.

Schneider, P., Bodmer, J. L., Thome, M., Hofmann, K., Holler, N., and Tschopp, J. (1997). Characterization of two receptors for TRAIL. *FEBS Lett.* **416,** 329–334.

Schreiber, R. D., Celada, A., and Buchmeier, N. (1986). The role of interferon-gamma in the induction of activated macrophages. *Ann. Inst. Pasteur Immunol.* **137C,** 203–206.

Sedger, L. M., Shows, D. M., Blanton, R. A., Peschon, J. J., Goodwin, R. G., Cosman, D., and Wiley, S. R. (1999). IFN-gamma mediates a novel antiviral activity through dynamic modulation of TRAIL and TRAIL receptor expression. *J. Immunol.* **163,** 920–926.

Seitz, S., Wassmuth, P., Fischer, J., Nothnagel, A., Jandrig, B., Schlag, P. M., and Scherneck, S. (2002). Mutation analysis and mRNA expression of trail-receptors in human breast cancer. *Int. J. Cancer* **102,** 117–128.

Sgadari, C., Angiolillo, A. L., Cherney, B. W., Pike, S. E., Farber, J. M., Koniaris, L. G., Vanguri, P., Burd, P. R., Sheikh, N., Gupta, G., Teruya-Feldstein, J., and Tosato, G. (1996). Interferon-inducible protein-10 identified as a mediator of tumor necrosis in vivo. *Proc. Natl. Acad. Sci. USA* **93,** 13791–13796.

Sgadari, C., Farber, J. M., Angiolillo, A. L., Liao, F., Teruya-Feldstein, J., Burd, P. R., Yao, L., Gupta, G., Kanegane, C., and Tosato, G. (1997). Mig, the monokine induced by interferon-gamma, promotes tumor necrosis in vivo. *Blood* **89,** 2635–2643.

Shankaran, V., Ikeda, H., Bruce, A. T., White, J. M., Swanson, P. E., Old, L. J., and Schreiber, R. D. (2001). IFNgamma and lymphocytes prevent primary tumour development and shape tumour immunogenicity. *Nature* **410,** 1107–1111.

Shimizu, S., Narita, M., and Tsujimoto, Y. (1999). Bcl-2 family proteins regulate the release of apoptogenic cytochrome c by the mitochondrial channel VDAC. *Nature* **399,** 483–487.

Shin, E. C., Ahn, J. M., Kim, C. H., Choi, Y., Ahn, Y. S., Kim, H., Kim, S. J., and Park, J. H. (2001a). IFN-gamma induces cell death in human hepatoma cells through a TRAIL/death receptor-mediated apoptotic pathway. *Int. J. Cancer* **93,** 262–268.

Shin, M. S., Kim, H. S., Lee, S. H., Park, W. S., Kim, S. Y., Park, J. Y., Lee, J. H., Lee, S. K., Lee, S. N., Jung, S. S., Han, J. Y., Kim, H., Lee, J. Y., and Yoo, N. J. (2001b). Mutations of tumor necrosis factor-related apoptosis-inducing ligand receptor 1 (TRAIL-R1) and receptor 2 (TRAIL-R2) genes in metastatic breast cancers. *Cancer Res.* **61,** 4942–4946.

Shyu, R. Y., Su, H. L., Yu, J. C., and Jiang, S. Y. (2000). Direct growth suppressive activity of interferon-alpha and -gamma on human gastric cancer cells. *J. Surg. Oncol.* **75,** 122–130.

Siegmund, D., Hadwiger, P., Pfizenmaier, K., Vornlocher, H. P., and Wajant, H. (2002). Selective inhibition of FLICE-like inhibitory protein expression with small interfering RNA oligonucleotides is sufficient to sensitize tumor cells for TRAIL-induced apoptosis. *Mol. Med.* **8,** 725–732.

Spets, H., Georgii-Hemming, P., Siljason, J., Nilsson, K., and Jernberg-Wiklund, H. (1998). Fas/APO-1 (CD95)-mediated apoptosis is activated by interferon-gamma and interferon-alpha interleukin-6 (IL-6)-dependent and IL-6-independent multiple myeloma cell lines. *Blood* **92,** 2914–2923.

Sprick, M. R., Weigand, M. A., Rieser, E., Rauch, C. T., Juo, P., Blenis, J., Krammer, P. H., and Walczak, H. (2000). FADD/MORT1 and caspase-8 are recruited to TRAIL receptors 1 and 2 and are essential for apoptosis mediated by TRAIL receptor 2. *Immunity* **12,** 599–609.

Stark, G. R., Kerr, I. M., Williams, B. R. G., Silverman, R. H., and Schreiber, R. D. (1998). How cells respond to interferons. *Annu. Rev. Biochem.* **67,** 227–264.

Steller, H. (1995). Mechanisms and genes of cellular suicide. *Science* **267,** 1445–1449.

Strander, H. (1986). Interferon treatment of human neoplasia. *Adv. Cancer Res.* **46,** 1–265.

Suk, K., Chang, I., Kim, Y. H., Kim, S., Kim, J. Y., Kim, H., and Lee, M. S. (2001). Interferon gamma (IFNgamma) and tumor necrosis factor alpha synergism in ME-180 cervical cancer cell apoptosis and necrosis. IFNgamma inhibits cytoprotective NF-kappa B through STAT1/IRF-1 pathways. *J. Biol. Chem.* **276**, 13153–13159.

Suliman, A., Lam, A., Datta, R., and Srivastava, R. K. (2001). Intracellular mechanisms of TRAIL: Apoptosis through mitochondrial-dependent and -independent pathways. *Oncogene* **20**, 2122–2133.

Sun, H., Gutierrez, P., Jackson, M. J., Kundu, N., and Fulton, A. M. (2000). Essential role of nitric oxide and interferon-gamma for tumor immunotherapy with interleukin-10. *J. Immunother.* **23**, 208–214.

Takeda, K., Hayakawa, Y., Smyth, M. J., Kayagaki, N., Yamaguchi, N., Kakuta, S., Iwakura, Y., Yagita, H., and Okumura, K. (2001a). Involvement of tumor necrosis factor-related apoptosis-inducing ligand in surveillance of tumor metastasis by liver natural killer cells. *Nat. Med.* **7**, 94–100.

Takeda, K., Smyth, M. J., Cretney, E., Hayakawa, Y., Yamaguchi, N., Yagita, H., and Okumura, K. (2001b). Involvement of tumor necrosis factor-related apoptosis-inducing ligand in NK cell-mediated and IFN-gamma-dependent suppression of subcutaneous tumor growth. *Cell. Immunol.* **214**, 194–200.

Takeda, K., Smyth, M. J., Cretney, E., Hayakawa, Y., Kayagaki, N., Yagita, H., and Okumura, K. (2002). Critical role for tumor necrosis factor-related apoptosis-inducing ligand in immune surveillance against tumor development. *J. Exp. Med.* **195**, 161–169.

Thomas, W. D., and Hersey, P. (1998). TNF-related apoptosis-inducing ligand (TRAIL) induces apoptosis in Fas ligand-resistant melanoma cells and mediates CD4 T cell killing of target cells. *J. Immunol.* **161**, 2195–2200.

Thome, M., Schneider, P., Hofmann, K., Fickenscher, H., Meinl, E., Neipel, F., Mattmann, C., Burns, K., Bodmer, J. L., Schroter, M., Scaffidi, C., Krammer, P. H., Peter, M. E., and Tschopp, J. (1997). Viral FLICE-inhibitory proteins (FLIPs) prevent apoptosis induced by death receptors. *Nature* **386**, 517–521.

Thompson, C. B. (1995). Apoptosis in the pathogenesis and treatment of disease. *Science* **267**, 1456–1462.

Tomita, Y., Bilim, V., Hara, N., Kasahara, T., and Takahashi, K. (2003). Role of IRF-1 and caspase-7 in IFN-gamma enhancement of Fas-mediated apoptosis in ACHN renal cell carcinoma cells. *Int. J. Cancer* **104**, 400–408.

Tran, S. E., Holmstrom, T. H., Ahonen, M., Kahari, V. M., and Eriksson, J. E. (2001). MAPK/ERK overrides the apoptotic signaling from Fas, TNF, and TRAIL receptors. *J. Biol. Chem.* **276**, 16484–16490.

Trinchieri, G. (1995). Interleukin-12: A proinflammatory cytokine with immunoregulatory functions that bridge innate resistance and antigen-specific adaptive immunity. *Annu. Rev. Immunol.* **13**, 251–276.

Varela, N., Muñoz-Pinedo, C., Ruiz-Ruiz, C., Robledo, G., Pedroso, M., and López-Rivas, A. (2001). Interferon-gamma sensitises human myeloid leukemia cells to death receptor-mediated apoptosis by a pleiotropic mechanism. *J. Biol. Chem.* **276**, 17779–17787.

Verhagen, A. M., Ekert, P. G., Pakusch, M., Silke, J., Connolly, L. M., Reid, G. E., Moritz, R. L., Simpson, R. J., and Vaux, D. L. (2000). Identification of DIABLO, a mammalian protein that promotes apoptosis by binding to and antagonizing IAP proteins. *Cell* **102**, 43–53.

von Mensdorff-Pouilly, S., Gourevitch, M. M., Kenemans, P., Verstraeten, A. A., Litvinov, S. V., van Kamp, G. J., Meijer, S., Vermorken, J., and Hilgers, J. (1996). Humoral immune response to polymorphic epithelial mucin (MUC-1) in patients with benign and malignant breast tumours. *Eur. J. Cancer* **32A**, 1325–1331.

Wadler, S., and Schwartz, E. L. (1990). Antineoplastic activity of the combination of interferon and cytotoxic agents against experimental and human malignancies: A review. *Cancer Res.* **50**, 3473–3486.

Walczak, H., Degli-Esposti, M. A., Johnson, R. S., Smolak, P. J., Waugh, J. Y., Boiani, N., Timour, M. S., Gerhart, M. J., Schooley, K. A., Smith, C. A., Goodwin, R. G., and Rauch, C. T. (1997). TRAIL-R2: A novel apoptosis-mediating receptor for TRAIL. *EMBO J.* **16,** 5386–5397.

Walczak, H., Miller, R. E., Ariail, K., Gliniak, B., Griffith, T. S., Kubin, M., Chin, W., Jones, J., Woodward, A., Le, T., Smith, C., Smolak, P., Goodwin, R. G., Rauch, C. T., Schuh, J. C. L., and Lynch, D. H. (1999). Tumoricidal activity of tumor necrosis factor-related apoptosis-inducing ligand *in vivo. Nat. Med.* **5,** 157–163.

Wang, C. Y., Mayo, M. W., Korneluk, R. G., Goeddel, D. V., and Baldwin, A. S., Jr. (1998). NF-kappaB antiapoptosis: Induction of TRAF1 and TRAF2 and c-IAP1 and c-IAP2 to suppress caspase-8 activation. *Science* **281,** 1680–1683.

Wang, Q., Ji, Y., Wang, X., and Evers, B. M. (2000). Isolation and molecular characterization of the 5′-upstream region of the human TRAIL gene. *Biochem. Biophys. Res. Commun.* **276,** 466–471.

Wesselborg, S., Engels, I. H., Rossmann, E., Los, M., and Schulze-Osthoff, K. (1999). Anticancer drugs induce caspase-8/FLICE activation and apoptosis in the absence of CD95 receptor/ligand interaction. *Blood* **93,** 3053–3063.

Wiley, S. R., Schooley, K., Smolak, P. J., Din, W. S., Huang, C. P., Nicholl, J. K., Sutherland, G. R., Smith, T. D., Rauch, C., Smith, C. A., and Goodwin, R. G. (1995). Identification and characterization of a new member of the TNF family that induces apoptosis. *Immunity* **3,** 673–682.

Wu, R. S., Kobie, J. J., Besselsen, D. G., Fong, T. C., Mack, V. D., McEarchern, J. A., and Akporiaye, E. T. (2001). Comparative analysis of IFN-gamma B7.1 and antisense TGF-beta gene transfer on the tumorigenicity of a poorly immunogenic metastatic mammary carcinoma. *Cancer Immunol. Immunother.* **50,** 229–240.

Yang, X., Merchant, M. S., Romero, M. E., Tsokos, M., Wexler, L. H., Kontny, U., Mackall, C. L., and Thiele, C. J. (2003). Induction of caspase 8 by interferon gamma renders some neuroblastoma (NB) cells sensitive to tumor necrosis factor-related apoptosis-inducing ligand (TRAIL) but reveals that a lack of membrane TR1/TR2 also contributes to TRAIL resistance in NB. *Cancer Res.* **63,** 1122–1129.

Yuan, X. J., and Whang, Y. E. (2002). PTEN sensitizes prostate cancer cells to death receptor-mediated and drug-induced apoptosis through a FADD-dependent pathway. *Oncogene* **21,** 319–327.

Zhang, X. D., Franco, A. V., Nguyen, T., Gray, C. P., and Hersey, P. (2000). Differential localization and regulation of death and decoy receptor for TNF-related apoptosis-inducing ligand (TRAIL) in human melanoma cells. *J. Immunol.* **164,** 3961–3970.

Zielinski, C. C., Muller, C., Kubista, E., Staffen, A., and Eibl, M. M. (1990). Effects of adjuvant chemotherapy on specific and non-specific immune mechanisms. *Acta Med. Austriaca* **17,** 11–14.

Zou, H., Li, Y., Liu, X., and Wang, X. (1999). An APAF-1-cytochrome c multimeric complex is a functional apoptosome that activates procaspase-9. *J. Biol. Chem.* **274,** 11549–11556.

Zuzak, T. J., Steinhoff, D. F., Sutton, L. N., Phillips, P. C., Eggert, A., and Grotzer, M. A. (2002). Loss of caspase-8 mRNA expression is common in childhood primitive neuroectodermal brain tumour/medulloblastoma. *Eur. J. Cancer* **38,** 83–91.

17

Retinoids and TRAIL: Two Cooperating Actors to Fight Against Cancer

Lucia Altucci* and Hinrich Gronemeyer[†]

*Dipartimento di Patologia Generale
Seconda Università degli Studi di Napoli
80138, Napoli, Italy
[†]Department of Cell Biology and Signal Transduction
Institut de Génétique et de Biologie Moléculaire et Cellulaire/
Centre National de la Recherche Scientifique/Institut National de la Santé et de la
Recherche Médicale/University Louis Pasteur, 67404 Illkirch Cedex
C. U. de Strasbourg, France

I. Introduction
II. Origin of Retinoids
III. Mechanism of Retinoid Action
IV. Anticancer Activity of Retinoic Acids
   A. RARα in Myeloid Leukemia
   B. RARβ: Marker for Tumor Progression or Tumor Suppressor?
   C. Blocking Tumor Promotion by Inhibiting AP1
   D. Rexinoids and Their Receptors are Not Silent
V. Retinoid Action on Human Cancers
   A. Molecular Genetics of APL
   B. Multiple Actions of Retinoic Acid in APL Cells
VI. APO2L/TRAIL and its Receptors
VII. Preventive and Therapeutic Potential of Retinoids: The TRAIL Connection
   A. Cancer Preventive Potential
   B. Anticancer Potential

C. How Does Retinoid Therapy of APL Work?
D. Atypical Retinoids and Death-Signaling Pathways
VIII. Future Directions
References

Multiple studies performed in *in vitro* and *in vivo* settings have confirmed the cancer therapeutic and cancer preventive capacity of retinoids and rexinoids. These compounds mediate their actions through the retinoid and rexinoid receptors, respectively, which exist in multiple isoforms and form a plethora of distinct heterodimers. Despite their apparent anticancer potential, with one exception the molecular basis of this activity has remained largely elusive. The exception concerns acute promyelocytic leukemia (APL), the prototype of retinoic acid-dependent differentiation therapy, for which both the molecular nature of the disease and the mechanism of action of retinoids are well understood. However, retinoids and rexinoids are active beyond the borderlines of the well-defined chromosomal translocation that gives rise to curable APL. In this context, particularly interesting is that retinoic acid induces a member of the tumor necrosis factor family, tumor necrosis factor–related apoptosis inducing ligand (TRAIL) or Apo2L. This ligand is exceptional in that it is capable of inducing apoptosis in cancer cells but not in normal cells. It is possible that this connection to the TRAIL signaling pathway contributes to the anti-tumor activity of retinoids and rexinoids. This review focuses on what is presently known about the regulation of cell life and death by the retinoid/rexinoid and TRAIL signaling pathways. © 2004 Elsevier Inc.

I. INTRODUCTION

Multicellular organisms require specific mechanisms of intercellular communication to organize their complex body plan during embryogenesis, and to maintain the physiologic functions of each organ throughout life. Retinoic acids (RAs) are signaling molecules that, together with their nuclear retinoid (RARα, β, and γ) and rexinoid (RXRα, β, and γ) receptors, establish genetic communication networks that are essential for embryonic development (Chytil, 1996; Clagett-Dame and DeLuca, 2002; Gavalas, 2002; Kastner *et al.*, 1995; Maden, 2002) and organogenesis (Gavalas, 2002; Niederreither *et al.*, 2001; Tulachan *et al.*, 2003). Furthermore, they also have important physiologic functions in the brain (Chiang *et al.*, 1998; de Urquiza *et al.*, 2000; Krezel *et al.*, 1998; Misner *et al.*, 2001) and the reproductive system (Kastner *et al.*, 1996); regulate phenomena such as organ "repair" (Massaro and Massaro, 2003); regeneration (Imai *et al.*, 2001b), and homeostasis (Imai *et al.*, 2001a); and, at the cellular level, trigger proliferation, differentiation, and death (Altucci *et al.*, 2001).

RARs and RXRs are transcription factors that act mainly, if not exclusively, as RAR–RXR heterodimers in which both ligands (agonists, antagonists, and several other types of ligands with defined characteristics) can modulate the generation of complexes that regulate transcription (Bourguet *et al.*, 2000; Germain *et al.*, 2002). These heterodimers have at least two distinct signaling options: (i) they modulate the frequency of transcription initiation of target genes after binding to RA response elements (RAREs) in their promoters, and recruit epigenetically active complexes and chromatin remodeling machines (for an *in vitro* study, see Dilworth *et al.*, 2000; for more general reviews, see Laudet and Gronemeyer, 2002; McKenna and O'Malley, 2002a,b; Westin *et al.*, 2000); and (ii) they affect the efficiency of other signaling pathways ('crosstalk') by mechanisms that have remained largely elusive. A well-established example is the mutual repression of transcriptional activation that is mediated by the AP1 (c-JUN–c-FOS) transcription factor and RARs (Chen *et al.*, 1995). In the case of a similar crosstalk between AP1 and the glucocorticoid receptor (reviewed by Gottlicher *et al.*, 1998), recent results indicate that the promoter and cell context define the nature of interactions in (some of) the transcriptional relevant complexes, thus determining if a given coactivator acts positively as coactivator, or negatively as corepressor (Rogatsky *et al.*, 2002). Thus, a picture emerges in which a complex pattern and the nature of transcription factor interactions determine the actual transcriptional outcome on a given target gene promoter. In addition post-translational modifications such as phosphorylation (Rochette-Egly, 2003) modulate the transcriptional capacity of the retinoid receptors. "Nongenomic effects" may contribute to (some of) the multiple physiologic actions of retinoids (Lopez-Carballo *et al.*, 2002) but more work is required to evaluate the impact of this signaling option.

Several observations over the past years have progressively solidified the reputation of retinoids and, more recently rexinoids (i.e., RXR-selective ligands), as promising anticancer agents. These are (i) the impressive success of the "differentiation therapy" of acute promyelocytic leukemia (APL) together with the elucidation of the molecular basis of the disease and the mechanism of action of the retinoid; (ii) the antitumor promotion activity of retinoids in models of chemical carcinogenesis; (iii) the epigenetic silencing or loss of the RARβ during tumor progression, suggesting that this isoform may be a tumor suppressor; and (iv) the recent discovery of a link between the retinoid and TRAIL signaling pathways, suggesting that the tumor-specific (and perhaps cancer preventive) effects of retinoids may be due to the activation of tumor-selective death signaling. These results have to be seen in the context of major advances in the design of synthetic retinoids that have provided molecules with entirely novel activities, such as partial, full, and inverse agonists; neutral antagonists; "dissociated" retinoids; isotype-selective and mixed-activity ligands that are, for example agonists for RXR and antagonists for RAR. This work has been facilitated by definition of the 3D structure of RAR and RXR ligand binding domain (hetero)dimers in presence and absence of various ligands (Bourguet *et al.*, 2000) and the availability of engineered ligand screening systems. While the molecular action of at least some ligands is reasonably well understood (Germain *et al.*, 2002) the pharmacologic exploitation of this toolkit is still in its infancy.

II. ORIGIN OF RETINOIDS

Dietary-derived all-*trans* RA (ATRA) is the main signaling retinoid in the body, and mediates its action through RAR–RXR heterodimers. Although 9-*cis* RA has been detected in humans, it remains to be shown that this isomer corresponds to a *bona fide* ligand for RARs and RXRs. In this context it is worth noting that in mouse brain a nonretinoid, docosahexanoic acid, has been identified as potential endogeneous ligand for RXR (de Urquiza *et al.*, 2000). Retro-retinoids represent another class of signaling vitamin A metabolites in humans but these molecules do not signal through the classical retinoid and rexinoid receptors and appear to act in a nongenomic manner (Hoyos *et al.*, 2000; Imam *et al.*, 2001). Numerous synthetic retinoids are presently available for experimental exploitation. Some of these compounds were generated with the aim of finding selective agonists and antagonists for each of the six receptors, which are known from gene-ablation studies to have distinct—albeit potentially redundant—functionalities. So, for nearly every receptor isotype, selective agonists or antagonists exist and furthermore, ligands have be synthesized that selectively activate a subset of the functions of the natural ligands. For example, ATRA can repress AP1-mediated signaling as well as directly activating RAR–RXR target genes, but ligands have been generated that perform only one of these functions (Chen *et al.*, 1995)

III. MECHANISM OF RETINOID ACTION

In the absence of the cognate agonists, or in the presence of certain antagonists (usually referred to as "inverse agonists"), the target genes of retinoic acid receptors are believed to be in a repressed state. This is apparently due to the (possibly gene-specific) recruitment of histone deacetylase (HDAC)–containing complexes that are tethered through co-repressors (CoRs)—such as nuclear receptor (NR) corepressor (NCoR) or silencing mediator for retinoid and thyroid hormone receptors (SMRT)—to the nonliganded (apo-) RAR–RXR dimer (Jepsen *et al.*, 2000). The consequence is local histone deacetylation, chromatin compaction, and silencing of target gene promoter/enhancer regions. It is not entirely clear if all or only a subset of RA-responsive genes can be silenced; note in this respect that ligand binding has been shown to be prerequisite for the receptor to interact *in vivo* with the RARβ2 responsive element (Chen *et al.*, 1996; Dey *et al.*, 1994). RAR agonist binding destabilizes the CoR-binding interface due to allosteric changes in the LBD. The most important conformational change is the movement of the carboxy terminal helix H12, which generates the interaction surface for coactivators (CoAs) (Bourguet *et al.*, 2000). It is possible that CoA interaction is the initiating event for the establishment of (at least) three types of complexes at the response-element-bound RAR–RXR heterodimer; however, it is unclear whether there is a precise order or requirement, cell or promoter selectivity of these events. For some nuclear receptor target genes a highly dynamic interaction of the enhancer/promoter regions with these (and additional) machineries has been observed (Burakov *et al.*, 2002; Reid *et al.*, 2003; Shang *et al.*, 2000). In any case one of the recruited complexes comprises the histone acetyltransferase (HAT) complexes—recruited by p160 CoAs—which acetylate histone tails and lead to chromatin decondensation over the target-gene promoter. The second machinery is the ATP-dependent chromatin remodeling complex (also referred to as SWI/SNF) (see, for example, Belandia *et al.*, 2002). A third type of complex, variously termed thyroid hormone receptor (TR)–associated protein (TRAP), vitamin D receptor–interacting protein (DRIP), or Srb and mediator protein–containing complex (SMCC), binds to the holo-receptor through one or more of its subunits and establishes contact with the basal transcription machinery. Consequently, transcription is activated by three principle mechanisms (Fig. 1): derepression, caused by chromatin decondensation, remodeling, and a receptor-dependent increase in the frequency of transcription initiation (Glass and Rosenfeld, 2000; Ito and Roeder, 2001; Rachez and Freedman, 2001). HATs can also acetylate proteins other than histones, such as p53, but the functional link with NR action, if it exists, is unclear. RAs do not act solely through the two subunits of the RAR–RXR heterodimer; indeed, RXR is a promiscuous heterodimerization partner for various NRs. Therefore, RXR ligands ("rexinoids") are subordinated to their partner ligands to avoid signaling conflicts (Germain *et al.*, 2002 and references therein). In adults, retinoids and/or their cognate receptors are required for the proper functioning of a number of organs, including the skin, lung, liver, neuronal and immune systems. Efforts are ongoing to

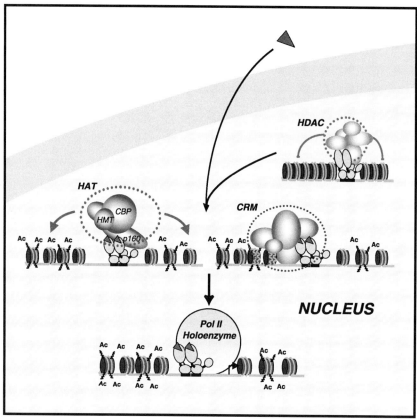

FIGURE 1. Machineries involved in nuclear receptor-mediated transcription activation. In the absence of ligand (triangle) several (but not all) nuclear receptors are believed to be bound to (some of) the regulatory regions of target genes as a corepressor or HDAC (histone deacetylase) complex. Histone deacetylation is generally believed to account for the gene silencing effect of apo receptors due to chromatin condensation. Ligand binding releases the HDAC complex and results in the recruitment of histone acetyltransferase (HAT) and chromatin-remodeling (CRM) compexes. The temporal order and requirement of (one or both of) these two complexes may occur in a receptor, target gene, and cell-specific fashion. As the last step the polymerase II holoenzyme, comprising in addition to the enzyme the TAF and MEDIATOR complexes, is recruited and increases the frequency of transcription initiation. While the temporal order of factor recruitment is being determined with increasing precision (Reid et al., 2003; Shang et al., 2000, 2002), complex composition and dynamics, the mechanistic basis for complex switching, and the receptor, promoter/enhancer, and cell specificities of these events are largely elusive.

improve our understanding of retinoid receptor function in adults using the CRE–LOX tissue-selective gene-ablation technology (for a recent review see Metzger and Chambon, 2001).

IV. ANTICANCER ACTIVITY OF RETINOIC ACIDS

Many factors contribute to tumorigenesis, including inherited and acquired genetic changes, chromosomal rearrangements, epigenetic phenomena, and chemical carcinogenesis. RAs can interfere with these events at several levels, their principal known actions being the induction of differentiation and/or apoptosis of tumor cells, and the inhibition of tumor promotion in chemically induced cancers. In keeping with their ability to regulate growth and induce differentiation throughout life, retinoids affect the growth of many tumor cell lines in culture. Given that there are six retinoid/rexinoid receptors, an important question is: which of the six RARs and RXRs are involved in the response to RA?

A. RARα IN MYELOID LEUKEMIA

APL can be effectively eradicated by retinoid signaling combined with chemotherapy, and therefore forms the prototype for retinoid-based therapies (Altucci and Gronemeyer, 2002; Altucci *et al.*, 2001; Fenaux *et al.*, 2001; Piazza *et al.*, 2001; Tenen, 2003). APL is caused by genetic translocations that lead to the production of fusion proteins containing RARα. These fusion proteins have maintained the ligand-responsive and DNA binding domains of RARα, producing a dysregulated but functional RAR (see later). APL is of prime importance for cancer research, not only because retinoid therapy is one of the rare success stories in the fight against cancer, but also because we understand most of the molecular details that explain the oncogenic properties of the fusion proteins, and the events that contribute to the differentiation and apoptosis of APL blasts.

B. RARβ: MARKER FOR TUMOR PROGRESSION OR TUMOR SUPPRESSOR?

In contrast to myeloid leukemias, a large body of evidence indicates an involvement of RARβ in a diverse range of solid tumors. Loss of RARβ expression is associated with human tumor progression; moreover, inducibility of the RARβ2 promoter by retinoic acid and/or expression of RARβ correlate with the growth-inhibitory effect of retinoic acid in solid tumors. Indeed, several ATRA-unresponsive tumor cell lines can be made ATRA responsive by introducing exogenous recombinant RARβ (for a recent summary see Altucci and Gronemeyer, 2001 and Table 2 in www-igbmc.u-strasbg.fr/Departments/Dep_I/Dep_IB/Publi/Paper1.html). In breast cancer cell lines, there is evidence that the RARβ2 promoter is silenced, and this is relieved by demethylation of DNA (Widschwendter *et al.*, 2000) or HDAC inhibition *in vitro*, as well as *in vivo* in xenograft tumors (Bovenzi and Momparler, 2001; Sirchia et al., 2002). Moreover, comparison of breast cancer biopsy specimens and non-neoplastic breast tissue indicated a correlation between RARβ2 silencing and methylation with tumor progression (Widschwendter et al., 2000). Recently an interesting link between the expression of an oncogenic transcription factor and methylation-dependent silencing of RARβ2 has been reported (Di Croce et al., 2002). The leukemogenic PML-RARα fusion protein induces RARβ2 hypermethylation and silencing by recruiting DNA methyltransferases to the promoter; this recruitment was apparently mediated by both the PML and (multimeric) RAR as no interaction of DNA methyltransferases was seen with (monomeric) RAR or RAR-RXR heterodimers. Retinoic acid treatment, known to relieve the differentiation block of APL cells, induces promoter demethylation and RARβ2 expression.

The gene program(s) that are regulated by RARβ are entirely unknown, but it is tempting to speculate that it may be related to two functionalities that distinguish RARβ from the other receptors: first, RARβ interacts only inefficiently with corepressors (Wong and Privalsky, 1998) resulting in an increased rexinoid responsiveness of RARβ–RXR heterodimers and second, RARβ constitutively represses AP1, in contrast to the ligand-dependent AP1 crosstalk of other retinoid receptors (Lin et al., 2000b). It is possible that these activities are unfavorable for rapid tumor growth, leading to the selection of cells in which RARβ is deleted or its expression or retinoid inducibility is impaired. A number of additional observations concerning RARβ2 regulation or action have been made. They comprise, for example, the report that expression of COUP-TF is required for RARβ2 promoter responsiveness to retinoic acid (Lin et al., 2000a) or the induction of RARβ2 by PPAR ligands (James et al., 2003), but it is unclear if and how these phenomena are linked to the tumor suppressive character of RARβ2.

C. BLOCKING TUMOR PROMOTION BY INHIBITING AP1

One of the models used to identify the functions of retinoid receptors that are required for the antitumor activity of retinoids is chemical carcinogenesis in skin. Chemical carcinogenesis is a multistep process comprising initiation, promotion, and progression. Typical skin carcinogenesis models combine two carcinogens—an initiator such as DMBA, and a promoter, which is often the phorbol ester TPA (Arora and Shukla, 2002; Chang et al., 2003). In a mouse epidermal cell line, retinoids block the promotion step, in which growth becomes anchorage independent, by inhibiting phorbol ester-induced AP1 activity (Dong et al., 1994). The antitumor activity of retinoids has also been demonstrated in experimental animals. Activation of *Hras1* is an early genetic change in mouse skin carcinogenesis that is associated with benign tumor formation (Greenhalgh et al., 1990). In a transgenic mouse that expresses oncogenic *Hras1*, phorbol ester–treated skin developed papillomas that increased in size and numbers and finally progressed with high frequency to invasive squamous cell carcinomas and underlying sarcomas (Leder *et al.*, 1990). Treatment with retinoic acid dramatically delayed, reduced, and often inhibited the appearance of these promoter-induced papillomas. Similarly, mammary carcinogenesis could be prevented by 9-*cis*-retinoic acid (Anzano *et al.*, 1994). It is tempting to speculate that repression of AP1 activity is responsible for this antioncogenic effect of the retinoid.

While the mechanistic basis of anti-AP1 activity of retinoids has remained elusive despite the proposal of several distinct mechanisms (critically reviewed by Gottlicher *et al.*, 1998; Resche-Rigon and Gronemeyer, 1998), the importance of this crosstalk for growth control is increasingly recognized. Particularly interesting are the observations that "dissociated" ligands (Chen *et al.*, 1995), which maintain the growth-inhibitory effects of ATRA, can block the growth-promoting action of the *Kras2* oncogene. These ligands do not (or at least do not efficiently) stimulate cognate gene transcription of RAR-responsive genes but are fully active in AP1 repression, suggesting that AP1 crosstalk accounts for the antitumor activity of retinoids in these models. The generation of "dissociated" and RAR/RXR isotype-selective ligands might lead to novel drugs with limited side effects.

D. REXINOIDS AND THEIR RECEPTORS ARE NOT SILENT

RXR is not naïve in APL cells and rexinoids can synergize with retinoids (Chen *et al.*, 1996) or exhibit RAR-independent activity (see subsequent discussion). In some AML cell lines, such as HL60, rexinoids are required in addition to retinoids to trigger post-maturation apoptosis (Nagy *et al.*, 1995; Rossin *et al.*, unpublished data).

Recent genetic data implicate RXRs in the chemopreventive activity of retinoic acid in experimental skin carcinogenesis. Conditional knockout mice lacking RXRα in their epidermis (Li *et al.*, 2000b) are hypersensitive to DMBA/TPA-induced skin tumorigenesis. Compared with wild-type littermates, RXR$\alpha^{-/-}$ mice have an increased frequency of papilloma formation, and these progress more rapidly to squamous, spinous, and basal cell carcinomas. Mice lacking epidermal RXRα do not respond to chemoprevention by retinoic acid (Chambon, personal communication). Whether this effect of RXR ablation results from a defect of autonomous RXR signaling (Benoit *et al.*, 2001) or from a lack of RXR signaling as a heterodimer with other nuclear receptors remains to be established.

Rexinoids are becoming increasingly appreciated not simply as silent heterodimerization partners of other nuclear receptors, but as active transducers of tumor suppressive signals. As well as being required for apoptosis induction following retinoid-induced differentiation (Nagy *et al.*, 1995) and synergizing with retinoids to inhibit growth (Chen *et al.*, 1996; Sun *et al.*, 2000b), under certain conditions they have potent effects independent of retinoids. These include induction of apoptosis *in vitro* (Benoit *et al.*, 2001) and in experimental animals the prevention of chemically induced mammary cancer, the regression of mammary carcinomas with signs of adipogenesis, and the inhibition of the growth of uterine leiomyoma (Bischoff *et al.*, 1998; Gamage *et al.*, 2000; Gottardis *et al.*, 1996). Notably, the RXR-selective compound (LGD1069, bexarotene, targretin) used in the latter studies did not manifest side effects that are normally associated with retinoid therapy. The minimal toxicity of oral bexarotene was recently confirmed in a multicenter phase II clinical study for patients with refractory metastatic breast cancer. For this group of patients, only limited efficacy of bexarotene was observed (Esteva *et al.*, 2003), while its safety and efficacy both as monotherapy and combination therapy is well established in the treatment of cutaneous T-cell lymphoma (Duvic *et al.*, 2001), possibly by inducing apoptosis, and may be effective for lymphomatoid papulosis (Krathen *et al.*, 2003).

V. RETINOID ACTION ON HUMAN CANCERS

A. MOLECULAR GENETICS OF APL

In the vast majority of cases, the origin of APL is a t(15;17)(22; q11.2-12) chromosomal translocation that fuses the promyelocytic leukemia gene *PML* and the *RAR*α gene (for a review, see Altucci and Gronemeyer, 2002). In rare cases, alternative chromosomal translocations generate RARα fusion proteins in which PML is replaced with PLZF, NUMA, NPM, or STAT5b (reviewed by Pandolfi, 2001). The generation of the leukemogenic PML–RARα has several molecular consequences. Compared to RARα, PML–RARα has gained the ability to form dimers or oligomers, which is both necessary and sufficient for its increased binding efficiency to corepressors/HDAC complexes. Indeed, RARα with a heterologous oligomerization domain recapitulates the properties of PML-RARα. This aberrant recruitment of corepressor-HDAC complexes accounts for the fact that nonphysiologic high doses of ATRA are needed to dissociate the HDAC-containing corepressor complex. The situation with PLZF–RARα is even worse: this fusion protein binds corepressors through both the apo-RARα and PLZF moieties, so ATRA cannot release HDACs at all. Consequently, PLZF-RARα remains a transcriptional repressor in the presence of ATRA. However, HDAC inhibitors (HDACi) can convert even PLZF-RARα into an activator of the retinoic acid signaling pathway (Grignani *et al.*, 1998; He *et al.*, 1998; Lin and Evans, 2000 and refs. cited therein; Lin *et al.*, 1998; Minucci *et al.*, 2000).

In addition, PML function is also disrupted by PML–RARα. PML is a multifunctional protein, involved in regulation of apoptosis, cell proliferation, and senescence (Salomoni and Pandolfi, 2002) acting as a transcriptional coactivator of p53 and regulates the p53 response to oncogenic signals (Guo *et al.*, 2000; Pearson *et al.*, 2000). PML is typically found in nuclear bodies (NBs)—macromolecular nuclear substructures that vary in number and size throughout the cell cycle and in response to stimuli such as interferon. PML–RARα interacts with the PML expressed from the intact allele, causing NB disintegration and relocalization of NB proteins in aberrant nuclear structures because PML is required for proper formation and stability of NBs. A large amount of data obtained from protein interaction, transcription and gene ablation studies has provided insight into some of the biological functions of PML. PML can suppress chemical carcinogenesis and is proapoptotic. $Pml^{-/-}$ mice are resistant to many apoptosis-inducing signals, whereas overexpression of *Pml* results in increased apoptosis (Zhong *et al.*, 2000) and senescence in a p53-dependent manner (Pearson *et al.*, 2000).

The final cell biologic outcome of these various aberrations that affect multiple signaling events by the formation of PML–RARα (or PLZF–RARα) is a block of differentiation at the promyelocytic stage. It is reasonable to assume that the altered functionality of PML in the fusion protein, such as loss of its proapoptogenic activity, or abrogation of corepressor function, adds to the growth potential/survival capacity of APL blasts, while HDAC-dependent silencing of "normal" retinoid signaling during myelopoiesis causes the differentiation block. However, there are still unexplained observations indicating that other genetic events, in addition to those generating the fusion protein, are involved in the etiology of APL (Piazza *et al.*, 2001).

B. MULTIPLE ACTIONS OF RETINOIC ACID IN APL CELLS

The genetics of nuclear receptors and APL explains the molecular basis of retinoid therapy: supraphysiologic concentrations of RA bind to the PML-RARα ligand-binding domain resulting in dissociation of the corepressor complex. This relieves the HDAC-dependent gene silencing, and block of differentiation and, through agonist-dependent recruitment of coactivator complexes, triggers the gene programs normally controlled by RARα-RXR. However, there are many more events induced by retinoic acid and we are just beginning to realize their importance in the APL and other cell systems. One example is the retinoic acid–induced proteolysis of PML-RARα and linked to this, the reorganization of the so-called "microspeckles" to intact nuclear bodies (Dyck *et al.*, 1994; Koken *et al.*, 1994; Weis *et al.*, 1994; Zhong *et al.*, 2000). This has led originally to the hypothesis that PML-RARα acts as a dominant-negative transcription factor whose degradation by retinoic acid results in reorganization of nuclear bodies, restoration of physiologic localization of PML and RXR, and return to "normal" retinoic acid–controlled gene programming but several arguments were not in support of this view (summarized by Grignani et al., 1999).

More recently it has been recognized that retinoic acid exerts a further activity in promyelocytic NB4 cells in vitro or APL patients' blasts ex vivo, namely the induction of post-maturation apoptosis (Altucci et al., 2001). The underlying molecular mechanism has been solved (Clarke and Gronemeyer, unpublished data) and leads to an exciting convergence of the anticancer activities of the retinoic acid and interferon signaling pathways on a tumor-selective death ligand, TRAIL.

VI. APO2L/TRAIL AND ITS RECEPTORS

Apo2L/TRAIL, one of the TNF superfamily members that induce apoptosis via recruitment of death receptors, was originally identified and cloned based on sequence homology to the Fas/Apo1 ligand (FasL) and TNF (Pitti et al., 1996; Wiley et al., 1995). Apo2L/TRAIL is unusual compared to any other cytokine because it interacts with a complex system of receptors: two proapoptotic death receptors and three antiapoptotic decoys (Fig. 2) (Almasan and Ashkenazi, 2003).

In fact, two of the receptors that bind Apo2L/TRAIL contain cytoplasmic "death domains" and signal apoptosis: death receptor 4 (DR4/TRAIL-R1) (Pan et al., 1997b) and death receptor 5 (DR5/TRAIL-R2) (MacFarlane et al., 1997; Pan et al., 1997a; Screaton et al., 1997; Sheridan et al., 1997; Walczak et al., 1997). The other three receptors appear to act as "decoys." Decoy receptor 1 (DcR1) (Degli-Esposti et al., 1997b; Mongkolsapaya et al., 1998; Pan et al., 1997b; Sheridan et al., 1997) and DcR2 (Degli-Esposti et al., 1997a; Marsters et al., 1997) have close homology to the extracellular domains of DR4 and DR5, but are incapable of transmiting an apoptotic signal given that DcR2 has a truncated, nonfunctional cytoplasmic death domain while DcR1 lacks a cytosolic region and is anchored to the membrane through a glycophospholipid moiety. The last receptor, the soluble TNFR family member osteoprotegerin (OPG) (Emery et al., 1998; Simonet et al., 1997) was discovered first to bind the TNF superfamily member RANKL, and later Apo2L/TRAIL. However, a biologic connection between OPG and Apo2L/TRAIL remains to be firmly established. OPG has a low affinity for Apo2L/TRAIL at physiologic temperature (Truneh et al., 2000). Apo2L/TRAIL is expressed as a type 2 transmembrane protein; however, its extracellular domain can be proteolytically cleaved from the cell surface. Like most other TNF family members, Apo2L/TRAIL forms a homotrimer (Fig. 3) that binds three

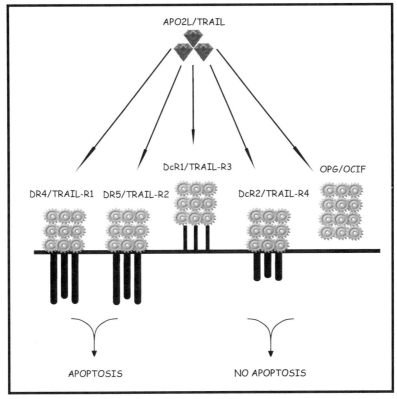

FIGURE 2. Apo2L/TRAIL and its receptors. TRAIL is a homotrimeric ligand that interacts with four highly related members of the TNFR superfamily. DR4/TRAIL-R1 and DR5/TRAIL-R2 contain a cytoplasmic death domain and signal apoptosis. DcR1/TRAIL-R3 is linked to the plasma membrane by a glycophosphatidylinositol moiety and lacks signaling activity. DcR2/TRAIL-R4 has a truncated, nonfunctional death domain. OPG/OCIF is a soluble more distantly related receptor capable of binding to TRAIL; the physiologic role of this possible interaction is unclear at present. Buttons illustrate the cystein-rich subdomains, termed CRDs, the horizontal line depicts the cytoplasmic membrane.

receptor molecules, each at the interface between two of its subunits (Hymowitz et al., 1999; Mongkolsapaya et al., 1999). A zinc atom bound by cysteines in the trimeric ligand is essential for trimer stability and optimal biologic activity (Bodmer et al., 2000; Hymowitz et al., 2000). Apo2L/TRAIL induces apoptosis by binding its death receptors and recruiting specific proteins such as the adaptor protein Fas–associated death-domain and caspase-8 and -10 to the intracellular death domain of the receptor, which form the death-inducing signaling complex (DISC) and consequently activation of the effector caspases such as caspase-3, -6, and -7 (illustrated in Fig. 4). Given that Apo2L/TRAIL induces apoptosis via both DR4 and 5, the specific function of these receptors remains

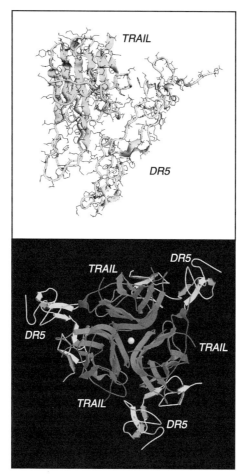

FIGURE 3. Crystal structure of the TRAIL-DR5 monomer and trimer. Top: Three-dimensional structure of the complex between human TRAIL (residues 91–281) and its cognate receptor, DR5 (residues 58–184), generated from the PDB file 1D4V (Mongkolsapaya *et al.*, 1999). Side chain residues are shown to illustrate the two binding interfaces. Bottom: Structure of the TRAIL-DR5 trimer generated from PDB file 1D0G (Hymowitz *et al.*, 1999). The central ball illustrates an essential zinc atom (Hymowitz *et al.*, 2000).

unclear. While apoptosis triggered by death ligands is considered a cell-extrinsic signaling pathway (LeBlanc and Ashkenazi, 2003), a cell-intrinsic death signaling activated by DNA damages exists that involves the activation and translocation of the Bcl-2 family member Bax to the mitochondria, dissipation of the mitochondrial potential, and cytochrome *c* release to the cytosol initiating the assembly of Apaf-1 and caspase-9 into the apoptosome.

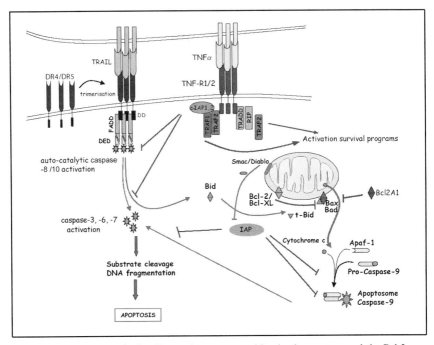

FIGURE 4. Apoptosis signaling pathways engaged by death receptors and the Bcl-2 gene superfamily. Engagement of the death receptors pathway by ApoL/TRAIL is sufficient to induce apoptosis in some cell types, while in others an amplification of the apoptotic signal through recruitment of the Bcl-2 family members is necessary for apoptosis. The crosstalk between the apoptosis pathways regulated by death receptors and Bcl-2 family members requires BID cleavage. Activation of caspase 3 is initiated by cleavage of caspase-8 and caspase-9. IAP family members can, on one hand, switch TNFα signaling into a survival pathway in presence of detectable levels of TRAF1, 2, and, on the other hand, block prodomain removal of caspase-3. Smac/Diablo can relieve this inhibition of apoptosis.

In some cell types (often referred to as "type 2 cells"), the mitochondrial death pathway is connected to Apo2L/TRAIL apoptosis via BID cleavage by caspase-8 and following activation of Bax and Bak, providing a mechanism of crosstalk between the extrinsic and the intrinsic pathway. Furthermore, it is possible that the requirement for the mitochondrial pathway may depend on the specific death receptor involved in the apoptotic signal and not—or not only—on the cell type. Therefore, it seems that in addition to promoting the activation of caspase-9, the mitochondrial pathway contributes to Apo2L/TRAIL-induced apoptosis by releasing Smac/DIABLO to the cytosol and relieving XIAP inhibition of caspase-3 (Du *et al.*, 2000; Ekert *et al.*, 2001). Finally, the finding of the FLICE inhibitory protein (FLIPs) that has high homology with caspase-8 and -10 lacking of protease activity, means that FLIP might be an inhibitor of caspase-8 activation at the Apo2L/TRAIL DISC (Irmler et al., 1997; Wang et al., 2000), but further observations are necessary for complete understanding of the role of FLIP at the DISC level.

VII. PREVENTIVE AND THERAPEUTIC POTENTIAL OF RETINOIDS: THE TRAIL CONNECTION

A. CANCER PREVENTIVE POTENTIAL

Much evidence supports the idea that retinoids can prevent cancer, inhibiting malignant progression. This notion is confirmed by i) epidemiologic studies and animal experiments indicating that retinoids reduce cancer risk; ii) inhibition of progression stage during chemical skin carcinogenesis; iii) efficiency in preneoplastic diseases such as leukoplakia, actinic keratosis, and cervical dysplasia; and iv) delay in the development of skin cancer in xeroderma pigmentosa patients. Increasing attention has been directed toward atypical retinoids (see subsequent discussion) because one of these (polyprenoic acid) is reported to prevent second primary hepatocarcinomas and a second one (4HPR) in combination with tamoxifen reduced the frequency of controlateral breast cancer in premenopausal women. Several observations indicate that classical and atypical retinoids may act via the TRAIL and TRAIL receptors pathway (see for example Sun et al., 2000b), but additional mechanisms have been proposed, such as the nuclear-to-mitochondria translocation of the TR3/NGFIB orphan nuclear receptor by CD437 (also termed "AHPN") and similar compounds (Dawson et al., 2001; Li et al., 2000a). It is tempting to speculate that cancer preventive activity of some retinoids might be linked to the induction of the cancer selective apoptogenic activity of TRAIL.

B. ANTICANCER POTENTIAL

Various cancers are already being treated with retinoid-based therapies and several are undergoing clinical evaluation. For several neoplastic diseases more than one retinoid is being evaluated and many clinical studies are underway to assess their efficacy (reviewed in Altucci and Gronemeyer, 2001). Retinoids are also being combined with other therapies, such as the combination of ret(x)inoids with chemotherapy (which can induce TRAIL receptor expression via p53), interferons (which activate the TRAIL signaling pathway), or antiestrogens in breast cancer treatment. The prototype of retinoid-based anticancer therapy is APL; for this leukemia, RA combined with chemotherapy is the first-line therapy.

C. HOW DOES RETINOID THERAPY OF APL WORK?

The molecular genetics of nuclear receptors and APL explain the basis of retinoid therapy: ATRA, RARα agonists, or RARα agonists/antagonists bind to the PML–RARα LBD resulting in dissociation of the corepressor complex. This relieves the HDAC-dependent block of differentiation and, through association of coactivator complexes, triggers the transcriptional programs normally controlled by RARα-RXR heterodimers. Eradication of PML-RARα by ATRA-induced proteolysis adds to the therapeutic effect, particularly in second-line therapy with As_2O_3 (Lallemand-Breitenbach *et al.*, 2001), but is not requisite for promyelocyte maturation (Benoit *et al.*, 1999).

In addition to the induction of antiapoptotic and survival programs, ATRA induces postmaturation apoptosis through the induction of TRAIL (Altucci *et al.*, 2001), a tumor-selective death ligand (French and Tschopp, 1999), and caspase-8, which mediates TRAIL action through the DR5/TRAIL-R2 receptor. This has several important implications. First, recombinant TRAIL, which is being developed as an anticancer agent, might prove useful in combination with lower doses of ATRA. This might reduce the adverse effects of ATRA, known as retinoic acid syndrome (Fenaux *et al.*, 2001). Second, it might be possible to substitute TRAIL—perhaps in combination with As_2O_3, for the chemotherapeutic agents that are currently used in combination with ATRA to reduce recurrence. Third, some chemotherapeutic drugs, as well as atypical retinoids, induce expression of the TRAIL receptors DR4/TRAIL-R1 and DR5/TRAIL-R2 (Bonavida *et al.*, 1999; Sun *et al.*, 2000b) so combining RARα-selective and atypical retinoids might improve treatment efficacy. Fourth, the known antitumor activity of retinoids might be linked to induction of TRAIL in other systems; indeed, our preliminary data reveal that TRAIL can also be induced by retinoids in other tumor cell lines (Jimenez-Lara and Clarke, unpublished data). Finally, it might be possible to use TRAIL induction as a surrogate marker for treatment efficacy, and as an assay for the development of novel retinoids that, for example, dissociate the apoptogenic activities of retinoid from their prosurvival effects.

D. ATYPICAL RETINOIDS AND DEATH-SIGNALING PATHWAYS

In addition to retinoids and rexinoids, the chemopreventive and chemotherapeutic potential of so-called atypical retinoids is being intensively investigated. The two main compounds in this class are 4-hydroxyphenylretinamide (4-HPR) and AHPN (also called CD437). These synthetic compounds are retinoids *in senso stricto* because they bind to and transactivate RARs, mainly RARγ and, more weakly, RARβ. However, this activity does not explain all their growth-inhibitory and apoptogenic effects because they are active in retinoid-resistant cells and retinoid antagonists do not completely block their activity (Holmes et al., 2000). Three novel apoptogenic signaling pathways have been described for AHPN. One involves the induction of three types of death receptors: Fas, which induces apoptosis upon binding Fas ligand (FasL); and DR4 and DR5, which are receptors for another apoptosis-inducing ligand, TRAIL (also called Apo2L) (Sun et al., 2000a,b,c). However, causal roles of increased death receptor expression in the induction of apoptosis were not demonstrated, but combined treatment of cells in culture with AHPN and TRAIL led to synergistic induction of cell death.

The second AHPN-induced death signaling involves the TR3/NGFIB/Nur77 orphan nuclear receptor, originally found to be necessary for activation-induced T cell apoptosis (which is negatively regulated by retinoids). In lung and prostate carcinoma cells, AHPN was reported to induce translocation of TR3 from the nucleus to mitochondria to induce cytochrome *c* release and apoptosis (Dawson et al., 2001; Li et al., 2000a). A third p53 APL signaling option is the AHPN-induced retinoid receptor-independent increase of *Egr1* and *TR3/NGFIB/Nur77* expression, which involves p38 MAP-kinase activation (Sakaue et al., 2001). Future research will have to clarify the contributions of the retinoid receptor, TR3 and Fas/TRAIL receptor–mediated activities to the growth-inhibitory and apoptogenic action of these compounds, as well as the mechanism of action of *retro*-retinoids, such as anhydroretinol, a physiologic metabolite of vitamin A that was reported to induce apoptosis in *in vitro* systems and prevent chemically induced mammary carcinogenesis in rats (reviewed in Chen et al., 1999).

VIII. FUTURE DIRECTIONS

As supported from the findings of the past few years, retinoids and rexinoids are a very promising class for cancer treatment and prevention. Different death signaling pathways have been found to be activated by retinoid treatment. One of the exciting objectives for future research is to combine our knowledge on the mechanisms that govern nuclear receptor action with the novel options for nuclear receptor–based drug design to interfere in a cancer-selective manner with aberrant cell growth. Because of its differential toxicity toward transformed and normal cells, the Apo2L/TRAIL signaling pathway shows promise both for the development of novel types of cancer therapies and as a tool for studying the molecular principles that underlie this apparent astounding selectivity. Is the anticancer activity of retinoids linked to TRAIL induction? The possibility of inducing expression of the TRAIL receptor DR5/TRAIL-R2 with atypical retinoids might pave the way for retinoid cocktail superinducers of apoptosis. Interferons (IFNs)-α and -β also induce TRAIL expression (Almasan and Ashkenazi, 2003; Ma *et al.*, 2001) and cotreatment of MCF-7 breast cancer cells with ATRA and IFN-β leads to upregulation of TRAIL, DR3, DR4/TRAIL-R1, DR5/TRAIL-R2, and Fas expression. Combinatorial treatments might lead to synergistic effects on growth control or induction of apoptosis, thereby allowing the use of lower concentrations as well as maintaining efficacy and reducing side effects. Several clinical trials of IFN–RA combinations are now in progress (Altucci and Gronemeyer, 2001), and it will be important to assess whether apoptogenic programs are induced in the tumor cells of patients treated with these combinations. Genetic alterations in mouse models, using tissue-selective conditional gene ablation technologies, should help to clarify the function of TNF family members and their receptors in retinoid signaling pathways. It has been demonstrated that Apo2L/TRAIL knockout mice have increased liver metastases, tumor growth of allografts in the mammary fat pad, and fibrosarcoma induction by methylcholantrene (Cretney *et al.*, 2002; Takeda *et al.*, 2002), and it will be interesting to test if ret(x)inods can have a therapeutic impact on the Apo2L/TRAIL pathways in these and related animal models. Mutations that distinguish between the ability of retinoid and rexinoid receptors to transactivate their own target genes and their ability to repress AP1 signaling and the use of "dissociated" retinoids (Chen *et al.*, 1995), should help to understand the contribution of AP1 crosstalk in retinoid-dependent cancer prevention in the skin carcinogenesis model and reveal its role in other growth-inhibitory and/or apoptogenic actions of retinoids. Moreover, isotype-selective simple or mixed activity agonists, antagonists, inverse agonists, or neutral antagonists have been synthesized as ligands for the retinoid/rexinoid receptor family. The potential of this enormous repertoire could be tested in genetically engineered cellular and animal models for defined growth regulatory capacities and for their capacity to activate Apo2L/TRAIL and Apo2L/TRAIL receptor pathway. On the receptor side, tissue-specific gene ablation(s) and mutant knock-ins might reveal why expression of RARβ is apparently incompatible with tumor progression (Sirchia *et al.*, 2002). One aspect to clarify is, for example, whether RARβ's ligand-independent ability to repress AP1 can contribute to what is often referred to as tumor suppressive action of RARβ.

Furthermore, defining the action spectra of HDACs, using genetics, antisense oligonucleotides, siRNAs and inhibitors, will provide the basis for targeted clinical use of potentially very powerful, isotype-selective, inhibitors. By combining these with retinoids or rexinoids, it might be possible to induce growth inhibitory ($p21^{WAF1}$), differentiative, and apoptogenic (TRAIL and/or TRAIL receptor induction; unblocking of RARβ inducibility) programs. In addition to HDAC inhibition, reversal of DNA hypermethylation by demethylating agents has been shown to restore

ATRA-mediated signaling in some leukemia and solid tumors. Defining and characterizing the players that modulate epigenetic events during tumorigenesis, and their effect on differentiation and apoptogenic pathways alone or in combination with other drugs such as retinoids might provide new tools to fight cancer.

Preliminary evidence indicates that specific p160 coactivators mediate the *in vivo* effects of a given nuclear receptor. These include TIF2 for RARs and AIB1 for estrogen-mediated growth of MCF7 breast cancer cells (List *et al.*, 2001). These studies should be extended to establish whether specific nuclear receptor–coregulator complexes mediate ligand-controlled growth in a (tumor) cell-type selective fashion.

Rexinoid signaling also deserves further analysis. If a rexinoid signaling activates a default death program (Benoit *et al.*, 2001), it will be important to define the endogenous rexinoids that trigger it, and the survival programs that sustain life.

The accumulated knowledge of the mechanistic, molecular, and pharmacologic actions of retinoids, together with the possibility of creating novel types of retinoid-related molecules with defined activities, is the basis for the development of efficient anticancer therapies, as is obvious from the success story of APL treatment with ATRA. Depending on the target cell and tumor type, we might, in the future, be able to choose combinations of drugs whose actions range from pleiotropic (as for atypical retinoids) to highly specific and tumor-selective (e.g., TRAIL-inducing RARα) ligands. Obviously, retinoids and rexinoids are a very promising class of compounds that will greatly enlarge our arsenal in the fight against cancer.

ACKNOWLEDGMENTS

Supported by the European Community (QLG1-CT-2000-01935 and QLK3-CT-2002-02029), the Institut Nationale de la Santé et de la Recherche Médicale, the Centre National de La Recherche Scientifique, the Hôpital Universitaire de Strasbourg, the Association for International Cancer Research, the Association pour le Recherche sur le Cancer, the Fondation de France, the Regione Campania L.41/94 annualità 2000, Ministero dell Salute R.F.02/184, French-Italian GALILEO project. We thank Sabrina Kammerer and Emmanuelle Wilhelm for their kind help in constructing Figs. 3 and 4, respectively.

REFERENCES

Almasan, A., and Ashkenazi, A. (2003). Apo2L/TRAIL: Apoptosis signaling, biology, and potential for cancer therapy. *Cytokine Growth Factor Rev.* **14,** 337–348.

Altucci, L., and Gronemeyer, H. (2001). The promise of retinoids to fight against cancer. *Nature Rev. Cancer* **1,** 181–193.

Altucci, L., and Gronemeyer, H. (2002). Decryption of the retinoid death code in leukemia. *J. Clin. Immunol.* **22,** 117–123.

Altucci, L., Rossin, A., Raffelsberger, W., Reitmair, A., Chomienne, C., and Gronemeyer, H. (2001). Retinoic acid-induced apoptosis in leukemia cells is mediated by paracrine action of tumor-selective death ligand TRAIL. *Nat. Med.* **7**, 680–686.

Anzano, M. A., Byers, S. W., Smith, J. M., Peer, C. W., Mullen, L. T., Brown, C. C., Roberts, A. B., and Sporn, M. B. (1994). Prevention of breast cancer in the rat with 9-cis-retinoic acid as a single agent and in combination with tamoxifen. *Cancer Res.* **54**, 4614–4617.

Arora, A., and Shukla, Y. (2002). Induction of apoptosis by diallyl sulfide in DMBA-induced mouse skin tumors. *Nutr. Cancer* **44**, 89–94.

Belandia, B., Orford, R. L., Hurst, H. C., and Parker, M. G. (2002). Targeting of SWI/SNF chromatin remodelling complexes to estrogen-responsive genes. *EMBO J.* **21**, 4094–4103.

Benoit, G., Altucci, L., Flexor, M., Ruchaud, S., Lillehaug, J., Raffelsberger, W., Gronemeyer, H., and Lanotte, M. (1999). RAR-independent RXR signaling induces t(15;17) leukemia cell maturation. *EMBO J.* **18**, 7011–7018.

Benoit, G. R., Flexor, M., Besancon, F., Altucci, L., Rossin, A., Hillion, J., Balajthy, Z., Legres, L., Segal-Bendirdjian, E., Gronemeyer, H. *et al.* (2001). Autonomous rexinoid death signaling is suppressed by converging signaling pathways in immature leukemia cells. *Mol. Endocrinol.* **15**, 1154–1169.

Bischoff, E. D., Gottardis, M. M., Moon, T. E., Heyman, R. A., and Lamph, W. W. (1998). Beyond tamoxifen: The retinoid X receptor-selective ligand LGD1069 (TARGRETIN) causes complete regression of mammary carcinoma. *Cancer Res.* **58**, 479–484.

Bodmer, J. L., Meier, P., Tschopp, J., and Schneider, P. (2000). Cysteine 230 is essential for the structure and activity of the cytotoxic ligand TRAIL. *J. Biol. Chem.* **275**, 20632–20637.

Bonavida, B., Ng, C. P., Jazirehi, A., Schiller, G., and Mizutani, Y. (1999). Selectivity of TRAIL-mediated apoptosis of cancer cells and synergy with drugs: The trail to non-toxic cancer therapeutics (review). *Int. J. Oncol.* **15**, 793–802.

Bourguet, W., Germain, P., and Gronemeyer, H. (2000). Nuclear receptor ligand-binding domains: Three-dimensional structures, molecular interactions and pharmacological implications. *Trends Pharmacol. Sci.* **21**, 381–388.

Bovenzi, V., and Momparler, R. L. (2001). Antineoplastic action of 5-aza-2′-deoxycytidine and histone deacetylase inhibitor and their effect on the expression of retinoic acid receptor beta and estrogen receptor alpha genes in breast carcinoma cells. *Cancer Chemother. Pharmacol.* **48**, 71–76.

Burakov, D., Crofts, L. A., Chang, C. P., and Freedman, L. P. (2002). Reciprocal recruitment of DRIP/mediator and p160 coactivator complexes in vivo by estrogen receptor. *J. Biol. Chem.* **277**, 14359–14362.

Chang, W. C., Jeng, J. H., Shieh, C. C., Tsai, Y. C., Ho, Y. S., Guo, H. R., Liu, H. I., Lee, C. C., Ho, S. Y., and Wang, Y. J. (2003). Skin tumor-promoting potential and systemic effects of pentachlorophenol and its major metabolite tetrachlorohydroquinone in CD-1 mice. *Mol. Carcinogen.* **36**, 161–170.

Chen, J. Y., Clifford, J., Zusi, C., Starrett, J., Tortolani, D., Ostrowski, J., Reczek, P. R., Chambon, P., and Gronemeyer, H. (1996). Two distinct actions of retinoid-receptor ligands. *Nature* **382**, 819–822.

Chen, J. Y., Penco, S., Ostrowski, J., Balaguer, P., Pons, M., Starrett, J. E., Reczek, P., Chambon, P., and Gronemeyer, H. (1995). RAR-specific agonist/antagonists which dissociate transactivation and AP1 transrepression inhibit anchorage-independent cell proliferation. *EMBO J.* **14**, 1187–1197.

Chen, Y., Buck, J., and Derguini, F. (1999). Anhydroretinol induces oxidative stress and cell death. *Cancer Res.* **59**, 3985–3990.

Chiang, M. Y., Misner, D., Kempermann, G., Schikorski, T., Giguere, V., Sucov, H. M., Gage, F. H., Stevens, C. F., and Evans, R. M. (1998). An essential role for retinoid receptors RARbeta and RXRgamma in long-term potentiation and depression. *Neuron* **21**, 1353–1361.

Chytil, F. (1996). Retinoids in lung development. *Faseb J.* **10**, 986–992.
Clagett-Dame, M., and DeLuca, H. F. (2002). The role of vitamin A in mammalian reproduction and embryonic development. *Annu. Rev. Nutr.* **22**, 347–381.
Cretney, E., Takeda, K., Yagita, H., Glaccum, M., Peschon, J. J., and Smyth, M. J. (2002). Increased susceptibility to tumor initiation and metastasis in TNF-related apoptosis-inducing ligand-deficient mice. *J. Immunol.* **168**, 1356–1361.
Dawson, M. I., Hobbs, P. D., Peterson, V. J., Leid, M., Lange, C. W., Feng, K. C., Chen, G., Gu, J., Li, H., Kolluri, S. K. *et al.* (2001). Apoptosis induction in cancer cells by a novel analogue of 6-[3-(1-adamantyl)-4-hydroxyphenyl]-2-naphthalenecarboxylic acid lacking retinoid receptor transcriptional activation activity. *Cancer Res.* **61**, 4723–4730.
de Urquiza, A. M., Liu, S., Sjoberg, M., Zetterstrom, R. H., Griffiths, W., Sjovall, J., and Perlmann, T. (2000). Docosahexaenoic acid, a ligand for the retinoid X receptor in mouse brain. *Science* **290**, 2140–2144.
Degli-Esposti, M. A., Dougall, W. C., Smolak, P. J., Waugh, J. Y., Smith, C. A., and Goodwin, R. G. (1997a). The novel receptor TRAIL-R4 induces NF-kappaB and protects against TRAIL-mediated apoptosis, yet retains an incomplete death domain. *Immunity* **7**, 813–820.
Degli-Esposti, M. A., Smolak, P. J., Walczak, H., Waugh, J., Huang, C. P., DuBose, R. F., Goodwin, R. G., and Smith, C. A. (1997b). Cloning and characterization of TRAIL-R3, a novel member of the emerging TRAIL receptor family. *J. Exp. Med.* **186**, 1165–1170.
Dey, A., Minucci, S., and Ozato, K. (1994). Ligand-dependent occupancy of the retinoic acid receptor beta 2 promoter in vivo. *Mol. Cell. Biol.* **14**, 8191–8201.
Di Croce, L., Raker, V. A., Corsaro, M., Fazi, F., Fanelli, M., Faretta, M., Fuks, F., Lo Coco, F., Kouzarides, T., Nervi, C. *et al.* (2002). Methyltransferase recruitment and DNA hypermethylation of target promoters by an oncogenic transcription factor. *Science* **295**, 1079–1082.
Dilworth, F. J., Fromental-Ramain, C., Yamamoto, K., and Chambon, P. (2000). ATP-Driven chromatin remodeling activity and histone acetyltransferases act sequentially during transactivation by RAR/RXR in vitro. *Mol. Cell* **6**, 1049–1058.
Dong, Z., Birrer, M. J., Watts, R. G., Matrisian, L. M., and Colburn, N. H. (1994). Blocking of tumor promoter-induced AP-1 activity inhibits induced transformation in JB6 mouse epidermal cells. *Proc. Natl. Acad. Sci. USA* **91**, 609–613.
Du, C., Fang, M., Li, Y., Li, L., and Wang, X. (2000). Smac, a mitochondrial protein that promotes cytochrome c-dependent caspase activation by eliminating IAP inhibition. *Cell* **102**, 33–42.
Duvic, M., Hymes, K., Heald, P., Breneman, D., Martin, A. G., Myskowski, P., Crowley, C., and Yocum, R. C. (2001). Bexarotene is effective and safe for treatment of refractory advanced-stage cutaneous T-cell lymphoma: Multinational phase II-III trial results. *J. Clin. Oncol.* **19**, 2456–2471.
Dyck, J. A., Maul, G. G., Miller, W. H., Jr., Chen, J. D., Kakizuka, A., and Evans, R. M. (1994). A novel macromolecular structure is a target of the promyelocyte-retinoic acid receptor oncoprotein. *Cell* **76**, 333–343.
Ekert, P. G., Silke, J., Hawkins, C. J., Verhagen, A. M., and Vaux, D. L. (2001). DIABLO promotes apoptosis by removing MIHA/XIAP from processed caspase 9. *J. Cell Biol.* **152**, 483–490.
Emery, J. G., McDonnell, P., Burke, M. B., Deen, K. C., Lyn, S., Silverman, C., Dul, E., Appelbaum, E. R., Eichman, C., DiPrinzio, R. *et al.* (1998). Osteoprotegerin is a receptor for the cytotoxic ligand TRAIL. *J. Biol. Chem.* **273**, 14363–14367.
Esteva, F. J., Glaspy, J., Baidas, S., Laufman, L., Hutchins, L., Dickler, M., Tripathy, D., Cohen, R., DeMichele, A., Yocum, R. C. *et al.* (2003). Multicenter phase II study of oral bexarotene for patients with metastatic breast cancer. *J. Clin. Oncol.* **21**, 999–1006.
Fenaux, P., Chomienne, C., and Degos, L. (2001). All-trans retinoic acid and chemotherapy in the treatment of acute promyelocytic leukemia. *Semin. Hematol.* **38**, 13–25.

French, L. E., and Tschopp, J. (1999). The TRAIL to selective tumor death [news; comment]. *Nat. Med.* **5,** 146–147.

Gamage, S. D., Bischoff, E. D., Burroughs, K. D., Lamph, W. W., Gottardis, M. M., Walker, C. L., and Fuchs-Young, R. (2000). Efficacy of LGD1069 (Targretin), a retinoid X receptor-selective ligand, for treatment of uterine leiomyoma. *J. Pharmacol. Exp. Ther.* **295,** 677–681.

Gavalas, A. (2002). ArRAnging the hindbrain. *Trends Neurosci.* **25,** 61–64.

Germain, P., Iyer, J., Zechel, C., and Gronemeyer, H. (2002). Coregulator recruitment and the mechanism of retinoic acid receptor synergy. *Nature* **415,** 187–192.

Glass, C. K., and Rosenfeld, M. G. (2000). The coregulator exchange in transcriptional functions of nuclear receptors. *Genes Dev.* **14,** 121–141.

Gottardis, M. M., Bischoff, E. D., Shirley, M. A., Wagoner, M. A., Lamph, W. W., and Heyman, R. A. (1996). Chemoprevention of mammary carcinoma by LGD1069 (Targretin): An RXR-selective ligand. *Cancer Res.* **56,** 5566–5570.

Gottlicher, M., Heck, S., and Herrlich, P. (1998). Transcriptional cross-talk, the second mode of steroid hormone receptor action. *J. Mol. Med.* **76,** 480–489.

Greenhalgh, D. A., Welty, D. J., Player, A., and Yuspa, S. H. (1990). Two oncogenes, v-fos and v-ras, cooperate to convert normal keratinocytes to squamous cell carcinoma. *Proc. Natl. Acad. Sci. USA* **87,** 643–647.

Grignani, F., De Matteis, S., Nervi, C., Tomassoni, L., Gelmetti, V., Cioce, M., Fanelli, M., Ruthardt, M., Ferrara, F. F., Zamir, I. *et al.* (1998). Fusion proteins of the retinoic acid receptor-alpha recruit histone deacetylase in promyelocytic leukaemia. *Nature* **391,** 815–818.

Grignani, F., Gelmetti, V., Fanelli, M., Rogaia, D., De Matteis, S., Ferrara, F. F., Bonci, D., Nervi, C., and Pelicci, P. G. (1999). Formation of PML/RAR alpha high molecular weight nuclear complexes through the PML coiled-coil region is essential for the PML/RAR alpha-mediated retinoic acid response. *Oncogene* **18,** 6313–6321.

Guo, A., Salomoni, P., Luo, J., Shih, A., Zhong, S., Gu, W., and Paolo Pandolfi, P. (2000). The function of PML in p53-dependent apoptosis. *Nat. Cell Biol.* **2,** 730–736.

He, L. Z., Guidez, F., Tribioli, C., Peruzzi, D., Ruthardt, M., Zelent, A., and Pandolfi, P. P. (1998). Distinct interactions of PML-RARalpha and PLZF-RARalpha with co-repressors determine differential responses to RA in APL. *Nat. Genet.* **18,** 126–135.

Holmes, W. F., Dawson, M. I., Soprano, R. D., and Soprano, K. J. (2000). Induction of apoptosis in ovarian carcinoma cells by AHPN/CD437 is mediated by retinoic acid receptors. *J. Cell. Physiol.* **185,** 61–67.

Hoyos, B., Imam, A., Chua, R., Swenson, C., Tong, G. X., Levi, E., Noy, N., and Hammerling, U. (2000). The cysteine-rich regions of the regulatory domains of Raf and protein kinase C as retinoid receptors. *J. Exp. Med.* **192,** 835–845.

Hymowitz, S. G., Christinger, H. W., Fuh, G., Ultsch, M., O'Connell, M., Kelley, R. F., Ashkenazi, A., and de Vos, A. M. (1999). Triggering cell death: The crystal structure of Apo2L/TRAIL in a complex with death receptor 5. *Mol. Cell* **4,** 563–571.

Hymowitz, S. G., O'Connell, M. P., Ultsch, M. H., Hurst, A., Totpal, K., Ashkenazi, A., de Vos, A. M., and Kelley, R. F. (2000). A unique zinc-binding site revealed by a high-resolution X-ray structure of homotrimeric Apo2L/TRAIL. *Biochemistry* **39,** 633–640.

Imai, T., Jiang, M., Chambon, P., and Metzger, D. (2001a). Impaired adipogenesis and lipolysis in the mouse upon selective ablation of the retinoid X receptor alpha mediated by a tamoxifen-inducible chimeric Cre recombinase (Cre-ERT2) in adipocytes. *Proc. Natl. Acad. Sci. USA* **98,** 224–228.

Imai, T., Jiang, M., Kastner, P., Chambon, P., and Metzger, D. (2001b). Selective ablation of retinoid X receptor alpha in hepatocytes impairs their lifespan and regenerative capacity. *Proc. Natl. Acad. Sci. USA* **98,** 4581–4586.

Imam, A., Hoyos, B., Swenson, C., Levi, E., Chua, R., Viriya, E., and Hammerling, U. (2001). Retinoids as ligands and coactivators of protein kinase C alpha. *Faseb J.* **15,** 28–30.

Irmler, M., Thome, M., Hahne, M., Schneider, P., Hofmann, K., Steiner, V., Bodmer, J. L., Schroter, M., Burns, K., Mattmann, C. *et al.* (1997). Inhibition of death receptor signals by cellular FLIP. *Nature* **388**, 190–195.

Ito, M., and Roeder, R. G. (2001). The TRAP/SMCC/Mediator complex and thyroid hormone receptor function. *Trends Endocrinol. Metab.* **12**, 127–134.

James, S. Y., Lin, F., Kolluri, S. K., Dawson, M. I., and Zhang, X. K. (2003). Regulation of retinoic acid receptor beta expression by peroxisome proliferator-activated receptor gamma ligands in cancer cells. *Cancer Res.* **63**, 3531–3538.

Jepsen, K., Hermanson, O., Onami, T. M., Gleiberman, A. S., Lunyak, V., McEvilly, R. J., Kurokawa, R., Kumar, V., Liu, F., Seto, E. *et al.* (2000). Combinatorial roles of the nuclear receptor corepressor in transcription and development. *Cell* **102**, 753–763.

Kastner, P., Mark, M., and Chambon, P. (1995). Nonsteroid nuclear receptors: What are genetic studies telling us about their role in real life? *Cell* **83**, 859–869.

Kastner, P., Mark, M., Leid, M., Gansmuller, A., Chin, W., Grondona, J. M., Decimo, D., Krezel, W., Dierich, A., and Chambon, P. (1996). Abnormal spermatogenesis in RXR beta mutant mice. *Genes Dev.* **10**, 80–92.

Koken, M. H., Puvion-Dutilleul, F., Guillemin, M. C., Viron, A., Linares-Cruz, G., Stuurman, N., de Jong, L., Szostecki, C., Calvo, F., Chomienne, C. *et al.* (1994). The t(15;17) translocation alters a nuclear body in a retinoic acid-reversible fashion. *EMBO J.* **13**, 1073–1083.

Krathen, R. A., Ward, S., and Duvic, M. (2003). Bexarotene is a new treatment option for lymphomatoid papulosis. *Dermatology* **206**, 142–147.

Krezel, W., Ghyselinck, N., Samad, T. A., Dupe, V., Kastner, P., Borrelli, E., and Chambon, P. (1998). Impaired locomotion and dopamine signaling in retinoid receptor mutant mice. *Science* **279**, 863–867.

Lallemand-Breitenbach, V., Zhu, J., Puvion, F., Koken, M., Honore, N., Doubeikovsky, A., Duprez, E., Pandolfi, P. P., Puvion, E., Freemont, P. *et al.* (2001). Role of promyelocytic leukemia (PML) sumolation in nuclear body formation, 11S proteasome recruitment, and As2O3-induced PML or PML/retinoic acid receptor alpha degradation. *J. Exp. Med.* **193**, 1361–1371.

Laudet, V. and Gronemeyer, H. (2002). The Nuclear Receptor Facts Book. Academic Press, San Diego.

LeBlanc, H. N., and Ashkenazi, A. (2003). Apo2L/TRAIL and its death and decoy receptors. *Cell Death Differ.* **10**, 66–75.

Leder, A., Kuo, A., Cardiff, R. D., Sinn, E., and Leder, P. (1990). v-Ha-ras transgene abrogates the initiation step in mouse skin tumorigenesis: Effects of phorbol esters and retinoic acid. *Proc. Natl. Acad. Sci. USA* **87**, 9178–9182.

Li, H., Kolluri, S. K., Gu, J., Dawson, M. I., Cao, X., Hobbs, P. D., Lin, B., Chen, G., Lu, J., Lin, F. *et al.* (2000a). Cytochrome c release and apoptosis induced by mitochondrial targeting of nuclear orphan receptor TR3. *Science* **289**, 1159–1164.

Li, M., Indra, A. K., Warot, X., Brocard, J., Messaddeq, N., Kato, S., Metzger, D., and Chambon, P. (2000b). Skin abnormalities generated by temporally controlled RXRalpha mutations in mouse epidermis. *Nature* **407**, 633–636.

Lin, B., Chen, G. Q., Xiao, D., Kolluri, S. K., Cao, X., Su, H., and Zhang, X. K. (2000a). Orphan receptor COUP-TF is required for induction of retinoic acid receptor beta, growth inhibition, and apoptosis by retinoic acid in cancer cells. *Mol. Cell. Biol.* **20**, 957–970.

Lin, F., Xiao, D., Kolluri, S. K., and Zhang, X. (2000b). Unique anti-activator protein-1 activity of retinoic acid receptor beta. *Cancer Res.* **60**, 3271–3280.

Lin, R. J., and Evans, R. M. (2000). Acquisition of oncogenic potential by RAR chimeras in acute promyelocytic leukemia through formation of homodimers. *Mol. Cell* **5**, 821–830.

Lin, R. J., Nagy, L., Inoue, S., Shao, W., Miller, W. H., Jr., and Evans, R. M. (1998). Role of the histone deacetylase complex in acute promyelocytic leukaemia. *Nature* **391**, 811–814.

List, H. J., Lauritsen, K. J., Reiter, R., Powers, C., Wellstein, A., and Riegel, A. T. (2001). Ribozyme targeting demonstrates that the nuclear receptor coactivator AIB1 is a rate-limiting factor for estrogen-dependent growth of human MCF-7 breast cancer cells. *J. Biol. Chem.* **276,** 23763–23768.

Lopez-Carballo, G., Moreno, L., Masia, S., Perez, P., and Barettino, D. (2002). Activation of the phosphatidylinositol 3-kinase/Akt signaling pathway by retinoic acid is required for neural differentiation of SH-SY5Y human neuroblastoma cells. *J. Biol. Chem.* **277,** 25297–25304.

Ma, X., Karra, S., Guo, W., Lindner, D. J., Hu, J., Angell, J. E., Hofmann, E. R., Reddy, S. P., and Kalvakolanu, D. V. (2001). Regulation of interferon and retinoic acid-induced cell death activation through thioredoxin reductase. *J. Biol. Chem.* **276,** 24843–24854.

MacFarlane, M., Ahmad, M., Srinivasula, S. M., Fernandes-Alnemri, T., Cohen, G. M., and Alnemri, E. S. (1997). Identification and molecular cloning of two novel receptors for the cytotoxic ligand TRAIL. *J. Biol. Chem.* **272,** 25417–25420.

Maden, M. (2002). Retinoid signalling in the development of the central nervous system. *Nat. Rev. Neurosci.* **3,** 843–853.

Marsters, S. A., Sheridan, J. P., Pitti, R. M., Huang, A., Skubatch, M., Baldwin, D., Yuan, J., Gurney, A., Goddard, A. D., Godowski, P. *et al.* (1997). A novel receptor for Apo2L/TRAIL contains a truncated death domain. *Curr. Biol.* **7,** 1003–1006.

Massaro, D., and Massaro, G. D. (2003). Retinoids, alveolus formation, and alveolar deficiency: Clinical implications. *Am. J. Respir. Cell Mol. Biol.* **28,** 271–274.

McKenna, N. J., and O'Malley, B. W. (2002a). Combinatorial control of gene expression by nuclear receptors and coregulators. *Cell* **108,** 465–474.

McKenna, N. J., and O'Malley, B. W. (2002b). Minireview: Nuclear receptor coactivators–an update. *Endocrinology* **143,** 2461–2465.

Metzger, D., and Chambon, P. (2001). Site- and time-specific gene targeting in the mouse. *Methods* **24,** 71–80.

Minucci, S., Maccarana, M., Cioce, M., De Luca, P., Gelmetti, V., Segalla, S., Di Croce, L., Giavara, S., Matteucci, C., Gobbi, A. *et al.* (2000). Oligomerization of RAR and AML1 transcription factors as a novel mechanism of oncogenic activation. *Mol. Cell* **5,** 811–820.

Misner, D. L., Jacobs, S., Shimizu, Y., de Urquiza, A. M., Solomin, L., Perlmann, T., De Luca, L. M., Stevens, C. F., and Evans, R. M. (2001). Vitamin A deprivation results in reversible loss of hippocampal long-term synaptic plasticity. *Proc. Natl. Acad. Sci. USA* **98,** 11714–11719.

Mongkolsapaya, J., Cowper, A. E., Xu, X. N., Morris, G., McMichael, A. J., Bell, J. I., and Screaton, G. R. (1998). Lymphocyte inhibitor of TRAIL (TNF-related apoptosis-inducing ligand): A new receptor protecting lymphocytes from the death ligand TRAIL. *J. Immunol.* **160,** 3–6.

Mongkolsapaya, J., Grimes, J. M., Chen, N., Xu, X. N., Stuart, D. I., Jones, E. Y., and Screaton, G. R. (1999). Structure of the TRAIL-DR5 complex reveals mechanisms conferring specificity in apoptotic initiation. *Nat. Struct. Biol.* **6,** 1048–1053.

Nagy, L., Thomazy, V. A., Shipley, G. L., Fesus, L., Lamph, W., Heyman, R. A., Chandraratna, R. A., and Davies, P. J. (1995). Activation of retinoid X receptors induces apoptosis in HL-60 cell lines. *Mol. Cell. Biol.* **15,** 3540–3551.

Niederreither, K., Vermot, J., Messaddeq, N., Schuhbaur, B., Chambon, P., and Dolle, P. (2001). Embryonic retinoic acid synthesis is essential for heart morphogenesis in the mouse. *Development* **128,** 1019–1031.

Pan, G., Ni, J., Wei, Y. F., Yu, G., Gentz, R., and Dixit, V. M. (1997a). An antagonist decoy receptor and a death domain-containing receptor for TRAIL. *Science* **277,** 815–818.

Pan, G., O'Rourke, K., Chinnaiyan, A. M., Gentz, R., Ebner, R., Ni, J., and Dixit, V. M. (1997b). The receptor for the cytotoxic ligand TRAIL. *Science* **276,** 111–113.

Pandolfi, P. P. (2001). Oncogenes and tumor suppressors in the molecular pathogenesis of acute promyelocytic leukemia. *Hum. Mol. Genet.* **10**, 769–775.

Pearson, M., Carbone, R., Sebastiani, C., Cioce, M., Fagioli, M., Saito, S., Higashimoto, Y., Appella, E., Minucci, S., Pandolfi, P. P. *et al.* (2000). PML regulates p53 acetylation and premature senescence induced by oncogenic Ras. *Nature* **406**, 207–210.

Piazza, F., Gurrieri, C., and Pandolfi, P. P. (2001). The theory of APL. *Oncogene* **20**, 7216–7222.

Pitti, R. M., Marsters, S. A., Ruppert, S., Donahue, C. J., Moore, A., and Ashkenazi, A. (1996). Induction of apoptosis by Apo-2 ligand, a new member of the tumor necrosis factor cytokine family. *J. Biol. Chem.* **271**, 12687–12690.

Rachez, C., and Freedman, L. P. (2001). Mediator complexes and transcription. *Curr. Opin. Cell Biol.* **13**, 274–280.

Reid, G., Hubner, M. R., Metivier, R., Brand, H., Denger, S., Manu, D., Beaudouin, J., Ellenberg, J., and Gannon, F. (2003). Cyclic, proteasome-mediated turnover of unliganded and liganded ERalpha on responsive promoters is an integral feature of estrogen signaling. *Mol. Cell* **11**, 695–707.

Resche-Rigon, M., and Gronemeyer, H. (1998). Therapeutic potential of selective modulators of nuclear receptor action. *Curr. Opin. Chem. Biol.* **2**, 501–507.

Rochette-Egly, C. (2003). Nuclear receptors: Integration of multiple signalling pathways through phosphorylation. *Cell Signal* **15**, 355–366.

Rogatsky, I., Luecke, H. F., Leitman, D. C., and Yamamoto, K. R. (2002). Alternate surfaces of transcriptional coregulator GRIP1 function in different glucocorticoid receptor activation and repression contexts. *Proc. Natl. Acad. Sci. USA* **99**, 16701–16706.

Sakaue, M., Adachi, H., Dawson, M., and Jetten, A. M. (2001). Induction of Egr-1 expression by the retinoid AHPN in human lung carcinoma cells is dependent on activated ERK1/2. *Cell Death Differ.* **8**, 411–424.

Salomoni, P., and Pandolfi, P. P. (2002). The role of PML in tumor suppression. *Cell* **108**, 165–170.

Screaton, G. R., Mongkolsapaya, J., Xu, X. N., Cowper, A. E., McMichael, A. J., and Bell, J. I. (1997). TRICK2, a new alternatively spliced receptor that transduces the cytotoxic signal from TRAIL. *Curr. Biol.* **7**, 693–696.

Shang, Y., Hu, X., DiRenzo, J., Lazar, M. A., and Brown, M. (2000). Cofactor dynamics and sufficiency in estrogen receptor-regulated transcription. *Cell* **103**, 843–852.

Shang, Y., Myers, M., and Brown, M. (2002). Formation of the androgen receptor transcription complex. *Mol. Cell* **9**, 601–610.

Sheridan, J. P., Marsters, S. A., Pitti, R. M., Gurney, A., Skubatch, M., Baldwin, D., Ramakrishnan, L., Gray, C. L., Baker, K., Wood, W. I. *et al.* (1997). Control of TRAIL-induced apoptosis by a family of signaling and decoy receptors. *Science* **277**, 818–821.

Simonet, W. S., Lacey, D. L., Dunstan, C. R., Kelley, M., Chang, M. S., Luthy, R., Nguyen, H. Q., Wooden, S., Bennett, L., Boone, T. *et al.* (1997). Osteoprotegerin: A novel secreted protein involved in the regulation of bone density. *Cell* **89**, 309–319.

Sirchia, S. M., Ren, M., Pili, R., Sironi, E., Somenzi, G., Ghidoni, R., Toma, S., Nicolo, G., and Sacchi, N. (2002). Endogenous reactivation of the RARbeta2 tumor suppressor gene epigenetically silenced in breast cancer. *Cancer Res.* **62**, 2455–2461.

Sun, S. Y., Yue, P., Chandraratna, R. A., Tesfaigzi, Y., Hong, W. K., and Lotan, R. (2000a). Dual mechanisms of action of the retinoid CD437: Nuclear retinoic acid receptor-mediated suppression of squamous differentiation and receptor-independent induction of apoptosis in UMSCC22B human head and neck squamous cell carcinoma cells. *Mol. Pharmacol.* **58**, 508–514.

Sun, S. Y., Yue, P., Hong, W. K., and Lotan, R. (2000b). Augmentation of tumor necrosis factor-related apoptosis-inducing ligand (TRAIL)-induced apoptosis by the synthetic retinoid 6-[3-(1-adamantyl)-4-hydroxyphenyl]-2-naphthalene carboxylic acid (CD437)

through up-regulation of TRAIL receptors in human lung cancer cells. *Cancer Res.* **60**, 7149–7155.
Sun, S. Y., Yue, P., Hong, W. K., and Lotan, R. (2000c). Induction of fas expression and augmentation of Fas/Fas ligand-mediated apoptosis by the synthetic retinoid CD437 in human lung cancer cells. *Cancer Res.* **60**, 6537–6543.
Takeda, K., Smyth, M. J., Cretney, E., Hayakawa, Y., Kayagaki, N., Yagita, H., and Okumura, K. (2002). Critical role for tumor necrosis factor-related apoptosis-inducing ligand in immune surveillance against tumor development. *J. Exp. Med.* **195**, 161–169.
Tenen, D. G. (2003). Disruption of differentiation in human cancer: AML shows the way. *Nat. Rev. Cancer* **3**, 89–101.
Truneh, A., Sharma, S., Silverman, C., Khandekar, S., Reddy, M. P., Deen, K. C., McLaughlin, M. M., Srinivasula, S. M., Livi, G. P., Marshall, L. A. *et al.* (2000). Temperature-sensitive differential affinity of TRAIL for its receptors. DR5 is the highest affinity receptor. *J. Biol. Chem.* **275**, 23319–23325.
Tulachan, S. S., Doi, R., Kawaguchi, Y., Tsuji, S., Nakajima, S., Masui, T., Koizumi, M., Toyoda, E., Mori, T., Ito, D. *et al.* (2003). All-trans retinoic acid induces differentiation of ducts and endocrine cells by mesenchymal/epithelial interactions in embryonic pancreas. *Diabetes* **52**, 76–84.
Walczak, H., Degli-Esposti, M. A., Johnson, R. S., Smolak, P. J., Waugh, J. Y., Boiani, N., Timour, M. S., Gerhart, M. J., Schooley, K. A., Smith, C. A. *et al.* (1997). TRAIL-R2: A novel apoptosis-mediating receptor for TRAIL. *EMBO J.* **16**, 5386–5397.
Wang, J., Lobito, A. A., Shen, F., Hornung, F., Winoto, A., and Lenardo, M. J. (2000). Inhibition of Fas-mediated apoptosis by the B cell antigen receptor through c-FLIP. *Eur. J. Immunol.* **30**, 155–163.
Weis, K., Rambaud, S., Lavau, C., Jansen, J., Carvalho, T., Carmo-Fonseca, M., Lamond, A., and Dejean, A. (1994). Retinoic acid regulates aberrant nuclear localization of PML-RAR alpha in acute promyelocytic leukemia cells. *Cell* **76**, 345–356.
Westin, S., Rosenfeld, M. G., and Glass, C. K. (2000). Nuclear receptor coactivators. *Adv. Pharmacol.* **47**, 89–112.
Widschwendter, M., Berger, J., Hermann, M., Muller, H. M., Amberger, A., Zeschnigk, M., Widschwendter, A., Abendstein, B., Zeimet, A. G., Daxenbichler, G. *et al.* (2000). Methylation and silencing of the retinoic acid receptor-beta2 gene in breast cancer. *J. Natl. Cancer Inst.* **92**, 826–832.
Wiley, S. R., Schooley, K., Smolak, P. J., Din, W. S., Huang, C. P., Nicholl, J. K., Sutherland, G. R., Smith, T. D., Rauch, C., Smith, C. A. *et al.* (1995). Identification and characterization of a new member of the TNF family that induces apoptosis. *Immunity* **3**, 673–682.
Wong, C. W., and Privalsky, M. L. (1998). Transcriptional silencing is defined by isoform- and heterodimer-specific interactions between nuclear hormone receptors and corepressors. *Mol. Cell. Biol.* **18**, 5724–5733.
Zhong, S., Salomoni, P., and Pandolfi, P. P. (2000). The transcriptional role of PML and the nuclear body. *Nat. Cell Biol.* **2**, E85–E90.

18

Potential for TRAIL as a Therapeutic Agent in Ovarian Cancer

Touraj Abdollahi

Department of Biochemistry and Molecular Pharmacology, Jefferson Medical College Thomas Jefferson University, Philadelphia, Pennsylvania 19107

    I. Introduction
   II. Ovarian Cancer
       A. *Ovarian Cancer Risk Factors*
       B. *Finding the Best Therapy for Ovarian Cancer*
  III. Tumor Necrosis Factor–Related Apoptosis-Inducing Ligand
       A. *TRAIL and Cancer*
       B. *TRAIL and Ovarian Cancer*
  IV. Interleukin-8
   V. Role of p38 MAPK in Apoptosis
       A. *Role of p38 MAPK in TRAIL-Induced Apoptosis*
  VI. Summary
      References

Tumor necrosis factor–related apoptosis-inducing ligand (TRAIL) is known to induce apoptosis, otherwise known as programmed cell death, in many malignant cells without any known detrimental effects to normal cells. These aspects of TRAIL indicate the potential of TRAIL as a therapeutic agent in cancer. Ovarian cancer remains the deadliest gynecologic malignancy and is the fourth leading cause of death due to cancer in women. However, it has been shown in studies that ovarian cancer cells are sensitive to TRAIL-induced cell death when treated with TRAIL alone or in combination with chemotherapeutic agents. TRAIL signals through two death receptors, TRAIL-R1 and TRAIL-R2, to induce apoptosis. TRAIL also binds to two other cell surface receptors, TRAIL-R3 and TRAIL-R4, which do not have intracellular death domains and therefore do not transmit the apoptotic signal upon ligation with TRAIL. It has been shown that a chemokine, interleukin-8 (IL-8), may play a role in ovarian tumor progression due to its elevated presence in the fluid surrounding ovarian cancer tissues. Possible roles for IL-8 in ovarian tumorigenesis include angiogenesis and metastasis. Because the mechanism of regulation for TRAIL-induced apoptosis needs to be clarified, the role of IL-8 in TRAIL-induced apoptosis of ovarian cancer cells was studied. Results showed that the presence of IL-8 regulates cell-surface expression of TRAIL receptors in ovarian cancer cell lines *in vitro*. There may be a role for the p38 mitogen-activated protein kinase (MAPK) pathway in TRAIL-induced apoptosis of ovarian cancer cell. © 2004 Elsevier Inc.

I. INTRODUCTION

Tumor necrosis factor–related apoptosis-inducing ligand (TRAIL) is a cytokine essentially present ubiquitously and exerts its effects via cell surface–specific receptors located at the surface of target tissues. Since cloning, there has been a considerable improvement in the knowledge of the mechanisms for TRAIL signaling. However, the mechanisms for TRAIL's ability to induce apoptosis in malignant cells and not normal cells remain largely unknown, as does the natural function and purpose of TRAIL in normal mammalian cells. In this review, the function of TRAIL, in particular with relation to ovarian cancer, the possible role of IL-8 in ovarian tumorigenesis, and possible role of p38 MAPK in ovarian apoptosis induced by TRAIL are presented.

II. OVARIAN CANCER

Ovarian cancer is the fourth leading cause of death from cancers among American women after lung, breast, and colon cancer. The epithelial cells of the ovary constitute a very small fraction ($<1\%$) of the total ovarian mass but constitute more than 90% of the ovarian neoplasms. Ovarian cancer of epithelial origin is the most lethal of gynecologic malignancies. The American Cancer Society estimates as many as 27,000 new cases and

14,000 deaths from ovarian cancer in the United States each year (Greenlee *et al.*, 2000). The overall 5-year survival rate for women with this cancer is only 30%; this is including women diagnosed in earlier stages (I and II) of the disease (Landis *et al.*, 1998). This poor prognosis is mainly due to the lack of sensitive tests for detection of the early stage of the disease, which is often asymptomatic. At the time of diagnosis, approximately 70% of patients are at stages III or IV, which is characterized by cancer cells spreading outside of the pelvis to the abdominal viscera or even further metastasis to the lungs and other organs (Landis *et al.*, 1998). When the disease is still confined to the ovary, surgery alone can be curative, with 5-year survival rates exceeding 90%. Currently, management of the disease requires histologic confirmation of the diagnosis, surgical staging, and aggressive surgical cytoreduction, followed by chemotherapy (Greenlee *et al.*, 2000).

A. OVARIAN CANCER RISK FACTORS

Commonly, epithelial ovarian cancers are thought to develop from the surface epithelium of the ovary, which shares the same developmental origin (coelomic epithelium) with the general pelvic and abdominal peritoneum. Laboratories have demonstrated that common epithelial ovarian cancer is from the same origin, suggesting that early stages of the disease can be detected if specific tumor markers expressed at early stages can be identified (Mok *et al.*, 1992; Tsao *et al.*, 1993). Because early stage disease is detected infrequently, very little is known about the molecular and biochemical events that cause transformation of normal ovarian epithelial cells from benign to malignant. Likewise, very little is known about the molecular and biochemical events regulating disease progression. Several lines of evidence suggest that the number of times a woman undergoes ovulation may play an important role in the development of epithelial ovarian cancer. First, ovarian cancer is exceedingly rare in women who do not ovulate because of gonadal dysgenesis. In addition, although epithelial ovarian cancer occurs rarely in most other animal species, it is common in hens, which like humans, are frequent ovulators. Finally, epidemiologic studies have shown that pregnancy, breast-feeding, and oral contraceptive pill use, all of which inhibit ovulation, are protective against the development of ovarian cancer (Whittemore *et al.*, 1992).

Although the relationship between ovulation and ovarian cancer is well accepted, the underlying causative mechanisms remain unclear. Several hypotheses have been advanced to explain this association. It has been suggested that exposure to high levels of steroid hormones at ovulation may facilitate transformation. In addition, ovulation leads to entrapment of epithelial cells in the underlying stroma with subsequent formation of inclusion cysts. These cysts could represent precursor lesions in which transformation is facilitated. Finally, proliferation of epithelial cells required to repair the disrupted ovarian surface after ovulation could contribute to carcinogenesis by increasing the likelihood of mutations due to spontaneous errors in DNA synthesis. If mutations that occur involve critical growth regulatory genes, this could facilitate clonal expansion of premalignant cells with an increased susceptibility to subsequently become fully transformed (Schildkraut et al., 1997). Therefore, women with uninterrupted ovulation appear to be at higher risk for malignant transformation of the ovarian surface epithelium. For instance, a higher risk of ovarian cancer is associated with a first birth after the age of 35 (Negri et al., 1991). Conversely, a lower risk is associated with childbirth, especially at age 25 or younger, or with the use of oral contraceptives (Negri et al., 1991). It has also been suggested that the progestin-dominant hormonal milieu associated with the pill and pregnancy increase apoptosis of ovarian epithelial cells (Marks et al., 1991). If this is true, then these factors may serve to "cleanse" the epithelium of accumulated genetic damage while also preventing the development of new mutations. However, a family history of ovarian cancer, especially if two or more first-degree relatives have been affected, appears to also remain as a very important risk factor (Lynch et al., 1992; Schildkraut et al., 1988).

B. FINDING THE BEST THERAPY FOR OVARIAN CANCER

Advancement in understanding the initiation and progression of ovarian carcinoma has been slow, mainly due to the lack of an appropriate experimental model. Despite its clinical importance, biology of the ovarian surface epithelium is poorly understood and evidence for its role in carcinogenesis is based almost entirely on morphologic and histologic examination of clinical tumor specimens and immortalized ovarian cancer cell lines. The direct progression of benign ovarian lesions to clinical carcinoma has not been clearly demonstrated (Feeley et al., 2001). Because early stage malignancy is infrequently detected in patients, the morphologic and genetic changes that occur as the benign epithelium becomes malignant are not well defined. At present, there is little evidence for a genetic model of multistep tumor progression in ovarian cancer, and there is speculation that ovarian carcinoma occurs without any precursor lesion (Bell et al., 1994). A suitable animal model, in which ovarian cancer could be predictably induced with defined genetic changes, may be crucial to our further understanding of the disease.

A vexing problem to approaches for treatment of ovarian cancer is that most tumors develop a broad cross-resistance to different chemotherapeutic agents and radiation therapy that they encounter during treatment. Drug resistance is thought to cause treatment failure and death in most patients with metastatic disease. The tumor microenvironment can play an important role in tumor-cell drug sensitivity. Historically, the main research seems to emphasize the tumor cell–specific mechanisms of drug resistance and, in particular, on those that influence drug-target interactions and subsequent cell damage. In addition, it has traditionally been hypothesized that drug resistance develops as a result of progressively acquired somatic mutations or epigenetic changes within tumor cells as they evolve over time (Nowell, *et al.*, 1976). However, it is important to consider that cytotoxic agents are primarily effective against proliferating cells and that, even in rapidly proliferating tumors, a significant proportion of cancer cells are in a quiescent state. These quiescent cells therefore show a degree of drug resistance relative to cycling cells (Shah *et al.*, 2001). So it seems like these phenomena may underlie clinical drug resistance in ovarian cancer.

These problems have led to a recent shift in emphasis away from drug-specific mechanisms of resistance to defects in the common apoptotic signaling and effector pathways downstream of drug-target interactions as the probable causes of resistance in clinical practice.

Numerous questions remain regarding the treatment of ovarian cancer. The challenge is to develop new approaches that will improve long-term survival. With a limited number of patient resources available, there is a definitive need for prioritization and international collaboration. The use of new agents based on evidence of *in vitro* activity and combinations that target specific intracellular processes may produce responses superior to those observed currently. Basic research efforts need to be continued to comprehend and ultimately manipulate the genetic and molecular mechanism connected with ovarian cancer initiation and evolution.

III. TUMOR NECROSIS FACTOR–RELATED APOPTOSIS-INDUCING LIGAND

Apoptosis is a cell-suicide mechanism that plays a crucial role in development and homeostasis, integral to the development and functioning of multicellular animals (Jacobson *et al.*, 1997; Steller, 1995). Apoptosis eliminates individual cells when they are no longer needed or have become seriously damaged. Cells appear to be programmed to die by default and execute apoptosis if they do not receive appropriate survival cues from their environment (Nagata, 1997). In addition, most cells have internal sensors for their well being that can initiate apoptosis if the cell is unable to repair defects such as DNA damage. Abnormal regulation of apoptosis has been implicated in cancer, autoimmune disease, and degenerative conditions. Therefore, an understanding of the cellular mediators of apoptosis is sure to provide valuable insights toward the development of new therapies for pathologic conditions resulting from inappropriate regulation of cell death.

Members of the TNF family of cytokines and receptors are critically involved in the apoptotic process (Amakawa et al., 1996; Cosman, 1994; Lynch, et al., 1995; Nagata et al., 1995; Smith et al., 1994a; Thompson, 1995; van Parijs et al., 1996). Among the ligands within this family, TNF, lymphotoxin α, Fas ligand FasL, Apo3L (Marsters et al., 1998), and TRAIL (Marsters et al., 1996; Pitti et al., 1996; Wiley et al., 1995) have been characterized as major mediators of apoptosis. In 1995, TRAIL was identified and characterized as a member of the TNF family of death-inducing ligands (Wiley et al., 1995). TRAIL is widely expressed in human tissues especially in the spleen, prostate gland, and lung, except for the brain, liver, or testis (Pitti et al., 1996; Rieger et al., 1999; Wiley et al., 1995). TRAIL, a characteristic type II transmembrane protein, shows the highest homology with FasL, sharing 28% identity in the extracellular receptor-binding motif (Wiley et al., 1995). Not only does TRAIL show high sequence homology with FasL, but in addition, TRAIL is capable of inducing rapid apoptosis in a variety of malignant cells (Jeremias et al., 1998; Pitti et al., 1996; Wiley et al., 1995).

To understand the function and regulation of the TRAIL apoptosis-inducing system, the TRAIL receptors were identified. TRAIL can bind two apoptosis-inducing receptors, TRAIL-R1 (DR4) and TRAIL-R2 (DR5); and two additional cell-bound receptors incapable of transmitting an apoptotic signal, TRAIL-R3 (DcR1) and TRAIL-R4 (DcR2) (Emery et al., 1998). The receptors, DR4 and DR5, have been shown to have an intracellular death domain to transmit the apoptotic signal. The DcR1 and DcR2 are known as "decoy receptors" because they lack a functional death domain to signal apoptosis. DcR2 has a truncated death domain and DcR1 is even more unique because it is devoid of any death domain (Marsters et al., 1997; Pan et al., 1997; Sheridan et al., 1997). Thus, the current hypothesis is that the nonsignaling receptors act as "decoys" to competitively bind TRAIL as a chief mechanism determining whether a cell is resistant or sensitive to TRAIL-induced death.

Two main signaling pathways initiate the apoptotic suicide machinery in mammalian cells. The cell-intrinsic pathway triggers apoptosis in response to DNA damage, defective cell cycle, detachment from the extracellular matrix, hypoxia, loss of survival factors, or other types of cell distress. This pathway generally involves activation of the proapoptotic members of the bcl-2 superfamily, which in turn, engages the mitochondria to cause the release of apoptotic factors, such as cytochrome *c*, into the cytosol (Adams et al., 1998; Green et al., 2000; Hunt et al., 2001). In the cytosol, cytochrome *c* binds an adaptor, APAF1, forming what is known as an "apoptosome" that activates the apoptosis-initiating protease caspase-9. In turn, caspase-9 activates downstream proteases caspase-3 and, -6, and -7 (Du et al., 2000;

Verhagen *et al.*, 2000). Most chemotherapy and irradiation triggers tumor cell apoptosis through the cell-intrinsic pathway, as an indirect consequence of causing cellular damage.

TRAIL is an example of ligand that initiates the extrinsic pathway of apoptosis. In response to engagement of death receptors by TRAIL, the extrinsic pathway is launched. This pathway stimulates the apoptotic caspase machinery. Ligand-induced activation of cell-surface death receptors leads to rapid assembly of a death-inducing signaling complex (DISC) and activation of the upstream caspases-8 and -10. These caspases, in turn, activate the same set of executioner caspases that are activated by the intrinsic pathway (Wallach *et al.*, 1999). The end result of these processes is the death of the target cells.

A. TRAIL AND CANCER

Advances in diagnosis, surgical techniques, radiation therapy, and chemotherapy have led to increased survival times for many cancer patients. This progress is not seen in relation to ovarian cancer patients due to late diagnosis and multidrug resistance. Acquired and innate resistance to chemotherapy and radiation therapy has been a major obstacle for clinical oncology. One potential addition to these conventional treatments is direct induction of cell death by activation of death receptor–mediated apoptosis. Ligation of death ligands with death receptors induces apoptosis by directly activating the caspase pathway and apparently bypassing the involvement of the primary mitochondrial damage caused by most chemotherapeutic agents and radiation. A number of *in vitro* studies have shown that many tumor cell lines of different origins including lung, breast, colon, and prostate gland cancers are sensitive to TRAIL-induced apoptosis (Pitti *et al.*, 1996; Rieger *et al.*, 1999; Sheridan *et al.*, 1997; Thomas and Hersey, 1998; Wiley *et al.*, 1995; Zhang *et al.*, 1999). The efficacy of the antitumor activity of TRAIL *in vivo* has been tested in animal xenograft models. Importantly, administration of human TRAIL has shown little toxicity in mice and nonhuman primates, with a very favorable regression of the tumors (Ashkenazi *et al.*, 1999; Nagane *et al.*, 2000; Walczak *et al.*, 1999). Treatment of mice inoculated with human breast or colon cancer cells with human cross-linked TRAIL powerfully suppressed the formation of xenografted tumors (Ashkenazi *et al.*, 1999; Walczak *et al.*, 1999).

Treatment with TRAIL also inhibited the growth of established xenografts in mice and even resulted in regression of the tumors by inducing apoptosis in tumor cells without causing generalized toxicity (Ashkenazi *et al.*, 1999; Walczak *et al.*, 1999). Furthermore, injection of TRAIL into tumors in mice lead to growth suppression and apoptosis of xenografts derived from TRAIL-sensitive human glioma cells and significant elongation of lifespan of the host mice (Roth *et al.*, 1999). Taken together, these results indicate substantial promise for TRAIL in human cancer therapy and clinical trials are eagerly awaited to see whether the positive results achieved in animal models will translate to a new treatment for cancer.

B. TRAIL AND OVARIAN CANCER

Despite the introduction of new drugs for the treatment of ovarian cancer, the overall survival of patients suffering from this malignancy is very poor. Therefore, the possibility of combining anticancer agents and compounds that can "reverse" the antiapoptotic factors in ovarian cancer may lead to an increase in the response of ovarian cancer cells to drug treatment. The major factor that limits the effectiveness of chemotherapy in patients with ovarian cancer is the acquisition of resistance. The biochemical and molecular mechanisms explaining this resistance are not completely known. It has been shown that in 12 chemoresistant ovarian epithelial cell lines some were sensitive to TRAIL alone. The other cell lines that were tested were resistant to TRAIL alone, however, in combination with chemotherapeutic drugs, cisplatin and paclitaxel, these cells were rendered sensitive to apoptosis (Cuello et al., 2001). These results indicate that for patients who may have advanced ovarian cancer that has become resistant to chemotherapy, may be made sensitive to drugs by the addition of another factor that can induce growth inhibition. Recently, one group determined the relative expression of TRAIL in 120 epithelial ovarian cancers from patients. This group has found that TRAIL expression in these tissues correlates with patient survival (Lancaster et al., 2003). In other words, the more TRAIL present in the tissues, then the more favorable survival.

IV. INTERLEUKIN-8

Interleukin-8 (IL-8) was initially described as a neutrophil chemoattractant (Matsushima et al., 1989). Later studies revealed that it is also an autocrine growth factor for keratinocytes, melanoma cells, and pancreatic cancer cells (Miyamoto et al., 1998; Schadendorf et al., 1994; Tuschil et al., 1992). IL-8 can also induce neovascularization (Koch et al., 1992) and modulate collagenase secretion (Luca et al., 1997), suggesting that it could modulate invasiveness and/or extracellular matrix remodeling in the tumor environment. Because cell proliferation, angiogenesis, migration, and invasion are all important components of the metastatic process, then IL-8 expression by tumor cells can influence their metastatic capacities (Singh et al., 1995). IL-8 secreted by tumor cells has been associated with progressive growth of bronchogenic carcinoma (Smith et al., 1994b), non–small-cell lung cancer (Arenberg et al., 1998), human colorectal carcinoma (Brew et al., 1996; Kitadai and Radinsky et al., 1996; Yatsunami et al.,

1997), human breast cancer (Miller *et al.*, 1998), human prostate gland cancer (Greene *et al.*, 1997), and human ovarian cancer (Harant *et al.*, 1995). Moreover, the expression of IL-8 has been shown to correlate with the metastatic potential of human melanoma (Scheibenbogen *et al.*, 1995; Singh *et al.*, 1995), ovarian cancer (Harant *et al.*, 1995), prostate gland cancer (Ferrer *et al.*, 1998; Miller *et al.*, 1998), human colon cancer (Kitadai *et al.*, 1996), and gastric carcinoma cells (Kitadai *et al.*, 1998). IL-8's role as a proinflammatory and proangiogenic factor may contribute to ovarian tumor development (Harada *et al.*, 1996; Koch *et al.*, 1992). Angiogenesis, the formation of new vessels from existing capillary beds, is a crucial process involved in the pathophysiologic condition of tumor growth. IL-8 has been shown to exert direct angiogenic effects on cells *in vitro* and *in vivo* (Hu *et al.*, 1993; Koch *et al.*, 1992; Strieter *et al.*, 1992).

Overexpression of IL-8 and/or its receptors has been shown in cancers, and IL-8 enhances tumor growth and angiogenesis, a critical step for tumor metastasis (Arenberg *et al.*, 1996; Kitadai *et al.*, 1999; Koch *et al.*, 1992; Merogi *et al.*, 1997; Radke *et al.*, 1996; Richards *et al.*, 1997; Singh *et al.*, 1994). In particular, high expression of IL-8 mRNA has been detected in clinical specimens of late-stage ovarian carcinomas (Merogi *et al.*, 1997; Yoneda *et al.*, 1998). Patients with ovarian cancer produce large amounts of ascitic fluid. The ascites fluid contains many growth factors (Mills *et al.*, 1988, 1990) and provides an excellent environment conducive to the growth of ovarian cancer cells (Berchuck *et al.*, 1997; Bookman *et al.*, 1998; Westermann *et al.*, 1997). The concentrations of various cytokines have been examined in the ascites from patients with ovarian cancer. Ascites fluid of patients with ovarian cancer contain significantly higher levels of IL-8 compared with those from patients with benign gynecologic disorders (Gawrychowski *et al.*, 1998; Ivarsson *et al.*, 1998).

I have shown that interleukin-8 (IL-8) may play an important role in blocking the TRAIL-induced apoptosis normally seen in the TRAIL-sensitive cell-line, OVCAR3 (Abdollahi *et al.*, 2003). In part it appears through the modulation of the TRAIL receptors. The inherent role that IL-8 plays in ovarian tumorigenesis is relatively unknown. My suggestion throughout this review has been the use of TRAIL as a potential therapeutic agent in the fight against ovarian cancer. However, it has emerged that IL-8 may be able to inhibit the positive effects seen by TRAIL on ovarian cancer cells. There is a lack of *in vivo* data as well as a very limited examination of IL-8's effects on multiple ovarian cancer cell lines to conclude a global ability of IL-8 to block TRAIL-induced apoptosis in ovarian carcinoma cells or any other malignant cells. However, these data seem to suggest that further investigation is necessary before TRAIL can effectively be used for therapy, and also the development of IL-8 inhibitors may be important in not only sensitizing ovarian cancer cells to TRAIL, but also to combat the possibly angiogenic and metastatic effects that IL-8 may have on ovarian cancer cells.

V. ROLE OF P38 MAPK IN APOPTOSIS

We have determined a possible role for the p38 mitogen-activated protein kinase (MAPK) pathway in TRAIL-induced apoptosis of ovarian cancer cells (Abdollahi *et al.*, 2003). MAPK signal transduction pathways are among the most widespread mechanisms of eukaryotic cell regulation (Waskiewicz *et al.*, 1995). Cells possess multiple MAPK pathways. They are activated through phosphorylation by upstream cascades of protein kinases, resulting in amplification of the signal. Once activated, MAP kinases phosphorylate a number of cytoplasmic and nuclear proteins, including transcription factors, therefore effecting changes on gene expression. Two important members of the human MAPK superfamily, other than p38, are the extracellular signal–regulated kinase (ERK) and c-Jun N-terminal kinase (JNK) (Davis *et al.*, 2000). Each of these three separate family members are preferentially recruited by distinct sets of stimuli.

The p38 MAPKs are a mammalian stress-activated MAPK family. In general, the stress-activated protein kinase, p38, is activated by noxious environmental stimuli such as ultraviolet radiation, osmotic stress, inflammatory cytokines, and inhibition of protein synthesis (Derijard *et al.*, 1994; Hibi *et al.*, 1993; Kyriakis *et al.*, 1994). There are five isoforms of the p38 group of MAP kinases that have been identified to date: p38α, p38β1, p38β2, p38γ, and p38δ (Hu *et al.*, 1999; Jiang *et al.*, 1996; Kumar *et al.*, 1997; Li *et al.*, 1996; Nick *et al.*, 1999). The p38s are strongly activated *in vivo* by environmental stresses and inflammatory cytokines (Kyriakis *et al.*, 1996). Also, the p38s are activated during ischemia and remain active (Bogoyevitch *et al.*, 1996; Kyriakis *et al.*, 1996; Pombo *et al.*, 1994).

Paclitaxel is a relatively new chemotherapeutic drug that has shown very encouraging effects in cancer chemotherapy-refractory ovarian cancer (McGuire *et al.*, 1989). Paclitaxel has exhibited significant antitumor activity against human ovarian cancer in a nude mouse model (McGuire *et al.*, 1989). Also, it has significantly inhibited the angiogenic response induced by tumor cells in mice (Derijard *et al.*, 1994; Hibi *et al.*, 1993). The significance of this to my review is that the p38 kinase pathway that can be activated by various stimuli is shown to be also activated to at least some degree in paclitaxel-treated human ovarian cancer cells (Kyriakis *et al.*, 1996; Raingeaud *et al.*, 1995).

A. ROLE OF P38 MAPK IN TRAIL-INDUCED APOPTOSIS

There is growing evidence for p38's role in apoptosis. For the purposes of this review, I report the evidence for p38's involvement in cell death induced in ovarian cancer or by the death receptors.

In previous studies it has been demonstrated that the p38 MAP kinase pathways and the mitochondrial death pathway play important roles in the induction of apoptosis of tumor cells (Ohtsuka *et al.*, 2002). The induction of apoptosis was observed as a result of the enhanced activation of p38 and induction of DR5-driven caspase activation by chemotherapy agents (Ohtsuka *et al.*, 2002). Further studies revealed that using agonistic antibodies to DR4- and DR5-induced apoptosis was a result of caspase processing and the activation of p38. Using chemotherapeutic agents, researchers were able to see an enhanced activation of caspase-8. When the p38 pathway was inhibited, not only was cell death blocked, but also the enhancing effect of chemotherapy agents on the caspase activation due to death-receptor induction (Ohtsuka *et al.*, 2003). Further evidence of p38's involvement in TRAIL-induced apoptosis comes from the treatment of adenocarcinoma HeLa cells with TRAIL. It has been shown that there is an accumulation of reactive oxygen species and activation of p38 MAPK when the HeLa cells are treated with TRAIL (Lee *et al.*, 2002). This is subsequently accompanied by caspase activation and apoptosis (Lee *et al.*, 2002).

There is an obvious need for further exploration of the role that the p38 MAPK pathway may play in TRAIL-induced apoptosis. The complex pathways of the MAPK are under very diverse regulation as is any complex process. The p38 pathway may be very important to TRAIL-induced apoptosis in various cells, or just a certain kind of cells, or it may be triggered indirectly by TRAIL, through other mechanisms when the cell undergoes apoptosis. The mechanism of TRAIL-induced apoptosis is slowly being deciphered, however the many facets of this cytokine are still unknown. The p38 MAPK pathway has been shown to play some kind of role in TRAIL-induced apoptosis, whether or not a large role is yet to be determined.

VI. SUMMARY

While the underlying causes of ovarian epithelial carcinogenesis remain unknown, we are beginning to define important clues that will ultimately lead us to effective treatments and future prevention. The extracellular matrix contains factors that regulate the growth and differentiation of epithelia and may determine some of the biologic properties of cancer cells.

Despite receiving the best possible chemotherapy, the majority of ovarian cancer patients eventually die of their disease. The management of recurrent ovarian cancer remains difficult, and innovative, effective therapies are needed to convert the high response rates into high cure rates. Treatment of ovarian cancer is both frustrating and encouraging. In view of its propensity to remain in the abdominal cavity and its responsiveness to chemotherapy, ovarian cancer would appear to be a curable disease. In reality, however, by the time the cancer is diagnosed, the tumor has progressed to a stage at which it is partially resistant to drug therapy. It is possible that the recognition of familial cancer syndromes and the use of improved methods of early detection will eventually permit diagnosis at an earlier, more limited stage of the disease, when drug resistance has not yet developed and the majority of patients can be cured with surgery and standard therapies.

Owing to its differential toxicity toward transformed versus normal cells, TRAIL shows potential for the treatment of ovarian cancer and possibly other cancers as long as the mechanisms become better understood, although further preclinical safety testing is important to allay concerns about possible hepatotoxicity. Much progress has been made on elucidating the endogenous biochemical pathway leading to TRAIL-induced apoptosis in cancer cells. Why normal cells generally resist TRAIL, how oncogenic transformation and perhaps infection makes them sensitive to this death ligand, and whether they respond to the ligand in other ways besides cell death is still not fully understood. Clarifying these issues will help advance a more in-depth biologic understanding of this ligand-receptor system and realize its exciting therapeutic potential. Thus, although ovarian cancer can be a particularly difficult disease to treat, we now have several promising strategies that may eventually result in improved survival for patients with this disease. However, it is likely that cure will result only from a combination of treatment strategies tailored to the needs of individual patients. The impact of these new treatment approaches on survival will be determined only through the continued research efforts and enrollment of patients in clinical trials.

REFERENCES

Abdollahi, T., Robertson, N., Abdollahi, A., and Litwack, G. (2003). Identification of Interleukin-8 as an Inhibitor of TRAIL-induced Apoptosis in the Ovarian Carcinoma Cell Line, OVCAR3. *Cancer Res.* **63**(15), 4521–4526.

Adams, J. M., and Cory, S. (1998). The Bcl-2 protein family: Arbiters of cell survival. *Science* **281**(5381), 1322–1326.

Amakawa, R., Hakem, A., Kundig, T. M., Matsuyama, T., Simard, J. J., Timms, E., Wakeham, A., Mittruecker, H. W., Griesser, H., Takimoto, H., Schmits, R., Shahinian, A., Ohashi, P., Penninger, J. M., and Mak, T. W. (1996). Impaired negative selection of T cells in Hodgkin's disease antigen CD30-deficient mice. *Cell* **84**(4), 551–562.

Arenberg, D. A., Kunkel, S. L., Polverini, P. J., Glass, M., Burdick, M. D., and Strieter, R. M. (1996). Inhibition of interleukin-8 correlates reduces tumorigenesis of human non-small cell lung cancer in SCID mice. *J. Clin. Invest.* **97**, 2792–2802.

Arenberg, D. A., Keane, M. P., DiGiovine, B., Kunkel, S. L., Morris, S. B., Xue, Y. Y., Burdick, M. D., Glass, M. C., Iannettoni, M. D., and Strieter, R. M. (1998). Epithelial-neutrophil activating peptide (ENA-78) is an important angiogenic factor in non-small cell lung cancer. *J. Clin. Invest.* **102**, 465–472.

Ashkenazi, A., Pai, R. C., Fong, S., Leung, S., Lawrence, D. A., Marsters, S. A., Blackie, C., Chang, L., McMurtrey, A. E., Hebert, A., DeForge, L., Koumenis, I. L., Lewis, D., Harris, L.,

Bussiere, J., Koeppen, H., Shahrokh, Z., and Schwall, R. H. (1999). Safety and anti-tumor activity of recombinant soluble Apo2 ligand. *J. Clin. Invest.* **104**(2), 155–162.

Bell, D. A., and Scully, R. E. (1994). Early de novo ovarian carcinoma. A study of fourteen cases. *Cancer* **73**, 1859–1864.

Berchuck, A., and Carney, M (1997). Human ovarian cancer of the surface epithelium. *Biochem. Pharmacal.* **54**, 541–544.

Bogoyevitch, M. A., Gillespie-Brown, J., Ketterman, A. J., Fuller, S. J., Ben-Levy, R., Ashworth, A., Marshall, C. J., and Sugden, P. H. (1996). Stimulation of the stress-activated mitogen-activated protein kinase subfamilies in perfused heart. p38/RK mitogen-activated protein kinases and c-Jun N-terminal kinases are activated by ischemia/reperfusion. *Circ. Res.* **79**, 162–173.

Bookman, M. (1998). Biological therapy of ovarian cancer: Current directions. *Semin. Oncol.* **25**, 381–396.

Brew, R., Southern, S. A., Flanagan, B. F., McDicken, I. W., and Christmas, S. E. (1996). Detection of interleukin-8 mRNA and protein in human colorectal carcinoma cells. *Eur. J. Cancer* **32A**, 2142–2147.

Cosman, D. (1994). A family of ligands for the TNF receptor superfamily. *Stem Cells* **12**(5), 440–455.

Cuello, M., Ettenberg, S. A., Nau, M. M., and Lipkowitz, S. (2001). Synergistic induction of apoptosis by the combination of trail and chemotherapy in chemoresistant ovarian cancer cells. *Gynecol. Oncol.* **81**(3), 380–390.

Davis, R. J. (2000). Signal transduction by the JNK group of MAP kinases. *Cell* **103**, 239–252.

Derijard, B., Hibi, M., Wu, I. H., Barrett, T., Su, B., Deng, T., Karin, M., and Davis, R. J. (1994). JNK1: A protein kinase stimulated by UV light and Ha-Ras that binds and phosphorylates the c-Jun activation domain. *Cell* **76**(6), 1025–1037.

Du, C., Fang, M., Li, Y., Li, L., and Wang, X. (2000). Smac, a mitochondrial protein that promotes cytochrome c-dependent caspase activation by eliminating IAP inhibition. *Cell* **102**(1), 33–42.

Emery, J. G., McDonnell, P., Burke, M. B., Deen, K. C., Lyn, S., Silverman, C., Dul, E., Appelbaum, E. R., Eichman, C., DiPrinzio, R., Dodds, R. A., James, I. E., Rosenberg, M., Lee, J. C., and Young, P. R. (1998). Osteoprotegerin is a receptor for the cytotoxic ligand TRAIL. *J. Biol. Chem.* **273**(23), 14363–14367.

Feeley, K. M., and Wells, M. (2001). Precursor lesions of ovarian epithelial malignancy. *Histopathology* **38**, 87–95.

Ferrer, F. A., Miller, L. J., Andrawis, R. I., Kurtzman, S. H., Albertsen, P. C., Laudone, V. P., and Kreutzer, D. L. (1998). Angiogenesis and prostate cancer: In vivo and in vitro expression of angiogenesis factors by prostate cancer cells. *Urology* **51**, 161–167.

Gawrychowski, K., Skopinska-Rozewska, E., Barcz, E., Sommer, E., Szaniawska, B., Roszkowska-Purska, K., Janik, P., and Zielinski, J. (1998). Angiogenic activity and interleukin-8 content of human ovarian cancer ascites. *Eur. J. Gynaecol. Oncol.* **19**, 262–264.

Green, D. R. (2000). Apoptotic pathways: Paper wraps stone blunts scissors. *Cell* **102**(1), 1–4.

Greene, G. F., Kitadai, Y., Pettaway, C. A., von Eschenbach, A. C., Bucana, C. D., and Fidler, I. J. (1997). Correlation of metastasis-related gene expression with metastatic potential in human prostate carcinoma cells implanted in nude mice using an in situ messenger RNA hybridization technique. *Am. J. Pathol.* **150**, 1571–1582.

Greenlee, R., Murray, T., Bolden, S., and Wingo, P. A. (2000). Cancer Statistics, 2000. *CA Cancer J. Clin.* **50**, 7–33.

Harada, A., Mukaida, N., and Matsushima, K. (1996). Interleukin 8 as a novel target for intervention therapy in acute inflammatory diseases. *Mol. Med. Today* **2**, 482–489.

Harant, H., Lindley, I., Uthman, A., Ballaun, C., Krupitza, G., Grunt, T., Huber, H., and Dittrich, C. (1995). Regulation of interleukin-8 gene expression by all-trans retinoic acid. *Biochem. Biophys. Res. Commun.* **210**, 898–906.

Hibi, M., Lin, A., Smeal, T., Minden, A., and Karin, M. (1993). Identification of an oncoprotein- and UV-responsive protein kinase that binds and potentiates the c-Jun activation domain. *Genes Dev.* **7**(11), 2135–2148.

Hu, D. E., Hori, Y., and Fan, T. P. (1993). Interleukin-8 stimulates angiogenesis in rats. *Inflammation* **17**(2), 135–143.

Hu, M. C., Wang, Y. P., and Mikhail, A. (1999). Murine p38-delta mitogen-activated protein kinase, a developmentally regulated protein kinase that is activated by stress and pro-inflammatory cytokines. *J. Biol. Chem.* **274**, 7095–7102.

Hunt, A., and Evan, G. (2001). Till death us do part. *Science* **293**, 1784–1785.

Ivarsson, K., Runesson, E., Sundfeldt, K., Haeger, M., Hedin, L., Janson, P. O., and Brannstrom, M. (1998). The chemotactic cytokine interleukin-8: A cyst fluid marker for malignant epithelial ovarian cancer? *Gynecol. Oncol.* **71**, 420–443.

Jacobson, M. D., Weil, M., and Raff, M. C. (1997). Programmed cell death in animal development. *Cell* **88**(3), 347–354.

Jeremias, I., Herr, I., Boehler, T., and Debatin, K. M. (1998). TRAIL/Apo-2-ligand-induced apoptosis in human T cells. *Eur. J. Immunol.* **28**(1), 143–152.

Jiang, Y., Chen, C., and Li, Z. (1996). Characterization of the structure and function of a new mitogen-activated protein kinase (p38beta). *J. Biol. Chem.* **271**, 17920–17926.

Kitadai, Y., Ellis, L. M., Tucker, S. L., Greene, G. F., Bucana, C. D., Cleary, K. R., Takahashi, Y., Tahara, E., and Fidler, I. J. (1996). Multiparametric in situ mRNA hybridization analysis to predict disease recurrence in patients with colon carcinoma. *Am. J. Pathol.* **149**, 1541–1551.

Kitadai, Y., Radinsky, R., Bucana, C. D., Takahashi, Y., Xie, K., Tahara, E., and Fidler, I. J. (1996). Regulation of carcinoembryonic antigen expression in human colon carcinoma cells by the organ microenvironment. *Am. J. Pathol.* **149**, 1157–1166.

Kitadai, Y., Haruma, K., Sumii, K., Yamamoto, S., Ue, T., Yokozaki, H., Yasui, W., Ohmoto, Y., Kajiyama, G., Fidler, I. J., and Tahara, E. (1998). Expression of interleukin-8 correlates with vascularity in human gastric carcinomas. *Am. J. Pathol.* **152**, 93–100.

Kitadai, Y., Takahashi, Y., Haruma, K., Naka, K., Sumii, K., Yokozaki, H., Yasui, W., Mukaida, N., Ohmoto, Y., Kajiyama, G., Fidler, I. J., and Tahara, E. (1999). Transfection of interleukin-8 increases angiogenesis and tumorigenesis of human gastric carcinoma cells in nude mice. *Br. J. Cancer* **81**, 647–653.

Koch, A. E., Polverini, P. J., Kunkel, S. L., Harlow, L. A., DiPietro, L. A., Elner, V. M., Elner, S. G., and Streiter, R. M. (1992). Interleukin-8 as a macrophage-derived mediator of angiogenesis. *Science* **258**, 1798–1801.

Kumar, S., McDonnell, P. C., and Gum, R. J. (1997). Novel homologues of CSBP/p38 MAP kinase: Activation, substrate specificity and sensitivity to inhibition by pyridinyl imidazoles. *Biochem. Biophys. Res. Commun.* **235**, 533–538.

Kyriakis, J. M., Banerjee, P., Nikolakaki, E., Dai, T., Rubie, E. A., Ahmad, M. F., Avruch, J., and Woodgett, J. R. (1994). The stress-activated protein kinase subfamily of c-Jun kinases. *Nature* **369**, 156–560.

Kyriakis, J. M., and Avruch, J. (1996). Sounding the alarm: Protein kinase cascades activated by stress and inflammation. *J. Biol. Chem.* **271**, 24313–24316.

Lancaster, J. M., Sayer, R., Blanchette, C., Calingaert, B., Whitaker, R., Schildkraut, J., Marks, J., and Berchuck, A. (2003). High expression of tumor necrosis factor apoptosis-inducing ligand is associated with favorable ovarian cancer survival. *Clin. Cancer Res.* **9**, 762–766.

Landis, S. H., Murray, T., Bolden, S., and Wingo, P. A. (1998). Cancer statistics, 1998. *CA Cancer J. Clin.* **48**, 6–29.

Lee, M. W., Park, S. C., Yang, Y. G., Yim, S. O., Chae, H. S., Bach, J. H., Lee, H. J., Kim, K. Y., Lee, W. B., and Kim, S. S. (2002). The involvement of reactive oxygen species (ROS) and p38 mitogen-activated protein (MAP) kinase in TRAIL/Apo2L-induced apoptosis. *FEBS Lett.* **512**(1–3), 313–318.

Li, Z., Jiang, Y., and Ulevitch, R. J. (1996). The primary structure of p38 gamma: A new member of p38 group of MAP kinases. *Biochem. Biophys. Res. Commun.* **228,** 334–340.

Luca, M., Huang, S., Gershenwald, J. E., Singh, R. K., Reich, R., and Bar-Eli, M. (1997). Expression of interleukin-8 by human melanoma cells up-regulates MMP-2 activity and increases tumor growth and metastasis. *Am. J. Pathol.* **151,** 1105–1113.

Lynch, D. H., Ramsdell, F., and Alderson, M. R. (1995). Fas and FasL in the homeostatic regulation of immune responses. *Immunol. Today* **16**(12), 569–574.

Lynch, H. T., and Lynch, J. F. (1992). Hereditary ovarian carcinoma. *Hematol. Oncol. Clin. North Am.* **6,** 783–811.

Marks, J. R., Davidoff, A. M., Kerns, B. J., Humphrey, P. A., Pence, J. C., Dodge, R. K., Clarke-Pearson, D. L., Iglehart, J. D., Bast, R. C., Jr., and Berchuck, A. (1991). Overexpression and mutation of p53 in epithelial ovarian cancer. *Cancer Res.* **51**(11), 2979–2984.

Marsters, S. A., Pitti, R. M., Donahue, C. J., Ruppert, S., Bauer, K. D., and Ashkenazi, A. (1996). Activation of apoptosis by Apo-2 ligand is independent of FADD but blocked by CrmA. *Curr. Biol.* **6**(6), 750–752.

Marsters, S. A., Sheridan, J. P., Pitti, R. M., Huang, A., Skubatch, M., Baldwin, D., Yuan, J., Gurney, A., Goddard, A. D., Godowski, P., and Ashkenazi, A. (1997). A novel receptor for Apo2L/TRAIL contains a truncated death domain. *Curr. Biol.* **7**(12), 1003–1006.

Marsters, S. A., Sheridan, J. P., Pitti, R. M., Brush, J., Goddard, A., and Ashkenazi, A. (1998). Identification of a ligand for the death-domain-containing receptor Apo3. *Curr. Biol.* **8**(9), 525–528.

Matsushima, K., and Oppenheim, J. J. (1989). Interleukin 8 and MCAF: Novel inflammatory cytokines inducible by IL 1 and TNF. *Cytokine* **1,** 2–13.

McGuire, W. P., Rowinsky, E. K., Rosenshein, N. B., Grumbine, F. C., Ettinger, D. S., Armstrong, D. K., and Donehower, R. C. (1989). Taxol: A unique antineoplastic agent with significant activity in advanced ovarian epithelial neoplasms. *Ann. Intern. Med.* **111**(4), 273–279.

Merogi, A. J., Marrogi, A. J., Ramesh, R., Robinson, W. R., Fermin, C. D., and Freeman, S. M. (1997). Tumor-host interaction: Analysis of cytokines, growth factors, and tumor-infiltrating lymphocytes in ovarian carcinomas. *Hum. Pathol.* **28,** 321–331.

Miller, L. J., Kurtzman, S. H., Wang, Y., Anderson, K. H., Lindquist, R. R., and Kreutzer, D. L. (1998). Expression of interleukin-8 receptors on tumor cells and vascular endothelial cells in human breast cancer tissue. *Anticancer Res.* **18,** 77–81.

Mills, G. B., May, C., McGill, M., and Roifman, C. M. (1988). A putative new growth factor in ascitic fluid from ovarian cancer patients: Identification, characterization, and mechanism of action. *Cancer Res.* **48,** 1066–1071.

Mills, G. B., May, C., Hill, M., Campbell, S., Shaw, P., and Marks, A. (1990). Ascitic fluid from human ovarian cancer patients contains growth factors necessary for intraperitoneal growth of human ovarian adenocarcinoma cells. *J. Clin. Invest.* **86,** 851–855.

Miyamoto, M., Shimizu, Y., Okada, K., Kashii, Y., Higuchi, K., and Watanabe, A. (1998). Effect of interleukin-8 on production of tumor-associated substances and autocrine growth of human liver and pancreatic cancer cells. *Cancer Immunol. Immunother.* **47,** 47–57.

Mok, C. H., Tsao, S. W., Knapp, R. C., Fishbaugh, P. M., and Lau, C. C. (1992). Unifocal origin of advanced human epithelial ovarian cancers. *Cancer Res.* **52**(18), 5119–5122.

Nagane, M., Pan, G., Weddle, J. J., Dixit, V. M., Cavenee, W. K., and Huang, H. J. (2000). Increased death receptor 5 expression by chemotherapeutic agents in human gliomas causes synergistic cytotoxicity with tumor necrosis factor-related apoptosis-inducing ligand in vitro and in vivo. *Cancer Res.* **60**(4), 847–853.

Nagata, S., and Golstein, P. (1995). The Fas death factor. *Science* **267**(5203), 1449–1456.

Nagata, S. (1997). Apoptosis by death factor. *Cell* **88,** 355–365.

Negri, E., Franceschi, S., and Tzonou, A. (1991). Pooled analysis of 3 European case-control studies. I. Reproductive factors and risk of epithelial ovarian cancer. *Int. J. Cancer* **49**, 50–56.

Nick, J. A., Avdi, N. J., and Young, S. K. (1999). Selective activation and functional significance of p38alpha mitogen-activated protein kinase in lipopolysaccharide-stimulated neutrophils. *J. Clin. Invest.* **103**, 851–858.

Nowell, P. C. (1976). The clonal evolution of tumor cell populations. *Science* **194**, 23–28.

Ohtsuka, T., and Zhou, T. (2002). Bisindolylmaleimide VIII enhances DR5-mediated apoptosis through the MKK4/JNK/p38 kinase and the mitochondrial pathways. *J. Biol. Chem.* **277**, 29294–29303.

Ohtsuka, T., Buchsbaum, D., Oliver, P., Makhija, S., Kimberly, R., and Zhou, T. (2003). Synergistic induction of tumor cell apoptosis by death receptor antibody and chemotherapy agent through JNK/p38 and mitochondrial death pathway. *Oncogene* **22**, 2034–2044.

Pan, G., Ni, J., Wei, Y. F., Yu, G., Gentz, R., and Dixit, V. M. (1997). An antagonist decoy receptor and a death domain-containing receptor for TRAIL. *Science* **277**(5327), 815–818.

Pitti, R. M., Marsters, S. A., Ruppert, S., Donahue, C. J., Moore, A., and Ashkenazi, A. (1996). Induction of apoptosis by Apo-2 ligand, a new member of the tumor necrosis factor cytokine family. *J. Biol. Chem.* **271**(22), 12687–12690.

Pombo, C. M., Bonventre, J. V., Avruch, J., Woodgett, J. R., Kyriakis, J. M., and Force, T. (1994). The stress-activated protein kinases are major c-Jun amino-terminal kinases activated by ischemia and reperfusion. *J. Biol. Chem.* **269**, 26546–26551.

Radke, J., Schmidt, D., Bohme, M., Schmidt, U., Weise, W., and Morenz, J. (1996). Zytokinspiegel im malignen Aszites und peripheren Blut von Patientinnen mit fortgeschrittenem Ovarialkarzinom. *Gubertshilfe Frauenheilkd* **56**, 83–87.

Raingeaud, J., Gupta, S., Rogers, J. S., Dickens, M., Han, J., Ulevitch, R. J., and Davis, R. J. (1995). Pro-inflammatory cytokines and environmental stress cause p38 mitogen-activated protein kinase activation by dual phosphorylation on tyrosine and threonine. *J. Biol. Chem.* **270**(13), 7420–7326.

Richards, B. L., Eisma, R. J., Spiro, J. D., Lindquist, R. L., and Kreutzer, D. L. (1997). Coexpression of interleukin-8 receptors in head and neck squamous cell carcinoma. *Am. J. Surg.* **174**, 507–512.

Rieger, J., Ohgaki, H., Kleihues, P., and Weller, M. (1999). Human astrocytic brain tumors express AP02L/TRAIL. *Acta Neuropathol.* **97**(1), 1–4.

Roth, W., Isenmann, S., Naumann, U., Kugler, S., Bahr, M., Dichgans, J., Ashkenazi, A., and Weller, M. (1999). Locoregional Apo2L/TRAIL eradicates intracranial human malignant glioma xenografts in athymic mice in the absence of neurotoxicity. *Biochem. Biophys. Res. Commun.* **265**(2), 479–483.

Schadendorf, D., Moller, A., Algermissen, B., Worm, M., Sticherling, M., and Czarnetzki, B. M. (1994). IL-8 produced by human malignant melanoma cells *in vitro* is an essential autocrine growth factor. *J. Immunol.* **151**, 2667–2675.

Scheibenbogen, C., Mohler, T., Haefele, J., Hunstein, W., and Keilholz, U. (1995). Serum interleukin-8 (IL-8) is elevated in patients with metastatic melanoma and correlates with tumor load. *Melanoma Res.* **5**, 179–181.

Schildkraut, J. M., and Thompson, W. D. (1988). Familial ovarian cancer: A population-based case-control study. *Am. J. Epidemiol.* **128**, 456–466.

Schildkraut, J. M., Bastos, E., and Berchuck, A. (1997). Relationship between lifetime ovulatory cycles and overexpression of mutant p53 in epithelial ovarian cancer. *J. Nat. Cancer Inst.* **89**(13), 932–938.

Shah, M. A., and Schwartz, G. K. (2001). Cell cycle-mediated drug resistance: An emerging concept in cancer therapy. *Clin. Cancer Res.* **7**, 2168–2181.

Sheridan, J. P., Marsters, S. A., Pitti, R. M., Gurney, A., Skubatch, M., Baldwin, D., Ramakrishnan, L., Gray, C. L., Baker, K., Wood, W. I., Goddard, A. D., Godowski, P., and Ashkenazi, A. (1997). Control of TRAIL-induced apoptosis by a family of signaling and decoy receptors. *Science* **277**(5327), 818–821.

Singh, R. K., Gutman, M., Radinsky, R., Bucana, C. D., and Fidler, I. J. (1994). Expression of interleukin-8 correlates with the metastatic potential of human melanoma cells in nude mice. *Cancer Res.* **54**, 3242–3247.

Singh, R. K., Gutman, M., Reich, R., and Bar-Eli, M. (1995). Ultraviolet B irradiation promotes tumorigenic and metastatic properties in primary cutaneous melanoma via induction of interleukin 8. *Cancer Res.* **55**, 3669–3674.

Smith, C. A., Farrah, T., and Goodwin, R. G. (1994a). The TNF receptor superfamily of cellular and viral proteins: Activation, costimulation, and death. *Cell* **76**(6), 959–962.

Smith, D. R., Polverini, P. J., Kunkel, S. L., Orringer, M. B., Whyte, R. I., Burdick, M. D., Wilke, C. A., and Strieter, R. M. (1994b). Inhibition of interleukin 8 attenuates angiogenesis in bronchogenic carcinoma. *J. Exp. Med.* **179**, 1409–1415.

Steller, H. (1995). Mechanisms and genes of cellular suicide. *Science* **267**(5203), 1445–1449.

Strieter, R. M., Kunkel, S. L., Elner, V. M., Martonyi, C. L., Koch, A. E., Polverini, P. J., and Elner, S. G. (1992). Interleukin-8, A corneal factor that induces neovascularization. *Am. J. Pathol.* **141**, 1279–1284.

Thomas, W. D., and Hersey, P. (1998). TNF-related apoptosis-inducing ligand (TRAIL) induces apoptosis in Fas ligand-resistant melanoma cells and mediates CD4 T cell killing of target cells. *J. Immunol.* **161**(5), 2195–2200.

Thompson, C. B. (1995). Apoptosis in the pathogenesis and treatment of disease. *Science* **267**(5203), 1456–1462.

Tsao, S. W., Mok, C. H., Knapp, R. C., Oike, K., Muto, M. G., Welch, W. R., Goodman, H. M., Sheets, E. E., Berkowitz, R. S., and Lau, C. C. (1993). Molecular genetic evidence of a unifocal origin for human serous ovarian carcinomas. *Gynecol. Oncol.* **48**(1), 5–10.

Tuschil, A., Lam, C., Haslberger, A., and Lindley, I. (1992). IL-8 stimulates calcium transients and promotes epidermal cell proliferation. *Invest. Dermatol.* **99**, 294–298.

Van Parijs, L., and Abbas, A. K. (1996). Role of Fas-mediated cell death in the regulation of immune responses. *Curr. Opin. Immunol.* **8**(3), 355–361.

Verhagen, A. M., Ekert, P. G., Pakusch, M., Silke, J., Connolly, L. M., Reid, G. E., Moritz, R. L., Simpson, R. J., and Vaux, D. L. (2000). Identification of DIABLO, a mammalian protein that promotes apoptosis by binding to and antagonizing IAP proteins. *Cell* **102**(1), 43–53.

Walczak, H., Miller, R. E., Ariail, K., Gliniak, B., Griffith, T. S., Kubin, M., Chin, W., Jones, J., Woodward, A., Le, T., Smith, C., Smolak, P., Goodwin, R. G., Rauch, C. T., and Lynch, D. H. (1999). Tumoricidal activity of tumor necrosis factor-related apoptosis-inducing ligand in vivo. *Nat. Med.* **5**(2), 157–163.

Wallach, D., Varfolomeev, E. E., Malinin, N. L., Goltsev, Y. V., Kovalenko, A. V., and Boldin, M. P. (1999). Tumor necrosis factor receptor and Fas signaling mechanisms. *Annu. Rev. Immunol.* **17**, 331–367.

Waskiewicz, A. J., and Cooper, J. A. (1995). Mitogen activated protein kinase (MAPK) signal transduction. *Curr. Opin. Cell Biol.* **7**, 798–805.

Westermann, A. M., Beijnen, J. H., Moolenaar, W. H., and Rodenhuis, S. (1997). Growth factors in human ovarian cancer. *Cancer Treat. Rev.* **23**, 113–131.

Whittemore, A. S., Harris, R., and Itnyre, J. (1992). Characteristics relating to ovarian cancer risk: Collaborative analysis of 12 US case-control studies. IV. The pathogenesis of epithelial ovarian cancer. Collaborative Ovarian Cancer Group. *Am. J. Epidemiol.* **136**(10), 1212–1220.

Wiley, S. R., Schooley, K., Smolak, P. J., Din, W. S., Huang, C. P., Nicholl, J. K., Sutherland, G. R., Smith, T. D., Rauch, C., Smith, C. A., and Goodwin, R. G. (1995). Identification and characterization of a new member of the TNF family that induces apoptosis. *Immunity* **3**(6), 673–682.

Yatsunami, J., Tsuruta, N., Ogata, K., Wakamatsu, K., Takayama, K., Kawasaki, M., Nakanishi, Y., Hara, N., and Hayashi, S. (1997). Interleukin-8 participates in angiogenesis in non-small cell, but not small cell carcinoma of the lung. *Cancer Lett.* **120,** 101–108.

Yoneda, J., Kuniyasu, H., Crispens, M. A., Price, J. E., Bucana, C. D., and Fidler, I. J. (1998). Expression of angiogenesis-related genes and progression of human ovarian carcinomas in nude mice. *J. Natl. Cancer Inst.* **90,** 447–454.

Zhang, X. D., Franco, A., Myers, K., Gray, C., Nguyen, T., and Hersey, P. (1999). Relation of TNF-related apoptosis-inducing ligand (TRAIL) receptor and FLICE-inhibitory protein expression to TRAIL-induced apoptosis of melanoma. *Cancer Res.* **59**(11), 2747–2753.

19

TRAIL AND CHEMOTHERAPEUTIC DRUGS IN CANCER THERAPY

Xiu-Xian Wu,[*] Osamu Ogawa,[†] and Yoshiyuki Kakehi[*]

[*]Department of Urology, Kagawa University, Kagawa 761-0793, Japan
[†]Department of Urology, Graduate School of Medicine, Kyoto University
Kyoto 606-8507, Japan

I. Introduction
II. Receptors for TRAIL
III. Signaling Pathways of TRAIL Receptors
    A. Recruitment of FADD and Caspase-8 and Caspase-8 Activation
    B. Activation of the Mitochondrial Pathway
    C. Regulation of Apoptotic Pathways
IV. Bioactivity of TRAIL
    A. Sensitivity of Tumor Cells to TRAIL
    B. In Vivo *Tumoricidal Activity of TRAIL*
    C. Involvement of TRAIL in Cell-Mediated Cytotoxicity
V. Synergistic Effect of TRAIL and Chemotherapeutic Drugs
VI. Molecular Mechanisms of the Synergistic Effect
VII. Conclusions and Prospects
    References

Tumor necrosis factor (TNF)–related apoptosis-inducing ligand (TRAIL/Apo2L) is a recently identified member of the TNF ligand family that selectively induces apoptosis in tumor cells *in vitro* and *in vivo* but not in most normal cells. Chemotherapeutic drugs induce apoptosis and the upregulation of death receptors or activation of intracellular signaling pathways of TRAIL. Numerous chemotherapeutic drugs have been shown to sensitize tumor cells to TRAIL-mediated apoptosis. Studies from our laboratory have also shown that TRAIL-resistant renal cell carcinoma, prostate gland cancer, and bladder cancer cells are sensitized by subtoxic concentrations of chemotherapeutic drugs including doxorubicin, epirubicin, pirarubicin, and cisplatin. TRAIL, particularly in combination with chemotherapeutic agents, is thus potentially promising in the treatment of cancer. This review addresses the putative role of TRAIL in cancer treatment and discusses the molecular basis of the synergistic effect of TRAIL and chemotherapeutic drugs. © 2004 Elsevier Inc.

I. INTRODUCTION

Tumor necrosis factor (TNF)–related apoptosis-inducing ligand (TRAIL), also known as Apo-2 ligand (Apo2L), is a recently identified member of the TNF family (Pitti *et al.*, 1996; Wiley *et al.*, 1995). Like other members of the TNF family, TRAIL is a type II transmembrane protein, but it is cleaved by cysteine proteases at the cell surface to yield a soluble form (Ashkenazi *et al.*, 1999a; Mariani *et al.*, 1998a). Unlike other members of the TNF family, TRAIL carries a zinc ion at the trimer interface, coordinated by the single unpaired cysteine residue (Cys230) of each monomer (Hymowitz *et al.*, 2000). The zinc ion is essential for the structural integrity of TRAIL, and the mutation of Cys230 to alanine or serine strongly inhibits its ability to induce apoptosis (Bodmer *et al.*, 2000a; Hymowitz *et al.*, 2000). Unlike other members of the TNF family whose expression is predominantly restricted to T cells and sites of immune privilege, mRNA for TRAIL is detected in a wide range of tissues including peripheral blood lymphocytes (PBLs), the spleen, lung, prostate gland, ovary, colon, and placenta, but not the brain, liver, or testis (Wiley *et al.*, 1995).

TRAIL induces apoptosis in a variety of tumor cells *in vitro* and *in vivo*, but shows little or no toxicity in most normal cells (Ashkenazi *et al.*, 1999a; Griffith *et al.*, 1998; Walczak *et al.*, 1999). Although other members of the TNF family such as TNF and FasL also strongly induce apoptosis in many types of tumor cells, administration of TNF causes a lethal inflammatory response that resembles septic shock (Havell *et al.*, 1988; Tartaglia *et al.*,

1992), and infusion of agonistic anti-Fas antibody causes lethal hepatotoxicity (Nagata, 1997; Ogasawara *et al.*, 1993). In contrast, TRAIL exhibits no detectable cytotoxicity in mice (Walczak *et al.*, 1999) or monkeys (Ashkenazi *et al.*, 1999a). These features have focused considerable attention on TRAIL as a potential candidate for anticancer therapeutics (Bonavida *et al.*, 1999; Nagane *et al.*, 2001). It was reported, however, that a histidine-tagged version of recombinant TRAIL induces apoptosis in normal human hepatocytes (Jo *et al.*, 2000). This apparent controversy was, however, recently resolved by findings demonstrating that, unlike the toxicity associated with the histidine-tagged recombinant version of the ligand, clinical-grade recombinant human TRAIL had minimal toxicity toward normal human hepatocytes (Lawrence *et al.*, 2001).

Many advances in cancer therapy have followed the introduction of chemotherapeutic drugs. A consequence of chemotherapy is, however, the development or acquisition of drug-resistant phenotypes. Cancer cell resistance to chemotherapeutic drugs thus remains a major obstacle and demonstrates an obvious need for alternatives. Combination treatment of biologic response modifiers with chemotherapeutic drugs has been considered a possible means to reverse drug resistance (Mizutani *et al.*, 1995; Wu *et al.*, 2000). Indeed, a number of studies, including ours, have shown that TRAIL works synergistically with chemotherapeutic drugs against most types of tumor cells and reversed resistance to such drugs (Wajant *et al.*, 2002; Wu *et al.*, 2002, 2003). These findings indicate that the combination of TRAIL and chemotherapeutic drugs is potentially promising in treating refractory cancers. In this review, we address the putative role of TRAIL in cancer treatment and discuss the molecular basis of the synergistic effect of TRAIL and chemotherapeutic drugs.

II. RECEPTORS FOR TRAIL

Of the five receptors for TRAIL, two DR4 (TRAIL-R1) and DR5 (TRAIL-R2/TRICK2/KILLER), contain a cytoplasmic death domain and can transmit an apoptotic signal upon stimulation by TRAIL (Chaudhary *et al.*, 1997; MacFarlane *et al.*, 1997; Pan *et al.*, 1997a,b; Screaton *et al.*, 1997; Sheridan *et al.*, 1997; Walczak *et al.*, 1997). Two decoy receptors, DcR1 (TRAIL-R3/TRID) and DcR2 (TRAIL-R4/TRUNDD/LIT), contain either no cytoplasmic death domain or a truncated death domain and compete with DR4 and DR5 for TRAIL binding, thereby acting as antagonists (Degli-Esposti *et al.*, 1997; MacFarlane *et al.*, 1997; Marsters *et al.*, 1997; Mongkolsapaya *et al.*, 1998; Pan *et al.*, 1997b; Sheridan *et al.*, 1997). DR4 and DR5 are expressed widely in both normal and malignant cells, while DcR1 and DcR2 are expressed preferentially in many normal tissues but in only a few transformed cells (Ashkenazi *et al.*, 1998, 1999b).

This suggests that TRAIL's low toxicity toward normal tissues is attributable to the expression of decoy receptors in normal tissues that protect normal cells from the induction of apoptosis (Ashkenazi et al., 1998, 1999b). The expression of decoy receptors, however, does not generally correlate with TRAIL sensitivity (Kim et al., 2000; Leverkus et al., 2000). Osteoprotegerin (OPG) is a soluble member of the TRAIL receptor family. Like other TRAIL decoy receptors, OPG also is capable of binding TRAIL and blocks TRAIL-mediated apoptosis (Emery et al., 1998), but its affinity to TRAIL is rather weak compared to that of other TRAIL receptors. The expression of TRAIL decoy receptors is therefore not the major factor in determining TRAIL sensitivity.

III. SIGNALING PATHWAYS OF TRAIL RECEPTORS

A. RECRUITMENT OF FADD AND CASPASE-8 AND CASPASE-8 ACTIVATION

Stimulation of death-domain–containing TRAIL receptors DR4 and DR5 results in recruitment of the cellular adaptor protein Fas-associated death domain (FADD) by interaction of the death domains of these molecules (French et al., 1999; Kuang et al., 2000; Wajant et al., 2002). FADD then recruits procaspase-8 in the death signaling complex of DR4 and DR5 through its death effector domain (Bodmer et al., 2000b; Kischkel et al., 2000; Sprick et al., 2000), where procaspase-8 undergoes autoactivation (Schneider et al., 2000). In accordance with the pivotal role of FADD and caspase-8 in TRAIL-mediated apoptosis, mice or cell lines deficient in these molecules are completely protected from the apoptotic action of TRAIL (Bodmer et al., 2000b; Kischkel et al., 2000; Kuang et al., 2000). Proapoptotic protein death-associated protein 3 (DAP3), originally found as a mediator of interferon-γ (IFN-γ)–induced cell death, was recently identified as an additional adaptor protein that binds to the death domain of death receptors and FADD and activates caspase-8 (Miyazaki et al., 2001).

B. ACTIVATION OF THE MITOCHONDRIAL PATHWAY

Activated caspase-8 triggers apoptosis, either by directly activating downstream executioner caspases such as caspase-3, -6, and -7 (Medema et al., 1997; Orth et al., 1996; Seol et al., 2001), or by cleaving Bid, a BH3 domain-containing proapoptotic member of the Bcl-2 family of proteins that activates the mitochondrial pathway to cell death (Yamada et al., 1999). Bid is cleaved by caspase-8 to a truncated version of the molecule tBid, which then translocates from cytosol to mitochondria and induces the release of cytochrome *c* (Li *et al.*, 1998; Luo *et al.*, 1998) and a second mitochondria-derived activator (Smac/Diablo) (Chai *et al.*, 2000; Verhagen *et al.*, 2000) from mitochondria Bax/Bak-dependently (Wei *et al.*, 2001). Cytochrome *c* induces formation of the apoptosome, a multimeric Apaf-1 and cytochrome *c* complex (Zou *et al.*, 1999). Caspase recruitment domains of Apaf-1 become exposed in the apoptosome, which subsequently facilitates the activation of caspase-9 (Wang, 2001). Smac also facilitates the activation of caspase-9 by blocking inhibitors of apoptosis (IAPs) and the x-linked inhibitor of apoptosis (xIAP) (Chai *et al.*, 2000; Verhagen *et al.*, 2000). Activated caspase-9 then activates downstream executioner caspases, such as caspase-3, which subsequently cleave many important intracellular proteins such as poly (ADP-ribose) polymerase and DNA fragmentation factor, resulting in characteristic morphologic changes in apoptosis (Hengartner, 2000) (Fig. 1). Postmitochondrial caspase-3 activation also completes the amplification loop by processing cytoplasmic procaspase-8 (Engels *et al.*, 2000; Tang *et al.*, 2000).

C. REGULATION OF APOPTOTIC PATHWAYS

Apoptotic pathways are regulated by most proteins including the caspase-8 homologue, FLICE-inhibitory protein (FLIP) (Irmler *et al.*, 1997; Kim *et al.*, 2000; Zhang *et al.*, 1999), and members of the IAP (Chai *et al.*, 2000) and Bcl-2 families (Fulda *et al.*, 2002; Munshi *et al.*, 2001; Sun *et al.*, 2001; Thomas *et al.*, 2000).

These signaling molecules involved in TRAIL-mediated apoptotic pathways become attractive therapeutic targets for controlling the sensitivity of tumor cells to TRAIL-mediated apoptosis.

IV. BIOACTIVITY OF TRAIL

A. SENSITIVITY OF TUMOR CELLS TO TRAIL

Since it was discovered in 1995 (Wiley *et al.*, 1995), the direct cytotoxic effect of soluble recombinant TRAIL on a variety of tumor cells has been widely studied. In numerous *in vitro* studies, certain tumor cell lines have been shown to be sensitive to TRAIL-induced apoptosis (Pitti *et al.*, 1996; Rieger *et al.*, 1998; Thomas *et al.*, 1998; Zhang *et al.*, 1999). It did not, however, induce apoptosis in most normal cells, including renal proximal tubule epithelial cells, lung fibroblasts, mammary epithelial cells, skeletal muscle cells, astrocytes, melanocytes, colon smooth muscle cells, or hepatocytes (Lawrence *et al.*, 2001; Walczak *et al.*, 1999). Several studies have also shown that although nontransformed cells such as keratinocytes (Kothny-Wilkes *et al.*, 1998) and PBL (Kayagaki *et al.*, 1999b;

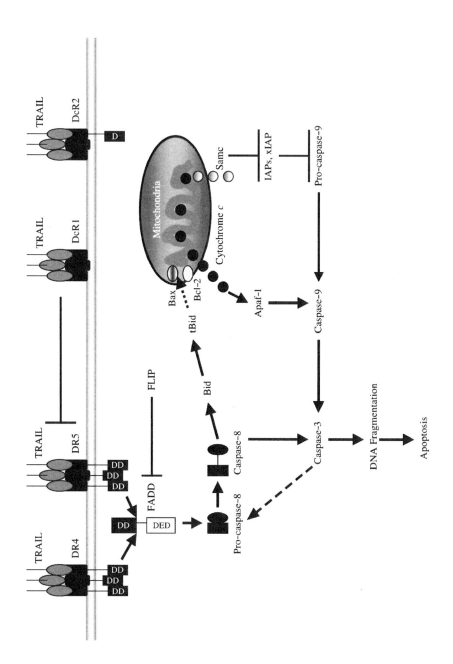

Screaton *et al.*, 1997) express TRAIL and TRAIL receptors, they are not sensitive to the cytotoxic effects of TRAIL.

B. *IN VIVO* TUMORICIDAL ACTIVITY OF TRAIL

The *in vivo* potential tumoricidal activity of TRAIL was recently demonstrated (Ashkenazi *et al.*, 1999a; Walczak *et al.*, 1999). Intraperitoneal injections of TRAIL effectively suppressed growth of TRAIL-sensitive human mammary adenocarcinoma cell line MDA-231 in CB.17SCID mice and prolonged murine survival with no detectable tumor mass in the majority of cases (Walczak *et al.*, 1999). TRAIL treatment of athymic nude mice bearing human colon carcinoma also induced apoptosis of tumor cells, suppressed tumor progression, and improved survival (Ashkenazi *et al.*, 1999a). These animal experiments also simultaneously showed that TRAIL has no systemic toxicity toward normal tissues. It was also recently demonstrated that combining TRAIL with either 5-fluorouracil (5-FU) or CPT-11 produced a greatly enhanced antitumor effect over that of either agent alone, with 50% of human colon cancer–implanted SCID mice achieving complete tumor regression with a combination of TRAIL and CPT-11 (Naka *et al.*, 2002).

C. INVOLVEMENT OF TRAIL IN CELL-MEDIATED CYTOTOXICITY

Constitutive or inducible TRAIL expression following T cell activation by calcium ionophore ionomycin, interleukin-2 (IL-2), IFN-α, β, or anti-CD3 monoclonal antibody (mAb) has been observed in CD4+ and CD8+ T cells (Kayagaki *et al.*, 1999a; Martinez-Lorenzo *et al.*, 1998). Expression of TRAIL was also seen on primary natural killer (NK) cells and in activated mouse and human T and B cells (Mariani *et al.*, 1998b; Zamai *et al.*, 1998). Freshly isolated murine NK cells from the liver were recently found to be responsible for spontaneous cytotoxicity against TRAIL-sensitive tumor cells *in vitro* along with perforin and the Fas ligand (Takeda *et al.*, 2001). The administration of neutralizing monoclonal antibody

---

FIGURE 1. Signaling pathways of TRAIL receptors. Stimulation of DR4 and DR5 results in recruiting procaspase-8 via interaction with FADD. Decoy receptors DcR1 and DcR2 inhibit the signal by competitive binding with TRAIL. Once procaspase-8 is recruited, it is autocleaved and activated. Activated caspase-8 further activates downstream executioner caspases such as caspase-3. Caspase-8 also cleaves the proapoptotic molecule Bid to tBid, which then translocates to mitochondria and promotes the release of cytochrome *c* and Smac. Cytochrome *c* triggers the activation of caspase-9 through Apaf-1 and Smac facilitates caspase-9 by blocking caspase inhibitory proteins IAPs and xIAP. The activation of caspase-9 activates caspase-3 and leads to DNA fragmentation and apoptosis. Activated caspase-3 also induces caspase-8 activation.

against TRAIL also significantly increased experimental liver metastases of several TRAIL-sensitive tumor cell lines (Takeda et al., 2001). These data suggest the involvement of TRAIL in T cell–mediated and NK cell–mediated cytotoxicity, which may complement the apoptotic activity of Fas and perforin/granzyme pathways (Johnsen et al., 1999; Kayagaki et al., 1999b; Monleon et al., 2001; Thomas et al., 1998).

V. SYNERGISTIC EFFECT OF TRAIL AND CHEMOTHERAPEUTIC DRUGS

The selective cytotoxicity of TRAIL on tumor cells but not most normal cells, its involvement in T cell–mediated and NK cell–mediated cytotoxicity, and the absence of toxic side effects upon *in vivo* administration have made TRAIL an attractive candidate for tumor treatment especially for tumors sensitive to TRAIL. A high percentage of tumor cell lines have, however, been relatively resistant to TRAIL-induced apoptosis (Grotzer *et al.*, 2000; Kagawa *et al.*, 2001; Wajant *et al.*, 2002). This led to numerous studies showing that TRAIL resistance is overcome by combined application with chemotherapeutic drugs (Table I). These data indicate that numerous chemotherapeutic drugs with diverse mechanisms of action, such as doxorubicin (DOX), cisplatin (CDDP), actinomycin D (Act D), and 5-FU, enhance TRAIL-mediated apoptosis in a variety of tumor cells, including those that are drug resistant.

VI. MOLECULAR MECHANISMS OF THE SYNERGISTIC EFFECT

Molecular mechanisms of the synergistic action of TRAIL and chemotherapy have yet to be clarified. Clearly, however, chemotherapeutic drugs and death receptor ligands induce apoptosis that converges at the level of the death receptor itself, the initial phase of caspase activation, or via activation of the mitochondrial pathway and effector caspases resulting in similar morphologic and biochemical features of cell death (Petak *et al.*, 2001). Chemotherapeutic drugs sensitize most tumor cells to TRAIL-mediated apoptosis by directly upregulating death domain–containing receptors DR4 and DR5, facilitating the activation of the mitochondrial pathway, or activating the caspase signaling cascade (Fig. 2). Paclitaxel, etoposide, 6-[3-(1-adamantyl)-4-hydroxyphenyl]-2-naphthalene carboxylic acid (CD437), DOX, and Ara-C, for example, upregulate the expression of DR4 and/or DR5 in several cancer cell lines (Nagane *et al.*, 2000; Nimmanapalli *et al.*, 2001; Sun *et al.*, 2000; Wen *et al.*, 2000). DOX and

TABLE I. Augmentation of TRAIL-Mediated Apoptosis by Chemotherapeutic Drugs

Cell types	Drugs	Modes of action	References
Prostate cancer lines	Paclitaxel	Induction of DR4 and DR5 expression	Nimmanapalli et al., 2001
	Etoposide		Munshi et al., 2002
	Doxorubicin		Voelkel-Johnson et al., 2002
	Doxorubicin	Cleavage of Bid	Wu et al., 2002
	Actinomycin D	Activation of caspase-8, -6, and -3	Ng et al., 2002
Breast cancer lines	Doxorubicin	Inhibition of xIAP and release of cytochrome c	Keane et al., 1999
	5-FU		Keane et al., 1999
	Herceptin	Inhibition of Akt kinase	Cuello et al., 2001a
	Etoposide	Induction of DR4 and DR5	Gibson et al., 2000
Lung cancer lines	CD437[a]	Induction of DR4 and DR5	Sun et al., 2000
Colorectal cancer cell lines	CPT-11		Gliniak et al., 1999
	Doxorubicin	Activation of caspase-8 and cleavage of Bid	Lacour et al., 2001
	Cisplatin	Activation of caspase-8 and cleavage of Bid	Lacour et al., 2001
	Sodium butyrate	Reduced FLIP expression	Hernandez et al., 2001b
	Sodium nitroprusside	Release of cytochrome c	Lee et al., 2001
	Actinomycin D	Reduced FLIP expression	Hernandez et al., 2001a
	Cycloheximide	Reduced FLIP expression	Hernandez et al., 2001a
Bladder cancer cells	Doxorubicin		Mizutani et al., 1999
	Cisplatin	Enhanced Bax expression	Mizutani et al., 2001
RCC cells	Topotecan		Dejosez et al., 2000
	Doxorubicin	Induction of DR4 and DR5 expression	Wu et al., 2003
	5-FU	Enhanced p53 expression	Mizutani et al., 2002
	Paclitaxel	Inhibition of Akt kinase	Asakuma et al., 2003
Ovary cancer cell lines	Herceptin	Inhibition of Akt kinase	Cuello et al., 2001a
	Cisplatin	Activation of caspase-3	Cuello et al., 2001b
	Doxorubicin	Activation of caspase-3	Cuello et al., 2001b
	Paclitaxel	Activation of caspase-3	Cuello et al., 2001b

(continues)

TABLE I. *(continued)*

Cell types	Drugs	Modes of action	References
Leukemia cell lines	Doxorubicin	Induction of DR5 expression	Wen et al., 2000
	Etoposide	Induction of DR5 expression	Wen et al., 2000
	Ara-C	Induction of DR5 expression	Wen et al., 2000
Heptacellular carcinoma cell lines	Actinomycin D		Yamanaka et al., 2000
	Doxorubicin		Yamanaka et al., 2000
	Camptothecin		Yamanaka et al., 2000
Glioma cell lines	Cisplatin	Induction of DR5 expression	Nagane et al., 2000
	Etoposide	Induction of DR5 expression	Nagane et al., 2000
Pancreatic cell lines	Actinomycin D		Matsuzaki et al., 2001
Multiple myeloma	Doxorubicin	Depolarization of mitochondrial membrane	Jazirehi et al., 2001
Osteogenic sarcoma cells	Doxorubicin		Evdokiou et al., 2002
	Cisplatin		Evdokiou et al., 2002
	Etoposide		Evdokiou et al., 2002
AIDS-Kaposi's sarcoma	Actinomycin D	Reduced Bcl-x_L expression	Mori et al., 1999
Melanoma cell lines	IFN-β	Depolarization of mitochondrial membrane	Chawla-Sarkar et al., 2002

[a]CD437, 6-[3-(1-adamantyl)-4-hydroxyphenyl]-2-naphthalene carboxylic acid.

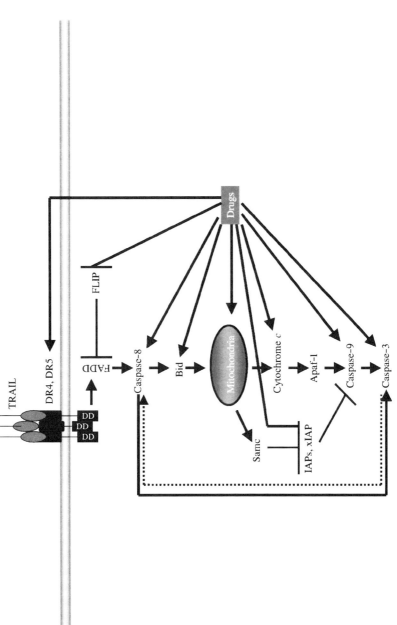

FIGURE 2. Enhancement of TRAIL-mediated apoptosis by chemotherapeutic drugs. Chemotherapeutic drugs directly up-regulate death domain-containing receptors DR4 and DR5, facilitate the activation of the mitochondrial pathway, or activate the caspase signaling cascade. Drugs also inhibit the expression of antiapoptotic molecules FLIP, IAPs, and xIAP.

CDDP induce the activation of caspase-8 and/or cleavage of Bid (Lacour et al., 2001; Voelkel-Johnson et al., 2002; Wu et al., 2002). DOX, Act D, sodium nitroprusside, and IFN-β cause the depolarization of mitochondria (Chawla-Sarkar et al., 2002; Jazirehi et al., 2001), and/or release of cytochrome c (Evdokiou et al., 2002; Lee et al., 2001). CDDP and 5-FU also increase proapoptotic Bax protein or p53 expression (Mizutani et al., 2001, 2002). Act D and sodium butyrate reduce antiapoptotic molecules Bcl-X_L (Hinz et al., 2000; Mori et al., 1999) and FLIP (Hernandez et al., 2001a,b). The augmentation of TRAIL-mediated apoptosis was also reported to be abrogated by overexpression of bcl-2 in human prostate cancer cell lines LNCaP and PC3 when used in combination with DOX, CDDP, or etoposide (Munshi et al., 2002).

We recently showed that subtoxic concentrations of DOX sensitized both established human renal cell carcinoma (RCC) cell lines and freshly patient-derived RCC cells to TRAIL-mediated apoptosis and resulted in a synergistic cytotoxic effect with TRAIL, but not in primarily isolated normal kidney cells (Wu et al., 2003). DOX significantly increased DR4 and DR5 expressions in RCC cells, but not in normal kidney cells. DOX also activated caspase cascade, including initiative caspases such as caspase-8 and -9 and effective caspases including caspase-6 and -3. The synergistic cytotoxicity of TRAIL and DOX was inhibited by DR4:Fc or DR5:Fc fusion proteins, which inhibit TRAIL-mediated apoptosis, and by general caspase inhibitor Z-VAD-FMK and specific caspase inhibitors for caspase-9, -8/6, and -3 (Wu et al., 2003). These results suggest that DOX sensitizes RCC cells to TRAIL-mediated apoptosis through the induction of DR4 and DR5 expression and activation of the caspase cascade.

VII. CONCLUSIONS AND PROSPECTS

Experimental evidence convincingly supports the concept that TRAIL shows no systemic cytotoxicity in clinical applications. Many tumor cells are only modestly sensitive to TRAIL-mediated apoptosis, but become highly sensitive after cotreatment with chemotherapeutic drugs. TRAIL, particularly in combination with chemotherapeutic agents, appears promising from a clinical perspective for use in cancer therapy. The mechanisms of the synergistic cytotoxicity of TRAIL and chemotherapeutic drugs is primarily through the induction of death domain–containing receptors DR4 and DR5, activation of mitochondria, or activating caspase cascades by chemotherapeutic drugs. The precise mechanism of action and the role of TRAIL in cancer treatment must, however, be further clarified in future clinical trials.

The possible side effects occurring by either TRAIL or cotreatment with chemotherapeutic agents may significantly limit the application of TRAIL in cancer therapy. Further study is thus needed to develop more sophisticated strategies that enable us to better exploit the therapeutic potential of death receptor–mediated apoptosis. A promising way in this respect is possibly the targeting-dependent selective action of death receptors at tumors. Indeed, it has been shown that killing activity of antihuman DR5 mAb against liver cancer cells, but not normal hepatocytes (Ichikawa et al., 2001), offers hope that rational engineering of recombinant molecules will provide new strategies for exploiting the TRAIL/TRAIL receptor pathway in cancer cells.

ACKNOWLEDGMENTS

Supported in part by Grant-in-Aid for Scientific Research (15591691) from the Ministry of Education, Science, Culture and Sports, Japan.

REFERENCES

Asakuma, J., Sumitomo, M., Asano, T., Asano, T., and Hayakawa, M. (2003). Selective Akt inactivation and tumor necrosis factor-related apoptosis-inducing ligand sensitization of renal cancer cells by low concentrations of paclitaxel. *Cancer Res.* **63,** 1365–1370.

Ashkenazi, A., and Dixit, V. M. (1998). Death receptors: Signaling and modulation. *Science* **281,** 1305–1308.

Ashkenazi, A., Pai, R. C., Fong, S., Leung, S., Lawrence, D. A., Marsters, S. A., Blackie, C., Chang, L., McMurtrey, A. E., Hebert, A., DeForge, L., Koumenis, I. L., Lewis, D., Harris, L., Bussiere, J., Koeppen, H., Shahrokh, Z., and Schwall, R. H. (1999a). Safety and antitumor activity of recombinant soluble Apo2 ligand. *J. Clin. Invest.* **104,** 155–162.

Ashkenazi, A., and Dixit, V. M. (1999b). Apoptosis control by death and decoy receptors. *Curr. Opin. Cell Biol.* **11,** 255–260.

Bodmer, J. L., Meier, P., Tschopp, J., and Schneider, P. (2000a). Cysteine 230 is essential for the structure and activity of the cytotoxic ligand TRAIL. *J. Biol. Chem.* **275,** 20632–20637.

Bodmer, J. L., Holler, N., Reynard, S., Vinciguerra, P., Schneider, P., Juo, P., Blenis, J., and Tschopp, J. (2000b). TRAIL receptor-2 signals apoptosis through FADD and caspase-8. *Nat. Cell Biol.* **2,** 241–243.

Bonavida, B., Ng, C. P., Jazirehi, A., Schiller, G., and Mizutani, Y. (1999). Selectivity of TRAIL-mediated apoptosis of cancer cells and synergy with drugs: The trail to non-toxic cancer therapeutics (review). *Int. J. Oncol.* **15,** 793–802.

Chai, J., Du, C., Wu, J. W., Kyin, S., Wang, X., and Shi, Y. (2000). Structural and biochemical basis of apoptotic activation by Smac/DIABLO. *Nature* **406,** 855–862.

Chaudhary, P. M., Eby, M., Jasmin, A., Bookwalter, A., Murray, J., and Hood, L. (1997). Death receptor 5, a new member of the TNFR family, and DR4 induce FADD-dependent apoptosis and activate the NF-kappaB pathway. *Immunity* **7,** 821–830.

Chawla-Sarkar, M., Leaman, D. W., Jacobs, B. S., and Borden, E. C. (2002). IFN-beta pretreatment sensitizes human melanoma cells to TRAIL/Apo2 ligand-induced apoptosis. *J. Immunol.* **169,** 847–855.

Cuello, M., Ettenberg, S. A., Clark, A. S., Keane, M. M., Posner, R. H., Nau, M. M., Dennis, P. A., and Lipkowitz, S. (2001a). Down-regulation of the erbB-2 receptor by trastuzumab (herceptin) enhances tumor necrosis factor-related apoptosis-inducing ligand-mediated apoptosis in breast and ovarian cancer cell lines that overexpress erbB-2. *Cancer Res.* **61**, 4892–4900.

Cuello, M., Ettenberg, S. A., Nau, M. M., and Lipkowitz, S. (2001b). Synergistic induction of apoptosis by the combination of trail and chemotherapy in chemoresistant ovarian cancer cells. *Gynecol. Oncol.* **81**, 380–390.

Degli-Esposti, M. A., Smolak, P. J., Walczak, H., Waugh, J., Huang, C. P., DuBose, R. F., Goodwin, R. G., and Smith, C. A. (1997). Cloning and characterization of TRAIL-R3, a novel member of the emerging TRAIL receptor family. *J. Exp. Med.* **186**, 1165–1170.

Dejosez, M., Ramp, U., Mahotka, C., Krieg, A., Walczak, H., Gabbert, H. E., and Gerharz, C. D. (2000). Sensitivity to TRAIL/APO-2L-mediated apoptosis in human renal cell carcinomas and its enhancement by topotecan. *Cell Death Differ.* **7**, 1127–1136.

Emery, J. G., McDonnell, P., Burke, M. B., Deen, K. C., Lyn, S., Silverman, C., Dul, E., Appelbaum, E. R., Eichman, C., DiPrinzio, R., Dodds, R. A., James, I. E., Rosenberg, M., Lee, J. C., and Young, P. R. (1998). Osteoprotegerin is a receptor for the cytotoxic ligand TRAIL. *J. Biol. Chem.* **273**, 14363–14367.

Engels, I. H., Stepczynska, A., Stroh, C., Lauber, K., Berg, C., Schwenzer, R., Wajant, H., Janicke, R. U., Porter, A. G., Belka, C., Gregor, M., Schulze-Osthoff, K., and Wesselborg, S. (2000). Caspase-8/FLICE functions as an executioner caspase in anticancer drug-induced apoptosis. *Oncogene* **19**, 4563–4573.

Evdokiou, A., Bouralexis, S., Atkins, G. J., Chai, F., Hay, S., Clayer, M., and Findlay, D. M. (2002). Chemotherapeutic agents sensitize osteogenic sarcoma cells, but not normal human bone cells, to Apo2L/TRAIL-induced apoptosis. *Int. J. Cancer* **99**, 491–504.

French, L. E., and Tschopp, J. (1999). The TRAIL to selective tumor death. *Nat. Med.* **5**, 146–147.

Fulda, S., Meyer, E., and Debatin, K. M. (2002). Inhibition of TRAIL-induced apoptosis by Bcl-2 overexpression. *Oncogene* **21**, 2283–2894.

Gibson, S. B., Oyer, R., Spalding, A. C., Anderson, S. M., and Johnson, G. L. (2000). Increased expression of death receptors 4 and 5 synergizes the apoptosis response to combined treatment with etoposide and TRAIL. *Mol. Cell. Biol.* **20**, 205–212.

Gliniak, B., and Le, T. (1999). Tumor necrosis factor-related apoptosis-inducing ligand's antitumor activity in vivo is enhanced by the chemotherapeutic agent CPT-11. *Cancer Res.* **59**, 6153–6158.

Griffith, T. S., Chin, W. A., Jackson, G. C., Lynch, D. H., and Kubin, M. Z. (1998). Intracellular regulation of TRAIL-induced apoptosis in human melanoma cells. *J. Immunol.* **161**, 2833–2840.

Grotzer, M. A., Eggert, A., Zuzak, T. J., Marwaha, S., Wiewrodt, B. R., Ikegaki, N., Brodeur, G. M., and Phillips, P. C. (2000). Resistance to TRAIL-induced apoptosis in primitive neuroectodermal brain tumor cells correlates with a loss of caspase-8 expression. *Oncogene* **19**, 4604–4610.

Havell, E. A., Fiers, W., and North, R. J. (1988). The antitumor function of tumor necrosis factor (TNF), I. Therapeutic action of TNF against an established murine sarcoma is indirect, immunologically dependent, and limited by severe toxicity. *J. Exp. Med.* **167**, 1067–1085.

Hengartner, M. O. (2000). The biochemistry of apoptosis. *Nature* **407**, 770–776.

Hernandez, A., Wang, Q. D., Schwartz, S. A., and Evers, B. M. (2001a). Sensitization of human colon cancer cells to TRAIL-mediated apoptosis. *J. Gastrointest. Surgery* **5**, 56–65.

Hernandez, A., Thomas, R., Smith, F., Sandberg, J., Kim, S., Chung, D. H., and Evers, B. M. (2001b). Butyrate sensitizes human colon cancer cells to TRAIL-mediated apoptosis. *Surgery* **130**, 265–272.

Hinz, S., Trauzold, A., Boenicke, L., Sandberg, C., Beckmann, S., Bayer, E., Walczak, H., Kalthoff, H., and Ungefroren, H. (2000). Bcl-XL protects pancreatic adenocarcinoma cells against CD95- and TRAIL-receptor-mediated apoptosis. *Oncogene* **19**, 5477–5486.

Hymowitz, S. G., O'Connell, M. P., Ultsch, M. H., Hurst, A., Totpal, K., Ashkenazi, A., de Vos, A. M., and Kelley, R. F. (2000). A unique zinc-binding site revealed by a high-resolution X-ray structure of homotrimeric Apo2L/TRAIL. *Biochemistry* **39,** 633–640.

Ichikawa, K., Liu, W., Zhao, L., Wang, Z., Liu, D., Ohtsuka, T., Zhang, H., Mountz, J. D., Koopman, W. J., Kimberly, R. P., and Zhou, T. (2001). Tumoricidal activity of a novel anti-human DR5 monoclonal antibody without hepatocyte cytotoxicity. *Nat. Med.* **7,** 954–960.

Irmler, M., Thome, M., Hahne, M., Schneider, P., Hofmann, K., Steiner, V., Bodmer, J. L., Schroter, M., Burns, K., Mattmann, C., Rimoldi, D., French, L. E., and Tschopp, J. (1997). Inhibition of death receptor signals by cellular FLIP. *Nature* **388,** 190–195.

Jazirehi, A. R., Ng, C. P., Gan, X. H., Schiller, G., and Bonavida, B. (2001). Adriamycin sensitizes the adriamycin-resistant 8226/Dox40 human multiple myeloma cells to Apo2L/ tumor necrosis factor-related apoptosis-inducing ligand-mediated (TRAIL) apoptosis. *Clin. Cancer Res.* **7,** 3874–3883.

Jo, M., Kim, T. H., Seol, D. W., Esplen, J. E., Dorko, K., Billiar, T. R., and Strom, S. C. (2000). Apoptosis induced in normal human hepatocytes by tumor necrosis factor-related apoptosis-inducing ligand. *Nat. Med.* **6,** 564–567.

Johnsen, A. C., Haux, J., Steinkjer, B., Nonstad, U., Egeberg, K., Sundan, A., Ashkenazi, A., and Espevik, T. (1999). Regulation of APO-2 ligand/trail expression in NK cells-involvement in NK cell-mediated cytotoxicity. *Cytokine* **11,** 664–672.

Kagawa, S., He, C., Gu, J., Koch, P., Rha, S. J., Roth, J. A., Curley, S. A., Stephens, L. C., and Fang, B. (2001). Antitumor activity and bystander effects of the tumor necrosis factor-related apoptosis-inducing ligand (TRAIL) gene. *Cancer Res.* **61,** 3330–3338.

Kayagaki, N., Yamaguchi, N., Nakayama, M., Eto, H., Okumura, K., and Yagita, H. (1999a). Type I interferons (IFNs) regulate tumor necrosis factor-related apoptosis-inducing ligand (TRAIL) expression on human T cells: A novel mechanism for the antitumor effects of type I IFNs. *J. Exp. Med.* **189,** 1451–1460.

Kayagaki, N., Yamaguchi, N., Nakayama, M., Kawasaki, A., Akiba, H., Okumura, K., and Yagita, H. (1999b). Involvement of TNF-related apoptosis-inducing ligand in human CD4+ T cell-mediated cytotoxicity. *J. Immunol.* **162,** 2639–2647.

Keane, M. M., Ettenberg, S. A., Nau, M. M., Russell, E. K., and Lipkowitz, S. (1999). Chemotherapy augments TRAIL-induced apoptosis in breast cell lines. *Cancer Res.* **59,** 734–741.

Kim, K., Fisher, M. J., Xu, S. Q., and el-Deiry, W. S. (2000). Molecular determinants of response to TRAIL in killing of normal and cancer cells. *Clin. Cancer Res.* **6,** 335–346.

Kischkel, F. C., Lawrence, D. A., Chuntharapai, A., Schow, P., Kim, K. J., and Ashkenazi, A. (2000). Apo2L/TRAIL-dependent recruitment of endogenous FADD and caspase-8 to death receptors 4 and 5. *Immunity* **12,** 611–620.

Kothny-Wilkes, G., Kulms, D., Poppelmann, B., Luger, T. A., Kubin, M., and Schwarz, T. (1998). Interleukin-1 protects transformed keratinocytes from tumor necrosis factor-related apoptosis-inducing ligand. *J. Biol. Chem.* **273,** 29247–29253.

Kuang, A. A., Diehl, G. E., Zhang, J., and Winoto, A. (2000). FADD is required for DR4- and DR5-mediated apoptosis: Lack of trail-induced apoptosis in FADD-deficient mouse embryonic fibroblasts. *J. Biol. Chem.* **275,** 25065–25068.

Lacour, S., Hammann, A., Wotawa, A., Corcos, L., Solary, E., and Dimanche-Boitrel, M. T. (2001). Anticancer agents sensitize tumor cells to tumor necrosis factor-related apoptosis-inducing ligand-mediated caspase-8 activation and apoptosis. *Cancer Res.* **61,** 1645–1651.

Lawrence, D., Shahrokh, Z., Marsters, S., Achilles, K., Shih, D., Mounho, B., Hillan, K., Totpal, K., DeForge, L., Schow, P., Hooley, J., Sherwood, S., Sherwood, S., Pai, R., Leung, S., Khan, L., Gliniak, B., Bussiere, J., Smith, C. A., Strom, S. S., Kelley, S., Fox, J. A., Thomas, D., and Ashkenazi, A. (2001). Differential hepatocyte toxicity of recombinant Apo2L/TRAIL versions. *Nat. Med.* **7,** 383–385.

Lee, Y. J., Lee, K. H., Kim, H. R., Jessup, J. M., Seol, D. W., Kim, T. H., Billiar, T. R., and Song, Y. K. (2001). Sodium nitroprusside enhances TRAIL-induced apoptosis via a mitochondria-dependent pathway in human colorectal carcinoma CX-1 cells. *Oncogene* **20,** 1476–1485.

Leverkus, M., Neumann, M., Mengling, T., Rauch, C. T., Brocker, E. B., Krammer, P. H., and Walczak, H. (2000). Regulation of tumor necrosis factor-related apoptosis-inducing ligand sensitivity in primary and transformed human keratinocytes. *Cancer Res.* **60,** 553–559.

Li, H., Zhu, H., Xu, C. J., and Yuan, J. (1998). Cleavage of BID by caspase 8 mediates the mitochondrial damage in the Fas pathway of apoptosis. *Cell* **94,** 491–501.

Luo, X., Budihardjo, I., Zou, H., Slaughter, C., and Wang, X. (1998). Bid, a Bcl2 interacting protein, mediates cytochrome c release from mitochondria in response to activation of cell surface death receptors. *Cell* **94,** 481–490.

MacFarlane, M., Ahmad, M., Srinivasula, S. M., Fernandes-Alnemri, T., Cohen, G. M., and Alnemri, E. S. (1997). Identification and molecular cloning of two novel receptors for the cytotoxic ligand TRAIL. *J. Biol. Chem.* **272,** 25417–25420.

Mariani, S. M., and Krammer, P. H. (1998a). Differential regulation of TRAIL and CD95 ligand in transformed cells of the T and B lymphocyte lineage. *Eur. J. Immunol.* **28,** 973–982.

Mariani, S. M., and Krammer, P. H. (1998b). Surface expression of TRAIL/Apo-2 ligand in activated mouse T and B cells. *Eur. J. Immunol.* **28,** 1492–1498.

Marsters, S. A., Sheridan, J. P., Pitti, R. M., Huang, A., Skubatch, M., Baldwin, D., Yuan, J., Gurney, A., Goddard, A. D., Godowski, P., and Ashkenazi, A. (1997). A novel receptor for Apo2L/TRAIL contains a truncated death domain. *Curr. Biol.* **7,** 1003–1006.

Martinez-Lorenzo, M. J., Alava, M. A., Gamen, S., Kim, K. J., Chuntharapai, A., Pineiro, A., Naval, J., and Anel, A. (1998). Involvement of APO2 ligand/TRAIL in activation-induced death of Jurkat and human peripheral blood T cells. *Eur. J. Immunol.* **28,** 2714–2725.

Matsuzaki, H., Schmied, B. M., Ulrich, A., Standop, J., Schneider, M. B., Batra, S. K., Picha, K. S., and Pour, P. M. (2001). Combination of tumor necrosis factor-related apoptosis-inducing ligand (TRAIL) and actinomycin D induces apoptosis even in TRAIL-resistant human pancreatic cancer cells. *Clin. Cancer Res.* **7,** 407–414.

Medema, J. P., Scaffidi, C., Kischkel, F. C., Shevchenko, A., Mann, M., Krammer, P. H., and Peter, M. E. (1997). FLICE is activated by association with the CD95 death-inducing signaling complex (DISC). *EMBO J.* **16,** 2794–2804.

Miyazaki, T., and Reed, J. C. (2001). A GTP-binding adapter protein couples TRAIL receptors to apoptosis-inducing proteins. *Nat. Immunol.* **2,** 493–500.

Mizutani, Y., Yoshida, O., Miki, T., and Bonavida, B. (1999). Synergistic cytotoxicity and apoptosis by Apo-2 ligand and adriamycin against bladder cancer cells. *Clin. Cancer Res.* **5,** 2605–2612.

Mizutani, Y., Bonavida, B., Koishihara, Y., Akamatsu, K., Ohsugi, Y., and Yoshida, O. (1995). Sensitization of human renal cell carcinoma cells to *cis*-diamminedichloroplatinum(II) by anti-interleukin 6 monoclonal antibody or anti-interleukin 6 receptor monoclonal antibody. *Cancer Res.* **55,** 590–596.

Mizutani, Y., Nakao, M., Ogawa, O., Yoshida, O., Bonavida, B., and Miki, T. (2001). Enhanced sensitivity of bladder cancer cells to tumor necrosis factor related apoptosis inducing ligand mediated apoptosis by cisplatin and carboplatin. *J. Urol.* **165,** 263–270.

Mizutani, Y., Nakanishi, H., Yoshida, O., Fukushima, M., Bonavida, B., and Miki, T. (2002). Potentiation of the sensitivity of renal cell carcinoma cells to TRAIL-mediated apoptosis by subtoxic concentrations of 5-fluorouracil. *Eur. J. Cancer* **38,** 167–176.

Mongkolsapaya, J., Cowper, A. E., Xu, X. N., Morris, G., McMichael, A. J., Bell, J. I., and Screaton, G. R. (1998). Lymphocyte inhibitor of TRAIL (TNF-related apoptosis-inducing ligand): A new receptor protecting lymphocytes from the death ligand TRAIL. *J. Immunol.* **160,** 3–6.

Monleon, I., Martinez-Lorenzo, M. J., Monteagudo, L., Lasierra, P., Taules, M., Iturralde, M., Pineiro, A., Larrad, L., Alava, M. A., Naval, J., and Anel, A. (2001). Differential secretion of Fas ligand- or APO2 ligand/TNF-related apoptosis-inducing ligand-carrying microvesicles during activation-induced death of human T cells. *J. Immunol.* **167,** 6736–6744.

Mori, S., Murakami-Mori, K., Nakamura, S., Ashkenazi, A., and Bonavida, B. (1999). Sensitization of AIDS-Kaposi's sarcoma cells to Apo-2 ligand-induced apoptosis by actinomycin D. *J. Immunol.* **162,** 5616–5623.

Munshi, A., Pappas, G., Honda, T., McDonnell, T. J., Younes, A., Li, Y., and Meyn, R. E. (2001). TRAIL (APO-2L) induces apoptosis in human prostate cancer cells that is inhibitable by Bcl-2. *Oncogene* **20,** 3757–3765.

Munshi, A., McDonnell, T. J., and Meyn, R. E. (2002). Chemotherapeutic agents enhance TRAIL-induced apoptosis in prostate cancer cells. *Cancer Chemother. Pharmacol.* **50,** 46–52.

Nagane, M., Pan, G., Weddle, J. J., Dixit, V. M., Cavenee, W. K., and Huang, H. J. (2000). Increased death receptor 5 expression by chemotherapeutic agents in human gliomas causes synergistic cytotoxicity with tumor necrosis factor-related apoptosis-inducing ligand in vitro and in vivo. *Cancer Res.* **60,** 847–853.

Nagane, M., Huang, H. J., and Cavenee, W. K. (2001). The potential of TRAIL for cancer chemotherapy. *Apoptosis* **6,** 191–197.

Nagata, S. (1997). Apoptosis by death factor. *Cell* **88,** 355–365.

Naka, T., Sugamura, K., Hylander, B. L., Widmer, M. B., Rustum, Y. M., and Repasky, E. A. (2002). Effects of tumor necrosis factor-related apoptosis-inducing ligand alone and in combination with chemotherapeutic agents on patients' colon tumors grown in SCID mice. *Cancer Res.* **62,** 5800–5806.

Ng, C. P., Zisman, A., and Bonavida, B. (2002). Synergy is achieved by complementation with Apo2L/TRAIL and actinomycin D in Apo2L/TRAIL-mediated apoptosis of prostate cancer cells: Role of XIAP in resistance. *Prostate* **53,** 286–299.

Nimmanapalli, R., Perkins, C. L., Orlando, M., O'Bryan, E., Nguyen, D., and Bhalla, K. N. (2001). Pretreatment with paclitaxel enhances apo-2 ligand/tumor necrosis factor-related apoptosis-inducing ligand-induced apoptosis of prostate cancer cells by inducing death receptors 4 and 5 protein levels. *Cancer Res.* **61,** 759–763.

Ogasawara, J., Watanabe-Fukunaga, R., Adachi, M., Matsuzawa, A., Kasugai, T., Kitamura, Y., Itoh, N., Suda, T., and Nagata, S. (1993). Lethal effect of the anti-Fas antibody in mice. *Nature* **364,** 806–809.

Orth, K., O'Rourke, K., Salvesen, G. S., and Dixit, V. M. (1996). Molecular ordering of apoptotic mammalian CED-3/ICE-like proteases. *J. Biol. Chem.* **271,** 20977–20980.

Pan, G., O'Rourke, K., Chinnaiyan, A. M., Gentz, R., Ebner, R., Ni, J., and Dixit, V. M. (1997a). The receptor for the cytotoxic ligand TRAIL. *Science* **276,** 111–113.

Pan, G., Ni, J., Wei, Y. F., Yu, G., Gentz, R., and Dixit, V. M. (1997b). An antagonist decoy receptor and a death domain-containing receptor for TRAIL. *Science* **277,** 815–818.

Petak, I., and Houghton, J. A. (2001). Shared pathways: Death receptors and cytotoxic drugs in cancer therapy. *Pathol. Oncol. Res.* **7,** 95–106.

Pitti, R. M., Marsters, S. A., Ruppert, S., Donahue, C. J., Moore, A., and Ashkenazi, A. (1996). Induction of apoptosis by Apo-2 ligand, a new member of the tumor necrosis factor cytokine family. *J. Biol. Chem.* **271,** 12687–12690.

Rieger, J., Naumann, U., Glaser, T., Ashkenazi, A., and Weller, M. (1998). APO2 ligand: A novel lethal weapon against malignant glioma? *FEBS Lett.* **427,** 124–128.

Schneider, P., and Tschopp, J. (2000). Apoptosis induced by death receptors. *Pharmacol. Acta Helv.* **74,** 281–286.

Screaton, G. R., Mongkolsapaya, J., Xu, X. N., Cowper, A. E., McMichael, A. J., and Bell, J. I. (1997). TRICK2, a new alternatively spliced receptor that transduces the cytotoxic signal from TRAIL. *Curr. Biol.* **7,** 693–696.

Seol, D. W., Li, J., Seol, M. H., Park, S. Y., Talanian, R. V., and Billiar, T. R. (2001). Signaling events triggered by tumor necrosis factor-related apoptosis-inducing ligand (TRAIL): Caspase-8 is required for TRAIL-induced apoptosis. *Cancer Res.* **61,** 1138–1143.

Sheridan, J. P., Marsters, S. A., Pitti, R. M., Gurney, A., Skubatch, M., Baldwin, D., Ramakrishnan, L., Gray, C. L., Baker, K., Wood, W. I., Goddard, A. D., Godowski, P., and Ashkenazi, A. (1997). Control of TRAIL-induced apoptosis by a family of signaling and decoy receptors. *Science* **277**, 818–821.

Sprick, M. R., Weigand, M. A., Rieser, E., Rauch, C. T., Juo, P., Blenis, J., Krammer, P. H., and Walczak, H. (2000). FADD/MORT1 and caspase-8 are recruited to TRAIL receptors 1 and 2 and are essential for apoptosis mediated by TRAIL receptor 2. *Immunity* **12**, 599–609.

Sun, S. Y., Yue, P., Hong, W. K., and Lotan, R. (2000). Augmentation of tumor necrosis factor-related apoptosis-inducing ligand (TRAIL)-induced apoptosis by the synthetic retinoid 6-[3-(1-adamantyl)-4-hydroxyphenyl]-2-naphthalene carboxylic acid (CD437) through up-regulation of TRAIL receptors in human lung cancer cells. *Cancer Res.* **60**, 7149–7155.

Sun, S. Y., Yue, P., Zhou, J. Y., Wang, Y., Choi Kim, H. R., Lotan, R., and Wu, G. S. (2001). Overexpression of BCL2 blocks TNF-related apoptosis-inducing ligand (TRAIL)-induced apoptosis in human lung cancer cells. *Biochem. Biophys. Res. Commun.* **280**, 788–797.

Takeda, K., Hayakawa, Y., Smyth, M. J., Kayagaki, N., Yamaguchi, N., Kakuta, S., Iwakura, Y., Yagita, H., and Okumura, K. (2001). Involvement of tumor necrosis factor-related apoptosis-inducing ligand in surveillance of tumor metastasis by liver natural killer cells. *Nat. Med.* **7**, 94–100.

Tang, D., Lahti, J. M., and Kidd, V. J. (2000). Caspase-8 activation and bid cleavage contribute to MCF7 cellular execution in a caspase-3-dependent manner during staurosporine-mediated apoptosis. *J. Biol. Chem.* **275**, 9303–9307.

Tartaglia, L. A., and Goeddel, D. V. (1992). Two TNF receptors. *Immunol. Today* **13**, 151–153.

Thomas, W. D., and Hersey, P. (1998). TNF-related apoptosis-inducing ligand (TRAIL) induces apoptosis in Fas ligand-resistant melanoma cells and mediates CD4 T cell killing of target cells. *J. Immunol.* **161**, 2195–2200.

Thomas, W. D., Zhang, X. D., Franco, A. V., Nguyen, T., and Hersey, P. (2000). TNF-related apoptosis-inducing ligand-induced apoptosis of melanoma is associated with changes in mitochondrial membrane potential and perinuclear clustering of mitochondria. *J. Immunol.* **165**, 5612–5620.

Verhagen, A. M., Ekert, P. G., Pakusch, M., Silke, J., Connolly, L. M., Reid, G. E., Moritz, R. L., Simpson, R. J., and Vaux, D. L. (2000). Identification of DIABLO, a mammalian protein that promotes apoptosis by binding to and antagonizing IAP proteins. *Cell* **102**, 43–53.

Voelkel-Johnson, C., King, D. L., and Norris, J. S. (2002). Resistance of prostate cancer cells to soluble TNF-related apoptosis-inducing ligand (TRAIL/Apo2L) can be overcome by doxorubicin or adenoviral delivery of full-length TRAIL. *Cancer Gene Ther.* **9**, 164–172.

Wajant, H., Pfizenmaier, K., and Scheurich, P. (2002). TNF-related apoptosis inducing ligand (TRAIL) and its receptors in tumor surveillance and cancer therapy. *Apoptosis* **7**, 449–459.

Walczak, H., Degli-Esposti, M. A., Johnson, R. S., Smolak, P. J., Waugh, J. Y., Boiani, N., Timour, M. S., Gerhart, M. J., Schooley, K. A., Smith, C. A., Goodwin, R. G., and Rauch, C. T. (1997). TRAIL-R2: A novel apoptosis-mediating receptor for TRAIL. *EMBO J.* **16**, 5386–5697.

Walczak, H., Miller, R. E., Ariail, K., Gliniak, B., Griffith, T. S., Kubin, M., Chin, W., Jones, J., Woodward, A., Le, T., Smith, C., Smolak, P., Goodwin, R. G., Rauch, C. T., Schuh, J. C., and Lynch, D. H. (1999). Tumoricidal activity of tumor necrosis factor-related apoptosis-inducing ligand in vivo. *Nat. Med.* **5**, 157–163.

Wang, X. (2001). The expanding role of mitochondria in apoptosis. *Genes Dev.* **15**, 2922–2933.

Wei, M. C., Zong, W. X., Cheng, E. H., Lindsten, T., Panoutsakopoulou, V., Ross, A. J., Roth, K. A., MacGregor, G. R., Thompson, C. B., and Korsmeyer, S. J. (2001). Proapoptotic BAX and BAK: A requisite gateway to mitochondrial dysfunction and death. *Science* **292**, 624–626.

Wen, J., Ramadevi, N., Nguyen, D., Perkins, C., Worthington, E., and Bhalla, K. (2000). Antileukemic drugs increase death receptor 5 levels and enhance Apo-2L-induced apoptosis of human acute leukemia cells. *Blood* **96,** 3900–3906.

Wiley, S. R., Schooley, K., Smolak, P. J., Din, W. S., Huang, C. P., Nicholl, J. K., Sutherland, G. R., Smith, T. D., Rauch, C., Smith, C. A., and Goodwin, R. G. (1995). Identification and characterization of a new member of the TNF family that induces apoptosis. *Immunity* **3,** 673–682.

Wu, X. X., Mizutani, Y., Kakehi, Y., Yoshida, O., and Ogawa, O. (2000). Enhancement of Fas-mediated apoptosis in renal cell carcinoma cells by adriamycin. *Cancer Res.* **60,** 2912–2918.

Wu, X. X., Kakehi, Y., Mizutani, Y., Kamoto, T., Kinoshita, H., Isogawa, Y., Terachi, T., and Ogawa, O. (2002). Doxorubicin enhances TRAIL-induced apoptosis in prostatic cancer. *Int. J. Oncol.* **20,** 949–954.

Wu, X. X., Kakehi, Y., Mizutani, Y., Nishiyama, H., Kamoto, T., Megumi, Y., Ito, N., and Ogawa, O. (2003). Enhancement of TRAIL/Apo2L-mediated apoptosis by adriamycin through inducing DR4 and DR5 in renal cell carcinoma cells. *Int. J. Cancer* **104,** 409–417.

Yamada, H., Tada-Oikawa, S., Uchida, A., and Kawanishi, S. (1999). TRAIL causes cleavage of bid by caspase-8 and loss of mitochondrial membrane potential resulting in apoptosis in BJAB cells. *Biochem. Biophys. Res. Commun.* **265,** 130–133.

Yamanaka, T., Shiraki, K., Sugimoto, K., Ito, T., Fujikawa, K., Ito, M., Takase, K., Moriyama, M., Nakano, T., and Suzuki, A. (2000). Chemotherapeutic agents augment TRAIL-induced apoptosis in human hepatocellular carcinoma cell lines. *Hepatology* **32,** 482–490.

Zamai, L., Ahmad, M., Bennett, I. M., Azzoni, L., Alnemri, E. S., and Perussia, B. (1998). Natural killer (NK) cell-mediated cytotoxicity: Differential use of TRAIL and Fas ligand by immature and mature primary human NK cells. *J. Exp. Med.* **188,** 2375–2380.

Zhang, X. D., Franco, A., Myers, K., Gray, C., Nguyen, T., and Hersey, P. (1999). Relation of TNF-related apoptosis-inducing ligand (TRAIL) receptor and FLICE-inhibitory protein expression to TRAIL-induced apoptosis of melanoma. *Cancer Res.* **59,** 2747–2753.

Zou, H., Li, Y., Liu, X., and Wang, X. (1999). An APAF-1.cytochrome c multimeric complex is a functional apoptosome that activates procaspase-9. *J. Biol. Chem.* **274,** 11549–11556.

20

Additive Effects of TRAIL and Paclitaxel on Cancer Cells: Implications for Advances in Cancer Therapy

Christine Odoux[*] and Andreas Albers[†]

[*]Division of Hematology/Oncology
[†]Department of Pathology, Otolaryngology, and Tumor Immunology
University of Pittsburgh Cancer Institute
Research Pavilion at The Hillman Cancer Center
University of Pittsburgh Medical School
Pittsburgh, Pennsylvania 15213

I. Introduction
II. Tumor Necrosis Factor–Related Apoptosis-Inducing Ligand
   A. TRAIL Molecule and its Biologic Effects
   B. Potential Clinical Value of TRAIL
III. Paclitaxel
   A. Paclitaxel Molecular Structure and Mechanisms of Action
   B. Paclitaxel Clinical and Biologic Characteristics
   C. Apoptotic Effects of Paclitaxel on Tumor Cells in Experimental Studies
IV. Biologic and Clinical Benefits of TRAIL and Paclitaxel Combination Therapy

A. Use of the Combination TRAIL and Paclitaxel in
    Experimental Studies
 B. In Vivo Relevance of Paclitaxel and TRAIL
    Combination Therapy
V. Conclusions
   References

In cancer therapy outgrowth of chemoresistant tumor cells is the most important factor that ultimately determines—apart from immediate adverse effects during treatment—the life span and prognosis of cancer patients. Despite many advances in cancer treatment and the integration of supportive medications, including new and better drugs for pain management, antiemesis, infection, and reconstitution of the hematopoietic system, both toxic effects and the development of resistance in response to the treatment remain a major problem. New treatment regimens have to be developed to target cancer more specifically using multiple cellular pathways. This will reduce toxic effects as well as the development of chemoresistance. Tumor necrosis factor–related apoptosis-inducing ligand (TRAIL) is the ligand for death receptors that belong to the TNF death receptor family. TRAIL triggers apoptosis *in vitro* in various cancer cell types. The antitumor drug, Paclitaxel (PA) was shown to increase the survival of patients with cancer. In *in vitro* experiments, PA also induces apoptosis in cancer cells. Together, PA and TRAIL lead to tumor regression in *in vivo* therapy and induce apoptosis through the interaction of TNF family death receptors, caspase activation, and/or cytochrome *c* release from mitochondria. PA and TRAIL complement each other using two distinct pathways that trigger apoptosis in addition to the anti-microtubule effect of PA. The combination of TRAIL and PA suppresses tumor growth that is otherwise resistant to treatment with either PA or TRAIL alone, by improving proapoptotic effects of the drugs. This observation support the use of the PA and TRAIL in future clinical trials. © 2004 Elsevier Inc.

I. INTRODUCTION

After its discovery, Paclitaxel (PA) raised interest in the scientific community due to its novel mechanism of cytotoxic action. PA alters microtubule assembly and activates proapoptotic signaling pathways. PA's broad spectrum of antitumor activity have earned it a unique place in experimental therapeutics, which was followed by a phase of extensive initial clinical trials in a variety of cancers. It now has established its place in many chemotherapeutic treatment regimens either as a single drug or in combination with other drugs. But still three and a half decades after its discovery, as many as 234 clinical trials are registered at the National Institutes of Health (NIH),[1] and of those, 158 are still actively recruiting patients, indicating that PA is still undergoing extensive clinical trials in a variety of cancers and other diseases, but also that there is still a great potential and need to improve current treatment schemas. Despite the potent efficacy of PA, anticancer treatment is frequently discontinued because of intolerable toxicity and/or the development of drug resistance. Thus, there is an urgent need for the development of new therapeutics and improved combinations with other drugs to reduce toxicity and to augment anticancer effects. A major mode of resistance to antitumor therapy is insensitivity to apoptosis induction. One possible new strategy in the cancer treatment is to combine chemotherapeutic agents with proapoptotic genes or molecules to enhance the apoptotic death of tumor cells.

Through many clinical trials, PA antitumor activity has been observed and improved by its combination with several known antitumor drugs, such as platinum, carboplatin, anthracyclins, doxorubicin, cisplatin, or docetaxel (Belani, 2000; Bunn, Jr. *et al.*, 2000; Nabholtz *et al.*, 2001; Novello *et al.*, 2001; Perez, 2001; Rosell *et al.*, 2001). New agents, such as antibodies, or modified genes are also presently investigated (Perez, 1999). In this perspective, as a proapoptotic molecule, we brought our interest on TRAIL, a TNF receptor member that is one of the key molecules inducing apoptotic cell death. TRAIL binds to its receptors (DR4 and DR5) and induces apoptosis through direct activation of the caspase cascade (Odoux *et al.*, 2002). The later pathway, so-called "death receptor pathway," is different from the mitochondrial pathway commonly used by conventional chemotherapeutic agents (Sprick *et al.*, 2000; Walczak *et al.*, 2000a). Thus, because of its different primary target, the caspases, when compared to the conventional drug primary target, the mitochondria and together with its ability to induce cell death exclusively in cancer cells, TRAIL was designated an attractive candidate for combination with conventional chemotherapeutic agents.

In this review, we summarize the biologic effects and clinical uses of PA and TRAIL. We also discuss the biologic effects induced by the combination of TRAIL and PA in *in vitro* experimentations and converse on the clinical benefit of a potential anticancer treatment using PA with TRAIL.

---

[1]Information on clinical trials for a wide range of diseases and conditions were obtained from the U.S. National Institutes of Health (NIH). Through its National Library of Medicine (NLM), NIH lists in a database studies that are conducted worldwide in about 90 countries. Access to this information is possible on internet at: www.ClinicalTrials.gov.

II. TUMOR NECROSIS FACTOR–RELATED APOPTOSIS-INDUCING LIGAND

A. TRAIL MOLECULE AND ITS BIOLOGIC EFFECTS

1. TRAIL Description

TRAIL, also called APO-2 ligand (Apo-2L), was identified in 1995 as apoptosis-inducing member of the TNF gene super family (Pitti *et al.*, 1996; Wiley *et al.*, 1995) on the basis of its sequence homology to the other members of the TNF family (Walczak *et al.*, 2000b).

TRAIL is a type II transmembrane protein consisting of 281 amino acids. TRAIL mRNA is expressed in a wide range of tissues including spleen, lung prostate ovary and bowel with little expression in testis, heart, skeletal muscle and pancreas. TRAIL protein is produced constitutively in many tissues.

2. TRAIL Receptors

TRAIL ligand has been intensively studied mainly for its ability to transduce an apoptotic death signal upon its binding to its death receptors in a wide variety of susceptible lymphoid and non-lymphoid tumor cell lines. Currently five TRAIL receptors are identified, including TRAIL-R1/DR4, TRAIL-R2/DR5/Killer, TRAIL-R3/DcR1/LIT, TRAIL-R4/DcR2/TRUNDD, and TRAIL-R5/osteoprotegerin/DcR3, a soluble receptor that binds to the osteoclast differentiation factor (Fig. 1).

TRAIL-R1 (DR4) and TRAIL-R2 (DR5) are type I transmembrane proteins that contain extracellular cysteine-rich domains and intracellular cytoplasmic domains. The intracellular cytoplasmic domains contain a conserved 80 amino acid sequence dubbed the death domain (DD) similar to that of the TNF family receptors. TRAIL receptors R1 and R2 have in fact 58% identical structure with the TNF family receptors. These two receptors, TRAIL-R1 and -R2, have been identified to mediate apoptosis (French *et al.*, 1999).

The other three receptors serve as "decoy" receptors (French *et al.*, 1999) and do not lead to apoptosis. It was observed that a cell expressing more decoy receptors than DR4 or DR5 is more likely to survive upon binding of the death ligand TRAIL (Griffith *et al.*, 1998b; Sheridan *et al.*, 1997).

TRAIL-R3/DcR1 and TRAIL-R4/DcR2 have a different structure than that seen in TRAIL-R1 or TRAIL-R2. TRAIL-R3/DcR1 contains only a partial nonfunctional cytoplasmic DD, while TRAIL-R4/DcR2 lacks any transmembrane component, and is instead attached to the cell surface by a glycosylphosphatidylinositol linker.

3. Cytotoxic and Apoptotic Effects of TRAIL

TRAIL is a cytokine that exerts cytotoxic and apoptotic effects on a wide variety of cancer cells derived from colon, ovary, prostate gland, lung, breast, kidney, brain, and skin cancer (Ashkenazi *et al.*, 1999; Lin *et al.*,

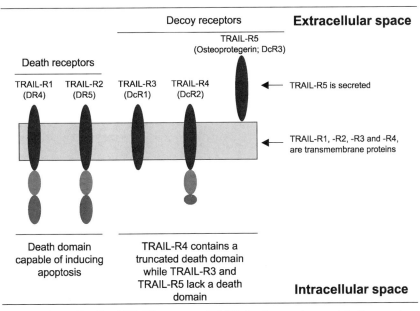

FIGURE 1. Family of TRAIL receptors. TRAIL family receptors contain five members. TRAIL-R1 and -R2 members are actively involved in TRAIL-induced apoptosis. The other three receptors, TRAIL-R3, -R4, and -R5, serve as "decoy" receptors and do not lead to apoptosis.

2003; Tomek et al., 2003; Walczak et al., 1999) and thus, has a potent anticancer activity (Ashkenazi et al., 1999). This cytokine in contrast to other members of the TNF family, seems to induce apoptosis only in cancer cells and not in normal tissue (Ashkenazi et al., 1999; Walczak et al., 1999). For instance, a high proportion of melanoma cell lines that were resistant to the other TNF family cytokine tested (TNF-alpha, FasL, or CD40L) could be killed by TRAIL (Griffith et al., 1998a; Thomas et al., 1998). TRAIL, and not FasL, was also found to mediate CD4 T cell killing of melanoma cells (Thomas et al., 1998). Thus, TRAIL may be a promising approach for selectively inducing apoptosis in tumor cells.

It has been hypothesized that physiologic mechanisms can protect many normal cell types from TRAIL-induced apoptosis. One such mechanism may involve the interaction of TRAIL with its three antagonistic decoy receptors. The selectivity occurs because TRAIL's death receptors (TRAIL-R1 and TRAIL-R2) are mainly expressed in transformed cells. Normal cells are resistant to TRAIL as they also express high levels of its decoy receptors (TRAIL-R3, TRAIL-R4, and TRAIL-R5) that antagonize TRAIL-induced apoptosis (Griffith et al., 1998b; Sheridan et al., 1997). More recently, it was shown that resistance to TRAIL on normal cells is controlled intracellularly rather than through the levels of the decoy receptors bound on the cellular surface (Leverkus et al., 2000). One important determinant of TRAIL activity is the level of the FLICE inhibitory protein, FLIP (Irmler et al., 1997), which interacts with and inhibits TRAIL signaling (Kim et al., 2000). For instance, in primary culture of normal keratinocytes versus transformed keratinocytes the levels of the cellular FLICE inhibitory protein, cFLIP inversely correlated with sensitivity to TRAIL and TRAIL-R1 and -R2–specific antibodies whereas TRAIL-R3 and -R4 were not involved (Leverkus et al., 2000). In cancer cells, apoptosis induced by TRAIL is sometimes inefficient also due to the development of cellular resistance to TRAIL. Despite the ubiquitous expression of TRAIL receptor, various cell lines originating from various cancer types demonstrate either partial or complete resistance to the proapoptotic effect of TRAIL. The level of intracellular FLIP is playing here again an important role. Cells expressing the long c-FLIP(L) and short c-FLIP(S) splice variants of the FLICE-like inhibitory protein (FLIP) were resistant to TRAIL despite the presence of DR4 (Tomek et al., 2003).

One possible way to decrease cancer cell resistance against TRAIL recombinant protein was suggested recently. Cancer cells that are resistant to the recombinant TRAIL protein in *in vitro* cultures are in fact more susceptible to TRAIL gene therapy. The transfection of an adenovector expressing the TRAIL or a green fluorescent protein, GFP-TRAIL-fusion gene driven by a human reverse transcriptase promoter (Ad/gTRAIL) (Lin et al., 2003) can induce apoptosis and apoptotic bystander effects in breast, lung, colon, ovary and prostate gland cancer cells (Griffith et al., 2002; Huang et al., 2002; Lin et al., 2002). Furthermore, apoptosis induction in Ad/gTRAIL transfected cells is very low in noncancerous cells, such as normal human fibroblasts, normal human hepatocytes, and normal human ovary and mammary epithelial cells (Lin et al., 2002). *In vivo*, the intratumoral injection of Ad/gTRAIL results in complete tumor regression in approximately 50% of mice bearing breast cancer xenografts (Lin et al., 2002). As well, Ad/gTRAIL inoculated intraperitoneally in nude mice with established abdominally spread tumors derived from ovarian cancer cell lines, significantly suppresses tumor growth, ascites formation, and prolongs animal survival (Huang et al., 2002).

Recent studies show that resistance of cancer cells to TRAIL could also be reversed by various combinations of TRAIL with chemotherapeutic agents, such as PA and cisplatin. As we will discuss further in this review, addition of chemotherapeutic drugs, such as PA, to TRAIL treatment produce in some cancer cells an increase in the expression of TRAIL receptors; DR4 and DR5, which consequently increases TRAIL-induced apoptosis (Griffith et al., 1998b; Nagane et al., 2000; Nimmanapalli et al., 2001).

4. Signaling Pathways Involved in TRAIL-Induced Apoptosis

TRAIL triggers apoptosis through the interaction of the two main death receptors DR4 (TRAIL-R1) and DR5 (TRAIL-R2) (Nimmanapalli et al., 2001). The DD is then responsible for the transduction of a death signal inside the cell by recruiting DD-containing adapter proteins (e.g., FADD, TRADD) to the receptor. The death effector domain (DED) containing the protein FLICE/caspase-8 joins this complex. FLICE, a member of the caspase family of cysteine proteases, is then activated by limited autoproteolysis. DR4 reportedly uses a FADD-independent pathway to induce apoptosis without activating the nuclear factor, NFκB pathway (MacFarlane et al., 1997).

Conversely, both DR4 and DR5 have been described to interact with FADD, caspase-8, TRADD, and RIP and to activate NFκB using a TRADD-dependent pathway (Schneider et al., 1997). The activated FLICE is responsible for the activation of downstream caspases (e.g. caspases-3, -6, -7, -10) that will execute the death sentence through the cleavage of cellular proteins (e.g. nuclear lamins, endonucleases, fodrin, poly [ADP-ribose] polymerase), resulting in the collapse of the cellular machinery and cell death. In the sequence of biochemical events that follow TRAIL stimulation, TRAIL-induced apoptosis involves late dissipation of the mitochondrial membrane potential and cytochrome c release that follows activation of caspase-8, caspase-3, and DNA fragmentation (Walczak et al., 2000b). The proapoptotic molecule Bid is also activated and contributes to the release of cytochrome c from mitochondria into the cytosol as well as the processing of procaspase-9 and -3 (Nimmanapalli et al., 2001).

In fact, two pathways for TRAIL apoptosis induction seems to exist, one that works via direct caspase activation operating independently from mitochondria, and another one that is dependent on mitochondrial apoptotic events and is, therefore inhibitable by antiapoptotic molecules, such as Bclx$_L$ or Bcl-2 (Walczak et al., 2000a).

B. POTENTIAL CLINICAL VALUE OF TRAIL

While the systemic effect of the two other members of the TNF family, CD95 ligand and TNF, have been proven to be disastrous (Ashkenazi et al., 1999; Mueller, 1998; Williamson et al., 1983), the ability of TRAIL to kill tumor cells more efficiently than normal cells has prompted great interest.

TRAIL has been demonstrated to kill a wide variety of tumor cells with minimal effects on normal cells. This selectivity makes TRAIL an attractive candidate for the development of novel cancer therapeutics. So far, the major concern about using TRAIL as a therapeutic is its toxicity to human hepatocytes. Although TRAIL has shown no effect on the normal murine hepatocytes, TRAIL can cause significant death of normal human hepatocytes. It is therefore important to investigate in greater detail the differences of receptor expression patterns and signaling pathways in normal cells derived from different tissues. This will lead to a better understanding of the mechanisms that support resistance and susceptibility in normal cells. This knowledge will prove to be invaluable to further increase selectivity of drug targeting and to reduce toxic effects for potential clinical applications. In the future, as we gain a better understanding on the regulation of receptor expression and signaling cascades, it may be possible to save entirely normal cells from the toxic effects of TRAIL. This could be either achieved by using more specific receptor agonists targeting tumor cells even more specifically. Another way could be to desensitize normal tissues, especially hepatocytes further by manipulating receptor expression or by inducing the expression of protective factors.

When TRAIL binds to death receptors, rapid cell death follows in various human cancer cell lines. TRAIL suppresses the growth of TRAIL-sensitive human cancers (including breast, colon, prostate gland, pancreas, glioblastoma, ovarian, and cervical cancers) implanted in mice (Huang *et al.*, 2002; Lin *et al.*, 2002, 2003). Together with a significant reduction of tumor growth, and improved survival, TRAIL does not induce cytotoxicity in tissues or organs of nonhuman primates (Ashkenazi *et al.*, 1999). Data suggest that TRAIL, either as a single agent or in combination with cytotoxic agents, might represent a new treatment option for advanced human soft tissue sarcoma (STS), which constitutes a largely chemotherapy-resistant disease (Tomek *et al.*, 2003).

Furthermore, in combination with some conventional chemotherapies, TRAIL suppresses tumor growth that otherwise resists treatment with either TRAIL or chemotherapy alone (Ashkenazi *et al.*, 1999; Lin *et al.*, 2003; Tomek *et al.*, 2003; Walczak *et al.*, 1999). This observation underscores the potential synergistic effects that derive from combining TRAIL with other chemotherapeutic agents. These will be discussed in more detail emphasizing a combination with PA in section IV of this review.

III. PACLITAXEL

A. PACLITAXEL MOLECULAR STRUCTURE AND MECHANISMS OF ACTION

1. Structure

PA, also known as taxol, belongs to the family of taxanes, one of the most important new classes of anticancer agents that emerged in the past decades. PA was isolated in 1971 from the bark of the Western Yew, *Taxus brevifolia* nut (Fuchs *et al.*, 1978). After the initial obstacle of a limited natural supply had been overcome through the development of novel synthetic methods and the identification of new sources for taxanes, a score of basic and clinical

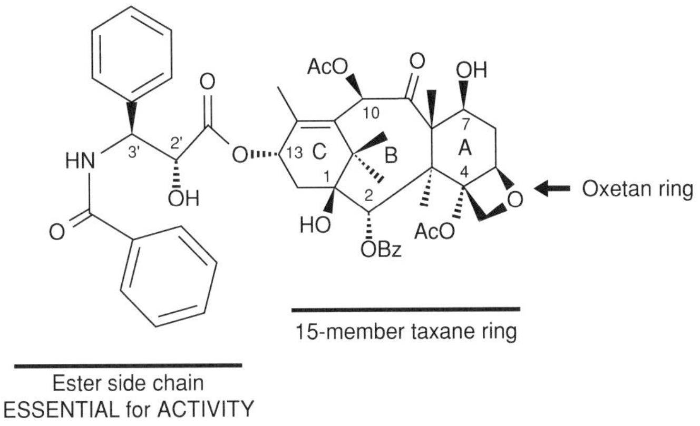

FIGURE 2. Paclitaxel molecular structure. Paclitaxel or Taxol ($C_{47}H_{51}NO_{14}$) (benzene-propanoic acid, beta-(benzoylamino))-alpha-hydroxy-6,12b-bis(acetyloxy)-12-(benzoloxy)-a,3,4,4a,5,6,9,10,12,12a-12b-dodecahydro-4,11-didihyroxy-4a,8,13,13-tetramethyl-5-oxo-7,11-ethano-1H-cyclodeca(3,4) benz (1,2-b)oxet-9-yl ester,(2aR-(2a-alpha, 4beta, 4a-beta, 6beta, 9alpha (alphaR*,betaS*),11alpha,1-2alpha,12a alpha,12b-alpha)-(9CI)). PA molecular structure includes a 15-member taxane ring linked to a four-member oxetan ring. The taxane ring of PA is linked to an ester side chain attached at the C-13 position of the ring, which is primordial for PA anti-tumor activity. AcO, Acetyl oxide function; Bz, benzene function.

studies have begun. Because of its unusual mechanism, PA has attracted much interest from basic scientists and clinicians. It has become a valuable tool for investigating microtubule function and its antitumor effects have earned it an important role in experimental and clinical use.

PA structure (Fig. 2) includes a 15-member taxane ring linked to a 4-member oxetan ring. The taxane ring of PA is linked to an ester side chain attached at the C-13 position of the ring, which is primordial for PA antitumor activity.

2. Mechanisms of Action

Among other antineoplastic drugs that interfere with microtubules, PA was found to be the most interesting cytotoxic agent due to its unique mechanism of action. PA binds with high affinity to the N-terminal 31 amino acids of the beta-tubulin subunit and induces a shift of the dynamic equilibrium between tubulin dimers and microtubules toward polymerization. The resulting effect is a stabilization of the microtubules followed by a perturbation of microtubule-dependent cytoplasmic structures that are required for cellular function, such as mitosis, maintenance of cellular morphology, shape change, locomotion, and secretion (Thuret-Carnahan et al., 1985). Cells treated with pharmacologic levels of PA (0.1–10 μM) are arrested in the G^2 and M phases of the cell cycle. Studies of the effects of low concentrations (0.25 μM) of PA on cultured cells, under conditions of minimal inhibition of DNA, RNA, and protein synthesis, demonstrated that PA specifically blocks progression of cells through the cell cycle in M phase, indicating that mitosis exhibits high sensitivity to PA (Mole-Bajer *et al.*, 1983). Higher concentrations of PA result in damage to interphase cells, with formation of microtubule bundles and loss of a variety of cell functions dependent on microtubules.

More recent studies suggest that PA may cause cytotoxicity through non-tubulin mechanisms. For example, PA was shown to activate a signaling pathway in a murine macrophage-like cell line by inducing phosphorylation of Shc, an adaptor protein that interacts with the intermediate signaling molecule Grb2 that can initiate activation of the Raf/MAPK cascade (Wolfson *et al.*, 1997). PA has also the ability to activate several intracellular caspases, allowing PA to kill cells through apoptosis (Alnemri *et al.*, 1996; Weigel *et al.*, 2000). Induction of apoptosis by PA, which is presented subsequently, was first described in 1993 in a human leukemia cell line, HL-60 cells (Bhalla *et al.*, 1993).

B. PACLITAXEL CLINICAL AND BIOLOGIC CHARACTERISTICS

1. Paclitaxel is a Chemotherapeutic Antitumor Drug

PA entered clinical phase I trials in 1983. In many trials, PA, as a single agent, induced acute hypersensitivity reactions that necessitated discontinuation of several clinical trials. However, concomitant administration of steroids, histamine H^1 and H^2 receptor antagonists, and prolonged infusion (6–24 h) have been used successfully. In early phase I trials, significant activity was demonstrated in refractory ovarian carcinoma, breast carcinoma, melanoma, non–small-cell lung carcinoma, and adenocarcinoma of unknown origin. Minor responses were observed in gastric, colon, and head and neck carcinomas as well as in lymphoblastic and myeloblastic leukemias (Rowinsky *et al.*, 1990). Phase II trials in refractory ovarian carcinoma have reported an objective response rate of 30% (Perez, 1999). PA has limited value in renal cell carcinoma, where no responses were observed in 18 patients receiving high doses of this drug (Einzig *et al.*, 1991).

2. Resistance to Paclitaxel

Resistance to PA has been studied on cells in *in vitro* cultures, and it appears in several forms. PA resistance is associated with altered expression (Haber *et al.*, 1995; Kavallaris *et al.*, 1997) or mutation (Giannakakou *et al.*, 1997) of tubulin isoforms; α- and β-tubulins. PA resistance was also shown to be associated with overexpression of P-glycoprotein (Pgp) and the multi-drug resistance (MDR) phenotype (Roy *et al.*, 1985). More recent studies revealed that selection for PA resistance led to expression of MDR1 and Pgp as well as decreased β-tubulin. Decreased β-tubulin expression is consistent with the finding that this subunit is a target for PA (Rao *et al.*, 1995). Finally, a recent study showed that caveolin-1, a major component of caveolae in membranes that may be involved in signal transduction, is upregulated in PA-resistant cells (Yang *et al.*, 1998), suggesting that factors other than tubulin may play a role in PA action and resistance.

3. Pharmacologic Parameters of Paclitaxel

Pharmacokinetic parameters for PA have been obtained in multiple clinical trials. PA is highly bound to plasma albumin, ranging from 88% to 98%. Renal elimination of PA is low, with only 2% to 10% recovered unchanged in urine. The elimination of PA in humans is saturable when administered as short infusions (<6 h) or when high dosage levels (≥ 300 mg/m^2) are administered over 24 hours. PA clearance has been shown to be variable (fivefold to sevenfold) between patients given the same dose. The primary route of elimination in both rodents and humans appears to involve hepatic metabolism and subsequent biliary elimination. When administered at very high dose levels (≥ 600 mg/m^2), PA is reported to cause acute encephalopathy, coma, and death.

C. APOPTOTIC EFFECTS OF PACLITAXEL ON TUMOR CELLS IN EXPERIMENTAL STUDIES

1. Paclitaxel Induces Tumor Cell Growth Arrest Through Apoptosis

PA is used in many *in vitro* experimental studies to control cell proliferation or to induce cell death of various types of cancer cells (Andre *et al.*, 2002; Guo *et al.*, 2002; Wang *et al.*, 1999; Zhou *et al.*, 2003). Apoptosis is a distinct form of programmed cell death and normally occurs in part to annihilate overgrowth of extra-numerous cells. Apoptosis plays a major role in chemotherapy-induced tumor cell killing. PA-induced apoptosis is associated with chromatin condensation, margination of the nuclear membrane, cytoskeletal alterations, membrane blebbing, and formation of apoptotic bodies (Majno *et al.*, 1995; Weigel *et al.*, 2000).

2. Mechanisms of Action Involved in Paclitaxel-Induced Apoptosis

PA-induced apoptosis may be triggered by two major intracellular signaling cascades, the mitochondrial pathway and the death receptor pathway, both leading to caspase activation and cleavage of specific cellular substrates. At least 14 separate caspases have been identified (Alnemri *et al.*, 1996). Amongst them, caspase-3 is one key member and is the last caspase to be activated in the caspase cascade before ensuing to the DNA cleavage. Caspase-3 was particularly shown by us to be activated in PA-induced apoptosis in human non–small-cell lung cancer cell lines (Weigel *et al.*, 2000), but also in several other cancer cell types, such as human breast cancer cell lines (von Haefen *et al.*, 2003), human erythroleukemia cells

(Brisdelli et al., 2003), thyroid cancer cells (Pan et al., 2001), acute myeloid leukemia cells (Ibrado et al., 1998), nasopharyngeal carcinoma cells (Tan et al., 2002), and lung adenocarcinoma cells (Oyaizu et al., 1999). PA also induces the activation of other caspase members; caspase-7, -8, and -9 in several cancer types (Goncalves et al., 2000; Pan et al., 2001; von Haefen et al., 2003). Depending on the cell line treated with PA, caspase-3 and -9 may not be activated (Ofir et al., 2002).

In other experimental and clinical studies, other mechanisms involved in PA-induced apoptosis that are not dependent on TNF death receptor family have been reported (Blagosklonny et al., 2002; Goncalves et al., 2000). Alternative pathways include mitochondrial permeability, release of cytochrome *c* from mitochondria and activation of proapoptotic molecules, such as Bad (Brisdelli et al., 2003; Tudor et al., 2000; von Haefen et al., 2003). PA enhances apoptosis through induction of the tumor suppressor gene *p53* (Ganansia-Leymarie et al., 2003; Tan et al., 2002) or through a p53-independent pathway that probably involves the proapoptotic gene, *Bax* (Gadducci et al., 2002).

3. Effects of Paclitaxel Combined with Other Antitumor Drugs

A major impediment to PA's cytotoxic effect is the establishment of multidrug resistance. Clinical studies have shown that regimens that use PA in combination with other antitumor drugs could improve the prognosis in terms of overall survival by 74% in patients with breast cancer (Moliterni et al., 1997). In the treatment of ovarian cancer, PA is being administered with cisplatin, carboplatin, or cyclophosphamide, as two-drug combinations (Harper, 2002). Since 1996, the combination of cisplatin and PA has been proven to prolong survival in comparison with older regimens containing both cisplatin and cyclophosphamide.

Significant antitumor activity with PA associated with cisplatin has been observed in a diverse range of tumor types that are generally refractory to conventional therapies, including non–small-cell and small-cell lung cancer, head and neck cancer, esophageal cancer, bladder cancer, germ cell malignancies, lymphoma and Kaposi's sarcoma (Dalpiaz et al., 2003; De Giorgi et al., 2003; Vansteenkiste et al., 2003). In addition, the introduction of carboplatin in combination with PA showed similar efficacy but preferable toxicity profiles when compared with cisplatin in combination with PA (Harper, 2002). Other studies report that combination of PA with doxorubicin increase regression rates in breast cancer compared to either drug used alone (Amadori et al., 1997; Moliterni et al., 1997). Combined therapy of PA with BR96-doxorubicin, an anticarcinoma immunoconjugate, improves efficacy relative to either agent alone at doses of conjugate in the range expected to be tolerated clinically (Moliterni et al., 1997). Lately, combination of PA with other drugs, such as the anthrapyrazole losoxantrone, significantly increased the response rate over PA alone (54% vs 15%).

The use of humanized anti-*HER-2* monoclonal antibody for HER-2 overexpressing breast carcinoma has also demonstrated some promise (Perez, 1999).

4. TRAIL Receptor Expression During Paclitaxel-Induced Apoptosis

One of the best characterized pathways for initiation of apoptosis involves the binding of extracellular death signal proteins (TNFα, Fas ligand [CD95L], TRAIL) to their cognate cell surface receptors; TNF receptors, Fas (CD95), DR4, and DR5 (Majno *et al.*, 1995). PA-induced apoptosis was particularly evocated through the activation of the Fas receptor (Pucci *et al.*, 1999). However, the involvement of TRAIL receptors in PA-induced apoptosis started to be described more recently in the literature. PA-induced apoptosis was shown to correlate with an upregulation of DR4 and DR5 proteins levels in prostate gland, renal cancer cells, and gliomas (Griffith *et al.*, 1998b; Nagane *et al.*, 2000; Nimmanapalli *et al.*, 2001). Conversely, PA-induced apoptosis did not correlate with an increase in DR4 or DR5 expression levels for breast cancer cells and non–small-cell lung cancer cells (Keane *et al.*, 1999; Odoux *et al.*, 2002).

These data fostered increasing interest in the research community to explore the effect of TRAIL in combination with PA both in experimental studies as well as for future clinical trials.

IV. BIOLOGIC AND CLINICAL BENEFITS OF TRAIL AND PACLITAXEL COMBINATION THERAPY

The efficacy of chemotherapy has been improved by regimens that combine several cytotoxic drugs with different mechanisms of action and/or different dose-limiting toxicities. Importantly, combination regimens with PA have shown an antitumor activity in both PA sensitive and insensitive tumors (Trail *et al.*, 1999). TRAIL increases cellular responses to cytotoxic drugs in different cancer types (Vignati *et al.*, 2002). In addition, TRAIL receptors are present in relative abundance on cancer cells, which is an important factor in terms of cellular response (Vignati *et al.*, 2002). Furthermore, recent *in vitro* studies showed that various chemotherapeutic agents could reverse resistance of cancer cells to TRAIL.

A. USE OF THE COMBINATION TRAIL AND PACLITAXEL IN EXPERIMENTAL STUDIES

1. TRAIL and Paclitaxel Combination Increases Apoptosis in Cancer Cells

In *in vitro* studies, triggering apoptosis in cancer cells at different levels and understanding the cellular mechanisms regulating apoptosis may offer a strong rationale for the combination of PA chemotherapy with the apoptotic death receptor ligand, TRAIL.

TRAIL is an attractive candidate to be used in combination with anticancer agents. Indeed, in addition to its potent induction of apoptosis in various cancer lines (Ashkenazi et al., 1999; Lin et al., 2003; Tomek et al., 2003; Walczak et al., 1999), TRAIL increases the cellular response to cytotoxic drugs in different cancer types. In in vitro cellular cultures, TRAIL and PA combination treatment remarkably increases the degree of apoptosis of either PA- or TRAIL-induced apoptosis in non–small-cell lung cancer cells (Frese et al., 2002; Odoux et al., 2002), in prostate gland cancer cells (Nimmanapalli et al., 2001), and in ovarian cancer cells (Vignati et al., 2002). For some chemoresistant cancer cells, TRAIL alone does not have any effect, but its combination with PA results in a significant growth inhibition (Cuello et al., 2001).

Because it is possible to reduce tumor growth using antibody treatment, we were also able to induce a significant rate of apoptosis in non–small-cell lung cancer cell lines in in vitro culture incubating cells with an anti-DR5 receptor antibody in combination with PA (Odoux et al., 2002). Possible mechanisms for this observation include a direct agonistic effect of the antibody on DR5, which can be potentiated by PA treatment or an increased availability of TRAIL by the antibody sequestration of soluble DR5 in the extracellular compartment. Strategies using antibodies directed to TRAIL receptors remain to be further investigated as they show promising results.

2. Proposed Mechanisms of Action

Only recent studies have described the details of the effect of TRAIL and PA combination therapy on cancer cells in culture. These mechanisms were particularly explored for prostate gland, renal, and breast cancer cells (Asakuma et al., 2003; Lin et al., 2003; Nimmanapalli et al., 2001). Enhancement of TRAIL's effect on the degree of PA-induced apoptosis is shown to be mediated through the increased expression of the TRAIL receptors DR4 and DR5 on prostate gland cancer cells (Nimmanapalli et al., 2001). It was associated with a greater processing of pro-caspase-8 and Bid, as well as caspase-3 and a greater cytosolic accumulation of cytochrome c (Nimmanapalli et al., 2001). However, it was not affecting the protein levels of decoy receptors, DcR1 and DcR2, or the TRAIL molecule itself or other TNF death receptor family members, such as Fas (Nimmanapalli et al., 2001).

Protein expression level of TRAIL receptors, DR4, and DR5 is playing an important role in TRAIL-induced apoptosis, and apoptosis resistance to TRAIL may be due to a lack of their expression. However, recent reports in which chemotherapeutic agents failed to show an increase in the expression of TRAIL receptors and thus the following increase in TRAIL sensitivity. These later results suggest that alternative mechanisms may be involved in PA-enhanced TRAIL cytotoxicity. For instance, apoptosis resistance to TRAIL correlates well with the level of phosphorylation of the well-known serine threonine kinase Akt/protein kinase B at serine 473 (Asakuma et al.,

2003). Recent studies on the apoptotic effect of TRAIL in combination with PA demonstrated that in fact, low concentrations of PA induce Akt inactivation through ceramide, a sphingolipid derived from sphingomyelin hydrolysis (Asakuma *et al.*, 2003).

Studies report that the combination of TRAIL with other chemotherapeutic drugs induces apoptosis through an increase in caspase activation, as well as an increase in JNK/p38 MAP kinase activation and in the release of cytochrome *c* from mitochondria (Ohtsuka *et al.*, 2003). This pathway could be further investigated in the context of TRAIL's combination with PA.

In the course of treatment, human malignancies often acquire resistance to chemotherapeutic agents or radiation resistance. Frequently overexpression of the antiapoptotic molecules Bcl-2 or Bcl-x_L in tumors is the leading reason for resistance (Hinz *et al.*, 2000). It is suggested that TRAIL can bypass the antiapoptotic effect of Bcl-2 or Bcl-x_L using the death receptor pathway versus the mitochondrial pathway as a way to induce apoptosis (Walczak *et al.*, 2000a). Therefore, TRAIL signaling and its downstream apoptotic effects are not blocked by these antiapoptotic molecules. This observation suggests a great potential for TRAIL against Bcl-2– or Bcl-x_L–overexpressing tumors.

Most favorable is a treatment that combines the selective cytotoxic effects of TRAIL with those of conventional chemotherapy. By attacking the tumor simultaneously from different angles using different cytotoxic mechanisms to kill the tumor cells, the chances for the tumor to develop resistant clones will diminish. Our data showed an additive effect of a combination of PA and TRAIL (Odoux *et al.*, 2002). These results may be due to the fact that TRAIL uses a more distinct pathway to trigger apoptosis than PA. TRAIL directly activates the caspase pathway and bypasses the mitochondrial pathway while PA acts like most other chemotherapeutic agents by primarily engaging the mitochondrial proapoptotic pathways.

B. *IN VIVO* RELEVANCE OF PACLITAXEL AND TRAIL COMBINATION THERAPY

A promising recent study examining the TRAIL-related novel monoclonal antibody TRA-8 combined with PA achieved together a high percentage of tumor regression in a mouse model of human breast cancer. This remission was observed in 38% of the animals treated and tumor regression was not observed with PA alone (Buschsboum *et al.*, 93rd annual meeting of the American Association for Cancer Research, April 2002, San Francisco, Abstract).

So far, no clinical trials have been conducted on cancer patients using combination PA and TRAIL therapy. However, Immunex Pharmaceutical Corporation constructed a soluble form of TRAIL that is nontoxic in animals and that stimulates tumor destruction from a variety of cancers.

TABLE I. Summary of the Clinical and Biologic Effects of PA, TRAIL, and PA + TRAIL Combination Treatment

Chemotherapeutic agent	Preclinical and clinical relevance	Experimental *in vitro* cytotoxic effects	Apoptotic mechanisms of action involved	Resistance to chemotherapy
PA	(1) Used as an antineoplastic drug in patients with cancer, notably, ovarian, breast, head and neck, and lung cancer (Rowinsky et al., 1990) (2) Active drug used in 234 clinical trials (3) Brings great interest for its unique cytotoxic mechanism of action (Fuchs and Johnson, 1978)	(1) Causes cytotoxicity through interference with microtubules (Thuret-Carnahan et al., 1985) (2) Induces apoptosis in various types of cancer (Andre et al., 2002; Guo et al., 2002; Vignati et al., 2002; Wang et al., 1999)	Sequential apoptotic events occur upon: (1) binding to TNF-related family receptors, (2) activation of the mitochondrial pathway, (3) DNA degradation, (4) late caspase activation (von Haefen et al., 2003)	PA resistance is associated with: (1) Altered expression (Haber et al., 1995) or mutation (Giannakakou et al., 1997) of tubulin isoforms; α- and β-tubulins, (2) Overexpression of P-glycoprotein (Pgp) (Roy and Horwitz, 1985) (3) Upregulation of caveolin-1, a major component of caveolae in membranes (Yang et al., 1998)
TRAIL	(1) Reduced tumor growth in mice implanted with human-derived tumor cells (Huang et al., 2002; Lin et al., 2002)	Induces apoptosis in various cancer cells, but not in normal cells (Tomek, Koestler et al., 2003)	(1) Interaction with TRAIL-R1 and -R2, and up-regulation of these receptors for some cancer cells (Nimmanapalli et al., 2001)	Resistance to TRAIL by normal cells or cancer cells is due to: (1) TRAIL interaction with its decoy receptors (Sheridan et al., 1997)

400

	(2) Improved survival of nonhuman primates with cancer, without inducing any detectable cytotoxicity in tissues or organs (Ashkenazi et al., 1999)		(2) Two pathways exist, one that works via direct caspase activation operating independently from mitochondria, and another one that is dependent on late mitochondrial apoptotic events (Walczak et al., 2000a)	(2) Inhibition of TRAIL signaling by FLIP protein expression (Leverkus et al., 2000) (3) Lack of DR4 and DR5 expression (Nimmanapalli et al., 2001)
PA + TRAIL	Suppressed tumor growth in mice implanted with human cancer cells that are resistant to chemotherapy or TRAIL alone (Lin et al., 2003; Walczak et al., 1999)	Trail and PA combination treatment remarkably increases the degree of apoptosis of either alone PA- or TRAIL-induced apoptosis in various types of cancer cells (Frese et al., 2002; Odoux et al. 2002)	(1) Increased expression of TRAIL receptors DR4 and DR5 (Nimmanapalli et al., 2001) (2) Increase in caspase activation and cytochrome c release (Nimmanapalli et al., 2001) (3) Bypass of the anti-apoptotic effect of Bcl-2 or Bcl-x_L by TRAIL (Walczak et al., 2000a)	(1) Bypass of the antiapoptotic molecule Bcl-2 and Bcl-x_L by TRAIL reducing resistance of cancer cells toward the combination TRAIL and PA (Walczak et al. 2000a) (2) Resistance decreases with inactivation of Akt, a serine threonine kinase that correlates with apoptosis resistance to TRAIL (Asakuma et al., 2003)

Immunex published in the journal *Nature Medicine* that the TRAIL molecule causes complete remission of tumors of mice injected with human breast cancer cells (Walczak *et al.*, 1999). TRAIL is thus a high priority predevelopment product for pharmaceutical companies that would allow an easier transition to study the combination effect of PA and TRAIL in clinical trials.

V. CONCLUSIONS

Until now, approaches applied to fight cancer are, apart from surgery, for the most part unspecific approaches: irradiation of the tumor and chemotherapy. Despite many advances, these treatments are still associated with strong, sometimes disastrous, side effects while long-term therapeutic success rates remain unsatisfyingly low.

The need for the development of novel therapeutic concepts is therefore eminent. Recent advances in recombinant technology have facilitated the expression of human proteins in mammalian cells enabling us to discover new cellular pathways. More sensitive and complete diagnostic tools on the molecular level will help us to predict susceptibilities and resistances to drugs and to develop new cancer classifications that are based on the molecular markers rather than on classical morphologic markers. Lately it has become apparent that pathologic changes of apoptosis play a major role in the development and maintenance of cancer. With the availability of recombinant apoptosis-inducing and -inhibiting molecules, apoptotic pathways can be probed and provide us with specific promising experimental molecules that could be used later as a basis for therapeutic agents.

With these new molecular tools a growing knowledge of the mechanisms of action and molecular pathways used by anticancer drugs will lead to a more rational development of therapy regimens. Factors that should be included in the design of therapies are the correct timing of application for each treatment component, and the combination of drugs based on molecular characteristics of the tumor targeting death-inducing pathways. This way, anticancer agents are more likely to fully unfold their actions toward their most synergistic and additive prospective while toxic effects and the development of resistance are reduced.

PA alone reduces the rate of cancer-related death in some cancers (Table I). TRAIL alone is a potent inducer of apoptosis of cancer cells (Table I). The currently available information on TRAIL and PA combination therapy suggests advances in the treatment of cancer, as its efficiency has been already shown in *in vitro* and *in vivo* studies (Table I), but it still remains to be further investigated. For proof of principle, new strategies and methods using a modified TRAIL gene, such as the *GFP-TRAIL* fusion gene, that are transfected into cancer cells in *in vitro* cultures resulted in great apoptotic response in various resistant cancer cells, such as breast, lung, colon, ovary, and prostate gland cancer cells. More recently, new molecules such as a TRAIL-related novel monoclonal antibody TRA-8 combined with PA have shown significant tumor regression in animals. Thus, the concept of triggering apoptosis at different levels and/or pathways using TRAIL-receptor agonists in combination with PA is very attractive. The future will show if the promises that the combination of these drugs hold *in vitro* and in animal studies will also hold in clinical trials for the benefit of cancer patients.

REFERENCES

Alnemri, E. S., Livingston, D. J., Nicholson, D. W., Salvesen, G., Thornberry, N. A., Wong, W. W., and Yuan, J. (1996). Human ICE/CED-3 protease nomenclature. *Cell* **87**, 171.

Amadori, D., Frassineti, G. L., Zoli, W., Milandri, C., Serra, P., Tienghi, A., Ravaioli, A., Gentile, A., and Salzano, E. (1997). Doxorubicin and paclitaxel (sequential combination) in the treatment of advanced breast cancer. *Oncol. (Huntingt)* **11**, 30–33.

Andre, N., Carre, M., Brasseur, G., Pourroy, B., Kovacic, H., Briand, C., and Braguer, D. (2002). Paclitaxel targets mitochondria upstream of caspase activation in intact human neuroblastoma cells. *FEBS Lett.* **532**, 256–260.

Asakuma, J., Sumitomo, M., Asano, T., Asano, T., and Hayakawa, M. (2003). Selective Akt inactivation and tumor necrosis factor-related apoptosis-inducing ligand sensitization of renal cancer cells by low concentrations of paclitaxel. *Cancer Res.* **63**, 1365–1370.

Ashkenazi, A., Pai, R. C., Fong, S., Leung, S., Lawrence, D. A., Marsters, S. A., Blackie, C., Chang, L., McMurtrey, A. E., Hebert, A., DeForge, L., Koumenis, I. L., Lewis, D., Harris, L., Bussiere, J., Koeppen, H., Shahrokh, Z., and Schwall, R. H. (1999). Safety and antitumor activity of recombinant soluble Apo2 ligand. *J. Clin. Invest.* **104**, 155–162.

Belani, C. P. (2000). Paclitaxel and docetaxel combinations in non-small cell lung cancer. *Chest* **117**, 144S–151S.

Bhalla, K., Ibrado, A. M., Tourkina, E., Tang, C., Mahoney, M. E., and Huang, Y. (1993). Taxol induces internucleosomal DNA fragmentation associated with programmed cell death in human myeloid leukemia cells. *Leukemia* **7**, 563–568.

Blagosklonny, M. V., Robey, R., Sheikh, M. S., and Fojo, T. (2002). Paclitaxel-induced FasL-independent apoptosis and slow (non-apoptotic) cell death. *Cancer Biol. Ther.* **1**, 113–117.

Brisdelli, F., Iorio, E., Knijn, A., Ferretti, A., Marcheggiani, D., Lenti, L., Strom, R., Podo, F., and Bozzi, A. (2003). Two-step formation of 1H NMR visible mobile lipids during apoptosis of paclitaxel-treated K562 cells. *Biochem. Pharmacol.* **65**, 1271–1280.

Bunn, P. A., Jr., and Kelly, K. (2000). New combinations in the treatment of lung cancer: A time for optimism. *Chest* **117**, 138S–143S.

Cuello, M., Ettenberg, S. A., Nau, M. M., and Lipkowitz, S. (2001). Synergistic induction of apoptosis by the combination of trail and chemotherapy in chemoresistant ovarian cancer cells. *Gynecol. Oncol.* **81**, 380–390.

Dalpiaz, O., al Rabi, N., Galfano, A., Martignoni, G., Ficarra, V., and Artibani, W. (2003). Small cell carcinoma of the bladder: A case report and a literature review. *Arch. Esp. Urol.* **56**, 197–202.

De Giorgi, U., Papiani, G., Severini, G., Fiorentini, G., Marangolo, M., and Rosti, G. (2003). High-dose chemotherapy in adult patients with germ cell tumors. *Cancer Control* **10**, 48–56.

Einzig, A. I., Gorowski, E., Sasloff, J., and Wiernik, P. H. (1991). Phase II trial of taxol in patients with metastatic renal cell carcinoma. *Cancer Invest.* **9**, 133–136.

French, L. E., and Tschopp, J. (1999). The TRAIL to selective tumor death. *Nat. Med.* **5,** 146–147.

Frese, S., Brunner, T., Gugger, M., Uduehi, A., and Schmid, R. A. (2002). Enhancement of Apo2L/TRAIL (tumor necrosis factor-related apoptosis-inducing ligand)-induced apoptosis in non-small cell lung cancer cell lines by chemotherapeutic agents without correlation to the expression level of cellular protease caspase-8 inhibitory protein. *J. Thorac. Cardiovasc. Surg.* **123,** 168–174.

Fuchs, D. A., and Johnson, R. K. (1978). Cytologic evidence that taxol, an antineoplastic agent from *Taxus brevifolia*, acts as a mitotic spindle poison. *Cancer Treat. Rep.* **62,** 1219–1222.

Gadducci, A., Cosio, S., Muraca, S., and Genazzani, A. R. (2002). Molecular mechanisms of apoptosis and chemosensitivity to platinum and paclitaxel in ovarian cancer: Biological data and clinical implications. *Eur. J. Gynaecol. Oncol.* **23,** 390–396.

Ganansia-Leymarie, V., Bischoff, P., Bergerat, J. P., and Holl, V. (2003). Signal transduction pathways of taxanes-induced apoptosis. *Curr. Med. Chem. Anti-Canc. Agents* **3,** 291–306.

Giannakakou, P., Sackett, D. L., Kang, Y. K., Zhan, Z., Buters, J. T., Fojo, T., and Poruchynsky, M. S. (1997). Paclitaxel-resistant human ovarian cancer cells have mutant beta-tubulins that exhibit impaired paclitaxel-driven polymerization. *J. Biol. Chem.* **272,** 17118–17125.

Goncalves, A., Braguer, D., Carles, G., Andre, N., Prevot, C., and Briand, C. (2000). Caspase-8 activation independent of CD95/CD95-L interaction during paclitaxel-induced apoptosis in human colon cancer cells (HT29-D4). *Biochem. Pharmacol.* **60,** 1579–1584.

Griffith, T. S., Chin, W. A., Jackson, G. C., Lynch, D. H., and Kubin, M. Z. (1998a). Intracellular regulation of TRAIL-induced apoptosis in human melanoma cells. *J. Immunol.* **161,** 2833–2840.

Griffith, T. S., and Lynch, D. H. (1998b). TRAIL: A molecule with multiple receptors and control mechanisms. *Curr. Opin. Immunol.* **10,** 559–563.

Griffith, T. S., Fialkov, J. M., Scott, D. L., Azuhata, T., Williams, R. D., Wall, N. R., Altieri, D. C., and Sandler, A. D. (2002). Induction and regulation of tumor necrosis factor-related apoptosis-inducing ligand/Apo-2 ligand-mediated apoptosis in renal cell carcinoma. *Cancer Res.* **62,** 3093–3099.

Guo, W., Zeng, C., Dong, F., and Lei, W. (2002). Paclitaxel-induced apoptosis in osteosarcoma cell line U-2 OS. *Chin. Med. J. (Engl.)* **115,** 1796–1801.

Haber, M., Burkhart, C. A., Regl, D. L., Madafiglio, J., Norris, M. D., and Horwitz, S. B. (1995). Altered expression of M beta 2, the class II beta-tubulin isotype, in a murine J774.2 cell line with a high level of taxol resistance. *J. Biol. Chem.* **270,** 31269–31275.

Harper, P. (2002). Current clinical practices for ovarian cancers. *Semin. Oncol.* **29,** 3–6.

Hinz, S., Trauzold, A., Boenicke, L., Sandberg, C., Beckmann, S., Bayer, E., Walczak, H., Kalthoff, H., and Ungefroren, H. (2000). Bcl-XL protects pancreatic adenocarcinoma cells against. *Oncogene* **19,** 5477–5486.

Huang, X., Lin, T., Gu, J., Zhang, L., Roth, J. A., Stephens, L. C., Yu, Y., Liu, J., and Fang, B. (2002). Combined TRAIL and Bax gene therapy prolonged survival in mice with ovarian cancer xenograft. *Gene Ther.* **9,** 1379–1386.

Ibrado, A. M., Kim, C. N., and Bhalla, K. (1998). Temporal relationship of CDK1 activation and mitotic arrest to cytosolic accumulation of cytochrome C and caspase-3 activity during Taxol-induced apoptosis of human AML HL-60 cells. *Leukemia* **12,** 1930–1936.

Irmler, M., Thome, M., Hahne, M., Schneider, P., Hofmann, K., Steiner, V., Bodmer, J. L., Schroter, M., Burns, K., Mattmann, C., Rimoldi, D., French, L. E., and Tschopp, J. (1997). Inhibition of death receptor signals by cellular FLIP. *Nature* **388,** 190–195.

Kavallaris, M., Kuo, D. Y., Burkhart, C. A., Regl, D. L., Norris, M. D., Haber, M., and Horwitz, S. B. (1997). Taxol-resistant epithelial ovarian tumors are associated with altered expression of specific beta-tubulin isotypes. *J. Clin. Invest.* **100,** 1282–1293.

Keane, M. M., Ettenberg, S. A., Nau, M. M., Russell, E. K., and Lipkowitz, S. (1999). Chemotherapy augments TRAIL-induced apoptosis in breast cell lines. *Cancer Res.* **59**, 734–741.

Kim, K., Fisher, M. J., Xu, S. Q., and el Deiry, W. S. (2000). Molecular determinants of response to TRAIL in killing of normal and cancer cells. *Clin. Cancer Res.* **6**, 335–346.

Leverkus, M., Neumann, M., Mengling, T., Rauch, C. T., Brocker, E. B., Krammer, P. H., and Walczak, H. (2000). Regulation of tumor necrosis factor-related apoptosis-inducing ligand sensitivity in primary and transformed human keratinocytes. *Cancer Res.* **60**, 553–559.

Lin, T., Huang, X., Gu, J., Zhang, L., Roth, J. A., Xiong, M., Curley, S. A., Yu, Y., Hunt, K. K., and Fang, B. (2002). Long-term tumor-free survival from treatment with the GFP-TRAIL fusion gene expressed from the hTERT promoter in breast cancer cells. *Oncogene* **21**, 8020–8028.

Lin, T., Zhang, L., Davis, J., Gu, J., Nishizaki, M., Ji, L., Roth, J. A., Xiong, M., and Fang, B. (2003). Combination of TRAIL gene therapy and chemotherapy enhances antitumor and antimetastasis effects in chemosensitive and chemoresistant breast cancers. *Mol. Ther.* **8**, 441–448.

MacFarlane, M., Ahmad, M., Srinivasula, S. M., Fernandes-Alnemri, T., Cohen, G. M., and Alnemri, E. S. (1997). Identification and molecular cloning of two novel receptors for the cytotoxic ligand TRAIL. *J. Biol. Chem.* **272**, 25417–25420.

Majno, G., and Joris, I. (1995). Apoptosis, oncosis, and necrosis. An overview of cell death. *Am. J. Pathol.* **146**, 3–15.

Mole-Bajer, J., and Bajer, A. S. (1983). Action of taxol on mitosis: Modification of microtubule arrangements and function of the mitotic spindle in *Haemanthus* endosperm. *J. Cell Biol.* **96**, 527–540.

Moliterni, A., Tarenzi, E., Capri, G., Terenziani, M., Bertuzzi, A., Grasselli, G., Agresti, R., Piotti, P., Greco, M., Salvadori, B., Pilotti, S., Lombardi, F., Valagussa, P., Bonadonna, G., and Gianni, L. (1997). Pilot study of primary chemotherapy with doxorubicin plus paclitaxel in women with locally advanced or operable breast cancer. *Semin. Oncol.* **24**, S17.

Mueller, H. (1998). Tumor necrosis factor as an antineoplastic agent: Pitfalls and promises. *Cell Mol. Life Sci.* **54**, 1291–1298.

Nabholtz, J. M., and Riva, A. (2001). Taxane/anthracycline combinations: Setting a new standard in breast cancer? *Oncologist* **6**(Suppl. 3), 5–12.

Nagane, M., Pan, G., Weddle, J. J., Dixit, V. M., Cavenee, W. K., and Huang, H. J. (2000). Increased death receptor 5 expression by chemotherapeutic agents in human gliomas causes synergistic cytotoxicity with tumor necrosis factor-related apoptosis-inducing ligand in vitro and in vivo. *Cancer Res.* **60**, 847–853.

Nimmanapalli, R., Perkins, C. L., Orlando, M., O'Bryan, E., Nguyen, D., and Bhalla, K. N. (2001). Pretreatment with paclitaxel enhances apo-2 ligand/tumor necrosis factor-related apoptosis-inducing ligand-induced apoptosis of prostate cancer cells by inducing death receptors 4 and 5 protein levels. *Cancer Res.* **61**, 759–763.

Novello, S., and Le Chevalier, T. (2001). European perspectives on paclitaxel/platinum-based therapy for advanced non-small cell lung cancer. *Semin. Oncol.* **28**, 3–9.

Odoux, C., Albers, A., Amoscato, A. A., Lotze, M. T., and Wong, M. K. (2002). TRAIL, FasL and a blocking anti-DR5 antibody augment paclitaxel-induced apoptosis in human non-small-cell lung cancer. *Int. J. Cancer* **97**, 458–465.

Ofir, R., Seidman, R., Rabinski, T., Krup, M., Yavelsky, V., Weinstein, Y., and Wolfson, M. (2002). Taxol-induced apoptosis in human SKOV3 ovarian and MCF7 breast carcinoma cells is caspase-3 and caspase-9 independent. *Cell Death Differ.* **9**, 636–642.

Ohtsuka, T., Buchsbaum, D., Oliver, P., Makhija, S., Kimberly, R., and Zhou, T. (2003). Synergistic induction of tumor cell apoptosis by death receptor antibody and chemotherapy agent through JNK/p38 and mitochondrial death pathway. *Oncogene* **22**, 2034–2044.

Oyaizu, H., Adachi, Y., Taketani, S., Tokunaga, R., Fukuhara, S., and Ikehara, S. (1999). A crucial role of caspase 3 and caspase 8 in paclitaxel-induced apoptosis. *Mol. Cell. Biol. Res. Commun.* **2**, 36–41.

Pan, J., Xu, G., and Yeung, S. C. (2001). Cytochrome c release is upstream to activation of caspase-9, caspase-8, and caspase-3 in the enhanced apoptosis of anaplastic thyroid cancer cells induced by manumycin and paclitaxel. *J. Clin. Endocrinol. Metab.* **86**, 4731–4740.

Perez, E. A. (1999). Paclitaxel plus nonanthracycline combinations in metastatic breast cancer. *Semin. Oncol.* **26**, 21–26.

Perez, E. A. (2001). Doxorubicin and paclitaxel in the treatment of advanced breast cancer: Efficacy and cardiac considerations. *Cancer Invest.* **19**, 155–164.

Pitti, R. M., Marsters, S. A., Ruppert, S., Donahue, C. J., Moore, A., and Ashkenazi, A. (1996). Induction of apoptosis by Apo-2 ligand, a new member of the tumor necrosis factor cytokine family. *J. Biol. Chem.* **271**, 12687–12690.

Pucci, B., Bellincampi, L., Tafani, M., Masciullo, V., Melino, G., and Giordano, A. (1999). Paclitaxel induces apoptosis in Saos-2 cells with CD95L upregulation and Bcl-2 phosphorylation. *Exp. Cell Res.* **252**, 134–143.

Rao, S., Orr, G. A., Chaudhary, A. G., Kingston, D. G., and Horwitz, S. B. (1995). Characterization of the taxol binding site on the microtubule. 2-(m-azidobenzoyl)taxol photolabels a peptide (amino acids 217-231) of beta-tubulin. *J. Biol. Chem.* **270**, 20235–20238.

Rosell, R., and Felip, E. (2001). Predicting response to paclitaxel/carboplatin-based therapy in non-small cell lung cancer. *Semin. Oncol.* **28**, 37–44.

Rowinsky, E. K., Cazenave, L. A., and Donehower, R. C. (1990). Taxol: A novel investigational antimicrotubule agent. *J. Natl. Cancer Inst.* **82**, 1247–1259.

Roy, S. N., and Horwitz, S. B. (1985). A phosphoglycoprotein associated with taxol resistance in J774.2 cells. *Cancer Res.* **45**, 3856–3863.

Schneider, P., Thome, M., Burns, K., Bodmer, J. L., Hofmann, K., Kataoka, T., Holler, N., and Tschopp, J. (1997). TRAIL receptors 1 (DR4) and 2 (DR5) signal FADD-dependent apoptosis and activate NF-kappaB. *Immunity* **7**, 831–836.

Sheridan, J. P., Marsters, S. A., Pitti, R. M., Gurney, A., Skubatch, M., Baldwin, D., Ramakrishnan, L., Gray, C. L., Baker, K., Wood, W. I., Goddard, A. D., Godowski, P., and Ashkenazi, A. (1997). Control of TRAIL-induced apoptosis by a family of signaling and decoy receptors. *Science* **277**, 818–821.

Sprick, M. R., Weigand, M. A., Rieser, E., Rauch, C. T., Juo, P., Blenis, J., Krammer, P. H., and Walczak, H. (2000). FADD/MORT1 and caspase-8 are recruited to TRAIL receptors 1 and 2 and are essential for apoptosis mediated by TRAIL receptor 2. *Immunity* **12**, 599–609.

Tan, G., Heqing, L., Jiangbo, C., Ming, J., Yanhong, M., Xianghe, L., Hong, S., and Li, G. (2002). Apoptosis induced by low-dose paclitaxel is associated with p53 upregulation in nasopharyngeal carcinoma cells. *Int. J. Cancer* **97**, 168–172.

Thomas, W. D., and Hersey, P. (1998). TNF-related apoptosis-inducing ligand (TRAIL) induces apoptosis in Fas ligand-resistant melanoma cells and mediates CD4 T cell killing of target cells. *J. Immunol.* **161**, 2195–2200.

Thuret-Carnahan, J., Bossu, J. L., Feltz, A., Langley, K., and Aunis, D. (1985). Effect of taxol on secretory cells: Functional, morphological, and electrophysiological correlates. *J. Cell Biol.* **100**, 1863–1874.

Tomek, S., Koestler, W., Horak, P., Grunt, T., Brodowicz, T., Pribill, I., Halaschek, J., Haller, G., Wiltschke, C., Zielinski, C. C., and Krainer, M. (2003). Trail-induced apoptosis and interaction with cytotoxic agents in soft tissue sarcoma cell lines. *Eur. J. Cancer* **39**, 1318–1329.

Trail, P. A., Willner, D., Bianchi, A. B., Henderson, A. J., TrailSmith, M. D., Girit, E., Lasch, S., Hellstrom, I., and Hellstrom, K. E. (1999). Enhanced antitumor activity of paclitaxel in combination with the anticarcinoma immunoconjugate BR96-doxorubicin. *Clin. Cancer Res.* **5,** 3632–3638.

Tudor, G., Aguilera, A., Halverson, D. O., Laing, N. D., and Sausville, E. A. (2000). Susceptibility to drug-induced apoptosis correlates with differential modulation of Bad, Bcl-2 and Bcl-xL protein levels. *Cell Death Differ.* **7,** 574–586.

Vansteenkiste, J., Vandebroek, J., Nackaerts, K., Dooms, C., Galdermans, D., Bosquee, L., Delobbe, A., Deschepper, K., Van Kerckhoven, W., Vandeurzen, K., Deman, R., D'Odemont, J. P., Siemons, L., Van den, B. P., and Dams, N. (2003). Influence of cisplatin-use, age, performance status and duration of chemotherapy on symptom control in advanced non-small cell lung cancer: Detailed symptom analysis of a randomised study comparing cisplatin-vindesine to gemcitabine. *Lung Cancer* **40,** 191–199.

Vignati, S., Codegoni, A., Polato, F., and Broggini, M. (2002). Trail activity in human ovarian cancer cells: Potentiation of the action of cytotoxic drugs. *Eur. J. Cancer* **38,** 177–183.

von Haefen, C., Wieder, T., Essmann, F., Schulze-Osthoff, K., Dorken, B., and Daniel, P. T. (2003). Paclitaxel-induced apoptosis in BJAB cells proceeds via a death receptor-independent, caspases-3/-8-driven mitochondrial amplification loop. *Oncogene* **22,** 2236–2247.

Walczak, H., Bouchon, A., Stahl, H., and Krammer, P. H. (2000a). Tumor necrosis factor-related apoptosis-inducing ligand retains its apoptosis-inducing capacity on Bcl-2- or Bcl-xL-overexpressing chemotherapy-resistant tumor cells. *Cancer Res.* **60,** 3051–3057.

Walczak, H., and Krammer, P. H. (2000b). The CD95 (APO-1/Fas) and the TRAIL (APO-2L) apoptosis systems. *Exp. Cell Res.* **256,** 58–66.

Walczak, H., Miller, R. E., Ariail, K., Gliniak, B., Griffith, T. S., Kubin, M., Chin, W., Jones, J., Woodward, A., Le, T., Smith, C., Smolak, P., Goodwin, R. G., Rauch, C. T., Schuh, J. C., and Lynch, D. H. (1999). Tumoricidal activity of tumor necrosis factor-related apoptosis-inducing ligand in vivo. *Nat. Med.* **5,** 157–163.

Wang, S., Lu, Y., and Ma, D. (1999). Paclitaxel-induced apoptosis in human ovarian cancer cell line COC1. *J. Tongji Med. Univ.* **19,** 124–126.

Weigel, T. L., Lotze, M. T., Kim, P. K., Amoscato, A. A., Luketich, J. D., and Odoux, C. (2000). Paclitaxel-induced apoptosis in non-small cell lung cancer cell lines is associated with increased caspase-3 activity. *J. Thorac. Cardiovasc. Surg.* **119,** 795–803.

Wiley, S. R., Schooley, K., Smolak, P. J., Din, W. S., Huang, C. P., Nicholl, J. K., Sutherland, G. R., Smith, T. D., Rauch, C., and Smith, C. A. (1995). Identification and characterization of a new member of the TNF family that induces apoptosis. *Immunity* **3,** 673–682.

Williamson, B. D., Carswell, E. A., Rubin, B. Y., Prendergast, J. S., and Old, L. J. (1983). Human tumor necrosis factor produced by human B-cell lines: Synergistic cytotoxic interaction with human interferon. *Proc. Natl. Acad. Sci. USA* **80,** 5397–5401.

Wolfson, M., Yang, C. P., and Horwitz, S. B. (1997). Taxol induces tyrosine phosphorylation of Shc and its association with Grb2 in murine RAW 264.7 cells. *Int. J. Cancer* **70,** 248–252.

Yang, C. P., Galbiati, F., Volonte, D., Horwitz, S. B., and Lisanti, M. P. (1998). Upregulation of caveolin-1 and caveolae organelles in Taxol-resistant A549 cells. *FEBS Lett.* **439,** 368–372.

Zhou, H. B., and Zhu, J. R. (2003). Paclitaxel induces apoptosis in human gastric carcinoma cells. *World J. Gastroenterol.* **9,** 442–445.

21

Regulation of Sensitivity to TRAIL by the PTEN Tumor Suppressor

Young E. Whang, Xiu-Juan Yuan,
Yuanbo Liu, Samarpan Majumder, and
Terrence D. Lewis

*Lineberger Comprehensive Cancer Center
Departments of Medicine and Pathology and Laboratory Medicine
University of North Carolina School of Medicine
Chapel Hill, North Carolina 27599-7295*

I. Introduction
II. Regulation of TRAIL-Induced Apoptosis
   A. Overview
   B. Decoy Receptor
   C. c-FLICE Inhibitory Protein
   D. Nuclear Factor-κB
   E. Other Factors
III. Regulation of the Phosphatidylinositol-3 Kinase Pathway by PTEN
   A. Phosphatidylinositol-3 Kinase/Akt Pathway
   B. PTEN as a Tumor Suppressor
IV. Modulation of TRAIL Sensitivity by the Phosphatidylinositol-3 Kinase/PTEN/Akt Pathway
   A. Activated Akt Resulting from PTEN Loss Protects Cells from TRAIL

B. Growth Factor–Induced Akt Activation Protects Cells from TRAIL
V. Downstream Targets of the Phosphatidylinositol-3 Kinase/PTEN/Akt Pathway
A. Overview
B. Nuclear Factor-κB
C. PED/PEA-15
VI. Strategies for Overcoming TRAIL Resistance
VII. Conclusion
References

The ability of tumor necrosis factor–related apoptosis-inducing ligand (TRAIL) to induce apoptosis preferentially in cancer cells is attractive for its development as a novel cancer therapeutic agent, but many cancer cell lines are resistant to TRAIL. While the molecular basis for TRAIL resistance is not always clear, a number of factors have been proposed to mediate TRAIL resistance, including decoy receptor, c-FLIP, nuclear factor (NF)-κB, and activation of antiapoptotic kinase signaling. Many growth factor receptors mediate their survival signals through the pathway involving recruitment and activation of phosphatidylinositol (PI) 3-kinase and the serine/threonine kinase Akt. The PTEN tumor suppressor is a phosphatase that dephosphorylates the phospholipids phosphorylated by PI-3 kinase, thereby opposing the action of PI 3-kinase, and acts as the primary negative regulator of the PI-3 kinase/Akt pathway in the cell. Loss of PTEN function occurs frequently in human tumors and leads to constitutive activation of Akt in cancer cells. Constitutively active Akt protects cells from TRAIL-induced apoptosis in multiple tumor types. Growth factors such as epidermal growth factor or insulin-like growth factor-1 also inhibit TRAIL-induced apoptosis through the Akt pathway. Akt exerts its antiapoptotic function by its ability to phosphorylate many key components of the cellular apoptotic regulatory circuit, such as BAD, MDM2, FOXO Forkhead transcription factors, and PED/PEA-15 as well as by its role in activating NF-κB. Because PTEN loss is common in tumors, strategies to inactivate Akt may be necessary to overcome TRAIL resistance and make TRAIL-based therapy more effective. © 2004 Elsevier Inc.

I. INTRODUCTION

Tumor necrosis factor (TNF)–related apoptosis-inducing ligand (TRAIL) belongs to the TNF superfamily of ligands that cause apoptosis in susceptible cells (Wiley et al., 1995). From its first report of cloning, its ability to induce apoptosis preferentially in transformed cells and not in normal cells as well as the relative lack of toxicity with systemic administration has raised hope that it may be developed as a novel therapeutic agent for cancer patients (Ashkenazi et al., 1999; Walczak et al., 1999). However, it quickly became apparent that there is a wide variability in the susceptibility of tumor cells to TRAIL-induced cytotoxicity and that many cancer cell lines are resistant to TRAIL. There is now emerging literature on mechanisms underlying resistance to TRAIL. This review will give a brief overview of the currently proposed molecular basis of TRAIL resistance and concentrate on the cell survival pathway regulated by the PTEN tumor suppressor and the Akt kinase as the regulator of sensitivity to TRAIL. We will also discuss several potential strategies to overcome resistance to TRAIL that may improve clinical efficacy of TRAIL as an anticancer agent.

II. REGULATION OF TRAIL-INDUCED APOPTOSIS

A. OVERVIEW

The general framework of the intracellular signaling pathway that results in the execution of the cell death program after exposure of susceptible cells to TRAIL is similar to that of other death receptors such as Fas and TNF receptor (Almasan and Ashkenazi, 2003). Engagement of the TRAIL receptor by the homotrimeric TRAIL ligand initiates the formation of intracellular molecular complexes termed death-inducing signaling complex (DISC), composed of the cytoplasmic death domain of the receptor, the adaptor protein FADD, and procaspase-8. FADD is composed of the death domain mediating the homotypic interaction with the receptor and the death effector domain capable of recruiting procaspase-8 that also contains the death effector domain at the N-terminus. The assembly of DISC promotes proteolytic autoactivation of procaspase-8 and subsequent proteolytic processing and activation of effector caspases such as caspase-3 or -7 directly in so-called type I cells or through steps involving the mitochondria and the release of cytochrome c in type II cells. In these type II cells, DISC-activated caspase-8 cleaves BID, a proapoptotic Bcl-2 family member, and truncated BID translocates to the mitochondria and triggers cytochrome c release.

Antiapoptotic Bcl-2 family members such as Bcl-2 or Bcl-X_L block cytochrome c release and thereby suppress the mitochondrial apoptotic pathway. Cytochrome c released into the cytosol results in the formation of a high molecular weight complex termed apoptosome, containing Apaf-1, procaspase-9, cytochrome c, and dATP, and promotes autoactivation by proteolysis of procaspase-9. This in turn leads to activation of downstream caspases such as caspase-3 and -7. Other mitochondrial factors released during apoptosis include Smac/DIABLO, which promotes caspase activation. TRAIL-induced apoptosis may be modulated at many points along this pathway.

B. DECOY RECEPTOR

Unique among death receptors is the complex system of TRAIL-binding receptors composed of two apoptosis-inducing receptors (TRAIL-R1/DR4 and TRAIL-R2/DR5) and three "decoy" receptors (TRAIL-R3/DcR1, TRAIL-R4/DcR2, and osteoprotegerin) (Ashkenazi and Dixit, 1999). Existence of decoy receptors lacking functional death domains and therefore incapable of transducing the death signal was initially proposed as the basis for resistance of normal cells to TRAIL (Pan et al., 1997; Sheridan et al., 1997). However, subsequent studies failed to show a good correlation between decoy receptor expression and resistance to TRAIL and the role of decoy receptors in mediating resistance to TRAIL is currently unclear (Hersey and Zhang, 2001; Leverkus et al., 2000; Zhang et al., 1999). However, the decoy receptor TRAIL-R4/DcR2 containing an incomplete death domain may signal to activate nuclear factor (NF)-κB and this may promote cell survival after TRAIL (Degli-Esposti et al., 1997).

C. c-FLICE INHIBITORY PROTEIN

Caspase activation by DISC formation could be modulated by c-FLIP (FLICE inhibitory protein), which shares homology with caspase-8 but lacks catalytic activity. c-FLIP prevents procaspase-8 activation by competing for binding to FADD. High levels of c-FLIP may confer cellular resistance to TRAIL, although not all studies show a connection between c-FLIP and TRAIL resistance (Burns and El-Deiry, 2001; Kim et al., 2000; Zhang et al., 1999). Interestingly, c-FLIP seems to be a labile protein and a potential mediator of pharmacologic agents (i.e., cycloheximide) that sensitize cells to TRAIL (Kim et al., 2002a; Kreuz et al., 2001; Leverkus et al., 2000; Sayers et al., 2003). Therefore, some investigators have targeted c-FLIP expression by antisense oligonucleotides or small interfering RNA and demonstrated sensitization of tumor cells to TRAIL (Mitsiades et al., 2002b; Siegmund et al., 2002). Taken together, high levels of c-FLIP expression may mediate TRAIL resistance in some cell types.

D. NUCLEAR FACTOR-κB

NF-κB, classically a heterodimeric transcription factor composed of p65 and p50 subunits, regulates inflammation and proliferation and survival in response to many cytokines and other stimuli (Orlowski and Baldwin, 2002). Signal-induced phosphorylation and degradation of IκB enable translocation of NF-κB to the nucleus where it transactivates antiapoptotic genes such as c-IAP, XIAP, c-FLIP, IEX-1L, and Bcl-X_L and protects cells from apoptosis induced by death ligands and chemotherapy and radiation therapy. TRAIL also serves as an activating signal for NF-κB and this activation is mediated by TRAIL-R1, -R2, and -R4 receptors (Chaudhary et al., 1997; Degli-Esposti et al., 1997; Hu et al., 1999; Jeremias et al., 1998; Schneider et al., 1997). Many reports now demonstrate that NF-κB activity, either induced by TRAIL or constitutive, protects cells from apoptotic effects of TRAIL and mediates TRAIL resistance. In melanoma cell lines, TRAIL-induced NF-κB activation correlated with TRAIL resistance whereas kidney cancer cell lines with constitutively active NF-κB were resistant to TRAIL (Franco et al., 2001; Oya et al., 2001). Tumor cells became sensitized to TRAIL when NF-κB activation was inhibited by blocking degradation of IκB through proteasome inhibition or overexpression of the nondegradable transdominant mutant form of IκB. In addition, the involvement of NF-κB in TRAIL resistance has been documented in leukemia, myeloma, pancreatic cancer, liver cancer, and other tumor types (Jeremias et al., 1998; Kim et al., 2002b; Mitsiades et al., 2001; Trauzold et al., 2001). Consistent with the postulated role of NF-κB on carcinogenesis, constitutively elevated NF-κB activity is commonly found in cancer cells and may play a critical role in enabling cells to escape from cytotoxic effects of TRAIL.

E. OTHER FACTORS

Many components of the intrinsic mitochondrial apoptotic pathway regulate TRAIL sensitivity. Deficiency of proapoptotic BAX and overexpression of antiapoptotic Bcl-2 family members prevent TRAIL-induced apoptosis (Burns and El-Deiry, 2001; Deng et al., 2002). The release of Smac/DIABLO from the mitochondria correlates with TRAIL sensitivity (Zhang et al., 2001). Kinase cascades that convey extracellular signals control virtually every aspect of cell behavior. Sensitivity to TRAIL has been shown to be modulated by many kinases such as protein kinase C, Erk, and casein kinase II (Hao et al., 2001; Ravi and Bedi, 2002; Tran et al., 2001; Trauzold et al., 2001; Zhang et al., 2003). Recently, kinases activated by lipids and a phosphatase PTEN have attracted attention as having a major role in regulating cell survival and proliferation as well as sensitivity to TRAIL.

III. REGULATION OF THE PHOSPHATIDYLINOSITOL-3 KINASE PATHWAY BY PTEN

A. PHOSPHATIDYLINOSITOL-3 KINASE/Akt PATHWAY

The phosphatidylinositols are phospholipids in the cell membrane that have recently emerged as an important second messenger linking growth factor receptors to intracellular pathways controlling cell survival and proliferation (Vivanco and Sawyers, 2002). Phosphatidylinositol (PI)-3 kinase is the key enzyme in this pathway because its activity is regulated by the activation status of growth factor receptor tyrosine kinases. When receptor tyrosine kinases bind to their cognate ligands, their intracellular tyrosine kinase domain becomes activated, leading to autophosphorylation of certain tyrosines and recruitment of proteins with affinity for phosphotyrosine residues. PI-3 kinase, through its p85 regulatory subunit, is translocated from the cytosol to the activated receptor tyrosine kinases and the activated PI-3 kinase phosphorylates its main substrate phosphatidylinositol-4,5 bisphosphate (PIP_2) to phosphatidylinositol-3,4,5 triphosphate (PIP_3) at the D3 position of the inositol ring. In quiescent cells, PIP_3 is present in vanishingly small quantities and when it is produced by PI-3 kinase, it mediates its action through its ability to bind to proteins containing the pleckstrin homology (PH) domain. There are many proteins with the PH domain that may be responsible for downstream effects of PI-3 kinase. However, much attention has been focused on the serine/threonine kinase Akt (also termed protein kinase B) as the critical downstream mediator of PI-3 kinase. Akt, the cellular homologue of the retroviral transforming oncogenes v-Akt, has three isoforms, Akt1, Akt2, and Akt3, and is composed of the PH domain and a kinase domain and a C-terminal regulatory tail. Akt is activated by PIP_3 in the membrane through several steps (Datta *et al.*, 1999). Akt is first translocated from the cytosol to the membrane due to the high affinity of its PH domain for PIP_3. Then, it is phosphorylated at Thr308 and Ser473 by the PDK1 kinase and an unidentified kinase. Membrane targeting of Akt (by fusing to the myristoylation signal sequence) results in constitutive activation of Akt and the myristoylated Akt mutant is frequently used to mimic activated Akt in many experiments. When it is fully activated, Akt appears to play a major role in enabling cells to become resistant to apoptosis and stimulating continued proliferation. Candidate downstream mediators of Akt will be discussed subsequently. Because proteins containing the PH domain are potentially regulated by phosphatidylinositols, it is likely that multiple proteins are responsible for the downstream actions of PI-3 kinase in the context of different signaling pathways and cell types. However, genetic evidence in *Drosophila* points to Akt as the major mediator of PIP_3 action as the Akt allele with the mutation in the PH domain is able to rescue flies from the lethal effects of elevated PIP_3 levels (Stocker *et al.*, 2002). Therefore, this review will primarily concentrate on Akt as the mediator of the PI-3 kinase signaling.

B. PTEN AS A TUMOR SUPPRESSOR

How is the PI-3 kinase pathway regulated? The major negative regulator of the PI-3 kinase pathway appears to be the PTEN phosphatase that dephosphorylates PIP_3 to PIP_2, thereby reversing the action of PI 3-kinase. PTEN (*p*hosphatase and *ten*sin homolog deleted on chromosome *ten*) was first identified as a tumor suppressor gene on the chromosome 10q23 locus deleted in tumor cells (Li *et al.*, 1997; Steck *et al.*, 1997). It was first thought to be a protein phosphatase, but Maehama and Dixon (1998) showed that it is a lipid phosphatase whose preferred *in vitro* substrate is PIP_3. The critical role of PTEN in the regulation of the PI-3 kinase pathway has been verified in many different experimental systems (Vivanco and Sawyers, 2002). Some investigators have used murine fibroblasts with targeted deletions of PTEN and have shown that loss of PTEN leads to the elevated levels of PIP_3 in cells and constitutive activation of Akt and increased resistance to apoptosis (Stambolic *et al.*, 1998). Others have shown that cancer cells lacking PTEN have constitutively activated Akt and reconstitution of PTEN leads to loss of tumorigenicity (Li and Sun, 1998; Wu *et al.*, 1998). The finding that loss of just one protein PTEN leads to constitutive activation of Akt serves as strong evidence that the physiologic determinant of relative levels of PIP_2 and PIP_3 is the balance between two opposing enzymatic activities on the D3 position of the inositol ring, PI-3 kinase being the kinase and PTEN being the phosphatase. The physiologic relevance of PTEN and the PI-3 kinase pathway for human cancer is supported by several lines of evidence. First, loss of expression of PTEN protein is very common in tumor cells occurring in a broad spectrum of human tumors and PTEN is among the most frequently mutated tumor suppressor in human tumors, perhaps second only to p53 (Ali *et al.*, 1999). Second, mice heterozygous for PTEN deletion develop tumors at a high frequency (Di Cristofano *et al.*, 1998). Third, patients with a germline mutation of PTEN are tumor prone (Liaw *et al.*, 1997). Aberrant activation of the PI-3 kinase/Akt pathway resulting from loss of PTEN or other mechanisms may result in protection of cancer cells from many apoptotic stimuli, including TRAIL-induced apoptosis.

IV. MODULATION OF TRAIL SENSITIVITY BY THE PHOSPHATIDYLINOSITOL-3 KINASE/ PTEN/Akt PATHWAY

A. ACTIVATED Akt RESULTING FROM PTEN LOSS PROTECTS CELLS FROM TRAIL

Several groups have recently made the observation linking constitutively active Akt resulting from PTEN loss to TRAIL resistance in a prostate cancer model (Chen *et al.*, 2001; Nesterov *et al.*, 2001; Yuan and Whang, 2002). A survey of prostate cancer cell lines indicated that LNCaP cells, which are PTEN-null due to mutations, were completely insensitive to TRAIL and had the highest level of Akt activation. Inactivation of Akt through treatment with PI-3 kinase inhibitors or expression of the dominant negative Akt or reintroduction of PTEN resulted in killing of cells by TRAIL. In untreated cells resistant to TRAIL, activation of caspase-8 after TRAIL treatment occurred normally but did not lead to the cleavage of BID whereas in cells treated with PI-3 kinase inhibitors, treatment with TRAIL led to caspase-8 activation, cleavage of BID, and the release of cytochrome *c* and apoptosis. Constitutively active myristoylated Akt introduced by an adenoviral vector also prevented apoptosis induced by a PI-3 kinase inhibitor and TRAIL and also prevented BID cleavage. These data suggest that Akt inhibits TRAIL-induced apoptosis at the point of BID cleavage. Interestingly, one of these groups also showed that PTEN reconstitution led to BID cleavage after chemotherapy treatment of cells at a dose too low to induce BID cleavage in control cells (Yuan and Whang, 2002). Taken together, these results imply that Akt inhibits apoptosis at the point of BID cleavage in both the intrinsic pathway after chemotherapy and the extrinsic pathway after TRAIL treatment. The next section will go over the potential mediators of Akt, but the action of Akt on preventing BID cleavage cannot be explained in terms of currently known substrates. BID is known to be phosphorylated on serine/threonine residues by casein kinases and phosphorylation regulates the cleavage of BID by caspases (Degli-Esposti *et al.*, 2003). Perhaps Akt plays a role in this process directly or indirectly, although no direct link between BID and Akt has been made. Experimental systems using leukemia, multiple myeloma, lung cancer, and thyroid cancer cell lines have also provided evidence supporting the role of the PI-3 kinase/PTEN/Akt pathway in regulating TRAIL sensitivity in tumor cells (Bortul *et al.*, 2003; Kandasamy and Srivastava, 2002; Mitsiades *et al.*, 2002a; Poulaki *et al.*, 2002).

B. GROWTH FACTOR–INDUCED Akt ACTIVATION PROTECTS CELLS FROM TRAIL

Because Akt is activated by the upstream signals emanating from growth factor receptors, it is reasonable to hypothesize that growth factors may affect TRAIL sensitivity. Poulaki et al. (2002) reported that growth factors such as insulin-like growth factor (IGF)-1, fibroblast growth factor, and epidermal growth factor (EGF) protected thyroid carcinoma cells from TRAIL and treatment with a PI-3 kinase inhibitor abrogated this protective effect. IGF-1 upregulated the expression of antiapoptotic proteins such as c-FLIP, c-IAP2, XIAP, and survivin. Another work performed with multiple myeloma cells similarly demonstrated that IGF-1 treatment led to Akt and NF-κB activation and made cells resistant to TRAIL, concomitantly with increased levels of c-FLIP, survivin, c-IAP2, XIAP, and A1/Bfl-1 (Mitsiades et al., 2002a). These effects were reversed not only by a PI-3 kinase inhibitor but also by an Akt kinase inhibitor, implicating the Akt pathway downstream of PI-3 kinase. Gibson et al. (2002) implicated the epidermal growth factor receptor (EGFR) pathway in TRAIL resistance. They showed that activation of Akt but not Erk after EGF treatment protected breast cancer cells and transformed human embryonic kidney 293 cells from TRAIL-induced apoptosis. Interestingly, EGF did not prevent caspase-8 activation or BID cleavage but prevented the release of cytochrome c from the mitochondria after TRAIL treatment in these cells.

V. DOWNSTREAM TARGETS OF THE PHOSPHATIDYLINOSITOL-3 KINASE/PTEN/ AKT PATHWAY

A. OVERVIEW

Activated Akt raises the cellular threshold to apoptosis after a variety of insults, such as deprivation of survival factors, stress (i.e., heat or osmotic shock), oncogenic stimulation (i.e., apoptosis induced by c-Myc), and DNA damage from chemotherapeutic agents and radiation. These agents induce apoptosis predominantly through a pathway involving the mitochondrial release of cytochrome c and subsequent caspase activation, sometimes referred to as the "intrinsic pathway." As discussed previously, Akt also suppresses apoptosis mediated by the "extrinsic pathway," resulting from activation of death receptors such as Fas ligand, TNF, and TRAIL. How Akt exerts its antiapoptotic function has been an area of active investigation and almost 30 putative substrates of Akt have been reported in the published literature so far. Although the physiologic relevance of some of these candidates is unclear, it seems likely that Akt influences multiple downstream pathways. Many Akt substrates are known to be a component of the cellular apoptotic machinery or to play a role in regulating sensitivity to apoptosis (Datta et al., 1999). For example, one of the first Akt substrate proteins to be identified is a proapoptotic Bcl-2 family member BAD, which heterodimerizes with and inactivates Bcl-X_L but phosphorylation by Akt prevents binding to Bcl-X$_L$. Another proposed substrate is caspase-9, which is activated by the release of cytochrome c. Phosphorylation of caspase-9 by Akt prevents its activation. However, whether caspase-9 is a physiologically relevant Akt target mediating resistance to apoptosis is unclear. Akt can also negatively modulate the p53-dependent apoptosis pathway by phosphorylating MDM2. Phosphorylated MDM2 translocates into the nucleus and binds to p53 and promotes ubiquitination and degradation of p53. Nuclear localization of the proapoptotic FOXO subfamily of Forkhead transcription factors is blocked by Akt-dependent phosphorylation, thereby preventing transcription of proapoptotic target genes. PTEN, through its ability to inactivate Akt, is expected to promote dephosphorylation of these substrates and many of these predictions have been verified in experimental models. Additional Akt substrates are detailed in a comprehensive review by Vivanco and Sawyers (Vivanco and Sawyers, 2002). In the following section, we will focus on Akt targets that have been described specifically in the context of modulating sensitivity to TRAIL.

B. NUCLEAR FACTOR-κB

The role of NF-κB activation in protecting cells from TRAIL-induced apoptosis is well established. NF-κB is a downstream target of Akt antiapoptotic signaling as NF-κB activity is stimulated by Akt. There are multiple mechanisms by which Akt activates NF-κB. Two groups initially reported that NF-κB activation by TNF or platelet-derived growth factor involves Akt and that Akt can phosphorylate and activate IKK (IκB kinase), resulting in IκB degradation and nuclear translocation of NF-κB (Ozes *et al.*, 1999; Romashkova and Makarov, 1999). However, other researchers have demonstrated that Akt stimulates NF-κB–dependent gene transcription by potentiating the transactivation function of the p65 subunit of NF-κB independent of nuclear translocation (Madrid *et al.*, 2000; Sizemore *et al.*, 1999). In support of the latter mechanism, Mayo *et al.* (2002) showed that in LNCaP prostate cancer cells with constitutively activated Akt by virtue of PTEN loss, reintroduction of PTEN suppressed Akt activation and inhibited NF-κB–dependent transcription, but did not block IκB degradation and nuclear translocation of NF-κB after TNF treatment. In terms of TRAIL resistance mediated by Akt and NF-κB, IGF-1 was shown to activate NF-κB in an Akt-dependent fashion and Akt-dependent NF-κB activation promoted cell survival after TRAIL treatment (Mitsiades *et al.*, 2002a). Regardless of the precise role of Akt in NF-κB activation, which may be stimulus- and cell type-dependent, it seems likely that Akt contributes to full activation of NF-κB in stimulation of antiapoptotic gene transcription.

C. PED/PEA-15

Another potential mediator downstream of Akt involved in regulating sensitivity to TRAIL is a small protein termed PED/PEA-15 (phosphoprotein enriched in diabetes/phosphoprotein enriched in astrocytes-15kDa). PED/PEA-15 inhibits apoptosis in response to multiple stimuli, including Fas ligand, TNF-α, and TRAIL (Condorelli *et al.*, 1999; Hao *et al.*, 2001). PED/PEA-15 contains death effector domains capable of interacting with death effector domains of FADD and caspase-8 and is recruited to the DISC of death receptors (Condorelli *et al.*, 1999). PED/PEA-15 inhibits association between FADD and caspase-8 and inhibits caspase activation. The ability of PED/PEA-15 to inhibit signaling by death receptors is regulated by phosphorylation by Akt (Trencia *et al.*, 2003). PED/PEA-15 contains an imperfect consensus Akt phosphorylation site at Ser116. Akt phosphorylates PED/PEA-15 at this site *in vitro* and in intact 293 cells. Furthermore, phosphorylation of PED/PEA-15 at this site increases the stability of protein and mutation of this serine residue leads to inhibition of its antiapoptotic function. In the U373MG glioma cells, inhibition of Akt activity by PI 3-kinase inhibitors leads to decreased phosphorylation of PED/PEA-15 and decreased PED/PEA-15 protein levels as well as increased induction of apoptosis by TRAIL. PI-3 kinase inhibitors also decrease the recruitment of PED/PEA-15 to TRAIL-induced DISC, consistent with the notion that the antiapoptotic function of PED/PEA-15 by inhibition of death receptor signaling is dependent on phosphorylation by Akt. There may be additional mechanisms mediating the antiapoptotic function of PED/PEA-15 involving phosphorylation by other kinases such as protein kinase C (Condorelli *et al.*, 1999; Hao *et al.*, 2001) and the relative contributions by these kinases are unknown.

VI. STRATEGIES FOR OVERCOMING TRAIL RESISTANCE

In cell culture model systems, resistance to TRAIL-induced apoptosis by the PI-3 kinase/PTEN/Akt pathway has been overcome by several approaches. The most commonly used strategy is to treat cells with PI-3 kinase inhibitors. Two widely used inhibitors are LY294002 and wortmannin. Although these agents reliably inactivate Akt *in vitro*, there are very few published reports of systemic administration of PI-3 kinase inhibitors in animal models, presumably because of severe side effects in intact animals. Wortmannin is associated with fatal systemic toxicity in rodents (Gunther *et al.*, 1989). The second approach is to express dominant negative mutant Akt protein by transfection or to reconstitute PTEN expression in

PTEN-null cells. Gene therapy is still in its infancy and many obstacles remain for this type of approach to be clinically applicable.

Combined administration of low doses of chemotherapeutic agents or ionizing radiation along with TRAIL is an approach that has been used successfully in animal models and may be translatable to clinic (Chinnaiyan *et al.*, 2000; Gliniak and Le, 1999). While the basis for synergy between TRAIL and chemotherapy or radiation is not entirely clear, the p53-dependent upregulation of the TRAIL-R2/DR5 receptor may contribute in this process.

A more specific approach to enhance efficacy of TRAIL is to inhibit kinase targets mediating TRAIL resistance. The previous sections touched on the role of the Akt kinase, growth factor receptors such as EGFR, and IGF-1 receptor, as well as Erk and other kinases, in mediating resistance to TRAIL. Because activation of these kinases is found commonly in many types of tumors and is thought to contribute to the process of oncogenesis, TRAIL resistance on the basis of activation of these kinases may be widespread and may limit the clinical effectiveness of TRAIL. Therefore, it may be necessary to inhibit these kinases along with TRAIL administration if optimal tumoricidal activity is to be achieved. Kinase inhibitors are rapidly entering the clinical arena with several agents already approved and many more in advanced clinical testing. Two recent publications highlight this approach of combining the inhibitors of EGFR or HER2 with TRAIL for optimizing their cell killing effect. Park and Seol (2002) showed that treating cells with small molecule inhibitors of the EGFR kinase enhanced the cytotoxic effect of TRAIL by inactivating the Akt pathway. Cuello *et al.* (2001) explored the effect of trastuzumab (Herceptin), an antibody against HER2, an EGFR family member, on modulating TRAIL sensitivity in breast cancer cell lines overexpressing HER2. Trastuzumab augmented the sensitivity of TRAIL in these cells and also led to Akt inactivation. The relevance of the Akt pathway in this system was supported by the finding that PI-3 kinase inhibitors sensitized cells to TRAIL and constitutively active Akt inhibited apoptosis induced by trastuzumab and TRAIL. Because trastuzumab is widely used currently to treat a subset of breast cancer patients and gefitinib (Iressa, ZD1839), an oral inhibitor of EGFR, has just received approval for use in lung cancer patients by the U.S. Food and Drug Administration, drugs are becoming available to test the strategy of the combination treatment of TRAIL and an inhibitor of growth factor receptors.

Given the pivotal importance of the PI-3 kinase/PTEN/Akt pathway in many aspects of oncogenesis, including TRAIL resistance, intense efforts are underway to develop inhibitors of PI-3 kinase or Akt with less systemic toxicity and better pharmacokinetic properties that may allow oral administration. These inhibitors may be effective alone or in combination with chemotherapy or radiation therapy or biologic therapy such as TRAIL administration or monoclonal antibodies. Preliminary results in this area have been published or appeared in abstract forms (Jin *et al.*, 2003; Lu *et al.*,

2003; Meuillet *et al.*, 2003). However, one caveat is that the strategy of combining TRAIL and kinase inhibitors for improving the tumoricidal activity of TRAIL is based on the *in vitro* cell culture experiments and has not yet been tested in relevant animal model systems. Therefore, further work will be necessary before we know the role of kinase inhibitors in TRAIL-based therapy.

Inhibition of NF-κB activity is another possible avenue of increasing TRAIL sensitivity. This could be accomplished by a cell-permeable peptide inhibitor of NF-κB or PS-341 proteasome inhibitor (Mitsiades *et al.*, 2002c; Sayers *et al.*, 2003). PS-341 (bortezomib) was recently approved by the FDA for treatment of multiple myeloma. Therefore, NF-κB inhibition is a clinically feasible approach for overcoming TRAIL resistance.

VII. CONCLUSION

Because of the relative lack of systemic toxicity as well as selectivity for tumor cells, TRAIL hold great promise as a novel cancer therapeutic agent. However, resistance to TRAIL in tumor cells by various signaling pathways that promote carcinogenesis may limit its clinical efficacy as a single agent. It is therefore important to understand the mechanistic basis for TRAIL resistance, both for selecting patients likely to respond to TRAIL and for devising strategies to overcome TRAIL resistance. Much evidence points to constitutively active Akt as an important mediator of TRAIL resistance in many tumor types. Constitutively active Akt occurs frequently in human tumors as a result of loss of function of the PTEN tumor suppressor, which negatively regulates the PI-3 kinase/Akt pathway through its lipid phosphatase activity. Growth factor–induced Akt activation also protects cells from TRAIL-induced apoptosis. Akt exerts its antiapoptotic function by its ability to phosphorylate many key components of the cellular apoptotic regulatory circuit, such as BAD, MDM2, FOXO Forkhead transcription factors, and PED/PEA-15 as well as by its role in activating NF-κB. It is interesting that many growth-promoting and antiapoptotic signaling pathways linked to tumorigenesis are also responsible for TRAIL resistance, suggesting that TRAIL resistance may not be an isolated phenomenon in the clinical application of TRAIL. Strategies to inactivate Akt may eventually be combined with TRAIL administration to enhance the clinical efficacy of TRAIL-based therapy for cancer patients.

ACKNOWLEDGMENTS

Y.E.W. was supported by grants from NIH (CA82399) and Department of Defense (DAMD17-00-1-0037). We thank Al Baldwin for helpful discussions about NF-κB. We apologize to many investigators whose significant contributions could not be cited due to space constraints.

REFERENCES

Ali, I. U., Schriml, L. M., and Dean, M. (1999). Mutational spectra of PTEN/MMAC1 gene: A tumor suppressor with lipid phosphatase activity. *J. Natl. Cancer Inst.* **91,** 1922–1932.

Almasan, A., and Ashkenazi, A. (2003). Apo2L/TRAIL: Apoptosis signaling, biology, and potential for cancer therapy. *Cytokine Growth Factor Rev.* **14,** 327–348.

Ashkenazi, A., and Dixit, V. M. (1999). Apoptosis control by death and decoy receptors. *Curr. Opin. Cell Biol.* **11,** 255–260.

Ashkenazi, A., Pai, R. C., Fong, S., Leung, S., Lawrence, D. A., Marsters, S. A., Blackie, C., Chang, L., McMurtrey, A. E., Hebert, A., DeForge, L., Koumenis, I. L., Lewis, D., Harris, L., Bussiere, J., Koeppen, H., Shahrokh, Z., and Schwall, R. H. (1999). Safety and antitumor activity of recombinant soluble Apo2 ligand. *J. Clin. Invest.* **104,** 155–162.

Bortul, R., Tazzari, P. L., Cappellini, A., Tabellini, G., Billi, A. M., Bareggi, R., Manzoli, L., Cocco, L., and Martelli, A. M. (2003). Constitutively active Akt1 protects HL60 leukemia cells from TRAIL-induced apoptosis through a mechanism involving NF-kappaB activation and cFLIP(L) up-regulation. *Leukemia* **17,** 379–389.

Burns, T. F., and El-Deiry, W. S. (2001). Identification of inhibitors of TRAIL-induced death (ITIDs) in the TRAIL-sensitive colon carcinoma cell line SW480 using a genetic approach. *J. Biol. Chem.* **276,** 37879–37886.

Chaudhary, P., Eby, M., Jasmin, A., Bookwaiter, A., Murray, J., and Hood, L. (1997). Death receptor 5, a new member of the TNFR family, and DR4 induce FADD-dependent apoptosis and activate the NF-kappaB pathway. *Immunity* **17,** 821–830.

Chen, X., Thakkar, H., Tyan, F., Gim, S., Robinson, H., Lee, C., Pandey, S. K., Nwokorie, C., Onwudiwe, N., and Srivastava, R. K. (2001). Constitutively active Akt is an important regulator of TRAIL sensitivity in prostate cancer. *Oncogene* **20,** 6073–6083.

Chinnaiyan, A. M., Prasad, U., Shankar, S., Hamstra, D. A., Shanaiah, M., Chenevert, T. L., Ross, B. D., and Rehemtulla, A. (2000). Combined effect of tumor necrosis factor-related apoptosis-inducing ligand and ionizing radiation in breast cancer therapy. *Proc. Natl. Acad. Sci. USA* **97,** 1754–1759.

Condorelli, G., Vigliotta, G., Cafieri, A., Trencia, A., Andalo, P., Oriente, F., Miele, C., Caruso, M., Formisano, P., and Beguinot, F. (1999). PED/PEA-15: An anti-apoptotic molecule that regulates FAS/TNFR1-induced apoptosis. *Oncogene* **18,** 4409–4415.

Cuello, M., Ettenberg, S. A., Clark, A. S., Keane, M. M., Posner, R. H., Nau, M. M., Dennis, P. A., and Lipkowitz, S. (2001). Down-regulation of the erbB-2 receptor by trastuzumab (herceptin) enhances tumor necrosis factor-related apoptosis-inducing ligand-mediated apoptosis in breast and ovarian cancer cell lines that overexpress erbB-2. *Cancer Res.* **61,** 4892–4900.

Datta, S. R., Brunet, A., and Greenberg, M. E. (1999). Cellular survival: A play in three Akts. *Genes Dev.* **13,** 2905–2927.

Degli-Esposti, M., Dougall, W., Smolak, P., Waugh, J., Smith, C., and Goodwin, R. (1997). The novel receptor TRAIL-R4 induces NF-kappaB and protects against TRAIL-mediated apoptosis, yet retains an incomplete death domain. *Immunity* **7,** 813–820.

Degli-Esposti, M., Ferry, G., Masdehors, P., Boutin, J. A., Hickman, J. A., and Dive, C. (2003). Post-translational modification of Bid has differential effects on its susceptibility to cleavage by caspase 8 or caspase 3. *J. Biol. Chem.* **278,** 15749–15757.

Deng, Y., Lin, Y., and Wu, X. (2002). TRAIL-induced apoptosis requires Bax-dependent mitochondrial release of Smac/DIABLO. *Genes Dev.* **16,** 33–45.

Di Cristofano, A., Pesce, B., Cordon-Cardo, C., and Pandolfi, P. P. (1998). Pten is essential for embryonic development and tumour suppression. *Nat. Genet.* **19,** 348–355.

Franco, A. V., Zhang, X. D., Van Berkel, E., Sanders, J. E., Zhang, X. Y., Thomas, W. D., Nguyen, T., and Hersey, P. (2001). The role of NF-kappaB in TNF-related apoptosis-inducing ligand (TRAIL)-induced apoptosis of melanoma cells. *J. Immunol.* **166,** 5337–5345.

Gibson, E. M., Henson, E. S., Haney, N., Villanueva, J., and Gibson, S. B. (2002). Epidermal growth factor protects epithelial-derived cells from tumor necrosis factor-related apoptosis-inducing ligand-induced apoptosis by inhibiting cytochrome c release. *Cancer Res.* **62,** 488–496.

Gliniak, B., and Le, T. (1999). Tumor necrosis factor-related apoptosis-inducing ligand's antitumor activity in vivo is enhanced by the chemotherapeutic agent CPT-11. *Cancer Res.* **59,** 6153–6158.

Gunther, R., Abbas, H. K., and Mirocha, C. J. (1989). Acute pathological effects on rats of orally administered wortmannin-containing preparations and purified wortmannin from Fusarium oxysporum. *Food Chem. Toxicol.* **27,** 173–179.

Hao, C., Beguinot, F., Condorelli, G., Trencia, A., Van Meir, E. G., Yong, V. W., Parney, I. F., Roa, W. H., and Petruk, K. C. (2001). Induction and intracellular regulation of tumor necrosis factor-related apoptosis-inducing ligand (TRAIL) mediated apotosis in human malignant glioma cells. *Cancer Res.* **61,** 1162–1170.

Hersey, P., and Zhang, X. (2001). How melanoma cells evade TRAIL-induced apoptosis. *Nat. Rev. Cancer* **1,** 142–150.

Hu, W.-H., Johnson, H., and Shu, H.-B. (1999). Tumor necrosis factor-related apoptosis-inducing ligand receptors signal NF-kappa B and JNK activation and apoptosis through distinct pathways. *J. Biol. Chem.* **274,** 30603–30610.

Jeremias, I., Kupatt, C., Baumann, B., Herr, I., Wirth, T., and Debatin, K. M. (1998). Inhibition of nuclear factor kappa B activation attenuates apoptosis resistance in lymphoid cells. *Blood* **91,** 4624–4631.

Jin, X., Gossett, D., Wang, S., Reynolds, K., and Lin, J. (2003). Inhibition of AKT survival pathway by a novel AKT-selective inhibitor in human endometrial cancer cells. *Proc. Am. Assoc. Cancer Res.* **44,** abstract R723.

Kandasamy, K., and Srivastava, R. K. (2002). Role of the phosphatidylinositol 3'-kinase/PTEN/Akt kinase pathway in tumor necrosis factor-related apoptosis-inducing ligand-induced apoptosis in non-small cell lung cancer cells. *Cancer Res.* **62,** 4929–4937.

Kim, K., Fisher, M. J., Xu, S.-Q., and El-Deiry, W. S. (2000). Molecular determinants of response to TRAIL in killing of normal and cancer cells. *Clin. Cancer Res.* **6,** 335–346.

Kim, Y., Suh, N., Sporn, M., and Reed, J. C. (2002a). An inducible pathway for degradation of FLIP protein sensitizes tumor cells to TRAIL-induced apoptosis. *J. Biol. Chem.* **277,** 22320–22329.

Kim, Y.-S., Schwabe, R. F., Qian, T., Lemasters, J. J., and Brenner, D. A. (2002b). TRAIL-mediated apoptosis requires NF-kappaB inhibition and the mitochondrial permeability transition in human hepatoma cells. *Hepatology* **36,** 1498–1508.

Kreuz, S., Siegmund, D., Scheurich, P., and Wajant, H. (2001). NF-kappaB inducers upregulate cFLIP, a cycloheximide-sensitive inhibitor of death receptor signaling. *Mol. Cell. Biol.* **21,** 3964–3973.

Leverkus, M., Neumann, M., Mengling, T., Rauch, C. T., Brocker, E.-B., Krammer, P. H., and Walczak, H. (2000). Regulation of tumor necrosis factor-related apoptosis-inducing ligand sensitivity in primary and transformed human keratinocytes. *Cancer Res.* **60,** 553–559.

Li, D. M., and Sun, H. (1998). PTEN/MMAC1/TEP1 suppresses the tumorigenicity and induces G1 cell cycle arrest in human glioblastoma cells. *Proc. Natl. Acad. Sci. USA* **95,** 15406–15411.

Li, J., Yen, C., Liaw, D., Podsypanina, K., Bose, S., Wang, S. I., Puc, J., Miliaresis, C., Rodgers, L., McCombie, R., Bigner, S. H., Giovanella, B. C., Ittmann, M., Tycko, B.,

Hibshoosh, H., Wigler, M. H., and Parsons, R. (1997). PTEN, a putative protein tyrosine phosphatase gene mutated in human brain, breast, and prostate cancer. *Science* **275,** 1943–1947.

Liaw, D., Marsh, D. J., Li, J., Dahia, P. L., Wang, S. I., Zheng, Z., Bose, S., Call, K. M., Tsou, H. C., Peacocke, M., Eng, C., and Parsons, R. (1997). Germline mutations of the PTEN gene in Cowden disease, an inherited breast and thyroid cancer syndrome. *Nat. Genet.* **16,** 64–67.

Lu, Y., Yu, Q., Hall, H., and Mills, G. B. (2003). Inhibition of PI3K-AKT pathway as a therapeutic target for breast cancer. *Proc. Am. Assoc. Cancer Res.* **44,** abstract 1000.

Madrid, L. V., Wang, C. Y., Guttridge, D. C., Schottelius, A. J., Baldwin, A. S., Jr., and Mayo, M. W. (2000). Akt suppresses apoptosis by stimulating the transactivation potential of the RelA/p65 subunit of NF-kappaB. *Mol. Cell. Biol.* **20,** 1626–1638.

Maehama, T., and Dixon, J. E. (1998). The tumor suppressor, PTEN/MMAC1, dephosphorylates the lipid second messenger, phosphatidylinositol 3,4,5-trisphosphate. *J. Biol. Chem.* **273,** 13375–13378.

Mayo, M. W., Madrid, L. V., Westerheide, S. D., Jones, D. R., Yuan, X.-J., Baldwin, A. S., Jr., and Whang, Y. E. (2002). PTEN blocks tumor necrosis factor-induced NF-kappa B-dependent transcription by inhibiting the transactivation potential of the p65 subunit. *J. Biol. Chem.* **277,** 11116–11125.

Meuillet, E. J., Mahadevan, D., Vankayalapati, H., Berggren, M., Williams, R., Coon, A., Kozikowski, A. P., and Powis, G. (2003). Specific inhibition of the Akt1 pleckstrin homology domain by D-3-deoxy-phosphatidyl-myo-inositol analogues. *Mol. Cancer Ther.* **2,** 389–399.

Mitsiades, C. S., Mitsiades, N., Poulaki, V., Schlossman, R., Akiyama, M., Chauhan, D., Hideshima, T., Treon, S., Munshi, N., Richardson, P., and Anderson, K. C. (2002a). Activation of NF-kappaB and upregulation of intracellular anti-apoptotic proteins via the IGF-1/Akt signaling in human multiple myeloma cells: Therapeutic implications. *Oncogene* **21,** 5673–5683.

Mitsiades, N., Mitsiades, C. S., Poulaki, V., Anderson, K. C., and Treon, S. P. (2002b). Intracellular regulation of tumor necrosis factor-related apoptosis-inducing ligand-induced apoptosis in human multiple myeloma cells. *Blood* **99,** 2162–2171.

Mitsiades, N., Mitsiades, C. S., Poulaki, V., Chauhan, D., Richardson, P. G., Hideshima, T., Munshi, N., Treon, S. P., and Anderson, K. C. (2002c). Biologic sequelae of nuclear factor-kappa B blockade in multiple myeloma: Therapeutic applications. *Blood* **99,** 4079–4086.

Mitsiades, C. S., Treon, S. P., Mitsiades, N., Shima, Y., Richardson, P., Schlossman, R., Hideshima, T., and Anderson, K. C. (2001). TRAIL/Apo2L ligand selectively induces apoptosis and overcomes drug resistance in multiple myeloma: Therapeutic applications. *Blood* **98,** 795–804.

Nesterov, A., Lu, X., Johnson, M., Miller, G. J., Ivashchenko, Y., and Kraft, A. S. (2001). Elevated AKT activity protects the prostate cancer cell line LNCaP from TRAIL-induced apoptosis. *J. Biol. Chem.* **276,** 10767–10774.

Orlowski, R. Z., and Baldwin, A. S. (2002). NF-kappaB as a therapeutic target in cancer. *Trends Mol. Med.* **8,** 385–389.

Oya, M., Ohtsubo, M., Takayanagi, A., Tachibana, M., Shimizu, N., and Murai, M. (2001). Constitutive activation of nuclear factor-kappaB prevents TRAIL-induced apoptosis in renal cancer cells. *Oncogene* **20,** 3888–3896.

Ozes, O. N., Mayo, L. D., Gustin, J. A., Pfeffer, S. R., Pfeffer, L. M., and Donner, D. B. (1999). NF-kappaB activation by tumour necrosis factor requires the Akt serine-threonine kinase. *Nature* **401,** 82–85.

Pan, G., Ni, J., Wei, Y.-F., Yu, G.-I., Gentz, R., and Dixit, V. M. (1997). An Antagonist Decoy Receptor and a Death Domain-Containing Receptor for TRAIL. *Science* **277,** 815–818.

Park, S.-Y., and Seol, D.-W. (2002). Regulation of Akt by EGF-R inhibitors, a possible mechanism of EGF-R inhibitor-enhanced TRAIL-induced apoptosis. *Biochem. Biophys. Res. Commun.* **295**, 515–518.

Poulaki, V., Mitsiades, C. S., Kotoula, V., Tseleni-Balafouta, S., Ashkenazi, A., Koutras, D. A., and Mitsiades, N. (2002). Regulation of Apo2L/tumor necrosis factor-related apoptosis-inducing ligand-induced apoptosis in thyroid carcinoma cells. *Am. J. Pathol.* **161**, 643–654.

Ravi, R., and Bedi, A. (2002). Sensitization of tumor cells to Apo2 ligand/TRAIL-induced apoptosis by inhibition of casein kinase II. *Cancer Res.* **62**, 4180–4185.

Romashkova, J. A., and Makarov, S. S. (1999). NF-kappaB is a target of AKT in anti-apoptotic PDGF signalling. *Nature* **401**, 86–90.

Sayers, T. J., Brooks, A. D., Koh, C. Y., Ma, W., Seki, N., Raziuddin, A., Blazar, B. R., Zhang, X., Elliott, P. J., and Murphy, W. J. (2003). The proteasome inhibitor PS-341 sensitizes neoplastic cells to TRAIL-mediated apoptosis by reducing levels of c-FLIP. *Blood* **102**, 303–310.

Schneider, P., Thome, M., Burns, K., Bodmer, J., Hofmann, K., Kataoka, T., Holler, N., and Tschopp, J. (1997). TRAIL receptors 1 (DR4) and 2 (DR5) signal FADD-dependent apoptosis and activate NF-kappaB. *Immunity* **7**, 831–836.

Sheridan, J. P., Marsters, S. A., Pitti, R. M., Gurney, A., Skubatch, M., Baldwin, D., Ramakrishnan, L., Gray, C. L., Baker, K., Wood, W. I., Goddard, A. D., Godowski, P., and Ashkenazi, A. (1997). Control of TRAIL-induced apoptosis by a family of signaling and decoy receptors. *Science* **277**, 818–821.

Siegmund, D., Hadwiger, P., Pfizenmaier, K., Vornlocher, H., and Wajant, H. (2002). Selective inhibition of FLICE-like inhibitory protein expression with small interfering RNA oligonucleotides is sufficient to sensitize tumor cells for TRAIL-induced apoptosis. *Mol. Med.* **2002**, 725–732.

Sizemore, N., Leung, S., and Stark, G. R. (1999). Activation of phosphatidylinositol 3-kinase in response to interleukin-1 leads to phosphorylation and activation of the NF-kappaB p65/RelA subunit. *Mol. Cell. Biol.* **19**, 4798–4805.

Stambolic, V., Suzuki, A., de la Pompa, J. L., Brothers, G. M., Mirtsos, C., Sasaki, T., Ruland, J., Penninger, J. M., Siderovski, D. P., and Mak, T. W. (1998). Negative regulation of PKB/Akt-dependent cell survival by the tumor suppressor PTEN. *Cell* **95**, 29–39.

Steck, P. A., Pershouse, M. A., Jasser, S. A., Yung, W. K., Lin, H., Ligon, A. H., Langford, L. A., Baumgard, M. L., Hattier, T., Davis, T., Frye, C., Hu, R., Swedlund, B., Teng, D. H., and Tavtigian, S. V. (1997). Identification of a candidate tumour suppressor gene, MMAC1, at chromosome 10q23.3 that is mutated in multiple advanced cancers. *Nat. Genet.* **15**, 356–362.

Stocker, H., Andjelkovic, M., Oldham, S., Laffargue, M., Wymann, M. P., Hemmings, B. A., and Hafen, E. (2002). Living with lethal PIP3 levels: Viability of flies lacking PTEN restored by a PH domain mutation in Akt/PKB. *Science* **295**, 2088–2091.

Tran, S. E. F., Holmstrom, T. H., Ahonen, M., Kahari, V.-M., and Eriksson, J. E. (2001). MAPK/ERK overrides the apoptotic signaling from Fas, TNF, and TRAIL receptors. *J. Biol. Chem.* **276**, 16484–16490.

Trauzold, A., Wermann, H., Arlt, A., Schutze, S., Schafer, H., Oestern, S., Roder, C., Underfroren, H., Lampe, E., Heinrich, M., Walczak, H., and Kalthoff, H. (2001). CD95 and TRAIL receptor-mediated activation of protein kinase C and NF-kappaB contributes to apoptosis resistance in ductal pancreatic adenocarcinoma cells. *Oncogene* **20**, 4258–4269.

Trencia, A., Perfetti, A., Cassese, A., Vigliotta, G., Miele, C., Oriente, F., Santopietro, S., Giacco, F., Condorelli, G., Formisano, P., and Beguinot, F. (2003). Protein kinase B/Akt binds and phosphorylates PED/PEA-15, stabilizing its antiapoptotic action. *Mol. Cell. Biol.* **23**, 4511–4521.

Vivanco, I., and Sawyers, C. L. (2002). The phosphatidylinositol 3-kinase AKT pathway in human cancer. *Nat. Rev. Cancer* **2**, 489–501.

Walczak, H., Miller, R., Ariail, K., Gliniak, B., Griffith, T., Kubun, M., Chin, W., Jones, J., Woodward, A., Le, T., Smith, C., Smolak, P., Goodwin, R., Rauch, C., Schuh, J., and Lynch, D. (1999). Tumoricidal activity of tumor necrosis factor-related apoptosis-inducing ligand in vivo. *Nat. Med.* **5,** 157–163.

Wiley, S. R., Schooley, K., Smolak, P. J., Din, W. S., Huang, C. P., Nicholl, J. K., Sutherland, G. R., Smith, T. D., Rauch, C., Smith, C. A., and Goodwin, R. G. (1995). Identification and characterization of a new member of the TNF family that induces apoptosis. *Immunity* **3,** 673–682.

Wu, X., Senechal, K., Neshat, M. S., Whang, Y. E., and Sawyers, C. L. (1998). The PTEN/MMAC1 tumor suppressor phosphatase functions as a negative regulator of the phosphoinositide 3-kinase/Akt pathway. *Proc. Natl. Acad. Sci. USA* **95,** 15587–15591.

Yuan, X. J., and Whang, Y. E. (2002). PTEN sensitizes prostate cancer cells to death receptor-mediated and drug-induced apoptosis through a FADD-dependent pathway. *Oncogene* **21,** 319–327.

Zhang, X., Borrow, J., Zhang, X., Nguyen, T., and Hersey, P. (2003). Activation of ERK1/2 protects melanoma cells from TRAIL-induced apoptosis by inhibiting Smac/DIABLO release from mitochondria. *Oncogene* **22,** 2869–2891.

Zhang, X. D., Franco, A., Myers, K., Gray, C., Nguyen, T., and Hersey, P. (1999). Relation of TNF-related apoptosis-inducing ligand (TRAIL) receptor and FLICE-inhibitory protein expression to TRAIL-induced apoptosis of melanoma. *Cancer Res.* **59,** 2747–2753.

Zhang, X. D., Zhang, X. Y., Gray, C. P., Nguyen, T., and Hersey, P. (2001). Tumor necrosis factor-related apoptosis-inducing ligand-induced apoptosis of human melanoma is regulated by Smac/DIABLO release from mitochondria. *Cancer Res.* **61,** 7339–7348.

22

TRAIL and Malignant Glioma

Christine J. Hawkins

Murdoch Children's Research Institute
Department of Haematology and Oncology, Royal Children's Hospital
Department of Paediatrics, University of Melbourne
Parkville, Victoria 3052, Australia

I. Malignant Glioma
   A. Classification
   B. Treatment
II. Apoptosis in Cancer
   A. Apoptosis in Glioma
   B. Cell Line Models for Glioma Apoptosis Research
III. Apoptosis Pathways Overview
   A. The Intrinsic Pathway
   B. The Extrinsic Pathway
   C. Type I/Type II Death Receptor Pathways
IV. TRAIL
   A. TRAIL Formulations
   B. TRAIL Receptors
V. TRAIL and Glioma
   A. Expression of TRAIL and its Receptors in Glioma
   B. TRAIL Signal Transduction in Sensitive Glioma Cells
   C. Nonapoptotic TRAIL Signaling in Glioma
   D. Synergy Between Chemotherapy and TRAIL
   E. Mechanisms of Resistance

F. Animal Experiments
G. Adenoviral TRAIL
VI. Future Directions
References

Encouragingly, some types of cancer can now be considered treatable, with patients reasonably expecting their disease to be cured. Chemotherapy and radiation therapy are effective against these cancers because they activate the so-called intrinsic apoptosis pathways within the cancer cells. Unfortunately currently available treatments are only effective against a subset of tumor types. In contrast, other cancers, such as malignant glioma, typically do not respond to currently available therapies. Some of this resistance can be attributed to these tumor cells failing to undergo apoptosis upon anticancer treatment. Recently, considerable research attention has focused on triggering apoptosis in chemotherapy- and radiation-therapy–resistant cancer cells via an alternative route—the "extrinsic" pathway, as a means of bypassing this block in apoptosis. Binding of members of the tumor necrosis factor-α (TNF-α) family of death ligands to their receptors on the cell surface triggers this pathway. Death ligands can kill some cancer cells that are resistant to the apoptotic pathway triggered by conventional anticancer treatments. Some death ligands, such as TNF-α and FasL, cause unacceptable toxicity to normal cells and are therefore not suitable anticancer agents. However another death ligand, TNF-related apoptosis-inducing ligand (TRAIL)/Apo-2L, and antibodies that emulate its actions, show greater promise as candidate anticancer drugs because they have negligible effects on normal cells. This review will discuss the ability of TRAIL to induce apoptosis in malignant glioma cells and the potential clinical applications of TRAIL-based agents for glioma treatment. © 2004 Elsevier Inc.

I. MALIGNANT GLIOMA

Gliomas account for approximately half of all malignant brain tumors. The three most frequent classes of glioma (astrocytomas, oligodendrogliomas, ependymomas) are named for their histologic resemblance to normal glial cell types (astrocytes, oligodendroglia, ependymal cells). Gliomas vary considerably in their age distribution, pathologic features, molecular profile, sensitivity to treatment, and prognosis. For example, astrocytomas tend to be resistant to chemotherapy whereas oligodendrogliomas are usually sensitive (Chinot, 2001). These differences may in part reflect underlying differences in the glial subsets from which each tumor type is derived (Nutt *et al.*, 2000), but considerable controversy surrounds the derivation of gliomas. Models for glioma oncogenesis include dedifferentiation of mature cells and oncogenic transformation of progenitor cells (Linskey, 2000; Noble *et al.*, 1995; Recht *et al.*, 2003).

A. CLASSIFICATION

Astrocytomas are the most common gliomas and are subdivided into four clinical grades based on histologic features and invasive behavior. Anaplastic astrocytoma (AA) and glioblastoma multiforme (GBM) are considered high-grade tumors. These cancers are often described as "malignant gliomas," a term which reflects their locally invasive behavior. GBMs and AAs only rarely metastasize to other sites within the body (Park *et al.*, 2000), at least within the patients' short post-diagnosis lifespans (Vertosick and Selker, 1990). Low-grade glioma can recur as high-grade tumors; so-called secondary glioblastomas (reviewed by Kleihues and Ohgaki, 1999).

Assignment of gliomas to distinct clinical grades is based on several histologic criteria, including cellularity, cell morphology, nuclear morphology, mitotic index, presence of necrotic foci, and degree of vasculature proliferation. GBMs are distinctive and their grading tends to be unambiguous, but lower grades are less consistently graded (Prados and Levin, 2000). Better criteria are needed that would permit more conclusive assessment of clinical stage. Certain genetic changes are commonly associated with various grades of gliomas. These characteristic molecular features may be exploited in the future for better grading of gliomas, as well as providing some insight about the biology underlying gliomagenesis. Such molecular alterations include amplification of the epidermal growth factor (EGF) receptor and/or CDK4; loss of heterozygosity of 10q, 1p, and/or 19q; and mutations in PTEN/MMAC, p53, INK4a-ARF, and/or CDKN2A (reviewed by Behin *et al.*, 2003; Rasheed *et al.*, 1999). Astrocytes can be immortalized through engineered overexpression of CDK4 or INK4a-ARF (Holland *et al.*, 1998), suggesting these molecular changes play a casual role in gliomagenesis. Global expression studies have recently been conducted to obtain molecular profiles of gliomas of various clinical stages (Benjamin *et al.*, 2003; Kim *et al.*, 2002a; Mischel *et al.*, 2003; Nutt *et al.*, 2003). Although an emerging methodology, the molecular models derived from these microarray surveys promise more accurate predictions of prognosis than standard histologic methods and may provide important clues to glioma biology.

B. TREATMENT

Surgery is mainly performed to debulk tumors, confirm diagnosis, and relieve symptoms associated with elevated intracranial pressure. Because high-grade gliomas are typically quite invasive, resection probably never removes all tumor cells, so adjuvant therapy is required to eliminate this residual tumor burden. Most studies support the postoperative treatment of high-grade glioma with external beam radiation therapy (Laperriere *et al.*, 2002). Most patients also receive systemic chemotherapy, such as the popular PCV combination (procarbazine, CCNU, and vincristine). A meta-analysis of 12 randomized trials suggested that chemotherapy provided survival benefits that are statistically significant but increases in mean survival were in the order of months, not years (Stewart, 2002). Even with surgery, radiation therapy, and chemotherapy, median survival for malignant glioma patients is still only 1 year after diagnosis (Behin *et al.*, 2003), a statistic that underscores the need for more effective treatments. New drugs being tested include temozolomide, a promising alkylating agent (Dinnes *et al.*, 2002). Other innovative approaches currently under evaluation include exploiting near tumor-specific expression of particular receptors such as those for EGF (Mamot *et al.*, 2003) or IL-13 (Joshi *et al.*, 2000) to target toxic treatments to glioma cells. Interstitial chemotherapy, in which wafers impregnated with chemotherapeutic drugs are placed into the tumor cavity after resection, is also being explored (Olivi *et al.*, 2003). This technique enables delivery of higher doses of anticancer drugs to glioma cells than would be tolerated if administered systemically.

II. APOPTOSIS IN CANCER

Dysregulated apoptosis can impinge on cancer in two general ways, contributing both to cancer development and resistance to treatment (reviewed by Reed, 1999). Defects in apoptotic pathways may confer a survival advantage to precancerous cells that then accumulate additional oncogenic mutations, leading to tumorigenesis. The earliest evidence of this link between cancer development and apoptosis inhibition was the finding that translocation-mediated overexpression of Bcl-2 contributes to follicular lymphoma development (Tsujimoto *et al.*, 1985) and the subsequent realization that Bcl-2 functions to block apoptosis (Vaux *et al.*, 1988). Many other apoptosis pathway defects have since been implicated in development of particular cancers (reviewed by Igney and Krammer, 2002). Such changes include overexpression of prosurvival molecules (including other Bcl-2 family members, inhibitors of apoptosis (IAPs), or death receptor inhibitors) or downregulation of proapoptotic proteins (for example, Apaf-1 or Bax).

One way in which anticancer treatments kill sensitive cells is by triggering apoptosis, so defects in apoptotic pathways may contribute to resistance to treatment. Countless cell line studies have demonstrated that engineered blockages to apoptosis pathways decrease the sensitivity of tumor cells to chemotherapeutic drugs and irradiation but logistical difficulties have, until recently, prevented a similarly direct analysis of this link in primary tumor cells. A recent study of primary human leukemias elegantly demonstrated that tumors that were resistant to anticancer therapies had blocks in apoptotic pathways (Schimmer *et al.*, 2003), verifying the results from the cell line studies.

A. APOPTOSIS IN GLIOMA

Cell death is a common feature of high-grade gliomas, but it is necrotic rather than apoptotic death. Necrosis is usually seen within the center of the tumor and probably results from insufficient nutrients and/or oxygen reaching the cells deep within these fast-growing tumors. Measures of apoptosis within *ex vivo* gliomas tend to suggest that the proportion of apoptotic cells correlates with clinical grade (reviewed by Bogler and Weller, 2002). However, assays typically used only detect a temporal window within the apoptotic process, before rapid engulfment of the dying cell by its neighbors. The small sections of tumor typically analyzed and possible post-resection apoptosis induction further complicate interpretation of these studies (Potten, 1996). Because surgical removal of gliomas precedes treatment, assessment of apoptosis induced by anticancer therapy in gliomas *in vivo* is impractical. Therefore, it is unfortunately necessary to rely on *in vitro* cell line–based studies or animal implantation models for clues as to the likely impact of anticancer therapies on glioma cell apoptosis in patients.

B. CELL LINE MODELS FOR GLIOMA APOPTOSIS RESEARCH

Much of the research into glioma biology, and glioma apoptosis in particular, has used human malignant glioma cell lines (Table I). While very reasonable feasibility considerations have necessitated the use of cell lines rather than fresh *ex vivo* tumor isolates, some important caveats apply to these studies. As with most tumor types, only a minority of glioma cells survive the transition from the *in vivo* environment to life in a tissue culture flask. It is not clear whether the cells that survive the crisis to yield the cell line are representative of the entire population of glioma cells *in vivo* or constitute a minority population especially amenable to *in vitro* conditions. Any selection process for cells that survive and proliferate *in vitro* may have particularly important ramifications for studying apoptosis in the resulting cell line because it is conceivable that subpopulations of cells with apoptotic pathway defects would be over-represented. Once established, only approximately two thirds of human glioma cell lines can form tumors upon intracranial implantation into immune-compromised animals (Bullard *et al.*, 1981; Van Meir *et al.*, 1994) (Table I).

The vast majority of gliomas acquire mesenchymal characteristics when cultured, a process that may be driven by ingredients in animal serum

TABLE I. Characteristics of Human Malignant Glioma Cell Lines

Cell line	Derivation	Tumorigenic?	Sensitivity	Sensitized by cycloheximide?	Receptor expression			
					DR4	DR5	DcR1	DcR2
A172	GBM[c]	No[c]	Sens[j], Int[j,k,l]	Yes[j,l]	No[j], Yes[k]	Yes[j,k]	No[j], Yes[k]	Yes[j], No[k]
D270	GBM[d]	Yes[d]	Int[m]	Yes[m]	No[m]	Yes[m]	Yes[m]	Yes[m]
D54	Mixed AA[e]	Yes[e]	Int[m]	Yes[m]	Yes[m]	Yes[m]	Yes[m]	Yes[m]
D645	GBM[f]	Yes[f]	Sens[m]	N/A[t]	No[m]	Yes[m]	Yes[m]	Yes[m]
LN-18	GBM[g]	Yes[h]	Sens[j,n,o,p], Int[q]	Yes[g]	Yes[j,n,o]	Yes[j,n,o]	Yes[j], No[n,o]	Yes[j,n,o]
LN-71	GBM[g]	No[g]	Sens[n]	N/A[t]	Yes[n]	Yes[n]	No[n]	Yes[n]
LN-215	GBM[g]	No[g]	Res[n,p]	No[n]	No[n]	Yes[n]	No[n]	No[n]
LN-229	GBM[g]	Very[g]	Res[j], Int[n,q,r,s]	Yes[j,n,q,r]	No[j,n]	Yes[j,n]	No[j,n]	Yes[j], No[n]
LN-308	GBM[g]	Very[h]	Res[j,n,o]	No[j,n]	No[j,n,o]	Yes[j,n,o]	No[j,n,o]	Yes[j], No[n,o]
LN-319	AA[g]	Yes[h]	Res[j]	Yes[j]	Yes[j]	Yes[j]	Yes[j]	Yes[j]
LN-428	GBM[g]	No[h]	Res[j]	Yes[j]	Yes[j]	Yes[j]	Yes[j]	Yes[j]
LN-751	GBM[g]	No[g]	Res[n]	No[n]	No[n]	Yes[n]	No[n]	No[n]
T98G	GBM[c]	No[h]	Int[j,n], Sens[k,o,p]	Yes[j,n]	No[j,k,n,o]	Yes[j,k,n,o]	No[j,n], Yes[k,o]	Yes[j,k,o], No[n]
U87MG	GBM[g]	Very[g]	Int[j,m,s], Res[j,n,m]	Yes[j,m], No[n]	No[j,m], Yes[n]	Yes[j,n,m]	No[j,m], Yes[n]	Yes[j,n], No[m]
U118MG[a]	GBM[g]	Yes[g]	Res[l], Int[n,m]	Yes[n,m]	No[n], Yes[m]	Yes[n,m]	No[n,m]	No[n,m], No[m]
U138MG[a]	GBM/AA[c,g]	No[g]	Res[j,n]	No[j], Yes[n]	No[j,n]	Yes[j,n]	Yes[j], No[n]	Yes[j], No[n]
U251MG[b]	GBM[c,g]	Yes[g]	Res[j,m]	Yes[j], No[m]	No[j,m]	Yes[j,m]	No[j], Yes[m]	Yes[j,m]
U343MG	AA[g]	Yes[g]	Sens[h,p]	Not reported	No[n]	Yes[n]	No[n]	No[n]
U373MG[b]	GBM[g]	Yes[g]	Res[j,n,m,s]	No[j,n,m]	No[j,n,m]	Yes[j,n,m]	Yes[j,n], No[n]	Yes[j,n], No[n]

Variation may reflect differences in techniques and reagents between laboratories, including: subline divergence, TRAIL formulations, apoptosis assay methodology, sensitivity and nature of receptor detection techniques (i.e., mRNA, total protein, cell surface protein).

[a] ATCC reports that U118MG and U138MG have identical VNTR and similar STR patterns and karyotype.
[b] The U373MG cell line available from ATCC is similar to U251MG and differs from the original Swedish isolates of U373MG.
[c] ATCC Catalog of Cell Lines and Hybridomas.
[d] Humphrey et al., 1988.

commonly used in tissue culture medium (reviewed by Noble and Mayer-Proschel, 1997). Using astrocyte conditioned medium rather than animal serum, Noble and colleagues have isolated and characterized a human glioma cell line (Hu-O-2A/Gb1), which more closely mimics key characteristics of glioma cells *in vivo* (Noble *et al.*, 1995). However, most glioma cell line studies, including those investigating cell death pathways, have used glioma cell lines established using animal serum.

III. APOPTOSIS PATHWAYS OVERVIEW

A. THE INTRINSIC PATHWAY

Numerous signals can tell a cell it is time to die. These include metabolic derangement within a cell, loss of growth factor signaling, abnormal transition through the cell cycle, viral infection, irradiation, or exposure to chemotherapeutic drugs. Although the upstream steps in molecular pathways triggered by various apoptotic stimuli differ, almost all apoptotic signals ultimately converge downstream, with the activation of a family of proteases termed caspases (see Hengartner, 2000 for a general review of apoptosis pathways). Caspases are translated as inactive precursors and must be activated before they can cleave substrates. The caspase family encompasses initiator and effector members. The upstream signals that promote the activation of the initiator caspases differ between apoptotic stimuli. Stimuli that provoke the "intrinsic" pathway include growth factor deprivation, irradiation, viral infection, and chemotherapeutic drugs. The death signal triggers alterations to mitochondrial membranes, leading to release of proapoptotic molecules including cytochrome *c* from mitochondria into the cytosol. Prosurvival members of the Bcl-2 family, such as Bcl-2 itself

---

[e]Bullard *et al.*, 1981.
[f]Rich *et al.*, 1999.
[g]Ishii *et al.*, 1999.
[h]Van Meir *et al.*, 1994.
[i]Pollack *et al.*, 2001.
[j]Rohn *et al.*, 2001.
[k]Shinohara *et al.*, 2001.
[l]Wu *et al.*, 2000.
[m]Knight *et al.*, 2001.
[n]Hao *et al.*, 2001.
[o]Dorr *et al.*, 2002b.
[p]Xiao *et al.*, 2002.
[q]Rieger *et al.*, 1998.
[r]Fulda *et al.*, 2002a.
[s]Roth *et al.*, 2000.
[t]Too sensitive to discern additional sensitization by cycloheximide.

and Bcl-X_L, can block this step. Cytosolic cytochrome c, in conjunction with ATP/dATP and an adaptor molecule (Apaf-1), promote activation of caspase-9, an initiator caspase. Once activated, caspase-9 cleaves and activates downstream caspases such as caspase-3, leading to the coordinated destruction of the cell via the cleavage of numerous cellular proteins (Fischer *et al.*, 2003). Active caspases can still be prevented from executing the cell by binding to members of the inhibitor of apoptosis protein (IAP) family. In cells expressing sufficient levels of these proteins, apoptotic signaling also requires antagonism of IAPs by molecules such as Smac/DIABLO and Omi/HtrA2, which are released from mitochondria following death stimuli.

B. THE EXTRINSIC PATHWAY

The "extrinsic" apoptotic pathway is triggered by "death ligands," including FasL (CD95L, Apo-1L) and TRAIL. These ligands bind cognate receptors, members of the TNF receptor family, causing their aggregation. The best characterized signaling process is that triggered by FasL. Ligation-induced clustering of the intracellular domains of Fas promotes recruitment of an adaptor molecule (FADD) and the initiator caspases-8 and/or -10. This complex has been termed the death-inducing signaling complex (DISC) (Kischkel *et al.*, 1995). Caspase-8 and -10 can proteolytically process and activate downstream caspases such as caspase-3, killing the cell. Caspase-8 and FADD were recently shown to also be important for apoptotic TNF-α signaling, but were found not to be complexed with the TNFR1 receptor (Harper *et al.*, 2003; Micheau and Tschopp, 2003). After some initial confusion, it now seems that the intracellular components of the FasL signaling pathway (FADD, caspase-8, caspase-10, and downstream caspases) are also important for TRAIL signaling (Bodmer *et al.*, 2000; Kischkel *et al.*, 2000, 2001; Sprick *et al.*, 2000, 2002; Werner *et al.*, 2002). At least three additional adaptors for TRAIL receptors have also been proposed (RIP, TRADD, and DAP3) (Chaudhary *et al.*, 1997; Miyazaki and Reed, 2001; Schneider *et al.*, 1997). The requirement for intrinsic amplification of TRAIL-mediated apoptotic signaling varies, with some cells rendered resistant to TRAIL-induced death by enforced expression of Bcl-2 while others remain sensitive despite overexpression of antiapoptotic Bcl-2 family members.

C. TYPE I/TYPE II DEATH RECEPTOR PATHWAYS

In most normal cell types (designated "type I"), the two pathways outlined in sections IIIA and B converge only at the level of downstream caspases. In these cells, blocking the instrinsic pathway, for example by overexpressing Bcl-2, has no effect on sensitivity to death ligands. However in some cell types, death ligand–induced apoptosis requires the mitochondrial membrane permeability changes associated with the intrinsic pathway

(Scaffidi et al., 1998). These cells are referred to as type II cells. Caspase-8–mediated cleavage of Bid represents the link between the two pathways in type II cells. Cleaved Bid translocates from the cytosol to the mitochondria and promotes release of the proapoptotic factors like cytochrome c and Smac/DIABLO, thus triggering the downstream apoptotic events. In type II cells, high levels of antiapoptotic Bcl-2 family members can block death receptor mediated apoptosis. Cells from a number of cancers such as colon (Deng et al., 2002), prostate gland (Munshi et al., 2001), pancreas (Hinz et al., 2000), lung (Sun et al., 2001), and some breast cancers (de Almodovar et al., 2001) have a type II phenotype. In contrast, the existence of nontransformed type II cells is controversial. Earlier studies designated hepatocytes as type II cells, because the hepatotoxicity associated with introduction of agonistic anti-Fas antibodies was ameliorated by enforced expression of Bcl-2 (Lacronique et al., 1996; Rodriguez et al., 1996). However when FasL was used rather than anti-Fas antibodies, Bcl-2 did not confer protection (Huang et al., 1999; Loo et al., 2003), indicating that like other normal cells, hepatocytes are probably type I.

IV. TRAIL

Relatively little is known of TRAIL's physiologic role, but natural killer cell–mediated antitumor and antimetastatic roles have been proposed (reviewed by Smyth et al., 2003). In most situations, TRAIL is a proapoptotic ligand, but like other death ligands it may also have nonapoptotic roles. TRAIL could promote survival and/or proliferation in cells resistant to TRAIL-induced apoptosis (Chaudhary et al., 1997; Degli-Esposti et al., 1997; Ehrhardt et al., 2003; Schneider et al., 1997) through pathways involving RIP and NF-κB but independent of caspase-8 and FADD (Lin et al., 2000; Muhlenbeck et al., 1998). However, most research attention has focused on the differential ability of TRAIL to induce apoptosis in normal versus transformed cells (Ashkenazi et al., 1999; Walczak et al., 1999), and the implications of these data for the development of TRAIL as an anticancer drug.

Cells expressing TRAIL on their surface can trigger apoptosis in cells they contact (Dorr et al., 2002b; Huang et al., 2003), indicating that membrane bound TRAIL is active. The extracellular portion of TRAIL can be released from the cell membrane, possibly by cysteine protease(s) (Mariani and Krammer, 1998). Unlike FasL, soluble TRAIL is active (Ashkenazi et al., 1999); however, enforced aggregation through crosslinking further enhances its activity (Schneider et al., 1998). Much remains to be established regarding the relative physiologic importance of membrane bound versus soluble endogenous TRAIL. Some evidence indicates that cross-linked TRAIL may signal differently from soluble TRAIL.

Muhlenbeck *et al.* have shown that one of the TRAIL receptors (DR4) could signal NF-κB activation and apoptosis upon binding either cross-linked or non–cross-linked soluble TRAIL, but another receptor (DR5) only responds to cross-linked TRAIL, signaling NF-κB activation, apoptosis, and JNK activation (Muhlenbeck *et al.*, 2000).

A. TRAIL FORMULATIONS

Hymowitz *et al.* solved the structure of soluble TRAIL complexed to the DR5 extracellular domain and found that a single zinc ion was coordinated by three cysteine residues (one from each of the three TRAIL molecules of a trimer) (Hymowitz *et al.*, 2000). Other groups have also published TRAIL structures that did not include metal ions (Cha *et al.*, 2000; Mongkolsapaya *et al.*, 1999), but this may be because the TRAIL used to derive these structures was refolded from inclusion bodies.

Many researchers initially used a histidine-tagged version of the extracellular domain of TRAIL in their apoptosis assays. These preparations were active, specifically killing many cell lines, but were unfortunately also toxic to hepatocytes (Jo *et al.*, 2000) and prostate gland epithelial cells (Nesterov *et al.*, 2002). Cross-linked TRAIL also killed hepatocytes (Ichikawa *et al.*, 2001) and cells from normal brain (Nitsch *et al.*, 2000). The addition of a polyhistidine tag removed the coordinated zinc ions from TRAIL and affected its activity (Lawrence *et al.*, 2001). Somewhat paradoxically, TRAIL lacking zinc was less toxic to tumor cells, but more toxic to hepatocytes than TRAIL trimers complexed with zinc (Lawrence *et al.*, 2001). This may be fortuitous in terms of the development of TRAIL as an anticancer drug, because these data predict that TRAIL complexed to zinc will exhibit considerable anti-tumor activity while minimizing side effects through negligible toxicity to normal cells (Ashkenazi *et al.*, 1999). Agonistic antibodies against TRAIL receptors have been generated and analyzed for suitability as anticancer agents (Ichikawa *et al.*, 2001, 2003; Ohtsuka *et al.*, 2003). At least one of these antibodies (TRA-8) was tumoricidal but not hepatotoxic (Ichikawa *et al.*, 2001) and therefore showed therapeutic promise.

B. TRAIL RECEPTORS

Five human receptors have been identified for TRAIL. Death receptor (DR) 4 (also known as TRAIL-R1) and DR5 (TRAIL-R2, Killer, TRICK2) can signal apoptosis upon TRAIL binding, with DR5 having the higher affinity for TRAIL at physiologic temperatures (Truneh *et al.*, 2000). Two "decoy" receptors (DcR) have been characterized: DcR1 (TRAIL-R3, TRID) and DcR2 (TRAIL-R4, TRUNDD). These receptors lack cytoplasmic domains that signal apoptosis and they bind TRAIL less strongly than the death receptors. While most frequently assumed to function merely by interfering with DR signal transduction, there is some evidence that DcR2 can signal NF-κB activation (Degli-Esposti *et al.*, 1997; Hu *et al.*, 1999). The lowest affinity TRAIL receptor (OPG) is a secreted protein. Overexpression of DcR1 or DcR2 rendered cells resistant to TRAIL-induced apoptosis, and higher expression of DcR1 and 2 was initially reported on cancerous cells compared to normal cells (reviewed by Sheridan *et al.*, 1997). However the majority of analyses have used reverse transcriptase–polymerase chain reaction (RT–PCR) to quantitate receptor expression. It is becoming increasingly clear that mRNA expression of TRAIL receptors is not a reliable indicator of the expression of the proteins on the cell surface (Lacour *et al.*, 2001; Zhang *et al.*, 2000), so the true importance of the decoy receptors in determining sensitivity is still unclear. In addition, without comparisons of the affinities of the reagents (antibodies or primers) used to detect the receptors, comparisons between the levels of each of the receptors on a particular cell population are impossible.

V. TRAIL AND GLIOMA

A. EXPRESSION OF TRAIL AND ITS RECEPTORS IN GLIOMA

The biology of TRAIL in human gliomas is an area of emerging interest. TRAIL expression was detected in human astrocyte cell lines (Rieger *et al.*, 1999) as well as in seven low grade astrocytomas, five AAs, and 11 GBMs (Rieger *et al.*, 1999). This finding contrasted with the lack of expression of TRAIL protein within normal brain cells. Hypertrophic cells observed following physical or chemical injury to the brain, known as reactive astrocytes (Wu and Schwartz, 1998), are the only brain cells in which TRAIL was detected (Dorr *et al.*, 2002a; Nakamura *et al.*, 2000; Rieger *et al.*, 1999). It is intriguing to speculate whether TRAIL upregulation might contribute to malignant transformation of glioma cells. Analogous to an acquired immune privlege model proposed to explain tumor cell expression of FasL (Maher *et al.*, 2002; Walker *et al.*, 1997), tumor cells expressing the TRAIL may kill activated T cells, protecting the tumor from immune destruction. Although controversial (Simon *et al.*, 2001), there is some evidence that TRAIL may play a role in activation-induced T cell death (Martinez-Lorenzo *et al.*, 1998). If this involvement is verified, TRAIL expression on tumor cells may reflect immune-counterattack.

Our knowledge of TRAIL receptor expression in *ex vivo* GBMs is based on RT-PCR analysis of a limited number of samples. As discussed previously, regulation of transcription is only one factor that determines the cell surface expression of TRAIL receptors so only limited conclusions can therefore be drawn from such investigations. The GBMs examined expressed DR4, DR5, and DcR1 but undetectable DcR2 mRNA (Frank et al., 1999). Many more studies have examined receptor levels in glioma cell lines, but it is impossible to know whether the levels of receptors detected in the cell lines mimic expression patterns *in vivo*. Most glioma cell lines express DR5 and some DR4, with fewer expressing detectable DcR1 and/or DcR2 (Dorr et al., 2002b; Hao et al., 2001; Knight et al., 2001; Rieger et al., 1998). Interestingly, freshly isolated melanoma cells upregulated their TRAIL receptor expression (chiefly DR5) and became more sensitive to TRAIL-induced apoptosis with increasing time in culture (Nguyen et al., 2001). Whether gliomas similarly change their sensitivity to TRAIL during *in vitro* culture is currently unknown—if they do, this would have major ramifications for predicting *in vivo* outcomes from cell line studies.

The data summarized in this section indicate that many glioma cell lines express both TRAIL death receptors and TRAIL. At first glance such a combination would seem incompatible with survival because the TRAIL expressed on one glioma cell would be expected to kill either itself or its neighbors by binding to DR4 and/or DR5. It is far from clear why this does not occur, especially because some glioma cell lines expressing both TRAIL and death receptors, including U343MG (Frank et al., 1999) and LN-18 (Rieger et al., 1999), are responsive to exogenous TRAIL (see Table I). A model of "intracellular/intragolgi engagement" has been proposed (Lee et al., 2002) which may explain this paradox. This model hypothesizes that the simultaneous translation of TRAIL and receptors permits their interaction before trafficking to the cell surface, interfering with downstream signaling.

B. TRAIL SIGNAL TRANSDUCTION IN SENSITIVE GLIOMA CELLS

Astrocytes, the likely nonneoplastic cell precursor of malignant glioma cells, are resistant to apoptosis induced by nonmultimerized TRAIL (Hao et al., 2001; Pollack et al., 2001), although sensitive to trimerized TRAIL (Walczak et al., 1999). Considerable heterogeneity has been found in glioma cell lines with respect to TRAIL sensitivity, as illustrated in Table I. TRAIL signaling in sensitive glioma cells appears to involve the same components identified in lymphoid cells. TRAIL binding to receptors DR4 or DR5 resulted in FADD, caspase-8, and caspase-10 recruitment to the cytoplasmic portion of the receptor(s) in sensitive glioma cells (Xiao et al., 2002). TRAIL signaling eventually resulted in activation of downstream caspases such as caspases-3 and -7, as evidenced by DEVDase activity in lysates of TRAIL-treated sensitive cells (Knight et al., 2001; Shinohara et al., 2001).

Inhibitors of candidate pathway components have been introduced into glioma cells to verify the obligate involvement of those proteins in TRAIL signaling. Enforced expression of a dominant negative FADD inhibited TRAIL-induced death in two glioma cell lines (Knight et al., 2001). Inhibition of caspase-8 with z-IETD-fmk halved the apoptosis seen in another line after TRAIL exposure and also decreased caspase-3 processing (Xiao et al., 2002). The remaining apoptosis signaling in those cells probably involved caspase-10 because addition of both z-IETD-fmk and z-AEVD-fmk abolished cleavage of caspases-3, -8, and -10 and blocked apoptosis (Xiao et al., 2002). Expression of the caspase-8 inhibitor CrmA (Knight et al., 2001; Rieger et al., 1998; Rohn et al., 2001) also protected glioma cells from TRAIL-mediated apoptosis, confirming the importance of caspase-8 in this pathway. The requisite involvement of downstream caspases was indicated, as glioma cells treated with a pan-caspase inhibitor (zVAD-fmk) or an inhibitor relatively specific for caspases-3 and -7 (z-DEVD-fmk) were somewhat protected from TRAIL-induced apoptosis (Hao et al., 2001; Pollack et al., 2001). Enforced overexpression of the broad-spectrum caspase inhibitor p35 had a similar protective effect (our unpublished results).

Glioma cells do not fall uniformly into either a type I or type II phenotype with respect to death receptor signaling. In at least one sensitive glioma cell line (D645), caspase-8 activation in the absence of mitochondrial involvement was sufficient for activation of downstream caspases and hence apoptosis (Knight et al., 2001). This is characteristic of a type I phenotype, as discussed previously. Although TRAIL treatment triggered mitochondrial release of proapoptotic molecules (Knight et al., in press), this was evidently not necessary for apoptosis to ensue because these mitochondrial changes could be suppressed by overexpression of Bcl-2 without inhibiting apoptosis. Other glioma lines such as D270, U87MG, LN-229, and LN-18 acted as type II cells with respect to TRAIL signaling (Fulda et al., 2002a,b; Knight et al., in press). Mitochondrial amplification of TRAIL signaling was required for apoptosis in these cells because overexpression of Bcl-2 protected them from TRAIL-induced death.

The primary distinguishing molecular features between type I and type II glioma cell lines are currently being investigated. Low caspase-8 levels influenced the necessity for mitochondrial amplification of TRAIL-mediated apoptotic signals in glioma cells (Knight et al., in press). The crucial factor(s) that must be released from the mitochondria for type II cells to undergo apoptosis in response to TRAIL treatment is also an area of intense investigation. Prime candidates for such factors would be the Apaf-1 cofactor cytochrome c or IAP antagonists (Smac/DIABLO and/or HtrA2). Caspase-9 activity was shown to be unnecessary for TRAIL-induced death of other type II cells (Deng et al., 2002), arguing against a vital role for cytochrome c in this process. In contrast, engineered neutralization of IAPs by enforced expression of cytosolic Smac/DIABLO sensitized many tumor cell lines to TRAIL (Deng et al., 2002; Kandasamy et al., 2003; Leverkus et al., 2003; Ng and Bonavida, 2002; Srinivasula et al., 2000), including glioma cell lines (Fulda et al., 2002b; Knight et al., in press). These data suggest that inhibition of caspases by IAP limits TRAIL-induced apoptosis. IAP antagonists such as Smac/DIABLO that are released from mitochondria following TRAIL exposure probably are important to relieve the IAP inhibition of caspases, hence allowing apoptosis to proceed.

1. TRAIL Killing of Glioma Cells by T Cells

Because T cells upregulated TRAIL after activation (Wendling et al., 2000), Dorr et al. assessed the ability of T cells to kill glioma cells in a TRAIL-dependent manner. The TRAIL-sensitive line LN-18 was efficiently killed by coculture with activated T cells (Dorr et al., 2002b). Using blocking reagents for TRAIL, FasL, or TNF-α, Dorr et al. determined that nine T cell lines of the 17 studied predominantly used a TRAIL-dependent killing mechanism. Addition of a semipermeable membrane ruled out the possibility that soluble TRAIL was responsible for the T cell killing in this context.

C. NONAPOPTOTIC TRAIL SIGNALING IN GLIOMA

As alluded to in section IV, TRAIL has been shown in nonglioma cell types to induce cellular responses other than apoptosis. TRAIL promoted IL-18 expression in two glioma lines (CRT-MG and U87MG) (Choi et al., 2002). Pathway analysis was undertaken using inhibitors of particular components. Global caspase inhibition or inhibition of caspase-1 and -8 prevented both apoptosis and IL-18 production, whereas inhibition of downstream caspases (using z-DEVD-fmk) only affected apoptosis. This study raises the intriguing question of what possible advantage glioma cells may derive *in vivo* from TRAIL induction of IL-18 expression. Although IL-18 is a neutrophil attractant, this would probably not be advantageous to the tumor and neutrophils are not commonly detected in gliomas (Van Meir et al., 1992). However, IL-18 is also angiogenic (Belperio et al., 2000), and tumor cells could conceivably derive significant advantage from this activity.

D. SYNERGY BETWEEN CHEMOTHERAPY AND TRAIL

TRAIL-induced apoptosis of glioma cells *in vitro* and *in vivo* could be substantially enhanced by treatment with chemotherapeutic drugs (Arizono et al., 2003; Nagane et al., 2000; Rohn et al., 2001; Shinohara et al., 2001). The lines that were sensitized to TRAIL by cycloheximide treatment tended to respond synergistically to chemotherapeutic drugs (Nagane et al., 2000; Rohn et al., 2001), but the mechanism underlying the synergy is still incompletely understood. DR5 is a p53-responsive gene (Wu *et al.*, 1997) and was upregulated in glioma cells treated with chemotherapy. Furthermore, enforced expression of a dominant negative p53 mutant modestly decreased the TRAIL sensitization by DNA damage (Arizono *et al.*, 2003). This would seem to imply that in cells with chemotherapy-mediated DNA damage, p53 induces upregulation of DR5, thus sensitizing the cells to TRAIL-induced apoptosis. However p53 status did not predict TRAIL sensitivity (Pollack *et al.*, 2001; Rieger *et al.*, 1998) and accumulating evidence points to the involvement of p53-independent mechanisms in sensitization to TRAIL by DNA damage. For example, some glioma lines that upregulated DR5 following DNA damage bore p53 mutations (Meng and El-Deiry, 2001; Nagane *et al.*, 2000) and engineered changes in p53 status did not affect the ability of chemotherapy drugs to sensitize glioma cell lines to TRAIL (Nagane *et al.*, 2000). Interestingly, p53-dependent induction of DR5 only occurred in cells where p53 caused apoptosis, not in cells where it promoted cell cycle arrest (Wu *et al.*, 1999a). Sensitization to TRAIL-induced apoptosis by chemotherapy has important therapeutic implications. Hopefully additional research will clarify the mechanism(s) accounting for this synergy.

E. MECHANISMS OF RESISTANCE

Inhibition, mutation, or absence of any component of the TRAIL signaling pathway could theoretically confer resistance to TRAIL induced apoptosis in glioma. Many resistant glioma cell lines can be sensitized by treatment with cycloheximide (which inhibits protein translation), implicating a labile inhibitor of TRAIL signaling in the resistance of those cells. Although differential expression levels of the "death" and "decoy" receptors were proposed to account for differences in sensitivity to TRAIL, more recent data suggest that this is unlikely to be a general mechanism through which sensitivity is determined, at least in glioma cells (Hao *et al.*, 2001; Knight *et al.*, 2001; Shinohara *et al.*, 2001). Mutations in DR4 and DR5 have been implicated in some breast, lung, and head and neck cancers (Fisher *et al.*, 2001; Shin *et al.*, 2001), but such mutations have not been reported to date in glioma. Variation in caspase-8 levels, achieved by gene deletion or methylation, have been associated with resistance to death ligands in other cancers, chiefly neuroblastoma (reviewed by Teitz *et al.*, 2001). Low caspase-8 levels appear to contribute to the resistance of one glioma cell line because artificial elevation of caspase-8 expression sensitized these cells to TRAIL (Knight *et al.*, 2001). The mechanism underlying the downregulation of caspase-8 in this line (U373MG) is not known. Modest comparative upregulation of an isoform of protein kinase C (ϵ) was implicated in the resistance of another resistant line, YKG-1 (Shinohara *et al.*, 2001).

Proteins that could interfere with DISC formation have been associated with resistance to death ligands. Enforced overexpression of cFLIP (a catalytically inactive relative of caspase-8) protected glioma cells from TRAIL-induced apoptosis (our unpublished results) and high endogenous cFLIP levels have been linked to resistance in other cell types (Bortul et al., 2003; Griffith et al., 1998; Hietakangas et al., 2003; Kim et al., 2002b; Leverkus et al., 2000; Siegmund et al., 2002). However alterations in cFLIP levels did not appear to represent a major resistance mechanism in glioma cell lines (Hao et al., 2001; Xiao et al., 2002). cFLIP was detected in the TRAIL DISC in glioma cells (Xiao et al., 2002), and its processing differed between sensitive and resistant lines, but this may reflect differential cFLIP cleavage by caspase-3 and/or -8, which were activated in the sensitive lines but not in the resistant ones (Xiao et al., 2002). PED/PEA-15 is another candidate inhibitor, which can bind to both FADD and caspase-8, potentially interfering with caspase-8 activation (Condorelli et al., 1999). In a small survey of glioma cell lines, higher levels of PEA-15 protein correlated with resistance to TRAIL-induced apoptosis (Hao et al., 2001), and its phosphorylation status also varied with sensitivity (Xiao et al., 2002). The involvement of doubly phosphorylated PEA-15 in the DISC was proposed to account for the resistance of the LN-215 cell line (Xiao et al., 2002).

Glioma cell lines vary considerably in their sensitivity to death ligands like TRAIL, as summarized in Table I. Perhaps it is not suprising, therefore, that the data published to date suggest heterogeneity in resistance mechanisms also, complicating the process of discerning the nature of signaling blockages in resistant gliomas.

F. ANIMAL EXPERIMENTS

No efficient means of triggering gliomagenesis in experimental animals exists, so animal models of glioma for use in apoptosis research necessarily involve intracranial implantation of glioma cell lines. Because TRAIL may be involved in immune system functioning, an immune-competent model would be preferable for these experiments. Immunocompetent rodent glioma models have been developed in which transplantable cell lines were derived from spontaneously arising or chemically induced tumors in mouse or rat strains (Barth, 1998; Serano et al., 1980). However, TRAIL signaling in murine cells may differ from that in human cells because only one TRAIL receptor has been identified in mice (Wu et al., 1999b). The *in vivo* anticancer properties of TRAIL have so far been evaluated using a single human glioma cell line (U87MG) implanted into nude mice. U87MG expresses wildtype p53 and has low to intermediate sensitivity to TRAIL *in vitro*.

TRAIL is an extremely effective antitumor agent in the U87MG/nude mouse model. Intratumor (Roth et al., 1999) or intraperitoneal (Pollack *et al.*, 2001) administration of TRAIL cured mice of intracranially implanted gliomas. Other researchers found that TRAIL alone was insufficient to cure animals implanted with U87MG cells (Nagane *et al.*, 2000). However, significant synergistic effects were observed when TRAIL was administered together with the chemotherapy drug cisplatin. This synergy occurred even with established tumors, replicating the *in vitro* results discussed previously. Peptides corresponding to the amino terminus of the IAP antagonist Smac/DIABLO mimicked the full-length protein in potentiating TRAIL killing *in vitro* (Fulda *et al.*, 2002b). Administration of these peptides to athymic mice implanted with U87MG cells also enhanced the antitumor effects of TRAIL *in vivo* (Fulda *et al.*, 2002b).

G. ADENOVIRAL TRAIL

Adenoviral administration represents one potential method for administering TRAIL clinically because adenoviruses infect both normal glial and neuronal cells *in vivo* (Dewey *et al.*, 1999). Lee *et al.* transduced adenoviruses expressing full-length human TRAIL into U87MG and T98 cells *in vitro*, killing 40% and 20% of cells, respectively (Lee *et al.*, 2002). Primary endothelial cells, fibroblasts, and normal human astrocytes could also be infected but were not killed. Infected cells (fibroblasts, differentiated neurons, or surviving infected glioma cells) could kill a proportion of cocultured glioma cells *in vitro*, but conditioned media from the infected cells was not toxic. In an established tumor model, U87MG cells were implanted intracranially and 5 days later infected by intratumor injection of adenoviral TRAIL. The animals were not cured but mice treated with the adenoviral TRAIL survived significantly longer than controls (Lee *et al.*, 2002). In direct contrast, however, Naumann *et al.* found that infection with a similar adenovirus induced some degree of apoptosis in the very TRAIL-sensitive line LN-18 but not in less sensitive lines including U87MG (Naumann *et al.*, 2003). Indeed, in this study, infection with the TRAIL adenovirus protected cells from death induced by soluble TRAIL. While such contradictory results do not inspire confidence in this therapeutic approach, it would be premature to discard adenoviral-based TRAIL delivery approaches until the reasons for these differences are more clear. Both studies used U87MG cells and expressed full-length human TRAIL from a CMV promoter. Interestingly, the cells infected by Naumann *et al.* secreted soluble ligand, unlike the cells infected by Lee *et al.* It is not clear why the secretion of soluble TRAIL would differ in the two experiments, but perhaps soluble TRAIL generated in this context was antagonistic, interfering with the ability of membrane-bound TRAIL to induce apoptosis.

VI. FUTURE DIRECTIONS

TRAIL is an exciting reagent for the treatment of malignant glioma, a disease that is currently incurable. Many glioma cell lines are efficiently killed by TRAIL *in vitro* and responses to TRAIL are often enhanced by treatment with chemotherapeutic drugs or an IAP antagonist. Promising antiglioma effects have also been reported *in vivo*, using nude mice intracranially implanted with a glioma cell line. The synergy observed *in vitro* between TRAIL and either chemotherapeutic drugs or IAP antagonism can be recapitulated in these animal glioma models.

Although these findings argue for the development of TRAIL as therapeutic agent for malignant glioma, several caveats apply to much of the research done to date. First, the vast majority of research into glioma responses to TRAIL has used glioma cell lines. Somewhat disturbingly, melanoma sensitivity to TRAIL increased following cell culture. If a similar phenomenon occurs in glioma, the cell line data could substantially overestimate the likely ability of TRAIL to kill glioma cells in patients. Second, while the murine glioma transplant model has yielded very encouraging results, these experiments have just used one cell line (U87MG), and should be expanded to encompass analyses of multiple glioma cell lines.

REFERENCES

Arizono, Y., Yoshikawa, H., Naganuma, H., Hamada, Y., Nakajima, Y., and Tasaka, K. (2003). A mechanism of resistance to TRAIL/Apo2L-induced apoptosis of newly established glioma cell line and sensitization to TRAIL by genotoxic agents. *Br. J. Cancer* **88**, 298–306.

Ashkenazi, A., Pai, R. C., Fong, S., Leung, S., Lawrence, D. A., Marsters, S. A., Blackie, C., Chang, L., McMurtrey, A. E., Hebert, A., DeForge, L., Koumenis, I. L., Lewis, D., Harris, L., Bussiere, J., Koeppen, H., Shahrokh, Z., and Schwall, R. H. (1999). Safety and antitumor activity of recombinant soluble Apo2 ligand. *J. Clin. Invest.* **104**, 155–162.

Barth, R. F. (1998). Rat brain tumor models in experimental neuro-oncology: The 9L, C6, T9, F98, RG2 (D74), RT-2 and CNS-1 gliomas. *J. Neurooncol.* **36**, 91–102.

Behin, A., Hoang-Xuan, K., Carpentier, A. F., and Delattre, J. Y. (2003). Primary brain tumours in adults. *Lancet* **361**, 323–331.

Belperio, J. A., Keane, M. P., Arenberg, D. A., Addison, C. L., Ehlert, J. E., Burdick, M. D., and Strieter, R. M. (2000). CXC chemokines in angiogenesis. *J. Leukoc. Biol.* **68**, 1–8.

Benjamin, R., Capparella, J., and Brown, A. (2003). Classification of glioblastoma multiforme in adults by molecular genetics. *Cancer J.* **9**, 82–90.

Bodmer, J. L., Holler, N., Reynard, S., Vinciguerra, P., Schneider, P., Juo, P., Blenis, J., and Tschopp, J. (2000). TRAIL receptor-2 signals apoptosis through FADD and caspase-8. *Nat. Cell. Biol.* **2**, 241–243.

Bogler, O., and Weller, M. (2002). Apoptosis in gliomas, and its role in their current and future treatment. *Front. Biosci.* **7**, E339–E353.

Bortul, R., Tazzari, P. L., Cappellini, A., Tabellini, G., Billi, A. M., Bareggi, R., Manzoli, L., Cocco, L., and Martelli, A. M. (2003). Constitutively active Akt1 protects HL60 leukemia cells from TRAIL-induced apoptosis through a mechanism involving NF-kappaB activation and cFLIP(L) up-regulation. *Leukemia* **17,** 379–389.

Bullard, D. E., Schold, S. C., Jr., Bigner, S. H., and Bigner, D. D. (1981). Growth and chemotherapeutic response in athymic mice of tumors arising from human glioma-derived cell lines. *J. Neuropathol. Exp. Neurol.* **40,** 410–427.

Cha, S. S., Sung, B. J., Kim, Y. A., Song, Y. L., Kim, H. J., Kim, S., Lee, M. S., and Oh, B. H. (2000). Crystal structure of TRAIL-DR5 complex identifies a critical role of the unique frame insertion in conferring recognition specificity. *J. Biol. Chem.* **275,** 31171–31177.

Chaudhary, P. M., Eby, M., Jasmin, A., Bookwalter, A., Murray, J., and Hood, L. (1997). Death receptor 5, a new member of the TNFR family, and DR4 induce FADD-dependent apoptosis and activate the NF-kappaB pathway. *Immunity* **7,** 821–830.

Chinot, O. (2001). Chemotherapy for the treatment of oligodendroglial tumors. *Semin. Oncol.* **28,** 13–18.

Choi, C., Kutsch, O., Park, J., Zhou, T., Seol, D. W., and Benveniste, E. N. (2002). Tumor necrosis factor-related apoptosis-inducing ligand induces caspase-dependent interleukin-8 expression and apoptosis in human astroglioma cells. *Mol. Cell. Biol.* **22,** 724–736.

Condorelli, G., Vigliotta, G., Cafieri, A., Trencia, A., Andalo, P., Oriente, F., Miele, C., Caruso, M., Formisano, P., and Beguinot, F. (1999). PED/PEA-15: An anti-apoptotic molecule that regulates FAS/TNFR1-induced apoptosis. *Oncogene* **18,** 4409–4415.

de Almodovar, C. R., Ruiz-Ruiz, C., Munoz-Pinedo, C., Robledo, G., and Lopez-Rivas, A. (2001). The differential sensitivity of Bcl-2-overexpressing human breast tumor cells to TRAIL or doxorubicin-induced apoptosis is dependent on Bcl-2 protein levels. *Oncogene* **20,** 7128–7133.

Degli-Esposti, M. A., Dougall, W. C., Smolak, P. J., Waugh, J. Y., Smith, C. A., and Goodwin, R. G. (1997). The novel receptor TRAIL-R4 induces NF-kappaB and protects against TRAIL-mediated apoptosis, yet retains an incomplete death domain. *Immunity* **7,** 813–820.

Deng, Y., Lin, Y., and Wu, X. (2002). TRAIL-induced apoptosis requires Bax-dependent mitochondrial release of Smac/DIABLO. *Genes Dev.* **16,** 33–45.

Dewey, R. A., Morrissey, G., Cowsill, C. M., Stone, D., Bolognani, F., Dodd, N. J., Southgate, T. D., Klatzmann, D., Lassmann, H., Castro, M. G., and Lowenstein, P. R. (1999). Chronic brain inflammation and persistent herpes simplex virus 1 thymidine kinase expression in survivors of syngeneic glioma treated by adenovirus-mediated gene therapy: Implications for clinical trials. *Nat. Med.* **5,** 1256–1263.

Dinnes, J., Cave, C., Huang, S., and Milne, R. (2002). A rapid and systematic review of the effectiveness of temozolomide for the treatment of recurrent malignant glioma. *Br. J. Cancer* **86,** 501–505.

Dorr, J., Bechmann, I., Waiczies, S., Aktas, O., Walczak, H., Krammer, P. H., Nitsch, R., and Zipp, F. (2002a). Lack of tumor necrosis factor-related apoptosis-inducing ligand but presence of its receptors in the human brain. *J. Neurosci.* **22,** RC209.

Dorr, J., Waiczies, S., Wendling, U., Seeger, B., and Zipp, F. (2002b). Induction of TRAIL-mediated glioma cell death by human T cells. *J. Neuroimmunol.* **122,** 117–124.

Ehrhardt, H., Fulda, S., Schmid, I., Hiscott, J., Debatin, K. M., and Jeremias, I. (2003). TRAIL induced survival and proliferation in cancer cells resistant towards TRAIL-induced apoptosis mediated by NF-kappaB. *Oncogene* **22,** 3842–3852.

Fischer, U., Janicke, R. U., and Schulze-Osthoff, K. (2003). Many cuts to ruin: A comprehensive update of caspase substrates. *Cell Death Differ.* **10,** 76–100.

Fisher, M. J., Virmani, A. K., Wu, L., Aplenc, R., Harper, J. C., Powell, S. M., Rebbeck, T. R., Sidransky, D., Gazdar, A. F., and El-Deiry, W. S. (2001). Nucleotide substitution in the ectodomain of trail receptor DR4 is associated with lung cancer and head and neck cancer. *Clin. Cancer Res.* **7,** 1688–1697.

Frank, S., Kohler, U., Schackert, G., and Schackert, H. K. (1999). Expression of TRAIL and its receptors in human brain tumors. *Biochem. Biophys. Res. Commun.* **257**, 454–459.

Fulda, S., Meyer, E., and Debatin, K. M. (2002a). Inhibition of TRAIL-induced apoptosis by Bcl-2 overexpression. *Oncogene* **21**, 2283–2294.

Fulda, S., Wick, W., Weller, M., and Debatin, K. M. (2002b). Smac agonists sensitize for Apo2L/TRAIL- or anticancer drug-induced apoptosis and induce regression of malignant glioma in vivo. *Nat. Med.* **8**, 808–815.

Griffith, T. S., Chin, W. A., Jackson, G. C., Lynch, D. H., and Kubin, M. Z. (1998). Intracellular regulation of TRAIL-induced apoptosis in human melanoma cells. *J. Immunol.* **161**, 2833–2840.

Hao, C., Beguinot, F., Condorelli, G., Trencia, A., Van Meir, E. G., Yong, V. W., Parney, I. F., Roa, W. H., and Petruk, K. C. (2001). Induction and intracellular regulation of tumor necrosis factor-related apoptosis-inducing ligand (TRAIL) mediated apoptosis in human malignant glioma cells. *Cancer Res.* **61**, 1162–1170.

Harper, N., Hughes, M., MacFarlane, M., and Cohen, G. M. (2003). Fas-associated death domain protein and caspase-8 are not recruited to the tumor necrosis factor receptor 1 signaling complex during tumor necrosis factor-induced apoptosis. *J. Biol. Chem.* **278**, 25534–25541.

Hengartner, M. O. (2000). The biochemistry of apoptosis. *Nature* **407**, 770–776.

Hietakangas, V., Poukkula, M., Heiskanen, K. M., Karvinen, J. T., Sistonen, L., and Eriksson, J. E. (2003). Erythroid differentiation sensitizes K562 leukemia cells to TRAIL-induced apoptosis by downregulation of c-FLIP. *Mol. Cell. Biol.* **23**, 1278–1291.

Hinz, S., Trauzold, A., Boenicke, L., Sandberg, C., Beckmann, S., Bayer, E., Walczak, H., Kalthoff, H., and Ungefroren, H. (2000). Bcl-XL protects pancreatic adenocarcinoma cells against CD95- and TRAIL-receptor-mediated apoptosis. *Oncogene* **19**, 5477–5486.

Holland, E. C., Hively, W. P., DePinho, R. A., and Varmus, H. E. (1998). A constitutively active epidermal growth factor receptor cooperates with disruption of G1 cell-cycle arrest pathways to induce glioma-like lesions in mice. *Genes Dev.* **12**, 3675–3685.

Hu, W. H., Johnson, H., and Shu, H. B. (1999). Tumor necrosis factor-related apoptosis-inducing ligand receptors signal NF-kappaB and JNK activation and apoptosis through distinct pathways. *J. Biol. Chem.* **274**, 30603–30610.

Huang, D., Hahne, M., Schroeter, M., Frei, K., Fontana, A., Villunger, A., Newton, K., Tschopp, J., and Strasser, A. (1999). Activation of Fas by FasL induces apoptosis by a mechanism that cannot be blocked by Bcl-2 or Bcl-XL. *Proc. Natl. Acad. Sci. USA* **96**, 14871–14876.

Huang, X., Lin, T., Gu, J., Zhang, L., Roth, J. A., Liu, J., and Fang, B. (2003). Cell to cell contact required for bystander effect of the TNF-related apoptosis-inducing ligand (TRAIL) gene. *Int. J. Oncol.* **22**, 1241–1245.

Humphrey, P. A., Wong, A. J., Vogelstein, B., Friedman, H. S., Werner, M. H., Bigner, D. D., and Bigner, S. H. (1988). Amplification and expression of the epidermal growth factor receptor gene in human glioma xenografts. *Cancer Res.* **48**, 2231–2238.

Hymowitz, S. G., O'Connell, M. P., Ultsch, M. H., Hurst, A., Totpal, K., Ashkenazi, A., de Vos, A. M., and Kelley, R. F. (2000). A unique zinc-binding site revealed by a high-resolution X-ray structure of homotrimeric Apo2L/TRAIL. *Biochemistry* **39**, 633–640.

Ichikawa, K., Liu, W., Zhao, L., Wang, Z., Liu, D., Ohtsuka, T., Zhang, H., Mountz, J. D., Koopman, W. J., Kimberly, R. P., and Zhou, T. (2001). Tumoricidal activity of a novel anti-human DR5 monoclonal antibody without hepatocyte cytotoxicity. *Nat. Med.* **7**, 954–960.

Ichikawa, K., Liu, W., Fleck, M., Zhang, H., Zhao, L., Ohtsuka, T., Wang, Z., Liu, D., Mountz, J. D., Ohtsuki, M., Koopman, W. J., Kimberly, R., and Zhou, T. (2003). TRAIL-R2 (DR5) mediates apoptosis of synovial fibroblasts in rheumatoid arthritis. *J. Immunol.* **171**, 1061–1069.

Igney, F. H., and Krammer, P. H. (2002). Death and anti-death: Tumour resistance to apoptosis. *Nat. Rev. Cancer* **2**, 277–288.

Ishii, N., Maier, D., Merlo, A., Tada, M., Sawamura, Y., Diserens, A. C., and Van Meir, E. G. (1999). Frequent co-alterations of TP53, p16/CDKN2A, p14ARF, PTEN tumor suppressor genes in human glioma cell lines. *Brain Pathol.* **9,** 469–479.

Jo, M., Kim, T. H., Seol, D. W., Esplen, J. E., Dorko, K., Billiar, T. R., and Strom, S. C. (2000). Apoptosis induced in normal human hepatocytes by tumor necrosis factor-related apoptosis-inducing ligand. *Nat. Med.* **6,** 564–567.

Joshi, B. H., Husain, S. R., and Puri, R. K. (2000). Preclinical studies with IL-13PE38QQR for therapy of malignant glioma. *Drug News Perspect.* **13,** 599–605.

Kandasamy, K., Srinivasula, S. M., Alnemri, E. S., Thompson, C. B., Korsmeyer, S. J., Bryant, J. L., and Srivastava, R. K. (2003). Involvement of proapoptotic molecules Bax and Bak in tumor necrosis factor-related apoptosis-inducing ligand (TRAIL)-induced mitochondrial disruption and apoptosis: Differential regulation of cytochrome c and Smac/DIABLO release. *Cancer Res.* **63,** 1712–1721.

Kim, S., Dougherty, E. R., Shmulevich, L., Hess, K. R., Hamilton, S. R., Trent, J. M., Fuller, G. N., and Zhang, W. (2002a). Identification of combination gene sets for glioma classification. *Mol. Cancer Ther.* **1,** 1229–1236.

Kim, Y., Suh, N., Sporn, M., and Reed, J. C. (2002b). An inducible pathway for degradation of FLIP protein sensitizes tumor cells to TRAIL-induced apoptosis. *J. Biol. Chem.* **277,** 22320–22329.

Kischkel, F. C., Hellbardt, S., Behrmann, I., Germer, M., Pawlita, M., Krammer, P. H., and Peter, M. E. (1995). Cytotoxicity-dependent APO-1 (Fas/CD95)-associated proteins form a death-inducing signaling complex (DISC) with the receptor. *EMBO J.* **14,** 5579–5588.

Kischkel, F. C., Lawrence, D. A., Chuntharapai, A., Schow, P., Kim, K. J., and Ashkenazi, A. (2000). Apo2L/TRAIL-dependent recruitment of endogenous FADD and caspase-8 to death receptors 4 and 5. *Immunity* **12,** 611–620.

Kischkel, F. C., Lawrence, D. A., Tinel, A., LeBlanc, H., Virmani, A., Schow, P., Gazdar, A., Blenis, J., Arnott, D., and Ashkenazi, A. (2001). Death receptor recruitment of endogenous caspase-10 and apoptosis initiation in the absence of caspase-8. *J. Biol. Chem.* **276,** 46639–46646.

Kleihues, P., and Ohgaki, H. (1999). Primary and secondary glioblastomas: From concept to clinical diagnosis. *Neurooncol.* **1,** 44–51.

Knight, M. J., Riffkin, C. D., Ekert, P. G., Ashley, D. M., and Hawkins, C. J. (2004). Caspase-8 levels affect necessity for mitochondrial amplification in death ligand-induced glioma cell apoptosis. *Mol. Carcinog.* (in press).

Knight, M. J., Riffkin, C. D., Muscat, A. M., Ashley, D. M., and Hawkins, C. J. (2001). Analysis of FasL and TRAIL induced apoptosis pathways in glioma cells. *Oncogene* **20,** 5789–5798.

Lacour, S., Hammann, A., Wotawa, A., Corcos, L., Solary, E., and Dimanche-Boitrel, M. T. (2001). Anticancer agents sensitize tumor cells to tumor necrosis factor-related apoptosis-inducing ligand-mediated caspase-8 activation and apoptosis. *Cancer Res.* **61,** 1645–1651.

Lacronique, V., Mignon, A., Fabre, M., Viollet, B., Rouquet, N., Molina, T., Porteu, A., Henrion, A., Bouscary, D., Varlet, P., Joulin, V., and Kahn, A. (1996). Bcl-2 protects from lethal hepatic apoptosis induced by an anti-fas antibody in mice. *Nat. Med.* **2,** 80–86.

Laperriere, N., Zuraw, L., and Cairncross, G. (2002). Radiotherapy for newly diagnosed malignant glioma in adults: A systematic review. *Radiother. Oncol.* **64,** 259–273.

Lawrence, D., Shahrokh, Z., Marsters, S., Achilles, K., Shih, D., Mounho, B., Hillan, K., Totpal, K., DeForge, L., Schow, P., Hooley, J., Sherwood, S., Pai, R., Leung, S., Khan, L., Gliniak, B., Bussiere, J., Smith, C. A., Strom, S. S., Kelley, S., Fox, J. A., Thomas, D., and Ashkenazi, A. (2001). Differential hepatocyte toxicity of recombinant Apo2L/TRAIL versions. *Nat. Med.* **7,** 383–385.

Lee, J., Hampl, M., Albert, P., and Fine, H. A. (2002). Antitumor activity and prolonged expression from a TRAIL-expressing adenoviral vector. *Neoplasia* **4,** 312–323.

Leverkus, M., Neumann, M., Mengling, T., Rauch, C. T., Brocker, E. B., Krammer, P. H., and Walczak, H. (2000). Regulation of tumor necrosis factor-related apoptosis-inducing ligand sensitivity in primary and transformed human keratinocytes. *Cancer Res.* **60**, 553–559.

Leverkus, M., Sprick, M. R., Wachter, T., Mengling, T., Baumann, B., Serfling, E., Brocker, E. B., Goebeler, M., Neumann, M., and Walczak, H. (2003). Proteasome inhibition results in TRAIL sensitization of primary keratinocytes by removing the resistance-mediating block of effector caspase maturation. *Mol. Cell. Biol.* **23**, 777–790.

Lin, Y., Devin, A., Cook, A., Keane, M. M., Kelliher, M., Lipkowitz, S., and Liu, Z. G. (2000). The death domain kinase RIP is essential for TRAIL (Apo2L)-induced activation of IkappaB kinase and c-Jun N-terminal kinase. *Mol. Cell. Biol.* **20**, 6638–6645.

Linskey, M. E. (2000). Developmental glial biology: The key to understanding glial tumors. *Clin. Neurosurg.* **47**, 46–71.

Loo, G., Lippens, S., Hahne, M., Matthijssens, F., Declercq, W., Saelens, X., and Vandenabeele, P. (2003). A Bcl-2 transgene expressed in hepatocytes does not protect mice from fulminant liver destruction induced by Fas ligand. *Cytokine* **22**, 62–70.

Maher, S., Toomey, D., Condron, C., and Bouchier-Hayes, D. (2002). Activation-induced cell death: The controversial role of Fas and Fas ligand in immune privilege and tumour counterattack. *Immunol. Cell Biol.* **80**, 131–137.

Mamot, C., Drummond, D. C., Greiser, U., Hong, K., Kirpotin, D. B., Marks, J. D., and Park, J. W. (2003). Epidermal growth factor receptor (EGFR)-targeted immunoliposomes mediate specific and efficient drug delivery to EGFR- and EGFRvIII-overexpressing tumor cells. *Cancer Res.* **63**, 3154–3161.

Mariani, S. M., and Krammer, P. H. (1998). Differential regulation of TRAIL and CD95 ligand in transformed cells of the T and B lymphocyte lineage. *Eur. J. Immunol.* **28**, 973–982.

Martinez-Lorenzo, M. J., Alava, M. A., Gamen, S., Kim, K. J., Chuntharapai, A., Pineiro, A., Naval, J., and Anel, A. (1998). Involvement of APO2 ligand/TRAIL in activation-induced death of Jurkat and human peripheral blood T cells. *Eur. J. Immunol.* **28**, 2714–2725.

Meng, R. D., and El-Deiry, W. S. (2001). p53-independent upregulation of KILLER/DR5 TRAIL receptor expression by glucocorticoids and interferon-gamma. *Exp. Cell Res.* **262**, 154–169.

Micheau, O., and Tschopp, J. (2003). Induction of TNF receptor I-mediated apoptosis via two sequential signaling complexes. *Cell* **114**, 181–190.

Mischel, P. S., Nelson, S. F., and Cloughesy, T. F. (2003). Molecular analysis of glioblastoma: Pathway profiling and its implications for patient therapy. *Cancer Biol. Ther.* **2**, 242–247.

Miyazaki, T., and Reed, J. C. (2001). A GTP-binding adapter protein couples TRAIL receptors to apoptosis-inducing proteins. *Nat. Immunol.* **2**, 493–500.

Mongkolsapaya, J., Grimes, J. M., Chen, N., Xu, X. N., Stuart, D. I., Jones, E. Y., and Screaton, G. R. (1999). Structure of the TRAIL-DR5 complex reveals mechanisms conferring specificity in apoptotic initiation. *Nat. Struct. Biol.* **6**, 1048–1053.

Muhlenbeck, F., Haas, E., Schwenzer, R., Schubert, G., Grell, M., Smith, C., Scheurich, P., and Wajant, H. (1998). TRAIL/Apo2L activates c-Jun NH2-terminal kinase (JNK) via caspase-dependent and caspase-independent pathways. *J. Biol. Chem.* **273**, 33091–33098.

Muhlenbeck, F., Schneider, P., Bodmer, J. L., Schwenzer, R., Hauser, A., Schubert, G., Scheurich, P., Moosmayer, D., Tschopp, J., and Wajant, H. (2000). The tumor necrosis factor-related apoptosis-inducing ligand receptors TRAIL-R1 and TRAIL-R2 have distinct cross-linking requirements for initiation of apoptosis and are non-redundant in JNK activation. *J. Biol. Chem.* **275**, 32208–32213.

Munshi, A., Pappas, G., Honda, T., McDonnell, T. J., Younes, A., Li, Y., and Meyn, R. E. (2001). TRAIL (APO-2L) induces apoptosis in human prostate cancer cells that is inhibitable by Bcl-2. *Oncogene* **20**, 3757–3765.

Nagane, M., Pan, G., Weddle, J. J., Dixit, V. M., Cavenee, W. K., and Huang, H. J. (2000). Increased death receptor 5 expression by chemotherapeutic agents in human gliomas causes synergistic cytotoxicity with tumor necrosis factor-related apoptosis-inducing ligand in vitro and in vivo. *Cancer Res.* **60,** 847–853.

Nakamura, M., Rieger, J., Weller, M., Kim, J., Kleihues, P., and Ohgaki, H. (2000). APO2L/TRAIL expression in human brain tumors. *Acta Neuropathol. (Berl.)* **99,** 1–6.

Naumann, U., Waltereit, R., Schulz, J. B., and Weller, M. (2003). Adenoviral (full-length) Apo2L/TRAIL gene transfer is an ineffective treatment strategy for malignant glioma. *J. Neurooncol.* **61,** 7–15.

Nesterov, A., Ivashchenko, Y., and Kraft, A. S. (2002). Tumor necrosis factor-related apoptosis-inducing ligand (TRAIL) triggers apoptosis in normal prostate epithelial cells. *Oncogene* **21,** 1135–1140.

Ng, C. P., and Bonavida, B. (2002). X-linked inhibitor of apoptosis (XIAP) blocks Apo2 ligand/tumor necrosis factor-related apoptosis-inducing ligand-mediated apoptosis of prostate cancer cells in the presence of mitochondrial activation: Sensitization by overexpression of second mitochondria-derived activator of caspase/direct IAP-binding protein with low pI (Smac/DIABLO). *Mol. Cancer Ther.* **1,** 1051–1058.

Nguyen, T., Zhang, X. D., and Hersey, P. (2001). Relative resistance of fresh isolates of melanoma to tumor necrosis factor-related apoptosis-inducing ligand (TRAIL)-induced apoptosis. *Clin. Cancer Res.* **7,** 966s–973s.

Nitsch, R., Bechmann, I., Deisz, R. A., Haas, D., Lehmann, T. N., Wendling, U., and Zipp, F. (2000). Human brain-cell death induced by tumour-necrosis-factor-related apoptosis-inducing ligand (TRAIL). *Lancet* **356,** 827–828.

Noble, M., Gutowski, N., Bevan, K., Engel, U., Linskey, M., Urenjak, J., Bhakoo, K., and Williams, S. (1995). From rodent glial precursor cell to human glial neoplasia in the oligodendrocyte-type-2 astrocyte lineage. *Glia* **15,** 222–230.

Noble, M., and Mayer-Proschel, M. (1997). Growth factors, glia and gliomas. *J. Neurooncol.* **35,** 193–209.

Nutt, C. L., Noble, M., Chambers, A. F., and Cairncross, J. G. (2000). Differential expression of drug resistance genes and chemosensitivity in glial cell lineages correlate with differential response of oligodendrogliomas and astrocytomas to chemotherapy. *Cancer Res.* **60,** 4812–4818.

Nutt, C. L., Mani, D. R., Betensky, R. A., Tamayo, P., Cairncross, J. G., Ladd, C., Pohl, U., Hartmann, C., McLaughlin, M. E., Batchelor, T. T., Black, P. M., von Deimling, A., Pomeroy, S. L., Golub, T. R., and Louis, D. N. (2003). Gene expression-based classification of malignant gliomas correlates better with survival than histological classification. *Cancer Res.* **63,** 1602–1607.

Ohtsuka, T., Buchsbaum, D., Oliver, P., Makhija, S., Kimberly, R., and Zhou, T. (2003). Synergistic induction of tumor cell apoptosis by death receptor antibody and chemotherapy agent through JNK/p38 and mitochondrial death pathway. *Oncogene* **22,** 2034–2044.

Olivi, A., Grossman, S. A., Tatter, S., Barker, F., Judy, K., Olsen, J., Bruce, J., Hilt, D., Fisher, J., and Piantadosi, S. (2003). Dose escalation of carmustine in surgically implanted polymers in patients with recurrent malignant glioma: A new approach to brain tumor therapy CNS consortium trial. *J. Clin. Oncol.* **21,** 1845–1849.

Park, C. C., Hartmann, C., Folkerth, R., Loeffler, J. S., Wen, P. Y., Fine, H. A., Black, P. M., Shafman, T., and Louis, D. N. (2000). Systemic metastasis in glioblastoma may represent the emergence of neoplastic subclones. *J. Neuropathol. Exp. Neurol.* **59,** 1044–1050.

Pollack, I. F., Erff, M., and Ashkenazi, A. (2001). Direct stimulation of apoptotic signaling by soluble Apo2l/tumor necrosis factor-related apoptosis-inducing ligand leads to selective killing of glioma cells. *Clin. Cancer Res.* **7,** 1362–1369.

Potten, C. S. (1996). What is an apoptotic index measuring? A commentary. *Br. J. Cancer* **74,** 1743–1748.

Prados, M. D., and Levin, V. (2000). Biology and treatment of malignant glioma. *Semin. Oncol.* **27,** 1–10.

Rasheed, B. K., Wiltshire, R. N., Bigner, S. H., and Bigner, D. D. (1999). Molecular pathogenesis of malignant gliomas. *Curr. Opin. Oncol.* **11**, 162–167.

Recht, L., Jang, T., Savarese, T., and Litofsky, N. S. (2003). Neural stem cells and neurooncology: Quo vadis? *J. Cell Biochem.* **88**, 11–19.

Reed, J. (1999). Dysregulation of apoptosis in cancer. *J. Clin. Oncol.* **17**, 2941–2953.

Rich, J. N., Zhang, M., Datto, M. B., Bigner, D. D., and Wang, X. F. (1999). Transforming growth factor-beta-mediated p15(INK4B) induction and growth inhibition in astrocytes is SMAD3-dependent and a pathway prominently altered in human glioma cell lines. *J. Biol. Chem.* **274**, 35053–35058.

Rieger, J., Naumann, U., Glaser, T., Ashkenazi, A., and Weller, M. (1998). APO2 ligand: A novel lethal weapon against malignant glioma? *FEBS Lett.* **427**, 124–128.

Rieger, J., Ohgaki, H., Kleihues, P., and Weller, M. (1999). Human astrocytic brain tumors express APO2L/TRAIL. *Acta Neuropathol. (Berl.)* **97**, 1–4.

Rodriguez, I., Matsuura, K., Khatib, K., Reed, J. C., Nagata, S., and Vassalli, P. (1996). A Bcl-2 transgene expressed in hepatocytes protects mice from fulminant liver destruction but not from rapid death induced by anti-Fas antibody injection. *J. Exp. Med.* **183**, 1031–1036.

Rohn, T. A., Wagenknecht, B., Roth, W., Naumann, U., Gulbins, E., Krammer, P. H., Walczak, H., and Weller, M. (2001). CCNU-dependent potentiation of TRAIL/Apo2L-induced apoptosis in human glioma cells is p53-independent but may involve enhanced cytochrome c release. *Oncogene* **20**, 4128–4137.

Roth, W., Isenmann, S., Naumann, U., Kugler, S., Bahr, M., Dichgans, J., Ashkenazi, A., and Weller, M. (1999). Locoregional Apo2L/TRAIL eradicates intracranial human malignant glioma xenografts in athymic mice in the absence of neurotoxicity. *Biochem. Biophys. Res. Commun.* **265**, 479–483.

Roth, W., Wild-Bode, C., Platten, M., Grimmel, C., Melkonyan, H. S., Dichgans, J., and Weller, M. (2000). Secreted frizzled-related proteins inhibit motility and promote growth of human malignant glioma cells. *Oncogene* **19**, 4210–4220.

Scaffidi, C., Fulda, S., Srinivasan, A., Friesen, C., Li, F., Tomaselli, K. J., Debatin, K. M., Krammer, P. H., and Peter, M. E. (1998). Two CD95 (APO-1/Fas) signaling pathways. *EMBO J.* **17**, 1675–1687.

Schimmer, A. D., Pedersen, I. M., Kitada, S., Eksioglu-Demiralp, E., Minden, M. D., Pinto, R., Mah, K., Andreeff, M., Kim, Y., Suh, W. S., and Reed, J. C. (2003). Functional blocks in caspase activation pathways are common in leukemia and predict patient response to induction chemotherapy. *Cancer Res.* **63**, 1242–1248.

Schneider, P., Thome, M., Burns, K., Bodmer, J. L., Hofmann, K., Kataoka, T., Holler, N., and Tschopp, J. (1997). TRAIL receptors 1 (DR4) and 2 (DR5) signal FADD-dependent apoptosis and activate NF-kappaB. *Immunity* **7**, 831–836.

Schneider, P., Holler, N., Bodmer, J., Hahne, M., Frei, K., Fontana, A., and Tschopp, J. (1998). Conversion of membrane-bound Fas(CD95) ligand to its soluble form is associated with downregulation of its proapoptotic activity and loss of liver toxicity. *J. Exp. Med.* **187**, 1205–1213.

Serano, R. D., Pegram, C. N., and Bigner, D. D. (1980). Tumorigenic cell culture lines from a spontaneous VM/Dk murine astrocytoma (SMA). *Acta Neuropathol. (Berl.)* **51**, 53–64.

Sheridan, J. P., Marsters, S. A., Pitti, R. M., Gurney, A., Skubatch, M., Baldwin, D., Ramakrishnan, L., Gray, C. L., Baker, K., Wood, W. I., Goddard, A. D., Godowski, P., and Ashkenazi, A. (1997). Control of TRAIL-induced apoptosis by a family of signaling and decoy receptors. *Science* **277**, 818–821.

Shin, M. S., Kim, H. S., Lee, S. H., Park, W. S., Kim, S. Y., Park, J. Y., Lee, J. H., Lee, S. K., Lee, S. N., Jung, S. S., Han, J. Y., Kim, H., Lee, J. Y., and Yoo, N. J. (2001). Mutations of tumor necrosis factor-related apoptosis-inducing ligand receptor 1 (TRAIL-R1) and receptor 2 (TRAIL-R2) genes in metastatic breast cancers. *Cancer Res.* **61**, 4942–4946.

Shinohara, H., Kayagaki, N., Yagita, H., Oyaizu, N., Ohba, M., Kuroki, T., and Ikawa, Y. (2001). A protective role of PKCepsilon against TNF-related apoptosis-inducing ligand (TRAIL)-induced apoptosis in glioma cells. *Biochem. Biophys. Res. Commun.* **284**, 1162–1167.

Siegmund, D., Hadwiger, P., Pfizenmaier, K., Vornlocher, H. P., and Wajant, H. (2002). Selective inhibition of FLICE-like inhibitory protein expression with small interfering RNA oligonucleotides is sufficient to sensitize tumor cells for TRAIL-induced apoptosis. *Mol. Med.* **8**, 725–732.

Simon, A. K., Williams, O., Mongkolsapaya, J., Jin, B., Xu, X. N., Walczak, H., and Screaton, G. R. (2001). Tumor necrosis factor-related apoptosis-inducing ligand in T cell development: Sensitivity of human thymocytes. *Proc. Natl. Acad. Sci. USA* **98**, 5158–5163.

Smyth, M. J., Takeda, K., Hayakawa, Y., Peschon, J. J., van den Brink, M. R., and Yagita, H. (2003). Nature's TRAIL–on a path to cancer immunotherapy. *Immunity* **18**, 1–6.

Sprick, M. R., Weigand, M. A., Rieser, E., Rauch, C. T., Juo, P., Blenis, J., Krammer, P. H., and Walczak, H. (2000). FADD/MORT1 and caspase-8 are recruited to TRAIL receptors 1 and 2 and are essential for apoptosis mediated by TRAIL receptor 2. *Immunity* **12**, 599–609.

Sprick, M. R., Rieser, E., Stahl, H., Grosse-Wilde, A., Weigand, M. A., and Walczak, H. (2002). Caspase-10 is recruited to and activated at the native TRAIL and CD95 death-inducing signalling complexes in a FADD-dependent manner but can not functionally substitute caspase-8. *EMBO J.* **21**, 4520–4530.

Srinivasula, S. M., Datta, P., Fan, X. J., Fernandes-Alnemri, T., Huang, Z., and Alnemri, E. S. (2000). Molecular determinants of the caspase-promoting activity of Smac/DIABLO and its role in the death receptor pathway. *J. Biol. Chem.* **275**, 36152–36157.

Stewart, L. A. (2002). Chemotherapy in adult high-grade glioma: A systematic review and meta-analysis of individual patient data from 12 randomised trials. *Lancet* **359**, 1011–1018.

Sun, S. Y., Yue, P., Zhou, J. Y., Wang, Y., Choi Kim, H. R., Lotan, R., and Wu, G. S. (2001). Overexpression of BCL2 blocks TNF-related apoptosis-inducing ligand (TRAIL)-induced apoptosis in human lung cancer cells. *Biochem. Biophys. Res. Commun.* **280**, 788–797.

Teitz, T., Lahti, J. M., and Kidd, V. J. (2001). Aggressive childhood neuroblastomas do not express caspase-8: An important component of programmed cell death. *J. Mol. Med.* **79**, 428–436.

Truneh, A., Sharma, S., Silverman, C., Khandekar, S., Reddy, M. P., Deen, K. C., McLaughlin, M. M., Srinivasula, S. M., Livi, G. P., Marshall, L. A., Alnemri, E. S., Williams, W. V., and Doyle, M. L. (2000). Temperature-sensitive differential affinity of TRAIL for its receptors. DR5 is the highest affinity receptor. *J. Biol. Chem.* **275**, 23319–23325.

Tsujimoto, Y., Cossman, J., Jaffe, E., and Croce, C. M. (1985). Involvement of the bcl-2 gene in human follicular lymphoma. *Science* **228**, 1440–1443.

Van Meir, E., Ceska, M., Effenberger, F., Walz, A., Grouzmann, E., Desbaillets, I., Frei, K., Fontana, A., and de Tribolet, N. (1992). Interleukin-8 is produced in neoplastic and infectious diseases of the human central nervous system. *Cancer Res.* **52**, 4297–4305.

Van Meir, E. G., Kikuchi, T., Tada, M., Li, H., Diserens, A. C., Wojcik, B. E., Huang, H. J., Friedmann, T., de Tribolet, N., and Cavenee, W. K. (1994). Analysis of the p53 gene and its expression in human glioblastoma cells. *Cancer Res.* **54**, 649–652.

Vaux, D. L., Cory, S., and Adams, J. M. (1988). Bcl-2 gene promotes haemopoietic cell survival and cooperates with c-myc to immortalize pre-B cells. *Nature* **335**, 440–442.

Vertosick, F. T., Jr., and Selker, R. G. (1990). Brain stem and spinal metastases of supratentorial glioblastoma multiforme: A clinical series. *Neurosurgery* **27**, 516–521 discussion 521–522.

Walczak, H., Miller, R. E., Ariail, K., Gliniak, B., Griffith, T. S., Kubin, M., Chin, W., Jones, J., Woodward, A., Le, T., Smith, C., Smolak, P., Goodwin, R. G., Rauch, C. T., Schuh, J. C., and Lynch, D. H. (1999). Tumoricidal activity of tumor necrosis factor-related apoptosis-inducing ligand in vivo. *Nat. Med.* **5**, 157–163.

Walker, P. R., Saas, P., and Dietrich, P. Y. (1997). Role of Fas ligand (CD95L) in immune escape: The tumor cell strikes back. *J. Immunol.* **158,** 4521–4524.

Wendling, U., Walczak, H., Dorr, J., Jaboci, C., Weller, M., Krammer, P. H., and Zipp, F. (2000). Expression of TRAIL receptors in human autoreactive and foreign antigen-specific T cells. *Cell Death Differ.* **7,** 637–644.

Werner, A. B., de Vries, E., Tait, S. W., Bontjer, I., and Borst, J. (2002). TRAIL receptor and CD95 signal to mitochondria via FADD, caspase-8/10, Bid, and Bax but differentially regulate events downstream from truncated Bid. *J. Biol. Chem.* **277,** 40760–40767.

Wu, G. S., Burns, T. F., McDonald, E. R., 3rd, Jiang, W., Meng, R., Krantz, I. D., Kao, G., Gan, D. D., Zhou, J. Y., Muschel, R., Hamilton, S. R., Spinner, N. B., Markowitz, S., Wu, G., and el-Deiry, W. S. (1997). KILLER/DR5 is a DNA damage-inducible p53-regulated death receptor gene. *Nat. Genet.* **17,** 141–143.

Wu, G. S., Burns, T. F., McDonald, E. R., 3rd, Meng, R. D., Kao, G., Muschel, R., Yen, T., and el-Deiry, W. S. (1999a). Induction of the TRAIL receptor KILLER/DR5 in p53-dependent apoptosis but not growth arrest. *Oncogene* **18,** 6411–6418.

Wu, G. S., Burns, T. F., Zhan, Y., Alnemri, E. S., and El-Deiry, W. S. (1999b). Molecular cloning and functional analysis of the mouse homologue of the KILLER/DR5 tumor necrosis factor-related apoptosis-inducing ligand (TRAIL) death receptor. *Cancer Res.* **59,** 2770–2775.

Wu, M., Das, A., Tan, Y., Zhu, C., Cui, T., and Wong, M. C. (2000). Induction of apoptosis in glioma cell lines by TRAIL/Apo-21. *J. Neurosci. Res.* **61,** 464–470.

Wu, V. W., and Schwartz, J. P. (1998). Cell culture models for reactive gliosis: New perspectives. *J. Neurosci. Res.* **51,** 675–681.

Xiao, C., Yang, B. F., Asadi, N., Beguinot, F., and Hao, C. (2002). Tumor necrosis factor-related apoptosis-inducing ligand-induced death-inducing signaling complex and its modulation by c-FLIP and PED/PEA-15 in glioma cells. *J. Biol. Chem.* **277,** 25020–25025.

Zhang, X. D., Franco, A. V., Nguyen, T., Gray, C. P., and Hersey, P. (2000). Differential localization and regulation of death and decoy receptors for TNF-related apoptosis-inducing ligand (TRAIL) in human melanoma cells. *J. Immunol.* **164,** 3961–3970.

23

REGULATION OF TRAIL-INDUCED APOPTOSIS BY ECTOPIC EXPRESSION OF ANTIAPOPTOTIC FACTORS

BHARAT B. AGGARWAL,
UDDALAK BHARDWAJ, AND YASUNARI TAKADA

*Cytokine Research Section, Department of Bioimmunotherapy
The University of Texas, M. D. Anderson Cancer Center, Houston, Texas 77030*

    I. Introduction
   II. Negative Regulation of TRAIL-Induced Apoptosis
      A. Decoy Receptors
      B. FLICE-Inhibitory Proteins
      C. Inhibitor of Apoptosis Proteins
      D. Bcl-2/Bcl-x_L
      E. Nuclear Factor-Kappa B
      F. Akt
      G. Other Mechanisms
  III. Inhibitors of Antiapoptotic Factors as Therapeutic Agents
  IV. Effect of TRAIL on Normal Cells
   V. Conclusions
      References

The discovery of an agent that selectively kills tumor cells and not normal cells is the dream of every cancer researcher. Tumor necrosis factor (TNF)–related apoptosis-inducing ligand (TRAIL), first discovered in 1995, was heralded as a selective killer of tumor cells, and its potential is still thought to be high. Almost immediately, broad efforts were made to understand its activity at the molecular level. TRAIL has been shown to interact with the cell surface through five distinct receptors, named death receptor (DR) 4, DR5, decoy receptor (Dc)R1, DcR2, and osteoprotegrin. It activates nuclear factor (NF)-κB, c-Jun N-terminal kinases, and apoptosis. The apoptotic signals are mediated through Fas-associated death domain protein (FADD)–mediated recruitment of caspase-8 and caspase-3. Additionally, caspase-8 can cleave Bcl-2 homology domain 3 (BH3)-interfering domain death agonist (Bid), and the cleaved Bid then causes the release of mitochondrial cytochrome c, leading to the activation of pro-caspase-9, which can then activate pro-caspase-3. TRAIL-induced apoptosis is negatively regulated by numerous cellular factors including decoy receptors, cellular FADD-like interleukin 1 β-converting enzyme (FLICE) interacting protein (cFLIP), cellular inhibitor of apoptosis protein (cIAP), X-linked IAP (XIAP), survivin, and NF-κB. Second mitochondria-derived activator of caspases (Smac)/direct IAP binding protein with low pI (DIABLO) mediates proapoptotic signals through inaction of IAP. How the TRAIL-induced apoptosis is downregulated by these factors is discussed in detail in this review. Whether TRAIL selectively kills tumor cells without harming normal cells is also discussed. © 2004 Elsevier Inc.

I. INTRODUCTION

The term apoptosis (Kerr *et al.*, 1972) was originally described as a genetically programmed cell death as opposed to random cell death and death caused by severe injury to the cell. Apoptosis is characterized by shrinkage, cytoplasmic membrane blebbing, chromatin condensation, DNA fragmentation, and finally destruction of the cell by its disassembly into fragments. Cell death by apoptosis differs from necrosis, which is characterized by mitochondrial swelling. Currently, the term apoptosis is used to describe any type of cell death induced by activation of cysteine-aspartic acid proteases, called caspases. Caspases are a novel family of proteases, so named because they contain cysteine at the active site and cleave the target protein following aspartic acid. It has been estimated that in humans there are 30×10^{12} cells (10 micron across) and 1×10^6 cells/sec (or 60×10^9 cells/day) that undergo apoptosis.

Tumor necrosis factor (TNF) was the first soluble factor to be isolated that could kill tumor cells *in vitro*. Since then, 19 different members of the TNF superfamily have been identified; they exhibit 15–20% amino acid sequence homology to each other (Aggarwal, 2003). The current review, however, focuses on only one of the TNF superfamily, TNF-related apoptosis-inducing ligand (TRAIL). This cytokine was first identified in

1995 by two independent investigators (Pitti *et al.*, 1996; Wiley *et al.*, 1995) and reported to selectively induce apoptosis of tumor cells. TRAIL is transmembrane protein expressed by natural killer (NK) cells, T lymphocytes, and dendritic cells.

Two major signaling pathways control apoptosis initiation in mammals. First, the cell-extrinsic pathway involves engagement of cell-surface death receptors by ligands that belong to the TNF superfamily and consequent activation of caspase-8. The second, cell-intrinsic pathway involves mitochondrial disruption by proapoptotic Bcl-2 family members and consequent release of factors such as cytochrome *c* that promote activation of caspase-9. Crosstalk between the death-receptor and mitochondrial pathways is mediated by caspase-8 cleavage of the protein Bcl-2 homology domain 3 (BH3)-interfering domain death agonist (Bid). Cleaved Bid can trigger mitochondrial cytochrome *c* release through the proapoptotic Bcl-2 homologs Bax and Bak (Fig. 1).

There is very little information about how TRAIL is released from its cell surface, but TRAIL is known to interact with five different receptors. Two are death receptors (DR4 and DR5), two are decoy receptors (DcR1 and DcR2) (Delgi-Esposti *et al.*, 1997a,b; MacFarlane *et al.*, 1997; Marsters *et al.*, 1997; Mongkolsapaya, 1998; Pan *et al.*, 1997a,b, 1998; Schneider *et al.*, 1997; Screaton *et al.*, 1997; Sheridan *et al.*, 1997; Walczak *et al.*, 1997), and the fifth is osteoprotegrin (OPG), a receptor without the transmembrane domain (Emery *et al.*, 1998). These receptors are present in a wide variety of normal tissues and in normal and tumor cells. By using overexpression systems, dominant-negative mutants, and coimmuno-precipitation, some investigators showed that TRAIL-induced apoptosis is dependent on Fas-associated death domain protein (FADD) (Chaudhary *et al.*, 1997; Schneider *et al.*, 1997; Wajant *et al.*, 1998), while others showed FADD-independent apoptosis (MacFarlane *et al.*, 1997; Marsters *et al.*, 1996; Pan *et al.*, 1997a,b).

FADD is a 208–amino acid protein that contains one death domain (DD) at its carboxyl-terminal end and one death effector domain (DED) at its amino terminus (Fig. 1). Fibroblasts from FADD-knockout mice were shown to undergo TRAIL-induced apoptosis, which suggests that FADD is not essential for TRAIL-induced apoptosis Yeh *et al.* (1998). Pan *et al.* (1997b) reported that DR4 did not bind FADD and that a dominant-negative FADD had little effect on DR4- or TRAIL-induced apoptosis. Similar reports were made for DR5 by MacFarlane *et al.* (1997). Some recent studies, however, indicate that TRAIL induces apoptosis by recruitment of FADD (Bodmer *et al.*, 2000; Kischkel *et al.*, 2000; Sprick *et al.*, 2002).

Why there are so many controversial reports about the role of FADD in TRAIL-induced apoptosis is not clear. Both the techniques and the cell type employed may explain some of these differences. Based on the paradigm for TNF and Fas signaling, it was concluded that FADD then binds to FADD-like interleukin-1β (IL-1β)–converting enzyme (FLICE), also called

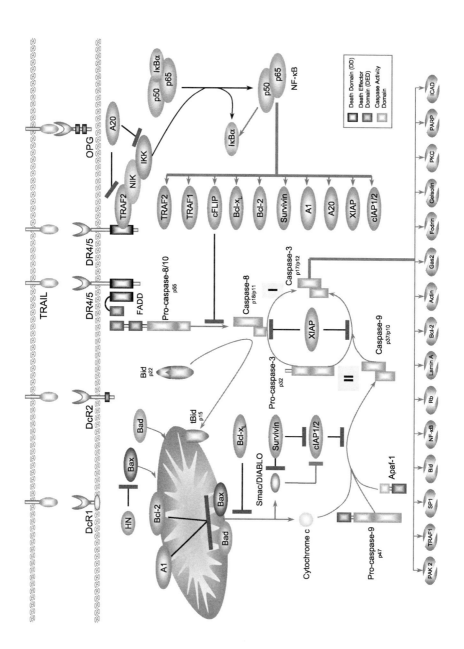

caspase-8. The latter is cleaved and then the activated enzyme causes either the activation of caspase-3 or the cleavage of Bid, a proapoptotic member of the Bcl-2 family of proteins, which leads to cytochrome *c* release from the mitochondria, resulting in caspase-9 activation (Li *et al.*, 1998b; Luo *et al.*, 1998).

Numerous caspases have been shown to be involved in TRAIL-induced apoptosis. While cleavage of caspase-8 has been shown following TRAIL stimulation (Kischkel *et al.*, 2000; Sprick *et al.*, 2002), the overexpression of dominant-negative mutant forms indicated the role of either caspase 10 alone (Pan *et al.*, 1997a) or a combination of caspase-8 and caspase-10 (MacFarlane *et al.*, 1997). Structurally, both pro–caspase-8 and pro–caspase-10 exhibit very similar architecture (Fig. 2), consisting of two DED toward the amino terminus and a caspase domain towards the carboxyl terminus. Cells expressing only one of the caspase undergoes ligand-induced apoptosis, indicating that each caspase can initiate apoptosis independently of the other. Thus, apoptosis signaling by TRAIL involves not only caspase-8 but also caspase-10, and both caspases may have equally important roles in initiation of apoptosis (Kischkel *et al.*, 2001).

Seol *et al.* (2000) showed that caspase-8 is required for TRAIL-induced apoptosis and that caspase-10 may play only a minor role, if any, in TRAIL-induced apoptosis. They showed that overexpression of a caspase-8 dominant-negative mutant inhibited apoptosis induced by TRAIL. Similarly, caspase-8–deficient Jurkat cells were found to be resistant to TRAIL-induced apoptosis, whereas wild-type Jurkat cells were susceptible to TRAIL-induced apoptosis. The caspase-8–reintroduced caspase-8–deficient Jurkat cells acquired normal susceptibility to TRAIL.

LeBlanc *et al.* (2002) showed that Bax is essential for TRAIL-induced apoptosis in cancer cells. Bax-deficient human colon carcinoma cells were resistant to TRAIL, whereas Bax-expressing sister clones were sensitive. Bax was dispensable for apical death-receptor signaling events including caspase-8 activation but crucial for mitochondrial changes and downstream caspase activation. Another report showed the role of a GTP-binding adaptor protein that links DR4 and DR5 to FADD, which leads to the activation of caspase-8 (Miyazaki and Reed, 2001).

FIGURE 1. Schematic representation of the mechanism of TRAIL-induced apoptosis and its regulation by intracellular proteins. DcR1 does not contain an intracellular domain, DcR2 contains a truncated death domain, and OPG contains two death domains but no transmembrane domain. The numbers below the proteins indicate molecular size in kilodaltons. Mitochondria-independent pathway is referred as pathway I, whereas mitochondria-dependent pathway is referred as pathway II. Various domains are shown in the box.

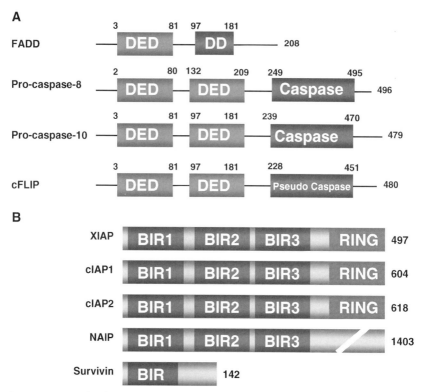

FIGURE 2. The domains structure of FADD, pro–caspase-8, pro–caspase-10, c-FLIP, XIAP, cIAP1, cIAP2, NAIP, and survivin. BIR is baculovirus inhibitory repeat domain.

II. NEGATIVE REGULATION OF TRAIL-INDUCED APOPTOSIS

A. DECOY RECEPTORS

Besides two proapoptotic receptors, TRAIL also binds to three different antiapoptotic receptors, which in turn suppresses TRAIL-induced apoptosis. These include DcR1 and DcR2, which are membrane bound receptors, whereas OPG is a secreted protein (Fig. 1). These receptors bind TRAIL but do not mediate signaling. Normal cells express all five of the receptors. The increased TRAIL sensitivity of tumor cells has been postulated to result from the lack of DcR expression. Van Noesel *et al.* (2002) showed that tumor-specific down-regulation of the TRAIL receptors DcR1 and DcR2, as well as DR4 and DR5, in a group of pediatric tumor cell lines (nine neuroblastoma and three peripheral primitive neuro-ectodermal tumors) and three cell lines from tumors in adults. Lack of expression of DcR1 and DcR2 was widespread (13 of 15 and 10 of 15 cell lines, respectively), in both the adult and the pediatric tumor lines. DR4 and DR5 were expressed in 8 of 15 and 12 of 15 cell lines, respectively. To understand the tumor-specific down-regulation of the TRAIL receptors, the promoter regions were studied for possible methylation changes in their CpG islands. All normal tissues were completely unmethylated, whereas in the tumor cell lines, the promoter was frequently hypermethylated. For DcR1 and DcR2, Van Noesel's group found dense hypermethylation in nine of 13 (69%) and nine of 10 (90%) nonexpressing cell lines. DR4 and DR5 were methylated in five of seven (71%) and two of three (67%) nonexpressing cell lines. Treatment with the demethylating agent 5-aza-2′deoxycytidine resulted in partial demethylation and restored mRNA expression. In addition, Van Noesel's group performed mutation analysis of the death domains (DD) of DR4 and DR5 by sequencing exon 9. Mutations were not present in any of the neuroblastoma or primitive neuro-ectodermal tumor cell lines. A panel of 28 fresh neuroblastoma tumor samples also lacked expression of DcR1 (85% of cases) and DcR2 (74%). DcR1 hypermethylation was observed in 6 (21%) of 28 cases and DcR2 hypermethylation in seven (25%). DR4 and DR5 were both expressed in 22 of 28 tumors, and no promoter methylation was observed. These data suggest that hypermethylation of the promoters of DcR1 and DcR2 is important in the down-regulation of expression in neuroblastoma and other tumor types.

In contrast to these studies, Zhang *et al.* found no correlation between expression of decoy receptors and susceptibility of human melanoma cells to TRAIL-induced apoptosis (1999). In view of this, they studied the localization of the receptors in melanoma cells by confocal microscopy to better understand their function (2000). They showed that DR4 and DR5 are located in the trans-Golgi network, whereas DcR1 and DcR2 are located in the nucleus. After exposure to TRAIL, DR4 and DR5 are internalized into endosomes, whereas DcR1 and DcR2 are relocated from the nucleus to the cytoplasm and cell membranes. Transfection of DcR1 and DcR2 increased resistance of the melanoma lines to TRAIL-induced apoptosis even in melanoma lines that naturally expressed these receptors. These results indicated that abnormalities in decoy receptor location or function may increase the sensitivity of melanoma to TRAIL-induced apoptosis.

OPG can also mediate resistance to TRAIL. Holen *et al.* (2002) demonstrated that high levels of OPG are produced by the hormone-insensitive prostate cancer cell lines PC3 and DU145, whereas the hormone-sensitive cell line LNCaP produced 10- to 20-fold less OPG under the same conditions. A strong negative correlation was observed between levels of endogenously produced OPG in the medium and the capacity of TRAIL to induce apoptosis in cells that produced high levels of OPG. The antiapoptotic effect of OPG was reversed by coadministration of 100-fold molar excess of receptor-activator of NF-κB ligand (RANKL), another protein that selectively binds OPG. These observations suggest that prostate cancer–derived

OPG may be an important survival factor in hormone-resistant prostate cancer cells.

B. FLICE-INHIBITORY PROTEINS

FLICE-inhibitory protein (FLIP) was first reported as a protein predominantly expressed in muscle and lymphoid tissues (Irmler et al., 1997). It is a 480-amino acid protein consisting of two DED and a pseudo-caspase domain (Fig. 2). The cellular short FLIP (cFLIP$_S$) contains two DED and is structurally related to the viral FLIP inhibitors of apoptosis (Thome et al., 1997), whereas the long form, FLIP$_L$, also contains a caspase-like domain in which a tyrosine residue is substituted for the active-center cysteine residue. cFLIP$_S$ and FLIP$_L$ interact with the adaptor protein FADD and the protease FLICE and potently inhibit apoptosis induced by all known human death receptors. FLIP$_L$ is expressed during the early stage of T-cell activation, but disappears when T-cells become susceptible to Fas ligand–mediated apoptosis. High levels of FLIP$_L$ protein are also detectable in melanoma cell lines and malignant melanoma tumors. Thus FLIP may be implicated in tissue homeostasis as an important regulator of apoptosis. vFLIP, which interferes with apoptosis signaled through DR, is present in several γ-herpes viruses (including Kaposi's-sarcoma–associated human herpesvirus-8), as well as in the tumorigenic human molluscipox virus (Thome et al., 1997). vFLIP contains two DED, which interact with the adaptor protein FADD, and this interaction inhibits the recruitment and activation of the protease FLICE by the death receptor. Cells expressing vFLIP are protected against apoptosis induced by TRAIL-R (TRAIL-receptors).

Griffith et al. (1998) examined the effect of TRAIL on melanoma cells and showed that FLIP expression was highest in TRAIL-resistant melanomas and low or undetectable in TRAIL-sensitive melanomas. Furthermore, addition of actinomycin D to TRAIL-resistant melanomas decreased intracellular concentrations of FLIP, a decrease that correlated with their acquisition of TRAIL sensitivity. Zhang et al. (1999), however, showed that the resistance of melanoma cell lines to TRAIL-induced apoptosis could not be correlated with the expression of either FLIP or the decoy receptors. Burns and El-Deiry (2001) used a genetic approach to identify TRAIL-resistant genes, and one of the genes identified by them was again cFLIP$_S$. Through this screen, they also identified Bcl-x$_L$. They showed that overexpression of cFLIP$_S$ and Bcl-x$_L$ were sufficient to completely inhibit TRAIL-induced apoptosis. Xiao and coworkers (2002) showed that caspase-8 and caspase-10 are recruited to the death-inducing signaling complex in both TRAIL-sensitive and TRAIL-resistant glioma cells, but caspase activation is inhibited by cFLIP in TRAIL-resistant cells.

Thus agents that can downregulate the expression of cFLIP can convert TRAIL-resistant cells to TRAIL-sensitive cells. A variety of natural and synthetic ligands of peroxisome proliferator-activated receptor γ (PPAR γ) have been shown to sensitize tumor but not normal cells to apoptosis induction by TRAIL by selectively reducing the levels of FLIP (Kim *et al.*, 2002b). The tumor suppressor gene p53 has also been shown to suppress the expression of cFLIP in human colon cancer cell lines expressing mutated and wild-type p53 (Fukazawa *et al.*, 2001). These workers showed that treating the cells with a specific inhibitor of the proteasome inhibited p53-induced decrease of cFLIP, suggesting that p53 enhances the degradation of cFLIP via an ubiquitin-proteasome pathway.

The expression of cFLIP was found to be regulated by the nuclear transcription factor NF-κB (Kreuz *et al.*, 2001). Treatment of SV80 cells with the proteasome inhibitor N-benzoyloxycarbonyl (Z)-Leu-leucinal (MG-132) or geldanamycin, a drug interfering with NF-κB activation, inhibited TNF-induced upregulation of cFLIP. Overexpression of a nondegradable IκBα mutant (IκBα-SR) or lack of IκB kinase-γ (IKK-γ) expression completely prevented phorbol myristate acetate-induced upregulation of cFLIP mRNA in Jurkat cells. These data point to an important role for NF-κB in the regulation of the cFLIP gene. SV80 cells normally show resistance to TRAIL and TNF because apoptosis can be induced only in the presence of low concentrations of cyclohexiamide (CHX).

However, overexpression of IκBα-SR rendered SV80 cells sensitive to TRAIL-induced apoptosis in the absence of CHX, and cFLIP expression reversed the proapoptotic effect of NF-κB inhibition. Western blot analysis further revealed that cFLIP, but not TRAF1, A20, and cIAP2, expression levels rapidly decrease upon CHX treatment. These data suggest a key role for cFLIP in the antiapoptotic response of NF-κB activation. Hietakangas *et al.* (2003) showed that erythroid differentiation sensitizes K562 leukemia cells to TRAIL-induced apoptosis by downregulating of cFLIP.

When phosphorylated cFLIP loses its ability to suppress apoptosis, Higuchi *et al.* (2003) showed that bile acids stimulate cFLIP phosphorylation, thus enhancing TRAIL-mediated apoptosis. Bile acid glycochenodeoxycholate stimulated phosphorylation of both cFLIP isoforms, which was associated with decreased binding to GST-FADD. The protein kinase C (PKC) antagonist chelerythrine prevented bile acid–stimulated cFLIP$_L$ and cFLIP$_S$ phosphorylation, restored cFLIP binding to GST-FADD, and attenuated bile acid ostentation of TRAIL-induced apoptosis. These results provide new insights into the mechanisms of bile acid cytotoxicity and the proapoptotic effects of cFLIP phosphorylation in TRAIL signaling. Mitsiades *et al.* (2002a) showed higher levels of expression for FLIP and lower procaspase-8 levels in TRAIL-resistant cells; sensitivity was restored by the protein synthesis inhibitor CHX and the PKC inhibitor bisindolylmaleimide (BIM), both of which lowered FLIP protein levels. Forced expression of procaspase-8 or FLIP antisense oligonucleotides also sensitized TRAIL-resistant cells to TRAIL. Finally, CHX, and BIM facilitated the cleavage and activation of pro–caspase-8 in TRAIL-resistant cells, confirming that inhibition of TRAIL-induced apoptosis occurs at this level and that these agents sensitize multiple myeloma cells by relieving this block. Kelly *et al.* (2002) showed that doxorubicin pretreatment sensitizes prostate cancer cell lines to TRAIL-induced apoptosis, in parallel with the loss of cFLIP expression. The decrease in cFLIP$_S$ correlated with onset and magnitude of caspase-8 and poly (ADP-ribose) polymerase (PARP) cleavage in PC3 cells. Selective inhibition of FLIP expression with small interfering RNA oligonucleotides has also been shown to sensitize tumor cells for TRAIL-induced apoptosis (Siegmund *et al.*, 2002).

There are other reports, however, that suggest that the levels of FLIP do not play a critical role in TRAIL-induced apoptosis in melanoma (Zhang *et al.*, 1999), multiple myeloma (Spencer *et al.*, 2002), or lung cancer (Frese *et al.*, 2002). It is quite likely that other mechanisms mediate resistance to TRAIL in these cells.

C. INHIBITOR OF APOPTOSIS PROTEINS

The IAP family of proteins, including XIAP, cIAP1, cIAP2, neuronal apoptosis inhibitor protein (NAIP), and survivin, are highly conserved through evolution (Fig. 2B). These proteins are characterized by the presence of baculoviral IAP repeat (BIR)-binding domains and RING zinc-finger domain. XIAP, cIAP1, and cIAP2 has been shown to bind specifically to the terminal effector cell death proteases caspases-3 and -7, but not to the proximal proteases, caspases-8, -1, or -6 (Deveraux *et al.*, 1997; Roy *et al.*, 1997). In contrast, NAIP failed to bind tightly to any of these proteases. The BIR domain-containing region of XIAP, cIAP1, and cIAP2 was sufficient for inhibition of these caspases, though proteins that retained the RING zinc-finger domain were somewhat more potent. Similar results were obtained in intact cells when XIAP, cIAP1, and cIAP2 were overexpressed by gene transfection, and apoptosis was induced by the anticancer drug etoposide. Other studies demonstrated that XIAP, cIAP1 and cIAP2 can prevent the proteolytic processing of pro–caspases-3, -6, and -7 by blocking the cytochrome c–induced activation of pro–caspase-9 (Deveraux *et al.*, 1998). In contrast, these IAP family proteins did not prevent caspase-8–induced proteolytic activation of pro–caspase-3; however, they subsequently inhibited caspase-3 directly, thus blocking downstream apoptotic events such as further activation of caspases. These findings demonstrate that IAPs can suppress different apoptotic pathways by inhibiting distinct caspases and identify pro–caspase-9 as a new target for IAP-mediated inhibition of apoptosis.

Suzuki *et al.* (2001b) found that XIAP promotes the degradation of active-form caspase-3, but not pro–caspase-3, in living cells. RING zinc-finger mutants of XIAP also could not promote the degradation of caspase-3. A proteasome inhibitor suppressed the degradation of caspase-3 by

XIAP, suggesting the involvement of a ubiquitin-proteasome pathway in the degradation. They also showed that XIAP acts as a ubiquitin-protein ligase for caspase-3. Caspase-3 was ubiquitinated in the presence of XIAP in living cells. Both the association of XIAP with caspase-3 and the RING zinc-finger domain of XIAP were essential for ubiquitination. Finally, the RING zinc-finger mutants of XIAP were less effective than wild-type XIAP at preventing apoptosis induced by overexpression of active-form caspase-3. These results demonstrate that the ubiquitin-protein ligase activity of XIAP promotes the degradation of caspase-3, which enhances its antiapoptotic effect.

Survivin is a structurally unique IAP apoptosis inhibitor that contains a single baculoviral IAP repeat and lacks a carboxyl-terminal RING zinc-finger (Fig. 2) (Ambrosini *et al.*, 1997). It was found to be present during fetal development but was undetectable in terminally differentiated adult tissues. However, survivin was found to be expressed in transformed cell lines and in all the most common human cancers of lung, colon, pancreas, prostate gland, and breast, *in vivo*. Survivin was also found in approximately 50% of high-grade non-Hodgkin's lymphomas (centroblastic, immunoblastic), but not in low-grade lymphomas (lymphocytic). Recombinant expression of survivin counteracted apoptosis of B lymphocyte precursors deprived of IL-3. These findings suggest that apoptosis inhibition may be a general feature of neoplasia and identified survivin as a potential new target for apoptosis-based therapy in cancer and lymphoma. Li *et al.* (1998a) showed that survivin is expressed in the G2/M phase of the cell cycle in a cycle-regulated manner. At the beginning of mitosis, survivin associates with microtubules of the mitotic spindle in a specific and saturable reaction that is regulated by microtubule dynamics. Disruption of survivin-microtubule interactions results in loss of survivin's anti-apoptosis function and increased caspase-3 activity, a mechanism involved in cell death, during mitosis. These results indicated that survivin may counteract a default induction of apoptosis in G2/M phase. The overexpression of survivin in cancer may overcome this apoptotic checkpoint and favor aberrant progression of transformed cells through mitosis.

TRAIL-induced apoptosis has been shown to be regulated by survivin (Griffith *et al.*, 2002). For instance, compared with TRAIL-sensitive renal cell carcinoma (RCC) cell lines (A-498, ACHN, and 769-P), the TRAIL-resistant RCC cell line 786-O expressed greater amounts of survivin. Incubation with actinomycin D decreased the intracellular levels of survivin, resulting in enhancing TRAIL-induced apoptotic death. The link between survivin and TRAIL regulation was confirmed when an increase in TRAIL resistance was observed after overexpression of survivin in the TRAIL-sensitive, survivin-negative cells.

Smac/DIABLO is a mitochondrial protein that is released along with cytochrome *c* during apoptosis and promotes cytochrome *c*–dependent caspase activation by neutralizing IAPs (Fig. 1). Srinivasula *et al.* (2000) showed that Smac/DIABLO functions at the levels of both the Apaf-1-caspase-9 apoptosome and effector caspases. The amino terminus of Smac/DIABLO is absolutely required for its ability to interact with the BIR3 of XIAP and to promote cytochrome c-dependent caspase activation. However, it is less critical for its ability to interact with BIR1/BIR2 of XIAP and to promote the activity of the effector caspases. Consistent with the ability of Smac/DIABLO to function at the level of the effector caspases, expression of a cytosolic Smac/DIABLO in cells allowed TRAIL to by-pass Bcl-x_L inhibition of death receptor-induced apoptosis. Combined, these data suggested that Smac/DIABLO plays a critical role in neutralizing IAPs inhibition of the effector caspases in the death receptor pathway of cells.

MacFarlane *et al.* (2002) showed that exposure of MCF-7 cells to TRAIL results in rapid Smac/DIABLO release from mitochondria, which occurs before or in parallel with loss of cytochrome *c*. Smac/DIABLO release is inhibited by Bcl-2/Bcl-x_L or by a pan-caspase inhibitor demonstrating that this event is caspase-dependent and modulated by Bcl-2 family members. Following release, Smac/DIABLO is rapidly degraded by the proteasome, an effect suppressed by cotreatment with a proteasome inhibitor. As the RING zinc-finger domain of XIAP possesses ubiquitin-protein ligase activity and XIAP binds tightly to mature Smac/DIABLO, an in vitro ubiquitination assay revealed that XIAP functions as a ubiquitin-protein ligase (E3) in the ubiquitination of Smac/DIABLO. Both the association of XIAP with Smac/DIABLO and the RING zinc-finger domain of XIAP are essential for ubiquitination, suggesting that the ubiquitin-protein ligase activity of XIAP may promote the rapid degradation of mitochondrial released Smac/DIABLO. Thus, in addition to its well-characterized role in inhibiting caspase activity, XIAP may also protect cells from inadvertent mitochondrial damage by targeting proapoptotic molecules for proteasomal degradation.

Ectopic overexpression of Smac/DIABLO or treatment with the N-terminus heptapeptide (Smac-7) or tetrapeptide (Smac-4) of Smac/DIABLO, while having no activity alone, significantly increased TRAIL-induced processing and PARP cleavage activity of caspase-3 (Guo *et al.*, 2002). This produced a significant increase in apoptosis of Jurkat cells. Increased apoptosis was also associated with the down-regulation of XIAP, cIAP1, and survivin. Along with the increased activity of caspase-3, ectopic over-expression of Smac/DIABLO or cotreatment with Smac-4 also increased TRAIL-induced processing of caspase-8 and Bid, resulting in enhanced cytosolic accumulation of cytochrome *c*. These findings demonstrate that cotreatment with the N-terminus Smac/DIABLO peptide is an effective strategy to enhance apoptosis triggered by the death receptor or mitochondrial pathway and may improve the antitumor activity of TRAIL. Smac/

DIABLO agonists sensitize cells for TRAIL- or anticancer drug-induced apoptosis and induce regression of malignant glioma *in vivo* (Fulda *et al.*, 2002b).

D. Bcl-2/Bcl-x_L

Bcl-2 (B-cell lymphoma-2), the gene that is linked to an immunoglobulin locus by chromosome translocation in follicular lymphoma, was found to inhibit cell death rather than promote proliferation (for a review, see Cory and Adams, 2002). In mammals, Bcl-2 has at least 20 relatives, all of which share at least one conserved Bcl-2 homology (BH) domain (Fig. 3). The clan includes four other antiapoptotic proteins, Bcl-x_L, Bcl-w, A1, and Mcl-1, and two groups of proteins that promote cell death: the Bax and the BH3-only families. Members of the Bax death family have sequences that are similar to those in Bcl-2, especially in the BH1, BH2, and BH3 regions, but the other proapoptotic proteins have only the short BH3 motif, an interaction domain that is both necessary and sufficient for their killing action. Both types of pro-apoptotic proteins are required to initiate apoptosis: the BH3-only proteins seem to act as damage sensors and direct antagonists of the pro-survival proteins, whereas the Bax-like proteins act further downstream, probably in mitochondrial disruption. Bcl-2 and its closest homologues, Bcl-x_L and Bcl-w, potently inhibit apoptosis in response to many, but not all, cytotoxic insults. Their hydrophobic carboxy-terminal domain helps target them to the cytoplasmic face of three intracellular membranes: the outer mitochondrial membrane, the endoplasmic reticulum, and the nuclear envelope. Bcl-2 is an integral membrane protein, even in healthy cells, whereas Bcl-w and Bcl-x_L only become tightly associated with the membrane after a cytotoxic signal.

Yang *et al.* (1997) showed that overexpression of Bcl-2 prevents cells from undergoing apoptosis in response to a variety of stimuli. Cytosolic cytochrome *c* is necessary for the initiation of the apoptotic program, suggesting a possible connection between Bcl-2 and cytochrome *c*, which is normally located in the mitochondrial intermembrane space. Cells undergoing apoptosis were found to have an elevation of cytochrome *c* in the cytosol and a corresponding decrease in the mitochondria. Overexpression of Bcl-2 prevented the efflux of cytochrome *c* from the mitochondria and the initiation of apoptosis. Thus, one possible role of Bcl-2 in prevention of apoptosis is to block cytochrome *c* release from mitochondria.

Interestingly, some proteins can be antiapoptotic to proapoptotic depending on their structure. Boise *et al.* (1993) reported the isolation of Bcl-x, a Bcl-2–related gene that can function as a Bcl-2–independent regulator of apoptosis. Alternative splicing results in two distinct Bcl-x mRNAs. The protein product of the larger mRNA, Bcl-x_L, is similar in size and predicted structure to Bcl-2. When stably transfected, Bcl-x_L inhibits cell death at least as

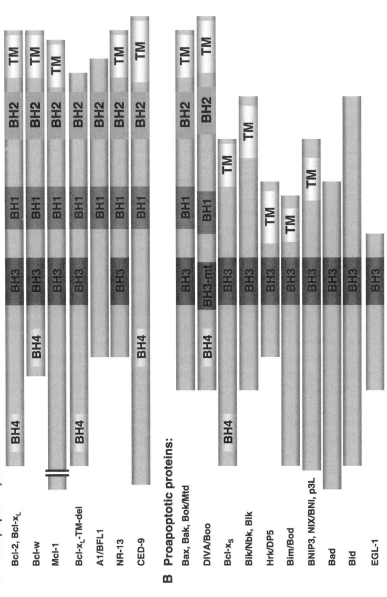

FIGURE 3. The domains structure of Bcl-2 family of proteins that inhibit TRAIL-induced apoptosis (panel A) and those that mediate apoptosis (panel B).

well as Bcl-2. Surprisingly, the second mRNA species, Bcl-x_S, encodes a protein that inhibits the ability of Bcl-2 to enhance the cell death. *In vivo*, Bcl-x_S mRNA is expressed at high levels in cells that undergo a high rate of turnover, such as developing lymphocytes. In contrast, Bcl-x_L is found in tissues containing long-lived postmitotic cells, such as adult brain. Like Bcl-2, Bcl-x_L has been shown to suppress apoptosis by inhibiting cytochrome *c* release without affecting mitochondrial depolarization (Johnson *et al.*, 2000).

Numerous reports indicate that overexpression of Bcl-2 or Bcl-x_L suppresses TRAIL-induced apoptosis (Fulda *et al.*, 2002a; Guo and Xu, 2001; Hinz *et al.*, 2002; Lamothe and Aggarwal, 2002; Munshi *et al.*, 2001; Rokhlin *et al.*, 2001; Sun *et al.*, 2001). The protection by Bcl-2/Bcl-x_L is not cell type specific because an effect was observed on a variety of cells including pancreatic adenocarcinoma cells (Hinz *et al.*, 2000), Jurkat T-cells (Guo and Xu, 2001), prostate cancer cells (Munshi *et al.*, 2001; Rokhlin *et al.*, 2001), lung cancer cells (Sun *et al.*, 2001), neuroblastoma, glioblastoma and breast carcinoma cell lines (Fulda *et al.*, 2002a), and acute myelogenous leukemia (Lamothe and Aggarwal, 2002). Fulda *et al.* (2002a) showed that Bcl-2 overexpression reduced TRAIL-induced cleavage of caspase-8 and Bid, indicating that caspase-8 was activated upstream and also downstream of mitochondria in a feedback amplification loop. Importantly, Bcl-2 blocked cleavage of caspases-9, -7, and -3 into active subunits. The cleavage of the caspase substrates DNA-fragmentation-factor 45 (DFF45), PARP, and XIAP was also blocked by Bcl-2. Lamothe and Aggarwal (2002) also showed that overexpression of Bcl-x_L or Bcl-2 inhibits TRAIL-induced activation of caspase-8, caspase-7, and caspase-3 and inhibits the cleavage of Bid and PARP proteins. In contrast to these studies, Srinivasan *et al.* (1998) showed that Bcl-x_L functions downstream of caspase-8 to inhibit death receptor-induced apoptosis of breast cancer cells. Another novel mechanism by which Bcl-x_L inhibits apoptosis is by maintaining efficient ATP/ADP exchange (Vander Heiden *et al.*, 1999). Kim *et al.* (2003) used a functional genetic screening to isolate genes interfering with TRAIL-induced apoptosis. They identified Bcl-x_L and FLIP from TRAIL-resistant clones and showed that increased expression of Bcl-x_L, but not Bcl-2, suppressed TRAIL-induced apoptosis in tumor cells.

While numerous reports suggest that TRAIL-induced apoptosis is suppressed by Bcl-2 and Bcl-x_L, others do not support this conclusion (Gazitt *et al.*, 1999; Keough *et al.*, 2000; Kim *et al.*, 2001; Rudner *et al.*, 2001; Walczak *et al.*, 2000). Gazitt *et al.* (1999) showed that TRAIL effectively induced apoptosis in RPMI 8226 and ARP-1 multiple myeloma cells, and Bcl-2-transfected RPMI 8226 and ARP-1 cells were equally sensitive to apoptosis induced by TRAIL. In addition to multiple myeloma cell lines, freshly isolated, flow-sorted myeloma cells from eight different multiple myeloma patients expressing variable levels of Bcl-2 were equally sensitive to TRAIL. Walczak *et al.* (2000) showed that tumor cells overexpressing Bcl-2 or Bcl-x_L became resistant to apoptosis induced by the chemotherapeutic drug etoposide but were sensitive to TRAIL-induced apoptosis; however, they showed a delay in TRAIL-induced mitochondrial permeability transition compared with control transfectants. This indicates that TRAIL-induced apoptosis depends on caspase-8 activation rather than on the disruption of mitochondrial integrity. Keogh *et al.* (2000) showed failure of Bcl-2 to block cytochrome *c* redistribution during TRAIL-induced apoptosis. These variations in results about the antiapoptotic effects of Bcl-2 and Bcl-x_L could be due either to differences in the cell type or other factors acting together with these proteins.

E. NUCLEAR FACTOR-KAPPA B

There are numerous reports that NF-κB activation can suppress apoptosis (Beg and Baltimore, 1996; Van Antwerp *et al.*, 1996; Wang *et al.*, 1996). NF-κB has been shown to regulate the expression of cFLIP, survivin, Bcl-x_L, A1/BFL1, cIAP, and XIAP (Chen *et al.*, 2000; Micheau *et al.*, 2001; Otaki *et al.*, 2000; Stehlik *et al.*, 1998; Wang *et al.*, 1998; Zong *et al.*, 1999). Several reports have demonstrated that NF-κB plays a critical role in TRAIL-induced apoptosis (Bernard *et al.*, 2001; Chen *et al.*, 2003; Delgi-Esposti *et al.*, 1997a; Eid *et al.*, 2002; Franco *et al.*, 2001; Ghosh *et al.*, 2003; Harper *et al.*, 2001; Hu *et al.*, 1999; Jeremias and Debatin, 1998; Keane *et al.*, 2000; Kim *et al.*, 2002a; Oya *et al.*, 2001; Southall *et al.*, 2001; Spalding *et al.*, 2002; Trauzold *et al.*, 2001). Both the TRAIL receptors DR4 and DR5 when overexpressed can by themselves induce apoptosis of cancer cells and activate NF-κB. DcR2 can also activate NF-κB and protect cells from TRAIL-induced apoptosis (Delgi-Esposti *et al.*, 1997a). Hu *et al.* (1999) showed that DR4, DR5, and DcR2–induced NF-κB activation is mediated by a TRAF2-NIK-IκB kinase α/β signaling cascade but is MEK kinase 1 (MEKK1) independent. They also showed that activation of NF-κB or overexpression of DcR2 does not protect against DR4-induced apoptosis. Moreover, inhibition of NF-κB by IκBα sensitizes cells to TNF- but not TRAIL-induced apoptosis. These findings suggest that TRAIL receptors induce apoptosis and NF-κB activation through distinct signaling pathways, and that activation of NF-κB is not sufficient for protecting cells from TRAIL-induced apoptosis. Bernard *et al.* (2001) showed that NF-κB protects against TRAIL-induced apoptosis by upregulating the TRAIL decoy receptor DcR1.

Chen *et al.* (2003) showed that overexpression of a transdominant-negative mutant of the IκBα results in downregulation of constitutively active NF-κB, induction of DR5, and tumor necrosis factor receptor (TNFR) 1–associated death domain (TRADD) expression and enhancement of TRAIL sensitivity. Overexpression of RelA or a transcription-deficient mutant of c-Rel inhibits TRAIL-induced apoptosis. Depletion of

RelA in mouse embryonic fibroblasts increases cytokine-induced apoptosis, whereas depletion of c-Rel blocks this process. Overexpression of the RelA subunit inhibits caspase-8 and DR4 and DR5 expression and enhances expression of cIAP1 and cIAP2 after TRAIL treatment. By comparison, overexpression of c-Rel enhances DR4, DR5, and Bcl-x_S and inhibits cIAP1, cIAP2, and survivin after TRAIL treatment. These results suggest that the RelA subunit acts as a survival factor by inhibiting expression of DR4/DR5 and caspase-8 and upregulating cIAP1 and cIAP2. The dual function of NF-κB, as an inhibitor or activator of apoptosis, depends on the relative levels of RelA and c-Rel subunits.

Similarly agents/conditions that activate NF-κB can induce resistance to TRAIL-induced apoptosis. For instance, treatment of keratinocytes with IL-1 has been shown to induce resistance to TRAIL, and MG132, which inhibits IL-1 induced NF-κB activation, suppressed the protective effects of IL-1 (Kothny-Wilkes *et al.*, 1998).

Harper *et al.* (2001) showed that TRAIL-induced NF-κB activation occurred in HeLa cells only upon pretreatment with the caspase inhibitor benzyloxycarbonyl-Val-Ala-Asp-(OMe) fluoromethyl ketone (z-VAD.fmk), indicating that this was due to a caspase-sensitive component of TRAIL-induced NF-κB activation. NF-κB activation was mediated by the death receptors DR4 and DR5, but not by DcR1 or DcR2 and was only observed in HeLa cells in the presence of z-VAD.fmk. Receptor-interacting protein (RIP), an obligatory component of TNF-induced NF-κB activation, was cleaved during TRAIL-induced apoptosis. Harper's group showed that RIP is recruited to the native TRAIL DISC (death-inducing signal complex) and that recruitment is enhanced in the presence of z-VAD.fmk, thus providing an explanation for the potentiation of TRAIL-induced NF-κB activation by z-VAD.fmk in TRAIL-sensitive cell lines. Examination of the TRAIL DISC in sensitive and resistant cells suggests that a high ratio of cFLIP to caspase-8 may partially explain cellular resistance to TRAIL-induced apoptosis. Sensitivity to TRAIL-induced apoptosis was also modulated by inhibition or activation of NF-κB. Thus, in some contexts, modulation of NF-κB activation, possibly at the level of apical caspase activation at the DISC, may be a key determinant of sensitivity to TRAIL-induced apoptosis.

There are also reports that NF-κB activation plays a proapoptotic role, possibly through the induction of TRAIL (Baetu *et al.*, 2001; Ravi *et al.*, 2001; Siegmund *et al.*, 2001) and TRAIL receptor expression (Ravi *et al.*, 2001). Ravi *et al.* (2001) showed that the c-Rel subunit of NF-κB induces expression of DR4 and DR5; conversely, a transdominant mutant of the inhibitory protein IκBα or a transactivation-deficient mutant of c-Rel reduces expression of either death receptor. Whereas NF-κB promotes death receptor expression, cytokine-mediated activation of the RelA subunit of NF-κB also increases expression of the apoptosis inhibitor Bcl-x_L and protects cells from TRAIL. Inhibition of NF-κB by blocking activation of the IKK complex reduces Bcl-x_L expression and sensitizes tumor cells to TRAIL-induced apoptosis. The ability to induce death receptors or Bcl-x_L may explain the dual roles of NF-κB as a mediator or inhibitor of cell death during immune and stress responses. That NF-κB can be proapoptotic is indicated by another recent report that showed that retinoic acid induces apoptosis in human acute promyelocytic leukemia cells through NF-κB–mediated induction of TRAIL (Altucci *et al.*, 2001).

There are also reports that NF-κB does not modulate sensitivity of renal carcinoma cells to TRAIL (Pawlowski *et al.*, 2000), while another report showed that constitutive activation of NF-κB prevents TRAIL-induced apoptosis in renal cancer cells (Oya *et al.*, 2001). Perhaps it is the balance between the proapoptotic and antiapoptotic effects of NF-κB that ultimately determines its effect, and this balance may vary in different cells.

F. Akt

There are numerous reports that suggest that activation of Akt inhibits TRAIL-induced apoptosis in a wide variety of tumor cells (Asakuma *et al.*, 2003; Bild *et al.*, 2002; Chen *et al.*, 2001; Kandasamy and Srivastava, 2002; Milani *et al.*, 2003; Mitsiades *et al.*, 2002b; Nam *et al.*, 2002; Nesterov *et al.*, 2001; Panka *et al.*, 2001; Park and Seol, 2002; Secchiero *et al.*, 2003; Wang *et al.*, 2002). The mechanism by which activation of Akt could suppress apoptosis may differ. Akt has been shown to induce the expression of cFLIP, which can block TRAIL-induced apoptosis (Nam *et al.*, 2002; Panka *et al.*, 2001). Akt can also induce NF-κB activation, which can then suppress apoptosis as indicated previously (Mitsiades *et al.*, 2002b). What may induce Akt activation leading to suppression of apoptosis induced by TRAIL may also vary. For instance some cells express constitutively active Akt as noted in prostate cancer cells (Chen *et al.*, 2001; Nesterov *et al.*, 2001) or growth factor such as EGF can activate Akt, inducing resistance to TRAIL (Park and Seol, 2002). Interestingly, however, in some cells such as primary human vascular endothelial cells, TRAIL itself can activate Akt, leading to survival and proliferation of these cells (Secchiero *et al.*, 2003). In human colon cancer cells, it was shown that the expression of TRAIL itself is regulated by activation of Akt (Wang *et al.*, 2002), suggesting an autocrine loop. Thus all these reports indicate that Akt plays an important role in TRAIL-induced apoptosis.

G. OTHER MECHANISMS

Besides FLIP, IAP, Bcl-2, NF-κB, and Akt, tumor cells may resist TRAIL-induced apoptosis through other mechanisms. These include loss of caspase-8 expression as observed in neuroblastoma cells (Eggert *et al.*, 2001). Shiiki *et al.* (2000) showed that granulocytic differentiation of cells by dimethyl sulfoxide induces resistance to TRAIL-induced apoptosis through expression of antagonistic decoy receptor DcR1 and DcR2. In addition, expression of Toso, a cell surface apoptosis regulator, seemed to block activation of caspase-8 by TRAIL via enhanced expression of $FLIP_L$ in granulocytic differentiated cells.

III. INHIBITORS OF ANTIAPOPTOTIC FACTORS AS THERAPEUTIC AGENTS

Agents that suppress the expression of FLIP, IAP, Bcl-2, NF-κB, and Akt have a potential to sensitize cells to TRAIL-induced apoptosis. Degterev et al. (2001) have identified a series of novel small molecules (BH3Is) that inhibit the binding of the Bak BH3 peptide to Bcl-x_L that target the BH3-binding pocket of Bcl-x_L. Inhibitors specifically block BH3-domain–mediated heterodimerization between Bcl-2 family members *in vitro* and *in vivo* and induce apoptosis. This small cell-permeable compound, BH3I-2′ (3-iodo-5-chloro-N-[2-chloro-5-([4-chlorophenyl]sulphonyl)phenyl]-2-hydroxybenzamide), was found to induce apoptosis by disrupting interactions mediated by the BH3 domain between proapoptotic and antiapoptotic members of the Bcl-2 family (Feng et al., 2003). This study found that BH3I-2′ induced cytochrome *c* release from the mitochondrial outer membrane in a Bax-dependent manner and that this correlated with the sensitivity of leukemic cells to apoptosis. Moreover, it also induced rapid damage to the inner mitochondrial membrane, represented by a rapid collapse of mitochondrial membrane potential (DeltaPsim), before cytochrome *c* release. This occurred both in whole cells and isolated mitochondria and was not associated with the sensitivity of cells to BH3I-2′–induced apoptosis. Exogenous Bcl-2 or Bcl-x_L neutralized BH3I-2′ *in vitro* and diminished its effect on the inner mitochondrial membrane.

Bax has been shown to play a critical role in TRAIL-induced apoptosis. Tumor cells that lack Bax expression are resistant to TRAIL. LeBlanc et al. (2002) have shown that chemotherapeutic agents can upregulate the expression of the TRAIL receptor DR5 and the Bax homolog Bak in $Bax^{-/-}$ cells and restore TRAIL sensitivity *in vitro* and *in vivo*. Bax mutation commonly seen in mismatch repair-deficient tumors can cause resistance to death receptor–targeted therapy, but preexposure to chemotherapy rescues tumor sensitivity.

Natural antagonists have also been identified that directly bind IAPs and inactivate their antiapoptotic activity (Verhagen et al., 2000). Smac/DIABLO is a novel protein that can bind XIAP and IAP. The N-terminally processed, IAP-interacting form of Smac/DIABLO is concentrated in membrane fractions in healthy cells but released into the XIAP-containing cytosolic fractions upon ultraviolet (UV) irradiation. As transfection of cells with Smac/DIABLO was able to counter the protection afforded by XIAP against UV irradiation, Smac/DIABLO may promote apoptosis by binding to IAPs and preventing them from inhibiting caspases. Another group reported (Du et al., 2000) that, Smac/DIABLO, which promotes caspase activation in the cytochrome c/Apaf-1/caspase-9 pathway, promotes caspase-9 activation by binding to inhibitor of apoptosis proteins, IAPs, and removing their inhibitory activity. Smac/DIABLO is normally a mitochondrial protein, but it is released into the cytosol when cells undergo apoptosis. Mitochondrial import and cleavage of its signal peptide are required for Smac/DIABLO to gain its apoptotic activity. Overexpression of Smac/DIABLO increases cells' sensitivity to apoptotic stimuli. Smac/DIABLO is the second mitochondrial protein, along with cytochrome c, that promotes apoptosis by activating caspases. Suzuki et al. (2001a) report that a serine protease called HtrA2/Omi is released from mitochondria and inhibits the function of XIAP by direct binding in a similar way to Smac/DIABLO. Moreover, when overexpressed extramitochondrially, HtrA2 induces atypical cell death, which is neither accompanied by a significant increase in caspase activity nor inhibited by caspase inhibitors, including XIAP. A catalytically inactive mutant of HtrA2, however, does not induce cell death. In short, HtrA2 is a Smac/DIABLO-like inhibitor of IAP activity with a serine protease-dependent cell death–inducing activity. The amino-terminal amino acids of Smac/DIABLO are indispensable for its function, and a seven-residue peptide derived from the amino terminus is sufficient to bind IAP and block their association with caspases, resulting in procaspase-3 activation (Chai et al., 2000). Huang et al. (2000) have identified a peptide as short as four amino acids long that reverses the effect of IAP.

For TRAIL-induced apoptosis, it was reported that Bax-dependent mitochondrial release of Smac/DIABLO was essential (Deng et al., 2002). Bax-null cancer cells are resistant to TRAIL-induced apoptosis. Bax deficiency had no effect on TRAIL-induced caspase-8 activation and subsequent cleavage of Bid; however, it resulted in an incomplete caspase-3 processing because of inhibition by XIAP. Release of Smac/DIABLO from mitochondria through the TRAIL-caspase-8 Bid-Bax cascade was required to remove the inhibitory effect of XIAP and allow apoptosis to proceed. Inhibition of caspase-9 activity had no effect on TRAIL-induced caspase-3 activation and cell death, whereas expression of the active form of Smac/DIABLO in the cytosol was sufficient to reconstitute TRAIL sensitivity in Bax-deficient cells. These results demonstrate that Bax-dependent release of Smac/DIABLO, not cytochrome c, from mitochondria mediates the contribution of the mitochondrial pathway to death receptor–mediated apoptosis.

The resistance of cancer cells to TRAIL could be reversed by various chemotherapeutic agents. This reversal may involve different mechanisms. Induction of death receptors for TRAIL has been described to be one of the mechanisms. Asakuma et al. (2003), however, showed that selective Akt inactivation sensitizes renal cancer cells to TRAIL. They showed that resistance to TRAIL-induced apoptosis correlates well with the level of Akt phosphorylation at Ser 473 and the sensitivity to TRAIL was altered by modulation of Akt activity, which increased the protein expression of cFLIP. Paclitaxel and cisplatin but not etoposide promoted TRAIL-induced apoptosis in cells, which was not mediated by increased TRAIL-receptor expression but by chemotherapeutic-induced Akt inactivation through ceramide formation derived from sphingomyelin hydrolysis. Similarly, regulation of Akt by EGFR inhibitors has also been shown to enhance TRAIL-induced apoptosis (Park and Seol, 2002).

IV. EFFECT OF TRAIL ON NORMAL CELLS

From the previous description, it is clear that TRAIL is a potent inducer of apoptosis and its activity is regulated by numerous factors. TRAIL has been shown to exert potent cytotoxic activity against many tumor cell lines but not against normal cells. It has been hypothesized that this difference in TRAIL sensitivity between normal and transformed cells might be due to the expression of the non–death-inducing TRAIL receptors DcR1 and DcR2, presumably by competition for limited amounts of TRAIL. However, no correlation has been found between the expression of decoy receptors and sensitivity to TRAIL in most tumor cells (Zhang et al., 2000). TRAIL was found to induce apoptosis in keratinocytes (Leverkus et al., 2000), fibroblasts (Ichikawa et al., 2003), primary plasma cells (Ursini-Siegel et al., 2002), hepatocytes (Lawrence et al., 2001), and thymocytes (Lamhamedi-Cherradi et al., 2003). TRAIL was also found to promote the survival and proliferation of primary human vascular endothelial cells by activating the Akt and extracellular signal–regulated kinase (ERK) pathways (Secchiero et al., 2003).

What is the true physiologic role of TRAIL? Song et al. (2000) reported that chronic blockade of TRAIL in mice exacerbated autoimmune arthritis and that intra-articular TRAIL gene transfer ameliorated the disease. In vivo, TRAIL blockade led to profound hyperproliferation of synovial cells and arthritogenic lymphocytes and heightened the production of cytokines and autoantibodies. In vitro, TRAIL inhibited DNA synthesis and prevented cell cycle progression of lymphocytes. Interestingly, TRAIL had no effect on apoptosis of inflammatory cells either in vivo or in vitro. Thus, unlike other members of the TNF superfamily, TRAIL is a prototype inhibitor protein that inhibits autoimmune inflammation by blocking cell cycle progression. Lamhamedi-Cherradi et al. (2003) showed that mice deficient in TRAIL have a severe defect in thymocyte apoptosis, so thymic deletion induced by T-cell receptor ligation is severely impaired. TRAIL-deficient mice are also hypersensitive to collagen-induced arthritis and streptozotocin-induced diabetes and develop heightened autoimmune responses. Thus, TRAIL mediates thymocyte apoptosis and is important in the induction of autoimmune diseases.

Mi *et al.* (2003) found that TRAIL blockade exacerbates the onset of type 1 diabetes in nonobese diabetics (NOD). In SCID recipients of transferred diabetogenic T-cells and in cyclophosphamide-treated NOD mice, TRAIL inhibits the proliferation of NOD diabetogenic T-cells by suppressing IL-2 production and cell cycle progression; this inhibition can be rescued by the addition of exogenous IL-2. cDNA array and Western blot analyses indicate that TRAIL upregulates the expression of the cdk inhibitor p27 (kip1). This data suggest that TRAIL is an important immune regulator of the development of type 1 diabetes.

V. CONCLUSIONS

Although gene-deletion studies indicate that TRAIL mediates thymocyte apoptosis and is important in the induction of autoimmune diseases, the studies described previously demonstrate that TRAIL is a potent inducer of apoptosis in most cells. However, its activity is negatively regulated by a wide variety of factors. TRAIL and agonistic antibodies against its receptor are currently in clinical trials for the treatment of various cancers. In order for TRAIL to be fully effective in suppressing the growth of tumor cells, one must find ways to inhibit negative regulatory signals as described here. Although much has been learned about this cytokine and its mechanism of action within the past few years, there are some important questions about the role of TRAIL that are still unanswered. For instance, why does TRAIL bind to five distinct receptors, two of which display agonistic effects and three of which mediate an antagonistic role? Irrespective of apoptosis, why does TRAIL activate NF-κB in some cells and not in others? Why are there so many different intracellular inhibitors of TRAIL-induced apoptosis? Why does TRAIL activate both apoptosis and antiapoptosis pathways (through activation of Akt and NF-κB)? Will administration of TRAIL to cancer patients enhance autoimmunity? Overall TRAIL is an exciting new cytokine of the TNF superfamily that is likely to demonstrate its clinical potential within the next few years.

ACKNOWLEDGMENTS

We thank Walter Pagel for his careful review of this manuscript. This work was supported by the Clayton Foundation for Research (to BBA), Department of Defense US Army Breast Cancer Research Program grant (BC010610, to BBA), a PO1 grant (CA91844) from the

National Institutes of Health on lung chemoprevention (to BBA) and a P50 Head and Neck SPORE grant from the National Institutes of Health (to BBA), and a Cancer Center Support Grant CA 16672. Dr. Aggarwal is a Ransom Horne, Jr. Distinguished Professor of Cancer Research.

REFERENCES

Aggarwal, B. B. (2003). Signalling pathways of the TNF superfamily: A double-edged sword. *Nat. Rev. Immunol.* **3,** 745–758.

Altucci, L., Rossin, A., Raffelsberger, W., Reitmair, A., Chomienne, C., and Gronemeyer, H. (2001). Retinoic acid-induced apoptosis in leukemia cells is mediated by paracrine action of tumor-selective death ligand TRAIL. *Nat. Med.* **7,** 680–686.

Ambrosini, G., Adida, C., and Altieri, D. C. (1997). A novel anti-apoptosis gene, survivin, expressed in cancer and lymphoma. *Nat. Med.* **3,** 917–921.

Asakuma, J., Sumitomo, M., Asano, T., and Hayakawa, M. (2003). Selective Akt inactivation and tumor necrosis factor-related apoptosis-inducing ligand sensitization of renal cancer cells by low concentrations of paclitaxel. *Cancer Res.* **63,** 1365–1370.

Baetu, T. M., Kwon, H. *et al.* (2001). Disruption of NF-kappaB signaling reveals a novel role for NF-kappaB in the regulation of TNF-related apoptosis-inducing ligand expression. *J. Immunol.* **167,** 3164–3173.

Beg, A. A., and Baltimore, D. (1996). An essential role for NF-kappaB in preventing TNF-alpha-induced cell death. *Science* **274,** 782–784.

Bernard, D., Quatannens, B., Vandenbunder, B., and Abbadie, C. (2001). Rel/NF-kappaB transcription factors protect against tumor necrosis factor (TNF)-related apoptosis-inducing ligand (TRAIL)-induced apoptosis by up-regulating the TRAIL decoy receptor DcR1. *J. Biol. Chem.* **276,** 27322–27328.

Bild, A. H., Mendoza, F. J., Gibson, E. M., Huang, M., Villanueva, J., Garrington, T. P., Jove, R., Johnson, G. L., and Gibson, S. B. (2002). MEKK1-induced apoptosis requires TRAIL death receptor activation and is inhibited by AKT/PKB through inhibition of MEKK1 cleavage. *Oncogene* **21,** 6649–6656.

Bodmer, J. L., Holler, N., Reynard, S., Vinciguerra, P., Schneider, P., Juo, P., Blenis, J., and Tschopp, J. (2000). TRAIL receptor-2 signals apoptosis through FADD and caspase-8. *Nat. Cell Biol.* **2,** 241–343.

Boise, L. H., Gonzalez-Garcia, M., Postema, C. E., Ding, L., Lindsten, T., Turka, L. A., Mao, X., Nunez, G., and Thompson, C. B. (1993). bcl-x, a bcl-2-related gene that functions as a dominant regulator of apoptotic cell death. *Cell* **74,** 597–608.

Burns, T. F., and El-Deiry, W. S. (2001). Identification of inhibitors of TRAIL-induced death (ITIDs) in the TRAIL-sensitive colon carcinoma cell line SW480 using a genetic approach. *J. Biol. Chem.* **276,** 37879–37886.

Chai, J., Du, C., Wu, J. W., Kyin, S., Wang, X., and Shi, Y. (2000). Structural and biochemical basis of apoptotic activation by Smac/DIABLO. *Nature* **406,** 855–862.

Chaudhary, P. M., Eby, M., Jasmin, A., Bookwalter, A., Murray, J., and Hood, L. (1997). Death receptor 5, a new member of the TNFR family, and DR4 induce FADD-dependent apoptosis and activate the NF-kappaB pathway. *Immunity* **7,** 821–830.

Chen, C., Edelstein, L. C., and Gelinas, C. (2000). The Rel/NF-kappaB family directly activates expression of the apoptosis inhibitor Bcl-x(L). *Mol. Cell. Biol.* **20,** 2687–2695.

Chen, X., Kandasamy, K., and Srivastava, R. K. (2003). Differential roles of RelA (p65) and c-Rel subunits of nuclear factor kappa B in tumor necrosis factor-related apoptosis-inducing ligand signaling. *Cancer Res.* **63,** 1059–1066.

Chen, X., Thakkar, H., Tyan, F., Gim, S., Robinson, H., Lee, C., Pandey, S. K., Nwokorie, C., Onwudiwe, N., and Srivastava, R. K. (2001). Constitutively active Akt is an important regulator of TRAIL sensitivity in prostate cancer. *Oncogene* **20**, 6073–6083.

Cory, S., and Adams, J. M. (2002). The Bcl2 family: Regulators of the cellular life-or-death switch. *Nat. Rev. Cancer* **2**, 647–656.

Degli-Esposti, M. A., Dougall, W. C., Smolak, P. J., Waugh, J. Y., Smith, C. A., and Goodwin, R. G. (1997a). The novel receptor TRAIL-R4 induces NF-kappaB and protects against TRAIL-mediated apoptosis, yet retains an incomplete death domain. *Immunity* **7**, 813–820.

Degli-Esposti, M. A., Smolak, P. J., Walczak, H., Waugh, J., Huang, C. P., DuBose, R. F., Goodwin, R. G., and Smith, C. A. (1997b). Cloning and characterization of TRAIL-R3, a novel member of the emerging TRAIL receptor family. *J. Exp. Med.* **186**, 1165–1170.

Degterev, A., Lugovskoy, A., Cardone, M., Mulley, B., Wagner, G., Mitchison, T., and Yuan, J. (2001). Identification of small-molecule inhibitors of interaction between the BH3 domain and Bcl-xL. *Nat. Cell Biol.* **3**, 173–182.

Deng, Y., Lin, Y., and Wu, X. (2002). TRAIL-induced apoptosis requires Bax-dependent mitochondrial release of Smac/DIABLO. *Genes Dev.* **16**, 33–45.

Deveraux, Q. L., Roy, N., Stennicke, H. R., Van Arsdale, T., Zhou, Q., Srinivasula, S. M., Alnemri, E. S., Salvesen, G. S., and Reed, J. C. (1998). IAPs block apoptotic events induced by caspase-8 and cytochrome c by direct inhibition of distinct caspases. *EMBO J.* **17**, 2215–2223.

Deveraux, Q. L., Takahashi, R., Salvesen, G. S., and Reed, J. C. (1997). X-linked IAP is a direct inhibitor of cell-death proteases. *Nature* **388**, 300–304.

Du, C., Fang, M., Li, Y., Li, L., and Wang, X. (2000). Smac, a mitochondrial protein that promotes cytochrome c-dependent caspase activation by eliminating IAP inhibition. *Cell* **102**, 33–42.

Eggert, A., Grotzer, M. A., Zuzak, T. J., Wiewrodt, B. R., Ho, R., Ikegaki, N., and Brodeur, G. M. (2001). Resistance to tumor necrosis factor-related apoptosis-inducing ligand (TRAIL)-induced apoptosis in neuroblastoma cells correlates with a loss of caspase-8 expression. *Cancer Res.* **61**, 1314–1319.

Eid, M. A., Lewis, R. W., Abdel-Mageed, A. B., and Kumar, M. V. (2002). Reduced response of prostate cancer cells to TRAIL is modulated by NFkappaB-mediated inhibition of caspases and Bid activation. *Int. J. Oncol.* **21**, 111–117.

Emery, J. G., McDonnell, P., Burke, M. B., Deen, K. C., Lyn, S., Silverman, C., Dul, E., Appelbaum, E. R., Eichman, C., DiPrinzio, R., Dodds, R. A., James, I. E., Rosenberg, M., Lee, J. C., and Young, P. R. (1998). Osteoprotegerin is a receptor for the cytotoxic ligand TRAIL. *J. Biol. Chem.* **273**, 14363–14367.

Feng, W. Y., Liu, F. T., Patwari, Y., Agrawal, S. G., Newland, A. C., and Jia, L. (2003). BH3-domain mimetic compound BH3I-2' induces rapid damage to the inner mitochondrial membrane prior to the cytochrome c release from mitochondria. *Br. J. Haematol.* **121**, 332–340.

Franco, A. V., Zhang, X. D., Van Berkel, E., Sanders, J. E., Zhang, X. Y., Thomas, W. D., Nguyen, T., and Hersey, P. (2001). The role of NF-kappa B in TNF-related apoptosis-inducing ligand (TRAIL)-induced apoptosis of melanoma cells. *J. Immunol.* **166**, 5337–5345.

Frese, S., Brunner, T., Gugger, M., Uduehi, A., and Schmid, R. A. (2002). Enhancement of Apo2L/TRAIL (tumor necrosis factor-related apoptosis-inducing ligand)-induced apoptosis in non-small cell lung cancer cell lines by chemotherapeutic agents without correlation to the expression level of cellular protease caspase-8 inhibitory protein. *J. Thorac. Cardiovasc. Surg.* **123**, 168–174.

Fukazawa, T., Fujiwara, T., Uno, F., Teraishi, F., Kadowaki, Y., Itoshima, T., Takata, Y., Kagawa, S., Roth, J. A., Tschopp, J., and Tanaka, N. (2001). Accelerated degradation of cellular FLIP protein through the ubiquitin-proteasome pathway in p53-mediated apoptosis of human cancer cells. *Oncogene* **20,** 5225–5231.

Fulda, S., Meyer, E., and Debatin, K. M. (2002a). Inhibition of TRAIL-induced apoptosis by Bcl-2 overexpression. *Oncogene* **21,** 2283–2294.

Fulda, S., Wick, W., Weller, M., and Debatin, K. M. (2002b). Smac agonists sensitize for Apo2L/TRAIL- or anticancer drug-induced apoptosis and induce regression of malignant glioma *in vivo. Nat. Med.* **8,** 808–815.

Gazitt, Y., Shaughnessy, P., and Montgomery, W. (1999). Apoptosis-induced by TRAIL and TNF-alpha in human multiple myeloma cells is not blocked by BCL-2. *Cytokine* **11,** 1010–1019.

Ghosh, S. K., Wood, C., Boise, L. H., Mian, A. M., Deyev, V. V., Feuer, G., Toomey, N. L., Shank, N. C., Cabral, L., Barber, G. N., and Harrington, W. J., Jr. (2003). Potentiation of TRAIL-induced apoptosis in primary effusion lymphoma through azidothymidine-mediated inhibition of NF-kappa B. *Blood* **101,** 2321–2327.

Griffith, T. S., Chin, W. A., Jackson, G. C., Lynch, D. H., and Kubin, M. Z. (1998). Intracellular regulation of TRAIL-induced apoptosis in human melanoma cells. *J. Immunol.* **161,** 2833–2840.

Griffith, T. S., Fialkov, J. M., Scott, D. L., Azuhata, T., Williams, R. D., Wall, N. R., Altieri, D. C., and Sandler, A. D. (2002). Induction and regulation of tumor necrosis factor-related apoptosis-inducing ligand/Apo-2 ligand-mediated apoptosis in renal cell carcinoma. *Cancer Res.* **62,** 3093–3099.

Guo, F., Nimmanapalli, R., Paranawithana, S., Wittman, S., Griffin, D., Bali, P., O'Bryan, E., Fumero, C., Wang, H. G., and Bhalla, K. (2002). Ectopic overexpression of second mitochondria-derived activator of caspases (Smac/DIABLO) or cotreatment with N-terminus of Smac/DIABLO peptide potentiates epothilone B derivative-(BMS 247550) and Apo-2L/TRAIL-induced apoptosis. *Blood* **99,** 3419–3426.

Guo, B. C., and Xu, Y. H. (2001). Bcl-2 over-expression and activation of protein kinase C suppress the trail-induced apoptosis in Jurkat T cells. *Cell Res.* **11,** 101–106.

Harper, N., Farrow, S. N., Kaptein, A., Cohen, G. M., and MacFarlane, M. (2001). Modulation of tumor necrosis factor apoptosis-inducing ligand-induced NF-kappaB activation by inhibition of apical caspases. *J. Biol. Chem.* **276,** 34743–34752.

Hietakangas, V., Poukkula, M., Heiskanen, K. M., Karvinen, J. T., Sistonen, L., and Eriksson, J. E. (2003). Erythroid differentiation sensitizes K562 leukemia cells to TRAIL-induced apoptosis by downregulation of c-FLIP. *Mol. Cell. Biol.* **23,** 1278–1291.

Higuchi, H., Yoon, J. H., Grambihler, A., Werneburg, N., Bronk, S. F., and Gores, G. J. (2003). Bile acids stimulate cFLIP phosphorylation enhancing TRAIL-mediated apoptosis. *J. Biol. Chem.* **278,** 454–461.

Hinz, S., Trauzold, A., Boenicke, L., Sandberg, C., Beckmann, S., Bayer, E., Walczak, H., Kalthoff, H., and Ungefroren, H. (2000). Bcl-XL protects pancreatic adenocarcinoma cells against CD95- and TRAIL-receptor-mediated apoptosis. *Oncogene* **19,** 5477–5486.

Holen, I., Croucher, P. I., Hamdy, F. C., and Eaton, C. L. (2002). Osteoprotegerin (OPG) is a survival factor for human prostate cancer cells. *Cancer Res.* **62,** 1619–1623.

Hu, W. H., Johnson, H., and Shu, H. B. (1999). Tumor necrosis factor-related apoptosis-inducing ligand receptors signal NF-kappaB and JNK activation and apoptosis through distinct pathways. *J. Biol. Chem.* **274,** 30603–30610.

Huang, Q., Deveraux, Q. L., Maeda, S., Salvesen, G. S., Stennicke, H. R., Hammock, B. D., and Reed, J. C. (2000). Evolutionary conservation of apoptosis mechanisms: Lepidopteran and baculoviral inhibitor of apoptosis proteins are inhibitors of mammalian caspase-9. *Proc. Natl. Acad. Sci. USA* **97,** 1427–1432.

Ichikawa, K., Liu, W., Fleck, M., Zhang, H., Zhao, L., Ohtsuka, T., Wang, Z., Liu, D., Mountz, J. D., Ohtsuki, M., Koopman, W. J., Kimberly, R., and Zhou, T. (2003). TRAIL-R2

(DR5) mediates apoptosis of synovial fibroblasts in rheumatoid arthritis. *J. Immunol.* **171**, 1061–1069.

Irmler, M., Thome, M., Hahne, M., Schneider, P., Hofmann, K., Steiner, V., Bodmer, J. L., Schroter, M., Burns, K., Mattmann, C., Rimoldi, D., French, L. E., and Tschopp, J. (1997). Inhibition of death receptor signals by cellular FLIP. *Nature* **388**, 190–195.

Jeremias, I., and Debatin, K. M. (1998). TRAIL induces apoptosis and activation of NFkappaB. *Eur. Cytokine Netw.* **9**, 687–688.

Johnson, B. W., Cepero, E., and Boise, L. H. (2000). Bcl-xL inhibits cytochrome c release but not mitochondrial depolarization during the activation of multiple death pathways by tumor necrosis factor-alpha. *J. Biol. Chem.* **275**, 31546–31553.

Kandasamy, K., and Srivastava, R. K. (2002). Role of the phosphatidylinositol 3'-kinase/PTEN/Akt kinase pathway in tumor necrosis factor-related apoptosis-inducing ligand-induced apoptosis in non-small cell lung cancer cells. *Cancer Res.* **62**, 4929–4937.

Keane, M. M., Rubinstein, Y., Cuello, M., Ettenberg, S. A., Banerjee, P., Nau, M. M., and Lipkowitz, S. (2000). Inhibition of NF-kappaB activity enhances TRAIL mediated apoptosis in breast cancer cell lines. *Breast Cancer Res. Treat.* **64**, 211–219.

Kelly, M. M., Hoel, B. D., and Voelkel-Johnson, C. (2002). Doxorubicin pretreatment sensitizes prostate cancer cell lines to TRAIL induced apoptosis which correlates with the loss of c-FLIP expression. *Cancer. Biol. Ther.* **1**, 520–527.

Keogh, S. A., Walczak, H., Bouchier-Hayes, L., and Martin, S. J. (2000). Failure of Bcl-2 to block cytochrome c redistribution during TRAIL-induced apoptosis. *FEBS Lett.* **471**, 93–98.

Kerr, J. F. R., Wyllie, A. H., and Searle, J. (1972). Apoptosis: A basic biological phenomenon with wide-ranging implications in tissue kinetics. *Br. J. Cancer* **26**, 239–257.

Kim, I. K., Jung, Y. K., Noh, D. Y., Song, Y. S., Choi, C. H., Oh, B. H., Masuda, E. S., and Jung, Y. K. (2003). Functional screening of genes suppressing TRAIL-induced apoptosis: Distinct inhibitory activities of Bcl-XL and Bcl-2. *Br. J. Cancer* **88**, 910–917.

Kim, Y. S., Schwabe, R. F., Qian, T., Lemasters, J. J., and Brenner, D. A. (2002a). TRAIL-mediated apoptosis requires NF-kappaB inhibition and the mitochondrial permeability transition in human hepatoma cells. *Hepatology* **36**, 1498–1508.

Kim, Y., Suh, N., Sporn, M., and Reed, J. C. (2002b). An inducible pathway for degradation of FLIP protein sensitizes tumor cells to TRAIL-induced apoptosis. *J. Biol. Chem.* **277**, 22320–22329.

Kim, E. J., Suliman, A., Lam, A., and Srivastava, R. K. (2001). Failure of Bcl-2 to block mitochondrial dysfunction during TRAIL-induced apoptosis. Tumor necrosis-related apoptosis-inducing ligand. *Int. J. Oncol.* **18**, 187–194.

Kischkel, F. C., Lawrence, D. A., Chuntharapai, A., Schow, P., Kim, K. J., and Ashkenazi, A. (2000). Apo2L/TRAIL-dependent recruitment of endogenous FADD and caspase-8 to death receptors 4 and 5. *Immunity* **12**, 611–620.

Kischkel, F. C., Lawrence, D. A., Tinel, A., LeBlanc, H., Virmani, A., Schow, P., Gazdar, A., Blenis, J., Arnott, D., and Ashkenazi, A. (2001). Death receptor recruitment of endogenous caspase-10 and apoptosis initiation in the absence of caspase-8. *J. Biol. Chem.* **7**, 46639–46646.

Kothny-Wilkes, G., Kulms, D., Poppelmann, B., Luger, T. A., Kubin, M., and Schwarz, T. (1998). Interleukin-1 protects transformed keratinocytes from tumor necrosis factor-related apoptosis-inducing ligand. *J. Biol. Chem.* **273**, 29247–29253.

Kreuz, S., Siegmund, D., Scheurich, P., and Wajant, H. (2001). NF-kappaB inducers upregulate cFLIP, a cycloheximide-sensitive inhibitor of death receptor signaling. *Mol. Cell. Biol.* **21**, 3964–3973.

Lamhamedi-Cherradi, S. E., Zheng, S. J., Maguschak, K. A., Peschon, J., and Chen, Y. H. (2003). Defective thymocyte apoptosis and accelerated autoimmune diseases in TRAIL−/− mice. *Nat. Immunol.* **4**, 255–260.

Lamothe, B., and Aggarwal, B. B. (2002). Ectopic expression of Bcl-2 and Bcl-xL inhibits apoptosis induced by TNF-related apoptosis-inducing ligand (TRAIL) through suppression of caspases-8, 7, and 3 and Bid cleavage in human acute myelogenous leukemia cell line HL-60. *J. Interferon Cytokine Res.* **22,** 269–279.

Lawrence, D., Shahrokh, Z., Marsters, S., Achilles, K., Shih, D., Mounho, B., Hillan, K., Totpal, K., DeForge, L., Schow, P., Hooley, J., Sherwood, S., Pai, R., Leung, S., Khan, L., Gliniak, B., Bussiere, J., Smith, C. A., Strom, S. S., Kelley, S., Fox, J. A., Thomas, D., and Ashkenazi, A. (2001). Differential hepatocyte toxicity of recombinant Apo2L/TRAIL versions. *Nat. Med.* **7,** 383–385.

LeBlanc, H., Lawrence, D., Varfolomeev, E., Totpal, K., Morlan, J., Schow, P., Fong, S., Schwall, R., Sinicropi, D., and Ashkenazi, A. (2002). Tumor-cell resistance to death receptor–induced apoptosis through mutational inactivation of the proapoptotic Bcl-2 homolog Bax. *Nat. Med.* **8,** 274–281.

Leverkus, M., Neumann, M., Mengling, T., Rauch, C. T., Brocker, E. B., Krammer, P. H., and Walczak, H. (2000). Regulation of tumor necrosis factor-related apoptosis-inducing ligand sensitivity in primary and transformed human keratinocytes. *Cancer Res.* **60,** 553–559.

Li, F., Ambrosini, G., Chu, E. Y., Plescia, J., Tognin, S., Marchisio, P. C., and Altieri, D. C. (1998a). Control of apoptosis and mitotic spindle checkpoint by survivin. *Nature* **396,** 580–584.

Li, H., Zhu, H., Xu, C. J., and Yuan, J. (1998b). Cleavage of Bid by caspase 8 mediates the mitochondrial damage in the Fas pathway of apoptosis. *Cell* **94,** 491–501.

Luo, X., Budihardjo, I., Zou, H., Slaughter, C., and Wang, X. (1998). Bid, a Bcl2 interacting protein, mediates cytochrome c release from mitochondria in response to activation of cell surface death receptors. *Cell* **94,** 481–490.

MacFarlane, M., Ahmad, M., Srinivasula, S. M., Fernandes-Alnemri, T., Cohen, G. M., and Alnemri, E. S. (1997). Identification and molecular cloning of two novel receptors for the cytotoxic ligand TRAIL. *J. Biol. Chem.* **272,** 25417–25420.

MacFarlane, M., Merrison, W., Bratton, S. B., and Cohen, G. M. (2002). Proteasome-mediated degradation of Smac during apoptosis: XIAP promotes Smac ubiquitination in vitro. *J. Biol. Chem.* **277,** 36611–36616.

Marsters, S. A., Pitti, R. M., Donahue, C. J., Ruppert, S., Bauer, K. D., and Ashkenazi, A. (1996). Activation of apoptosis by Apo-2 ligand is independent of FADD but blocked by CrmA. *Cur. Biol.* **6,** 750–752.

Marsters, S. A., Sheridan, J. P., Pitti, R. M., Huang, A., Skubatch, M., Baldwin, D., Yuan, J., Gurney, A., Goddard, A. D., Godowski, P., and Ashkenazi, A. (1997). A novel receptor for Apo2L/TRAIL contains a truncated death domain. *Cur. Biology* **7,** 1003–1006.

Mi, Q. S., Ly, D., Lamhamedi-Cherradi, S. E., Salojin, K. V., Zhou, L., Grattan, M., Meagher, C., Zucker, P., Chen, Y. H., Nagle, J., Taub, D., and Delovitch, T. L. (2003). Blockade of tumor necrosis factor-related apoptosis-inducing ligand exacerbates type 1 diabetes in NOD mice. *Diabetes* **52,** 1967–1975.

Micheau, O., Lens, S., Gaide, O., Alevizopoulos, K., and Tschopp, J. (2001). NF-kappaB signals induce the expression of c-FLIP. *Mol. Cell. Biol.* **21,** 5299–5305.

Milani, D., Zauli, G., Rimondi, E., Celeghini, C., Marmiroli, S., Narducci, P., Capitani, S., and Secchiero, P. (2003). Tumor necrosis factor-related apoptosis-inducing ligand sequentially activates pro-survival and pro-apoptotic pathways in SK-N-MC neuronal cells. *J. Neurochem.* **86,** 126–135.

Mitsiades, N., Mitsiades, C. S., Poulaki, V., Anderson, K. C., and Treon, S. P. (2002a). Intracellular regulation of tumor necrosis factor-related apoptosis-inducing ligand-induced apoptosis in human multiple myeloma cells. *Blood* **99,** 2162–2171.

Mitsiades, C. S., Mitsiades, N., Poulaki, V., Schlossman, R., Akiyama, M., Chauhan, D., Hideshima, T., Treon, S. P., Munshi, N. C., Richardson, P. G., and Anderson, K. C. (2002b). Activation of NF-kappaB and upregulation of intracellular anti-apoptotic proteins via the IGF-1/Akt signaling in human multiple myeloma cells: Therapeutic implications. *Oncogene* **21,** 5673–5683.

Miyazaki, T., and Reed, J. C. (2001). A GTP-binding adapter protein couples TRAIL receptors to apoptosis-inducing proteins. *Nat. Immunol.* **2,** 493–500.

Mongkolsapaya, J., Cowper, A. E., Xu, X. N., Morris, G., McMichael, A. J., Bell, J. I., and Screaton, G. R. (1998). Lymphocyte inhibitor of TRAIL (TNF-related apoptosis-inducing ligand): A new receptor protecting lymphocytes from the death ligand TRAIL. *J. Immunol.* **160,** 3–6.

Munshi, A., Pappas, G., Honda, T., McDonnell, T. J., Younes, A., Li, Y., and Meyn, R. E. (2001). TRAIL (APO-2L) induces apoptosis in human prostate cancer cells that is inhibitable by Bcl-2. *Oncogene* **20,** 3757–3765.

Nam, S. Y., Amoscato, A. A., and Lee, Y. J. (2002). Low glucose-enhanced TRAIL cytotoxicity is mediated through the ceramide-Akt-FLIP pathway. *Oncogene* **21,** 337–346.

Nesterov, A., Lu, X., Johnson, M., Miller, G. J., Ivashchenko, Y., and Kraft, A. S. (2001). Elevated AKT activity protects the prostate cancer cell line LNCaP from TRAIL-induced apoptosis. *J. Biol. Chem.* **276,** 10767–10774.

Otaki, M., Hatano, M., Kobayashi, K., Ogasawara, T., Kuriyama, T., and Tokuhisa, T. (2000). Cell cycle-dependent regulation of TIAP/m-survivin expression. *Biochim. Biophys. Acta* **1493,** 188–194.

Oya, M., Ohtsubo, M., Takayanagi, A., Tachibana, M., Shimizu, N., and Murai, M. (2001). Constitutive activation of nuclear factor-kappaB prevents TRAIL-induced apoptosis in renal cancer cells. *Oncogene* **20,** 3888–3896.

Pan, G., Ni, J., Yu, G., Wei, Y. F., and Dixit, V. M. (1998). TRUNDD, a new member of the TRAIL receptor family that antagonizes TRAIL signalling. *FEBS Lett.* **424,** 41–45.

Pan, G., Ni, J., Wei, Y. F., Yum, G., Gentz, R., and Dixit, V. M. (1997a). An antagonist decoy receptor and a death domain-containing receptor for TRAIL. *Science* **277,** 815–818.

Pan, G., O'Rourke, K., Chinnaiyan, A. M., Gentz, R., Ebner, R., Ni, J., and Dixit, V. M. (1997b). The receptor for the cytotoxic ligand TRAIL. *Science* **276,** 111–113.

Panka, D. J., Mano, T., Suhara, T., Walsh, K., and Mier, J. W. (2001). Phosphatidylinositol 3-kinase/Akt activity regulates c-FLIP expression in tumor cells. *J. Biol. Chem.* **276,** 6893–6896.

Park, S. Y., and Seol, D. W. (2002). Regulation of Akt by EGF-R inhibitors, a possible mechanism of EGF-R inhibitor-enhanced TRAIL-induced apoptosis. *Biochem. Biophys. Res. Commun.* **295,** 515–518.

Pawlowski, J. E., Nesterov, A., Scheinman, R. I., Johnson, T. R., and Kraft, A. S. (2000). NF-kB does not modulate sensitivity of renal carcinoma cells to TNF alpha-related apoptosis-inducing ligand (TRAIL). *Anticancer Res.* **20,** 4243–4255.

Pitti, R. M., Marsters, S. A., Ruppert, S., Donahue, C. J., Moore, A., and Ashkenazi, A. (1996). Induction of apoptosis by Apo-2 ligand, a new member of the tumor necrosis factor cytokine family. *J. Biol. Chem.* **271,** 12687–12690.

Ravi, R., Bedi, G. C., Engstrom, L. W., Zeng, Q., Mookerjee, B., Gelinas, C., Fuchs, E. J., and Bedi, A. (2001). Regulation of death receptor expression and TRAIL/Apo2L-induced apoptosis by NF-kappaB. *Nat. Cell Biol.* **3,** 409–416.

Rokhlin, O. W., Guseva, N., Tagiyev, A., Knudson, C. M., and Cohen, M. B. (2001). Bcl-2 oncoprotein protects the human prostatic carcinoma cell line PC3 from TRAIL-mediated apoptosis. *Oncogene* **20,** 2836–2843.

Roy, N., Deveraux, Q. L., Takahashi, R., Salvesen, G. S., and Reed, J. C. (1997). The c-IAP-1 and c-IAP-2 proteins are direct inhibitors of specific caspases. *EMBO J.* **16,** 6914–6925.

Rudner, J., Lepple-Wienhues, A., Budach, W., Berschauer, J., Friedrich, B., Wesselborg, S., Schulze-Osthoff, K., and Belka, C. (2001). Wild-type, mitochondrial and ER-restricted Bcl-2 inhibit DNA damage-induced apoptosis but do not affect death receptor-induced apoptosis. *J. Cell Sci.* **114,** 4161–4172.

Schneider, P., Thome, M., Burns, K., Bodmer, J. L., Hofmann, K., Kataoka, T., Holler, N., and Tschopp, J. (1997). TRAIL receptors 1 (DR4) and 2 (DR5) signal FADD-dependent apoptosis and activate NF-kappaB. *Immunity* **7**, 831–5397.

Screaton, G. R., Mongkolsapaya, J., Xu, X. N., Cowper, A. E., McMichael, A. J., and Bell, J. I. (1997). TRICK2, a new alternatively spliced receptor that transduces the cytotoxic signal from TRAIL. *Current Biol.* **7**, 693–696.

Secchiero, P., Gonelli, A., Carnevale, E., Milani, D., Pandolfi, A., Zella, D., and Zauli, G. (2003). TRAIL Promotes the Survival and Proliferation of Primary Human Vascular Endothelial Cells by Activating the Akt and ERK Pathways. *Circulation* **107**, 2250–2256.

Seol, D. W., Li, J., Seol, M. H., Park, S. Y., Talanian, R. V., and Billiar, T. R. (2000). Signaling events triggered by tumor necrosis factor-related apoptosis-inducing ligand (TRAIL): Caspase-8 is required for TRAIL-induced apoptosis. *Cancer Res.* **60**, 3152–3154.

Sheridan, J. P., Marsters, S. A., Pitti, R. M., Gurney, A., Skubatch, M., Baldwin, D., Ramakrishnan, L., Gray, C. L., Baker, K., Wood, W. I., Goddard, A. D., Godowski, P., and Ashkenazi, A. (1997). Control of TRAIL-induced apoptosis by a family of signaling and decoy receptors. *Science* **277**, 818–821.

Shiiki, K., Yoshikawa, H., Kinoshita, H., Takeda, M., Ueno, A., Nakajima, Y., and Tasaka, K. (2000). Potential mechanisms of resistance to TRAIL/Apo2L-induced apoptosis in human promyelocytic leukemia HL-60 cells during granulocytic differentiation. *Cell Death Diff.* **7**, 939–946.

Siegmund, D., Hadwiger, P., Pfizenmaier, K., Vornlocher, H. P., and Wajant, H. (2002). Selective inhibition of FLICE-like inhibitory protein expression with small interfering RNA oligonucleotides is sufficient to sensitize tumor cells for TRAIL-induced apoptosis. *Mol. Med.* **8**, 725–732.

Siegmund, D., Hausser, A., Peters, N., Scheurich, P., and Wajant, H. (2001). Tumor necrosis factor (TNF) and phorbol ester induce TNF-related apoptosis-inducing ligand (TRAIL) under critical involvement of NF-kappaB essential modulator (NEMO)/IKKgamma. *J. Biol. Chem.* **276**, 43708–43712.

Song, K., Chen, Y., Goke, R., Wilmen, A., Seidel, C., Goke, A., Hilliard, B., and Chen, Y. (2000). Tumor necrosis factor-related apoptosis-inducing ligand (TRAIL) is an inhibitor of autoimmune inflammation and cell cycle progression. *J. Exp. Med.* **191**, 1095–1104.

Southall, M. D., Isenberg, J. S., Nakshatri, H., Yi, Q., Pei, Y., Spandau, D. F., and Travers, J. B. (2001). The platelet-activating factor receptor protects epidermal cells from tumor necrosis factor (TNF) alpha and TNF-related apoptosis-inducing ligand-induced apoptosis through an NF-kappa B-dependent process. *J. Biol. Chem.* **276**, 45548–45554.

Spalding, A. C., Jotte, R. M., Scheinman, R. I., Geraci, M. W., Clarke, P., Tyler, K. L., and Johnson, G. L. (2002). TRAIL and inhibitors of apoptosis are opposing determinants for NF-kappaB-dependent, genotoxin-induced apoptosis of cancer cells. *Oncogene* **21**, 260–271.

Spencer, A., Yeh, S. L., Koutrevelis, K., and Baulch-Brown, C. (2002). TRAIL-induced apoptosis of authentic myeloma cells does not correlate with the procaspase-8/cFLIP ratio. *Blood* **100**, 3049; author reply 3050–3051.

Sprick, M. R., Weigand, M. A., Rieser, E., Rauch, C. T., Juo, P., Blenis, J., Krammer, P. H., and Walczak, H. (2002). FADD/MORT1 and caspase-8 are recruited to TRAIL receptors 1 and 2 and are essential for apoptosis mediated by TRAIL receptor 2. *Immunity* **12**, 599–609.

Srinivasan, A., Li, F., Wong, A., Kodandapani, L., Smidt, R., Jr., Krebs, J. F., Fritz, L. C., Wu, J. C., and Tomaselli, K. J. (1998). Bcl-xL functions downstream of caspase-8 to inhibit Fas and tumor necrosis factor receptor 1-induced apoptosis of MCF7 breast carcinoma cells. *J. Biol. Chem.* **273**, 4523–4529.

Srinivasula, S. M., Datta, P., Fan, X. J., Fernandes-Alnemri, T., Huang, Z., and Alnemri, E. S. (2000). Molecular determinants of the caspase-promoting activity of Smac/DIABLO and its role in the death receptor pathway. *J. Biol. Chem.* **275**, 36152–36157.

Stehlik, C., de Martin, R., Kumabashiri, I., Schmid, J. A., Binder, B. R., and Lipp, J. (1998). Nuclear factor (NF)-kappaB-regulated X-chromosome-linked iap gene expression protects endothelial cells from tumor necrosis factor alpha-induced apoptosis. *J. Exp. Med.* **188,** 211–216.

Sun, S. Y., Yue, P., Zhou, J. Y., Wang, Y., Choi Kim, H. R., Lotan, R., and Wu, G. S. (2001). Overexpression of BCL2 blocks TNF-related apoptosis-inducing ligand (TRAIL)-induced apoptosis in human lung cancer cells. *Biochem. Biophys. Res. Commun.* **280,** 788–797.

Suzuki, Y., Imai, Y., Nakayama, H., Takahashi, K., Takio, K., and Takahashi, R. (2001a). A serine protease, HtrA2, is released from the mitochondria and interacts with XIAP, inducing cell death. *Mol. Cell* **8,** 613–621.

Suzuki, Y., Nakabayashi, Y., and Takahashi, R. (2001b). Ubiquitin-protein ligase activity of X-linked inhibitor of apoptosis protein promotes proteasomal degradation of caspase-3 and enhances its anti-apoptotic effect in Fas-induced cell death. *Proc. Natl. Acad. Sci. USA* **98,** 8662–8667.

Thome, M., Schneider, P., Hofmann, K., Fickenscher, H., Meinl, E., Neipel, F., Mattmann, C., Burns, K., Bodmer, J. L., Schroter, M., Scaffidi, C., Krammer, P. H., Peter, M. E., and Tschopp, J. (1997). Viral FLICE-inhibitory proteins (FLIPs) prevent apoptosis induced by death receptors. *Nature* **386,** 517–521.

Trauzold, A., Wermann, H., Arlt, A., Schutze, S., Schafer, H., Oestern, S., Roder, C., Ungefroren, H., Lampe, E., Heinrich, M., Walczak, H., and Kalthoff, H. (2001). CD95 and TRAIL receptor-mediated activation of protein kinase C and NF-kappaB contributes to apoptosis resistance in ductal pancreatic adenocarcinoma cells. *Oncogene* **20,** 4258–4269.

Ursini-Siegel, J., Zhang, W., Altmeyer, A., Hatada, E. N., Do, R. K., Yagita, H., and Chen-Kiang, S. (2002). TRAIL/Apo-2 ligand induces primary plasma cell apoptosis. *J. Immunol.* **169,** 5505–5513.

Van Antwerp, D. J., Martin, S. J., Kafri, T., Green, D. R., and Verma, I. M. (1996). Suppression of TNF-alpha-induced apoptosis by NF-kappaB. *Science* **274,** 787–789.

Van Noesel, M. M., Van Bezouw, S., Salomons, G. S., Voute, P. A., Pieters, R., Baylin, S. B., Herman, J. G., and Versteeg, R. (2002). Tumor-specific down-regulation of the tumor necrosis factor-related apoptosis-inducing ligand decoy receptors DcR1 and DcR2 is associated with dense promoter hypermethylation. *Cancer Res.* **62,** 2157–2161.

Vander Heiden, M. G., Chandel, N. S., Schumacker, P. T., and Thompson, C. B. (1999). Bcl-xL prevents cell death following growth factor withdrawal by facilitating mitochondrial ATP/ADP exchange. *Mol. Cell* **3,** 159–167.

Verhagen, A. M., Ekert, P. G., Pakusch, M., Silke, J., Connolly, L. M., Reid, G. E., Moritz, R. L., Simpson, R. J., and Vaux, D. L. (2000). Identification of DIABLO, a mammalian protein that promotes apoptosis by binding to and antagonizing IAP proteins. *Cell* **102,** 43–53.

Wajant, H., Johannes, F. J., Haas, E., Siemienski, K., Schwenzer, R., Schubert, G., Weiss, T., Grell, M., and Scheurich, P. (1998). Dominant-negative FADD inhibits TNFR60-, Fas/Apo1- and TRAIL-R/Apo2-mediated cell death but not gene induction. *Cur. Biol.* **8,** 113–116.

Walczak, H., Bouchon, A., Stahl, H., and Krammer, P. H. (2000). Tumor necrosis factor-related apoptosis-inducing ligand retains its apoptosis-inducing capacity on Bcl-2- or Bcl-xL-overexpressing chemotherapy-resistant tumor cells. *Cancer Res.* **60,** 3051–3057.

Walczak, H., Degli-Esposti, M. A., Johnson, R. S., Smolak, P. J., Waugh, J. Y., Boiani, N., Timour, M. S., Gerhart, M. J., Schooley, K. A., Smith, C. A., Goodwin, R. G., and Rauch, C. T. (1997). TRAIL-R2: A novel apoptosis-mediating receptor for TRAIL. *EMBO J.* **16,** 5386–5397.

Wang, C. Y., Mayo, M. W., and Baldwin, A. S., Jr. (1996). TNF- and cancer therapy-induced apoptosis: Potentiation by inhibition of NF-kappaB. *Science* **274,** 784–787.

Wang, C. Y., Mayo, M. W., Korneluk, R. G., Goeddel, D. V., and Baldwin, A. S., Jr. (1998). NF-kappaB antiapoptosis: Induction of TRAF1 and TRAF2 and c-IAP1 and c-IAP2 to suppress caspase-8 activation. *Science* **281,** 1680–1683.

Wang, Q., Wang, X., Hernandez, A., Hellmich, M. R., Gatalica, Z., and Evers, B. M. (2002). Regulation of TRAIL expression by the phosphatidylinositol 3-kinase/Akt/GSK-3 pathway in human colon cancer cells. *J. Biol. Chem.* **277,** 36602–36610.

Wiley, S. R., Schooley, K., Smolak, P. J., Din, W. S., Huang, C. P., Nicholl, J. K., Sutherland, G. R., Smith, T. D., Rauch, C., Smith, C. A. *et al.* (1995). Identification and characterization of a new member of the TNF family that induces apoptosis. *Immunity* **3,** 673–682.

Xiao, C., Yang, B. F., Asadi, N., Beguinot, F., and Hao, C. (2002). Tumor necrosis factor-related apoptosis-inducing ligand-induced death-inducing signaling complex and its modulation by c-FLIP and PED/PEA-15 in glioma cells. *J. Biol. Chem.* **277,** 25020–25025.

Yang, J., Liu, X., Bhalla, K., Kim, C. N., Ibrado, A. M., Cai, J., Peng, T. I., Jones, D. P., and Wang, X. (1997). Prevention of apoptosis by Bcl-2: Release of cytochrome c from mitochondria blocked. *Science* **275,** 1129–1132.

Yeh, W. C., Pompa, J. L., McCurrach, M. E., Shu, H. B., Elia, A. J., Shahinian, A., Ng, M., Wakeham, A., Khoo, W., Mitchell, K., El-Deiry, W. S., Lowe, S. W., Goeddel, D. V., and Mak, T. W. (1998). FADD: Essential for embryo development and signaling from some, but not all, inducers of apoptosis. *Science* **279,** 1954–1958.

Zhang, X. D., Franco, A., Myers, K., Gray, C., Nguyen, T., and Hersey, P. (1999). Relation of TNF-related apoptosis-inducing ligand (TRAIL) receptor and FLICE-inhibitory protein expression to TRAIL-induced apoptosis of melanoma. *Cancer Res.* **59,** 2747–2753.

Zhang, X. D., Franco, A. V., Nguyen, T., Gray, C. P., and Hersey, P. (2000). Differential localization and regulation of death and decoy receptors for TNF-related apoptosis-inducing ligand (TRAIL) in human melanoma cells. *J. Immunol.* **164,** 3961–3970.

Zong, W. X., Edelstein, L. C., Chen, C., Bash, J., and Gelinas, C. (1999). The prosurvival Bcl-2 homolog Bfl-1/A1 is a direct transcriptional target of NF-kappaB that blocks TNFalpha-induced apoptosis. *Genes Dev.* **13,** 382–387.

INDEX

Page numbers followed by f and t indicate figures and tables, respectively.

A

AAs. *See* Astrocytomas
Acid
  lysophosphatidic (LPA), 209
  response elements, retinoic
    (RAREs), 321
  retinoic (RA), 320, 325–330
  6-3-(1-adamantyl)-4-hydroxyphenyl-2-
    naphthalene carboxylic
    (CD437), 372
  SMase (A-SMase), 236
Actinomycin D (Act D), 372
Actinomycins, 199
Agonists, 322
AIF. *See* Apoptosis
Airways, 149
  TRAIL pathways in, 150–153
  TRAIL's effect on, 153–162, 156f,
    157f, 158f
American Cancer
  Society, 348
Amino acid(s), 8, 12
  domains, 102
  Exon codes for, 56
  KDEL, 171
  peptides, 22
  sequences, 53f, 56, 83
  substitutions, 29
  TRAIL and, 82
Animal experiments, 442–443

Ankyrin repeat domain (ARD),
  103, 105
Antagonists, 322
Antibodies, monoclonal, 216
  binding of, 73–74, 74t
  immunoassays and, 68t–69t, 70–73, 71t,
    72t, 73t
  TRAIL and, 65–76, 68t–69t, 71t, 72t,
    73t, 74t
APL. *See* Leukemia
Apoptosis, 2, 44, 298f
  breast cancer and, 303–308
  cancer and, 277, 430–431
  CARD and, 133–143, 136f
  CARDINAL and, 140
  ceramides in, 230, 241
  ER Ca^{2+} and, 171–183, 174f, 177f,
    178f, 179f, 180f, 181f, 182f,
    183f, 184f
  in glioma, 431–433, 432t
  IAPs in, 175, 215, 233, 280, 369,
    462–465
  inducing factor (AIF),
    116, 277
  initiation, 82
  p38 MAPK in, 356–357
  PA and, 395–397
  pathways, 369, 433–435
  process, 258–259, 275–277, 292–293,
    351–353

Apoptosis (*continued*)
  proteins, 4, 280–281
  TRAIL and, 4, 52, 53*f*, 54–55, 57–59, 264–265, 300, 453–475, 456*f*, 458*f*, 466*f*
Apoptosis, 2 ligand (Apo2L). *See* Ligand, TNF-related apoptosis-inducing
Apoptosis, TRAIL-induced, 4, 52, 53*f*, 54–55, 57–59
  DISC and, 85–86, 86*f*, 91
  mechanisms of, 110–113, 110*f*, 112*f*
  NFkB and, 113–121
ARD. *See* Ankyrin repeat domain
A-SMase. *See* Acid
Assays
  DGK, 238–239
  enzyme-linked immunosorbent (ELISAs), 67, 70–73, 71*t*, 72*t*, 74*t*
Asthma, 72*t*, 75, 150
  TRAIL in, 151–152, 154–156, 159–162
Astrocytes, 369, 438–439
Astrocytomas, 428–429
  anaplastic (AAs), 429

B

Baculoviral IAP repeat (BIR), 114
BAFF/TALL-1/BlyS, 2–3
  in complex, 9–12, 13
BAL. *See* Bronchoalveolar lavage
BH. *See* Homology
Binding
  cell, 73–74, 74*t*
  cytokine, 209
  DNA sites, 102
  of monoclonal antibodies, 73–74, 74*t*
  peptide, 23
  RANKL, 26–27, 27*f*, 84
  sites, 37, 38*f*, 302
  TRAIL, 73–74, 74*t*
  Zn^{2+}, 5–8, 6*f*, 366
BIR. *See* Baculoviral IAP repeat
Bisindolylmaleimides, 199
Bone
  disorders, 20, 30
  metabolism, 19–23, 29–30
  OPG and, 21–22, 29–30
  RANK and, 21, 23, 29–30
  RANKL and, 21–23, 29–30
  remodeling, 20
  resorption, 110
Brains, 320–321, 428

Bronchoalveolar lavage (BAL), 152
Bronchoconstriction, 152
BTK. *See* Kinases

C

Calcium (Ca^{2+}), 169
  apoptosis and, 171–183, 174*f*, 177*f*, 178*f*, 179*f*, 180*f*, 181*f*, 182*f*, 183*f*, 184*f*
  ER, 170–172, 172*f*, 175–176
  homeostasis, 175–176, 178–183, 178*f*, 179*f*, 180*f*, 181*f*, 182*f*, 183*f*, 184*f*
  TRAIL/receptors and, 176–183, 177*f*, 178*f*, 179*f*, 180*f*, 181*f*, 182*f*, 183*f*, 184*f*
Cambridge Antibody Technology, 76
CaMKII. *See* Protein kinases II, calcium/calmodulin-dependent
Cancer
  apoptosis and, 277, 430–431
  breast, 216, 296–297, 303–308
  cells, 2, 36, 55, 91–93, 275–276, 304, 385–386, 397–398, 459
  chemicals and, 326–327
  counter-attack hypothesis, 196
  defense against, 72*t*, 74–75, 74*t*
  drugs, 43, 45, 45*f*, 92, 196, 215–216
  EGF and, 216–217, 221*f*
  immune evasion and, 195–196
  retinoids and human, 328–330, 334–336
  skin, 326–327
  TRAIL and, 91–93, 218–219, 275–284, 353–354
  types, 72*t*, 428
Cancer, ovarian
  definition of, 348–349
  risk factors of, 349–350
  therapy, 350–351
  TRAIL and, 347–358
Cancer therapy, 43, 45, 45*f*, 65, 92, 107
  advances in, 367
  chemotherapy as, 43, 366, 372–377, 373*t*–374*t*, 375*f*, 428, 430, 440–444
  combination, 282–283
  PA/TRAIL in, 385–403, 389*f*, 393*f*, 400*t*–401*t*
  radiation as, 428
  TRAIL signaling for, 275–284
CAPK. *See* Kinases
Carboplatins, 396
Carcinogens, 326–327. *See also* Cancer
Carcinoma, renal cell (RCC), 376
CARD. *See* Domain

INDEX 487

Cardiovascular system, 194–195
Caspase(s), 36
  activation of, 277
  CARDINAL proteins and, 140–142
  DFF45/ICAD, 259
  effector, 86*f*
  -8, 85–86, 111–113, 112*f*, 179–180, 180*f*, 305–307, 368
  -11, 142
  -5, 142
  inhibitors, 114–115, 120, 215, 376
  -9, 112, 179–180, 180*f*, 218, 277, 369, 370*f*, 418
  -1, 140–142
  -10, 85–86
  -3, 112, 114, 179–180, 180*f*, 305, 462
Catabolism, 237–238
CC. *See* Coiled-coil motif
CD437. *See* Acid
CDDP. *See* Cisplatin
Cell(s), 2, 473–474. *See also* Apoptosis
  APL, 321, 329–330, 335
  B, 73–74, 74*t*, 371
  BAL, 161–162
  BAX, 181–183, 181*f*, 182*f*, 183*f*, 184*f*, 471–472
  binding of, 73–74, 74*t*
  breast cancer, 216, 296–297, 303–308
  cancer/melanoma, 2, 36, 55, 91–93, 275–276, 304, 385–386, 397–398, 459
  cycle arrest, 240
  cytokines and, 240
  cytotoxic, 258, 262, 268
  degradation of host, 259
  dendritic (DC), 73–74, 74*t*, 150, 161, 230, 455
  endothelial, 150–151, 154
  epithelial, 150–151, 153–154, 208, 349–350, 369
  fate/regulation, 239–241
  glioma, 439–440
  Hela, 296, 468
  hepatoma, 304
  human airway smooth muscle (HASM), 154–156, 156*f*, 157*f*, 158*f*, 159
  hybridoma, 67
  inflammatory, 150–151, 154–155, 159–162
  leukemia, 196, 304, 321
  lines, 38–39
  MCF-7, 305
  MEF, 214
  natural killer (NK), 73–74, 74*t*, 150, 230–231, 276, 284, 293, 295, 371–372, 455
  necrosis, 88
  nonstimulated, 103
  NSO plasmacytoma, 67
  phagocytic, 297
  proliferation, 87–90
  protein, 89–90
  skeletal, 369
  smooth muscle, 150–151, 154–156, 156*f*, 157*f*, 158*f*, 159, 369
  spleen, 67
  survival, 207
  T, 73–74, 74*t*, 118, 150, 161, 262, 276, 293, 371–372, 440
  tumor, 4, 92–93, 197–199, 276–277, 283, 300, 395
  types, 114, 150–151, 261, 265
  virus-infected, 262–263
Cells, human umbilical vein endothelial (HUVEC), 57
Ceramide(s)
  analysis, 238–239
  in apoptosis, 230, 241
  catabolism, 237–238
  cell fate/regulation, 239–241
  formation, 238
  sphingolipids and, 234–242, 235*f*, 243*f*
  synthesis of, 236–237
  targets, 242
  TRAIL and, 229–244, 235*f*, 243*f*
  use of, 237–242
C-FLIP. *See* Proteins, FLICE-inhibitory
Chemokines, 160
Chemotherapy, 43, 366, 372–377, 373*t*–374*t*, 375*f*, 428, 430, 440–444
Chromatography, liquid, 238
Chromosomes, 53, 54*f*
Cisplatin (CDDP), 372, 390, 396
Clontech. *See* Human Promoter Finder DNA Walking Kit
CMV. *See* Viruses
Co-activators (CoAs), 323
Coiled-coil (CC) motif, 137
Colony-stimulating factor-1 (CSF-1), 20
Computer programs, 302
Contraceptives, oral, 350
Co-repressors (CoRs), 323
CRD. *See* Cysteine-rich domains

CRE-LOX tissue-selective-gene-ablation
  technology, 324
CSF-1. *See* Colony-stimulating factor-1
Cyclic AMP response element
  (CAMP), 172
Cycloheximides, 199
Cyclosphosphamides, 396
Cys230. *See* Cysteine
Cysteine residue (Cys230), 5, 22
Cysteine-rich domains (CRD), 2
  1/2, 13
  1-4, 22
  extracellular, 3, 21
  of TNF, 53*f*, 56
Cysts, 349–350
Cytochromes, 87, 175, 369, 412
Cytokine(s), 4. *See also* Ligand, TNF-related
    apoptosis-inducing
  activity of, 13, 150, 155, 219
  binding, 209
  cells and, 240
  pro-inflammatory, 151
  roles/systems, 160
  TNF-related activation-induced
    (TRANCE), 22
Cytotoxicity, 231–232, 233–234, 371–372,
  388–390

D

DAG. *See* Diacylglycerol
DC. *See* Cells
Death domains, 232, 455
  Fas-associated (FADD), 83, 85, 119–121,
    232–233, 261, 265, 279, 368, 411,
    455, 456*f*
  TNF receptor 1-associated (TRADD), 85,
    119, 468
  TNFR 1, 83
Death-inducing signaling complex (DISC),
    85–86, 86*f*, 88–90, 89*f*
  analysis, 264–265
  formation of, 192, 192*f*, 233
  TRAIL and, 85–86, 91, 110–111, 217,
    261–262, 331, 333*f*, 411–412
DED. *See* Domain
Diacylglycerol (DAG), 171
DIG. *See* Digitoxins
Digitoxins (DIG), 70
Dimerizations, 102
DISC. *See* Death-inducing signaling complex
Diseases, 65. *See also specific diseases*

AIDS/HIV-related, 72*t*, 75, 265–266
autoimmune, 72*t*
conditions of, 75
inflammatory, 107
malignant, 266–268
Niemann-Pick, 236
Domain
  caspase recruitment (CARD), 114,
    133–143, 136*f*
  death effector (DED), 191, 232–233
  nucelotide-binding (NBD), 136–138, 136*f*
  pyrin (PYD), 137
  silencer of death (SODD), 233
Doxorubicon (DOX), 199, 372, 396
DR. *See* Receptors, death
DRIP. *See* Proteins
Drugs, 36. *See also specific drugs*
  5-FU, 372
  anti-tumor, 396–397
  Ara-C, 372
  C225, 216
  cancer, 43, 45, 45*f*, 92, 196, 215–216
  chemotherapeutic, 366, 370*f*, 372–377,
    373*t*–374*t*, 375*f*, 390, 440–444
  resistance to, 350–351
  ZD1839 (Iressa), 216–217

E

EGF. *See* Receptors, epidermal
    growth factor
ELISAs. *See* Assays
Encephalomyelitis, 75
Endoplasmic reticulum (ER), 169, 236
  apoptosis and, 171–183, 174*f*, 177*f*, 178*f*,
    179*f*, 180*f*, 181*f*, 182*f*, 183*f*, 184*f*
  CA^{2+}, 170–172, 172*f*, 175–176
  roles of, 170–171
  rough (RER), 170
  smooth (SER), 170
Endothelium, 150–151, 154
Eosinophils, 155, 160–161
Eosinopoietins, 160
Epithelium, 150–151, 153–154, 208,
    349–350, 369
ER. *See* Endoplasmic reticulum
ERK. *See* Kinases
EST. *See* Expressed sequence tag
Etoposides, 372
Exons, 56, 89
Experiments, animal, 442–443
Expressed sequence tag (EST), 56

F

FADD. *See* Death domains
FDR. *See* Receptors, decoy
Fibroblasts, 150–151, 153–154
    lung, 369
    murine embryonic (MEFs), 241
    Niemann-Pick, 236, 241
FLIP-L. *See* Proteins, FLICE-inhibitory
FLIP-S. *See* Proteins, FLICE-inhibitory
Flow cytometry, 155

G

GBM. *See* Glioblastoma multiforme
GC. *See* Glucocorticoid
Gene(s), 36, 102
    ancestral, 13
    carboxy-terminal really interesting new (RING), 114
    cathepsin-D, 296
    cell-cycle arrest, 40, 240
    expression, 214
    IAP family, 114–116
    Myeloid differentiation primary response gene 88 (MYD88-s), 109
    PML, 328–329
    pro-apoptotic, 40
    responsive, 327
    thioredoxin, 296
Glioblastoma multiforme (GBM), 429
Glioma, malignant, 427–428
    apoptosis in, 431–433, 432*t*
    cells, 439–440
    classification of, 429
    lines, 440
    TRAIL and, 427–428, 432*t*, 433–444
    treatment of, 429–430
Gliomagenesis, 442
Glucocorticoid (GC), 155, 156, 158*f*, 159
    response element (GRE), 159
Glucosylceramides, 237
GR. *See* Receptors
GRE. *See* Glucocorticoid
Growth factors, 160, 208–209, 416–417. *See also* Receptors, epidermal growth factor; Tumor growth factor-beta
    platelet-derived (PDGFs), 239

H

HASM. *See* Cells
HAT. *See* Histone acetyltransferase
HCMV. *See* Viruses
HDACi. *See* Inhibitors
HDACS. *See* Histone deacetylases
Health, 65
Hepatocytes, human, 7
Heterotrimers, 55
Histamines, 394
Histone acetyltransferase (HAT), 323
Histone deacetylases (HDACs), 159, 323
Homeostasis, 2, 43
    Ca^{2+}, 175–176, 178–183, 178*f*, 179*f*, 180*f*, 181*f*, 182*f*, 183*f*, 184*f*
Homology, 230
    Bcl-2 (BH), 174, 174*f*, 455, 456*f*, 465–468, 466*f*
    plecksrin (PH), 210–211
    region (RHR), 102
Homotrimers, 52, 83
Hormones, 20
Human Genome Sciences, 76
Human Promoter Finder DNA Walking Kit (Clontech), 58
HUVEC. *See* Cells, human umbilical vein endothelial
Hypermethylation, 36, 38–39

I

IAPs. *See* Inhibitors
IFN. *See* Interferon
IFN-c. *See* Interferon-gamma
IKK. *See* Inhibitors
IL. *See* Interleukin
Immune systems, 2, 195–196, 258
Immunosuppression, 265–266
Inflammasomes, 141–143
Inflammation, 72*t*, 75, 107–109, 150–151, 154–155, 159–162
Inhibitors, 441, 471–473
    of apoptosis (IAPs), 175, 215, 233, 280, 369, 462–465
    of apoptosis, X-linked (XIAPs), 115, 233, 333, 369, 462–465
    caspase, 114–115, 120, 215, 376
    death, 233
    of EGFR/HER2, 420
    HDAC (HDACi), 328
    of κB kinase (IKK), 88, 105–109, 108*f*, 119–121, 134
    of metalloproteinase-1, tissue (TIMP-1), 153
    plasminogen activator, -1 (PAI-1), 153
    PS-341, 199

Inhibitors (*continued*)
    TRAIL-R3, 57
Initiators, 326
Interferon (IFN), 230–231, 262
Interferon-gamma (IFN-c), 291
    breast cancer and, 296–297, 303–308
    effects of, 295–296
    regulation by, 303–308
    roles of, 293–295, 294f
    sensitization by, 303–306
    signaling, 294f, 295
    TRAIL and, 291–309, 294f, 298f, 302f
Interleukin (IL)
    -8, 348, 354–55
    -11, 20
    -1, 105, 134
    1 receptor-associated kinase-M
        (IRAK-M), 109
    -6, 20
IP$_3$R. *See* Receptors
IRAK-M. *See* Interleukin
Iressa. *See* Drugs
Irradiation, 283

K

Keratinocytes, 369
KILLER/D5. *See* Ligand-receptor2,
    TNF-related apoptosis-inducing
Kinase suppressor of RAS (KSR), 242
Kinases, 413
    AKT, 210–213, 410, 416–421, 469–470
    Bruton's tyrosine (BTK), 233
    ceramide activated protein (CAPK), 242
    extracellular regulated (ERK),
        90, 473
    IKK, 88, 105–109, 108f, 119–121, 134
    JAK, 213
    MEKK1, 219
    mitogen-activated protein (MAPK),
        171–172
    NFκB-inducing (NIK), 107
    p38 MAPK, 356–357
    phosphoinositide-dependent, -1
        (PDK-1), 244
    phosphotidylinositol-3 (P13-K), 171,
        209–211, 210f
    proline-rich tyrosine, 2 (PYK2), 172
    protein (PKR), 87–88, 307
    protein C (PKC), 90, 172
    ribosomal S6 (RSK), 213
    serine, 105–107

KSR. *See* Kinase suppressor of RAS

L

Leucine-rich repeats (LRRs), 136, 136f
Leukemia
    acute promyelocytic (APL), 321,
        329–330, 335
    cells, 193, 304, 321
    myeloid, 325
Ligand(s)
    CD95, 292
    cognate, 2, 21
    death, 82, 189–192
    osteoprotegerin (OPGL), 22
    rexinoid, 323
    systems, 260–262
Ligand, receptor activator of NF-κB
    (RANKL), 19
    binding, 26–27, 27f, 84
    bone metabolism and, 21–23, 29–30
    interaction of, 27–29, 27f
    mutagenesis and, 29
    residue specific to, 28
    structure of, 24f, 25–30, 27f
    TNF and, 22–23, 110
    TRAIL and, 20–21, 55
Ligand, TNF-related apoptosis-inducing
    (TRAIL), 1, 473–474
    AA" loop of, 8, 9f, 10f
    activation, 152–153
    adenoviral, 443
    administration, 281–284
    in airways, 149–162, 156f, 157f, 158f
    amino acids and, 82
    antitumor activity of, 282
    apoptosis and, 4, 52, 53f, 54–55, 57–59,
        264–265, 300, 453–474, 456f, 458f,
        466f
    in asthma, 151–152, 154–156, 159–162
    binding, 73–74, 74t
    bioactivity of, 369–372, 388
    breast cancer and, 303–308
    cancer, 91–93, 218–219, 275–284, 353–354
    ceramides and, 229–244, 235f, 243f
    chemotherapeutic drugs and, 366–377,
        370f, 373t–374t, 375f, 428, 430,
        440–444
    clinical value of, 391–392
    crossactivity of, 3, 28
    cytotoxic effect of, 231–232, 233–234,
        371–372, 388–390

INDEX   491

death receptors and, 83, 110–113, 112f,
    217–218
decoy receptors, 52, 54–55, 59, 82, 84,
    456f, 458–460
description, 388
DISC and, 85–86, 91, 110–111, 217,
    261–262, 331, 337f, 411–412
EGF and, 207–222, 210f, 220f, 221f
ER Ca^{2+} and, 176–183, 177f, 178f,
    179f, 180f, 181f, 182f,
    183f, 184f
expressions, 38–39, 150–152, 263
FADD and, 232–233, 368
fibroblasts and, 150–151, 153–154
FLIP and, 197–199, 460–462
formulations, 436
functions of, 3, 65–66, 91, 352
future of, 76
Gln-205 in, 28
IFN-c and, 291–309, 294f, 298f, 302f
immunoassays and, 68t–69t, 70–73, 71t,
    72t, 73t
in immunosuppression, 265–266
induced Bid cleavage, 86f,
    87, 416
induced cell proliferation, 87–90
induced c-FLIP recruitment, 88–90, 89f,
    116–117, 279–280, 412–413,
    460–461
induced death signal, 232–233
induction, 231
ligation, 217
malignant diseases and, 266–268
malignant glioma and, 427–428, 432t,
    433–444
monoclonal antibodies against, 65–76,
    68t–69t, 71t, 72t, 73t, 74t
mRNA, 155, 156, 158f
NF-κB signaling and, 101–121, 104f, 108f,
    110f, 112f, 418, 468–470
OPG and, 109–110, 110f
other factors of, 413, 470–471
ovarian cancer and, 347–358
PA and, 385–403, 389f, 393f, 400t–401t
PED recruitment and, 89f, 90–91
physiologic roles of, 301, 302f
promoter, 151, 156–159, 158f
RANKL and, 20–21
receptors, 3, 36–37, 52–53, 53f, 54f,
    109–110, 110f, 260–261, 278–279,
    330–334, 331f, 332f, 333f, 367–368,
    388, 389f, 436–437

recombinant, 91–93
regulation of, 352, 409–421
resistance, 419–421, 441–442
retinoids and, 320–338, 324f, 331f,
    332f, 333f
sensitivity, 232, 263, 409–421
signaling, 81–92, 86f, 89f, 150–153,
    261–262, 275–284, 298f, 299–300,
    368–369, 370f, 391
significance of, 72t, 74–75, 74t
Smac/DIABLO and, 234, 463–465, 471
smooth muscle and, 154–156, 156f, 157f,
    158f, 159
soluble (sTRAIL), 66–67, 68t–69t, 70–73,
    71t, 72t, 73t
sphingolipids and, 234–242, 235f, 243f
structure of, 5, 6f, 9f, 12–14
synergistic effects of, 372–376,
    373t–374t, 375f
system, 297, 298f
transcriptional regulation of, 301–302
transduction, 438–440
tumors and, 90–91
vessels and, 154
viral infections and, 257–269
in vivo activity of, 371, 399–402
Zn^{2+} and, 5–8, 6f
Ligand-receptor1, TNF-related apoptosis-
    inducing (TRAIL-R1), 36–37
comparisons of, 39–40, 41f
domains, 53f, 56, 109–110, 110f, 176–177,
    177f, 260–261
expressions of, 38–39, 52, 53f, 152–153
promoter, 57
Ligand-receptor2, TNF-related apoptosis-
    inducing (TRAIL-R2), 35–37
cancer therapy of, 43, 45, 45f
comparisons of, 39–40, 41f, 42f
domains, 53f, 56, 109–110, 110f, 176–177,
    177f, 260–261
ER Ca^{2+} and, 43
expressions of, 38–39, 52–55, 53f, 54f,
    152–153
hypermethylation of, 36, 38–39
p53 and, 40–43, 418
promoter, 37–45, 38f, 39f, 41f, 42f, 57
regulator of, 40–43
structure of, 37–40, 38f, 39f, 44t
Ligand-receptor3, TNF-related apoptosis-
    inducing (TRAIL-R3), 36–37
decoy receptors and, 52, 54–55, 59
domains, 109–110, 110f, 176, 260–261

Ligand-receptor3, TNF-related apoptosis-
inducing (TRAIL-R3) (*continued*)
  expressions of, 38–39, 52–55, 53*f*, 54*f*
  identification of, 53*f*, 54*f*, 56
  as inhibitor, 57
  promoter, 54*f*, 57–58
  transcriptional regulation of, 51–60, 53*f*,
    54*f*, 59*f*
Ligand-receptor4, TNF-related apoptosis-
inducing (TRAIL-R4), 36–37
  domains, 109–110, 110*f*,
    176, 260–261
  expressions of, 38–39, 53–55, 53*f*
Lipid charring, 238
Lipopolysaccharides (LPS), 105
Loops, AA″
  DE, 27–28, 29
  length of, 26
  TRAIL, 8, 9*f*, 10*f*
LPA. *See* Acid
LPS. *See* Lipopolysaccharides
LRRs. *See* Leucine-rich repeats
Lymphocytes, 266, 293, 455
  peripheral blood (PBLs), 265, 366

M

Macrophages, 150, 161, 266, 295
MAPK. *See* Kinases
M-CSF. *See* Colony-stimulating factor-1
Mediators, 154
MEFs. *See* Fibroblasts
Melanocytes, 369
Mitochondria, 173, 241, 368–369
Mitochondrial permeability transition
  (MPT), 241
MK. *See* Mouse homologue of
  TRAIL-R2
Molecules
  adhesion, 155
  effector, 134
  ICAM-1, 155
  lipid, 234
  VCAM, 155
Monocytes, 73–74, 74*t*, 75, 150, 161
Mouse homologue of TRAIL-R2 (MK),
  40, 42*f*
MPT. *See* Mitochondrial permeability
  transition
Mutagenesis, 29, 114
Mutations, 441
MYD88-s. *See* Genes

N

National Institutes of Health (NIH), 387
National Library of Medicine (NLM), 387
NBD. *See* Domain
NCoRs. *See* Corepressors
NDV. *See* Viruses
NIH. *See* National Institutes of Health
NIK. *See* Kinases
NK. *See* Cells
NLM. *See* National Library of Medicine
NR. *See* Receptors
Nuclear factor-kappa B (NF-κB), 134, 410
  activation of, 87–88, 89*f*, 119–121,
    134–135, 193
  CARD and, 133–143, 136*f*
  cellular sensitivity and, 113–116
  EGF and, 214
  IKK complex and, 105–109, 108*f*, 119–121
  proteins and, 102–105, 104*f*
  regulation by, 113–119, 134–135
  signaling, 101–121, 104*f*, 108*f*, 110*f*, 112*f*
  TRAIL and, 101–121, 104*f*, 108*f*, 110*f*,
    112*f*, 418, 468–470
  transcription factors, 102–103, 104*f*,
    281, 413
Nuclear localization sequence (NLS),
  102–103, 193
Nuclear receptor co-repressors (NCoRs), 323

O

OCIF. *See* Osteoclastogenesis
  inhibitory factor
ODF. *See* Osteoclast differentiation factor
Oncogenesis, 107
Open reading frames (ORF), 56
OPG. *See* Osteoprotegerin
OPGL. *See* Ligands
ORF. *See* Open reading frames
Organ repairs, 321
Organogenesis, 2
Osteoblasts, 20–21
Osteoclast differentiation factor (ODF),
  22, 110
Osteoclastogenesis inhibitory factor (OCIF),
  21, 29
Osteoclasts, 20–21
Osteoporosis, 20, 30
Osteoprotegerin (OPG), 3, 20. *See also*
  Ligand, osteoprotegerin; Receptors
  bone metabolism and, 21–22, 29–30
  residues of, 28

soluble, 66, 84
  TNF and, 21–22, 52
  TRAIL and, 109–110, 110f
Ovulation, 350

P

Paclitaxel (PA), 372
  apoptotic effects of, 395–397
  characteristics of, 394–395
  discovery of, 386–387
  mechanisms of, 393–394
  parameters of, 395
  resistance to, 394
  structure of, 392–393, 393f
  TRAIL and, 385–403, 389f, 393f, 400t–401t
  in vivo relevance of, 399–402
PAI-1. See Inhibitors
PBLs. See Lymphocytes
PDB. See Protein Data Bank
PDGFs. See Growth factors
PDI. See Proteins
PDK-1. See Kinases
PED. See Phosphoprotein enriched in diabetes
Peptides
  amino acid, 22
  binding, 23
  signal, 83
Perforin/Granzyme pathway, 197
PEST. See Proline, glutamic acid, serine and threonine domain
PH. See Homology
Phosphatidylinositol-3 kinase/AKT pathway, 414
  PTEN and, 415–421
  targets of, 417–421
  TRAIL and, 416–417
Phospholipase C (PLC), 171–172
  Phospatidylinositol-specific C (PI-PLC), 57
Phospholipids, 210–211
Phosphoprotein enriched in diabetes (PED), 90–91
Phosphorylation, 105–107, 211–212
PI-PLC. See Phospholipase
PKC. See Kinases
PKR. See Kinases
PLC. See Phospholipase
PNETs. See Tumors, primitive neuroectodermal

Proline, glutamic acid, serine and threonine (PEST) domain, 105
Promoter(s), 35
  CAAT, 301
  Gc-box-containing, 36
  regulation, 37
  structure, 57–58
  TATA-box, 36, 37, 57, 301
  TRAIL, 151, 156–159, 158f
  TRAIL-R1, 57
  TRAIL-R2, 37–45, 38f, 39f, 41f, 42f, 57
  TRAIL-R3, 54f, 57–58
Prostaglandin E2, 20
Protein(s)
  activator, 137–138
  adaptor, 23–25, 36, 111, 137–138
  antiapoptotic, 107–109, 115–116, 211, 214–216, 220, 453–454
  apoptosis, 4, 280–281
  ASC, 142
  Bcl-12 family, 215–216
  Bid, 86f, 87, 112, 116, 180, 221, 277, 368–369, 416
  CATERPILLER, 137
  cell, 89–90
  DEDD, 193
  disulfide isomerase (PDI), 171
  DRAL, 139
  FADD, 36, 89, 119–121, 217, 411, 455, 456f
  FAP-1, 233
  functions, 135–138, 136f
  HtrA2/Omi, 115
  IAP family, 114–116, 215
  ICEBERG, 140–141
  IκB, 103–105
  NAC/NALP1, 142
  NALP, 137
  NBD-CARD, 142
  NF-κB, 102–103, 104f
  Nod, 136–138, 136f, 142
  PED/PEA-15, 410, 419
  PP1, 243–244
  proinflammatory, 107–109
  pseudoICE, 140–141
  RAS, 212–213
  receptor-interacting (RIP), 87–88, 89f, 119–121, 468
  release, 214–215
  RICK, 141
  T-Bid, 217–218
  thyroid hormone receptor-associated (TRAP), 323

Protein(s) (continued)
    type I transmembrane, 2, 52, 82, 84
    type II transmembrane, 52
    vitamin D receptor-interacting (DRIP), 323
    XIAP, 115, 233, 333
Protein Data Bank (PDB), 25
Protein kinases II, calcium/calmodulin-dependent (CaMKII), 90
Proteins, BcL-2 family, 4, 112, 116, 221
    anti-apoptotic subfamily of, 174, 174f, 211, 412
    BH3-only subfamily of, 175
    Ca^{2+} homeostasis and, 175–176, 178–183, 178f, 179f, 180f, 181f, 182f, 183f, 184f
    domains, 174, 174f, 455, 456f, 465–468, 466f
    pro-apoptotic subfamily of, 174
    roles of, 261–262, 277, 280, 368–369
Proteins, CARDINAL (CARD inhibitor of NFκB activating ligands), 135–137, 136f
    apoptosis and, 140
    caspases and, 140–142
    contradictions, 142–143
    functions of, 138
    as regulator, 138–139
Proteins, FLICE-inhibitory (FLIP), 4, 410, 455
    v (vFLIP), 264–265
    action mechanisms of, 191–192, 192f, 193f
    alternative mechanisms of, 193
    c (c-FLIP), 88–90, 89f, 116–117, 279–280, 412–413, 460–461
    expression, 194
    functions, 194–196, 218
    long (FLIP-L), 116–117, 218
    short (FLIP-S), 116–117, 218
    structure of, 190–191, 191f
    synonyms for, 190
    TRAIL and, 197–199, 459–462
Proteolysis, 105
P13-K. *See* Kinases
PYK2. *See* Kinases

R

RA. *See* Acid
Radiolabeling, 238
RANKL. *See* Ligand, receptor activator of NF-κB
RAREs. *See* Acid
RCC. *See* Carcinoma, renal cell
Reactive oxygen species (ROS), 241
Receptors, 21. *See also* Osteoprotegerin
    AP1, 321, 326–327
    BAFF-R, 3
    BCMA, 3
    CD3/T cell (TCR), 265
    CD95L, 54
    ErbB1/B4, 214
    GC (GR), 159
    G-coupled, 209
    inositol-1,4,5-triphosphate (IP_3R), 3, 171
    internalization and degradation (RID), 264
    non-tyrosine, 209
    nuclear (NR), 323
    PDGF, 217
    proapoptotic, 52, 54–55
    of rexinoids, 327–328
    ryanodine (RYRs), 171
    thyroid hormone (TR), 323
    TRAIL, 3, 36–37, 52–53, 53f, 54f, 109–110, 110f, 260–261, 278–279, 330–334, 331f, 332f, 333f, 367–368, 388, 389f, 436–437
Receptors, death (DR), 82, 189–190. *See also* Ligand-receptor1, TNF-related apoptosis-inducing; Ligand-receptor2, TNF-related apoptosis-inducing 4, 3, 66, 83, 85–87, 152–153, 176–177, 177f, 219, 222, 229, 5, 2–3, 36, 66, 83, 85–87, 152–153, 176–177, 177f, 219, 222, 229
    Fas, 213–214
    ligand systems and, 260–262
    pathways, 434–435
    TRAIL and, 83, 110–113, 112f, 217–218
Receptors, decoy, 3, 21–22, 36–37, 412. *See also* Ligand-receptor3, TNF-related apoptosis-inducing; Ligand-receptor4, TNF-related apoptosis-inducing
    DcR1/DcR2, 3, 66, 84, 152
    DcR3, 54
    Fas/CD95 decoy (FDR), 54
    TRAIL, 52, 54–55, 59, 82, 84, 456f, 457–459
Receptors, epidermal growth factor (EGF), 429
    cancer and, 216–217, 221f
    family, 208–209
    NF-κB and, 214
    signaling pathways and, 209–213, 210f, 219–222, 220f, 221f
    TRAIL and, 207–222, 210f, 220f, 221f

transcription factors' activation by, 213–214
Receptors, RANK, 20, 84
  bone metabolism and, 21, 23, 29–30
  residues of, 28
Receptors, tumor necrosis factor (TNFR), 1–3, 83, 212
Reproductive systems, 321
RER. See Endoplasmic reticulum
Retinoids
  APL and, 335
  atypical, 335–236
  future of, 336–238
  human cancers and, 328–330, 334–336
  mechanism of, 322–324, 324f
  nuclear (RARs), 320–322, 325–326
  origin of, 322
  preventive/therapeutic potential of, 334–336
Reverse transcriptase-polymerase chain reaction (RT-PCR), 155, 156f
Rexinoids (RXRs), 320–322
  future of, 336–338
  ligand, 323
  myeloid leukemia and, 325
  receptors of, 327–328
RHR. See Homology
RID. See Receptors
RING. See Genes
RIP. See Proteins
ROS. See Reactive oxygen species
RSK. See Kinases
RT-PCR. See Reverse transcriptase-polymerase chain reaction
RYRs. See Receptors

S

Sarcoplasmic/endoplasmic reticulum calcium-transporting ATPases (SERCAs), 171, 173
SER. See Endoplasmic reticulum
SERCAs. See Sarcoplasmic/endoplasmic reticulum calcium-transporting ATPases
Signaling pathways, 150–153, 352
  death, 335–336
  EGF, 209–213, 210f, 219–222, 220f, 221f
  IFN-c, 294f, 295
  JAK/STAT, 208, 209–213, 210f
  NF-κB, 101–121, 104f, 108f, 110f, 112f
  P13K/AKT, 208, 209–213, 210f, 219

Ras/MAPK, 208, 209–213, 210f, 219
TRAIL, 81–92, 86f, 89f, 150–153, 261–262, 275–284, 298f, 299–300, 368–269, 370f, 391
Smac/DIABLO (second mitochondria-derived activator of caspase/direct inhibitor of apoptosis binding protein), 87, 114–116, 175, 218, 277
  TRAIL and, 234, 463–465, 471
SMCC. See Srb and mediator protein-containing complex
SODD. See Domain
Sodium butyrates, 199
Son of sevenless (SOS), 212
Spectrometry, mass, 238–239
Sphingolipids, 234–242, 235f, 243f. See also Ceramides
Sphingomyelinases, 235–236, 237f
Sphingomyelins, 235–236, 237, 237f
Srb and mediator protein-containing complex (SMCC), 323
sTRAIL. See Ligand, TNF-related apoptosis-inducing
Surgery, 429
Survivin, 458f, 463

T

TCR. See Receptors
TFSEARCH, 37, 58
TG. See Thapsigargin
TGF-b. See Tumor growth factor-beta
Thapsigargin (TG), 43, 173
  ER Ca^{2+} and, 178–183, 178f, 179f, 180f, 181f, 182f, 183f, 184f
TIMP. See Inhibitors
TNF receptor-associate factor (TRAF), 23–25
  -5, 120
  -1, 26
  -2, 120
TNF. See Tumor necrosis factor
TNFR. See Receptors, tumor necrosis factor
TR. See Receptors
TRADD. See Death domains
TRAF. See TNF receptor-associate factor
TRAIL receptor inducer of cell killing 2 (TRICK2), 36. See also Ligand-receptor2, TNF-related apoptosis-inducing

TRAIL. *See* Ligand, TNF-related apoptosis-inducing
TRAIL with N-terminal polyhistidine tag (TRAIL-His), 7
TRAIL-R1. *See* Ligand-receptor1, TNF-related apoptosis-inducing
TRAIL-R2. *See* Ligand-receptor2, TNF-related apoptosis-inducing
TRAIl-R3. *See* Ligand-receptor3, TNF-related apoptosis-inducing
TRAIL-R4. *See* Ligand-receptor4, TNF-related apoptosis-inducing
TRAIL:sDR5 Complex, 1, 26
structure of, 9–12, 9f, 11f, 12–14
TRANCE. *See* Cytokines
Transcription factors. *See also* Retinoids; Rexinoids
  CREB, 211–212
  EGF and, 213–214
  forkhead, 211–212, 213, 410, 418
  NF-κB, 102–103, 104f, 281
  STAT, 213
Translational initiation codon ATG, 36–37, 40–43
TRAP. *See* Proteins
TRICK2. *See* TRAIL receptor inducer of cell killing 2
Tumor
  anti, 282, 396–397
  brain, 428–433
  cells, 4, 92–93, 197–199, 276–277, 283, 300, 395
  progression, 325–326
  PTEN suppressor, 410, 415–421
  suppressor p53, 41–44, 418
  suppressors, 325–326, 415
  surveillance, 91, 284
  TRAIL and, 90–91
Tumor growth factor-beta (TGF-β), 115, 151
Tumor necrosis factor (TNF). *See also* Receptors, tumor necrosis factor
  -b, 2, 5, 12, 24f, 25–26
  CRD of, 53f, 56
  death ligands of, 82, 189–192
  family, 8, 9f, 10f, 21–22, 25, 52–53, 53f, 258–259, 351–353
  OPG and, 21–22, 52
  -R55, 2, 12–13
  RANKL and, 22–23, 110
  roles of, 2–3, 82, 134, 454–455
Tumorigenesis, 325
Tumors, primitive neuroectodermal (PNETs), 39

U

Untranslated regions (UTRs), 3′/5′, 56
U.S. Food and Drug Administration (USDA), 420
UTRs. *See* Untranslated regions

V

vFLIP. *See* Proteins, FLICE-inhibitory
Virus(es), 72t
  adeno, 262
  cytomegalo (CMV), 231
  encephalomyocarditis, 262
  encodes, 265
  human cytomegalo (HCMV), 262–263, 266
  infected cells, 262–263
  infections/TRAIL, 257–269
  malignant diseases and, 266–268
  measle, 266
  Newcastle disease (NDV), 267
  reo, 262–263
Vitamins
  D3, 20

X

XIAPs. *See* Inhibitors; Proteins
X-ray crystallography, 23

Z

Zinc binding (Zn^{2+}), 5–8, 6f, 366